Case Studies in Infectious Disease

Second Edition

Peter M Lydyard
Michael F Cole
John Holton
William L Irving
Nina Porakishvili
Pradhib Venkatesan
Katherine N Ward

CRC Press

Taylor & Francis Group

Boca Raton London New York

CRC Press is an imprint of the
Taylor & Francis Group, an **informa** business

Second edition published 2023
by CRC Press
6000 Broken Sound Parkway NW, Suite 300, Boca Raton, FL 33487-2742

and by CRC Press
4 Park Square, Milton Park, Abingdon, Oxon, OX14 4RN

CRC Press is an imprint of Taylor & Francis Group, LLC

ISBN: 978-0-367-72587-7 (hbk)
ISBN: 978-0-367-69639-9 (pbk)
ISBN: 978-1-003-15544-7 (ebk)

DOI: 437827

Typesetting by Evolution Design and Digital Ltd

Instructor & Student Resources can be accessed at:
www.routledge.com/cw/lydyard

Table of Contents

Additional cases are available for download at **www.routledge.com/cw/lydyard**

Preface

Following the SARS-CoV-2 pandemic, microbiology has never been such an important topic to both students and indeed the general public. For this edition, we have updated the 40 cases which appeared in the first edition (2010), having acquired much more information in relation to the five core questions that the students are asked for each case. Again, each case follows the natural history of the infection from point of entry of the pathogen through pathogenesis, clinical presentation, diagnosis, and treatment. All features of this edition are identical to those of the first. There are 48 cases, 8 of which are new, and the remainder thoroughly updated. Because of the size limitation of the book there are 40 cases in the book and 8 available on the website, where they will be regularly updated. As for the previous book, we consider this new edition to provide invaluable key information for both science courses and particularly for medical students who need instant access to specific medically important organisms.

We hope that you enjoy the new edition

The authors

Preface to the First Edition

The idea for this book came from a successful course in a medical school setting. Each of the forty cases has been selected by the authors as being those that cause the most morbidity and mortality worldwide. The cases themselves follow the natural history of infection from point of entry of the pathogen through pathogenesis, clinical presentation, diagnosis, and treatment. We believe that this approach provides the reader with a logical basis for understanding these diverse medically important organisms.

Following the description of a case history, the same five sets of core questions are asked to encourage the student to think about infections in a common sequence. The initial set concerns the nature of the infectious agent, how it gains access to the body, what cells are infected, and how the organism spreads; the second set asks about host defense mechanisms against the agent and how disease is caused; the third set enquires about the clinical manifestations of the infection and the complications that can occur; the fourth set is related to how the infection is diagnosed, and what is the differential diagnosis; and the final set asks how the infection is managed, and what preventative measures can be taken to avoid the infection.

In order to facilitate the learning process, each case includes summary bullet points, a reference list, a further reading list and some relevant reliable websites. Some of the websites contain images that are referred to in the text. Each chapter concludes with multiple-choice questions for self-testing with the answers given online.

In the contents section, diseases are listed alphabetically under the causative agent. A separate table categorizes the pathogens as bacterial, viral, protozoal/worm/fungal and acts as a guide to the relative involvement of each body system affected. Finally, there is a comprehensive glossary to allow rapid access to microbiology and medical terms highlighted in bold in the text. All figures are available in JPEG and PowerPoint® format at www.routledge.com/cw/lydyard.

We believe that this book would be an excellent textbook for any course in microbiology and in particular for medical students who need instant access to key information about specific infections.

Happy learning!!

The authors
March, 2009

Author Affiliations

Peter M Lydyard, Emeritus Professor of Immunology, University College Medical School, London, UK, Honorary Professor of Immunology, School of Life Sciences, University of Westminster, London, UK and Professor of Immunology, University of Georgia, Tbilisi, Georgia.

Michael F Cole, Emeritus Professor of Microbiology and Immunology, Georgetown University Department of Microbiology & Immunology, Washington DC, USA.

John Holton, Department of Natural Sciences, University of Middlesex, London, UK and Department of Pathology, Darrent Valley Hospital, Dartford, UK.

William L Irving, Professor and Honorary Consultant in Virology, University of Nottingham and Nottingham University Hospitals NHS Trust, UK.

Nina Porakishvili, Principal Lecturer in Life Sciences, University of Westminster, London, UK.

Pradhib Venkatesan, Consultant in Infectious Diseases, Department of Infectious Diseases, Nottingham University Hospitals City Campus, Nottingham, UK.

Katherine N Ward, Honorary Professor, Division of Infection and Immunity, University College London, London, UK.

Acknowledgements

In writing this book we have benefited greatly from the advice of many microbiologists and immunologists. We would like to thank the following for their suggestions in preparing this edition.

William R. Abrams (New York University College of Dentistry, USA); Abhijit M. Bal (Crosshouse Hospital, UK); Keith Bodger (University of Liverpool, UK); Carolyn Hovde Bohach (University of Idaho, USA); Robert H. Bonneau (The Pennsylvania State University College of Medicine, USA); Dov L. Boros (Wayne State University, USA); Thomas J. Braciale (University of Virginia Health Systems, USA); Stephen M. Brecher (VA Boston Healthcare System, USA); Patrick J. Brennan (Colorado State University, USA); Christine M. Budke (Texas A&M University, USA); Neal R. Chamberlain (A.T. Still University of Health Sciences/KCOM, USA); Dorothy H. Crawford (University of Edinburgh, UK); Jeremy Derrick (University of Manchester, UK); Joanne Dobbins (Bellarmine University, USA); Michael P. Doyle (University of Georgia, USA); Sean Doyle (National University of Ireland); Gary A. Dykes (Food Science Australia); Stacey Efstathiou (University of Cambridge, UK); Roger Evans (Raigmore Hospital, UK); Ferric C. Fang (University of Washington School of Medicine, USA); Robert William Finberg (University of Massachusetts Medical School, USA); Joanne Flynn (University of Pittsburgh School of Medicine, USA); Scott G. Franzblau (University of Illinois at Chicago, USA); Caroline Attardo Genco (Boston University School of Medicine, USA); Geraldo Gileno de Sá Oliveira (Oswaldo Cruz Foundation, Brazil); John W. Gow (Glasgow Caledonian University, UK); Carlos A. Guerra (University of Oxford, UK); Paul Hagan (University of Glasgow, UK); Anders P. Hakansson (SUNY at Buffalo, USA); Tim J. Harrison (University College London, UK); Robert S. Heyderman (Liverpool School of Tropical Medicine, UK); Geoff Hide (University of Salford, UK); Stuart Hill (Northern Illinois University, USA); Stephen Hogg (University of Newcastle, UK); Malcolm J. Horsburgh (University of Liverpool, UK); Michael Hudson (University of North Carolina at Charlotte, USA); Karsten Hueffer (University of Alaska Fairbanks, USA); Paul Humphreys (University of Huddersfield, UK); Ruth Frances Itzhaki (University of Manchester, UK); Prof G. Janossy (University College, London, UK); Aras Kadioglu (University of Leicester, UK); A. V. Karlyshev (Kingston University, UK); Ruth A. Karron (Johns Hopkins University, USA); Stephanie M. Karst (Louisiana State University Health Sciences Center, USA); C. M. Anjam Khan (University of Newcastle, UK); Peter G.E. Kennedy (University of Glasgow, UK); Martin Kenny (University of Bristol, UK); H. Nina Kim (University of Washington, USA); George Kinghorn (Royal Hallamshire Hospital, UK); Michael Klemba (Virginia Polytechnic Institute and State University, USA); Brent E. Korba (Georgetown University Medical Center, USA); Awewura Kwara (Warren Alpert Medical School of Brown University, USA); Jerika T. Lam (Loma Linda University, USA); Robert A. Lamb (Northwestern University, USA); Audrey Lenhart (Liverpool School of Tropical Medicine, UK); Michael D. Libman (McGill University, Canada); David Lindsay (Virginia Technical University, USA); Dennis Linton (University of Manchester, UK); Martin Llewelyn (Brighton and Sussex Medical School, UK); Diana Lockwood (London School of Hygiene & Tropical Medicine, UK); Francesco A. Mauri (Imperial College, UK); Don McManus (Queensland Institute of Medical Research, Australia); Keith R. Matthews (University of Edinburgh, UK); Ernest Alan Meyer (Oregon Health and Science University, USA); Manuel H. Moro (National Institutes of Health, USA); Kristy Murray (The University of Texas Health Science Center, USA); Tim Paget (The Universities of Kent and Greenwich at Medway, UK); Andrew Pekosz (Johns Hopkins University, USA); Lennart Philipson (Karolinska Institute, Sweden); Gordon Ramage (University of Glasgow, UK); Julie A. Ribes (University of Kentucky, USA); Alan Bernard Rickinson (University of Birmingham, UK); Adam P. Roberts (University College London, UK); Nina Salama (Fred Hutchinson Cancer Research Center and University of Washington, USA); John W. Sixbey (Louisiana State University Health Sciences Center-Shreveport, USA); Deborah F. Smith (York Medical School University of York, UK); John S. Spencer (Colorado State University, USA); Richard Stabler (London School of Hygiene & Tropical Medicine, UK); Catherine H. Strohbehn (Iowa State University, USA); Sankar Swaminathan (University of Florida Shands Cancer Center, USA); Clive Sweet (University of Birmingham, UK); Clarence C. Tam (London School of Hygiene & Tropical Medicine, UK); Mark J. Taylor (Liverpool School of Tropical Medicine, UK); Yasmin Thanavala (Roswell Park Cancer Institute, USA); Christian Tschudi (Yale University, USA); Mathew Upton (University of Manchester, UK); Juerg Utzinger (Swiss Tropical Institute, Switzerland); Julio A. Vázquez (National Institute of Microbiology, Institute of Health Carlos III, Spain); Joseph M. Vinetz (University of California, San Diego, USA); J. Scott Weese (University of Guelph, Canada); Lee Wetzler (Boston University School of Medicine, USA); Peter Williams (University of Leicester, UK); Robert Paul Yeo (Durham University, UK); Qijing Zhang (Iowa State University, USA); Shanta M. Zimmer (Emory University School of Medicine, USA).

Guide to the relative involvement of each body system affected by the infectious organisms described in this book: the organisms are categorized into bacteria, viruses, and protozoa/fungi/worms.

Organism Bacteria	Resp	MS	GI	H/B	GU	CNS	CV	Skin	Syst	L/H
Borrelia burgdorferi						4+		1+	1+	
Campylobacter jejuni			4+			2+				
Chlamydia trachomatis				2+	4+	2+				
Clostridium difficile			4+							
Escherichia coli	4+				4+	4+			4+	
Helicobacter pylori			4+							
Leptospira spp.				4+	4+	4+				
Listeria monocytogenes			2+			4+	2+		4+	
Mycobacterium leprae						2+		4+		
Mycobacterium tuberculosis	4+								2+	
Neisseria gonorrhoeae					4+				2+	
Neisseria meningitidis						4+			4+	
Rickettsia spp.							4+	4+	4+	
Salmonella typhi			4+						4+	
Staphylococcus aureus	1+	1+	2+				1+	4+	1+	
Streptococcus pneumoniae	4+					4+				
Streptococcus pyogenes		3+						4+	3+	
Viruses										
Coxsackie B virus	1+	1+				4+		1+		
Cytomegalovirus	2+		3+	2+		1+				2+
Enteroviruses	1+	1+				4+	1+	3+		
Epstein-Barr virus									2+	4+
Hepatitis B virus				4+						
Hepatitis C virus				4+						
Herpes simplex virus					4+	4+		4+		
Human immunodeficiency virus			2+						2+	4+
Influenza virus	4+					1+	1+			
Norovirus			4+							
Parvovirus	2+	3+						4+	2+	
Respiratory syncytial virus	4+									
Varicella-zoster virus	2+					2+		4+		

Protozoa/Fungi/Worms	Resp	MS	GI	H/B	GU	CNS	CV	Skin	Svst	L/H
Aspergillus fumigatus	4+							1+	2+	
Candida albicans	1+		2+		2+		1+	4+*	1+	
Echinococcus spp.	2+			4+						
Giardia intestinalis			4+							
Histoplasma capsulatum	3+					1+			4+	
Leishmania spp.								4+	4+	
Plasmodium spp.			4+							4+
Schistosoma spp.			4+	4+	4+					
Trichophyton spp.								4+		
Toxoplasma gondii						2+				4+
Trypanosoma spp.			4+			4+			4+	
Wuchereria bancrofti										4+

Online Cases

Organism	Resp	MS	GI	H/B	GU	CNS	CV	Skin	Svst	L/H
Bartonella bacilliformis							3+	3+		
Brucella				2+		1+				
Coxiella burnetti	4+						4+			
Dengue virus		2+		1+			2+	3+	2+	3+
Enterococcus faecalis and E. faecium					3+		1+	1+		
Mycobacterium abscessus	3+	2+								
SARS-CoV-2	4+	1+	1+			1+				
Streptococcus mitis						1+	4+			

The rating system (+4 the strongest, +1 the weakest) indicates the greater to lesser involvement of the body system.

KEY:

Resp: Respiratory; MS: Musculoskeletal; GI: Gastrointestinal;

H/B: Hepatobiliary; GU: Genitourinary; CNS: Central Nervous System;

CV: Cardiovascular; Skin: Dermatological; Syst: Systemic; L/H: Lymphatic-Hematological;

*: + mucosa

Digital Resources

There are three digital resources that accompany this book.

- **Eight additional case studies.** So that they can be regularly updated, and to keep the size of the book reasonable, eight case studies are available to students as a digital resource, which can be downloaded at www.routledge.com/cw/lydyard.
  - *Bartonella bacilliformis*
  - *Brucella*
  - *Coxiella burnetti*
  - Dengue virus
  - *Enterococcus faecalis* and *E. faecium*
  - *Mycobacterium abscessus*
  - SARS-CoV-2
  - *Streptococcus mitis*

- **Multiple choice questions.** Each case study has a range of interactive MCQs to test understanding of the case. Answers are also provided enabling the reader to self-test. Students can access these MCQs by visiting www.routledge.com/cw/lydyard.

- **Figure slides** – A library of all the figures from the book for instructors to use in presentations and lectures. Figures are available in two formats: PowerPoint and PDF. Instructors wishing to access this library of figures must register at the Instructor Resources Download Hub: www.routledgetextbooks.com/textbooks/instructor_downloads/.

Aspergillus fumigatus

1

A 68-year-old Caucasian man was diagnosed with B-cell chronic lymphocytic **leukemia** (B-CLL) and received chemo- and immuno-therapy and attended the CLL clinic regularly. Ten years later, the patient presented with **pneumonia** symptoms and was examined by chest **CT scan**. The results were suggestive of aspergillosis and additional laboratory tests were performed. Positive *Aspergillus* serology allowed the doctors in the clinic to give a diagnosis of *A. fumigatus* pneumonia. The patient was not neutropenic and his condition improved following an 8-month course of itraconazole followed by voriconazole for 6 months.

Two years later, the patient was diagnosed with pulmonary aspergillosis. The diagnosis was based on a CT scan, cytology results, and a history of prior infection (Figure 1.1). He was treated with amphotericin B, monitored by radiography, but he died 2 days later. An autopsy was performed and the diagnosis of invasive pulmonary aspergillosis was confirmed.

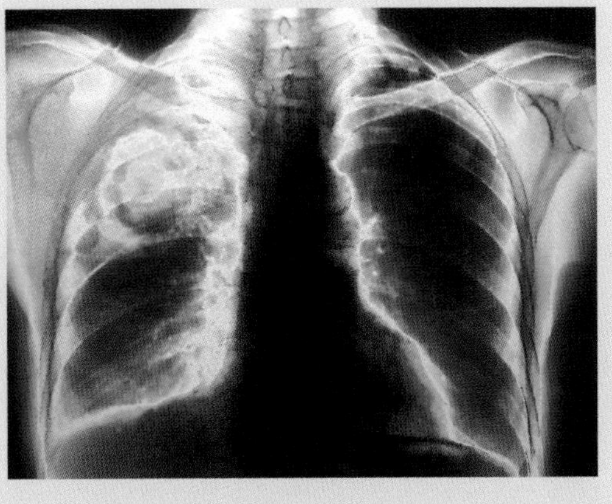

Figure 1.1 Chest X-ray showing that the fungus has invaded the lung tissue. There is a large cavity in the upper left lobe of the lung, with a fungus ball within the cavity. *From Sovereign, ISM / Science Photo Library. With permission.*

1. WHAT IS THE CAUSATIVE AGENT, HOW DOES IT ENTER THE BODY AND HOW DOES IT SPREAD A) WITHIN THE BODY AND B) FROM PERSON TO PERSON?

CAUSATIVE AGENT

Aspergillosis is caused by *Aspergillus*, a saprophytic, filamentous fungus found in soil, decaying vegetation, hay, stored grain, compost piles, mulches, sewage facilities, and bird excreta. It is also found in water storage tanks (for example in hospitals), fire-proofing materials, bedding, pillows, ventilation and air conditioning, and computer fans. It is a frequent contaminant of laboratory media and clinical specimens and can even grow in disinfectants!

Although *Aspergillus* is not the most abundant fungus in the world, it is one of the most ubiquitous. There are more than 100 species of *Aspergillus*. Although about 10 000 genes have been identified in the *Aspergillus* genome, none of the gene sets is shared with other fungal pathogens.

The cell wall of *A. fumigatus* contains various polysaccharides (Figure 1.2). Newly synthesized β(1–3)-glucans are modified and associated to the other cell wall polysaccharides (chitin, galactomannan (GM), and β(1–3)-,

β(1–4)-glucan) leading to the establishment of a rigid cell wall. Glycosyltransferases bound to the membrane by a glycosylphosphatidyl inositol (GPI) anchor play a major role in the biosynthesis of the cell wall. Fungal cell composition affects its **virulence** and susceptibility to immune responses.

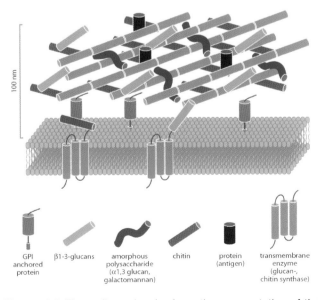

100 nm

| GPI anchored protein | β1-3-glucans | amorphous polysaccharide (α1,3 glucan, galactomannan) | chitin | protein (antigen) | transmembrane enzyme (glucan-, chitin synthase) |

Figure 1.2 Three-dimensional schematic representation of the *Aspergillus fumigatus* cell wall.

Of more than 100 species of *Aspergillus*, only a few are pathogenic, most of all *A. fumigatus*, but other species, including *A. niger*, *A. terreus*, *A. flavus*, *A. clavatus*, *A. nidulans*, and *A. oryzae* might also contribute to pulmonary allergic disorders. However, other than *A. fumigatus*, *Aspergillus* species do not normally grow at temperatures higher than 37°C, and therefore do not cause invasive disease (see Section 3). *A. nidulans* can cause occasional infections in children with **chronic granulomatous disease**.

A. fumigatus is a primary pathogen of man and animals. It is characterized by thermotolerance: ability to grow at temperatures ranging from 15°C to 55°C, it can even survive temperatures of up to 75°C. This is a key feature for *A. fumigatus*, which allows it to grow over other aspergilli species and within the mammalian respiratory system.

A. fumigatus is a fast grower. It reproduces by tiny spores known as conidia, formed on specialized conidiophores. *A. fumigatus* sporulates abundantly, with every conidial head producing thousands of conidia. The conidia released into the atmosphere ranges from 2–5µm in diameter and are small enough to fit in the lung alveoli. The spores are released through disturbances of the environment and air currents and become airborne both indoors and outdoors since their small size makes them buoyant.

A. fumigatus can be identified by the morphology of the conidia and conidiophores: green-blue echinulate conidia are produced in chains basipetally from greenish phialides (Figure 1.3). A few isolates are nonpigmented and produce white conidia. *A. fumigatus* strains with colored conidia indicate the presence of accumulated metabolites and are more **virulent** than nonpigmented strains since they give the fungus an advantage to adapt to its environment.

ENTRY AND SPREAD WITHIN THE BODY

We normally inhale 100–200 *Aspergillus* conidial spores daily, but only susceptible individuals develop a clinical condition. The spores enter the body via the respiratory tract and lodge in the lungs or sinuses. Once inhaled, spores can reach distal areas of the lung due to their small size. Very rarely other sites of primary infection have been described such as the skin, peritoneum, kidneys, bones, eyes, and gastrointestinal (GI) tract, but these are not clinically important. Usually the invasion of other organs by *A. fumigatus* is secondary and follows its spread from the respiratory tract.

When *Aspergillus* conidia enter the human respiratory system at body temperature they swell and develop into a different form, thread-like hyphae, which absorb nutrients required for the growth of the fungus. Some enzymes, particularly proteases, are essential for this fungal pathogen to invade the host tissue. Proteases are involved in the digestion of the lung matrix composed of elastin and collagen. In the case of infection of respiratory tissues, this contributes to the pathogenesis (see Section 3). Together, the hyphae can form a dense mycelium in the lungs. However, in the case of healthy immunocompetent individuals, the spores are prevented

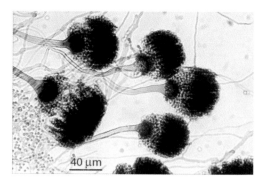

Figure 1.3 Microscopic morphology of *Aspergillus fumigatus* showing typical conidial heads. Conidiophores are short, smooth-walled, and have conical-shaped terminal vesicles, which support a single row of phialides. Conidia are produced in basipetal succession forming long chains. They are green and rough-walled to echinulate. Chains of spores can be seen emerging from phialides surrounding the head. *Courtesy of Dr David Ellis, University of Adelaide, Australia.*

from reaching this stage due to optimal immune responses (see Section 2), and there is some colonization but limited pathology.

Due to the increase in aggressive immunosuppressive therapies and AIDS, the number of immunocompromised patients developing aspergillosis has grown significantly. Thus, severe and often fatal invasive infections with *A. fumigatus* have increased several times in developed countries and it has become the dominant airborne fungal pathogen. Since *Aspergillus* species are essentially ubiquitous in the environment worldwide, no geographic preference of the exposure to airborne conidia or spores has been reported.

PERSON-TO-PERSON SPREAD

This organism is not spread from person to person.

2. WHAT IS THE HOST RESPONSE TO THE INFECTION AND WHAT IS THE DISEASE PATHOGENESIS?

When the fungi colonize the respiratory tract, the clinical manifestation and severity of the disease depend on the efficiency of the immune responses, genetic predisposition and genetic polymorphism. In susceptible hosts, *Aspergillus* conidia germinate to form swollen conidia and then progress to hyphae, its invasive form. The main goal of the immune system is to recognize and kill *Aspergillus* conidia and to prevent its transition to the hyphal form.

MECHANICAL BARRIERS AND INNATE IMMUNITY

In immunocompetent hosts, innate immunity to the inhaled spores begins with the epithelium of the respiratory tract. In larger airways, the epithelial layer contains goblet cells and ciliated cells. Goblet cells produce mucus which traps fungal spores, and ciliated cells move the mucus up the airways where it can be coughed up or swallowed. The majority of the conidia are normally removed from the lungs through the ciliary action. However, *A. fumigatus* can produce toxic metabolites such as gliotoxin, which inhibit ciliary activity, and proteases which can damage the epithelial tissue. The mucociliary function is impaired in individuals with **cystic fibrosis (CF)** or asthma. In addition, increased viscosity of mucus in patients with CF severely affects ciliary movements. In the smaller airways, Type-II pneumocytes secrete surfactant proteins A and D. These act as opsonins that bind *Aspergillus* conidia to be targeted by neutrophils and macrophages.

Because of the site of the infection by *A. fumigatus*, bronchoalveolar macrophages – the resident phagocytic cells of the lung, together with recruited neutrophils – are the major cells involved in the **phagocytosis** of *A. fumigatus*. While macrophages mostly attack conidia, neutrophils are more important for elimination of developing hyphae.

Bronchoalveolar macrophages sense *A. fumigatus* through **pathogen-associated membrane patterns (PAMPs)** such as β-glucan, chitin, and GM, the conidia with their pattern recognition receptors (PRRs) including membrane-associated **Toll-like receptors (TLR)** TLR2 and TLR4. Endosomal TLR3 and TLR9 are also involved, hence their agonists are considered as adjuvants for Allergic Bronchopulmonary Aspergillosis (ABPA) therapy (see later).

Inhaled conidia via GM bind some soluble receptors such as pentraxin-3 and lung surfactant protein D. This enhances phagocytosis and inflammatory responses. Phagosomes containing conidia fuse with endosomes followed by activation of **NADPH oxidase-dependent killing** and production of **reactive oxygen species (ROS)**. Nonoxidative mechanisms are also essential for the digestion of phagocytosed conidia by macrophages. Swelling of the conidia inside the macrophage or in the bronchoalveolar space appears to be a prerequisite for fungal killing. Conidial swelling alters cell-wall composition and exposes fungal β-glucan. This further triggers fungicidal responses via mammalian β-glucan receptor dectin-1.

In the cases when resident bronchoalveolar macrophages fail to control the fungus, conidia germinate into hyphae. Neutrophils and monocytes are then recruited from the circulation to phagocytose and kill hyphae.

Neutrophils adhere to the surface of the hyphae, since hyphae are too large to be engulfed. They are often seen clustered around fungal hyphae. This triggers a respiratory burst, secretion of ROS, release of lysozyme, neutrophil cationic peptides, and degranulation of neutrophils. NADPH oxidase-independent killing through the release of lactoferrin, an iron-sequestering molecule, is important. In contrast to the slow and subefficient killing of conidia by macrophages, hyphal damage by neutrophils is rapid, possibly through a release of fungal cell-wall glycoproteins with the help of polysaccharide hydrolases. **Defensins** may also play a role in responses to *A. fumigatus* hyphae. The importance of neutrophils in protection against *Aspergillus* is illustrated by a development of invasive aspergillosis (IA) in immunodeficient patients with chemotherapy-induced **neutropenia**. Corticosteroid-based treatment, purine analogs (fludarabine) and some monoclonal antibody treatment (Campath 1H – anti-CD52) and immunosuppression lead to neutropenia and/or neutrophil dysfunction. **Corticosteroids** reduce the oxidative burst and superoxide anion release by neutrophils, thereby inhibiting hyphal killing.

Platelets also play some role in protection against *Aspergillus*, by attaching to the cell walls of invasive hyphae and becoming activated, thus enhancing direct cell-wall damage of *A. fumigatus* and neutrophil-mediated fungicidal effect. Hence **thrombocytopenia**, which is associated with prolonged neutropenia during chemotherapy, increases the risk of infection by *A. fumigatus*.

Invasion with *A. fumigatus* enhances the levels of serum fibrinogen, C-reactive protein, and other **acute-phase proteins**. Resting conidia activate the **alternative complement pathway** and induce neutrophil chemotaxis and deposition of complement components on the fungal surface. Antigen-presenting **dendritic cells (DC)** exposed to hyphae in the lung migrate to the spleen and draining lymph nodes where they launch peripheral T helper (Th)-cell responses.

ADAPTIVE IMMUNITY

T-Cell Responses

T-cell responses against *A. fumigatus* are mostly confined to the **CD4+ T cells**. To a certain extent, their efficiency resides in the ability of **Th1** cells to further enhance neutrophil-mediated killing. A Th1 response, associated with a strong cellular immune component and increased levels of **IFN-γ**, granulocyte and granulocyte-macrophage colony-stimulating factors (**G-CSF** and **GM-CSF**), tumor necrosis factor-alpha (**TNF-α**), **interleukin-1 (IL-1)**, IL-6, IL-12, and IL-18, provides resistance to mycotic disease. Th1 pro-inflammatory signals recruit neutrophils into sites of infection. TNF-α enhances the capacity of neutrophils to damage hyphae; G-CSF, GM-CSF, and especially IFN-γ enhance monocyte and neutrophil activity against hyphae; while IL-15 enhances hyphal damage and IL-8 release by neutrophils. IL-8 recruits more neutrophils to sites of inflammation and mediates release of **antimicrobial peptides**.

On the other hand, a **Th2** response, which is associated with a minimal cellular component and an increase in antibody production, and secretion of IL-4, IL-5, and IL-10, appears to facilitate fungal invasion. Production of IL-4 by CD4+ T lymphocytes impairs neutrophil antifungal activity and IL-10 suppresses oxidative burst. Pathogenicity of ABPA, (see Section 3) is associated with a pulmonary **eosinophilia** – the result of production of Th2 **cytokines**. *Aspergillus*-specific CD4+ T-cell clones isolated from ABPA patients have a Th2 phenotype.

It appears that a balance between beneficial Th1 and damaging Th2 types of immune responses is dependent upon the nature of antigens that prime DCs. Exposure of conidia to DCs leads to the activation of Th1 CD4+ T cells, while priming of DCs with hyphae enhances Th2-mediated pathways of CD4+ T-cell responses. **Regulatory T cells** may also be involved in determination of the Th1/Th2 bias.

A role of **CD8+ T cells** in the resistance to *A. fumigatus* was found to be very limited since fungal gliotoxin suppresses granule **exocytosis**-associated cellular cytotoxicity.

Antibody Responses

Humoral immunity to *Aspergillus* species is poorly characterized. Although even in severely immunocompromised patients the production of specific antibodies has been described, their protective role, if any, remains unclear. The **antibody isotypes** produced are **IgG1**, IgG2, and IgA (particularly in bronchial lavage). Immune serum did not enhance phagocytosis of conidia *in vitro* but did induce macrophage-mediated killing. Neutralizing antibodies to proteases or toxins may also be beneficial to the host. Serum antibodies to *A. fumigatus* are often found in the absence of disease as a result of environmental exposure.

Serum samples from patients with ABPA (see Section 3) contain elevated levels of antigen-specific circulating antibodies, mainly of IgG and **IgE** isotypes, which participate in pathogenesis of ABPA. B cells secrete IgE spontaneously as a result of IL-4 production, while IL-5 recruits eosinophils. Eosinophilic infiltration and basophil and mast-cell degranulation in response to *A. fumigatus* antigens and IgE complex releases **pro-inflammatory mediators**. This leads to further chemotaxis of eosinophils and activated CD4+ lymphocytes to the site of the infection.

Patients with **aspergilloma** (see Section 3), particularly those who recover from **granulocytopenia**, have increased levels of specific IgG and **IgM**, mostly against fungal carbohydrates and glycoproteins.

Generally, the efficiency of host immune responses to *A. fumigatus* is a result of a dynamic interaction between fungal cell wall components and immune cells. Redundancy of host defense mechanisms may lead to the tissue-damaging inflammation favoring the invasive potential of the fungal cells and development of aspergillosis.

Pathogenesis

The pathogenicity of *A. fumigatus* is usually based on (a) a direct attack to the host cells with the pathogen's virulent factors and toxins; (b) evasion of the immune responses; (c) the hypersensitivity of innate and adaptive responses of patients.

Toxins produced by *A. fumigatus* often represent secondary metabolites of fungi. They can affect the synthesis of DNA, RNA, and proteins, impair cellular functions of the host. Inhibiting phagocytic function of innate cells is critical for the survival of the fungus. Gliotoxin is the most potent toxin produced by *A. fumigatus* which can suppress T-cell responses, inhibit phagocytosis by macrophages and, most importantly – neutrophils by inhibiting NADPH and ROS production.

A. fumigatus is often able to block phagocytosis by producing oxidoreductases and hydrophobic pigments – melanins such

as conidial dihydroxynaphthalene-melanin. Melanins are expressed on the conidial surface and protect the pathogen by quenching ROS.

Another strategy of evading innate immune responses by *A. fumigatus* is production of a specific lipophilic inhibitor of the alternative complement pathway which enhances proteolytic cleavage of C3 complement component bound to conidia.

Some of the host defense – such as recently discovered innate lymphoid cells (ILCs), might be involved in pathogenesis of ABPA (see Section 3) by enhancing **pulmonary eosinophilia** and mucus hypersecretion through the release of Th2 cytokines IL-5 and IL-13. ILC inhibition might reduce *Aspergillus*-induced inflammation.

In patients with ABPA, immediate skin reactions are mediated mainly by **type I hypersensitivity** and IgE antibody. The late reaction (**Arthus reaction**) to *Aspergillus* antigens is the result of IgE-mediated mast-cell activation or immune complex formation (**type III hypersensitivity**) with complement activation. **Immune complexes** of specific IgG and *A. fumigatus* antigens trigger the generation of leukotriene C4 by mast cells which, in turn, promotes mucus production, bronchial constriction, **hyperemia**, and **edema**. **Granuloma** formation in the lung has also been reported since some patients have granulomatous **bronchiolitis**.

3. WHAT IS THE TYPICAL CLINICAL PRESENTATION AND WHAT COMPLICATIONS CAN OCCUR?

The spectrum of pulmonary diseases caused by *A. fumigatus* is grouped under the name of aspergillosis. These conditions vary in the severity of the course, pathology, and outcome and can be classified according to the site of the disease within the respiratory tract, the extent of fungal invasion or colonization, and the immunological competence of the host. There are four main clinical types of pulmonary aspergillosis: ABPA, chronic necrotizing *Aspergillus* pneumonia (CNPA), IA, and pulmonary aspergilloma (Figure 1.4).

ALLERGIC BRONCHOPULMONARY ASPERGILLOSIS (ABPA)

ABPA develops as a result of a Type-I hypersensitivity reaction to *A. fumigatus* colonization of the tracheobronchial tree. Estimating the frequency of ABPA is difficult due to the lack of standard diagnostic criteria (see Section 4). It often appears not as a primary pathology, but as a complication of other chronic lung diseases such as **atopic asthma**, **CF**, and **sinusitis**. Depending on definition criteria, ABPA occurs in approximately 2–19% of CF patients. The estimated prevalence of ABPA in asthmatic patients varies between 0.7% and 22%, and is likely to affect 4.8 million patients globally, while its prevalence in severe asthma is as high as 70%.

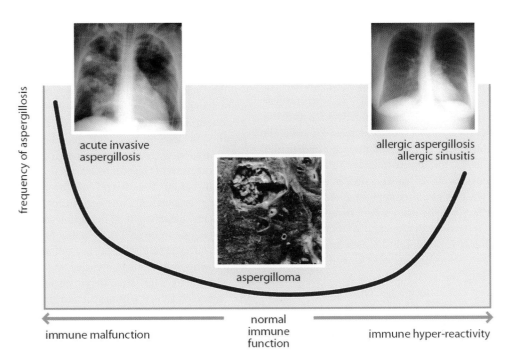

Figure 1.4 Relative risk of Aspergillus infection. The two chest X-rays show examples of acute invasive and allergic pulmonary aspergillosis. The fungal ball (aspergilloma) that was removed from a lung and measures about 6cm in diameter is also shown. Hypersensitivity accompanies development of allergic aspergillosis, immunodeficiency leads to invasive aspergillosis, while aspergilloma can be observed in immunocompetent individuals. *From the Aspergillus Website, https://www.aspergillus.org.uk and Dr Jennifer Bartholomew, School of Medicine, University of Manchester. With permission.*

The clinical course often follows as classic asthma but can also lead to a fatal destruction of the lungs. IgE- and IgG-mediated Type I hypersensitivity and Type III hypersensitivity are the leading causes of pathology (see Section 2).

Since ABPA presents as a bronchial asthma, the symptoms are similar to poorly controlled asthma and include wheezing, cough, fever, malaise, hemoptysis and weight loss. Additional symptoms include recurrent pneumonia, release of brownish mucoid plugs with fungal hyphae, and recurrent lung obstruction. In the case of secondary ABPA, unexplained worsening of asthma and CF is observed. It is essential to diagnose and treat ABPA at the onset of the disease, which can be traced to early childhood or even infancy. ABPA should be suspected in children with a history of recurrent wheezing and pulmonary infiltrates.

The outcome of the disease depends on asthma control, presence of widespread **bronchiectasis**, and resultant chronic fibrosis of the lungs (Figure 1.5). Respiratory failure and fatalities can occur in patients in the third or fourth decade of life. The long-term prognosis of ABPA beyond 5 years is currently not well defined.

CHRONIC NECROTIZING PULMONARY ASPERGILLOSIS (CNPA)

CNPA is a subacute condition mostly developing in mildly immunocompromised patients and is commonly associated with underlying lung disease such as steroid-dependent **chronic obstructive pulmonary disease (COPD)**, interstitial lung disease, previous thoracic surgery, chronic corticosteroid therapy or alcoholism.

CNPA presents as a subacute pneumonia unresponsive to antibiotic therapy, which progresses and results in cavity formations over weeks or months. Symptoms include fever, cough, night sweats, and weight loss. Because it is uncommon, CNPA often remains unrecognized for weeks or months and causes a progressive cavitatory pulmonary infiltrate. It is often found at autopsy. The reported mortality rate for CNPA is 10–40% or higher if it remains undiagnosed.

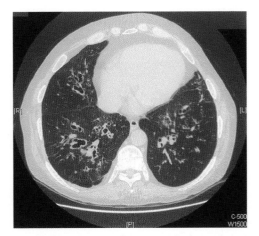

Figure 1.5 High-resolution CT scan of chest demonstrating remarkable bronchial wall thickening in the context of longstanding ABPA. *From the Aspergillus Website, https://www.aspergillus.org.uk and Dr Jennifer Bartholomew, School of Medicine, University of Manchester. With permission.*

INVASIVE ASPERGILLOSIS (IA)

Exposure to *A. fumigatus* in immunocompromised individuals can lead to IA, which is the most serious, life-threatening condition. Due to the development of immunosuppression in transplantation and anticancer chemotherapy leading to severe immunodeficiency and the AIDS pandemic, the incidence of IA has increased. Leukemia or bone marrow transplant (BMT) patients are at particular risk. IA is responsible for approximately 30% of fungal infections in patients dying of cancer, and it is estimated that IA occurs in 10–25% of all leukemia patients, in whom the mortality rate is 80–90%, even when treated. It occurs in 5–10% of cases following allogeneic BMT and in 0.5–5% after **cytotoxic therapy** of blood diseases or autologous BMT. In solid-organ transplantation, IA is diagnosed in 19–26% of heart-lung transplant patients and in 1–10% of liver, heart, lung, and kidney recipients. Other patients at risk include those with chronic granulomatous disease (25–40%), neutropenic patients with leukemia (5–25%), and patients with AIDS, **multiple myeloma**, and **severe combined immunodeficiency** (about 4%). Drugs such as antimicrobial agents and steroids can predispose the patient to colonization with *A. fumigatus* and invasive disease.

Four types of IA have been described. Clinical symptoms of the different types of IA depend on the organ localization and the underlying disease.

- Acute or chronic pulmonary aspergillosis (lungs).

- Tracheobronchitis and obstructive bronchial disease (bronchial mucosa and cartilage).

- Acute invasive rhinosinusitis (sinuses).

- Disseminated disease (brain, skin, kidneys, heart, eyes).

IA starts with pneumonia, and then the fungus usually disseminates to various organs causing **endocarditis**, **osteomyelitis**, **otomycosis**, **meningitis**, vision obstruction, and cutaneous infection (Figure 1.6). *Aspergillus*-related endocarditis and wound infections may occur through cardiac surgery. In the developing world, infection with Aspergillus can cause **keratitis** – a unilateral blindness.

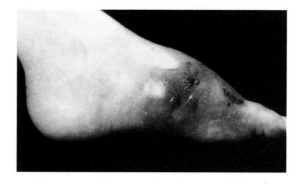

Figure 1.6 Bone infection caused by invasive aspergillosis. *From the Aspergillus Website, http://www.aspergillus.org.uk and Dr Jennifer Bartholomew, School of Medicine, University of Manchester. With permission.*

Symptoms are usually variable and nonspecific: fever and chills, weakness, unexplained weight loss, chest pain, **dyspnea**, headaches, bone pain, a heart murmur, decreased **diuresis**, blood in the urine or abnormal urine color, and straight, narrow red lines of broken blood vessels under the nails. Patients develop **tachypnea** and progressive worsening **hypoxemia**. IA is accompanied by increased sputum production (sometimes with blood), sinusitis, and acute inflammation with areas of **ischemic necrosis**, **thrombosis**, and **infarction** of the organs involved.

Even before the arrival of the COVID-19 pandemic, a strong association between influenza infection and IA had been established. The latter contributed to further complications in up to 23% of patients with severe influenza. A high variability in clinical presentation was seen, ranging from tracheobronchitis to invasive and angioinvasive disease.

With the COVID-19 pandemic throughout 2020, fewer than 200 cases of COVID-19-associated pulmonary aspergillosis (CAPA) were reported worldwide, mostly associated with underlying pulmonary disease, and characterized by high mortality. If not treated in a timely manner, these cases quickly developed into invasive forms. It must also be noted that patients with SARS-CoV-2 often develop other fungal infections, particularly candidemia following a prolonged stay in the ICU.

ASPERGILLOMA

An aspergilloma, also known as a mycetoma or fungus ball, is a clump of fungus which populates a lung cavity. It occurs in 10–15% of patients with preexisting lung cavities due to conditions such as tuberculosis, CF, lung abscess, **sarcoidosis**, emphysematous bullae, and chronically obstructed paranasal sinuses. Although *Aspergillus* species are the most common, some *Zygomycetes* and *Fusarium* may also form mycetomas. In patients with AIDS, aspergilloma may occur in cystic areas resulting from prior *Pneumocystis jiroveci* pneumonia infection. The fungus invades, settles, and multiplies in a cavity mostly outside the reach of the immune system. The growth results in the formation of a ball shaped like a half-moon (crescent). It consists of a mass of hyphae surrounded by a proteinaceous matrix, which incorporates dead tissue and mucus with sporulating structures at the periphery. Some cavities may contain multiple aspergilloma (Figure 1.7).

Patients with aspergilloma do not manifest many related symptoms, and the condition may go on for many years undiagnosed, often discovered incidentally by chest X-ray or by CT scans. The most common, but still rare, symptom is **hemoptysis**. The patients may cough up the fungus elements, and sometimes chains of conidia can be seen in the sputum. Aspergilloma can lead to **pleural thickening**.

Rarely, in immunocompromised individuals, aspergillomas can be formed in other body cavities. They may cause abscesses in the brain, or populate various face sinuses, ear canals, kidneys, urinary tracts, and even heart valves. Secondary aspergillomas may occur as a result of IA when

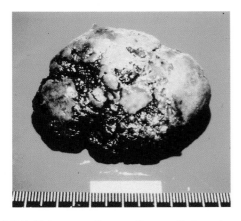

Figure 1.7 Multiple aspergillomas. Gross pathology showing three fungus balls in one cavity. *From the Aspergillus Website, http://www. aspergillus.org.uk and Dr Jennifer Bartholomew, School of Medicine, University of Manchester. With permission.*

a solid lesion of IA erodes to the surface of the lung. These lesions can be detected by a chest CT scan and must be taken into account when further immunosuppressive therapy for relapsed IA is prescribed.

4. HOW IS THE DISEASE DIAGNOSED AND WHAT IS THE DIFFERENTIAL DIAGNOSIS?

DIAGNOSIS OF ABPA

ABPA is a particularly difficult syndrome to diagnose since the symptoms are nonspecific. The current diagnostic approaches are known as the Rosenberg–Patterson criteria. The disease presents with bronchial asthma with transient pulmonary infiltrates and, at later stages, proximal bronchiectasis and lung **fibrosis**. Chest radiography is also not specific and shows various transient abnormalities: consolidation or collapse, thickened bronchial wall, and/or peripheral shadows. The following criteria are used for diagnosis of ABPA: asthma, a history of pulmonary infiltrates, and central bronchiectasis.

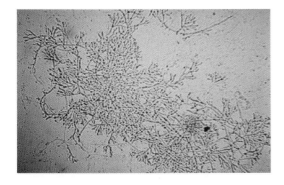

Figure 1.8 Microscopy of sputum. A typical example of a wet mount of a sputum sample from a patient with ABPA. *From the Aspergillus Website, http://www.aspergillus.org.uk and Dr Jennifer Bartholomew, School of Medicine, University of Manchester. With permission.*

The radiological diagnosis is supplemented by laboratory tests: peripheral blood eosinophilia (>10% or 1000 mm^{-3}), immediate skin reactivity to *A. fumigatus* antigenic extracts within 15 ± 5 min, detection of precipitating IgG and IgM antibodies in >90% of cases, and elevated levels of total IgE in serum (>1000 ng/ml). Specific IgE antibodies against *A. fumigatus* are also normally measured.

Isolation of *A. fumigatus* from sputum (Figure 1.8), expectoration of brown plugs containing eosinophils and **Charcot–Leyden crystals**, and a skin reaction occurring 6 ± 2 h after the application of antigen are used as a complementary diagnosis.

DIAGNOSIS OF IA

IA in the early stages is also difficult to diagnose. A safe diagnosis can only be made at autopsy with the histopathological evidence of mycelial growth in tissue. Differential diagnosis from the invasion of hyphae of other filamentous fungi such as *Fusarium* or *Pseudallescheria* is often difficult and requires **immunohistochemical** staining or *in situ* **hybridization** techniques. Clinical symptoms are usually nonspecific and require further laboratory tests.

Criteria currently used for the diagnosis of IA are: a positive CT scan (see Figure 1.1), culture and/or microscopic evaluation, and the detection of *Aspergillus* antigens in the serum.

A CT scan can demonstrate single or multifocal nodules, with and without cavitation, or widespread, often bilateral, infiltrates. However, the appearances are heterogeneous throughout the course of the disease, with the most specific being at the early stages and presenting a "halo" of hemorrhagic necrosis surrounding the fungal lesion or pleura-based lesions. In non-pulmonary forms of the disease such as cerebral aspergillosis, a CT scan, together with brain **magnetic resonance imaging** (**MRI**), can detect the extent of the disease and the bone invasion.

The diagnostic value of the microscopic examination of sputum is limited due to the presence of airborne conidia of *Aspergillus* and the possibility of accidental contamination. However, in neutropenic or BMT patients, the predictive value of a sputum culture positive for *A. fumigatus* exceeds 70%. The presence of *A. fumigatus* in **bronchoalveolar lavage** fluid (BAL) samples from patients with leukemia and BMT is found in 50–100% of those who have definitive or probable aspergillosis. Nasal swabs of patients also have diagnostic value, although **bronchoscopy** is preferable due to the sterility of the clinical sample. For the same reason, percutaneous lung biopsy or aspirated material are the specimens of choice. However, invasive procedures in immunocompromised patients require careful consideration.

CELL CULTURE

After the microscopic identification of *A. fumigatus*, cell culture may be critical in supporting the diagnosis of aspergillosis. The specimen is usually inoculated onto a plate with Sabouraud glucose agar, inhibitory mold agar (IMA) or other appropriate medium with antibiotics – gentamicin or chloramphenicol. The plates are incubated at 30°C for up to 6 weeks with the cultures being examined at 3-day intervals.

Antigen Detection

A highly specific (99.6%) and sensitive (1 ng ml–1) test for detection of *Aspergillus* GM has been developed for screening and early diagnosis of IA in serum, BAL, and cerebrospinal fluid (CSF). GM is a part of the *Aspergillus* cell wall (see Section 1), and can often be released into the patient's bodily fluids. The detection of *A. fumigatus* GM by **Enzyme-Linked Immunosorbent Assay** (**ELISA**) becomes possible at an early stage of infection thus allowing timely initiation of therapy. In 65.2% of patients, GM can be detected in serum 5–8 days before the development of IA symptoms. In addition, ELISA can be used for monitoring the disease treatment. Positive results in two consecutive serum samples allows the diagnosis of IA. However, in some cases, false positive reactions can be observed due to cross-reactivity with nonspecific antigens derived from other fungi such as *Rhodotorula rubra*, *Paecilomyces varioti*, *Penicillium chrysogenum* and *P. digitatum*.

DIAGNOSIS OF ASPERGILLOMA

A definitive diagnosis of aspergilloma requires bronchoscopy, lung biopsy or resection, but this is rare. The diagnosis is usually made accidentally or specifically by chest radiography. A pulmonary aspergilloma appears as a solid ball of water density, sometimes mobile, within a spherical or ovoid cavity. It is separated from the wall of the cavity by the air space. Pleural thickening is also characteristic. A chest CT scan can sometimes detect aspergilloma with a negative chest

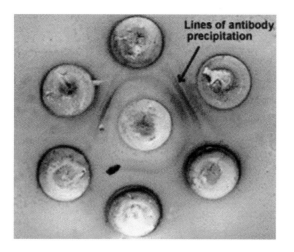

Figure 1.9 Double diffusion test for aspergillosis. The central well contains *A. fumigatus* antigen and wells at the top and bottom contain control antiserum. The three peripheral wells with precipitin bands contain sera from patients with *A. fumigatus* aspergilloma. More bands present in the upper right case are characteristic of aspergilloma. The well in the bottom left position is negative. *From the Aspergillus Website, http://www.aspergillus.org.uk and Dr Jennifer Bartholomew, School of Medicine, University of Manchester. With permission.*

radiograph. The radiographic picture must be differentiated from other conditions such as cavitating neoplasm, blood clot, disintegrating **hydatid cyst**, and pulmonary **abscess** with necrosis.

Clinical analysis should be coupled with serologic tests since a number of other fungi such as *Candida*, *Torulopsis*, *Petriellidium*, *Sporotrichum*, and *Streptomyces* can lead to the development of mycetoma. Since aspergilloma is a condition often observed in immunocompetent individuals, a laboratory diagnosis based on a humoral response is feasible. The most commonly used methods in clinical diagnosis are double immunodiffusion (Figure 1.9), immunoprecipitation, and counter-immunoelectrophoresis because they are simple, cheap, and easy to perform. It must be noted that patients undergoing corticosteroid treatment may become seronegative.

For the **polymerase chain reaction** (**PCR**) analysis the most reliable and well-characterized antigens of *A. fumigatus* are RNase, catalase, dipeptidylpeptidase V, and the GM.

5. HOW IS THE DISEASE MANAGED AND PREVENTED?

ABPA

Although ABPA is a chronic condition, acute corticosteroid-responsive asthma can occur and lead to fibrotic end-stage lung disease. The aim of treatment of ABPA is to suppress the immune reaction to the fungus and to control bronchospasm. For this, high doses of oral corticosteroids are used: 30–45 mg/day of prednisolone or prednisone in acute phase and a lower maintenance dose of 5–10 mg/day. Oral triazole antifungal drugs are also effective against *A. fumigatus*. These include first-generation drugs such as itraconazole and fluconazole and second-generation antifungals – voriconazole, posaconazole, and isavuconazole.

The administration of medications can be combined with removal of mucus plugs by bronchoscopic aspiration. Regular monitoring by X-rays, pulmonary function tests, and serum IgE levels are essential. Successive control of ABPA leads to a drop in IgE levels, while their increase indicates relapse.

Novel treatments include immunotherapy, although these have not been studied in large-scale randomized trials. The following antibodies have been proposed: anti-IgE Omalizumab, Ligelizumab and Quilizumab; anti-IL-5 Mepolizumab and Resilizumab; anti-IL-5 receptor Benralizumab, and others. TLR agonists are also considered as adjuvants in allergen immunotherapy. Currently, the development of nanoparticle-based therapies is underway.

INVASIVE ASPERGILLOSIS

It is difficult to achieve a timely diagnosis of IA due to the rapid progression (1–2 weeks from onset to death) and waiting for the confirmation of diagnosis would put the patients at a greater risk of untreatable fungal burden. The decision to start antifungal treatment has therefore to be empirical and based more on the presence of risk factors such as long-term (12–15 days) severe neutropenia (< 100–500 mm^{-3}).

The antifungal regimen for the treatment of IA includes voriconazole, amphotericin B (deoxycholate and lipid preparations), and itraconazole. Voriconazole is particularly effective against IA and in reducing mortality. In addition, voriconazole, itraconazole, and amphotericin B exhibit a broad-spectrum activity against *Aspergillus* and the related hyaline molds. In the early treatment of CAPA, the most effect was achieved with isavuconazole demonstrating broad-spectrum, predictable pharmacokinetics, high dose in lungs, few side-effects and few interactions with other drugs.

Amphotericin B (AmB) has been used in antifungal therapy for more than 30 years, mostly under the name of Fungizone®, and remains a first-line drug, although the overall success rate of AmB therapy for IA is only 34%. Despite some progress in antifungal therapy, mortality rates from IA remain very high: 86%, 66%, and 99% for pulmonary, sinus, and cerebral aspergillosis, respectively. CAPA cases were associated with high mortality. Few patients with severe persistent neutropenia survive, and neutrophil recovery is usually followed by resolution of aspergillosis. The duration of neutropenia is therefore an indication of a possible outcome of IA.

ASPERGILLOMA

Aspergilloma can be prevented by timely and effective management of diseases that increase the risk of its development, such as tuberculosis. Complication of aspergilloma by severe hemoptysis is an indication for surgery, although it is associated with high risks of morbidity and mortality. Surgical resection of aspergilloma is one of the most complex procedures in thoracic surgery, since prolonged chronic infection and inflammation lead to thickened fibrotic tissue, induration of the hilar structures, and obliteration of the pleural space. Postoperative complications include hemorrhage, bronchopleural fistula, a residual pleural space, and **empyema**. Surgery is restricted to patients with severe hemoptysis and adequate pulmonary function. On the other hand, in the absence of surgical intervention, the course of the disease remains unpredictable.

An alternative treatment in patients with severe pulmonary dysfunction comprises topical therapy. This includes intracavitary instillation of an antifungal drug such as AmB, percutaneous injection into aspergilloma cavities of sodium and potassium bromide, and endobronchial instillation of ketoconazole via fiberoptic bronchoscopy. However, topical therapy is labor-intensive and nonapplicable for cases with multiple aspergilloma.

SUMMARY

1. WHAT IS THE CAUSATIVE AGENT, HOW DOES IT ENTER THE BODY, AND HOW DOES IT SPREAD A) WITHIN THE BODY AND B) FROM PERSON TO PERSON?

- *Aspergillus* is a saprophytic fungus, and one of the most ubiquitous. Its natural habitat is soil, but it is also present around construction sites and indoors in water storage tanks, fire-proofing materials, bedding, and ventilation and air-conditioning systems.
- The cell wall of *A. fumigatus* contains various polysaccharides including a galactomannan core.
- Of more than 100 species of Aspergillus, only a few are pathogenic: *A. fumigatus, A. flavus, A. terreus, A. clavatus,* and *A. niger.*
- *A. fumigatus* is a primary pathogen of man and animals and causes all manifestations of aspergillosis. It is thermotolerant and grows at temperatures ranging from 15°C to 55°C; it can even survive temperatures of up to 75°C. This is a key feature that allows it to grow over other aspergilli species and within the mammalian respiratory system.
- *A. fumigatus* sporulates abundantly, with every conidial head producing thousands of conidia. The conidia released into the atmosphere range from 2.5–3.0 μm in diameter. We normally inhale 100–200 spores daily, but only susceptible individuals develop a clinical condition. The spores enter the body via the respiratory tract and lodge in the lungs or sinuses.
- Once inhaled, spores at body temperature develop into a different form – thread-like hyphae, which invade the host tissue. Together, the hyphae can form a dense mycelium in lungs. However, in the case of healthy immunocompetent individuals, the spores are prevented from reaching this stage due to the optimal immune responses, and there is some colonization but limited pathology.

2. WHAT IS THE HOST RESPONSE TO THE INFECTION AND WHAT IS THE DISEASE PATHOGENESIS?

- In immunocompetent hosts, fungal conidia may be cleared by ciliated epithelium of the terminal bronchioles and ingested by tissue macrophages or alveolar macrophages. While macrophages mostly attack conidia, neutrophils are more important for elimination of the next, hyphal form of the fungus.
- Neutrophils adhere to the surface of the hyphae and trigger a respiratory burst, secretion of NADPH and reactive oxygen species (ROS), release of lysozyme, neutrophil cationic peptides, and lactoferrin.
- In immunodeficient, particularly IA patients, corticosteroid-based treatment, purine analogs (fludarabine) and some monoclonal antibody treatment lead to neutropenia and/or neutrophil dysfunction. Corticosteroids reduce oxidative burst and superoxide anion release by neutrophils, thereby inhibiting hyphal killing.
- Th1 cytokines are important in neutrophil-mediated killing of hyphae. The Th2 response, which is associated with an increase in antibody production, seems to facilitate fungal invasion rather than protect.

Continued...

....continued

- ABPA pathogenesis is associated with elevated levels of antigen-specific circulating IgG and IgE. B cells secrete IgE spontaneously as a result of IL-4 production, while IL-5 recruits eosinophils. Development of Types I and II hypersensitivity leads to inflammation and tissue damage.
- Patients with aspergilloma have increased levels of specific IgG and IgM, mostly against Aspergillus carbohydrates and glycoproteins. The pathogenicity of *A. fumigatus* is usually based on (a) a direct attack to the host cells with the pathogen's virulent factors and toxins; (b) evasion of the immune responses; (c) the hypersensitivity of innate and adaptive responses of patients.

3. WHAT IS THE TYPICAL CLINICAL PRESENTATION AND WHAT COMPLICATIONS CAN OCCUR?

- Pulmonary diseases caused by *A. fumigatus* are classified into four main clinical types: allergic bronchopulmonary aspergillosis, chronic necrotizing Aspergillus pneumonia, invasive aspergillosis, and pulmonary aspergilloma.
- ABPA is the result of a hypersensitivity reaction to *A. fumigatus* colonization of the tracheobronchial tree. It often appears as a complication of other chronic lung diseases such as atopic asthma (0.5–2%), cystic fibrosis (7–35%), and sinusitis. ABPA occurs in up to 15% of asthmatic patients sensitized to *A. fumigatus*.
- ABPA symptoms are similar to asthma and include wheezing, cough, fever, malaise, and weight loss. Additional symptoms include recurrent pneumonia, release of brownish mucoid plugs with fungal hyphae, and recurrent lung obstruction. In secondary ABPA, there is unexplained worsening of asthma and cystic fibrosis.
- Chronic obstructive pulmonary disease (COPD) presents as a subacute pneumonia that is unresponsive to antibiotic therapy. Symptoms include fever, cough, night sweats, and weight loss. It develops mostly in mildly immunocompromised patients and is commonly associated with underlying lung disease.
- Invasive aspergillosis (IA) is the most serious form of aspergillosis and normally occurs in immunocompromised individuals. There are four types of IA: acute or chronic pulmonary aspergillosis (lungs); tracheobronchitis and obstructive bronchial disease (bronchial mucosa and cartilage); acute invasive rhinosinusitis (sinuses); and disseminated disease (brain, skin, kidneys, heart, eyes).
- IA symptoms are variable and nonspecific: fever and chills, weakness, unexplained weight loss, chest pain, shortness of breath, headaches, bone pain, a heart murmur, blood in the urine, decreased diuresis, and straight, narrow red lines of broken blood vessels under the nails. IA is accompanied by increased sputum production (sometimes with hemoptysis), sinusitis, and acute inflammation with ischemic necrosis, thrombosis, and infarction of the organs.
- IA often accompanies severe influenza-based pulmonary distress. Since the beginning of the COVID-19 pandemic, COVID-associated pulmonary aspergillosis has been identified which quickly develops into invasive forms and requires early treatment.
- Aspergilloma occurs in 10–15% of patients with cavitating lung diseases such as tuberculosis, sarcoidosis, lung abscess, emphysematous bullae, cystic fibrosis, and paranasal sinuses. It is often discovered incidentally by chest X-ray or by computed tomography (CT) scans. The most common symptom is hemoptysis.

4. HOW IS THE DISEASE DIAGNOSED, AND WHAT IS THE DIFFERENTIAL DIAGNOSIS?

- ABPA diagnostic criteria include: asthma, a history of pulmonary infiltrates, and central bronchiectasis supplemented by the laboratory tests: peripheral blood eosinophilia, immediate skin reactivity to *A. fumigatus*, precipitating IgG and IgM, elevated levels of total IgE in serum and specific IgE against *A. fumigatus*. Diagnosis of IA is based on a positive CT scan, cell culture and/or microscopic evaluation, and detection of Aspergillus antigens in serum.
- Since IA is life-threatening, early diagnosis using ELISA and PCR-based techniques is essential. Reliable antigens of *A. fumigatus* include RNase, catalase, dipeptidylpeptidase V, and the GM. GM detection by ELISA has become the basis for the most popular *A. fumigatus* diagnostic tests.
- A definitive diagnosis of aspergilloma requires bronchoscopy, lung biopsy or resection. The diagnosis is usually made accidentally or specifically by chest radiography. Clinical analysis should be coupled with serologic tests.

5. HOW IS THE DISEASE MANAGED AND PREVENTED?

- The aim of the treatment of ABPA is to suppress the immune reaction to the fungus and to control bronchospasm. High doses of oral corticosteroids are used. Antifungal drugs such as itraconazole are sometimes used. Regular monitoring by X-rays, pulmonary function tests, and serum IgE levels is essential.
- Since it is difficult to achieve timely diagnosis of IA, the decision to start antifungal treatment has to be empirical and based on the presence of risk factors, particularly prolonged neutropenia. Antifungal treatments include voriconazole (particularly effective), amphotericin B (AmB), and itraconazole, which exhibit a broad-spectrum activity against Aspergillus and the related hyaline molds. Itraconazole is prescribed to patients with AmB-induced nephrotoxicity. Isavuconazole appears to be advantageous for the CAPA treatment.
- Aspergilloma can be prevented by the treatment of diseases that increase its risk, such as tuberculosis. In the absence of surgical intervention, the course of the disease remains unpredictable. Itraconazole for 6–18 months is recommended for the oral treatment.

FURTHER READING

Barnes PD, Marr KA. Aspergillosis: spectrum of disease, diagnosis and treatment. Infect Dis Clin N Am, 20: 545–561, 2006.

Thompson GRA, Young J-AH. Aspergillus Infections. NEJM, 385: 1496–1509, 2021.

REFERENCES

Alastruey-Izquierdoa A, Cadranel J, Flick H, et al. Treatment of Chronic Pulmonary Aspergillosis: Current Standards and Future Perspectives. Respiration, 96: 159–170, 2018.

Asano K, Hebisawa A, Ishiguro T, et al. New Clinical Diagnostic Criteria for Allergic Bronchopulmonary Aspergillosis/Mycosis and its Validation. J Allergy Clin Immunol, 147: 1261–1268, 2020.

Chumbita M, Puerta-Alcalde P, Garcia-Pouton N, Garcia-Vidal C. COVID-19 and Fungal Infections: Etiopathogenesis and Therapeutic Implications. Rev Esp Quimioter, 34: 72–75, 2021.

Gerson SL, Talbot GH, Hurwitz S, et al. Prolonged Granulocytopenia: The Major Risk Factor for Invasive Pulmonary Aspergillosis in Patients with Acute Leukemia. Ann Intern Med, 100: 345–351, 1984.

Gu X, Hua Y-H, Zhang Y-D, et al. The Pathogenesis of Aspergillus fumigatus, Host Defense Mechanisms, and the Development of AFMP4 Antigen as a Vaccine. Polish J Microbiol, 70: 3–11, 2021.

Herbrecht R, Letscher-Bru V, Oprea C, et al. Aspergillus Galactomannan Detection in the Diagnosis of Invasive Aspergillosis in Cancer Patients. J Clin Oncol, 20: 1898–1906, 2002.

Hohl TM, Feldmesser M. Aspergillus fumigatus: Principles of Pathogenesis and Host Defense. Eukaryot Cell, 6: 1953–1963, 2007.

Koutsokera A, Corriveau S, Sykes J, et al. Omalizumab for Asthma and Allergic Bronchopulmonary Aspergillosis in Adults with Cystic Fibrosis. J Cyst Fibros, 19: 119–124, 2020.

Latgé J-P, Chamilos G. Aspergillus fumigatus and Aspergillosis in 2019. Clin Microbiol Rev, 33: e00140–18, 2019.

Lewington-Gower E, Chan L, Shah A. Review of Current and Future Therapeutics in ABPA. Ther Adv Chronic Dis, 12: 1–17, 2021.

Moss RB. Pathophysiology and Immunology of Allergic Bronchopulmonary Aspergillosis. Med Mycol, 43: S203–S206, 2005.

Tekaia F, Latgé J-P. Aspergillus fumigatus: Saprophyte or Pathogen? Curr Opin Microbiol, 8: 385–392, 2005.

Vanderbeke L, Spriet I, Breynaert C, et al. Invasive Pulmonary Aspergillosis Complicating Severe Influenza: Epidemiology, Diagnosis and Treatment. Curr Opin Infect Dis, 31: 471–480, 2018.

Walsh TJ, Anaissie EJ, Denning DW, et al. Treatment of Aspergillosis: Clinical Practice Guidelines of the Infectious Diseases Society of America. Clin Infect Dis, 46: 327–360, 2008.

Wassano NS, Goldman GH, Damasio A. Aspergillus fumigatus. Trends Microbiol, 28: 594–595, 2020.

WEBSITES

American Lung Association, Aspergillosis Symptoms and Diagnosis, 2021: https://www.lung.org/lung-health-diseases/lung-disease-lookup/aspergillosis/symptoms-diagnosis

Aspergillus Website, The Fungal Research Trust, Copyright ©, 2007: http://www.aspergillus.org.uk

John Hopkins Medicine, Shmuel Shoham, Aspergillus, 2020: https://www.hopkinsguides.com/hopkins/view/Johns_Hopkins_ABX_Guide/540036/all/Aspergillus

Mayo Clinic, Aspergillosis, 2022: https://www.mayoclinic.org/diseases-conditions/aspergillosis/symptoms-causes/syc-203696198

Utah State University, Intermountain Herbarium: http://herbarium.usu.edu

Students can test their knowledge of this case study by visiting the Instructor and Student Resources: [www.routledge.com/cw/lydyard] where several multiple choice questions can be found.

Borrelia burgdorferi and related species

2

A 45-year-old woman was on vacation in Cape Cod, Massachusetts and decided to attend an outdoor music festival. In order to get there, she walked 3 miles each way at night, through a dark, wooded, grassy area. Shortly thereafter, she developed nonspecific symptoms that included fever, headache, muscle aches, mild neck stiffness, and joint pain. She also noticed an oval "bull's eye" **rash** on her right arm, which increased in size and cleared in the center. Over the ensuing month, she felt increasingly fatigued and developed facial paralysis (**Bell's palsy**) (Figure 2.1), which precipitated a visit to her family physician. She couldn't remember being bitten by a tick but based on her description of the rash and her other symptoms, her doctor suspected Lyme disease and took a blood sample for **serology**. **Enzyme immunoassay** and **Western blot** confirmed the presence of *Borrelia burgdorferi*-reactive antibodies. After confirming that the patient was not pregnant, she was prescribed doxycycline 100 mg twice daily for 30 days.

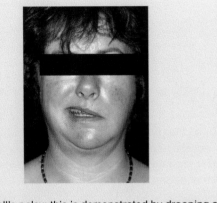

Figure 2.1 Bell's palsy: this is demonstrated by drooping at the left corner of the mouth, loss of the left naso-labial fold, and inability to completely close the left eye (not shown in image). *Reprint permission kindly granted by Dr Charles Goldberg, MD, and Regents of the University of California.*

1. WHAT IS THE CAUSATIVE AGENT, HOW DOES IT ENTER THE BODY AND HOW DOES IT SPREAD A) WITHIN THE BODY AND B) FROM PERSON TO PERSON?

CAUSATIVE AGENT

The patient has Lyme disease. The causative agent of Lyme disease is *Borrelia burgdorferi sensu lato* (meaning in the broad sense). There are at least 20 genospecies within the *Borrelia burgdorferi* complex (*B. burgdorferi sensu lato*) worldwide. The three main pathogenic genospecies comprising this group are *B. burgdorferi sensu stricto*, *B. garinii*, and *B. afzelii*. Strains found in North America belong to *B. burgdorferi sensu stricto* whereas all three species are found in Europe and Asia. Borreliae are microaerophilic **spirochetes** (Figure 2.2) that are extremely difficult to culture because of their complex nutrient requirements. Thus, they are usually detected by the immune response they induce in blood of the infected person (see above and Section 4). The bacteria have a gram-negative wall structure and have a spiral mode of motility produced by axial filaments termed endoflagella. In contrast to the usual type of flagella exhibited by gram-negative bacteria that are anchored in the cytoplasmic membrane and extend through the cell wall into the external environment of the cell, the endoflagella of borreliae are found within the periplasmic space contained between a semi-rigid peptidoglycan layer and a multi-layer, flexible outer-membrane sheath. Rotation of the endoflagella within the periplasmic space causes the borreliae to move in a cork-screw fashion (Figure 2.2). Borreliae lack lipopolysaccharide in their outer membrane (OM); instead the OM contains phospho- and glycolipids. Importantly, OM lipoproteins are phase variable during the infection

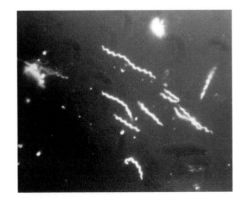

Figure 2.2 *Borrelia burgdorferi* is a spirochete: it is 0.2–0.3 micrometers (μm) wide and its length may exceed 15–20 μm. *Courtesy of the Centers for Disease Control, Atlanta, Georgia. Image is found in the Public Health Image Library #6631.*

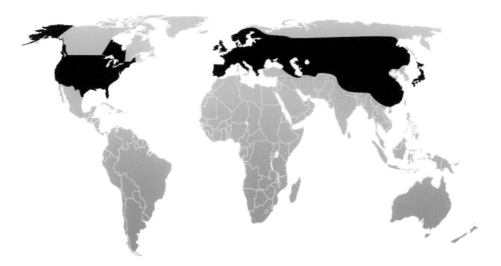

Figure 2.3 The global distribution of *Ixodes* spp. ticks able to transmit the agent of Lyme disease, *Borrelia burgdorferi*. *Modified from http://geo.arc.nasa.gov/sge/health/sensor/disease/lyme.html*

cycle. In addition, *Borrelia* species, instead of having circular chromosomes, have linear chromosomes and contain circular and linear plasmids, with some species containing more than 20 different plasmids.

TICKS – THE VECTORS OF
B. BURGDORFERI

The bacteria are maintained in an enzootic cycle involving hard-bodied ticks belonging to the *Ixodes ricinus* species complex and a wide range of reservoir vertebrate hosts. The global distribution of *Ixodes* species is shown in Figure 2.3. In the eastern US, the vector is primarily *Ixodes scapularis* and in the western US it is *I. pacificus*. *I. ricinus* and *I. persulcatus* are the vectors in Europe and Eurasia, respectively. These ticks have a 2-year life cycle (Figure 2.4). Adult ticks feed and mate on large animals, especially white-tailed deer, in the autumn and early spring. However, white-tailed deer are not considered reservoirs of *B. burgdorferi* because they do not support a sufficiently high level of spirochetes in their blood to infect ticks. Nevertheless, deer are important in tick reproduction and serve to increase tick numbers in an area and spread ticks into new areas. Female ticks then drop off the animals and lay eggs on the ground. By summer, the eggs hatch into larvae, which feed on mice and other small mammals and birds through to early autumn; then they become inactive until the following spring when they molt into nymphs. Nymphs feed on small rodents and other small mammals and birds during the late spring and summer and molt into adults in the autumn, completing the 2-year life cycle. Larvae and nymphs typically become infected with borreliae when they feed on infected small animals, particularly the white-footed mouse and chipmunks. The tick remains infected with the borreliae as it matures from larva to nymph or from nymph to adult. Infected nymphs and adult ticks then bite and transmit the bacteria to other small rodents, other animals, or humans

in the course of their normal feeding behavior. The ticks are slow feeders, requiring several days to complete a blood meal. Transfer of the borreliae from the infected tick to a vertebrate host probably does not occur unless the tick has been attached to the body for 36 hours or so.

Ticks transmit Lyme disease to humans generally during the nymph stage, probably because nymphs are more likely to feed on a person and are rarely noticed because of their small size (< 2 mm) (Figure 2.5). Although tick larvae are smaller than nymphs, they rarely carry borreliae at the time of feeding

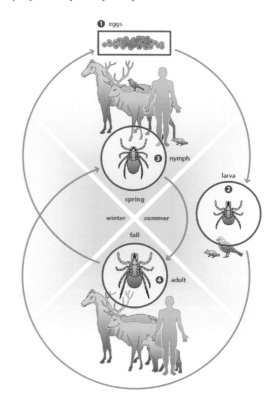

Figure 2.4 Tick life cycle. *Courtesy of Carolyn Klass, Cornell University.*

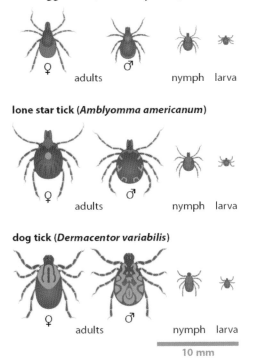

blacklegged tick (*Ixodes scapularis*)

adults ♀ ♂   nymph   larva

lone star tick (*Amblyomma americanum*)

adults ♀ ♂   nymph   larva

dog tick (*Dermacentor variabilis*)

adults ♀ ♂   nymph   larva

10 mm

Figure 2.5 Appearance and relative sizes of adult male and female, nymph, and larval ticks including deer ticks (*Ixodes scapularis*), lone star ticks (*Amblyomma americanum*), and dog ticks (*Dermacentor variabilis*). Of those pictured, only the *I. scapularis* ticks are known to transmit Lyme disease. *Courtesy of the Centers for Disease Control, Division of Vector-Borne Infectious Diseases, Fort Collins, Colorado.*

move toward the salivary glands. Once the borreliae enter the vertebrate host, they down-regulate OspA and express OspC, DbpA, and BBK32. The environmental cues for up- and down-regulation of Osps include nutrient availability, oxygen tension, temperature, and pH. The sequence of OspC is strain-specific, so the population of *B. burgdorferi* injected by the tick expresses a spectrum of antigenically distinct OspC proteins.

Several salivary gland proteins are induced during tick feeding and one of these, Salp15, has immunosuppressive activity where the tick saliva is deposited in the skin. It has been shown that there is a specific interaction between tick Salp15 and OspC, both *in vitro* and *in vivo*.

ENTRY AND SPREAD WITHIN THE BODY

B. burgdorferi sensu lato spirochetes enter the tissues while the tick takes a blood meal. The bacteria may establish a localized infection in the skin at the site of the tick bite, producing a characteristic skin lesion known as **erythema migrans (EM)**. In addition, they may disseminate via the bloodstream and/ or lymphatics. The organism demonstrates a tropism for the central nervous system (CNS), heart, joints, and eyes, all of which may become chronically infected, giving rise to neurologic disease, **carditis**, **arthritis**, and **conjunctivitis** (see Section 3). Even in the absence of systemic symptoms, it appears that as many as half of persons with EM have evidence of borreliae in the blood or cerebrospinal fluid (CSF). Furthermore, borreliae can also persist in skin and perhaps the CNS for years without causing symptoms.

PERSON-TO-PERSON SPREAD

Lyme disease is not spread from person to person. It is possible in a woman who contracts Lyme disease during pregnancy for the borreliae to cross the placenta leading to infection of the fetus, but this rarely occurs. The consequence of fetal infection remains unclear. For this reason, the Centers for Disease Control and Prevention (CDC) maintains a registry of pregnant women with Lyme disease to accrue data on the effects of Lyme disease on the developing fetus.

EPIDEMIOLOGY

Lyme disease is the most common tick-borne disease in North America and Europe. Although Lyme disease has now been reported in all 50 states in the US, almost all reported cases are confined to New England (Connecticut, Maine, Massachusetts, New Hampshire), the Mid-Atlantic region (Delaware, Maryland, New Jersey, Pennsylvania), the East-North Central region (Wisconsin), and the West North-Central region (Minnesota). A 2021 report, based on the use of commercial claims data for evaluating trends in Lyme disease diagnoses in the US between 2010–2018, gave an annual incidence of Lyme disease diagnoses per 100 000 enrollees of 49 to 88. However, all of these diagnoses may not have been proven to be Lyme disease. Although Lyme disease is common

and are probably not important in the transmission of Lyme disease to humans. While adult ticks can transmit borreliae they are less likely to do so than nymphs. This is because their larger size means that they are more likely to be noticed and removed from a person's body within a few hours and they are most active during the cooler months of the year, when outdoor activity is limited. It should be noted that dogs, horses, cattle, deer, and other animals are also susceptible to Lyme disease.

Borreliae express a number of outer-surface lipoproteins termed Osps (outer surface proteins) and they play an important role in the life cycle of the spirochete by interacting with intercellular and cellular components of its arthropod and vertebrate hosts. They are also important in the evasion of the host immune system (see Section 3). *B. burgdorferi* selectively expresses specific Osps during distinct phases of its life cycle and in specific tissue locations. OspA and OspB are expressed on spirochetes in unfed nymphs and adult ticks. Both OspA and OspB mediate adherence of the spirochetes to the cells of the tick mid-gut, which allows them to avoid endocytosis by tick enterocytes during digestion of the blood meal and subsequently allows their detachment when the tick takes a second blood meal so that the bacteria can enter the vertebrate host. In the mid-gut, during tick feeding, the bacteria up-regulate expression of OspC, which presages their in the US and Scandinavia and has been reported in other countries in Western and Eastern Europe, Japan, China, and Australia, it is not a common disease in the UK, with about 3000 cases per year being reported in England and Wales in the most recent data. The incidence of Lyme disease in Europe is estimated to be a little over 232 000 cases.

2. WHAT IS THE HOST RESPONSE TO THE INFECTION AND WHAT IS THE DISEASE PATHOGENESIS?

Because the spirochete is delivered to the host via the bite of a tick, it bypasses the physical barrier of the intact skin and the antimicrobial factors in sweat.

INNATE IMMUNITY

Components of the innate immune system are brought into play to combat the spirochetes at the site of their delivery by the interaction of **microbe-associated molecular patterns** (**MAMPs**) on the surface of the borreliae with **Toll-like receptors** (**TLRs**) on epithelium and leukocytes. *B. burgdorferi* activates both the innate and classical pathways of the complement cascade but is resistant to complement-mediated lysis because OspE and other proteins on its surface bind the complement control glycoprotein, factor H, which inactivates complement factor C3b. The **membrane attack complex** (**MAC**) can also be inactivated in a similar manner. Resident macrophages in the area of the inoculum are able to bind, phagocytose, and kill borreliae without the need for **opsonization** by complement or antibody. Binding may be mediated by the mannose-binding receptor and/or the MAC-1 receptor. Polymorphonuclear leukocytes (PMNs) can also kill borreliae without opsonization.

ADAPTIVE IMMUNITY

There is little doubt that **IgM** and **IgG** antibodies play the principal role in the clearance of *B. burgdorferi*. Furthermore, these antibodies do not need to be complement fixing. Murine IgG and IgM **monoclonal antibodies** have been developed that are bactericidal for borreliae in the absence of complement. The finding that mice deficient in α/β T cells or deficient in α/β- and γ/δ T cells can clear spirochetemia indicates that T cells are not required for spirochete clearance. Lyme arthritis appears to result from a constellation of factors that include production of pro-inflammatory **cytokines** and **immune complexes**, and arthritis is linked to HLA-DR4 and HLA-DR2.

How do the borreliae evade these host defense mechanisms? First of all, as mentioned earlier, Salp15 salivary gland protein induced during tick feeding has immunosuppressive activity where the tick saliva containing the borreliae is deposited in the skin. Salp15 inhibits the IgG antibody response by blocking **CD4+** T-cell activation and so may protect *B. burgdorferi* by suppressing production of neutralizing antibodies. This mechanism may be particularly important in Lyme-endemic areas where infected ticks frequently feed on their primary hosts that may possess preexisting antibodies against *B. burgdorferi*.

Furthermore, borreliae can stimulate **interleukin**-10 (IL-10) production by macrophages, mast cells, and **Th2** CD4+ T cells. IL-10 inhibits synthesis of pro-inflammatory cytokines and suppresses antigen presentation to CD4+ helper T cells by antigen-presenting cells.

B. burgdorferi undergoes **antigenic variation** and modulates the expression of Osps on the cell surface during infection. VlsE is an example of an Osp that undergoes antigenic variation. The vls locus of *B. burgdorferi* is located on a 28-kb linear plasmid and consists of an expression site (vlsE) and 15 silent vls cassettes. Another candidate may be OspE, since the gene for this Osp has two hypervariable domains and repeat regions that allow recombination with other genes, which may result in the formation of new antigens.

PATHOGENESIS

The tissue injury in Lyme disease is mediated by inflammation induced by *B. burgdorferi*. The manner in which the bacterium induces inflammation in the host is not fully understood. Spirochetemia results in the invasion of tissues such as the heart and joints and the host responds with a vigorous inflammatory response. The role of *B. burgdorferi* MAMPs–TLR interactions in the induction of pro-inflammatory cytokines has been questioned by results of knockout experiments, which indicate that TLR-2-deficient mice or mice deficient in TLR signaling molecules develop arthritis and have a much larger burden of *B. burgdorferi*. It has been suggested that **chemokines** produced at the site of infection may be more important in the influx of inflammatory cells to the site of infection.

3. WHAT IS THE TYPICAL CLINICAL PRESENTATION AND WHAT COMPLICATIONS CAN OCCUR?

Whether the disease develops following *B. burgdorferi* infection depends on the balance between the pathogen and the host's immune response. There are three potential outcomes of the borrelia–host interaction. The spirochete may be cleared without any manifestations of disease, the only indicator of infection being that the individual is **seropositive**. Alternatively, the spirochete establishes in the skin and, after a variable incubation period ranging from a few days to a month, produces a characteristic spreading rash termed EM. The rash begins as a small **macule** (a visible change in the color of the skin that cannot be felt) or **papule** (a small, solid and usually conical elevation of the skin), which then expands, ranging in diameter to between 5 and 50 cm (**Figure 2.6**). The rash has a flat border and central clearing so that it resembles a "bull's-eye". EM is probably caused by the inflammatory response to the spirochete infection. From the initial focus of infection in the skin, the spirochete spreads throughout the body. Systemic spread of the spirochete results in malaise, headaches, chills, joint pain, **myalgia**, **lymphadenopathy**, and severe fatigue. This phase may last for up to one month. Unless treated, over two-thirds of infected individuals manifest neurologic and cardiac symptoms. These manifestations may occur as early as one month or as late as 2 years or more post-infection. Neurologic sequelae include **meningitis**, **encephalitis**, and **peripheral nerve neuropathy**, particularly seventh cranial nerve palsy (Bell's palsy). Cardiac sequelae include heart block, **myopericarditis**, and congestive heart failure. Neurologic and cardiac sequelae may be followed by **arthralgia** and arthritis. About two-thirds of patients with untreated infection will experience intermittent bouts of arthritis, with severe joint pain and swelling. Large joints are most often affected, particularly the knees. These manifestations may last for months to years with little evidence of bacterial invasion. The manifestations of Lyme disease are related to the particular genospecies of *Borrelia* involved. In Europe *B. garinii* is associated with neurologic disease, while *B. afzelii* is associated with a dermatologic manifestation known as **acrodermatitis chronica atrophicans**, a progressive fibrosing skin process.

The existence of an entity termed "chronic Lyme disease" or post-treatment Lyme disease is controversial. The term chronic Lyme disease is used in North America and in Europe as a diagnosis for patients with persistent pain, neurocognitive

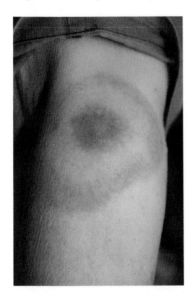

symptoms, fatigue, either separately or together, with or without clinical or serologic evidence of previous early or late Lyme disease.

However, persistent joint swelling lasting as long as several years is seen in about 10% of adult patients with Lyme arthritis following appropriate antibiotic therapy. The inability to detect borreliae in joint aspirates or tissues using **polymerase chain reaction** (**PCR**) has led to a proposed autoimmune etiology.

OTHER DISEASES TRANSMITTED BY HARD-BODIED TICKS

Hard-bodied ticks belonging to the *I. ricinus* species complex are also vectors for the infectious agents *Ehrlichia phagocytophila*, *Babesia microti*, and tick-borne encephalitis (TBE) virus. *E. phagocytophila* is a small intracellular gram-negative coccobacillus that parasitizes neutrophils (human granulocytic ehrlichiosis – HGE). Once taken up into phagosomes, the bacteria prevent fusion with lysosomes and replicate forming membrane-enclosed masses termed morulae. Disease presents as a flu-like illness with **leukopenia** and **thrombocytopenia**. Most infected individuals require hospitalization and severe complications are not infrequent. *B. microti* is an intracellular sporozoan parasite that causes babesiosis. The infectious stage, termed pyriform (pear-shaped) bodies, enter the bloodstream and infect erythrocytes. Within the erythrocyte, trophozoites replicate by binary fission forming tetrads. The erythrocytes lyse releasing merozoites, which may infect new red blood cells. Disease presents as a flu-like illness leading to **hemolytic anemia** and renal failure. **Hepatomegaly** and **splenomegaly** may be observed in advanced disease. TBE virus is a member of the Flaviviridae. The disease spectrum ranges from a mild **febrile** illness to meningitis, encephalitis or **meningoencephalitis**. Chronic or permanent neuropsychiatric sequelae are observed in as many as 20% of infected patients.

4. HOW IS THE DISEASE DIAGNOSED, AND WHAT IS THE DIFFERENTIAL DIAGNOSIS?

In patients with signs and symptoms consistent with Lyme disease, a diagnosis is confirmed by antibody detection tests. The current recommendation from the US Centers for Disease Control and Prevention (CDC) (https://www.cdc.gov/lyme/diagnosistesting/) is for a two-test approach consisting of a sensitive enzyme immunoassay (EIA) or **immunofluorescence assay** (**IFA**) followed by a Western immunoblot. Similar tests are used worldwide. All specimens positive or equivocal by the EIA or IFA should be tested by a standardized Western immunoblot. Specimens negative by the EIA or IFA do not require further testing unless clinically indicated. The EIA or IFA can be performed either as a total Lyme titer or as separate IgG and IgM titers. If a Western immunoblot is performed during the first 4 weeks of disease onset, both IgM and IgG immunoblots should be performed. A positive IgM test result alone is not recommended for use in determining active disease in persons with illness of greater than one month's duration because the likelihood of a false-positive test result for a current infection is high for these persons. If a patient with suspected early Lyme disease has a negative serology, serologic evidence of infection is best obtained by testing paired acute- and convalescent-phase serum samples. Serum samples from persons with disseminated or late-stage Lyme disease almost always have a strong IgG response to *B. burgdorferi* antigens. For an IgM immunoblot to be considered positive, two of the following three bands must be present: 24 or 21 kDa (OspC) (the apparent molecular mass of OspC is dependent on the strain of *B. burgdorferi* being tested), 39 kDa (BmpA), 41 kDa (Fla). For an IgG immunoblot to be considered positive, five of the following 10 bands must be present: 18 kDa, 21 or 24 kDa (OspC, see above), 28 kDa, 30 kDa, 39 kDa (BmpA), 41 kDa (Fla), 45 kDa, 58 kDa (not GroEL), 66 kDa, and 93 kDa. The different genospecies of *B. burgdorferi sensu lato* in Europe make the serologic diagnosis of Lyme disease more complex and the standardization of tests, particularly the Western blot, more difficult.

DIFFERENTIAL DIAGNOSIS

The following conditions should be considered in the differential diagnosis: calcium pyrophosphate deposition disease, **fibromyalgia**, gonococcal arthritis, **gout**, meningitis, **psoriatic arthritis**, **rheumatoid arthritis**, **syncope**, **systemic lupus erythematosus** (**SLE**), and **urticaria**.

5. HOW IS THE DISEASE MANAGED AND PREVENTED?

MANAGEMENT

Routine use of antibiotics or serologic testing after a tick bite is not recommended. Persons who remove attached ticks (see below) should be monitored for up to one month for signs and symptoms of Lyme and other tick-borne diseases. Persons developing EM or other illness should seek medical attention.

The recommended antimicrobial regimens and therapy for patients with Lyme disease is shown in Table 2.1. More information on antibiotic therapy can be found at https://www.cdc.gov/lyme/treatment/index.html and https://www.idsociety.org/practice-guideline/lyme-disease/.

PREVENTION

The best method of preventing Lyme disease is to avoid tick-infested areas. If this is not feasible then the following are recommended:

- wear light-colored clothing so that ticks can be more easily spotted;
- tuck trouser cuffs into socks or boots and tuck shirts into trousers so that ticks cannot crawl under clothing;
- wear a long-sleeved shirt and hat;
- use DEET (meta-N,N-diethyl toluamide) tick repellent on skin;
- spray clothing and boots with permethrin;

Table 2.1 Recommended treatment for adults with Lyme borreliosis

Manifestation	Antibiotic	Treatment duration (days)
Erythema migrans, borrelial lymphocytoma or acrodermatitis chronica atrophicans[a]	Doxycycline	10
	Amoxicillin	14
	Cefuroxime axetil	14
	Phenoxymethylpenicillin	14
	Azithromycin[b]	5–10
Lyme meningitis, Cranial neuropathy or radiuclopathy	Doxycycline[c]	14
	Ceftriaxone[d]	14
Lyme encephalomyelitis	Ceftriaxone	14–28
Cardiac Lyme disease	Doxycycline	14–21
	Amoxicillin	14–21
	Cefuroxime axetil	14–21
	Ceftriaxone	14–21
Lyme arthritis	Doxycycline	28
	Amoxicillin	28
	Cefuroxime axetil	28
	Ceftriaxone[e]	14–28

[a] Treatment duration for borrelial lymphocytoma is 14 days for β-lactam and tetracycline antibiotics; the duration for acrodermatitis chronica atrophicans is 21–28 days. [b] For patients who are unable to take β-lactams or tetracyclines. [c] For ambulatory patients. [d] For hospitalized patients. [e] Used when there is only a minimal response to oral antibiotics. Adapted from the Infectious Diseases Society of America (IDSA) treatment guidelines. (From Steere A, Strle F, Wormser G *et al.* (2016) *Nat Rev Dis Primers* 2, 16090. https://doi.org/10.1038/nrdp.2016.90. With permission from Nature Springer.)

- check the entire body for ticks every day;
- remove any attached ticks as soon as possible since it takes about 36 hours of attachment before borreliae are transmitted.

VACCINES

Currently, no vaccine is available for the prevention of Lyme disease. However, in December 1998, the US Food and Drug Administration licensed the LYMErix™ vaccine against Lyme disease for human use. In February 2002, the vaccine was withdrawn from the market, reportedly because of poor sales but more likely because of the fear of inducing autoimmunity. LYMErix™ contained lipidated recombinant OspA from

B. burgdorferi sensu stricto. The vaccine was targeted for use in persons aged 15–70 years at high risk of exposure to infected ticks. Interestingly, OspA is not expressed by spirochetes in infected humans, but anti-OspA IgG antibodies in human blood are taken up by the infected tick during feeding and kill the borreliae in the tick hind-gut, preventing transmission. Despite withdrawal from the market, research on modified OspA-based vaccines and other spirochaete protein-based vaccines has continued. With modified multivalent OspA-based vaccines lacking a T-cell epitope thought to have autoreactive potential. Phase I and phase II trials have confirmed that the vaccine is safe and induces significant antibody responses but no attempt has been made to bring the vaccine to market.

SUMMARY

1. WHAT IS THE CAUSATIVE AGENT, HOW DOES IT ENTER THE BODY, AND HOW DOES IT SPREAD A) WITHIN THE BODY AND B) FROM PERSON TO PERSON?

- Lyme disease is the most common tick-borne disease in North America and Europe.
- There are three genospecies within the *Borrelia burgdorferi* complex: *B. burgdorferi* (*sensu lato*), *B. burgdorferi* (*sensu stricto*), *B. garinii*, and *B. afzelii*.
- Strains found in North America belong to *B. burgdorferi sensu stricto*, whereas the other two species are found in Europe and Asia.
- The bacteria are maintained in an enzootic cycle involving hard-bodied ticks belonging to the *Ixodes ricinus* species complex and a wide range of reservoir vertebrate hosts.
- Because the ticks are slow feeders, transfer of the borreliae from the infected tick to a vertebrate host probably does not occur unless the tick has been attached to the body for about 36 hours.
- Ticks transmit Lyme disease to humans generally during the nymph stage.
- Borreliae are microaerophilic spirochetes that are extremely difficult to culture because of their complex nutrient requirements.
- *B. burgdorferi sensu lato* are detected by the immune response that they induce in blood of the infected person.
- The bacteria have a gram-negative wall structure and have endoflagella within the periplasmic space.
- *Borrelia* species have linear chromosomes and contain circular and linear plasmids with some species containing more than 20 different plasmids.
- Borreliae are adept at evading host immunity by varying the lipoproteins on their outer surface.
- The Osps play an important role in the life cycle of the spirochete by interacting with intercellular and cellular components of its arthropod and vertebrate hosts.

- *B. burgdorferi* selectively expresses specific Osps during distinct phases of its life cycle and in specific tissue locations.
- *B. burgdorferi sensu lato* establishes a localized infection in the skin at the site of the tick bite, producing a characteristic skin lesion known as erythema migrans (EM).
- In addition, it may disseminate via the bloodstream and/or lymphatics to the CNS, heart, joints, and eyes, all of which may become chronically infected.
- Lyme disease is not spread person to person.
- Rarely, borreliae may cross the placenta leading to infection of the fetus.

2. WHAT IS THE HOST RESPONSE TO THE INFECTION AND WHAT IS THE DISEASE PATHOGENESIS?

- Because the spirochete is delivered to the host via the bite of a tick, it bypasses the physical barrier of the intact skin and the antimicrobial factors in sweat.
- The innate pathways of the complement cascade, macrophages, and **dendritic cells** are the first line of defense in the skin.
- *B. burgdorferi* activates both the innate and classical pathways of the complement cascade but is resistant to complement-mediated lysis.
- Macrophages and polymorphonuclear leukocytes can phagocytose and kill borreliae without the need for opsonization by complement or antibody.
- IgM and IgG antibodies play the principal role in the clearance of *B. burgdorferi*.
- *B. burgdorferi* undergoes antigenic variation and modulates the expression of Osps on the cell surface during infection.
- The pathogenesis of Lyme disease is not well understood.
- Tissue injury results from the inflammatory response mounted against the borreliae.
- Chemokines may be the principal mediators of inflammation in Lyme disease.

Continued...

...continued

3. WHAT IS THE TYPICAL CLINICAL PRESENTATION AND WHAT COMPLICATIONS CAN OCCUR?

- Lyme disease can be divided into three stages: (1) erythema migrans and some associated symptoms; (2) intermittent arthritis, cranial nerve palsies and nerve pain, atrioventricular node block, and severe malaise and fatigue; (3) prolonged arthritis, chronic encephalitis, myelitis, and parapareses (partial paralysis of the lower limbs).
- The manifestations of Lyme disease are related to the particular genospecies of *Borrelia* involved.

4. HOW IS THE DISEASE DIAGNOSED, AND WHAT IS THE DIFFERENTIAL DIAGNOSIS?

- In patients with signs and symptoms consistent with Lyme disease, a diagnosis is confirmed by antibody detection tests comprising an enzyme immunoassay (EIA) or immunofluorescence assay (IFA) followed by a Western immunoblot.

- All specimens positive or equivocal by the EIA or IFA should be tested by a standardized Western immunoblot. Specimens negative by the EIA or IFA do not require further testing.
- The following conditions should be considered in the differential diagnosis: calcium pyrophosphate deposition disease, fibromyalgia, gonococcal arthritis, gout, meningitis, psoriatic arthritis, rheumatoid arthritis, syncope, systemic lupus erythematosus, and urticaria.

5. HOW IS THE DISEASE MANAGED AND PREVENTED?

- The preferred oral regimens for erythema migrans and early disease are amoxicillin 500 mg three times daily, doxycycline 100 mg twice daily or cefuroxime 500 mg twice daily for 14 days.
- Meningitis or radiculopathy require parenteral ceftriaxone 2 g IV for 14 days.
- Late disease can be treated for 28 days with the oral regimen, except for central or peripheral nervous system disease, which should be treated using the parenteral regimen.

FURTHER READING

Goering R, Dockrell H, Zuckerman M, Chiodini PL. Mims' Medical Microbiology and Immunology, 6th edition. Elsevier, Philadelphia, 2018.

Murray PR, Rosenthal KS, Kobayashi GS, Pfaller MA. Medical Microbiology, 9th edition. Elsevier, Philadelphia, 2021.

REFERENCES

Hyde JA. Borrelia burgdorferi Keeps Moving and Carries on: A Review of Borrelial Dissemination and Invasion. Front Immunol, 8: 114, 2017.

Kullberg, BJ, Vrijmoeth HD, van de Schoor F, Hovius JW. Lyme Borreliosis: Diagnosis and Management. BMJ, 369: m1041, 2020.

Kurokawa C, Lynn GE, Pedra JHF, et al. Interactions Between *Borrelia burgdorferi* and Ticks. Nat Rev Microbiol, 18: 587–600, 2020.

Margos G, Fingerle V, Reynolds S. Borrelia bavariensis: Vector Switch, Niche Invasion, and Geographical Spread of a Tick-Borne Bacterial Parasite. Front Ecol Evol, 7: 401, 2019.

Ozdenerol E. GIS and Remote Sensing Use in the Exploration of Lyme Disease Epidemiology. Int J Environ Res Public Health, 12: 15182–15203, 2015.

Schwartz AM, Kugeler KJ, Nelson CA, et al. Use of Commercial Claims Data for Evaluating Trends in Lyme Disease Diagnoses, United States, 2010–2018. Emerg Infect Dis, 27: 499–507, 2021.

Steere AC, Strle F, Wormser GP, et al. Lyme Borreliosis. Nat Rev Dis Primers, 2: 16090, 2016.

WEBSITES

European Centre for Disease Prevention and Control, Tick maps: https://www.ecdc.europa.eu/en/disease-vectors/surveillance-and-disease-data/tick-maps

Lyme Disease Association, Inc.: http://www.lymedisease association.org

WHO, Vector-borne diseases: https://www.who.int/news-room/fact-sheets/detail/vector-borne-diseases

Students can test their knowledge of this case study by visiting the Instructor and Student Resources: [www.routledge.com/cw/lydyard] where several multiple choice questions can be found.

Campylobacter jejuni

3

A 35-year-old man had been feeling unwell for a few days with nonspecific aches and pains in his joints and a slight headache. He put this down to a barbecue he had attended a few days previously, where he had also consumed a considerable amount of alcohol.

The following day, he felt rather worse with severe colicky abdominal pain and he developed bloody diarrhea, going to the lavatory 10 times during the day. This persisted overnight and he attended his local hospital's Emergency Department.

On examination, he was dehydrated and rather pale. He was admitted to hospital for intravenous (IV) rehydration and blood and feces samples were sent for culture. He was started on antibiotics and over the subsequent few days he improved with lessening of the symptoms and was discharged home.

Some weeks later, he began to develop weakness in his feet, which gradually affected his legs. He contacted his primary care physician who admitted him to hospital once again. Over the subsequent few days, the paralysis affected his upper leg muscles, and gradually over the ensuing weeks slowly resolved with treatment.

1. WHAT IS THE CAUSATIVE AGENT, HOW DOES IT ENTER THE BODY AND HOW DOES IT SPREAD A) WITHIN THE BODY AND B) FROM PERSON TO PERSON?

CAUSATIVE AGENT

Campylobacter jejuni is a slender, motile non-spore-forming curved gram-negative bacterium (Figure 3.1) measuring 0.2–0.5 μm wide by 0.5–5.0 μm long, with a single unsheathed polar flagellum. It is resistant to complement. *C. jejuni* also produces superoxide dismutase (*sodB*), an enzyme which catalyzes the breakdown of superoxide radicals, thus protecting the organism from an inflammatory response. It has a complex respiratory chain and can use other electron acceptors instead of oxygen giving it phenotypic versatility. Its genome has been sequenced. Several species of *Campylobacter* exist, two of which cause the majority of human disease: *C. jejuni* and *C. coli*. A small number of other "emerging" campylobacters have also been linked to human disease (Table 3.1). Two closely related genera are: *Arcobacter* and *Helicobacter* (see Case study 12).

C. jejuni has a non-typical gram-negative cell wall structure with lipo-oligosaccharide (LOS) rather than lipopolysaccharide (LPS). *C. jejuni* is subdivided into Penner serogroups based on the **antigenic variation** of the heat stable capsular polysaccharide antigens (HS) or Lior serogroups based on the heat labile capsular and LOS antigens (HL). Although serogrouping is being replaced by molecular typing methods, a distribution of the serotypes is valuable for developing a vaccine and as certain Penner/Lior groups are associated with immune-mediated diseases (see below). *Campylobacter* can be typed by molecular methods such as **restriction fragment length polymorphism** (**RFLP**), **multilocus enzyme electrophoresis** (**MLEE**) and whole genome sequencing (WGS), although the problem with this emerging technology is understanding the natural genome diversity of campylobacter species.

The organism is microaerophilic, requiring an atmosphere of 5–10% oxygen and 10% CO_2, growing on campylobacter

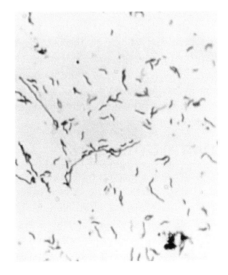

Figure 3.1 *Campylobacter jejuni*: a silver stain showing the "seagull" morphology of *Campylobacter* spp. *Courtesy of the Centers for Disease Control, Atlanta, Georgia. Image is found in the Public Health Image Library #6654. Additional photographic credit is given to Robert Weaver, PhD, who took the photo in 1980.*

selective agar (Brucella blood agar with added trimethoprim, vancomycin and polymyxin B) within 2–5 days.

The two main human pathogens are thermotolerant and can grow at 42°C, which is used as a selective condition for clinical isolates. Other campylobacters have been linked with human disease. *C. fetus* can cause intestinal or systemic infections in immunocompromised individuals or those who work with animals. *C. concisus* and *C. showae*, part of the oral flora, have been associated with inflammatory bowel disease, although not proven. *C. curvus* is associated with gastroenteritis, abscesses, and bacteremias. *C. hyointestinalis* found in the intestine of pork, is an opportunist pathogen in humans causing diarrheal disease. *C. sputorum* has been isolated from gastroenteritis, abscesses, and blood (Table 3.1).

C. jejuni is found widely in the animal kingdom and is part of the normal flora of many food-source animals, for example, poultry, lamb, and pig. The commonest meat source is poultry. Campylobacteriosis is thus a **zoonosis**, with infection being acquired principally from eating poorly cooked meats. Each year in the EU approximately 23 million people become ill from eating contaminated food. Infection can also be acquired from raw milk contaminated at source, contaminated water sources, or from close contact with animals such as in children's zoos, infected dogs or bird excreta. Sexual contact is also a risk factor since MSM have 14 times the chance of acquiring *Campylobacter* compared to controls. Those who were positive for *Campylobacter* were also 74 times more likely to have *Shigella*.

ENTRY INTO THE BODY

Entry of the organism is via the mucosal surfaces of the intestine where it mostly remains. The infectious dose is about 500–1000 organisms and colonization affects the small and large intestine, with the terminal ileum and colon affected most often. Mesenteric **adenitis** regularly occurs. The shape and motility of the organisms enable them to penetrate the mucus layer where they adhere to the enterocytes and are internalized utilizing microfilament/microtubule-dependent mechanisms. Important in the internalization is the campylobacter invasion antigen (Cia), which is secreted by a flagellar secretion apparatus (the evolutionary forerunner of the **type III secretion apparatus**).

SPREAD WITHIN THE BODY

C. jejuni rarely causes a **bacteremia**, and remains in the gastrointestinal (GI) tract, causing a **terminal ileitis** and **colitis**. Systemic infections are more commonly associated with *Campylobacter fetus*, which principally causes diseases of the reproductive system in sheep or cattle and infections in immunocompromised humans. *C. fetus* possesses an S-layer (protein microcapsule), which is anti-phagocytic and may explain its tendency to cause **septicemia**.

PERSON-TO-PERSON SPREAD

Direct person-to-person spread is uncommon.

Table 3.1 Some species of Campylobacter associated with human disease

Campylobacter	
	C. jejuni
	C. coli
	C. curvus
	C. lari
	C. fetus
	C. conciscus
	C. showae
	C. hyointestinalis
	C. sputorum
	C. upsaliensis

EPIDEMIOLOGY

Campylobacteriosis is the most common bacterial GI infectious disease globally. The true global incidence is unknown but, in industrialized countries, the annual incidence is up to 9.3/100 000 population with all ages being susceptible. Globally, the three common Penner types are HS4c, HS2, and HS1/4cc. In Europe, HS4c and HS2 are equally common (17.3% and 15.3%), in Asia 8.9% and 11.8%, in N. America 23.5% and 10.7%, in Oceania 17.4% and 18.2%, and in Africa 7.0 % and 6.2% respectively. In all regions, HS1/44c is the least common and varies between 10.5% and 4.2%. Globally, eight serotypes represented over 50% of the isolates.

In the UK in 2017, the number of reported cases was 56 729 (96.57/100 000) increasing from 49 891 (90.97/100 000) in 2008 with a peak in 2012 of 65 044 (114.88/100 000). In the EU, it has been the most reported GI disease since 2005 affecting 220 000 individuals in 2019. There is a seasonality to infection, with most cases in different countries occurring in the spring–summer and declining in the winter months. In the UK, the highest rate of laboratory-confirmed cases occurs in June/July with the highest number of cases in the 50–59 age group. Confirmed disease is notifiable in Austria, Denmark, Finland, Germany, Italy, Sweden, and Norway. In the UK, a case of food poisoning or a case of diarrhea with blood are notifiable; both can relate to campylobacter but also to other infections. Isolation of campylobacter is notifiable by the laboratory.

2. WHAT IS THE HOST RESPONSE TO THE INFECTION AND WHAT IS THE DISEASE PATHOGENESIS?

INNATE IMMUNITY

Following invasion of the enterocytes, an acute IL-8 induced inflammatory reaction occurs with infiltration of the **lamina propria** by granulocytes. A variety of pathogen recognition receptors on and in the enterocytes are triggered, for example TLRs, and this leads to the production of pro-inflammatory cytokines that play a role in generating the acute inflammation reaction.

ADAPTIVE IMMUNITY

Soon after infection, the host mounts an antibody response, which peaks at about 2 weeks and declines over the following weeks. The main antigens are flagella, outer-membrane proteins (OMPs), and LOS. Since antibodies to OMP and LOS cross-react with myelin components, this can lead to the development of **Guillain-Barré** or the **Miller Fisher syndromes** (see below).

Little is known about the role of the cell-mediated response. However, CD4+ T-helper cells are involved in the production of **IgG** and **IgA** antibodies to the bacterial antigens. Secretory IgA in both the intestinal secretions and in maternal milk are important in protection against further infection and against children in early life getting severe disease.

PATHOGENESIS

Following infection, ulceration of the epithelium with formation of crypt **abscesses** occurs. Motility is essential for initial colonization of the GI tract and a variety of surface proteins are important for: adhesion – CadF, FlpA which bind to fibronectin cell; invasion – CapC an autotransported protein involved in adhesion and inflammation; ciaB (a protein similar to a Type III secretion protein) which is internalized by the gastric cell; iamA (invasion-associated marker, function unknown) and Cas9 which appears to regulate the campylobacter virulence factors. The bacteria also produce a cytolethal distending toxin (*cdtA, cdtB, cdtC*) which is present in most strains of *C. jejuni* and a similar toxin is found in other bacteria. The toxin induces cell-cycle arrest leading to **apoptosis**. The cdtA and cdtC components form the binding subunit which attaches to lipid rafts in the cell membrane. The cdtB unit is the active toxic component having DNAse and phosphatidylinositol-3, 4, 5-triphosphate (PIP3) phosphatase activity which, after internalization by endocytosis, enters the nucleus and causes DNA damage, cell-cycle arrest, cell distention, and apoptosis. The toxin also affects cell signaling and induces a pro-inflammatory response by up-regulating cytokines IL-6 IL-8. *C. jejuni* also has a Type VI secretion system (T6SS) with a single hemolysin-coregulated protein (Hcp) the possession of which is associated with bloody diarrhea. The protein is important in adhesion and invasion of intestinal cells.

In animal studies, IFNγ producing innate lymphoid cells (ILCs) induce intestinal pathology caused by *C. jejuni*. This distinct population of cells NK1.1-T-bet+ ILCs accumulate in the intestine after infection with campylobacter. They develop from RORγt+ progenitors and promote the pro-inflammatory response.

Neurologic complications (e.g. Guillain-Barré syndrome, Miller-Fisher syndrome) may occur, caused by antibodies to the LOS and cross-reacting with GM1 gangliosides at the nodes of Ranvier, affecting Schwann cells and ion channels leading to a conduction block (Figure 3.2).

The cause of the fluid loss is due to a direct effect of the organism affecting the sodium channel proteins and indirectly due to the immune response generated by the organism.

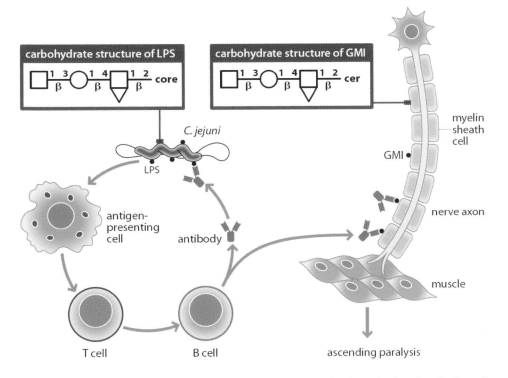

Figure 3.2 **Putative mechanism for development of Guillain-Barré syndrome.** Antibodies raised against the lipo-oligosaccharide (LOS) component of the cell wall of some *Campylobacter* strains cross-react with GM1 gangliosides in the myelin sheath of the nerve, due to a structural similarity, leading to damage, loss of nerve conduction, and paralysis.

Campylobacter inhibits expression of the β and γ subunits of the sodium channel in the colonic epithelial cells leading to sodium malabsorption. In addition, the organism causes the redistribution of tight junction proteins. Occludin is cleaved by HtrA protease (high temperature requirement protein A) and is displaced to the cytoplasm with ZO1 to the lateral plasma membrane along with claudins 1, 3, 4, 5, and 8. The organism gains access to the basolateral membrane and entry into the epithelial cells via the fibronectin receptor inducing cellular apoptosis leading to epithelial barrier porosity for fluids.

3. WHAT IS THE TYPICAL CLINICAL PRESENTATION AND WHAT COMPLICATIONS CAN OCCUR?

The incubation period is commonly 1–3 days but can be as long as 8 days. Infection may be symptomless in a proportion of individuals. A typical clinical presentation is one of generalized systemic upset with fever and **myalgia** accompanied by abdominal pain and diarrhea. Diarrhea may range from a few loose motions to profuse and watery or grossly bloody. The symptoms may last from one or two days to over a week in about 20% of patients and are generally self-limiting, although relapses may occur in about 10% of patients.

In severe cases of **enterocolitis**, **toxic megacolon** may occur and rare cases of **cholecystitis** or **pancreatitis** have been reported. Three major extraintestinal complications are recognized. The commonest complication is reactive **arthritis**, which occurs in about 1% of cases and occurs several weeks after the initial infection. It is more common in subjects who are HLA B27. The **hemolytic uremic syndrome** may also follow infection and usually comes on within a few days. Infection may be followed by the Guillain-Barré or the Miller Fisher syndrome variant. Although a rare complication, because campylobacteriosis is so common, an infection with *C. jejuni* is the cause of these immune pathologies in about 50% of cases. Other extraintestinal complications that have occasionally been reported are **IgA nephropathy** and **interstitial nephritis**.

4. HOW IS THIS DISEASE DIAGNOSED AND WHAT IS THE DIFFERENTIAL DIAGNOSIS?

Campylobacter enteritis is confirmed by the isolation of the organism from the feces using a charcoal-based selective medium and culturing under microaerobic conditions at 42°C. Both a **polymerase chain reaction** (**PCR**) and an immunochromatographic assay for antigen detection are also available. In resource-limited countries, a rapid means of provisional diagnosis is to look at a wet preparation of feces under the microscope, where the rapid motion of the *Campylobacter* can be seen – in contrast to *Shigella* dysentery where there is no obvious microbial movement as *Shigella* are nonmotile. **Pus** cells are also prominent. *C. jejuni* may also be detected using the PCR and immunochromatographic assays which can give a rapid (15 min) turnaround time.

DIFFERENTIAL DIAGNOSIS

It is not possible to make a microbiological diagnosis from the clinical presentation, so all the other causes of gastroenteritis are in the differential diagnosis such as *Shigella*, *Salmonella*, *Escherichia coli*, and so forth. Additionally, as the presentation may be severe with bloody diarrhea, **inflammatory bowel disease** may also be confused with campylobacteriosis.

5. HOW IS THE DISEASE MANAGED AND PREVENTED?

MANAGEMENT

The disease is self-limiting and, in most cases, no therapy is required. In severe cases, particularly if it lasts for several days, the patient may have to be admitted to hospital for rehydration. Antibiotics should also be given in severe cases. Erythromycin or ciprofloxacin are the two most frequently used, although for both agents, antibiotic-resistant isolates are increasing in prevalence due to overuse of antibiotics in both human and veterinary practice. In the presence of macrolide- or quinolone-resistant isolates, alternative agents such as tetracycline (except in children) or clindamycin may be used. *Campylobacter* is not susceptible to β-lactams, with the exception of co-amoxiclav. If systemic spread occurs, gentamicin can be used but is not useful for the enteric infection.

PREVENTION

Preventive measures rely mainly on adequate food hygiene and establishing disease-free animal stocks. In order to reduce contamination of chickens, a variety of approaches have been adopted such as acidification of feed water, feed supplements such as prebiotics and bacteriophages and biosecurity measures.

Important areas of cross contamination in kitchens are sinks, after washing chickens in running water, lack of hand washing, and cutting boards.

An established vaccine does not yet exist, although several trial vaccines for chickens and humans are in trial. The most effective chicken vaccines are based on either total outer-membrane proteins or cysteine ABC transporter substrate binding protein, but a capsule-conjugated vaccine seems to be promising. The two leading contenders for a human vaccine are a capsule-conjugate vaccine and a DNA-based vaccine.

SUMMARY

1. WHAT IS THE CAUSATIVE AGENT, HOW DOES IT ENTER THE BODY, AND HOW DOES IT SPREAD A) WITHIN THE BODY AND B) FROM PERSON TO PERSON?

- *Campylobacter* is microaerophilic.
- The cell wall has an atypical gram-negative structure.
- The antigenic variation of the capsular polysaccharide is the basis for the Penner groups.
- It is one of the commonest causes of gastroenteritis globally.
- Infection occurs at any age, with peaks in children and young adults.
- Infection is a zoonosis.
- The commonest food source is poultry.
- The organism is located in the small and large intestine.

2. WHAT IS THE HOST RESPONSE TO THE INFECTION AND WHAT IS THE DISEASE PATHOGENESIS?

- Motility and the organism's spiral shape are important in colonization.
- The infectious dose is about 500–1000 organisms.
- Cell invasion involves *Campylobacter* invasion antigens.
- Invasion involves the microtubules.
- Innate immunity leads to pro-inflammatory cytokines and acute inflammation.
- *Campylobacter* produces a cytolethal distending toxin, which causes cell-cycle arrest and apoptosis.

- Histologically, there is epithelial ulceration and infiltration by granulocytes.
- Antibodies produced in response to the organism can be protective but some can cross-react with myelin.

3. WHAT IS THE TYPICAL CLINICAL PRESENTATION AND WHAT COMPLICATIONS CAN OCCUR?

- The incubation period is 1–8 days, usually 1–3 days.
- Presents with constitutional symptoms.
- Gastrointestinal symptoms are colicky abdominal pain and diarrhea, which may contain blood.
- Complications include megacolon, reactive arthritis, hemolytic uremic syndrome, and Guillain-Barré syndrome.

4. HOW IS THIS DISEASE DIAGNOSED, AND WHAT IS THE DIFFERENTIAL DIAGNOSIS?

- Diagnosis is by culture on selective medium incubated microaerobically.
- Differential diagnosis is other bacterial causes of diarrhea and dysentery including *Shigella*, *Salmonella*, *E. coli*, and inflammatory bowel disease.

5. HOW IS THE DISEASE MANAGED AND PREVENTED?

- Most infections require no treatment.
- In severe cases, rehydration and antibiotics may be required.
- Prevention is largely by adequate food processing and hygiene.

FURTHER READING

Fischer GH, Paterek E. Campylobacter. StatPearls, Treasure Island, 2022.

Ketley JM, Konkel ME. Campylobacter: Molecular & Cellular Biology. Garland Science, Abingdon, 2005.

Nachamkin I, Blaser MJ. Campylobacter. Blackwell Publishing, Oxford, 2000.

REFERENCES

Bucker R, Krug SM, Moos V, et al. Campylobacter jejuni Impairs Sodium Transport and Epithelial Barrier Function via Cytokine Release in Human Colon. Mucosal Immunol, 11: 474–485, 2018.

Cardoso MJ, Ferreira V, Truninger, M, et al. Cross-Contamination Events of Campylobacter spp. in Domestic Kitchens Associated with Consumer Handling Practices of Raw Poultry. Int J Food Microbiol, 338: 108984, 2021.

Desvaux M, Hebraud M, Henderson IR, Pallen MJ. Type III Secretion: What's in a Name? Trends Microbiol, 14: 157–160, 2006.

Dingle KE, Colles FM, Wareing DRA, et al. Multilocus Sequence Typing for Campylobacter jejuni. J Clin Microbiol, 39: 14–23, 2001.

Elgamoudi BA, Andrianova EP, Shewell LK, et al. The Campylobacter jejuni Chemoreceptor Tlp10 has a Bimodal Ligand-Binding Domain and Specificity or Multiple Classes of Chemo-Effectors. Sci Signal, 14: eabc8521, 2021.

Harrer A, Bücker R, Boehm M, et al. Campylobacter jejuni Enters Gut Epithelial Cells and Impairs Intestinal Barrier Function Through Cleavage of Occludin by Serine Protease HtrA. Gut Pathol, 11: 4, 2019.

Karenlampi R, Rautelin H, Hanninen ML. Evaluation of Genetic Markers and Molecular Typing Methods for Prediction of Sources of Campylobacter jejuni and C. coli Infections. Appl Environ Microbiol, 73: 1683–1685, 2007.

Konkel ME, Monteville MR, Rivera-Amill V, Joens LA. The Pathogenesis of Campylobacter jejuni-Mediated Enteritis. Curr Issues Intest Microbiol, 2: 55–71, 2001.

Lu T, Marmion M, Ferone M, et al. On Farm Interventions to Minimise Campylobacter spp. Contamination in Chicken. Br Poult Sci, 62: 53–67, 2021.

Muraoka WT, Korchagina AA, Xia Q, et al. Campylobacter Infection Promotes IFNγ-Dependent Intestinal Pathology via

ILC3 to ILC1 Conversion. Mucosal Immunol, 14: 703–716, 2021.

Nylen G, Dunstan F, Palmer SR, et al. The Seasonal Distribution of Campylobacter Infection in Nine European Countries and New Zealand. Epidemiol Infect, 128: 383–390, 2002.

Parkhill J, Wren BM, Mungall E, et al. The Genome Sequence of the Food Borne Pathogen Campylobacter jejuni Reveals Hypervariable Sequences. Nature, 403: 665–668, 2000.

Payot S, Bolla JM, Corcoran D, et al. Mechanisms of Fluoroquinolone and Macrolide Resistance in Campylobacter spp. Microb Infect, 8: 1967–1971, 2006.

Prokhorova TA, Nielsen PN, Petersen J, et al. Novel Surface Polypeptides of Campylobacter jejuni as Travellers Diarrhoea Vaccine Candidates Discovered by Proteomics. Vaccine, 24: 6446–6455, 2006.

Solomon T, Willison H. Infectious Causes of Acute Flaccid Paralysis. Curr Opin Infect Dis, 16: 375–381, 2005.

WEBSITES

Centers for Disease Control and Prevention (CDC), Morbidity and Mortality Weekly Report, 2007: http://www.cdc.gov/mmwr/preview/mmwrhtml/mm5614a4.htm

Centre for Infections, Health Protection Agency, HPA Copyright, 2008: http://www.hpa.org.uk/infections/topics_az/campy/menu.htm

Students can test their knowledge of this case study by visiting the Instructor and Student Resources: [www.routledge.com/cw/lydyard] where several multiple choice questions can be found.

Candida albicans

4

A 38-year-old man presented to his GP complaining of acute pain, numbness and tingling of the lower extremities and evidence of ischemia. There were no recent flare ups of fever, chills, sweating, respiratory, gastroenteritis, or urinary tract infection. The patient's notes, however, revealed that as a result of endocarditis, he had previously suffered from a mycotic aneurysm and underwent aortic valve replacement 17 months prior to the visit. A second episode of endocarditis occurred post operation, but since no organism was recovered and the patient was treated empirically with a combination of vancomycin and ceftriaxone. The patient was also on long-term anticoagulation therapy (warfarin) due to associated deep vein thrombosis (DVT). Laboratory tests showed elevated white blood cell (WBC), neutrophilia with hypersegmented neutrophils and anemia. Clinical chemistry results revealed decreased serum iron, increased ferritin, increased lactate dehydrogenase and abnormal coagulation. The patient was referred to his local hospital. Surgical intervention to correct occlusion of the distal aortae revealed white debris identified as *Candida albicans*. Aortic vegetation was determined to be the source of the infection. The patient was successfully treated with antifungal therapy: combination of amphotericin B and flucytosine followed by prolonged suppression with fluconazole.

1. WHAT IS THE CAUSATIVE AGENT, HOW DOES IT ENTER THE BODY AND HOW DOES IT SPREAD A) WITHIN THE BODY AND B) FROM PERSON TO PERSON?

CAUSATIVE AGENT

Candida is a commensal fungus found in the gastrointestinal (GI) tract, genitourinary tract, mouth, and skin. In healthy individuals, it does no harm. However, in immunocompromised people *Candida* can cause a range of pathologic conditions from mild superficial infections to life-threatening invasive candidiasis (IC) which comprises candidemia (bloodstream infection), disseminated and deep-seated candidiasis.

The *Candida* species most frequently isolated from severe IC are *C. albicans*, *C. glabrata*, *C. parapsilosis*, and *C. tropicalis*. *C. auris* has emerged as a global pathogen over the past decade, with the majority of cases associated with high mortality. The shift toward other non-*albicans Candida* species is thought to be the result of selective pressure by first-line drugs effectively used against invasive infections caused by *C. albicans*, in particular, fluconazole and echinocandins (see Section 5). However, *C. albicans* remains the most frequent clinically isolated species, arguably the most successful, and the focus of this case.

In response to environmental pressure, *C. albicans* appears in two reversible morphologic forms: unicellular budding yeast and a filamentous form (hyphae and pseudohyphae). Yeast cells have a round-to-oval morphology, while hyphae consist of firmly attached branched tubular cells. Pseudohyphae resemble both yeasts and hyphae (Figure 4.1). Under certain conditions, *C. albicans* can undergo additional morphologic transitions to tochlamydospores, gray cells, white/opaque cells and a particular GUT phenotype characteristic of GI transition.

Among the factors regulating the morphogenesis between yeast and hyphal forms are temperature and pH. Higher temperatures (37°C) and neutral pH favor hyphal morphology, while the yeast form is supported by lower temperatures (less than 30°C) and acidic pH (<6.0). The hyphal form is required at the early stage of infection, penetrating the tissues and

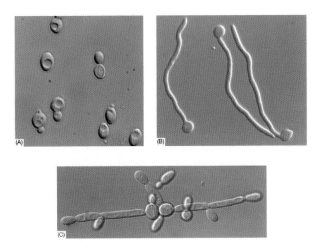

Figure 4.1 Morphologic forms of *C. albicans*: unicellular budding yeast (A); filamentous hyphae (B) and pseudohyphae which resemble both yeasts and hyphae (C). Scale bars represent 5 μm. Thanks to Peter Sudbery for these images. *From Sudbery PE (2011) Nat Rev Micro 9:737–748. With permission from Springer Nature.*

developing resistance to phagocytosis. The yeast form is needed for adhesion to endothelial cells and dissemination into the bloodstream.

The cell wall helps to protect *C. albicans* from the microenvironment within each of the organs and from host immune responses. The primary pathogen-associated molecular patterns (PAMPs) that elicit immune responses to fungal pathogens are chitin and β-glucans. However, in *C. albicans*, the β-glucans are mostly found in the inner cell wall layer and are masked by mannan fibrils located in the outer layer. This masking is a major virulence factor of *C. albicans*.

ENTRY INTO THE BODY

Although a part of the normal microbiota, *Candida* spp. are commonly associated with yeast infections. *C. albicans* can cause a spectrum of pathologies ranging from mucosal to life-threatening systemic infections. Being both a commensal and opportunistic fungal pathogen of humans, *C. albicans* is present in our bodies almost from birth. It is able to colonize most areas of the body, including skin, mucosal membranes, and internal organs. Colonization of the skin and GI tract is thought to occur during passage through the birth canal, where *C. albicans* is a natural resident. As an opportunistic infection, *C. albicans* is capable of rapidly colonizing almost all anatomic sites in the body, developing resistance to antifungal drugs and forming biofilms on medical devices. Biofilms are a collection of one or more types of microorganisms that can grow together on many different surfaces.

Both yeast and hyphal forms of *C. albicans* can create biofilms through a polysaccharide matrix, and they have an important role in drug resistance.

The balance between commensal colonization and opportunistic infection is defined by the innate immune system and resident microbiota. During immune suppression, *C. albicans* can overcome the innate immune system and switch from a harmless commensal to a pathogenic growth, such as seen in superficial mucosal infections (oral and genital thrush), subcutaneous infections and systemic disease (see Section 3).

In order to reach internal organs, *Candida* (the yeast form) must cross the endothelium lining of the blood vessels. *Candida* adheres to endothelial cells by expressing a range of adhesins: αMβ2-like adhesin, which interacts with endothelial intracellular adhesion molecule (ICAM)-1 and ICAM-2; αvβ3-like adhesin, which interacts with platelet endothelial cell adhesion molecule (PECAM)-1; agglutinin-like protein 3, which interacts with N-cadherin. The transmigration of *C. albicans* across endothelial cells might be facilitated by the endocytosis of *Candida*, mediated by the endothelial cell receptor N-cadherin.

SPREAD WITHIN THE BODY

The yeast-to-hyphal transition of *C. albicans* is essential for the pathogen's further spread and is highly dependent on the complex interaction between internal signaling pathways including protein kinases (PK) such as cAMP-PKA, MAP kinase, PKC and external environmental conditions. Subsequent hyphal development requires mechanisms of polarized growth ensuring hyphal elongation and the formation of long tubular cells. This includes continuous delivery of membrane-bound secretory vesicles, along the actin cables.

During infection of the host, *C. albicans* uses several sources of iron via reduction of extracellular ferric chelates and production of siderophores with high iron-binding affinity. In order to obtain heme iron, *C. albicans* hemolyzes erythrocytes. The limited iron conditions in anatomic niches leads to a decrease in mitochondrial activity of the fungus and subsequent changes in cellular metabolism resulting in an increase in lactate production. Lactate production is reduced in a high iron-containing environment.

When colonizing the abdomen, vagina, and GI tract, *C. albicans* experiences a varied pH. The fungus preferably grows in an acidic atmosphere.

PERSON-TO-PERSON SPREAD

Most infections arise from the endogenous flora of patients who have risk factors (see Section 3) following disruption of skin and mucosal barriers. Less commonly, *Candida* can be transmitted via healthcare workers' hands or contaminated medical devices (fomites). A few outbreaks of candidemia have been linked to healthcare workers and hence hand hygiene is essential for preventing the spread of infection. There is also a possibility of the sexual transmission of *Candida*.

EPIDEMIOLOGY

As an opportunistic fungal pathogen, *C. albicans* causes a range of conditions from mild infections to life-threatening candidemia. The use of broad-spectrum antibiotics, immunosuppressive therapies such as for cancer chemotherapy and organ transplantation, increases the probability of the development of IC. Currently *C. albicans* is a leading cause of hospital-acquired infections (nosocomial infections, 15%) with 47% mortality in intensive care unit (ICU) patients.

The exact number of cases of oral (mouth and throat) and esophageal candidiasis (see Section 3) is difficult to determine due to the lack of global surveillance. The risk of these infections correlates with underlying medical conditions. Although uncommon in healthy adults, it is present in almost 30% of people living with HIV/AIDS.

In contrast, vaginal candidiasis (see Section 3) is quite common. In the US, it is the second most frequent vaginal infection after bacterial infection, with around 1.4 million outpatient visits annually.

The average incidence of candidemia, the most common bloodstream infection and IC, is 3.88/100 000 population/year. However, it varies in different geographic regions from 1.0 to 10.4 per 100 000. Importantly, there was an increasing trend from the 1990s (median ratio 2.18) up to 3.22 in the last decade. This trend can be explained by the prolonged survival of critically ill patients, an aging population and increasing usage of long-term intravenous (IV) devices: tunneled intravascular and peripherally inserted central catheters (PICC).

The average incidence of candidemia in the US from 2013–2017, was approximately 9/100 000 population with 25 000 cases recorded annually. Among all age groups, candidemia rates are approximately twice as high in Black people compared with other races/ethnicities. This might be due to differences in underlying conditions, socioeconomic status, healthcare access, etc. As a leading cause of bloodstream infections in US hospitals and its associated long hospital stay (including in intensive care), invasive *Candida* infections are costly for patients and the healthcare system.

Although the prevalence of nosocomial candidemia is high, the occurrence of fungal endocarditis is rare (approximately 1.3–6% of all cases) and mainly affects younger patients with an average age of 44.5 years (see the Case). Fungal endocarditis is most frequently caused by *C. albicans* (30.3%) and *Aspergillus fumigatus* (28%). Endocarditis with *C. albicans* has a better prognosis than *A. fumigatus*, although it has a morbidity and mortality rate of 50%.

2. WHAT IS THE HOST RESPONSE TO THE INFECTION AND WHAT IS THE DISEASE PATHOGENESIS?

The immune response to *Candida* infections includes mucocutaneous protective barriers, innate immunity, and adaptive immunity to a lesser extent.

INNATE IMMUNITY

Innate immunity is essential at preventing opportunistic invasion of *C. albicans*. The phagocytosis of *C. albicans* by macrophages and polymorphonuclear neutrophils (PMNs) is critical for the effective clearance of *C. albicans* from the host tissue. Fungal β-glucan is recognized by Dectin-1, a C-type lectin pathogen recognition receptor (PRR) expressed by phagocytic cells, leading to the activation of phagocytosis and the release of pro-inflammatory cytokines, ROS and reactive nitrogen species (RNS). Infected patients usually have impaired C-type lectin receptor (CLR) activity that are often caused by mutations. Co-stimulation of CLRs and Toll-like receptors (TLRs) is essential for the inflammatory response.

TLR4, mannose receptor, and β-glucan receptor complex Dectin 1/TLR2 recognize N-mannan and O-mannan which are the main components of the *Candida* cell wall.

Dectin-1 increases the production of TLR2 and TLR4-induced cytokines, such as tumor necrosis factor (TNF). INF-γ and TNF are also produced through the co-operation of TLR4 with the mannose receptor. Human keratinocytes, which express a mannose receptor, exhibit anti-*Candida* activity through the production of nitric oxide.

IL-1 plays a key role in immune responses to *Candida* in mucosal surfaces. IL-1R and IL-36R are required for PMN recruitment and protective Th17 responses (see below).

PMNs react to *C. albicans* early in infection by forming neutrophil extracellular traps (NETs) comprised of DNA and other antimicrobial components. These are released during so called NETosis, often stimulated through autophagy, and ROS leading to damage of the invading fungi. Importantly, NETs are more harmful to hyphae forms than to yeast forms of *C. albicans*.

ADAPTIVE IMMUNITY

The adaptive immune response against fungal commensals in the normal skin and gut is mainly generated through T-helper (Th)17 cells (Figure 4.2). A small proportion of fungus-specific Th1 and Th2 cells are also involved. As mentioned above, IL-17 production in response to commensal fungi may regulate epithelial barrier integrity and support the equilibrium between bacterial and fungal commensals. Interestingly, Th17 cells generated against *C. albicans* may modulate neutrophil-mediated immunity against systemic infection with *Streptococcus aureus*.

Stimulation of Th1 cytokine synthesis is thought to be mediated by mannose receptors. It has been shown that during systemic disease, IL-18 drives protective Th1 responses, while IL-33 promotes Th2 and suppresses Th1 immunity.

Differently from phagocytic cells and T cells, the role of B cells and the protective function of antibodies are less well characterized. Specific antibodies directed against fungal cell wall components increase upon fungal infection. It has been shown, *in vitro*, that human serum antibodies impair the adherence and inhibit the invasion of human oral epithelial cells by *C. albicans* through blocking induced endocytosis.

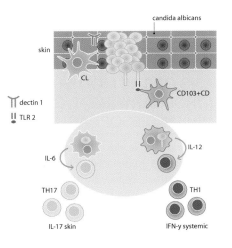

Figure 4.2 Dectin 1 (blue) and Toll-like receptor 2 (TLR 2, green) receptors co-expressed by phagocytic cells (yellow) recognize the main components of the *Candida albicans* (red yeasts) cell wall and release interleukin (IL)-6 and IL-12. IL-6 triggers differentiation of naïve CD4+T cells (white) into Th17+ cells, while IL-12 – into Th1+ cells. Th17+ cells then produce IL-17, and Th1+ cells – interferon gamma (IFN-γ) which represent the leading components of adaptive T-cell immune responses to *C. albicans*.

However, currently the antibody response to *C. albicans* is mostly used for diagnosis purposes (see Section 4).

PATHOGENESIS

Inflammation and epithelial barrier disruption of the skin (such as atopic dermatitis) or in the gut (such as inflammatory bowel disease) might be also triggered by commensal fungi-specific Th17 cells. Serum amyloid A (SAA) could be one of the regulators of Th17 cell function in the gut. *C. albicans*-specific Th17 cells may also cross-react with other fungi, such as *A. fumigatus* in the inflamed lung that leads to chronic obstructive pulmonary disease (COPD) or synergize with *A. fumigatus*-specific Th2 cells causing allergic bronchopulmonary aspergillosis (ABPA - see case 1).

Thrombocytopenia is a known risk factor for invasive fungal infections, including candidiasis. However, the role of platelets and thrombosis in *Candida* infection is poorly understood and is yet to be determined.

In order to evade immune responses, *C. albicans* can adapt to hypoxia and anoxia (absence of oxygen) by repressing the transcription factor of filamentous growth Efg1. Since *C. albicans* is often exposed to fluctuations in pH in different body systems and organs such as the digestive tract, vagina, oral cavity, and blood, it has adapted by modulating the extracellular pH. As a result, alkalinization of the fungus microenvironment would counter macrophage phagosome acidification thus supporting *C. albicans* survival in the macrophage.

3. WHAT IS THE TYPICAL CLINICAL PRESENTATION AND WHAT COMPLICATIONS CAN OCCUR?

In immunocompromised individuals *C. albicans* can infect and colonize large numbers of organs and tissues (Figure 4.3) causing a spectrum of conditions from superficial to life-threatening.

CANDIDIASIS OF THE MOUTH AND THROAT (OROPHARYNGEAL CANDIDIASIS OR THRUSH) AND ESOPHAGUS

This form of candidiasis is uncommon in healthy adults. However, babies have a higher risk for getting candidiasis in the mouth and throat, especially those younger than 1 month of age. For adults, the risk factors include wearing dentures, smoking, taking antibiotics or corticosteroids, such as inhaled corticosteroids for asthma. Any medication that causes dry mouth is considered to be a risk factor. However, in the majority of cases, it is immunodeficiency that is the major risk factor for candidiasis of the mouth, throat and esophagus such as patients with HIV/AIDS, cancer (particularly while undergoing cancer treatment), and diabetes.

Patients usually present with white patches or sores on the inner cheeks, tongue, roof of the mouth, and throat. When the corners of the mouth are red (inflamed), eroded and cracked because of a *Candida* infection, the condition is called angular cheilitis or perleche. This is accompanied by a cotton-like feeling in the mouth, loss of taste, and pain while eating. Symptoms of candidiasis in the esophagus include pain and difficulty in swallowing. Nails might undergo discoloration, onycholysis, and dystrophy.

VAGINAL CANDIDIASIS

This condition is also called "vulvovaginal candidiasis," or "candidal vaginitis." It often appears in pregnant women, women with diabetes, immunodeficiency, and those who take hormonal contraceptives and antibiotics. Vaginal candidiasis usually first presents as a thick white or yellow vaginal discharge (leukorrhea) with itching and redness of the female genitalia (vagina and vulva), pain when urinating and during sexual intercourse. However, about 20% of women who have vaginal *Candida* are asymptomatic. Despite being categorized as a mild condition, vulvovaginal candidiasis can have a debilitating impact on quality of life, and the infection is often recurrent.

Infection of the tip of the penis with *Candida* (Glans Penis) is less common than vulvovaginal candidiasis. This infection may be seen in men whose sexual partners have vulvovaginal candidiasis and in men with diabetes. The lesion appearance is similar to that described above.

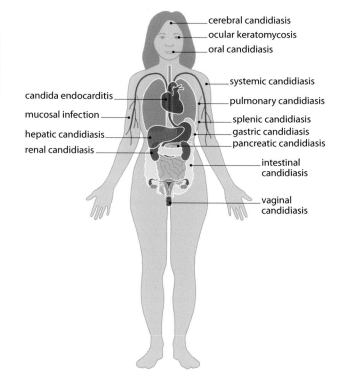

Figure 4.3 In patients with primary or secondary (acquired) immunodeficiency *C. albicans* can infect and colonize almost all organs and tissues in the body causing mild (oral and vaginal candidiasis) or severe life-threatening conditions – blood candidemia and invasive candidiasis of internal organs.

INVASIVE CANDIDIASIS (IC)

The most common form of IC is candidemia, a bloodstream infection, but it does not represent all forms of IC. As systemic candidiasis, a very serious condition, infection can occur in the heart, kidney, bones, and other internal organs without even being detected in the blood. In fact, the number of cases of IC might be twice as high as the estimate for candidemia. On the other hand, 5–20% of such cases of primary deep-seated candidiasis can lead to secondary candidemia.

The risk factors include: primary or acquired immunodeficiency due to various underlying conditions, neutropenia, stem-cell transplantation, prolonged ICU stay, central venous catheters, parenteral nutrition, being on hemodialysis due to the renal failure, and taking broad-spectrum antibiotics. Pre-term and very low birth weight infants are also at risk.

Candidemia and deep-seated candidiasis may occur concurrently or independently. Primary candidemia most often occurs from GI tract transition of commensal *Candida* into the blood or colonization of an IV catheter. Hematogenous seeding of *Candida* accounts for approximately 50% of secondary deep-seated candidiasis. However, the latter may also be caused by the non-hematogenous introduction of *Candida* into sterile sites, particularly into the abdominal cavity following GI tract perforation or via an infected peritoneal catheter.

The most common symptoms of IC are fever and chills that do not improve after antibiotic treatment for suspected bacterial infections. Other symptoms can develop if the infection spreads to other parts of the body, such as the heart, brain, eyes, bones, or joints as systemic candidiasis. Inflammation of the membrane lining the heart (endocarditis, see the Case at the start of the chapter), the membrane lining the skull (meningitis), or rarely inflammation of the bone (osteomyelitis) may also occur.

IC in neonatal intensive care units (NICUs) remains a leading cause of newborn mortality and morbidity, and particularly in those infants requiring surgery.

IC is often associated with high rates of morbidity and mortality, as well as an increased length of hospital stay. In hospital, mortality among patients with candidemia is approximately 19–24%. However, because people who develop IC typically suffer from other severe medical conditions, it can be difficult to determine the proportion of deaths directly attributable to *Candida* infection.

4. HOW IS THE DISEASE DIAGNOSED, AND WHAT IS THE DIFFERENTIAL DIAGNOSIS?

SAMPLE COLLECTION

Candidiasis of the mouth or throat is usually diagnosed simply by observation, although sometimes a swab might be taken from the mouth or throat for laboratory testing. Candidiasis in the esophagus often requires endoscopic examination of the digestive tract. In order to diagnose vaginal candidiasis, a small sample of vaginal discharge is usually collected for microscopy or used for fungal culture. The results should be interpreted with caution since a positive fungal culture does not always mean symptomatic *Candida* infection.

In the case of IC, identification of *C. albicans* is required to be made in the isolate from a clinical specimen. Fungal cultures can be grown from blood or from tissue and sterile body fluids collected from deep sites of infection. They require invasive procedures for sample collection and might be too risky for further spreading of the fungal infection.

GOLD STANDARD: FUNGAL CULTURE

Blood cultures are still the gold standard method for the diagnosis of candidemia despite the variable sensitivity (21–71%) in the first stages of infection. Blood cultures can take around 2–5 days to achieve conclusive results and may not become positive until late in the disease course. This might lead to delayed therapeutic interventions and poor disease outcome.

Usually, a concentrated sample of blood is directly inoculated onto solid media, or a non-concentrated blood is first incubated in broth media. After incubation at 37°C on Sabouraud Dextrose Agar, *Candida* species appear as creamy white colonies with a smooth and waxy surface. At the base of the colony, there are usually tiny projections, known as feet.

In cell cultures, identification of *C. albicans* from other *Candida* species can be achieved by incubating in serum or plasma at 37°C for 3 hours. Under these conditions, only *C. albicans* and *C. dubliniensis* produce germ tubes, the initial stage of hypha formation in the yeast cell. Cornmeal agar induces spore formation and is important for further identification. *C. albicans* at 42°C produces thick-walled spores, while *C. dubliniensis* does not.

Cell culture is sensitive for detecting viable *Candida*. Blood cultures are positive in most patients if samples are collected during active candidemia, but only in approximately 40% of patients with concomitant deep-seated infection. Blood cultures are often negative during deep-seated IC without candidemia. Across the spectrum of IC, the sensitivity of blood cultures is therefore approximately 50%. Moreover, the collection of deep-tissue cultures requires invasive procedures that may be contraindicated in patients at risk of *Candida* infections.

HISTOPATHOLOGY

Histopathologic detection of *Candida* in biopsy specimens allows the confirmation of IC and assessment of the level of tissue invasion and inflammation, although its downside is the invasiveness of the procedure and its low sensitivity. It is contraindicated in patients with severe thrombocytopenia and coagulopathies.

Serum Assays

Serum assays have been developed for *Candida* antigens and anti-*Candida* antibodies. These include detection of the most abundant cell-wall components mannan and 1,3-β-D-glucan (BDG). However, antigen-based assays are limited due to decreased serum concentrations and rapid clearance of antigens from the blood as well as low sensitivity in immunosuppressed hosts. The major concern for the BDG essay is false positivity. Antibody based assays have also been used to detect serum IgG specific to mannan, rather than IgM. In addition, *C. albicans* germ tube antibody (CAGTA) assays for IA infection are employed in the EU, but not in the US since they are not cleared by the US Food and Drug Administration (FDA).

Sensitivity and specificity of the combined mannan/antimannan assay can reach 85%, although reports of individual assay specificity from different laboratories and the types of IC tested vary dramatically from 5% to 100%.

POLYMERASE CHAIN REACTION (PCR)-BASED ASSAYS

Various sources of *Candida* DNA can be used for PCR: blood, serum, plasma, cerebrospinal fluid (CSF) and different tissues. The DNA targets include multi-copy broad-range panfungal genes (such as *5.8S, 18S* or *28S* ribosomal sequences; or the internal transcribed spacer (ITS) 1 or ITS2 regions within the *Rrna* gene as well as *Candida*-specific genes, such as *ERG11, HSP90, SAP1-6, CHS1* or *ACT1*. Some commercial broader multiplex PCR-based assays for bacterial and fungal pathogens might also be of use. These assays target the five most common pathogenic *Candida* species (*C. albicans, C. glabrata, C. parapsilosis, C. tropicalis,* and *C. krusei*), which account for >95% of IC. However, PCR essays can yield false-negative results in the case of a low number of fungal cells available in the sample as well as problems with DNA extraction and false positivity – due to sample contamination. There are currently no FDA-cleared PCR assays for *Candida*.

T2 Candida Nanodiagnostic Panel

The FDA-approved automated T2 Candida nanodiagnostic panel assesses *Candida* directly in whole blood using magnetic resonance technology. The method is applicable for *C. albicans/C. tropicalis, C. glabrata/C. krusei,* and *C. parapsilosis*.

Fluorescence in situ hybridization assay using peptide nucleic acid probes (PNA FISH) targets species-specific RNA in yeast and can identify *C. albicans* in a blood culture in three hours.

A *latex agglutination test* is based on the detection of heat-labile *Candida* protein antigens in the serum such as a 47-kDa fragment of Hsp90, Eno1, Mp65 or Sap1/2. It has a low sensitivity since, as mentioned above, *Candida* antigens rapidly disappear from the bloodstream. Candida mannans can also be measured using this test.

Loop-mediated isothermal amplification (**LAMP**) is a relatively new technique based on the usage of specific primers for the yeast DNA targeting the ITS2 gene of *C. albicans*. This is a highly sensitive method allowing one to differentiate *C. albicans* samples from other yeast species. The LAMP method is as specific and precise as common diagnostic methods, but is faster, easier deployable, more sensitive, and can be used in the field.

Differential diagnosis of candidiasis include the following:

* Cutaneous candidiasis - Dermatitis (contact, allergic), folliculitis.

* GI tract candidiasis – Esophagitis due to herpes simplex virus, herpes zoster, induced by radiation, gastroesophageal reflux disease.

* Respiratory candidiasis – Bacterial pneumonia, viral pneumonia, tracheitis, Aspergillus pneumonia, pulmonary cryptococcosis.

* For IC and/or candidemia, differential diagnosis should be considered from abdominal abscess, bacterial sepsis, and septic shock.

5. HOW IS THE DISEASE MANAGED AND PREVENTED?

MANAGEMENT

Three classes of antifungal drugs are used as first-line treatment of IC: polyenes (amphotericin B), azoles (fluconazole), and echinocandins (anidulafungin, caspofungin, and micafungin). Amphotericin B and fluconazole target fungal sterol ergosterol, while anidulafungin affects the fungal cell wall through inhibition of β-1,3-glucan synthase. Due to the superior fungicidal activity and positive safety record, the echinocandins are now the first-line antifungal treatment for IC in both neutropenic and non-neutropenic critical patients.

For most adults with IC, the initial recommended antifungal treatment is an echinocandin given IV. Fluconazole can be used as an alternative as initial therapy for those patients who are not critically ill and who are considered unlikely to have a fluconazole-resistant *Candida* infection. Other treatments include voriconazole and amphotericin B formulations. The treatment should continue for 2 weeks after clearance of *Candida* from the bloodstream and resolution of symptoms.

For neonatal candidiasis, the recommended primary treatment is amphotericin B deoxycholate or fluconazole for

2 weeks after clearance of *Candida* from the bloodstream and resolution of symptoms. Other forms of IC, such as infections in the bones, joints, heart, or central nervous system (CNS), usually need to be treated for a longer period of time.

However, some types of *Candida* are becoming increasingly resistant to the first-line and second-line treatment with fluconazole and the echinocandins. Therefore, new anti-fungal drugs have recently been introduced such as tetrazoles, encochleated amphotericin B, terpenoid derivative ibrexafungerp, rezafungin, the novel echinocandin with an enhanced half-life, and other drugs which are in the process of becoming licensed.

For candidiasis of the mouth, throat, or esophagus, clotrimazole, miconazole, or nystatin are usually applied to the inside of the mouth for 7 to 14 days. For severe infections, fluconazole is taken orally or IV. The treatment for candidiasis in the esophagus is usually fluconazole.

For vaginal candidiasis, an antifungal medicine is applied inside the vagina or a single dose of fluconazole is taken orally. For unresponsive or more severe infections, the doses of oral fluconazole can be increased or boric acid, nystatin, or flucytosine applied inside the vagina.

PREVENTION

Candidiasis in the mouth and throat can be prevented through maintaining good oral health, rinsing the mouth or brushing teeth after using inhaled corticosteroids, particularly for immunocompromised individuals. Wearing cotton underwear might help reduce the chances of getting Candida infection as well as taking antibiotics strictly as prescribed.

In hospital environments, it is important to adhere to hand hygiene, and follow recommendations for placement and maintenance of central venous catheters.

Patients who may benefit from antifungal prophylaxis are solid-organ transplant recipients, stem-cell transplant recipients with neutropenia, high-risk ICU patients, and those with chemotherapy-induced neutropenia.

There are currently no vaccines available for the prevention of Candidiasis. However, there are some encouraging results. Two anti-*Candida* vaccines have reached the stage of clinical trials.

- PEV7 contains recombinant aspartyl-proteinase 2 (Sap2), (an enzyme secreted by *C. albicans*), delivered via virosomes.

- NDV-3 contains the recombinant N terminus of *C. albicans* agglutinin-like sequence 3 protein Als3p, which is a cell-surface adhesin and invasin. Aluminum hydroxide (Alum) is used as adjuvant.

An alternative strategy would be to produce a vaccine based on common antigens expressed by multiple genera of fungi in order to target a broad range of mycoses.

SUMMARY

1. WHAT IS THE CAUSATIVE AGENT, HOW DOES IT ENTER THE BODY, AND HOW DOES IT SPREAD A) WITHIN THE BODY AND B) FROM PERSON TO PERSON?

- *Candida* is a commensal fungus found in the GI tract, genitourinary tract, mouth, and skin. In healthy individuals, it does no harm. However, in immunocompromised people *Candida* can cause a range of pathologic conditions from mild superficial infections to life-threatening invasive candidiasis (IC).

- *C. albicans* appears in two reversible morphologic forms: a unicellular budding yeast and filamentous form (hyphae and pseudohyphae).

- As an opportunistic infection, *C. albicans* is capable of rapidly colonizing almost all anatomic sites in the body. In order to reach internal organs, *Candida* yeast adheres to endothelial cells by expressing a range of adhesion molecules. The yeast-to-hyphal transition of *C. albicans* is essential for the pathogen's further spread.

- The use of broad-spectrum antibiotics, immunosuppressive therapies such as cancer chemotherapy, and organ transplantation, increases the probability of the development of invasive candidiasis.

- Currently, *C. albicans* is a leading cause of hospital-acquired infections with 47% mortality in intensive care unit (ICU) patients.

2. WHAT IS THE HOST RESPONSE TO THE INFECTION AND WHAT IS THE DISEASE PATHOGENESIS?

- Innate immunity is essential for preventing infection by the opportunist *C. albicans*. The phagocytosis of *C. albicans* by macrophages and polymorphonuclear neutrophils (PMNs) is critical for the effective clearance of *C. albicans* from the host tissues.

- Fungal β-glucan is recognized by Dectin-1, a C-type lectin pathogen recognition receptor (PRR) expressed by phagocytic cells, leading to the activation of phagocytosis and the release of pro-inflammatory cytokines, reactive oxygen species (ROS) and reactive nitrogen (RNS) species.

- The adaptive immune response against fungal commensals in the normal skin and gut is mainly through Th17 cells and through a small proportion of fungus-specific Th1 and Th2 cells.

Continued...

...continued

- Specific antibodies directed against fungal cell-wall components increase upon fungal infection and can impair the adherence and inhibit invasion of human oral epithelial cells by *C. albicans* through blocking induced endocytosis.

3. WHAT IS THE TYPICAL CLINICAL PRESENTATION AND WHAT COMPLICATIONS CAN OCCUR?

- Candidiasis of the mouth and throat (oropharyngeal candidiasis or thrush) and esophagus presents as white patches or sores on the inner cheeks, tongue, roof of the mouth, and throat, accompanied by a cotton-like feeling in the mouth, loss of taste, and pain while eating.
- Symptoms of candidiasis in the esophagus include pain and difficulty in swallowing.
- Vaginal candidiasis usually first presents as a thick white or yellow vaginal discharge (leukorrhea) with itching and redness of the female genitalia (vagina and vulva), pain when urinating and during sexual intercourse.
- The most common form of invasive candidiasis (IC) is candidemia, a bloodstream infection.
- Systemic candidiasis infection can occur in the heart, kidney, bones, and other internal organs. The risk factors include: primary or acquired immunodeficiency, neutropenia, stem-cell transplantation, prolonged intensive care unit stay, central venous catheters, parenteral nutrition, being on hemodialysis due to the renal failure, and taking broad-spectrum antibiotics.
- The most common symptoms of IC are fever and chills that don't improve after antibiotic treatment for suspected bacterial infections. IC is often associated with high rates of morbidity and mortality, as well as increased length of hospital stays.

4. HOW IS THE DISEASE DIAGNOSED, AND WHAT IS THE DIFFERENTIAL DIAGNOSIS?

- Blood cultures are still the gold standard method for the diagnosis of candidemia despite the variable sensitivity (21–71%) in the first stages of infection.

- Serum assays include detection of the most abundant cell-wall components mannan and 1,3-β-D-glucan (BDG). Antibody based assays have also been used for detecting specific serum IgG antibody levels to mannan. The sensitivity and specificity of the combined mannan/antimannan assay can reach 85%.
- Various sources of *Candida* DNA can be used for PCR: blood, serum, plasma, cerebrospinal fluid and different tissues. The DNA targets include multi-copy broad-range panfungal genes as well as *Candida*-specific genes. However, PCR essays can yield false-negative results in the case of a low number of fungal cells in the sample.
- The FDA approved automated T2 Candida nanodiagnostic panel assesses *Candida* directly in whole blood using magnetic resonance technology.
- Loop-mediated isothermal amplification (LAMP) is a highly sensitive method allowing one to differentiate *C. albicans* samples from other yeast species.

5. HOW IS THE DISEASE MANAGED AND PREVENTED?

- For most adults with IC, the initial recommended antifungal treatment is an echinocandin given intravenously. Fluconazole can be used as an alternative as initial therapy for those patients who are not critically ill and who are considered unlikely to have a fluconazole-resistant *Candida* infection. Other treatments include voriconazole and amphotericin B formulations. The treatment should continue for 2 weeks after clearance of *Candida* from the bloodstream and resolution of symptoms.
- For candidiasis of the mouth, throat, or esophagus, clotrimazole, miconazole, or nystatin are usually applied to the inside of the mouth for 7 to 14 days. For severe infections, fluconazole is taken orally or intravenously.
- For vaginal candidiasis, an antifungal medicine is applied inside the vagina or a single dose of fluconazole is taken orally.
- There are currently no vaccines available for the prevention of Candidiasis.

FURTHER READING

Ostrosky-Zeichner L (ed). Fungal Infections, An Issue of Infectious Disease Clinics of North America. Elsevier, Philadelphia, 2021.

Pappas PG, Lionakis MS, Arendup MC, et al. Invasive Candidiasis. Nat Rev Disease Primers, 4: 18026, 2018.

REFERENCES

Bojang E, Ghuman H, Kumwenda P, Hall RA. Immune Sensing of Candida albicans. J Fungi, 7: 119, 2021.

Chow EWL, Pang LM, Wang Y. From Jekyll to Hyde: The Yeast–Hyphal Transition of Candida albicans. Pathogens, 10: 859, 2021.

Clancy CJ, Nguyen MH. Diagnosing Invasive Candidiasis. J Clin Microbiol, 56: e01909–17, 2018.

Griffiths JS, Camilli G, Kotowicz NK, et al. Role for IL-1 Family Cytokines in Fungal Infections. Front Microbiol, 12: 633047, 2021.

Hameed S, Hans S, Singh S, et al. Revisiting the Vital Drivers and Mechanisms of β-Glucan Masking in Human Fungal Pathogen, Candida albicans. Pathogens, 10: 942, 2021.

Murphy SE, Bicanic T. Drug Resistance and Novel Therapeutic Approaches in Invasive Candidiasis. Front Cell Infect Microbiol, 11: 759408, 2021.

Oliveira LVN, Wang R, Specht CA, Levitz SM. Vaccines for Human Fungal Diseases: Close but still a Long Way to Go. NPJ Vaccines, 6: 33, 2021.

Scheffold A, Bacher P, LeibundGut-Landmann S. T-cell Immunity to Commensal Fungi. Curr Opin Microbiol, 58: 116–123, 2020.

Tortorano AM, Prigitan A, Morroni G, et al. Candidemia: Evolution of Drug Resistance and Novel Therapeutic Approaches. Infect Drug Resist, 14: 5543–5553, 2021.

Wich M, Greim S, Ferreira-Gomes M, et al. Functionality of the Human Antibody Response to Candida albicans. Virulence, 12: 3137–3148, 2021.

WEBSITES

Centers for Disease Control and Prevention, Fungal Diseases, 2021: https://www.cdc.gov/fungal/diseases/candidiasis/thrush/index.html

Web MD, What is Candidiasis? 2021: https://www.webmd.com/skin-problems-and-treatments/guide/what-is-candidiasis-yeast-infection

> Students can test their knowledge of this case study by visiting the Instructor and Student Resources: [www.routledge.com/cw/lydyard] where several multiple choice questions can be found.

Chlamydia trachomatis

5

A 19-year-old woman was seen by her doctor for a routine gynecologic examination and complained about some mid-cycle bleeding. She had been with her current boyfriend for a year, had a pregnancy termination 2 years previously, and was taking birth control pills.

Internal examination revealed a **mucopurulent** discharge at the external cervical os (**Figure 5.1**). The cervix was friable and bled easily. The doctor suspected chlamydial infection and collected an endocervical swab specimen for a *Chlamydia* test. The woman returned for the results and was told that the test for *Chlamydia* was positive.

The doctor prescribed a course of doxycycline for 1 week and explained to the patient that this treatment is sufficient and effective in more than 95% of cases, and no repeat testing to prove the eradication of this infection is necessary. However, due to the high risk of re-infection in sexually active young adults, the doctor recommended her to return for a follow-up visit in 6 months.

He further stated that a timely cleared chlamydial infection does not normally lead to infertility, although around a 10% probability of ectopic pregnancy remains. The woman was given a leaflet on chlamydial infection and on other sexually transmitted infections (STIs).

The patient was counseled regarding safe-sex practices. The doctor also advised her to contact the local genitourinary clinic to be tested for other STIs including HIV and made all necessary arrangements according to the national guidelines.

As a part of the disease management, the doctor carried out all the appropriate actions recommended in the national guidelines to notify the patient's partner and to advise him to visit a genitourinary clinic or to see his doctor for *Chlamydia* and other STI tests. The couple were advised to abstain from sex until both were cleared of the infection.

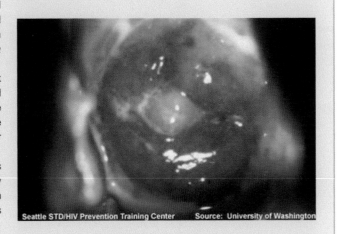

Seattle STD/HIV Prevention Training Center   Source: University of Washington

Figure 5.1 Mucopurulent cervical discharge caused by chlamydial infection, showing ectopy and edema. *Courtesy of the Seattle STD/HIV Prevention Training Center, University of Washington, Seattle.*

1. WHAT IS THE CAUSATIVE AGENT, HOW DOES IT ENTER THE BODY AND HOW DOES IT SPREAD A) WITHIN THE BODY AND B) FROM PERSON TO PERSON?

CAUSATIVE AGENT

This patient was infected with *Chlamydia trachomatis*, which belongs to the Family Chlamydiaceae of the Order Chlamydiales. The Chlamydiaceae Family consists of two genuses – *Chlamydia* and *Chlamydophila*. Bacteria of this Family are obligate intracellular human and animal pathogens. The term "Chlamydia" is derived from the word "chlamys," which means cloak (Khlamus) in Greek, an appropriate name reflecting the cloak-like chlamydial inclusion around the host cell nucleus (see below).

Chlamydiaceae are some of the most widespread bacterial pathogens in the world and there are several species that infect a variety of hosts based on a wide range of tissue tropism. Two species, *Chlamydia trachomatis* and *Chlamydophila pneumoniae*, are human pathogens and are responsible for various diseases that represent a significant economic burden. *Chlamydophila psittaci* and *C. pecorum* are mainly bird/animal pathogens, although **zoonotic** transmission of the former to humans can occur resulting in the disease psittacosis.

Chlamydiaceae species share some important structural features. Like other gram-negative bacteria, they have inner and outer membranes, but have a specific **lipopolysaccharide** (**LPS**) that differs from that of other bacteria. The extracellular osmotic stability of Chlamydiaceae is provided by several complex disulfide cross-linked membrane proteins, the main ones being a 40 kDa major outer-membrane protein (MOMP, a product of the *ompA* gene); a hydrophilic cysteine-rich 60 kDa protein (OmcA); and a low molecular weight cysteine-rich lipoprotein (OmcB). The Chlamydiaceae are thought to have little or no muramic acid, the hallmark constituent of peptidoglycan (PG). No transfer RNAs were identified in chlamydial cells, thus confirming the parasitism of this bacterium.

There are two human biological variants (biovars) of *C. trachomatis*: **trachoma** and lymphogranuloma venereum (LGV), and one biovar *C. pneumoniae* infecting mice, and causing mouse pneumonitis (which will not be discussed here).

Fifteen serologic variants (**serovars**) have been identified in the trachoma biovar (A–K, Ba, Da, Ia, and Ja). These mostly infect columnar and squamo-columnar epithelial cells of mucous membranes (see below). Serovars D–K, Da, Ia, and Ja typically infect genitourinary tissues, but were also found in the mucous membranes of the eye conjunctiva and epithelial tissues in the neonatal lung. Serovars A, B, Ba, and C generally infect the conjunctiva and cause trachoma. The LGV biovar consists of four serovars (L1, L2, L2a, and L3), which predominantly infect monocytes and macrophages passing through the epithelial surface to regional lymphoid tissue. Proteomic analysis of the pathogen is very difficult since Chlamydiaceae species are all obligate intracellular parasites and cannot grow in a cell-free system. As a result, most of the data obtained on the structural and functional proteins and

biochemical pathways utilized by Chlamydiaceae are derived from gene sequencing and indirect evidence. The genome of *C. trachomatis* (serovars A, D, and L2) has been fully sequenced.

Chlamydiaceae species have a more complicated biphasic developmental life cycle (Figure 5.2) than other bacteria in that they have two different forms, a metabolically inert infectious elementary body (EB) and a larger noninfectious reticulate body (RB). Interestingly, the term "elementary body" belongs to the virology world and is derived from the time when Chlamydiaceae were initially considered to be viruses. The EB form of the bacterium survives outside the host cell whereas the RB form lives and replicates in a specialized vacuole of the host cell called an inclusion (another virology-derived term).

ENTRY INTO THE BODY

The infectious EB form of the majority of *Chlamydia* strains is typically 0.2–0.3 μm in diameter. MOMP makes up 60% of its cell wall. Due to their rigid outer membrane, EBs are able to survive outside the eukaryotic host cells. *C. trachomatis* infecting genital tissue usually does so through small abrasions in the mucosal surfaces. The EBs infect nonciliated columnar, cuboidal or transitional epithelial cells, but can also infect macrophages. The infection process is multivalent. EBs bind to the epithelial cells directly via cellular proteoglycans

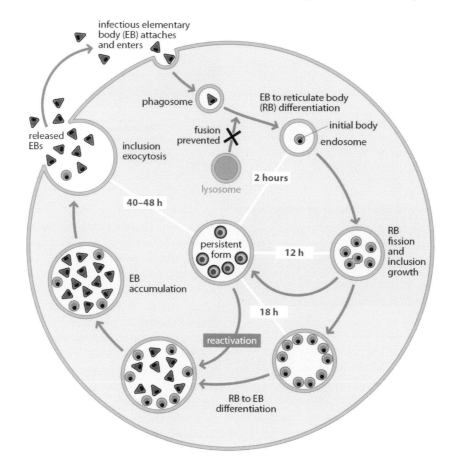

Figure 5.2 Biphasic developmental cycle of *C. trachomatis*. *From Brunham R & Rey-Ladino J (2005) Nat Rev Immunol 5:149-161. https://doi. org/10.1038/nri1551. With permission from Springer Nature.*

with heparan sulfate (HS) moieties (HSPGs) mainly through electrostatic interactions. Depending on the serovar, the degree of HS involvement varies. Additionally, *C. trachomatis* EB form is able to adhere to and enter cells indirectly through binding with fibroblast growth factor 2 (FGF2). This complex then interacts with the FGF2 receptor, which mediates EB internalization into cells. The EB-FGF2 complex may involve synergistic interactions with the EB membrane bacterial protein OmcB, which, in turn, interacts with HSPGs. This facilitates chlamydial entry through **phagocytosis**, receptor-mediated **endocytosis**, and **pinocytosis**.

Early intracellular phase (0–2 hours after infection). Once inside the cell, EB-containing vacuoles move toward the perinuclear region. The vacuole membrane phospholipids promote homotypic fusion of EB vacuoles with each other, but not with **lysosomes** – an important feature that helps the pathogen to avoid intracellular destruction. The homotypic fusion is specific to *C. trachomatis* only and not to other *Chlamydia* species. This results in the formation of a single fusion vacuole (a nascent inclusion) containing several EBs. Some of these vacuoles may contain different serovars such as F and E, leading to the possibility for genetic exchange to occur.

The inclusions then move toward the microtubule organization center where they are supplied with nutrients via the host-cell Golgi apparatus. Bacterial proteins are directly secreted into the host cell cytosol.

Inclusion development (2–40 hours after infection). The EB forms now undergo a lengthy and complex development using host-cell ATP and nutrients as a source of energy. With a reduced genome, *C. trachomatis* is dependent on its host for survival and hijacks host-cell metabolism particularly glycolytic enzymes, aldolase A, pyruvate kinase, and lactate dehydrogenase which are enriched at the *C. trachomatis* inclusion membrane during infection. The increased requirement for glutamine, important for the growth of *C. trachomatis* in infected cells is achieved by reprogramming the glutamine metabolism.

Still remaining in the inclusion, the EB forms now transform into RB forms, which are typically 0.8–1.0 μm in diameter. RBs multiply by binary fission so that the resulting inclusions may contain 500–1000 progeny RBs and occupy up to 90% of the cell cytoplasm. After several rounds of replication, RB forms revert to the infectious EB forms. In tissue culture, the productive infectious cycle of *Chlamydia* lasts about 48–72 hours depending on the serovar. Eventually, the EBs are released to infect other adjacent cells. *C. trachomatis* is able to regulate host cell **apoptosis** throughout the early and productive growth stages. Nonreplicating forms of *Chlamydia* inhibit apoptosis through interfering with the *TP53* tumor suppressor gene. However, late in the life cycle, the pathogen produces a caspase-independent pro-apoptotic Bax protein, which facilitates apoptosis of the host cell thus freeing the secondary EB forms.

Under conditions unfavorable for a pathogen, such as the presence of **interferon-γ** (**IFN-γ**), lack of nutrients or drug treatment (see Section 5) *Chlamydia* may enter a nonreplicating mode called persistence (Figure 5.2). During persistent infection, the developmental cycle is lengthened or aborted and RB forms are produced that do not divide or differentiate back into the EB forms. Persistent infection with *C. trachomatis* may lead to serious clinical conditions that are difficult to treat. However, dormant forms can revert to metabolically active forms if the unfavorable conditions are removed.

SPREAD WITHIN THE BODY

EBs of *C. trachomatis* infect cervical columnar epithelial cells, but the bacteria can spread by ascending into the endometrium and the fallopian tubes, causing **pelvic inflammatory disease** (**PID**), ectopic pregnancy, and infertility (see Section 3). Sexually transmitted *C. trachomatis* serovars D–K can also lead to conjunctivitis through autoinoculation or ocular–genital contact.

PERSON-TO-PERSON SPREAD

Genital Tract Infections

C. trachomatis (serovars D–K) is sexually transmitted in vaginal fluid or semen containing the EB form, through vaginal intercourse but occasionally via oral and anal sex. These serovars can also be vertically transmitted from mother to child during birth through an infected birth canal, causing conjunctivitis (**ophthalmia neonatorum**) or chlamydial **pneumonia**.

C. trachomatis (biovar LGV) is also sexually transmitted. In some parts of Africa, Asia, South America, and the Caribbean it is largely found in heterosexuals. In outbreaks in industrialized countries, the cases are mostly confined to men who have sex with men (MSM) with multiple sexual partners.

Ocular Infections

C. trachomatis (serovars A–C) is found predominantly in areas of poverty and overcrowding. Infection can be transmitted from eye-to-eye by fingers, shared cloths or towels, by eye-seeking flies, and by droplets (coughing or sneezing). The latter route is possible because *C. trachomatis* can exist in the nasopharynx and external nasal exudates of children with trachoma.

Importantly, the undiagnosed and untreated children can contribute to a so-called "age-reservoir effect" responsible for the continuous transmission within the community. These must be identified and treated to prevent further spread of the pathogen in communities.

EPIDEMIOLOGY

C. trachomatis (serovars D–K) is the leading bacterial cause of STIs, with over 50 million new cases occurring yearly worldwide and 4 million new cases each year in the US. The highest infection rates are detected in African Americans, American Indian/Alaska Natives, and Hispanics. Two-thirds of new chlamydial infections occur among youth aged 15–24

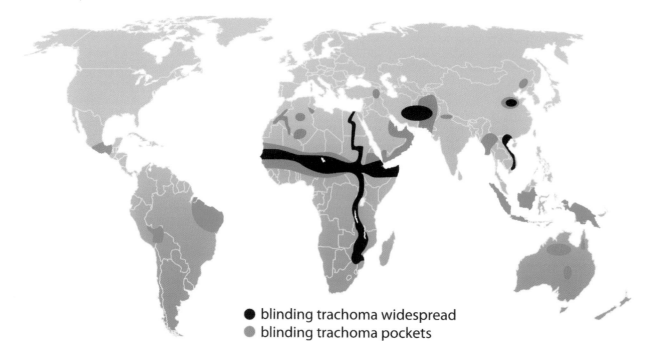

Figure 5.3 **World distribution of trachoma according to the WHO.** *C. trachomatis* (biovar trachoma) is endemic in large areas of Africa and the Middle East, and focal areas of disease are found in India, South-East Asia, and Latin America. As per 2020 WHO data, trachoma affects some 140 million people worldwide, and about 1.9 million people suffer visual loss and blindness. *From Belland R, Ojcius D & Byrne G (2004) Focus: Chlamydia Nat Rev Microbiol 2:530. https:doi.org/10.1038/nrmicro931. With permission from Springer Nature.*

years. In general practice, around 1 in 20 sexually active women aged less than 25 years may be infected. Among MSM worldwide incidence of rectal chlamydial infection range from 3.0% to 10.5%, and pharyngeal chlamydial infection from 0.5% to 2.3%.

Interestingly, there was a 30% decline in Chlamydia cases as well as in all major STIs in England in 2020 compared to 2019, most likely due to the COVID-19 lockdowns.

Every year *C. trachomatis* (serovars A–C) is a major cause of 500 000 cases of trachoma worldwide (Figure 5.3). Active trachoma affects some 85 million people, more than 10 million have **trichiasis** (turned-in eyelashes that touch the eye globe, Figure 5.4), and about 6 million people suffer visual loss and

blindness. Active disease is most commonly seen in children and, in adults, the prevalence of trichiasis is about three times higher in women than in men. Trachoma is **endemic** in large areas of Africa and the Middle East, and focal areas of disease are found in India, South-West Asia, Latin America, and Aboriginal communities in Australia. The disease is generally found in clusters in certain communities or even households, indicating the existence of local risk factors in addition to the generally accepted poverty and lack of water and sanitation.

C. trachomatis (biovar LGV – serotypes L1–L3) causes sexually transmitted disease that is prevalent in parts of Africa, Asia, South America, the Caribbean, and increasingly in Europe and the US. Humans are the only natural host, with MSM being the major reservoir of the disease. In the US, the incidence is 300–500 cases per year.

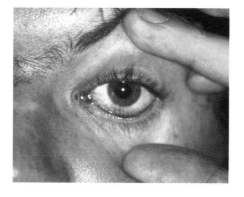

Figure 5.4 **Trachomatous trichiasis: at least one eyelash rubs on the eyeball.** *Courtesy of the Centers for Disease Control, Atlanta, Georgia. Image is found in the Public Health Image Library #4076. Additional photographic credit is given to Joe Miller, PhD, who took the photo in 1976.*

2. WHAT IS THE HOST RESPONSE TO THE INFECTION AND WHAT IS THE DISEASE PATHOGENESIS?

INNATE IMMUNITY

Both innate and adaptive immune responses are induced during *C. trachomatis* infection. However, these responses are often insufficient, providing only partial clearance of the pathogen from the body. This can lead to protracted chronic infections and chronic inflammation contributing to pathogenesis.

Innate immune responses. Chlamydial infection initially induces an influx of polymorphonuclear cells and macrophages into the infection site as a part of acute inflammation. This is facilitated by the release by the infected epithelial cells of **cytokines** and **chemokines**, particularly **interleukin-8 (IL-8)**, which is a powerful neutrophil attractant. Infiltration with neutrophils and macrophages is followed by accumulation of B cells, T cells, and **dendritic cells (DCs)** in submucosal areas launching T-cell responses and antibody production.

Chlamydial PAMPs (pathogen-associated molecular patterns) are recognized by multiple PRRs (pattern recognition receptors), particularly TLR2 which was shown to be activated by MOMP and heat shock protein 60 (Hsp60), during the productive infection, although the direct ligand for TLR2 in this infection is still unknown.

Antibody responses. A humoral response is invoked resulting in production of mucosal secretory **IgA** and circulating **IgM** and **IgG** antibodies. These antibodies are mostly specific for MOMP and Hsp60. Anti-chlamydial IgG antibodies are also found in ocular secretions in trachoma patients. The effectiveness of these antibodies in damaging or blocking further entry of EBs to the cells adjacent to the infected ones is unclear. However, it appears, that IgG antibodies bound to the bacterial MOMP invoke antibody-dependent cellular cytotoxicity (ADCC) by **natural killer (NK)** cells.

Cellular responses. It is likely that at least some professional antigen-presenting cells (APCs) at the site of infection engulf EBs, process and present Chlamydia peptides via the major histocompatibility complex (MHC) class II-mediated pathway leading to the activation of CD4+ T cells. DCs in response to the infection produce IL-12 and drive Th1-cell development and hence **production of interferon γ (IFN-γ)** – the major inhibitory cytokine for *Chlamydia* (see below). Recent data suggests that in order to counteract this, *C. trachomatis* up-regulates expression of PD-L1 on the DCs present in the uterus. Its interaction with PD-1 receptor on T cells might lead to blocking of T-cell responses.

That both *Chlamydia*-specific CD4+ and CD8+ T cells are involved in controlling *C. trachomatis* infection is indicated by their expansion with memory phenotype in the endocervix during the infection. Patients with *C. trachomatis* infection had the highest levels of T-cell recruiting cytokines out of major STI, a factor associated with higher risk of HIV co-infection which may occur in patients with untreated Chlamydia infection. In epithelial cells, which do not express MHC class II molecules, recognition of *Chlamydia* peptides may occur via the MHC class I presentation pathway, which activates CD8+ T cells. These can recognize proteins present in the host-cell cytosol or cytosolic domains of membrane proteins. MOMP is one of the potential antigenic targets for CD8+ CTLs. However, no evidence has been found so far of CD8+ T-cell-mediated killing of the infected cells that would disrupt pathogen replication and intracellular survival.

It appears, therefore, that the production of the **Th1** cytokine IFN-γ is one of the most powerful anti-chlamydial

T-cell-mediated immune mechanisms. IFN-γ inhibits the growth of *Chlamydia* in cell culture and, in experimental models, disruption of its production enhances host susceptibility to *Chlamydia* infection. It is thought that IFN-γ can potentially limit *C. trachomatis* infection by the following mechanisms:

- activation of macrophages and their phagocytic potential;
- up-regulation of the expression of MHC molecules by professional and nonprofessional APCs;
- expression of **indoleamine 2,3-dioxygenase (IDO)**, a host enzyme that degrades intracellular tryptophan essential for *C. trachomatis* growth;
- up-regulation of **inducible nitric oxide synthase (iNOS)**, which catalyzes the production of nitric oxide (NO) and other reactive nitrogen intermediates, that can enhance damage to intracellular pathogens;
- down-regulation of transferrin receptor on infected cells, resulting in an intracellular iron deficiency that may limit *C. trachomatis* replication.

Even although IFN-γ plays a role in protective immunity to *C. trachomatis* infection, its overall effect is limited. Infection does not stimulate long-lasting immunity and repeated re-infections are common, which results in a prolonged inflammatory response and subsequent tissue damage (see Section 3). One of the reasons for poor anti-chlamydial immunity is the ability of the pathogen to evade immune responses.

HOW DOES *C. TRACHOMATIS* EVADE THE HOST IMMUNE RESPONSES?

This bacterium employs several strategies to evade the host immune response.

- Its intracellular location protects it from antibodies and complement.
- It down-regulates the expression of MHC class I molecules on the surface of infected cells, blocking recognition and MHC class I-restricted CD8+ T-cell-mediated cytotoxicity. The pathogen uses a protease-like activity factor (CPAF) able to degrade host transcription factors required for MHC gene activation.
- Fusion of the pathogen-containing **phagosome** with host cell **lysosomes** is prevented.
- *Chlamydia*-infected macrophages induce apoptosis of T cells, by paracrine effects and **tumor necrosis factor-α (TNF-α)**.

PATHOGENESIS

Tissue damage is the result of the host inflammatory response to the persistent infection as well as direct damage to infected cells by the bacteria. Various species of *Chlamydia* produce cytotoxins that can deliver immediate cytotoxicity of host cells if infected with large doses of the pathogen.

With unsuccessful initial elimination by the innate and adaptive immune systems leading to persistence of the pathogen, the site of infection becomes infiltrated with macrophages, plasma cells, and eosinophils. The continuous production of cytokines and chemokines results in development of lymphoid follicles and tissue scarring due to **fibrosis**. It is believed that IFN-γ production by T cells at the site of infection is reduced as the bacterial load decreases. This, in turn, supports replication of *C. trachomatis* and the inflammatory process resumes, entering into a vicious circle. Chronic inflammation leads to many of the clinical symptoms seen with *C. trachomatis*. Trachoma is characterized by the conjunctival lymphoid follicle formation, which contain germinal centers consisting predominantly of B lymphocytes, with CD8+ T lymphocytes in the parafollicular region. The inflammatory infiltrate contains plasma cells, DCs, macrophages, and polymorphonuclear leucocytes. Inflammatory infiltrates taken from patients with scar tissue are characterized by the expansion of CD4+ T lymphocytes.

3. WHAT IS THE TYPICAL CLINICAL PRESENTATION AND WHAT COMPLICATIONS CAN OCCUR?

Chlamydia is known as the "silent epidemic", since it may not cause any symptoms sometimes for months or years before being discovered. There are several conditions caused by various biovars and serovars of *C. trachomatis* (Table 5.1).

UROGENITAL INFECTIONS

Most people with *C. trachomatis* infection do not have any symptoms and are unaware of the infection. However, when symptoms develop, treatment is urgently required to prevent complications.

Males. When the symptoms develop, the patient may suffer from nongonococcal **urethritis** (**NGU**), which can result in discharge from the penis or pain and burning sensation when urinating. If not treated, it can lead to inflammation near the testicles with considerable pain. Spread to the testicles may cause **epididymitis** and, rarely, sterility. *Chlamydia* causes more than 250 000 cases of epididymitis in the US each year. Post-gonococcal urethritis may occur in men infected with both *Neisseria gonorrhoeae* and *C. trachomatis* who receive antibiotic treatment effective solely for gonorrhoea. MSM men could also develop rectal infection.

Females. The incubation period is usually 1–3 weeks after which the symptoms of urethritis and **cervicitis** may develop: **dysuria** and **pyuria**, cervical discharge or vaginal spotting, and lower abdominal pain. Physical examination reveals yellow or cloudy mucoid discharge from the os (see Figure 5.1). If untreated, *Chlamydia* may spread through the uterus to the fallopian tubes, causing salpingitis. In the US, chlamydial infection is the leading cause of first trimester pregnancy-related deaths. Women infected with *Chlamydia* have a three-to five-fold increased probability of acquiring HIV due to the increased behavioral risk.

Complications. More than 4 billion US dollars are spent annually on the treatment of the most common and severe complication of the sexually transmitted *C. trachomatis* PID. Over 95% of women with uncomplicated and effectively treated chlamydial infection will not develop tubal infertility. Chlamydia can also cause subclinical inflammation of the upper genital tract, so called "subclinical PID". Some patients develop perihepatitis, or "Fitz-Hugh-Curtis Syndrome", an inflammation of the liver capsule and surrounding peritoneum, causing pain in the right upper quadrant.

More than 85% of women with PID remain fertile. However, an approximate 10% risk of ectopic pregnancy – a potentially life-threatening condition – remains after both clinical and subclinical PID due to the permanent damage to

Table 5.1 *C. trachomatis* serovars and associated human diseases

Serovars	Human disease	Method of spread	Pathology
A, B, Ba, and C	Ocular trachoma	Hand to eye, fomites, and eye-seeking flies	Conjunctivitis and conjunctival and corneal scarring
D, Da, E, F, G, H, I, Ia, J, Ja, and K	Oculogenital disease	Sexual and perinatal	Cervicitis, urethritis, endometritis, pelvic inflammatory disease, tubal infertility, ectopic pregnancy, neonatal conjunctivitis, and infant pneumonia
L1, L2, and L3	Lymphogranuloma venereum	Sexual	Submucosa and lymph node invasion, with necrotizing granuloma and fibrosis

Reproduced with permission from Brunham RC and Rey-Ladino J. Immunology of Chlamydia infection: implications for Chlamydia trachomatis vaccine. (2005). Nature Reviews Immunology, 5: 149–161.

the fallopian tubes, uterus, and surrounding tissues. Women with PID often suffer later from abdominal pain and may require a hysterectomy. It is difficult to give a precise prognosis of infertility until a patient tries to have a child.

In both sexes, an asymptomatic infection may be present in either the throat or rectum if the patient has had oral and/ or anal intercourse.

INCLUSION CONJUNCTIVITIS

This condition is caused by *C. trachomatis* (serovars D–K) being associated with genital infections. It is often transferred from the genital tract to the eye by contaminated hands. The main symptom is a sensation of a foreign body in the eye, redness, and irritation. Other symptoms include mucosal discharge later replaced by purulent discharge, large lymphoid follicles, and papillary hyperplasia of conjunctiva, corneal infiltrates, and vascularization. Corneal scarring is rare and happens mostly in the chronic stage followed by epithelial **keratitis**. Ear infection and **rhinitis** can accompany the ocular disease.

INFANT PNEUMONIA

Infants vertically infected with *C. trachomatis* (serovars D–K) from their mother at birth can develop pneumonia presented by staccato cough and **tachypnea** often preceded by conjunctivitis.

LYMPHOGRANULOMA VENEREUM (LGV)

The causative agent is *C. trachomatis* biovar LGV. The first symptom of the infection is the development of a primary lesion – a small painless **papule** or ulcer at the site of infection, often the penis or vagina. Several weeks after the primary lesion, patients develop painful inguinal and/or femoral **lymphadenopathy**. In the case of extragenital infection, the lymphadenopathy can occur in the cervix. Patients develop a fever, headache, and **myalgia** followed by inflammation of the draining lymph nodes. As a result, the lymph nodes become enlarged and painful and may eventually rupture. **Elephantiasis** of the genitalia, more often in women, can develop due to obstruction of the lymphatics. In females, lymphatic drainage occurs usually in perianal sites and can involve **proctitis** and **recto-vaginal fistulae**. In males, proctitis develops from anal intercourse or from lymphatic spread from the urethra.

OCULAR INFECTIONS

Trachoma is the most serious of the eye infections caused by *C. trachomatis*. The word "trachoma" in Greek means rough (trakhus) and reflects the roughened appearance of the conjunctiva. Repeated re-infection with the ocular serovars A, B, Ba, and C results in chronic **keratoconjunctivitis**. Following infection there is an incubation period of 5–12 days, after which the symptoms start to appear and include a

mild conjunctivitis and eye discharge. Initial infection is often self-limiting and heals spontaneously. A repeated infection, however, leads to the development of chronic inflammation (see Section 2), characterized by swollen eyelids and swelling of lymph nodes in front of the ears. Years of re-infection and chronic inflammation may result in fibrosis and in scarring in the upper subtarsal conjunctiva. Scarring is more frequent in young adults, particularly in women. The progress of scarring over many years causes distortion of the lid margin and the lashes turn inward and rub against the cornea. This is called trichiasis (Figure 5.4). If untreated, persistent trauma can result in ulceration of the cornea, corneal opacity, and blindness.

Inflammation and scarring in the eye may block the natural flow of tears, which represent an important first line of defense against bacteria. This can facilitate secondary re-infection with *C. trachomatis* as well as infection with other bacteria or fungi.

OCULAR LYMPHOGRANULOMA VENEREUM

Ocular infection with *C. trachomatis* biovar LGV can lead to conjunctivitis and **preauricular lymphadenopathy**.

REACTIVE ARTHRITIS

This can occur in men and women following symptomatic or asymptomatic chlamydial infection, sometimes coupled with urethritis and conjunctivitis and painless mucocutaneous lesions. Formerly, this complication was referred to as Reiter's Syndrome.

4. HOW IS THE DISEASE DIAGNOSED, AND WHAT IS THE DIFFERENTIAL DIAGNOSIS?

CLINICAL DIAGNOSIS OF GENITAL INFECTIONS

The following clinical indicators are used for the diagnosis and screening of chlamydial infection in women:

- less than 25 years of age, sexually active;
- more than one sexual partner;
- mucopurulent vaginal discharge (Figure 5.1);
- burning sensation when passing urine;
- friable cervix or bleeding after sex or between menstrual periods;
- lower abdominal pain, or pain during sexual intercourse.

Because of the possibility of multiple STIs, all patients with any STI should be evaluated for chlamydial infection and offered an HIV test.

CLINICAL DIAGNOSIS OF OCULAR INFECTIONS

Examination of an eye for the clinical signs of trachoma involves careful inspection of the lashes and cornea, then aversion of the upper lid and inspection of the upper tarsal conjunctiva using binocular loupes. Clinical diagnosis is best made based on investigation of the history of living in a trachoma-endemic environment, in combination with clinical signs.

CLINICAL DIAGNOSIS OF LYMPHOGRANULOMA VENEREUM

The clinical symptoms may initially be unclear since they overlap with those of other STIs. Some men may have had treatment for a range of conditions including inflammatory bowel disease, **Crohn's disease**, and so forth. Diagnosis is largely based on the history of the disease, physical examination, and laboratory tests. The clinical course of LGV is divided into three stages.

1. Primary painless lesion, which develops after incubation of 3–30 days. A papule or ulcer can be found on the genitalia (glans of the penis, vaginal wall, labia, or cervix) or, in some cases, in the oral cavity.
2. Secondary lesion/**lymphadenitis**. It is a regional dissemination causing inguinal and femoral lymphadenopathy and possibly bubo formation that ulcerates and discharges pus. Lymphadenopathy is usually unilateral involving the retroperitoneal lymph nodes in women and the inguinal lymph nodes in men.
3. Tertiary stage/genito-ano-rectal syndrome. The majority of patients will recover from the second stage without sequelae. However, a few may develop proctitis and fibrosis that may result in chronic genital ulcers or fistulas, rectal strictures, and genital elephantiasis. Early symptoms of LGV **proctocolitis** include anal **pruritus** and discharge, fever, rectal pain, and **tenesmus**.

LABORATORY DIAGNOSIS

Sample collection. For genital infections, swabs are collected from the cervix or vagina of women or the urethra of men. Self-collection of swabs should be taken from the throat or rectum if there is a possibility of infection there.

With the use of **nucleic acid amplification tests** (**NAATs**) (see below) a noninvasive urine test can be used for screening for *Chlamydia* with sufficient sensitivity instead of swabs. It is particularly useful in asymptomatic cases where genital examination and sampling may not be justified.

For suspected LGV infections in the primary stage, a swab of the lesion can be taken. In the secondary stage, bubo pus, saline aspirates of the bubo, swabs of the rectum, vagina, urethra, urine, serum, or biopsy specimens of the lower gastrointestinal (GI) tract are used.

For conjunctival specimens, epithelial cells are collected by rubbing a dry swab over the everted palpebral conjunctiva.

The following tests can be made on the samples collected.

Cytology is used to detect the inclusion bodies in stained cell scrapings (Figure 5.5), but this method lacks sensitivity and is time-consuming.

Cell culture. For many years, cell culture has been the gold standard for the diagnosis of *C. trachomatis* and is very specific for this pathogen, but NAATs are more sensitive and represent a new gold standard (see below). Cultured cells are stained with Giemsa or iodine, or by fluorescent antibodies and examined for the presence of iodine-staining inclusion bodies. Iodine stains glycogen, which is only found in the inclusions of *C. trachomatis*, not in other *Chlamydia* species.

NAATs are now recommended to replace culture techniques. They include first of all the **polymerase chain reaction** (**PCR**), but also transcription-mediated amplification (TMA). NAATs are characterized by high sensitivity (> 89%) and specificity (95–99.5%) for endocervical, urethral, and conjunctival samples. NAATs are the only testing techniques currently recommended by the English National Chlamydia Screening Programme. Certain NAAT test platforms have been cleared by FDA for non-genital sites. NAATs can be used for detection of the bacterium as well as for the quantification of the bacterial load. The wider use of NAATs worldwide in screening strategies is, however, limited by the costs and complexity of the equipment required.

Direct hybridization probe tests have been used in the past, but now are largely replaced by NAATs. They include Gen-Probe PACE 2 test, which uses a synthetic single-stranded DNA probe complementary to the chlamydial rRNA region, and Digene HC-II-CT-II probe based on **enzyme immunoassay** (**EIA**) for the detection of the RNA probe binding to the single-stranded bacterial DNA.

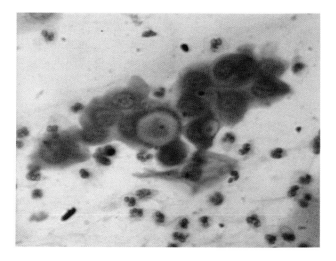

Figure 5.5 Chlamydial inclusions, which may contain 100–500 RB progeny. *From Mabey DCW, Solomon AW & Foster A (2003) The Lancet 362: 223-229. https://doi.org/10.1016/S0140-6736(03)13914-1. With permission from Elsevier.*

Antibody-based laboratory techniques include direct fluorescent antibody (DFA) detection and **enzyme-linked immunosorbent assay (ELISA)**. The latter can be used for determination of the serovars. Usually diagnostic antibodies target group-specific LPS or serovar-specific outer-membrane proteins (OMPs). However, ELISA-based technology is widely used in screening programs and is of great help in reducing the overall chlamydial infection and incidence of PID and NGU.

DIFFERENTIAL DIAGNOSIS

Differential diagnosis of genital chlamydial infection includes gonorrhea, *Ureaplasmaurealyticum*, *Mycoplasma genitalium*, *M. hominis*, and *Trichomonas vaginalis*, as well as urinary tract infections and bacterial vaginosis. Some other conditions such as periurethral **abscess**, **endometriosis**, urethral/vaginal foreign body, other causes of PID, prostatitis, and **epididymo-orchitis** must also be ruled out.

In the case of trachoma, a differential diagnosis can be made with the following conditions: adult inclusion conjunctivitis; other bacterial infections, especially with *Moraxella* species and *Streptococcus pneumoniae*; viral infections including adenovirus, herpes simplex virus, and molluscum contagiosum; *Pediculosis palpebrarum*; toxic conjunctivitis, which is secondary to topical drugs or eye cosmetics; Axenfeld's follicular conjunctivitis; Parinaud's oculoglandular syndrome; and vernal conjunctivitis.

For LGV and inguinal **adenopathy**, differential diagnosis of the genital ulcer includes chancroid, herpes, syphilis, and donovanosis. For proctitis, the differential diagnosis includes inflammatory bowel disease.

The differential diagnosis is mostly based on the laboratory findings, particularly PCR analysis, supplemented by clinical observations.

5. HOW IS THE DISEASE MANAGED AND PREVENTED?

MANAGEMENT

In 2021, the CDC introduced new guidelines for the treatment of all STIs.

Urogenital Infections

For urogenital infections, the antibiotic doxycycline is the drug of choice and 100 mg is taken twice daily for 7 days. This is effective in 95% of cases but can interfere with the contraceptive pill and can cause photosensitivity and stomach irritation. Doxycycline is also available in a delayed-release 200-mg tablet formulation, which requires once-daily dosing for 7 days.

An alternative CDC-recommended regimen includes azithromycin, 1 g orally in a single dose or levofloxacin 500 mg

orally once daily for 7 days. Azithromycin is a derivative of erythromycin and is characterized by improved bioavailability and ability to maintain high tissue concentrations, particularly at sites of inflammation. Azithromycin has also been proven to be neonatally safe. Erythromycin is no longer recommended due to the GI side effect. In the past, it was given to the infected woman if pregnant or lactating, as 500 mg four times a day for 7 days or twice daily for 14 days. Pregnant women are given a single 1 g dose of azithromycin orally or amoxicillin 500 mg orally 3 times a day for 7 days. Neonates – erythromycin base or ethylsuccinate, 50 mg/kg body weight/day orally, divided into 4 doses daily for 14 days. The same regimen is used for infants. However, in addition, they can be prescribed an azithromycin suspension, 20 mg/kg body weight/day orally, 1 dose daily for 3 days.

Since genital chlamydial infections often have no symptoms, particularly in women, the sexual partner could have become infected months ago. In case of multiple sexual partners, all of them should be tested for chlamydial infection and treated if positive. The patients should also be tested for other STIs. Having sex is not recommended during treatment and for at least a week after the completion of the treatment, particularly if both partners are infected (see Case Study).

A high prevalence of *C. trachomatis* infection has been observed post-treatment, due to re-infection caused by failure of sex partners to receive treatment or initiation of sexual activity with a new infected partner, indicating a need for improved education and treatment of sex partners. Repeat infections confer an elevated risk for PID and other complications among women. As a preventive measure, men and women who have been treated for chlamydia should be retested approximately 3 months after treatment, or whenever persons next seek medical care during the first year after initial treatment.

Trachoma

For the treatment of active trachoma, two antibiotic regimens are currently recommended: tetracycline ointment applied twice daily for 6 weeks or one 20 mg kg^{-1} dose of azithromycin can be used instead. It must be noted that the application of the ointment is inconvenient in children. Azithromycin is also effective for treating extraocular reservoirs of chlamydial infection, although antibiotic resistance could eventually develop.

Currently, annual mass treatment for 3 years is recommended by the WHO in districts and communities where the prevalence of follicular trachoma in children aged 1–9 years is equal to or greater than 10%. Surgical correction of trichiasis to fix eyelid deformities and to prevent vision loss, coupled with post-operational one dose of azithromycin, is applied. Due to the severe damage of eyes in trichiasis, cornea transplantation is not considered.

PREVENTION

Prevention of genital chlamydial infections involves safe sexual practices, using a condom during sexual intercourse, and prompt treatment of infected patients and their sexual partners. It is recommended to have a *Chlamydia* test under the following conditions:

* after sex with a new or casual partner;
* immediately if symptoms occur (see Section 3);
* if a sexual partner has *Chlamydia* or symptoms of *Chlamydia*.

The best way to avoid becoming infected with *Chlamydia* and other STIs is to have sexual contact or to be in a long-term, mutually monogamous relationship with a partner who is not infected.

Flies acting as physical vectors for transmission of *C. trachomatis* transmit ocular infection, particularly in children in areas with poor sanitation. The presence of cattle pens has been associated with trachoma in some African countries. Crowded living conditions in the family unit constitute another factor increasing the risk of trachoma. Therefore, special socioeconomic measures have been recommended to reduce the risk of transmission of ocular *C. trachomatis* infection. These include increasing access to water, fly control interventions, education, and improved living conditions.

VACCINE

Due to its high incidence, relapses, prolonged bouts of infection, and significant morbidity, *C. trachomatis* is an important target for vaccine development. A safe vaccine administered prior to adolescence and effective through child-bearing age would have a significant impact on the spread of the disease. The attempts of using inoculation of whole inactivated bacteria, however, led to the exacerbation of inflammatory disease and subsequent re-infection. This shifted the focus of vaccine development toward MOMP-based subunit vaccines. **Monoclonal antibodies** to MOMP neutralize *C. trachomatis* infection *in vitro* and *in vivo*.

The results of clinical trials were reported in 2019 on a multivalent vaccine, CTH522 incorporating MOMP proteins. The vaccine gave promising results by significantly increasing the levels of antigen-specific mucosal IgG and IgA, antibody neutralization, and the production of IFNγ. Further human clinical trials are required to fine-tune this vaccine, which is likely to include recombinant protein combination with adjuvants.

Since 2013, vaccine development against *C. trachomatis* has been intensified and various delivery platforms have been suggested, such as nanoparticles, Hepatitis B core antigen and *Neisseria lactamicaporin* B as career molecules, adenoviral and mRNA vectors.

SUMMARY

1. WHAT IS THE CAUSATIVE AGENT, HOW DOES IT ENTER THE BODY, AND HOW DOES IT SPREAD A) WITHIN THE BODY AND B) FROM PERSON TO PERSON?

* *Chlamydia trachomatis* belongs to the family Chlamydiaceae, which contains two genuses: *Chlamydia* and *Chlamydophila*. Bacteria of this family are obligatory intracellular human and animal pathogens.
* *Chlamydia trachomatis* and *Chlamydophila pneumoniae* are human pathogens. *C. trachomatis* can cause sexually transmitted urogenital infections, neonatal conjunctivitis and pneumonia, ocular trachoma, and urogenital infection associated with lymphadenopathy and lymphadenitis (LGV). *C. pneumoniae* can cause bronchitis, sinusitis, and pneumonia, and accelerates atherosclerosis.
* All Chlamydiaceae species have specific lipopolysaccharides and envelope proteins: 40 kDa major outer-membrane protein (MOMP), a hydrophilic cysteine-rich 60 kDa protein, and a low molecular weight cysteine-rich lipoprotein.
* *C. trachomatis* contains two human biological variants (biovars): trachoma and LGV, and one biovar infecting mice. Fifteen serologic variants (serovars) have been identified in trachoma biovar. MOMP confers serovar specificity.
* Chlamydiaceae species are characterized by a biphasic developmental cycle. The infectious EB form is typically 0.2–0.6 μm in diameter and resistant to unfavorable environmental conditions outside of their eukaryotic host cells.
* EBs bind to the epithelial cells directly via cellular proteoglycans with heparan sulfate (HS) moieties (HSPGs). Additionally, *C. trachomatis* EB form is able to adhere to and enter cells indirectly through binding with fibroblast growth factor 2 (FGF2). This complex then interacts with the FGF2 receptor, which mediates EB internalization into cells.
* EB entry is followed by translocation of EB-containing endosomes to the perinuclear region and their homotypic fusion with each other forming a fusion vacuole – a nascent inclusion. EBs accumulate in an inclusion where they use a supply of nutrients via Golgi apparatus and mature into the RB form.
* *C. trachomatis* is dependent on its host for survival and hijacks host-cell metabolism particularly glycolytic enzymes, aldolase A, pyruvate kinase and lactate dehydrogenase, which are enriched at the *C. trachomatis* inclusion membrane during infection. The increased requirement for glutamine, important for the growth of *C. trachomatis* in infected cells is achieved by reprogramming the glutamine metabolism.
* The RB form (up to 1.5 μm in diameter) multiplies by binary fission. After several rounds of replication during 48–72 hours, the RB form reverts to the infectious EB form. Eventually, the EBs are released through the extrusion of the inclusions by reverse endocytosis or through apoptosis or are released after cell lysis.

Continued...

...continued

- *C. trachomatis* (serovars D–K) is sexually transmitted in the EB form, usually through vaginal intercourse but occasionally can be transmitted by oral and anal sex. Spread to the fallopian tubes in females, it can lead to pelvic inflammatory disease (PID). It can also lead to conjunctivitis through autoinoculation. Genital infection of *C. trachomatis* (serovars D–K) is the leading bacterial cause of sexually transmitted infection with over 50 million new cases occurring yearly worldwide.

- Ocular infection of *C. trachomatis* (serovars A–C) is spread in areas of poverty and overcrowding. Every year *C. trachomatis* is a major cause of 500 000 cases of trachoma worldwide, approximately 85 million people worldwide have active trachoma, more than 10 million have trichiasis (inturned eyelashes that touch the globe), and about 6 million people suffer visual loss and blindness.

- Infection can be transmitted from eye-to-eye by fingers, shared cloths or towels, by eye-seeking flies, and by droplets (coughing or sneezing).

- *C. trachomatis* (biovar LGV) causes sexually transmitted disease that is prevalent in Africa, Asia, and South America. Humans are the only natural host.

2. WHAT IS THE HOST RESPONSE TO THE INFECTION AND WHAT IS THE DISEASE PATHOGENESIS?

- Both innate and adaptive immune responses are induced during *C. trachomatis* infection, but they are usually inefficient at controlling the infection. This may lead to the partial clearance of the pathogen from the body resulting in bacterial persistence, chronic inflammation, tissue damage, and severe clinical symptoms.

- Chlamydial PAMPs are recognized by multiple PRRs, particularly TLR2 which was shown to be activated by MOMPs and heat shock protein 60 (Hsp60), during the productive infection. Infection with *Chlamydia* induces production of secretory IgA and circulatory IgM and IgG antibodies mostly directed to MOMP and Hsp60. IgG antibodies bound to the bacterial MOMP invoke antibody-dependent cellular cytotoxicity (ADCC) by **natural killer (NK)** cells.

- Dendritic cells (DCs) in response to the infection release IL-12 and drive Th1 cell development and hence **production of interferon γ (IFN-γ)** – the major inhibitory cytokine for *Chlamydia* (see below). Recent data suggests that to counteract this, *C. trachomatis* up-regulates expression of PD-L1 on the DCs in the uterus. Its interaction with PD-1 receptor on T cells might lead to blocking of T-cell responses.

- One of the most powerful anti-chlamydial T-cell-mediated immune mechanisms is production of the Th1 cytokine IFN-γ. IFN-γ can limit *C. trachomatis* infection by activating macrophages, up-regulating expression of MHC, suppressing tryptophan production in the host cells, enhancing production of nitric oxide, and down-regulating transferrin receptor on the host cells.

- Infection does not stimulate long-lasting immunity and repeated episodes of infection are common. Re-infection results in an inflammatory response and subsequent tissue damage.

- One of the reasons for poor anti-chlamydial immunity is the ability of the pathogen to evade or block host immune responses by intracellular location, down-regulation of MHC class I molecules, prevention of the formation of phagolysosome, and induction of apoptosis of cytotoxic T cells.

- Pathogenesis is a result of a direct killing of infected cells and of chronic inflammation. The chronic inflammation induced by *C. trachomatis* infection leads to episodes of PID or conjunctivitis resulting in tubal infertility or blindness, respectively.

3. WHAT IS THE TYPICAL CLINICAL PRESENTATION AND WHAT COMPLICATIONS CAN OCCUR?

- Urogenital infection in men causes urethritis, which can produce a discharge from the penis or pain and burning sensation when urinating. If not treated, it can lead to epididymitis and, rarely, sterility.

- Urogenital infection in women leads to urethral discharge, dysuria and pyuria, urethritis, and cervical discharge and friability. If untreated, the chlamydial infection may spread through the uterus to the fallopian tubes, causing salpingitis, infertility, or ectopic pregnancy.

- A severe complication of the sexually transmitted *C. trachomatis* is PID.

- Inclusion conjunctivitis is associated with genital infections transferred from the genital tract to the eye by contaminated hands. The main symptom is a sensation of a foreign body in the eye, redness, irritation.

- Infants infected with *C. trachomatis* vertically from their mother at birth can develop pneumonia, which presents as a tachypnea, staccato cough and is often preceded by conjunctivitis.

- The first symptom of LGV is the development of a primary lesion, often in the penis or vagina. Several weeks after the primary lesion, patients develop painful inguinal and/or femoral lymphadenopathy, fever, headache, and myalgia followed by inflammation of the draining lymph nodes.

4. HOW IS THE DISEASE DIAGNOSED, AND WHAT IS THE DIFFERENTIAL DIAGNOSIS?

- Clinical diagnosis of genital chlamydial infection in women is based on age, sexual activity, more than one sexual partner, mucopurulent vaginal discharge, burning when passing urine, friable cervix or bleeding after sex or between menstrual periods, lower abdominal pain, or pain during sexual intercourse.

- Clinical diagnosis of LGV is based on the clinical presentation and the course of the disease assisted by differential diagnostic and laboratory tests.

continued...

...continued

- Eye examination for the clinical signs of trachoma involves careful inspection of the lashes and cornea, then reversion of the upper lid and inspection of the upper tarsal conjunctiva. A grading system is used by the WHO for the assessment of the prevalence and severity of trachoma.
- Nucleic acid amplification tests (NAATs), including polymerase chain reaction (PCR), are now increasingly used for diagnosis and differential diagnosis of chlamydial infections, although their wider use in screening strategies worldwide is limited by the costs and complexity of the equipment required.
- Cell culture has traditionally been the gold standard for the diagnosis of *C. trachomatis* and is specific for this pathogen. However, it is now replaced by NAATs.
- Antibody-based laboratory techniques – immunofluorescence and enzyme-linked immunosorbent assay (ELISA) – are widely used for determining the serovars in screening programs.

5. HOW IS THE DISEASE MANAGED AND PREVENTED?

- For urogenital infections, the antibiotic doxycycline (Doryx®, Vibramycin®) is given 100 mg twice daily for 7 days. Alternative CDC-recommended regimen include azithromycin, 1 g orally in a single dose or levofloxacin. 500 mg orally once daily for 7 days. Pregnant women are given a single 1 g dose of azithromycin orally or amoxicillin 500 mg orally 3 times/day for 7 days. Neonates –

erythromycin base or ethylsuccinate 50 mg/kg body weight/day orally, divided into 4 doses daily for 14 days. The same regimen is used for infants. In addition, they can be prescribed an azithromycin suspension, 20 mg/kg body weight/day orally, 1 dose daily for 3 days.

- All sexual partners should be tested for chlamydial infection and treated if positive. Patients should also be tested for other sexually transmitted infections.
- Environmental improvement includes removing the risk factors for transmission of infection such as flies, the presence of cattle pens, and crowded living conditions.
- Safe sexual practices, using a condom during sexual intercourse, and prompt treatment of infected patients and their sexual partners can prevent genital infections with *C. trachomatis*.
- Repeat infections confer an elevated risk for PID and other complications among women. As a preventive measure, men and women who have been treated for chlamydia should be retested approximately 3 months after treatment, or whenever persons next seek medical care during the first year after initial treatment.
- The major target for vaccine development currently is the multivalent MOMP-based vaccine in combination with adjuvants. For the delivery platforms nanoparticles, Hepatitis B core antigen and Neisseria lactamicaporin B as career molecules, adenoviral and mRNA vectors are considered.

FURTHER READING

Goering RV, Dockrell H, Zuckerman M, Chiodini P. Mims' Medical Microbiology and Immunology, 6th edition. Mosby, Edinburgh, 2018.

Murray PR, Rosenthal KS, Pfaller MA. Medical Microbiology, 9th edition. Elsevier Mosby, Philadelphia, 2021.

REFERENCES

Corsaro D, Greub G. Pathogenic Potential of Novel Chlamydiae and Diagnostic Approaches to Infections due to these Obligate Intracellular Bacteria. Clin Microbiol Rev, 19: 283–297, 2006.

Dautry-Varsat A, Subtil A, Hackstadt T. Recent Insights into the Mechanisms of Chlamydia Entry. Cell Microbiol, 7: 1714–1722, 2005.

Ende RJ, Derre I. Host and Bacterial Glycolysis During Chlamydia trachomatis Infection. Infect Immune, 88: e00545–20, 2020.

Gambhir M, Basáñez M-G, Turner F, et al. Trachoma: Transmission, Infection, and Control. Lancet Infect Dis, 7: 420–427, 2007.

McClarty G, Caldwell HD, Nelson DE. Chlamydial Interferon Gamma Immune Evasion Influences Infection Tropism. Curr Opin Microbiol, 10: 47–51, 2007.

Murray SM, McKay PF. Chlamydia trachomatis: Cell Biology, Immunology and Vaccination. Vaccine, 39: 2965–2975, 2021.

Starnbach MN. The Well Evolved Pathogen. Curr Opin Microbiol, 54: 33–36, 2020.

West SK. Milestones in the Fight to Eliminate Trachoma. Ophthalmic Physiol Opt, 40: 66–74, 2020.

Wright HR, Taylor HR. Clinical Examination and Laboratory Tests for Estimation of Trachoma Prevalence in a Remote Setting: What are they Really Telling us? Lancet Infect Dis, 5: 313–320, 2005.

WEBSITES

Centers for Disease Control and Prevention, Atlanta, 2022: https://www.cdc.gov/std/chlamydia/STDFact-Chlamydia.htm

National Institutes of Health, Department of Health and Human Services, USA: www.nih.com

UK Health Security Agency, English National Chlamydia Screening Programme, 2022: https://assets.publishing.service.

gov.uk/government/uploads/system/uploads/attachment_data/file/759846/NCSP_Standards_7th_edition_update_November_2018.pdf

Wikipedia – The Free Encyclopedia, Chlamydia: https://en.wikipedia.org/wiki/Chlamydia

World Health Organization, WHO guidelines for the treatment of Chlamydia trachomatis, 2016: https://www.who.int/publications/i/item/978-92-4-154971-4

Students can test their knowledge of this case study by visiting the Instructor and Student Resources: [www.routledge.com/cw/lydyard] where several multiple choice questions can be found.

Clostridioides difficile

6

An elderly woman, of no fixed abode, arrived at the hospital's emergency department after having fallen. She was admitted to hospital for fixation of a fracture of the hip. Shortly after admission, she developed signs of a chest infection and was started on a cephalosporin, which she remained on for a week. Subsequently, she developed profuse watery diarrhea and abdominal pain. A fecal sample was sent to the laboratory to test for the toxins of *Clostridioides difficile*, which proved positive and she was commenced on oral vancomycin. Despite treatment, the diarrhea persisted and it also failed to respond to a course of metronidazole. The condition of the patient worsened and a **sigmoidoscopy** was performed, revealing that she had **pseudomembranous colitis** (Figure 6.1). Her clinical condition deteriorated and she developed **toxic megacolon** and an emergency **colectomy** was performed. The patient died shortly after the operation.

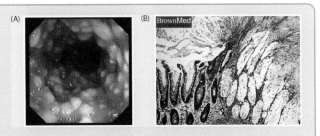

Figure 6.1 Endoscopic view (A) of a patient with pseudomembranous colitis showing plaques of pseudomembranes. Histologically (B) the pseudomembranes are composed of mushroom-shaped collections of neutrophils, cellular debris, and fibrin. *From David M Martin, MD / Science Photo Library. With permission. Courtesy of Jae Lim, Calvin Oyer and Murray Resnick, editors and photographic contributors to the Digital Pathology website associated with Brown Medical School, Providence, Rhode Island.*

1. WHAT IS THE CAUSATIVE AGENT, HOW DOES IT ENTER THE BODY AND HOW DOES IT SPREAD A) WITHIN THE BODY AND B) FROM PERSON TO PERSON?

CAUSATIVE AGENT

Clostridioides difficile (previously *Clostridium difficile* [CD]) is an anaerobic spore-forming motile gram-positive rod measuring $0.5–1.0 \times 3.0–16.0\,\mu m$ (Figure 6.2). It produces irregular white colonies on blood agar (Figure 6.3). Toxigenic strains of CD produce three toxins, TcdA, TcdB, and CDT. TcdA and B are carried on a pathogenicity locus (PaLoc) which is a mobile element and can be transferred to non-toxigenic strains converting them to toxigenic strains. Both TcdA and B are glycosyl transferases. Regulation of the production of the toxins is by TcdR (Group V sigma factor) and inhibition is by TcdC (an anti-sigma factor) which acts directly on TcdR. TcdR also activates its own production in a positive feedback loop. A further protein, TcdE, is found on the PaLoc which is involved with secretion of the toxins (Figure 6.4) TcdC hyper-virulent mutants have been described. These strains produce increased amounts of toxins A and B and are caused by a single nucleotide deletion. Toxin production is influenced by several environmental factors such as sugars, amino

acids, and sporulation. Toxin A and B production occurs maximally in stationary phase and during nutrient limitation. Carbohydrate limitation results in a switch to amino-acid fermentation which, in turn, results in a shortage of amino acids and toxin production. Low levels of biotin also increase toxin production. Additionally, *C. difficile* has a thiolactone quorum-signaling system, which regulates toxin expression independent of TcdC.

Figure 6.2 Gram stain of *C. difficile* showing that it is a gram-positive rod. The spore can be clearly seen as a subterminal clear area in most of the bacilli. *Courtesy of the Centers for Disease Control, Atlanta, Georgia. Image is found in the Public Health Image Library #3876. Additional photographic credit is given to Dr Gilda Jones who took the photo in 1980.*

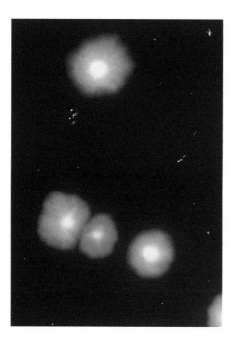

Figure 6.3 Colonies of *C. difficile* showing the typical appearance of white colonies with crenated edges. Under UV light, the colonies fluoresce a greenish-yellow color. *Courtesy of the Centers for Disease Control, Atlanta, Georgia. Image is found in the Public Health Image Library #3876. Additional photographic credit is given to Dr Gilda Jones who took the photo in 1980.*

The binary toxin CDT is found at a different location on the genome. The toxin consists of two genes *cdtA* and *ctdB* and is an ADP-ribosyltranferase. Production of the toxin is regulated by cdtR, which is one of the common strains of CD (027) also up-regulated TcdA and B.

Several methods of typing *C. difficile* are available, the most frequently used being **pulse field gel electrophoresis**, and **ribotyping** (Figure 6.5), although **multilocus enzyme** electrophoresis (**MLEE**) and analysis of variable number of tandem repeats (VNTR) may also be used.

Phylogenomic studies of *C. difficile* using micro-arrays demonstrate five clades: a hypervirulent **clade**, a defective trehalose metabolism clade, a TcdA-negative/TcdB-positive clade, and two other human and animal clades. Differences between the clades are related to virulence, ecology, antibiotic resistance, motility, adhesion, and metabolism.

ENTRY INTO THE BODY

Following the ingestion of spores from contaminated feces or a contaminated environment, the organism colonizes mainly the large intestine of the GI tract. The dynamics of the GI microbiome is important in *C. difficile* infections as a dysbiosis caused by antibiotics predisposes to colonization. Treatment of *C. difficile* disease can therefore be ameliorated by fecal transplants. Eradication of colonization by *C. difficile* relies on a normal intestinal microbiota. Colonization resistance is provided by bacteria in the Ruminococcaceae and Lachnospiraceae which produce short-chain fatty acids (SCFA) such as butyric acid. Many studies have demonstrated that the microflora in an individual is rather stable, although variations have been shown in relation to diet and stress. *C. difficile* is present in the intestines of 2–3% of asymptomatic adults, but may increase to 30% in hospital, and in between 60% and 70% of neonates. The reason why neonates do not often get disease is unclear but may be related to the nature of the neonate microbiome or toxin not binding to cells.

SPREAD WITHIN THE BODY

The organism remains within the GI tract and only rarely is extra-intestinal disease reported.

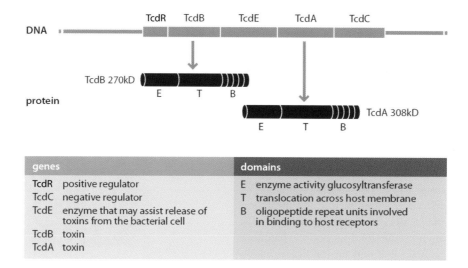

genes		domains	
TcdR	positive regulator	E	enzyme activity glucosyltransferase
TcdC	negative regulator	T	translocation across host membrane
TcdE	enzyme that may assist release of toxins from the bacterial cell	B	oligopeptide repeat units involved in binding to host receptors
TcdB	toxin		
TcdA	toxin		

Figure 6.4 This figure shows the arrangement of the operon for toxins A and B of *C. difficile*. The expression of the toxin genes (*TcdA* and *TcdB*) is controlled by positive (*TcdR*) and negative (*TcdC*) regulators, which respond to changes in the environment. The operon also carries a gene, *TcdE*, which produces a protein thought to assist in the release of the toxins across the bacterial cell wall. Toxin A is a 308 kDa protein and toxin B is a 270 kDa protein. Each protein has an enzyme active domain E (glucosyltransferase); a domain responsible for binding B, which consists of multiple oligopeptide repeat units; and a domain T that is involved in translocation of the toxin across the host cell membrane.

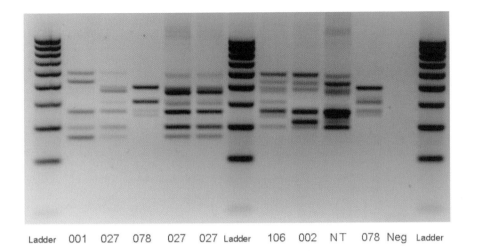

Ladder   001   027   078   027   027   Ladder   106   002   N T   078   Neg   Ladder

Figure 6.5 Ribotyping of *Clostridium difficile* – photograph of a gel showing some different common ribotypes of *C. difficile* (001, 027, etc.). The pattern of bands produced by the restriction enzyme is characteristic for the different ribotypes and the patterns can be used to determine if there is cross infection or a common source outbreak where the bands would be identical. Because some ribotypes are currently very common, e.g. 027, a more discriminatory typing system, e.g. variable number of tandem repeats (VNTR) may be used to subdivide the common ribotypes. *Courtesy of the Windeyer Institute of Medical Sciences, University College London, UK.*

PERSON-TO-PERSON SPREAD

Patients with diarrhea, especially if it is severe or accompanied by incontinence, may unintentionally spread the infection to other patients in hospital. In addition, the ability of *C. difficile* to form spores enables it to survive for long periods in the environment, for example, on floors and around toilets, and protects it against heat and chemical disinfectants. The spores can survive for 5 months on inanimate surfaces and are resistant to most disinfectants. Acquisition of *C. difficile* is probably as a spore rather than the vegetative organism. Colonization in hospitals is directly related to the floor area in single rooms: for every 15 m², there is an OR of 3.0. Effective agents for whole room decontamination are UV light and H_2O_2 mist.

EPIDEMIOLOGY

C. difficile is a well-recognized hospital "superbug" that has received considerable media attention and generated much public concern. In the US, there were around 500 000 cases annually with 15 000 deaths at a cost to the healthcare system of $4.8 billion (2015 data). In the UK, cases have been subject to mandatory reporting since 2007 when the infection rate was 110/100 000. In 2018, the infection rate was 20/100 000 recording all cases over the age of 2 years. This represented 13 286 cases of which 4739 were acquired in hospital.

The standard typing method is PCR sequencing of the 16-23S inter-spacer region (ribotype, RT).

Although important in typing *C. Difficile*, more discrimination can be achieved with multilocus variable number tandem repeats (MLVA) and whole genome sequencing (WGS). Investigation of outbreaks by specific ribotypes has indicated widely separate subtypes or, in some cases, closely related subtypes. In the UK, the two most prevalent ribotypes

are RT002 and RT015. The hypervirulent ribotype RT027 was very common in 2008–2009 (40%) but by 2018 had decreased about 1% of the total and it remains uncommon. RT220 is another more virulent isolate that is linked to high C-reactive protein levels and a high mortality but is not a common isolate. Some types are particularly associated with outbreaks such as RT001 and TR106 but again their prevalence has decreased since 2008. Certain ribotypes are also geographically linked. In Europe (2012–2013), the common ribotypes were RT027, RT 001/072, RT014/020. A study in 2018 using WGS demonstrated specific country clades: RT356 and RT018 were found in Italy, RT176 in the Czech Republic and Germany, RT001/072 in Spain, Slovakia, and Germany, RT027 in Poland, Romania, Hungary, Italy, and Germany. Other ribotypes did not have specific country clades: RT078, RT015, RT002, RT014 and RT020. Overall, two distinct routes of spread have been identified, hospital-acquired and non-hospital acquired, the latter probably transmitted by several routes such as food or feco-oral.

Recognizing a "One Health" approach, a study on prevalence of *C. difficile* in dogs and Zoo animals demonstrated 17% of dog feces, 56% for dog paws and 22% of environmental samples from Parks were positive for toxigenic *C. difficile* found associated with humans. Eighteen percent of Zoo animals were also colonized.

2. WHAT IS THE HOST RESPONSE TO THE INFECTION AND WHAT IS THE DISEASE PATHOGENESIS?

IMMUNE RESPONSES

Colonization is initially with a spore. Once germinated, the organism expresses a number of factors that stimulate the

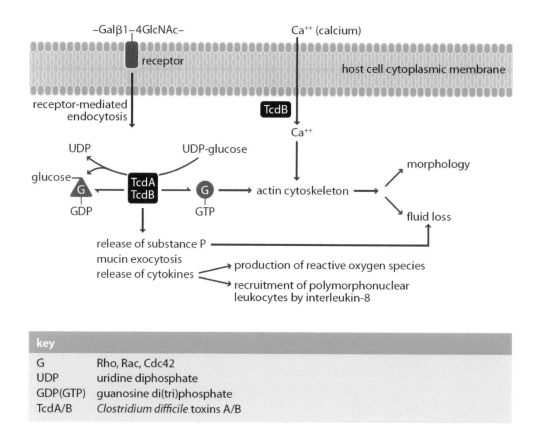

Figure 6.6 Mode of action of *C. difficile* toxins A/B. This figure shows the action of the toxins A and B on the host cell. The toxins are taken up by receptor-mediated endocytosis after binding to specific glycoprotein receptors. The toxins inhibit the normal functioning of G proteins Rho, Rac, and Cdc42 by glucosylating them. This inactivation affects the actin cytoskeleton which, in turn, affects cell morphology and the integrity of the tight junctions between the cells, thereby increasing fluid loss. The toxins also stimulate the release of cytokines, e.g. IL-8, which recruits PMNs and establishes an inflammatory response and the production of reactive oxygen radicals. The toxins increase the release of substance P, a peptide that enhances fluid loss from the intestines and decreases the exocytosis of the protective mucus barrier.

immune system such as the flagella and surface proteins. The organism also produces a number of toxins (TcdA, TcdB, CDT) that are also immunogenic. Some surface proteins are pathogen-associated molecular patterns (PAMPs) and trigger a pro-inflammatory response along with an inhibitory response (IL-1β, IL-6, TNF-α, IL-10, and IL-12, IL-23). Both T and B cells are generated. Immunoglobulin A (IgA) is produced along with IgM and IgG. High levels of both IgA and IgM at the initial infection are related to a low risk of recurrence and a low level of IgA is also linked to prolonged disease as well as a high risk of recurrence. High levels of IgG protect against *C. difficile* disease but if diarrhea develops, high levels of IgG at day 12 are linked to a low risk of recurrence. Specific deficiencies of IgG2 and 3 are also a risk factor for recurrences. Studies on IgG and IgA antibodies against TcdA and TcdB are contradictory, some showing a neutralizing effect, and others not. Despite this, humanized monoclonal antibodies directed against the toxins TcdA and TcdB have been developed as an effective immunotherapy.

PATHOGENESIS

The toxins TcdA and TcdB are A-B type toxins (Figure 6.6). Both toxins consist of a glucosyltransferase domain (GTD), an auto-processing domain (APD) a translocation domain (TD), and a C-terminal repetitive binding domain (CROP). The CROP of TcdA is homologous to the carbohydrate binding region of streptococcal glycosyltransferase and binds to various carbohydrate moieties on the cell membrane. The binding of TcdB is to the poliovirus receptor-like 3 (PVRL3) expressed on colonic epithelial cells. The toxins are internalized by receptor-mediated endocytosis, but with slightly different mechanisms. TcdA uptake is independent of clathrin whereas TcdB is mediated by clathrin. In the endosome, the toxins undergo a conformational change caused by the low pH environment. The proteins from the TD of the toxins insert into the membrane forming a pore through which the active enzyme component (glucosyltransferase) and the auto-processing component enter the cytosol. The glucosyltransferase enzyme is released from rest of the toxin by proteolytic cleavage by the auto-processing component. The GT enzyme then inactivates the GTPases Rho and Ras which leads to disruption of the actin cytoskeleton, apoptosis or necrosis (Figure 6.6).

Both TcdA and B toxins induce a strong inflammatory response involving a number of factors including **interleukin** (IL)-8, IL-6, **tumor necrosis factor**-α (**TNF**-α), leptin, and **substance P**. TcdA stimulates the vanilloid receptor, VR1, which leads to the release of substance P from sensory neurons (and increases the expression of its receptor – neurokinin 1-R) which, in turn, leads to the release of **neurotensin** and mast-cell degranulation. TcdA also increases the release of endocannabinoids, which increase the release of more substance P via stimulation of VR1. Both toxins also induce an inflammatory response with infiltration of polymorphonuclear leukocytes (PMNs) into the **lamina propria** by up-regulation of IL-8 secretion from sensory neurons and colonocytes. The polymorphism within the IL-8 gene may be related to the presentation of more severe disease.

The role of biofilm development in the GI tract, along with its role in pathogenesis and antibiotic resistance, is an area of further study.

3. WHAT IS THE TYPICAL CLINICAL PRESENTATION AND WHAT COMPLICATIONS CAN OCCUR?

C. difficile is associated with antibiotic-associated diarrhea (CDAD) at one extreme to pseudomembranous colitis (PMC) at the other. The latter has a significant mortality. *C. difficile* is largely linked to hospital admission and has been associated with hospital outbreaks. Infection has usually occurred in those aged over 65 years and usually in association with antibiotics, bowel surgery or chemotherapeutic agents. The illness may follow antibiotics taken briefly or even several weeks previously. Recently, community cases of *C. difficile*-associated disease have been reported in a younger age group who have not necessarily been taking antibiotics. These cases have often been caused by hypervirulent strains such as ribotype 027.

Patients present with colicky abdominal pain, bloating, and watery diarrhea with an almost characteristic smell for *C. difficile*. In severe cases (PMC), there may be blood in the feces and the colon may perforate. Toxic megacolon may occur with ileus, and colectomy may be required. Thus *C. difficile* with little or no diarrhea, and pain predominating, may be an ominous sign.

Severe damage to the large bowel can result in rupture or perforation.

The rare patients with extra-intestinal disease almost always have long-term carriage of the organism or severe GI disease. Diseases reported include **bacteremia**, **abscess**, and a prosthetic hip infection. The significance of the isolate is questionable as, in abscesses, they are often part of a mixed infection.

4. HOW IS THE DISEASE DIAGNOSED, AND WHAT IS THE DIFFERENTIAL DIAGNOSIS?

The routine diagnosis of *C. difficile*-associated disease is the detection of toxins in the feces by **enzyme immunoassays** (**ELISA**). An additional useful assay is the detection of glutamate dehydrogenase in the feces as a surrogate marker for the organism. Culture is also available and is important for the investigation of suspected outbreaks where the isolates can be ribotyped. The organism may be grown on cycloserine/cefoxitin/egg yolk agar and a variety of other more sensitive media, always anaerobically producing 2–3 mm gray-white colonies. **Polymerase chain reaction** (**PCR**) for toxin-related genes is also available. PMC is diagnosed by colonoscopy where raised yellowish plaques may be seen on the mucosa. If severe disease is suspected, limited **sigmoidoscopy** may be used as colonoscopy may result in perforation.

The differential diagnosis includes other infective causes of diarrhea, inflammatory bowel disease, and **diverticulitis**.

5. HOW IS THE DISEASE MANAGED AND PREVENTED?

MANAGEMENT

For patients who are on antibiotics and develop diarrhea, one should stop the antibiotics if at all possible. Such a decision is on a case-by-case basis. Mild disease is not associated with a raised white cell count, and increased frequency of type 5–7 stools. Moderate disease is associated with a raised white cell count but less than 15×10^9. Severe disease has a white cell count of greater than 15×10^9, a temperature of 38.5°C or greater, and a rising creatinine with clinical evidence of colitis such as pain and or ileus.

For a first attack, mild-to-moderate disease is treated with oral metronidazole 400 mg TDS for 10–14 days. Severe disease is treated with oral vancomycin 125 mg QDS for 10–14 days or Fidaxomicin 200 mg BD for 10–14 days. Fidaxomicin reduces the episode of recurrence but is expensive and the decision to use Fidaxomicin should be based on a cost-benefit analysis.

If there is no response, and colitis is suspected (and the initial treatment was with vancomycin), one can use vancomycin 500 mg QDS with IVI metronidazole 500 mg TDS or Fidaxomicin 200 mg BD. With a worsening situation, intracolonic vancomycin 500 mg 4 hourly and IVI immunoglobulin 400 mg/kg.

Irrespective of the antibiotic treatment, there was a similar reduction in the alpha diversity, (richness of the microbiome in relation to the number of different bacterial species) however, the beta diversity (comparative richness of the microbiome) differed according to the antibiotic used. Each antibiotic (with the exception of fidaxomicin) was associated with an increase in pathogenic fungi and a functional increase in xenobiotic metabolism which may perpetuate a dysbiosis.

If a surgical resolution is needed, a total colectomy or loop ileostomy with lavage have been performed.

Recurrence of *C. difficile*-associated disease occurs in about 25% of cases and, if not severe, is treated with Fidaxomicin 200 mg BD. Alternative treatments are a tapering course of oral vancomycin such as 125 mg qds week 1, 125 mg tds week 2,

125 mg week 3 , 125 mg week 4, 125 mg on alternate days week 5, 125 mg every third day week 6, although other schedules have also been used. Similarly, many other treatments have been tried such as probiotics, rifamycin with little evidence of success. Fecal transplant (from a close family member) seems to be very effective in cases of multiple recurrences. Currently, several fecal microbial transplant (FMT) biotics are available which require assessment for cost-benefit analysis. Actoxumab (a human monoclonal antibody to toxin A) and bezlotoxumab (a human monoclonal antibody to toxin B), are two mAb that reduce recurrence by binding to the CROP domain and inhibiting binding of the toxins. The use of a triple mutant of adeno-associated virus (AAV6.2FF) that expresses actoxumab or bezlotoxumab may prevent relapses of the disease.

PREVENTION

In hospitals and nursing homes, prevention is by strict control-of-infection procedures. Contact precautions for any patients in whom CDAD is suspected are essential. The patient should be isolated in a single room with dedicated toilet facilities and the use of personal protective equipment (PPE), that is gloves and apron. If more than one patient has the illness, the patients should be nursed in a dedicated ward. A decision whether to close the ward/unit must be made on an individual basis but if there is evidence of widespread cross-infection then the unit should be closed. Horizontal and "high touch" surfaces and other items within the immediate environment, particularly toilet facilities, should be regularly cleaned with a hypochlorite disinfectant. All items and rooms after discharge of the patient(s) should be cleaned in the same manner. Attention to hand-washing practices is important and should be undertaken with soap and water rather than alcohol, as the latter has no effect on the spores of *C. difficile*. In fact, the spores of *C. difficile* are highly resistant to many of the generic disinfectants.

There is currently no vaccine although research is ongoing.

SUMMARY

1. WHAT IS THE CAUSATIVE AGENT, HOW DOES IT ENTER THE BODY, AND HOW DOES IT SPREAD A) WITHIN THE BODY AND B) FROM PERSON TO PERSON?

- *Clostridium difficile* is a gram-positive sporulating anaerobe.
- Several typing methods are available including ribotyping and PFGE.
- The spore is acquired by **feco-oral** spread or from the environment and infects colonocytes of the host.
- Up to 70% of neonates may be colonized without disease.
- Asymptomatic colonization may occur in up to 30% of patients in hospital.
- A hypervirulent strain (ribotype 027) first isolated in Canada produces large amounts of toxin and has spread across Europe, North America, and globally.
- *C. difficile* produces three toxins: toxins A and B and sometimes a binary toxin.
- Important risk factors for disease are increased age (> 60 years) and taking antibiotics.
- The organism only rarely causes extra-intestinal disease.

2. WHAT IS THE HOST RESPONSE TO THE INFECTION AND WHAT IS THE DISEASE PATHOGENESIS?

- Antibodies to toxins A and B are produced following *C. difficile* diarrhea, some of which are neutralizing.
- The toxins bind to host receptors on colonocytes and are taken up into the cell.
- Toxins A/B induce disaggregation of the actin cytoskeleton by inactivating GTPase proteins.
- Impairment of tight junctions and cell death occurs leading to fluid loss and a localized inflammatory response.

- Both the toxins and the inflammatory response itself are thought to be involved in tissue damage.

3. WHAT IS THE TYPICAL CLINICAL PRESENTATION AND WHAT COMPLICATIONS CAN OCCUR?

- The patient presents with abdominal pain and diarrhea.
- Frequently, the patient is aged > 60 years, in hospital, and is on antibiotics.
- Community-acquired disease in younger patients who have not been on antibiotics is relatively rare and associated with hypervirulent strains.
- The disease may be mild-to-moderate CDAD or severe PMC.
- Complications include perforation or toxic megacolon.

4. HOW IS THE DISEASE DIAGNOSED, AND WHAT IS THE DIFFERENTIAL DIAGNOSIS?

- The disease is routinely diagnosed by detection of the toxins in the feces using enzyme immunoassays.
- Culture is important for typing the organism in identifying and managing outbreaks.
- PMC is diagnosed by colonoscopy or sigmoidoscopy.
- Differential diagnosis includes other causes of gastroenteritis, diverticulitis and inflammatory bowel disease, and antibiotic-associated diarrhea not due to *C. difficile*.

5. HOW IS THE DISEASE MANAGED AND PREVENTED?

- The first-line treatment is oral metronidazole or vancomycin for 14 days.
- Alternative treatments include agents to bind the toxin, probiotics, or fecal enema.
- In hospital, prevention is by standard control of infection methods.

FURTHER READING

Cimolai N. Laboratory Diagnosis of Bacterial Infections. CRC Press, New York, 2001.

Mandell GL, Bennet JE, Dolin R. Principles & Practice of Infectious Diseases, 6th edition, Vol 3. Elsevier/Churchill Livingstone, Edinburgh, 1249–1259, 2005.

Mayhall GC. Hospital Epidemiology & Infection Control, 3rd edition. Lippincott Williams & Wilkins, Philadelphia, 623–631, 2004.

Murphy K, Weaver C. Janeway's Immunobiology, 9th edition. Garland Science, New York/London, 2016.

REFERENCES

Alam MJ, McPherson J, Miranda J, et al. Molecular Epidemiology of Clostridioides difficile in Domestic Dogs and Zoo Animals. Anaerobe, 59: 107–111, 2019.

Ausiello CM, Cerquetti M, Fedele G, et al. Surface Layer Proteins from Clostridium difficile Induce Inflammatory and Regulatory Cytokines in Human Monocytes and Dendritic Cells. Microbes Infect, 8: 2640–2646, 2006.

Balaji A, Ozer EA, Kociolek LK. Clostridioides difficile Whole-Genome Sequencing Reveals Limited Within-Host Genetic Diversity in a Pediatric Cohort. J Clin Micro, 57: e00559–19, 2019.

Bruxelle J-F, Péchiné S, Collignon A. Immunization Strategies Against Clostridium difficile. Adv Exp Med Biol, 1050: 197–225, 2018.

Ciftci Y, Girinathan BP, Dhungel BA, et al. Clostridioides difficile SinR' Regulates Toxin, Sporulation and Motility Through Protein-Protein Interaction with SinR. Anaerobe, 59: 1–7, 2019.

Durai R. Epidemiology, Pathogenesis and Management of Clostridium difficile Infection. Dig Dis Sci, 52: 2958–2962, 2007.

Fawley J, Napolitano LM. Vancomycin Enema in the Treatment of Clostridium difficile Infection. Surg Infect, 20: 311–316, 2019.

Khoruts A, Staley C, Sadowsky MJ, et al. Faecal Microbiota Transplantation for Clostridioides difficile: Mechanisms and Pharmacology. Nat Rev Gastroenterol Hepatol, 18: 67–80, 2021.

Lamendella R, Wright JR, Hackman J, et al. Antibiotic Treatments for Clostridium difficile Infection are Associated with Distinct Bacterial and Fungal Community Structures. mSphere, 3: e00572–17, 2018.

Matamouros S, England P, Dupuy B. Clostridium difficile Toxin Expression Is Inhibited by the Novel Regulator TvdC. Mol Microbiol, 64: 1274–1288, 2007.

McMaster-Baxter NL, Musher DM. Clostridium difficile: Recent Epidemiological Findings and Advances in Therapy. Pharmacotherapy, 27: 1029–1039, 2007.

Mounsey A, Lacy Smith K, Reddy VC, Nickolich S. Clostridioides difficile Infection: Update on Management. Am Fam Physician, 101: 168–175, 2020.

Owens RC. Clostridium difficile Associated Disease: Changing Epidemiology and Implications for Management. Drugs, 67: 487–502, 2007.

Peled N, Pitlik S, Samra Z, et al. Predicting Clostridium difficile Toxin in Hospitalized Patients with Antibiotic Associated Diarrhoea. Infect Control Hosp Epidemiol, 28: 377–381, 2007.

Smith JA, Cooke DL, Hyde S, et al. Clostridium difficile Toxin Binding to Human Intestinal Epithelial Cells. J Med Microbiol, 46: 953–958, 1997.

Summers BB, Yates M, Cleveland KO, et al. Fidaxomicin Compared with Oral Vancomycin for the Treatment of Severe Clostridium difficile-Associated Diarrhea: A Retrospective Review. J Hosp Pharm, 55: 268–272, 2020.

Voth DE, Ballard JD. Clostridium difficile Toxins: Mechanism of Action and Role in Disease. Clin Microbiol Rev, 18: 247–263, 2005.

Vuotto C, Donelli G, Buckley A. Clostridium difficile Biofilm. Adv Exp Med Biol, 1050: 97–115, 2018.

Wetzel D, McBride SM. The Impact of pH on Clostridioides difficile Sporulation and Physiology. Appl Environ Microbiol, 86: e02706–19, 2020.

WEBSITES

GOV.UK, Clostridioides difficile: guidance, data and analysis, 2014: http://www.hpa.org.uk/infections/topics_az/clostridium_difficile/C_diff_faqs.htm

Students can test their knowledge of this case study by visiting the Instructor and Student Resources: [www.routledge.com/cw/lydyard] where several multiple choice questions can be found.

Cytomegalovirus

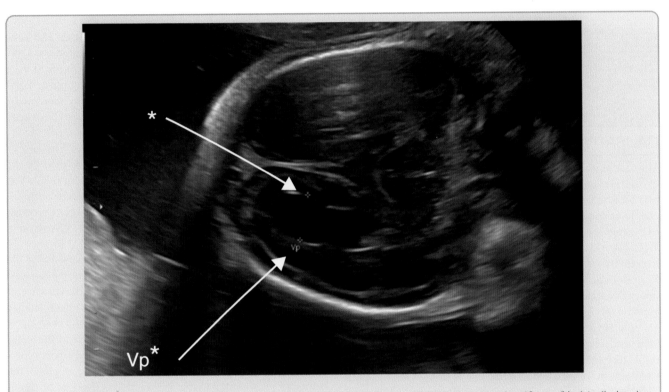

7

Figure 7.1 Fetal hydrops noted at growth scan (30 wk 6d). Lateral ventricles of fetal brain measuring 12 mm (Vp * to *) showing ventriculomegaly.

A 26-year-old healthy primigravida woman was booked into hospital early in the second trimester. At her first trimester antenatal screen, she was Rhesus positive, rubella immune, and negative for HIV, hepatitis B and C and syphilis. She underwent a routine morphology scan at 19-weeks' gestation. The biparietal diameter (BPD) and head circumference (HC) were on the 5th centile, the abdominal circumference (AC) on the 57th centile with an estimated fetal weight (EFW) of 304 g, on the 40th centile. Fetal anatomy appeared normal except for low-grade echogenic bowel. In view of these results, she was advised to have interval growth scans at 28 and 34 weeks. The planned third trimester growth scan was performed at 30 weeks 6 days, and showed the BPD on the 35th centile, HC on the 44th centile, AC >95th centile, femur length (FL) on the 10th centile and the EFW was estimated as >95th centile (~2218 g). Several abnormalities were noted. The fetus had developed gross hydrops; gross ascites, pericardial effusions of 6mm on both sides of the fetal cardia, and ventriculomegaly

of 12mm was noted (Figure 7.1). A Doppler scan of the middle cerebral artery suggested moderate to severe fetal anemia.

At 31-weeks' gestation, various possibilities were discussed with the patient, including aneuploidy, structural abnormalities, viral infection, or a genetic syndrome. Urgent tests were organized for CMV antibody, and amniocentesis to test for CMV DNA by polymerase chain reaction (PCR) and chromosomal abnormalities by microarray. Because of the possibility of severe fetal anemia, cordocentesis and intrauterine fetal transfusion were also performed. Pre-transfusion full blood count revealed a hemoglobin of 79 g/dl, a hematocrit of 0.25% and a platelet count of 29.9 × 10⁹/L (thrombocytopenia). After intrauterine transfusion, the hemoglobin increased to 100 g/dl, the hematocrit to 0.30 and the platelet count was 30 × 10⁹/L. Results of requested tests were positive CMV IgG and IgM with high IgG avidity, and on amniotic fluid no chromosomal abnormalities but a high CMV DNA load. The patient was told that primary CMV infection was the most likely diagnosis.

Continued...

...continued

Given the severe hydrops, ascites, pericardial effusion, and pancytopenia with probable bone-marrow involvement, as well as ventriculomegaly, there was a poor perinatal prognosis with a high chance of early neonatal death or significant long-term neurologic impact. The late termination was discussed but the patient wished for all reasonable measures for perinatal resuscitation to be taken. Use of CMV-specific immunoglobulin and antivirals were also considered, but at 32 weeks' gestation, the patient spontaneously ruptured membranes, went into labor and the baby was delivered by cesarean section. The neonate was admitted into the neonatal intensive care unit for ongoing supportive care. On day four of life, a cranial ultrasound scan revealed severe periventricular leukomalacia, ventriculitis and diffuse calcifications. The poor long-term outlook was explained to the parents and the decision was taken to withdraw care.

Adapted with permission from Mathias CR & Joung SJS. Diagnostic challenges in congenital cytomegalovirus infection in pregnancy: A case report. Case Rep Women's Health, 2019, 22: e00119. doi: 10.1016/j.crwh.2019.e00119

1. WHAT IS THE CAUSATIVE AGENT, HOW DOES IT ENTER THE BODY AND HOW DOES IT SPREAD A) WITHIN THE BODY AND B) FROM PERSON TO PERSON?

CAUSATIVE AGENT

The case above is an example of cytomegalic inclusion disease of the newborn. Cytomegalovirus (CMV), also known as human cytomegalovirus (HCMV), is so named because of the characteristic owl's eye or cytomegalic inclusions it produces in the nuclei of infected cells (Figure 7.2).

CMV is a beta herpesvirus, one of eight human herpesviruses (see Table 7.1, Epstein-Barr virus case). The *Herpesviridae* are characterized by a double-stranded (ds) linear DNA genome. CMV has the largest human herpesvirus genome of 236 kb encoding more than 200 proteins. The genome structure is similar to herpes simplex virus in that long and short unique sequences are bound by left and right terminal repetitive sequences. Individual genes of CMV are numbered sequentially in the unique long, unique short or terminal repeat regions. For example, UL128 refers to the 128th gene in the unique long region and pUL128 to its protein product (see Table 7.1 for other genes and gene products). Using next-generation sequencing, extensive CMV genome-wide variability has recently been recognized even within a single individual such as congenitally infected and immunocompromised patients. This variability potentially affects virus pathogenesis, but no definitive studies have been reported.

Virus structure is that of a typical herpesvirus (Figure 7.3). The space between the capsid and envelope is filled by the tegument or matrix, of which a major component in CMV is pp65 but there are dozens of other viral tegument proteins. Embedded in the CMV envelope are many viral glycoproteins including those important for cell entry, namely glycoprotein B (gB), gH, gL, gO, pUL128, pUL130, and pUL131A.

In common with all herpesviruses, CMV persists for life after primary infection. In latency, the viral genome remains in the cell's nucleus as a circular episome in the absence of lytic replication and virus production. The site of latency is primarily CD34+ hematopoietic progenitor cells. Upon cellular differentiation into monocytes and dendritic cells, the virus reactivates. Given the diversity of CMV genotypes, re-infection does occur but, in clinical practice, given the difficulty of distinguishing between reactivation of latent virus (endogenous) and re-infection (exogenous), the term CMV reactivation is applied to both scenarios.

VIRUS REPLICATION

CMV can infect a remarkably broad cell range within its host, including parenchymal and connective tissue cells of virtually any organ, as well as hematopoietic and myeloid cells. Lytic virus replication occurs predominantly in epithelial cells, endothelial cells, fibroblasts, and smooth muscle cells.

Host-cell entry is initiated when gB attaches to heparin sulphate. Binding to various cellular receptors by one of two gH/gL complexes on the virion is critical. The trimer gH/gL/gO mediates entry into fibroblasts and gB fusion at the plasma membrane. Whereas the pentamer gH/gL/pUL128/pUL130/pUL131A mediates entry into epithelial/endothelial and myeloid cells, and gB fusion at the endocytic membrane. Tegument proteins are released upon entry into the cytoplasm and function to ensure successful infection by modulating both the initial host response to infection and transcription of immediate early genes. The nucleocapsid is transported to

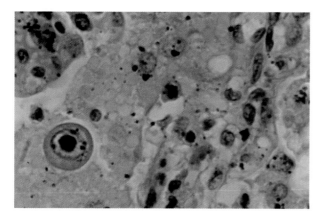

Figure 7.2 Photomicrograph of a section of lung tissue from an infant with congenital CMV infection. A large cytomegalic inclusion cell is shown on the lower left. *Courtesy of the Centers for Disease Control, Atlanta, Georgia. Image is found in the Public Health Image Library #22212. Additional photographic credit is given to Roger A. Feldman, who took the photo in 1976.*

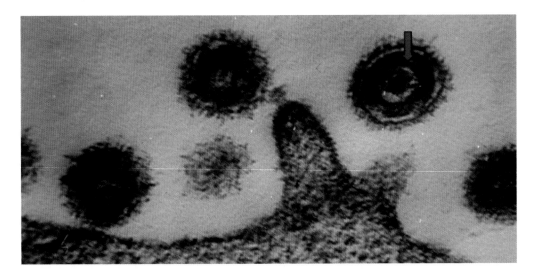

Figure 7.3 Thin section electron micrograph of herpesviruses budding after replication in an infected cell. Lipid virus envelope with embedded glycoproteins surrounds the tegument (red arrow), within which is the nucleocapsid.

Table 7.1 Selected CMV genes and gene products

Gene	Gene product
Nucleic acid metabolism	
UL97	Protein kinase confers ganciclovir sensitivity
UL122	Immediate early protein – IE2, responsible for regulation of viral gene expression and virus replication
UL54	DNA polymerase
UL51/UL56/UL89	Terminase complex enables viral DNA packaging
Structural proteins	
UL83	Phosphoprotein 65 (pp65), major tegument protein, also known as ppUL83 or lower matrix protein. Provides scaffold for optimal tegumentation. See also immunomodulatory molecules below
UL86	Major capsid protein
Virus attachment/membrane fusion proteins	
UL55	Glycoprotein B (gB), fusion protein, major target of neutralizing antibodies
UL75	Glycoprotein H (gH), target of neutralizing antibodies
UL115	Glycoprotein L (gL)
UL74	Glycoprotein O (gO), unique to CMV, forms a trimer with gH/gL critical for cell entry to fibroblasts
UL128/UL130/UL131A	pUL128/pUL130/pUL131A, unique to CMV, forms a pentamer with gH & gL critical for entry to epithelial/endothelial and myeloid cells
Immunomodulatory molecules	
US2	gpUS2 inserts into ER membrane and redirects human leukocyte antigen (HLA) class I to the cytosol for proteasomal degradation
US3	gpUS3 inhibits maturation and transport of peptide complexes with HLA class I
US6	gpUS6 inhibits TAP-mediated ER peptide transport
US10	gpUS10 degrades HLA G
US11	gpUS11 dislocates and degrades HLA class I from the ER
UL83	pp65 suppresses interferon response by inhibiting DNA sensing pathways, e.g. cGAS, IFI16
UL111A	Viral homolog of interleukin (IL)-10 regulates CD4+ T cell response

ER = endoplasmic reticulum

the nuclear pore and the genome enters the nucleus. Gene expression occurs sequentially in immediate early, early, and late phases. The linear viral DNA is circularized and copied by rolling circle replication to produce long concatemers consisting of linearly linked copies of the viral genome. The concatemer is cut into separate genome copies by the viral terminase complex and packaged into a newly formed capsid.

ENTRY AND SPREAD WITHIN THE BODY

In primary infection, CMV usually enters the body by replication in oropharyngeal or genital mucosal epithelia. Infection spreads to endothelial cells and leukocytes. Viremia facilitates systemic infection and presumably seeds secretory organs such as salivary glands and kidneys as well as the bone marrow where the virus becomes latent in CD34+ cells. Infection of ubiquitous cell types such as fibroblasts and smooth muscle cells provides the platform for virus proliferation.

PERSON-TO-PERSON SPREAD

After primary infection, CMV is shed asymptomatically in saliva and urine for many months in children and for a shorter time in adults. Notably, children with congenital CMV infection whether symptomatic or asymptomatic shed large amounts of virus of the order of 10^6 genomes/mL of urine or saliva for years. Shedding recurs periodically in all infected persons especially those with immunodeficiency or immunosuppression. Virus is also found in breast milk, cervical secretions, blood, and semen.

In childhood, CMV is acquired by contact with saliva or urine from infected children or on fomites. Children in day care nurseries transmit the virus to each other and to susceptible adult carers. Later in life, sexual transmission is a major route of infection; seroprevalence approaches 100% in men who have sex with men and sex workers. Another major route of transmission is from mother to child both prenatally when virus infects and crosses the placenta and perinatally through cervical secretions, breast milk, and saliva.

CMV can also be transmitted in leukocytes by blood transfusion from seropositive donors with a risk of 2.5%/unit of blood. Solid-organ and hematopoietic cell transplant recipients can acquire CMV from seropositive donors.

EPIDEMIOLOGY OF CMV INFECTION AND CONGENITAL CMV INFECTION

In developing countries, CMV is usually acquired in childhood and nearly 100% of young adults are seropositive. In developed countries, seroconversion progresses with age, but seroprevalence is higher in lower socioeconomic groups; overall about 40% of adolescents have been infected and seroprevalence increases by approximately 1% per year thereafter.

More than one million congenital CMV infections are predicted to occur each year worldwide but detailed data from developing countries is lacking. The best population-based estimates of congenital CMV infection come from the US, where 0.4%–0.7% of babies (approximately 28 000) are born each year with congenital CMV infection. Of these, 12.7% may be symptomatic at birth, and 13.5% of these developed symptoms on follow-up, making congenital CMV infection the most common viral cause of sensorineural hearing loss and neurodevelopmental delay. The prevalence of congenital CMV infection in Europe is similar.

2. WHAT IS THE HOST RESPONSE TO THE INFECTION AND WHAT IS THE DISEASE PATHOGENESIS?

HOST RESPONSE

Mammalian cytomegaloviruses are species specific and so human CMV infection cannot be studied in animals. Additionally, the host response has been difficult to dissect since human infection is mostly asymptomatic. Much of the information comes from murine CMV infections but many mechanisms have also been confirmed with human CMV. On first encounter with the virus, both innate and adaptive immunity responses play a role in restricting viral replication. Adaptive immunity in healthy individuals controls viremia, and productive infection in the face of virus reactivation or re-infection.

INNATE IMMUNITY

CMV infection of host cells triggers a rapid innate response through recognition of viral proteins and DNA by a variety of pattern recognition receptors (PRRs). These include **Toll-like receptor (TLR) 2** on the surface of antigen-presenting cells such as dendritic cells and macrophages. TLR2 binds CMV envelope proteins, gB and gH, and activates the NF-κB pathway that directs the production of inflammatory cytokines, such as interleukin (IL)-6, IL-8, IL-12, and IFN-β. TLR3 and TLR9 recognize CMV DNA in the endosome and nucleotide-binding oligomerization domain (NOD)-like receptors (NLRs) perform the same action in the cytoplasm.

Type I interferons (IFN-α/β), together with natural killer (NK) cells, are very important in controlling primary infection with herpesviruses; this is amply demonstrated in patients with NK-cell deficiencies who suffer severe herpesvirus infections. NK cells target virus infected cells by recognizing reduced levels of major histocompatibility complex (MHC) class I, together with elevated levels of infection-induced stress ligands; they do not express either immunoglobulin or T-cell receptor genes. These cells have been shown to inhibit human CMV transmission in fibroblasts, endothelial and epithelial cells by production of IFN-γ, and induction of IFN-β in infected cells.

Notably, although NK cells are part of the innate immune response, under certain circumstances they exhibit immunologic memory, a feature usually confined to adaptive responses. Studies of the antiviral response to human CMV have demonstrated clonally expanded NK cells based on their avidity to certain antigens, which leads to lifelong memory and readiness to respond to re-infection or reactivation.

Murine CMV experiments have shown that other innate effector cells are also important in controlling viral infection. Mast cells are activated and produce chemokines that recruit CD8+ T cells (see later) to infection sites. Neutrophils are recruited by NK-cell produced IL-22. Finally, invariant NKT (iNKT) cells help to control murine CMV infection and therefore may be active against human CMV. In humans, iNKT cells express an invariant T-cell receptor that recognizes glycolipids presented by the non-classical MHC class I molecule, CD1d.

ADAPTIC IMMUNITY – ANTIBODY RESPONSES

When assessed using cultures of human fibroblasts, gB and gH were found to be the targets of neutralizing antibodies directed against human CMV infection. However, it is now clear that the most important neutralizing antibodies are those targeted against the pentameric complex which block virus replication in endothelial/epithelial cells but not in fibroblasts.

ADAPTIVE IMMUNITY – T-CELL RESPONSES

MHC class I restricted CD8+ cytotoxic cells (T-cell receptor composed of α and β chains) are the single most important immune defense to give lifelong control over human CMV. Both CMV-specific CD4+ and CD8+ T cells dominate the memory compartments of seropositive subjects (often exceeding 10%). CD4+ T cells that recognize epitopes of CMV in the context of MHC class II are also important for helping generate cytotoxic CD8+ T cells but also some are cytotoxic for virus-infected cells themselves. Lastly, it is likely that γδ T cells are also involved; unlike αβ T cells, these cells have a receptor composed of γ and δ chains that can recognize antigens directly without the need for antigen processing and resemble antibodies in recognizing whole unprocessed molecules.

HUMAN CMV IMMUNE EVASION MECHANISMS

Remarkably, each CMV species appears to have coevolved within the mammalian species with which it is associated today. Such species-specific evolution produced different solutions to the problem of evading host immune effector mechanisms. Human CMV encodes more genes for immune evasion than it does to produce the virion. Mechanisms by which human CMV evades both primary and secondary immune responses are mainly the result of proteins coded for by the viral genome and microRNAs. These include the following:

1. Latency. Maintenance of the latent viral genome in the absence of lytic replication and virus production allows evasion or delaying of the immune response.
2. The interferon response is disabled at multiple points in the viral life cycle. For example, pp65 (Table 7.1) interferes with DNA sensing pathways.
3. Regulation of NK-cell activation. Some UL genes of CMV have been identified that suppress NK-cell recognition and activation through up-regulation of NK-cell inhibitory receptors and inactivating NK-cell signaling. In addition, by suppressing co-stimulatory molecules expression, they inhibit both CD8+ T-cell and NK-cell mediated antibody dependent cellular cytotoxicity.
4. Targeting antigen presentation by classical and nonclassical human leukocyte antigen (HLA) class I pathways. Several human CMV coded glycoproteins (gpUS2, gpUS3, gpUS6, gpUS10, and gpUS11) down-regulate these pathways using several different mechanisms to suppress the action of CD8+ T cells (Table 7.1).
5. Viral homolog of IL-10 (Table 7.1). This promotes down-regulation of HLA class II molecules and reduces presentation of viral epitopes to CD4+ T cells and consequent activation.

PATHOGENESIS

Mammalian cytomegaloviruses are species specific and so CMV cannot be studied in animal models. Pathology is presumably produced by direct viral cytopathic effects, although indirect effects produced by viral-encoded proteins or the host immune response remain a possibility.

CMV primary infection or reactivation in a healthy individual is rarely a cause of morbidity; subclinical infection occurs frequently in the normal host but is controlled by a robust immune response. Thus, a surfeit of virus-encoded immune evasion mechanisms is not sufficient to completely evade immunosurveillance. Uncontrolled viral replication and disease occurs both in the setting of impaired T-cell function and when CMV crosses the placenta and infects the fetus.

Primary infection in pregnancy carries a 40% risk of the virus crossing the placenta and infecting the fetus; virus transmission is highest in the late 2nd and 3rd trimester. Conversely, intrauterine transmission in the late 1st trimester and early 2nd trimester is associated with a greater likelihood of clinical abnormalities at birth and long-term neurodevelopmental abnormalities. Notably, the incidence of symptomatic congenital infection is much lower, although not absent, in maternal CMV reactivation or re-infection. Although maternal preexisting neutralizing antibody in

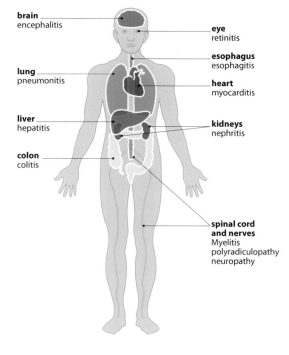

brain
encephalitis

eye
retinitis

esophagus
esophagitis

lung
pneumonitis

heart
myocarditis

liver
hepatitis

kidneys
nephritis

colon
colitis

spinal cord
and nerves
Myelitis
polyradiculopathy
neuropathy

Figure 7.4 Sites of CMV organ disease.

seropositive women can reduce intrauterine transmission, the neurodevelopmental consequences of infected infants are strikingly similar to those in infants infected in utero after primary maternal infection.

3. WHAT IS THE TYPICAL CLINICAL PRESENTATION AND WHAT COMPLICATIONS CAN OCCUR?

Severe CMV disease mostly occurs in the immunocompromised, namely those with advanced HIV disease (AIDS), solid-organ and hematopoietic stem-cell transplants, and in congenital CMV infection. Because of its broad cell tropism, CMV infection can affect many different organs (Figure 7.4). The disease pattern varies according to the patient's condition (see below).

CMV MONONUCLEOSIS

Although primary infection is usually clinically silent in the immunocompetent, it may present as a mononucleosis syndrome in a young adult. Classically, the syndrome comprises any combination of fever, tiredness, sore throat, lymphadenopathy, hepatosplenomegaly, and rash, accompanied by lymphocytosis and altered liver function tests. The disease is clinically indistinguishable from infectious mononucleosis caused by primary Epstein-Barr virus (EBV) infection (see Section 4, differential diagnosis).

Complications include CMV pneumonitis, which is usually mild with no treatment necessary. Guillain-Barré syndrome, associated with the emergence of anti-ganglioside antibodies, is rare. There is ascending muscle weakness in the legs and loss of tendon reflexes, accompanied by varying extents of sensory loss.

AIDS AND CMV

CMV is the most frequent viral opportunistic infection in patients with AIDS. Disease is usually due to virus reactivation as CMV seropositivity has been noted in more than 90% of HIV-infected men who have sex with men. CMV retinitis (Figure 7.5) is, by far, the most common outcome usually occurring when the CD4 cell count falls to less than 50cells/mm^3. The peripheral retina first shows retinal necrosis, which progresses to more central areas. By the time the patient notices visual disturbance, the retinal infection may be advanced. Blindness occurs within 4 to 6 months.

Gastrointestinal (GI) disease is also common. Esophageal CMV infection can cause painful ulceration with dysphagia whereas CMV colitis causes diarrhea. Less frequently, the nerve roots emerging from the spinal cord can become inflamed with a polyradiculopathy. Notably, pneumonitis is not seen; when CD4 counts are below 50 cells/mm^3, virus may be detectable in lung washings, but there are no pulmonary infiltrates. Nowadays, CMV end-organ disease is relatively infrequent and observed only in untreated HIV patients or those with poorly controlled HIV infection.

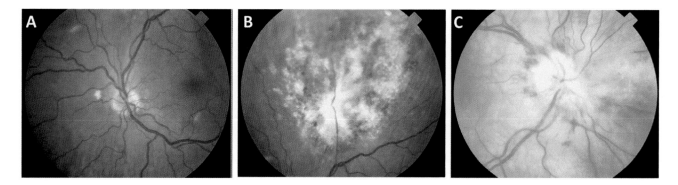

Figure 7.5 Cytomegalovirus (CMV) retinitis. (A) Early disease with retinal involvement along blood vessels. (B) Extensive retinal damage, and retinal hemorrhages. (C) CMV retinitis with papillitis. *From Cytomegalovirus (CMV) in: Bennett JE, Dolin R, Blaser MJ, eds. Mandell, Douglas and Bennett's Principles and Practice of Infectious Diseases, 8th Edition. Elsevier Saunders, 2015. With permission from Elsevier.*

SOLID-ORGAN TRANSPLANT RECIPIENTS

Patients receiving a solid-organ transplant are heavily immunosuppressed to prevent organ rejection. CMV may be transmitted from infected leukocytes in blood perfusing the donor organ, but endothelial cells may also be a source of latent or persistent infection. The risk of CMV disease is greatest in a seronegative recipient (R-) receiving a graft from a seropositive donor (D+) as compared with a seropositive recipient (R+) with re-infection or reactivation of latent virus.

The clinical spectrum of CMV infection in the early posttransplant period ranges from the CMV syndrome to organ- and life-threatening end-organ disease. The CMV syndrome includes fever, fatigue, low white blood counts, reactive lymphocytes, elevated hepatic transaminases, and CMV viremia. End-organ disease includes esophagitis, colitis, hepatis, and rarely interstitial pneumonitis or CMV retinitis. Disease varies according to the organ transplanted, for example, nephritis in renal transplant recipients, pancreatitis in pancreas transplants, hepatitis after liver transplant, pneumonitis in lung and heart-lung transplants, myocarditis in heart transplants.

HEMATOPOIETIC STEM-CELL TRANSPLANT RECIPIENTS

Patients receiving an allogeneic hematopoietic stem-cell transplant have their hematopoietic cells ablated before receiving donor (D) cells to reconstitute their immune system which takes at least several months. In contrast to solid-organ transplant recipients (R), those at highest risk of disease are the D-R+ cohort; on receipt of a transplant, the patient's endogenous CMV reactivates as the donor cells lack immune experience of CMV. Transmission from a D+ stem-cell donor to an R- recipient is an intermediate risk because few cells, <0.01%, will be latently infected with CMV. Additionally, the graft will transfer CMV cell-mediated immunity to the recipient.

As regards disease, hematopoietic stem-cell transplant patients are at high risk of CMV pneumonitis with a high mortality rate even with aggressive antiviral therapy. Infiltrates can be seen in the interstitial tissue of the lung rather than the airways on chest radiography (Figure 7.6). The pathology involves recruitment of lymphocytes increasing the distance that oxygen travels between the alveoli and blood vessels. Apart from pneumonitis, GI disease is also common and often concurrent with acute GI graft versus host disease making the assessment of each complication's relative contribution to symptoms difficult. CMV retinitis is a late complication after cessation of antiviral prophylaxis. Rarely, hepatitis or encephalitis may occur.

In contrast to an allogeneic transplant, an autologous stem-cell transplant has a much lower incidence of CMV pneumonitis. Instead, reactivation of the patient's endogenous virus or re-infection are a frequent cause of fever.

CONGENITAL CMV INFECTION

Approximately 15% of congenitally infected babies show signs and symptoms of severe generalized infection at birth. Clinical features of congenital CMV are maldevelopment of the brain with microcephaly and calcification in affected areas, chorioretinitis of the eye, sensorineural deafness, hepatitis with jaundice and hepatosplenomegaly, anemia, and thrombocytopenia. Such classical cytomegalic inclusion disease has a high mortality; 80% of all infants symptomatic at birth who survive have serious sequelae such as cognitive, visual, and hearing impairment. Of the approximately 85% of asymptomatic neonates, about 10–15% will have long-term neurodevelopmental abnormalities, principally sensorineural deafness.

PERINATAL OR POST-NATAL CMV INFECTION

Infection acquired during birth or early in the perinatal period almost never results in clinical symptoms and long-term sequelae in term infants. In contrast, premature infants can develop severe infections with multiorgan disease, including hepatitis, pneumonitis, and sepsis-like syndromes; however, linking such early CMV infection to long-term neurodevelopmental outcomes has been confounded by the impact of prematurity on neurodevelopment.

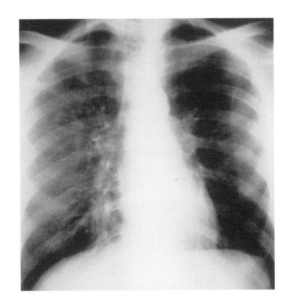

Figure 7.6 Bilateral interstitial pneumonitis caused by cytomegalovirus in a bone marrow transplant recipient. *From Cytomegalovirus (CMV) in: Bennett JE, Dolin R, Blaser MJ, eds. Mandell, Douglas and Bennett's Principles and Practice of Infectious Diseases, 8th Edition. Elsevier Saunders, 2015. With permission from Elsevier.*

Table 7.2 Methods for diagnosis of CMV infection

	Gene product	Sample Tested	Comment
Virus detection	Virus isolation	Blood, urine, throat, and bronchoalveolar washings	Tissue culture; virus grown in fibroblasts
	Shell vial (DEAFF)	Blood, urine, throat, and bronchoalveolar washings	Detection of CMV early antigens in infected fibroblasts using fluorescent labeled monoclonal antibody
	Antigenemia	Blood	Detection of pp65 in neutrophils using a monoclonal antibody; semiquantitative
	Nucleic acid amplification	Blood, cerebrospinal fluid	Quantitation of viral DNA by real-time polymerase chain reaction using primers from a conserved region of a CMV gene, e.g. immediate early or DNA polymerase
		Tissue biopsy	Reverse transcription polymerase chain reaction to detect mRNA
	Histopathology	Tissue biopsy	Identification of CMV infection in cells with typical owl eye inclusions; now superseded by immunohistochemistry or *in situ* hybridization to demonstrate productively infected cells
Antibody detetion	CMV-specific IgM	Serum or plasma	Identification of recent primary infection
	CMV-specific IgG	Serum or plasma	Identification of past infection or IgG seroconversion
	CMV-specific IgG avidity	Serum or plasma	Low avidity indicates recent primary infection

DEAFF= Detection of antigen fluorescent foci

4. HOW IS THE DISEASE DIAGNOSED, AND WHAT IS THE DIFFERENTIAL DIAGNOSIS?

DIAGNOSTIC TESTS

Laboratory diagnosis of CMV infection depends on virus isolation, detection of viral antigens or viral nucleic acid, and antibody tests. The various available tests are summarized in Table 7.2.

CMV grows very slowly in culture, and it may take several weeks before a cytopathic effect develops. With the advent of antiviral therapy, it became necessary to detect CMV infection before disease and treat early enough to reduce the incidence of CMV-related morbidity and mortality in the immunocompromised. In response to this need, a shell vial test was developed to detect virus infection, within 24 to 48 hours. This was superseded by the antigenemia test (Figure 7.7) as this gives semiquantitative results in blood and a good positive predictive value for development of disease; neutrophils make the largest contribution to overall virus infection in blood and pp65 can be readily detected in these cells by immunofluorescence or immunoperoxidase.

Currently, the most used test for CMV viremia is a real-time PCR. Accurate quantitative results can be achieved rapidly and serial measurements of viral DNA have largely replaced other assays for virologic monitoring of patients at risk of invasive CMV infection.

CMV-specific IgG tests are used to determine prior infection with CMV, i.e. seropositive or seronegative. IgM and IgG assays are important to diagnose primary from past infection in the immunocompetent but specific IgM can persist for months following first exposure in some individuals. Testing for IgG avidity depends on the principle that antibody avidity increases progressively with time from low to high after exposure to an immunogen. Testing involves modification of standard assays for anti-CMV IgG antibodies by the inclusion of chaotropes, such as urea, to disrupt IgG antibody binding to CMV antigens.

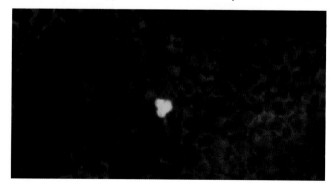

Figure 7.7 Immunofluorescence test for antigenemia showing pp65 in a neutrophil nucleus.

Examples of the appropriate use of the above tests to diagnose different conditions follows.

CMV MONONUCLEOSIS

The clinical diagnosis is confirmed by CMV IgG seroconversion or the presence of IgM.

CONGENITAL CMV INFECTION

If congenital infection is suspected before delivery, evidence of primary maternal CMV infection should be sought by antibody tests. It is often necessary to test for CMV IgG avidity as the timing of unsuspected asymptomatic primary infection is unknown and coincident serum samples unavailable. If primary infection is confirmed, amniocentesis should be done at least 7 weeks after the presumed time of maternal infection, and after 21 weeks of gestation, to ensure that amniotic fluid contains a large proportion of fetal urine. The sample should be tested for CMV by quantitative PCR as the higher the viral load the more likely severe symptomatic infection. Diagnosis of congenital infection at birth relies on CMV DNA detection in urine during the first 3 weeks of life; testing later cannot distinguish between congenital infection and an infection acquired during or after delivery. Other samples that can be tested are blood or saliva but urine is preferred.

CMV END-ORGAN DIEASE AFTER ORGAN TRANSPLANT

IgG antibody testing of allograft donors and recipients pretransplant is used to assess risk of CMV infection. After transplant, antibody testing is inappropriate and testing for virus is required. CMV pneumonitis causes the pulmonary infiltrates seen on chest radiography (Figure 7.6), and CMV DNA will be present in bronchoalveolar lavage (lung washings) obtained by bronchoscopy. For other end-organ disease, the clinical manifestations are difficult to differentiate from organ rejection and tissue from the affected organ should be tested for CMV infection. Diagnosis requires symptoms at the affected site plus tests on a biopsy to demonstrate productively infected cells; evidence should be sought by culture, immunochemistry, *in situ* hybridization or reverse transcription PCR for mRNA. Detection of viral DNA in tissue is insufficient as this may reflect viral load in blood rather than tissue damage due to viral replication.

CMV retinitis cannot be predicted by the presence of viral DNA in blood. Rather than testing for virus as described above, it is diagnosed on its clinical appearance, characteristically a white fluffy retinal infiltrate together with several areas of hemorrhage (Figure 7.5).

DIFFERENTIAL DIAGNOSIS

CMV-induced infectious mononucleosis is clinically indistinguishable from that induced by EBV. Seventy-nine percent of infectious mononucleosis is estimated to be caused by EBV; most of the other 21% is caused by primary CMV infection. Heterophil agglutinin test results are negative in CMV mononucleosis and usually positive in EBV mononucleosis. Another distinguishing feature is a sore throat with enlarged exudate-covered tonsils, which is more common with EBV infection. Whereas with CMV, fever and tiredness can predominate over other features and the age range of CMV-infected patients is 18–66 years, with a median age of 29 years, higher than in those with EBV-induced mononucleosis.

Cytomegalic inclusion disease of the newborn has features that may be caused by congenital infection with rubella virus, herpes simplex virus, and *Toxoplasma gondii*. All of these infections can cause microcephaly. In 2016, Zika virus was recognized as a cause of severe congenital infection with microcephaly after its epidemic spread in South America. Rubella, prior to widespread vaccination, was a prime cause of sensorineural deafness.

5. HOW IS THE DISEASE MANAGED AND PREVENTED?

MANAGEMENT

Antiviral agents were first used to control CMV infections in allograft recipients and individuals with HIV/AIDS. With the advent of highly active antiretroviral therapy, the number of AIDS patients fell significantly as did those with invasive CMV infections. More recently, treatment of newborn infants with symptomatic congenital CMV infections has been shown to improve outcomes in infants with symptomatic infections, principally protection against further hearing loss.

The antiviral agents approved for CMV therapy are listed in Table 7.3 together with their modes of action and side effects. Ganciclovir was licensed in 1989 and remains the most widely used drug. In allograft patients, resistance to ganciclovir develops at different rates depending on the clinical setting and duration of therapy such that it can develop in up to 10% of high-risk solid-organ transplant recipients; resistance is due to mutations in UL97 or UL54. Foscarnet is used as a second-line drug in the face of ganciclovir resistance, and cidofovir in the case of failure with the other anti-CMV drugs. Letermovir is a recently developed drug licensed for CMV prophylaxis (see below). The use of fomivirsen is limited to local therapy of CMV retinitis.

PREVENTION

CMV Infection Acquired from Blood Transfusion

Seronegative patients at risk of CMV disease can be given seronegative or leukocyte depleted blood.

Prophylaxis and Pre-emptive Therapy in Allograft Recipients

Since combined therapy with ganciclovir and intravenous (IV) immunoglobulin proved insufficient to prevent mortality

Table 7.3 Antiviral drugs for CMV and their side effects

	Mode of action	Side effects
Inhibitors of pUL54 DNA polymerase		
Ganciclovir (valganciclovir, oral formulation)	Deoxyguanosine analog phosphorylated by UL97 and cellular kinases to the triphosphate which inhibits DNA chain elongation	Thrombocytopenia and leukopenia
Cidofovir	Deoxycytidine monophosphate analog further phosphorylated to the triphosphate by host cell kinases inhibits DNA chain elongation	Irreversible renal toxicity
Foscarnet	Pyrophosphate analog. Non-competitive inhibitor of polymerase stops incorporation of nucleoside triphosphates into viral DNA	Renal toxicity
Inhibitor of viral DNA terminase complex		
Letermovir	Interferes with cytomegalovirus genome formation and virion maturation	Diarrhea, nausea, vomiting
Antisense inhibitor		
Fomivirsen	A phosphorothioate oligonucleotide complementary to a sequence in mRNA transcripts of the major immediate early region 2 (IE2) resulting in inhibition of IE2 protein synthesis and CMV replication	

due to CMV pneumonitis, two different approaches were developed for allograft recipients at risk. First, **prophylaxis** in which antiviral drugs, ganciclovir or valganciclovir, are given immediately after transplant and continued for the period of most significant immunosuppression. Alternatively, **preemptive therapy** in which transplant recipients are monitored intensively after transplant and treatment initiated when quantitative CMV DNA assays indicate increasing virus DNA in blood. The superiority of one strategy over the other for improving outcomes in allograft recipients remains to be determined.

One of the problems with CMV prophylaxis after hematopoietic stem-cell transplant is deciding when to start therapy. Whereas prophylaxis in patients who have received solid-organ transplants can be initiated almost immediately posttransplant, in patients who have undergone stem-cell transplantation, prophylaxis is initiated after engraftment (recovery of neutrophil count in blood) because of the bone marrow toxicity of ganciclovir and valganciclovir. Letermovir has recently been shown of value in this situation due to both its antiviral activity and, importantly, its lack of the myelosuppression associated with ganciclovir therapy.

Vaccine

An effective and safe vaccine is needed to immunize CMV-seronegative adolescent females prior to pregnancy and thus prevent congenital CMV infection with its accompanying sensorineural hearing loss and neurodevelopmental delay. Another urgent target is seronegative recipients of organ transplants. However, there are challenges facing vaccine development including the many mechanisms CMV has for immune evasion, and the fact that natural immunity does not prevent CMV re-infection.

Although some vaccines have demonstrated encouraging results in clinical trials, none have so far progressed to phase III. Two replication deficient vaccines have recently entered phase II studies. The first is a lymphocytic choriomeningitis virus vector-based vaccine that can express either gB or pp65. The second is a genetically inactivated CMV strain modified to express the pentameric complex. In the meantime, studies of passive immunotherapy using monoclonal antibodies with defined activity against specific CMV proteins should help to delineate immune correlates of protection.

SUMMARY

1. WHAT IS THE CAUSATIVE AGENT, HOW DOES IT ENTER THE BODY, AND HOW DOES IT SPREAD A) WITHIN THE BODY AND B) FROM PERSON TO PERSON?

- Cytomegalovirus (CMV), also known as human cytomegalovirus (HCMV) is a beta herpesvirus with an outer envelope with embedded viral glycoproteins surrounding an icosahedral capsid which encloses the linear ds DNA genome. The space between the capsid and envelope is filled by the tegument, of which a major component is phosphoprotein 65 (pp65).

- CMV persists for life after primary infection. The site of latency is CD34+ hematopoietic progenitor cells, in which the virus is a circular episome with limited gene expression and no virus production. Virus reactivation occurs upon cellular differentiation into monocytes and dendritic cells. Given the diversity of CMV genotypes, re-infection can also occur.

- CMV infects a broad cell range within its host, including parenchymal and connective tissue cells, as well as hematopoietic and myeloid cells. Lytic virus replication occurs predominantly in fibroblasts, epithelial, endothelial, and smooth muscle cells.

- There are two routes of cell entry: the trimer gH/gL/gO mediates entry into fibroblasts whereas the pentamer gH/gL/pUL128/pUL130/pUL131A mediates entry into epithelial/endothelial and myeloid cells.

- In virus replication, gene expression occurs sequentially in immediate early, early, and late phases. The linear genome is circularized and copied by rolling circle replication. The DNA product is cut into separate genome copies by the viral terminase complex and packaged into a newly formed capsid.

- CMV usually enters the body via oropharyngeal or genital mucosal epithelia. Infection spreads to endothelial cells and leukocytes facilitating systemic infection.

- After primary infection, virus is shed periodically in saliva and urine. Virus is also found in breast milk, cervical secretions, blood, and semen.

- In childhood, CMV is acquired by contact with saliva or urine from infected children or on fomites. Later in life, sexual transmission is a major route of infection. CMV passes from mother to child both prenatally when virus infects and crosses the placenta and perinatally through cervical secretions, breast milk, and saliva.

- In developing countries, CMV is usually acquired in childhood. In developed countries, about 40% of adolescents are seropositive and seroprevalence increases by approximately 1% per year thereafter.

- More than one million new congenital CMV infections are predicted to occur each year worldwide. Congenital CMV infection is the most common viral cause of sensorineural hearing loss and neurodevelopmental delay in children.

2. WHAT IS THE HOST RESPONSE TO THE INFECTION AND WHAT IS THE DISEASE PATHOGENESIS?

- CMV infection of host cells triggers a rapid innate response through recognition of viral proteins and DNA by a variety of pattern recognition receptors (PRRs).

- Type I interferons, together with NK cells, are very important in controlling primary infection. Patients with NK-cell deficiencies suffer severe herpesvirus infections. Invariant NKT (iNKT) cells may also be active against human CMV.

- Adaptive immunity in healthy individuals controls viremia, and productive infection in the face of virus reactivation or re-infection.

- Neutralizing antibodies include those directed against gB and gH which block infection of human fibroblasts. In contrast, neutralizing antibodies that block infection of endothelial/epithelial cells are thought to be directed against the pentamer complex and are not able to block fibroblast infection.

- MHC class I restricted CD8+ cytotoxic cells are the single most important immune defense to give lifelong control over CMV. Both CMV specific CD4+ and CD8+ T cells dominate the memory compartments of seropositive subjects (often exceeding 10%).

- CMV encodes more genes for immune evasion than it does to produce the virion.

- Evasion mechanisms include viral latency, disabling the interferon response, regulation of NK-cell activation, down-regulating the HLA class I pathway to suppress the action of CD8+ T cells and a viral homolog of IL-10, an immunomodulator.

- Mammalian cytomegaloviruses are species specific and human CMV cannot be studied in animals. Pathology is presumably produced by direct viral cytopathic effects, although indirect effects produced by viral-encoded proteins, or the host immune response remain a possibility.

3. WHAT IS THE TYPICAL CLINICAL PRESENTATION AND WHAT COMPLICATIONS CAN OCCUR?

- Severe CMV disease is rare in the immunocompetent but occurs in the immunocompromised with impaired T-cell function. Due to CMV's broad tropism, organ-specific disease may involve the nervous system, eyes, lungs, heart, liver, kidneys, and gastrointestinal tract.

- Immunocompetent adults with primary CMV infection sometimes present with a mononucleosis syndrome, with fevers, sore throat, and lymphadenopathy. A rare complication is Guillain-Barré syndrome.

- CMV reactivation is the most frequent viral opportunistic infection in AIDS. CMV retinitis is by far the most common outcome. Esophagitis and colitis are also common.

Continued...

...continued

- Solid-organ transplants are at risk of primary CMV infection acquired from a CMV seropositive donor. End-organ disease includes esophagitis, colitis, hepatitis, and rarely interstitial pneumonitis or CMV retinitis. The pattern of disease varies according to the organ transplanted.
- Allogeneic hematopoietic stem-cell transplant recipients are at highest risk of CMV disease if they are CMV seropositive but receive an organ from a CMV seronegative donor; upon transplant, the patient's endogenous CMV reactivates as the donor cells lack immune experience of CMV. CMV pneumonitis has a high morbidity and mortality. Gastrointestinal disease is also common. CMV retinitis is a late complication.
- Congenital CMV infection may cause a severe systemic illness with jaundice, hepatosplenomegaly, bleeding manifested as petechiae in the skin, and associated with thrombocytopenia and anemia. Organ-specific damage includes maldevelopment of the brain with microcephaly. There is a high mortality and those who survive have cognitive, visual, and hearing impairment.
- Of neonates infected with CMV but asymptomatic, 10–15% will have long-term neurodevelopmental abnormalities principally sensorineural deafness.
- CMV infection acquired by premature infants during birth or early in the perinatal period may develop severe infections with multiorgan disease, including hepatitis, pneumonitis, and sepsis-like syndromes.

4. HOW IS THE DISEASE DIAGNOSED, AND WHAT IS THE DIFFERENTIAL DIAGNOSIS?

- Laboratory diagnosis of CMV infection depends on virus isolation, detection of viral antigens or viral nucleic acid, and antibody tests.
- Viremia is diagnosed by identification of CMV DNA in blood by real-time polymerase chain reaction. Accurate quantitative tests have largely replaced other assays for virologic monitoring of allogeneic transplant recipients at risk of invasive CMV infection.
- Other uses of CMV DNA tests include testing of amniotic fluid if congenital CMV infection is suspected, or of infant's urine within the first 3 weeks of life. Bronchoalveolar fluid is tested for confirmation of a radiologic diagnosis of CMV pneumonitis in allogeneic transplant recipients.
- Diagnosis of end-organ disease requires symptoms at the affected site, together with tests on a biopsy, to demonstrate

productively infected cells; evidence should be sought by culture, immunochemistry, *in situ* hybridization or reverse transcription PCR for mRNA. CMV retinitis is an exception as it is diagnosed on its clinical appearance alone.

- CMV-specific IgG antibody tests are used to stratify risk of CMV infection in allograft donors and recipients. Tests for IgM, IgG, and antibody avidity are used to diagnose CMV mononucleosis and infection in pregnant women.
- The main differential diagnosis of infectious mononucleosis is that due to EBV infection, but the second most likely cause is CMV.
- The features of systemic congenital infection may arise not only due to CMV but also *Toxoplasma gondii*, rubella, and herpes simplex viruses. These pathogens can all cause microcephaly, as can Zika virus. Rubella, prior to widespread vaccination, was a prime cause of congenital sensorineural deafness.

5. HOW IS THE DISEASE MANAGED AND PREVENTED?

- Antiviral agents were first used to control CMV infections in allograft recipients and individuals with HIV/AIDS but, more recently, with some success in infants with congenital CMV.
- Ganciclovir and its oral formulation valganciclovir are the most widely used drugs. Foscarnet is used in the face of ganciclovir resistance, and cidofovir in the case of failure with the other anti-CMV drugs. Letermovir is a recently developed drug licensed for CMV prophylaxis. Fomivirsen is limited to local therapy of CMV retinitis.
- CMV infection acquired from blood transfusion can be prevented by using seronegative or leukocyte depleted blood. Allograft recipients can be protected from CMV infection by prophylaxis or preemptive therapy when treatment with these drugs is initiated at onset of viremia.
- An important priority is for an effective and safe vaccine to reduce congenital CMV infection, but no candidate vaccine has progressed to phase III trials.
- Two replication deficient vaccines have recently entered phase II studies, one can express gB or pp65 and the second the pentameric complex.
- Studies of passive immunotherapy using monoclonal antibodies with defined activity against specific CMV proteins should help to delineate immune correlates of protection.

FURTHER READING

Crumpacker II CS. Cytomegalovirus (CMV). In: Bennett JE, Dolin R, Blaser MJ, editors. Mandell, Douglas and Bennett's Principles and Practice of Infectious Diseases, 8th edition. Elsevier Saunders, New York, 2015.

Goering RV, Dockrell HM, Zuckerman M, Chiodini PL. Medical Microbiology and Immunology, 6th edition. Elsevier, Philadelphia, 2019.

Goodrum F, Britt W, Mocarski ES. Cytomegalovirus. In: Howley PM and Knipe DM, editors. Fields Virology, 7th edition. Wolters Kluwer, 2022.

Sissons JGP. Herpesviruses (excluding Epstein-Barr virus). In: Firth J, Conlon C, Cox T, editors. Oxford Textbook of Medicine, 6th edition. Oxford University Press, Oxford, 2020.

REFERENCES

Biolatti M, Gugliesi F, Dell'Oste V, Landolfo S. Modulation of the Innate Immune Response by Human Cytomegalovirus. Infect Genet Evol, 64: 105–114, 2018.

Biron CA, Byron KS, Sullivan JL. Severe Herpesvirus Infections in an Adolescent Without Natural Killer Cells. N Engl J Med, 320: 1731–1735, 1989.

Cheeran MC-J, Lokensgard JR, Schleiss MR. Neuropathogenesis of Congenital Cytomegalovirus Infection: Disease Mechanisms and Prospects for Intervention. Clin Microbiol Rev, 22: 99–126, 2009.

Dollard SC, Grosse SD, Ross DS. New Estimates of the Prevalence of Neurological and Sensory Sequelae and Mortality Associated with Congenital Cytomegalovirus Infection. Rev Med Virol, 17: 355–363, 2007.

Griffiths P, Plotkin S, Mocarski E, et al. Desirability and Feasibility of a Vaccine Against Cytomegalovirus. Vaccine, 31: B197–203, 2013.

Griffiths P, Reeves M. Pathogenesis of Human Cytomegalovirus in the Immunocompromised Host. Nat Rev Microbiol, 19: 759–773, 2021.

Kimberlin DW, Jester PM, Sanchez PJ, et al. Valganciclovir for Symptomatic Congenital Cytomegalovirus Disease. N Engl J Med, 372: 933–943, 2015.

Liu Y, Freed DC, Li, L, et al. A Replication-Defective Human Cytomegalovirus Vaccine Elicits Humoral Immune Responses Analogous to Those with Natural Infection. J Virol, 93: e00747–19, 2019.

Manandhar T, Hò G-GT, Pump WC, et al. Battle Between Host Immune Cellular Responses and CMV Immune Evasion. Int J Mol Sci, 20: 3626, 2019.

Mathias CR, Joung SJS. Diagnostic Challenges in Congenital Cytomegalovirus Infection in Pregnancy: A Case Report. Case Rep Womens Health, 22: e00119, 2019.

Schwendinger M, Thiry G, De Vos B, et al. A Randomized Dose-Escalating Phase I Trial of a Replication-Deficient Lymphocytic Choriomeningitis Virus Vector-Based Vaccine Against Human Cytomegalovirus. J Infect Dis, 225: 1399–1410, 2020.

WEBSITES

Centers for Disease Control and Prevention (CDC), Cytomegalovirus (CMV) and Congenital CMV Infection, 2020: https://www.cdc.gov/cmv/index.html

Students can test their knowledge of this case study by visiting the Instructor and Student Resources: [www.routledge.com/cw/lydyard] where several multiple choice questions can be found.

8

Echinococcus spp.

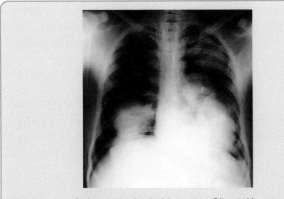

Figure 8.1 Pulmonary hydatid cysts. Chest X-ray showing hydatid cysts (opaque areas at center right and center left) in a patient's lungs. The cysts are caused by the parasitic tapeworm *Echinococcus* spp. Humans are not a natural host for the parasite but may become infected from ingesting eggs shed in the feces of an infected dog or other canids. Cysts may be formed in the liver, lungs or any other organ in the body. Large pulmonary cysts may cause symptoms, including shortness of breath, chest discomfort and sometimes wheeze. Treatment involves giving anti-parasitic drugs and careful surgical resection when possible. *From Zephyr / Science Photo Library, with permission.*

A 28-year-old Kurdish refugee complained of upper abdominal pain for 4 months. He had recently noticed some pain on the right side of his chest. He had not noticed any fever and his weight was stable. His doctor requested a chest X-ray. This showed a mass lesion in the right lung. He was referred to the Thoracic Surgical Department at the local hospital. A **CT scan** of his chest and abdomen was requested. This demonstrated a cystic structure in his liver, with other **cysts** in his chest (Figure 8.1). The surgeons excised the right lung lesion. The histopathologists reported that it was a **hydatid cyst**. The patient was referred to the Infectious Diseases Department. A serologic test was positive for *Echinococcus*. He was given treatment with albendazole and his remaining cysts were initially monitored serially on scans.

1. WHAT IS THE CAUSATIVE AGENT, HOW DOES IT ENTER THE BODY AND HOW DOES IT SPREAD A) WITHIN THE BODY AND B) FROM PERSON TO PERSON?

CAUSATIVE AGENT

Echinococcus species are tapeworms, also known as cestodes. Adults are only 2–7 mm in length. Two species are responsible for the majority of human infections. *Echinococcus granulosus* arises from dogs and other canids and has been described worldwide. It causes cystic echinococcosis. *Echinococcus multilocularis* arises from arctic or red foxes and is restricted to the Northern Hemisphere. It causes alveolar echinococcosis. Infection with the latter seems to be spreading as red-fox populations have been growing. Adult tapeworms, numbering in the thousands, live in the intestines of the dog or wild canid hosts. They attach to the mucosa with suckers and hooks. There are different parts to the structure of the tapeworm, and these are shown in Figure 8.2. The adult tapeworms lay eggs, which pass out in the feces and contaminate the soil. Unless dessicated, eggs may survive in the soil for one year. These may be swallowed by an intermediate host such as sheep for *E. granulosus* and small mammals for *E. multilocularis*.

ENTRY AND SPREAD WITHIN THE BODY

Humans are infected if they accidentally swallow eggs present in soil, water or food. The eggs have a tough shell that breaks down and releases a larval form called the oncosphere. The oncosphere burrows into the intestinal wall and then spreads to other parts of the body along blood vessels or lymphatics. The liver is the commonest site of infection as most of the intestinal drainage of blood is along the portal vein to the liver. However, infection could arise anywhere in the body.

When the oncosphere lodges in an organ it develops into a vesicle. This progressively grows into a cyst. The cyst wall includes an inner germinal layer. From this buds a stage called a protoscolex within brood capsules. Around the germinal layer is an acellular, laminated layer. For *E. granulosus* this cyst is unilocular and known as a hydatid cyst. For *E. multilocularis*, the cyst is multilocular or referred to as alveolar. Occasionally,

sucker   uterus with eggs

Echinococcus granulosus

Figure 8.2 An adult tapeworm of *Echinococcus granulosus*. *(From www.healthatoz.com/healthatoz/Atoz/images/)*

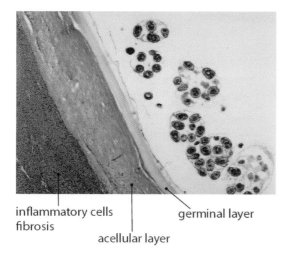

inflammatory cells
fibrosis

germinal layer

acellular layer

Figure 8.3 Histopathology of the structure of a hydatid cyst. Round protoscoleces are seen on the right. They arise from a thin germinal layer, which is surrounded by an acellular layer; the host inflammatory cells can be seen to the left of this. *Courtesy of the Centers for Disease Control, Atlanta, Georgia. Image is found in the Public Health Image Library #910. Additional photographic credit is given to Dr Peter Schantz who took the photo in 1975.*

cysts can rupture giving rise to secondary, "daughter" cysts. The structure of a hydatid cyst due to *E. granulosus* is shown in Figure 8.3.

PERSON-TO-PERSON SPREAD

Humans do not spread *Echinococcus* spp. from person to person. Humans are an end host, with occasional exceptions.

INTERMEDIATE HOSTS

Animals such as sheep or cattle may ingest *E. granulosus* eggs shed in dog feces. They can act as intermediate hosts. When sheep or cattle are killed, dogs may eat their meat or offal. When the dogs swallow cysts within infected tissues the protoscoleces are released. They attach to the dogs' intestinal mucosa to mature into the adult tapeworms. Thus, the life cycle is completed (Figure 8.4). Dogs are the definitive hosts. This life cycle for *E. granulosus* occurs on farms but for *E. multilocularis* there is a rural, or **sylvatic**, life cycle with small mammals serving as the intermediate host and arctic or red foxes as the definitive host.

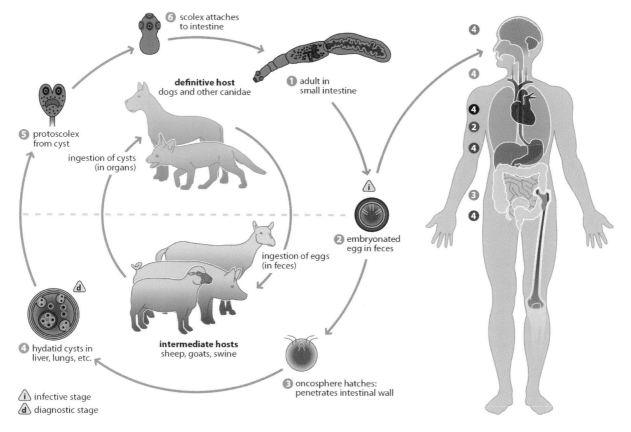

Figure 8.4 Life cycle of *Echinococcus granulosus*. Adult worms living in the dog intestine shed eggs in dog feces. These contaminate the environment and may be ingested by intermediate hosts such as sheep. In the intestine, oncospheres hatch from the eggs and penetrate the intestinal wall. The oncosphere spreads to other organs and eventually matures into a cyst. Within the cyst, protoscolices develop. When a cyst is ingested by a dog the protoscolices are released, attach to the intestinal mucosa as a scolex, and mature into adult worms. *Courtesy of the Centers for Disease Control, Atlanta, Georgia. Image is found in the Laboratory Identification of Parasites of Public Health Concern.*

EPIDEMIOLOGY

Given the life cycle, hydatid disease is most common where dogs and animals such as sheep and cattle mix. For *E. granulosus*, worldwide rates are highest in South America, Africa, Asia, and parts of Europe. In some rural regions close to 10% of the population may be infected. *E. multilocularis* is found in northern regions of China, Mongolia, Russia, Europe, and North America, with geographic spread relating to growing fox populations.

2. WHAT IS THE HOST RESPONSE TO THE INFECTION AND WHAT IS THE DISEASE PATHOGENESIS?

Potentially, the host may mount a response to the oncospheres, after they have entered the intestine, or against the cysts, which become surrounded by the acellular, laminated layer. There may also be a host response to fluid that leaks from a ruptured cyst.

Little is known of the host response to the oncospheres. A single or small number of oncospheres may not pose a sufficient antigenic stimulus to trigger an adaptive immune response. The innate immune response, utilizing phagocytic cells or complement, may attack them after invasion of the intestinal mucosa. Once a hydatid cyst has developed, an antibody response occurs that is initially **IgM** followed by an **IgG** and **IgE** response. If these antibodies also recognize the oncospheres, they could play a part in preventing further ingested oncospheres from causing additional infection.

An inflammatory response occurs around the maturing cyst. This is triggered by complement activation. There is an infiltration of macrophages and lymphocytes and these include both **CD4+** and **CD8+** T lymphocytes. Both **Th1**- and **Th2**-type **cytokines** are found, such as **interferon (IFN)-γ, interleukin** (IL)-4, and interleukin-10. Eventually, a host-derived, fibrous layer forms around the cyst. Complement activation and inflammation subside as the acellular layer accumulates. It is thought that there is local immunomodulation with impaired macrophage responses, impaired **dendritic cell** differentiation, and T-regulatory cells. Cyst-derived antigens, including an antigen B, affect the maturation and function of dendritic cells, such that when they present antigen to lymphocytes the immune response skews to a Th2 pattern. The latter is characterized by IL-4 production and increased IgG and IgE levels.

PATHOGENESIS

Large cysts cause pressure and may cause pain or dysfunction of the organ affected. If a hydatid cyst ruptures its proteinaceous contents leak, with a sudden, large antigenic stimulus. This can cause an anaphylactic reaction if IgE antibodies trigger the activation of mast cells and the systemic release of histamine and other mediators (**type 1 hypersensitivity**).

3. WHAT IS THE TYPICAL CLINICAL PRESENTATION AND WHAT COMPLICATIONS CAN OCCUR?

Hydatid cysts grow gradually. They may not cause any physical problems and may remain asymptomatic. Just over half of infected individuals are asymptomatic. When the cyst is large, pressure effects may cause pain or dysfunction of the organ affected. If a cyst ruptures, daughter cysts may cause additional problems and cystic fluid can trigger life-threatening **anaphylaxis**.

Hydatid cysts may be found in any part of the body but about 90% occur in the liver or lung or both. Large cysts in the liver can cause upper abdominal pain. The pain may localize more to the right side of the abdomen, as the right lobe of the liver is most frequently affected. The liver can enlarge and become palpable. The cyst may become secondarily infected with bacteria. If the cyst presses on the biliary tree, obstructive **jaundice** can develop. Rupture into the biliary tree may lead to daughter cysts in the common bile duct. Again, these can cause an obstructive jaundice. In the lung, large cysts may cause pain. This may be felt as a central chest ache. When cysts abut the chest wall, they cause localized pain at these sites. Very large cysts may make individuals short of breath and cause a cough.

Hydatid disease due to *E. granulosus* has a case fatality rate of about 2%. Infection due to *E. multilocularis* has a higher mortality rate. It can cause any of the clinical features associated with hydatid disease but can have a progressive course with growing cysts that enlarge the liver and compromise its function.

If cysts rupture, an abrupt antigenic challenge may trigger anaphylaxis. A classical anaphylactic reaction comprises swelling of lips, tongue, vocal chords, and bronchial airway mucosa; urticarial reactions in the skin; and a fall in blood pressure. **Eosinophilia** may be observed if there is cyst leakage but is not uniformly present with intact cysts.

4. HOW IS THE DISEASE DIAGNOSED, AND WHAT IS THE DIFFERENTIAL DIAGNOSIS?

The case history illustrates that sometimes the diagnosis is made after surgical removal of a lesion and histopathologic examination. However, it is preferable to make a diagnosis before surgery as accidental incision of a cyst and leakage of its contents may trigger anaphylaxis.

Either ultrasound of the liver or a CT scan of thorax or abdomen may reveal a cystic structure. Some features are highly suggestive of a hydatid cyst. There is a thick cyst wall. Septa may be seen within the cyst. If the germinal layer detaches from the cyst wall, the inner septa may float in the fluid like a "water lily". Mature, degenerating cysts may become calcified. Cystic appearances determine treatment strategies.

Serologic tests are available for *Echinococcus* antibodies. Methods used include **enzyme-linked immunosorbent assay (ELISA)**, indirect hemagglutination assays, and latex agglutination assays. The antigens used in assays are derived from hydatid cyst fluid. There may be cross-reactions with other tapeworm infections affecting specificity. Initial serology may be followed by immunoblotting for more specific results. Sensitivity may be as low as 60% for infection in the lung, but more than 80% for liver.

DIFFERENTIAL DIAGNOSIS

The differential diagnosis is from other causes of cystic structures. The range of possibilities is larger for the lung than for the liver. In the liver, benign cysts with thin walls can occur. A liver abscess may cause a different overall clinical picture and aspiration may yield pus. Liver tumors may have the same shape but are more solid structures. The more extensive liver changes in *E. multilocularis* infection may appear like nodular cirrhosis.

5. HOW IS THE DISEASE MANAGED AND PREVENTED?

MANAGEMENT

Treatment options depend on the size of the cyst (a viable cyst <5 cm may be treated with chemotherapy alone), whether there is a single or multiple compartments (a viable cyst >5 cm with a single compartment may be aspirated alongside chemotherapy) and whether there are features of viability (an inactive cyst may be observed alone). Surgical removal may be required for troublesome cysts, if technically feasible.

Albendazole is the chemotherapeutic agent of choice and may be used continuously for months. Cysts reduce in size in about 70% of cases. The addition of praziquantel to albendazole may help prevent the development of daughter cysts if there is cyst rupture. Chemotherapy is used as an adjunct to surgery and aspiration. Albendazole may be used one week to one month prior to surgery and for at least one month afterward, reducing cyst viability and making surgery safer. At surgery, 20% hypertonic saline may also be injected into a cyst to reduce viability prior to excision.

Another approach to the management of some cases of cystic echinococcosis has been aspiration of accessible cysts with a needle and injection of an agent such as 95% ethanol to kill the protoscoleces inside the cyst. The ethanol is then aspirated back after 20 minutes. Microscopy of the aspirate shows free protoscoleces, which are sometimes referred to as "hydatid sand". This method is referred to as **PAIR**, which stands for **P**uncture, **A**spiration, **I**njection, **R**easpiration. The key risk is anaphylaxis if cyst fluid leaks out but is rare in experienced hands.

PREVENTION

Humans need to maintain high standards of washing hands and fresh food to avoid swallowing eggs. Prevention of *Echinococcus* infection requires breaking the natural life cycle. Dogs have been treated with arecoline or praziquantel. Praziquantel baiting has been used to treat red foxes. On farms, dogs are kept away from offal.

Vaccines have been trialed in sheep and dogs. A recombinant DNA vaccine containing the gene for an oncosphere antigen, and designated EG95, has shown >90% protective efficacy in sheep.

Despite efforts, *Echinococcus* infection continues to be a problem in many countries.

SUMMARY

1. WHAT IS THE CAUSATIVE AGENT, HOW DOES IT ENTER THE BODY, AND HOW DOES IT SPREAD A) WITHIN THE BODY AND B) FROM PERSON TO PERSON?

- The causative agent is *Echinococcus granulosus* or *Echinococcus multilocularis*. These are tapeworms, or cestodes.
- Adult Echinococcus tapeworms live in the intestine of dogs or foxes and shed eggs in the feces. Dogs and foxes are the definitive hosts.
- The eggs are accidentally swallowed by other species. Once swallowed, oncospheres emerge from the eggs, penetrate the intestinal wall, and spread to other parts of the body where they form cysts.
- Dogs are infected when they eat meat or offal containing cysts. This happens for instance on farms, where they might eat offal from sheep or cattle.
- Species such as sheep or cows are therefore referred to as intermediate hosts. Humans are usually an end host.

2. WHAT IS THE HOST RESPONSE TO THE INFECTION AND WHAT IS THE DISEASE PATHOGENESIS?

- Little is known about the host response to the oncospheres when they penetrate the intestinal wall.
- Once the oncospheres develop into cysts in tissues, there is a surrounding inflammatory response triggered by complement activation, involving macrophages and then a CD4+ and CD8+ T-lymphocyte response.
- Eventually, an acellular and fibrous layer develops around the cyst with down-regulation of the immune and inflammatory response.
- The pathogenesis of disease depends on the size of cysts. Large cysts can pose a physical problem in the infected organ.
- If cysts rupture and leak, an IgE-mediated response can be triggered.

Continued...

...continued

3. WHAT IS THE TYPICAL CLINICAL PRESENTATION AND WHAT COMPLICATIONS CAN OCCUR?

- Small cysts may remain asymptomatic.
- Hydatid cysts due to *E. granulosus* are mostly found in the liver or lung or both. The cysts of *E. multilocularis* are primarily found in the liver.
- Large hydatid cysts in the liver can cause pain, may obstruct the flow of bile, and may rupture into the biliary tree.
- Disease due to *E. multilocularis* can be steadily progressive.
- Cyst rupture and leakage of proteinaceous contents can trigger an anaphylactic reaction.

4. HOW IS THE DISEASE DIAGNOSED, AND WHAT IS THE DIFFERENTIAL DIAGNOSIS?

- Diagnosis is based on imaging such as ultrasound or CT scan. Typical radiologic features may be present to diagnose hydatid disease.

- Additional tests include serology.
- Cross-reactions with other tapeworm infections affect the specificity of serology. Sensitivity ranges from 60% to 80%.
- The differential diagnosis is from other causes of cystic lesions.

5. HOW IS THE DISEASE MANAGED AND PREVENTED?

- Treatment options depend on cyst size, cyst compartmentalization, and features of cyst viability.
- The main chemotherapeutic agent is albendazole.
- Surgical resection may be performed if necessary and if feasible.
- Some experienced centers also undertake a procedure called PAIR, which involves percutaneous puncture of cysts, aspiration of contents, injection of ethanol, and then reaspiration.
- Prevention requires breaking the life cycle by treating dogs and preventing the consumption of offal by dogs. To avoid accidental infection, humans must maintain high standards of hand hygiene.

FURTHER READING

Murphy K, Weaver C. Janeway's Immunobiology, 9th edition. Garland Science, New York/London, 2016.

Stojkovic M, Gottstein B, Junghanns T. Echinococcosis. In: Farrar J, Hotez P, Junghanns T, et al. (eds). Manson's Tropical Diseases, 23rd edition. Elsevier Saunders, London, 795–819, 2014.

REFERENCES

Bakhtiar NM, Spotin A, Mahami-Oskouei M, et al. Recent Advances on Innate Immune Pathways Related to Host-Parasite Cross-Talk in Cystic and Alveolar Echinococcosis. Parasit Vectors, 3: 232, 2020.

Brunetti E, Kern P, Vuitton DA. Writing Panel for the WHO-IWGE. Expert Consensus for the Diagnosis and Treatment of Cystic and Alveolar Echinococcosis in Humans. Acta Trop, 114: 1–16, 2010.

McManus DP, Zhang W, Li J, Bartley PB. Echinococcosis. Lancet, 362: 1295–1304, 2003.

Neumayr A, Toia G, de Bernardis C, et al. Justified Concern or Exaggerated Fear: The Risk of Anaphylaxis in Percutaneous Treatment of Cystic Echinococcosis – a Systematic Literature Review. PLos Negl Trop Dis, 5: e1154, 2011.

Zhang W, Wen H, Li J, et al. Immunology and Immunodiagnosis of Cystic Echinococcosis: An Update. Clin Dev Immunol, 2012: 101895, 2012.

WEBSITES

Centers for Disease Control and Prevention (CDC), Parasites – Echinococcosis, 2020: https://www.cdc.gov/parasites/echinococcosis/index.html

World Health Organization, Echinococcosis, 2012: https://www.who.int/health-topics/echinococcosis

Students can test their knowledge of this case study by visiting the Instructor and Student Resources: [www.routledge.com/cw/lydyard] where several multiple choice questions can be found.

9

Enteroviruses

A 22-year-old medical student walked into an Emergency Department complaining of a headache that had begun 3 hours before and seemed to be getting worse. He felt generally unwell, but without any other specific symptoms. On examination, he was febrile (38°C), and was noted to have neck stiffness. He complained when a bright light was shone into his eyes. The casualty officer made a diagnosis of acute meningitis and admitted him to the medical ward.

A lumbar puncture was performed and 2 hours later the following results were received:

- cerebrospinal fluid (CSF) – clear and colorless;
- white cell count $120 \times 10^6/l$, 75% mononuclear cells, no red cells;
- CSF protein 4.2 g/l (normal 1.5–4.0 g/l);
- CSF glucose 4.3 mmol/l, blood glucose 5.7 mmol/l;
- Gram stain – no organisms seen.

A throat swab and a sample of feces were also sent to the laboratory. The following day, the ward received a telephone call from the microbiology laboratory informing them that the CSF and throat swab were positive for an enterovirus.

Three days later, the patient was feeling much better and was ready to be discharged home. Two weeks later, a report from the reference laboratory confirmed that the virus present in the CSF was coxsackie B4.

1. WHAT IS THE CAUSATIVE AGENT, HOW DOES IT ENTER THE BODY, AND HOW DOES IT SPREAD A) WITHIN THE BODY AND B) FROM PERSON TO PERSON?

CAUSATIVE AGENT

Coxsackie viruses belong to the *Enterovirus* genus within the family of viruses known as the *Picornaviridae*. "Pico" means small, so this translates as the small RNA viruses. The *Picornaviridae* carry a genome of positive sense single-stranded RNA in an unenveloped viral particle. The other important human pathogens within this family are the rhinoviruses, the commonest cause of the common cold, the parechoviruses and hepatitis A virus.

Enteroviruses are small, icosahedral in shape, and approximately 25–30 nm in diameter. They are non-enveloped, and the virions are relatively simple, consisting of a protein capsid surrounding a single-stranded, positive sense RNA genome of around 7500 bases. The enterovirus genus contains a large number of human pathogens. Historically, their nomenclature has been somewhat eclectic, with a number of polioviruses, Coxsackie A and B viruses (named after the place of first isolation and split into two because of different effects in experimental mice), and ECHO viruses (enteric cytopathic human orphan viruses). More recently, the genus has been divided on the basis of genetic phylogeny based on the VP1 capsid region into 10 species, of which seven – Enterovirus A, B, C, D (containing over 100 serotypes) and Rhinovirus A, B, C – (with over 150 serotypes) are known to infect humans.

Any new isolates are designated as EV- (or RV) followed by the letter denoting the species, and the next consecutive number. All these viruses are serologically distinct, that is infection with any one enterovirus induces antibody production that does not cross-neutralize any of the other enteroviruses. The more common/important human enteroviruses are listed in Table 9.1.

ENTRY AND SPREAD WITHIN THE BODY

The term enterovirus reflects the fact that these viruses enter and leave the human host via the enteric tract (otherwise known as the gastrointestinal (GI) tract). Thus, they are swallowed in contaminated food or water, being able to survive the highly acidic Ph of the stomach (non-enveloped viruses are in general more hardy than enveloped ones). Initial replication in the small bowel wall results in viral excretion in the feces, which may persist for several days. At some stage, virus passes through the small bowel wall and into the host bloodstream. This **viremic** stage allows access of the virus to many different cells and tissues of the body, resulting in a number of different clinical manifestations (see below). In addition, some enteroviruses, for example EV-D68, appear to infect via the respiratory tract, while EV-D70 is found almost exclusively in conjunctival and throat swabs.

PERSON-TO-PERSON SPREAD

Spread between individuals mostly arises through contamination of food and water, for example, through inadequate hand hygiene after defecation. Enterovirus contamination of water supplies through inadequate disposal

Table 9.1 The human enteroviruses

Old classification:	
Polioviruses (serotypes 1, 2, and 3)	
Coxsackie A viruses (23 serotypes, numbered A1–A24, as coxsackie A23 is the same virus as echovirus 9)	
Coxsackie B viruses (serotypes B1–B6)	
Echoviruses (31 serotypes, numbered 1–33, as echoviruses 10 and 28 have been reclassified into other virus families)	
Enterovirus serotypes (EV 68–71)	
New classification:	
Enterovirus species	Types known to infect humans
Enterovirus A	Coxsackievirus A16
	EV-A71
Enterovirus B	Coxsackievirus A9
	Coxsackie B viruses 1-6
	All echoviruses
Enterovirus C	Coxsackievirus A1
	Polioviruses 1-3
	EV-C95 et seq
Enterovirus D	EV-D68

of sewage can give rise to outbreaks of infection. Outbreaks may also occur in closed communities such as boarding schools or households.

EPIDEMIOLOGY

Enteroviral meningitis may arise at any age but is most common in infants and young children. There is a seasonality to enterovirus infections, most arising in the summer months in temperate climates. Enterovirus infections are common worldwide. The current epidemiology of poliovirus infection is described below under Section 5 (Prevention).

2. WHAT IS THE HOST RESPONSE TO THE INFECTION AND WHAT IS THE DISEASE PATHOGENESIS?

IMMUNE RESPONSES

Innate immune responses are crucial in determining the fate of many virus infections. Initial recognition of viral entry is via pattern recognition receptors (PRRs), which bind pathogen-associated molecular patterns (PAMPs). There are five PRRs, including Toll-like receptors (TLRs), retinoic acid-inducible gene 1 (RIG-I)-like receptors, and nucleotide-binding and oligomerization domain (NOD)-like receptors (NLRs). Binding of these receptor families to fragments of viral DNA, RNA or proteins triggers signaling pathways leading ultimately to activation of NF-κB and AP-1 transcription factors which, in turn, generate rapid pro-inflammatory responses, including type-1 interferon production.

In order to evade these initial antiviral mechanisms, viruses have evolved a number of countermeasures, and enteroviruses are no exception. The enteroviral genome encodes a number of non-structural proteins, including at least two with protease function – proteases 2A and 3C. These have been shown in *in-vitro* systems, and in animal models, to mediate a number of cleavages of key host antiviral proteins – the EV-71 2A protein cleaves melanoma differentiation-associated protein 5 (MDA5), RIG-I, and the mitochondrial antiviral signaling protein (MAVS), all of which are integral to the signaling pathways leading to pro-inflammatory and interferon responses. It further reduces the effect of any type-1 interferon that is generated by cleaving the type-1 interferon receptor, thereby blocking the expression of interferon-stimulated genes. More recently, EV-71 proteases C3 and 2A have been shown to antagonize the activation of the NOD-like receptor family pyrin domain containing 3 (NLRP3) inflammasome which is responsible for IL-1β production in response to infection. Study of the interactions between enteroviruses and these complex innate immune recognition signaling pathways is not only generating new knowledge about the function and mechanisms of those pathways but may also identify new targets for antiviral drugs.

Natural killer (NK) cells appear to be important in the early response to coxsackie B viruses, not through cytotoxicity but rather through their production of interferon (IFN)-γ. In fact, mice lacking the IFN-g receptor die quickly after challenge with coxsackie B viruses.

Antibody responses against enteroviruses appear 7–10 days after infection. Those directed against the capsid proteins may be neutralizing.

T-cell responses are also made but the humoral immune response appears to be more important for this group of viruses than the cell-mediated response, as evidenced by the following:

1. In patients with antibody deficiency, enterovirus infection may become chronic, for example, persistence of live poliovirus vaccines given to patients with **hypogammaglobulinemia**, with a particular risk of serious CNS manifestations, for example, chronic enteroviral **encephalitis**.

2. Enteroviruses are not recognized as particularly serious pathogens in the context of patients with cell-mediated immune deficiencies, for example, those with HIV infection, or transplant recipients.

IgA might play a role in protection by preventing entry via the intestinal epithelium.

Coxsackie B viruses and other enteroviruses have a number of strategies to prevent immune responses. These include: down-regulation of surface MHC class I molecules to thwart **CD8**+ T-cell responses and inhibition of **dendritic cell** function (seen with echoviruses but not coxsackie B viruses).

PATHOGENESIS

Evidence has been accumulating that coxsackie B viruses and other enteroviruses may play a role in autoimmune disease. Enteroviruses have been implicated in the etiology of type-I diabetes through "molecular mimicry". There is some evidence that antibodies to coxsackie viruses cross-react with human islet cell antigens and it has been reported that some epitopes of coxsackie B viruses cross-react at the T-cell level with glutamic acid decarboxylase (GAD).

Enteroviruses are thought to be the main etiologic agents of viral **myocarditis**, which is a common cause of idiopathic dilated cardiomyopathy, a severe pathologic condition that often requires heart transplantation.

3. WHAT IS THE TYPICAL CLINICAL PRESENTATION AND WHAT COMPLICATIONS CAN OCCUR?

As mentioned above, the clinical manifestations of enteroviral infection are **protean**, although, despite the name enterovirus, these viruses do not cause gastroenteritis. The most common outcome is probably no disease at all, that is, asymptomatic infection. Febrile illnesses with nonspecific **rashes** are common in children, and some at least are due to enterovirus infections. Coxsackie A viruses may give rise in children to **herpangina**, or hand, foot, and mouth disease (**vesicles** in the mouth and on the hands and feet). **Conjunctivitis** is particularly associated with EV70.

The disease poliomyelitis has been known for centuries and arises from infection with one of the three serotypes of poliovirus. Most infections with polioviruses are asymptomatic; a small percentage (up to 10%) may present with a febrile illness with, in some patients, evidence of CNS involvement such as headache, photophobia, or neck stiffness. Paralytic poliomyelitis occurs in around 1 in 1000 children or 1 in 75 adults. Infection of nerve cells supplying muscle tissue (collectively referred to as lower motor neurons, whose cell bodies lie within the anterior horn of the spinal cord) results in death of those cells and flaccid paralysis of the relevant muscle (Figure 9.1). This is potentially life-threatening, especially if muscles involved with respiration are affected. Very occasionally, poliomyelitis-like illness can arise from infections with other enteroviruses, for example, coxsackie viruses, or even with completely unrelated viruses, for example, Japanese encephalitis or West Nile viruses.

MENINGITIS

As in our specific case, meningitis typically presents with a fever, irritability and/or lethargy, especially in young children, neck stiffness (although the absence of this symptom/sign does not exclude a diagnosis of meningitis, especially in young children), and an aversion to bright light – **photophobia**.

Complications of enteroviral meningitis are unusual in an immunocompetent child or adult. Spread of virus from the **meninges** to the brain, resulting in **meningoencephalitis**, may occur rarely. This may be heralded by an abrupt deterioration in mental state, or the onset of **seizures**. In patients with immunodeficiencies, particularly those associated with impaired antibody production, meningoencephalitis is much more common, and may become chronic.

Figure 9.1 This image shows muscle wasting (right lower limb) arising from poliomyelitis. *From Science Photo Library, with permission.*

MUSCLE INFECTIONS

Coxsackie viruses can also infect muscle tissue itself, giving rise to **Bornholm's disease** (with involvement of chest muscles) or **myocarditis**. Enteroviruses are also the commonest cause of viral **pericarditis**.

RESPIRATORY INFECTIONS

Within the respiratory system, enteroviruses occasionally cause upper respiratory tract manifestations such as the common cold or, especially in young babies, can involve the lower respiratory tract, presenting with **bronchiolitis** or even **pneumonia**.

NEONATAL INFECTIONS

Enteroviruses are particularly feared pathogens in neonates, in whom they may cause devastating disseminated infections, resulting in multisystem life-threatening disease including myocarditis, **hepatitis**, and encephalitis. Furthermore, patient-to-patient spread within a neonatal ward has been reported on a number of occasions.

4. HOW IS THE DISEASE DIAGNOSED, AND WHAT IS THE DIFFERENTIAL DIAGNOSIS?

DIAGNOSIS

Meningitis is a medical emergency, and appropriate diagnostic tests must be initiated as soon as possible to expedite effective therapy (if available). Initial investigations should include blood cultures and a lumbar puncture (although this is contraindicated if there is evidence of raised intracranial pressure) to obtain a sample of CSF. Laboratory examination of the CSF usually allows distinction between bacterial and viral meningitis. In the former, CSF protein is grossly raised, CSF glucose is reduced to less than half of the blood sugar level, and the predominant cellular infiltrate is with polymorphonuclear leucocytes (PMNs) (Figure 9.2A). Gram staining of centrifuged CSF, that is the pellet, will often reveal the presence of bacteria. In viral meningitis, CSF protein is only marginally raised, CSF glucose is usually normal, and the predominant cellular infiltrate is with mononuclear cells, that is lymphocytes and monocytes (Figure 9.2B) (although an early predominance of PMNs can sometimes be seen very early in the course of viral meningitis). A Gram stain will be negative.

Mumps virus used to be the commonest cause of viral meningitis, but with the introduction of effective mumps vaccines (given universally precisely because of this manifestation of infection) enteroviruses now account for the majority of cases of viral meningitis. Herpes simplex virus (HSV) can also cause meningitis, usually in young women with extensive genital herpes infection. Confirmation that viral meningitis is due to an enterovirus, as opposed to other

viral causes, can be made in a number of ways. The most sensitive approach is to detect the presence of enteroviral RNA in the CSF sample by a genome amplification technique such as the **polymerase chain reaction** (**PCR**). By sequencing the PCR product, it is possible to determine which particular enterovirus is the culprit – this is essential if patients present with flaccid paralysis, in order to demonstrate or rule out infection with polioviruses. However, genome detection assays may miss infection with novel enteroviruses if there is sequence divergence at the oligonucleotide primer binding sites. Many (but not all) enteroviruses may also be isolated from CSF in routine tissue culture. If CSF is not available, then alternative samples are a throat swab or a sample of feces. Detection of enterovirus in feces may be possible up to 2 weeks after initial clinical presentation.

Serologic assays are available for detection of an antibody response to an enterovirus. **IgM** anti-enterovirus antibodies are indicative of a recent infection, but this test is usually only available within a reference laboratory. An alternative approach is to demonstrate a rise in **IgG** anti-enterovirus antibody titers in a sample taken 1 or 2 weeks after the initial presentation (known as a convalescent sample) compared with those in a sample taken at the time of acute illness. Thus, this is not a technique that allows the diagnosis to be made when the patient first presents acutely ill.

Other clinical manifestations may be evident on examination of the patient, depending on the particular causative agent, for example, meningococcal meningitis is

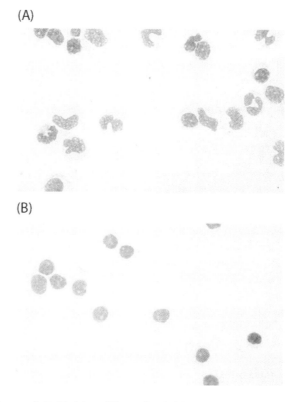

(A)

(B)

Figure 9.2 Staining (Giemsa) of CSF deposit showing: (A) polymorphonuclear cells typical of bacterial meningitis; (B) mononuclear cells typical of viral meningitis.

often associated with a spreading, nonblanching **purpuric rash** (see Case 25).

DIFFERENTIAL DIAGNOSIS

Other causes of viral meningitis include mumps virus, HSV, and some arboviruses.

5. HOW IS THE DISEASE MANAGED AND PREVENTED?

MANAGEMENT

Enteroviral meningitis will usually resolve spontaneously, and the vast majority of patients will make a complete recovery. There are no licensed antiviral drugs with activity against enteroviruses, so management is supportive only.

PREVENTION

There are two types of vaccines available for the prevention of poliovirus infection (but neither offers any protection against infection with other enteroviruses): (a) live attenuated vaccines, known as Sabin vaccines after the person who first developed them, which may contain one, two or all three poliovirus serotypes. These are administered orally (i.e. mimicking the natural route of infection), which therefore induces immunity within the gut mucosa, as well as within the bloodstream. While these are effective and inexpensive, the two main drawbacks are the necessity of maintaining a cold chain so that the viruses remain viable, and the potential for these viruses to revert to pathogenicity as they are excreted from the bowel in feces. Such vaccine-derived polioviruses (VDPV) can circulate and cause disease; (b) inactivated vaccines (generically referred to as Salk vaccines), again containing one, two or all three poliovirus serotypes. These are administered by intramuscular (IM) injection and may therefore not induce as good a mucosal immune response as Sabin vaccines but are safer as there is no risk of reversion to virulence. The World Health Organization (WHO) is currently co-ordinating a campaign for the elimination of poliovirus infection, which has resulted in a decrease of over 99% in the numbers of cases from an estimated 350 000 in 1988 to 175 in 2019. Wild type polioviruses 2 and 3 have officially been certified as eradicated. Wild type poliovirus 1 currently is only circulating in Pakistan and Afghanistan. However, VDPV type 2 viruses are circulating in 30 countries, including the UK as of August 2022, giving rise to over 400 cases of polio in 2021. The dramatic changes in poliovirus circulation and epidemiology has led to the modification of the traditional trivalent forms of polio vaccines, most notably with the development of mono- and bi-valent vaccines such as a more genetically stable novel OPV2 vaccine (nOPV2) designed to combat the circulating VDPV2.

SUMMARY

1. WHAT IS THE CAUSATIVE AGENT, HOW DOES IT ENTER THE BODY, AND HOW DOES IT SPREAD A) WITHIN THE BODY AND B) FROM PERSON TO PERSON?

- The enteroviruses are non-enveloped positive single-stranded RNA viruses belonging to the family *Picornaviridae*.
- The enterovirus genus is now classified into 10 species, seven of which (four enterovirus and three rhinovirus) contain types which infect humans. The EV types (A, B, C and D) contain over 100 individual enterovirus types, including poliovirus types 1–3, coxsackie A viruses (23 types), coxsackie B viruses (6 types), echoviruses (31 types), and enteroviruses numbered from 68 to over 100.
- Virus entry is usually via the gastrointestinal tract. Initial replication occurs within the small bowel, associated with excretion of virus in the feces.
- This is followed by viremic spread to various target organs.
- Human-to-human spread is via the **feco–oral route**.

2. WHAT IS THE HOST RESPONSE TO THE INFECTION AND WHAT IS THE DISEASE PATHOGENESIS?

- Enteroviruses have evolved several mechanisms for evasion of the innate immune response.
- Antibody responses are essential to control enterovirus infections.

- In antibody-deficient individuals, infection may become chronic and can be life-threatening.

3. WHAT IS THE TYPICAL CLINICAL PRESENTATION AND WHAT COMPLICATIONS CAN OCCUR?

Enteroviruses cause a wide range of diseases, including the following:
- Febrile illness with or without skin rashes.
- Respiratory tract infections (e.g. EV-D68).
- Herpangina (usually coxsackie A viruses).
- Hand, foot, and mouth disease (usually coxsackie A viruses).
- Conjunctivitis (e.g. EV-D70).
- Meningitis.
- Encephalitis (in antibody-deficient patients).
- Myocarditis and pericarditis.
- Bornholm's disease (epidemic pleurodynia).
- Disseminated multisystem disease in the newborn.
- Poliomyelitis.

4. HOW IS THE DISEASE DIAGNOSED, AND WHAT IS THE DIFFERENTIAL DIAGNOSIS?

Enteroviral meningitis is diagnosed on the basis of typical CSF findings (mononuclear cell infiltrate, protein raised only marginally, glucose normal, gram-stain negative) plus any of the following:
- Genome amplification (e.g. PCR) within CSF.
- Isolation of the virus in tissue culture from CSF, throat swab or feces. *Continued...*

...continued

- IgM antibodies, or rising titers of IgG antibodies.
- Other causes of viral meningitis include mumps virus, herpes simplex virus, some arboviruses.

5. HOW IS THE DISEASE MANAGED AND PREVENTED?

- Enterovirus infections, including meningitis, are almost always self-limiting in immunocompetent hosts and resolve spontaneously.

- There are no licensed antiviral drugs with proven activity against enteroviruses.
- Poliovirus infection can be prevented by use of either live attenuated (Sabin) or inactivated (Salk) vaccines. The WHO poliovirus eradication campaign has achieved considerable success, although wild-type poliovirus 1 still circulates in Pakistan and Afghanistan, while vaccine-derived polioviruses also continue to circulate.

FURTHER READING

Barer M, Irving W, Swann A, Perera N. Medical Microbiology, 19th edition. Elsevier, Philadelphia, 2019.

Murphy K, Weaver C. Janeway's Immunobiology, 9th edition. Garland Science, New York/London, 2016.

Richman DD, Whitley RJ, Hayden FG. Clinical Virology, 4th edition. ASM Press, Washington, DC, 2016.

REFERENCES

Autore G, Bernardi L, Perrone S, Esposito S. Update on Viral Infections Involving the Central Nervous System in Paediatric Patients. Children, 8: 782, 2021.

Brown DM, Zhang H, Scheuermann RH. Epidemiology and Sequence-Based Evolutionary Analysis of Circulating Non-Polio Enteroviruses. Microorganisms, 8(12): 1856, 2020.

Chumakov K, Ehrenfeld E, Agol VI, Wimmer E. Polio Eradication at the Crossroads. Lancet Global Health, 9: e1172–1175, 2021.

Elrick MJ, Pekosz A, Duggal P. Enterovirus D68 Molecular and Cellular Biology and Pathogenesis. J Biol Chem, 296: 100317, 2021.

Zhang Y, Li J, Li Q. Immune Evasion of Enteroviruses Under Innate Immune Monitoring. Front Microbiol, 9: 1866, 2018.

WEBSITES

All the Virology on the WWW Website, Tulane University, 1995: http://www.virology.net/garryfavweb13.html#entero

Centers for Disease Control and Prevention, USA: http://www.cdc.gov/ncidod/dvrd/revb/enterovirus/non-polio_entero.html

GOV.UK, Polio: the green book, chapter 26, 2013: https://www.gov.uk/government/publications/polio-the-green-book-chapter-26

GOV.UK, Polio: guidance, data and analysis, 2013: https://www.gov.uk/government/collections/polio-guidance-data-and-analysis

Virology Online, Enteroviruses Slide Set: http://virology-online.com/viruses/Enteroviruses.htm

Students can test their knowledge of this case study by visiting the Instructor and Student Resources: [www.routledge.com/cw/lydyard] where several multiple choice questions can be found.

Epstein-Barr virus

10

A 20-year-old woman went to her doctor complaining of a sore throat. It started 4 days previously with associated episodes of fever and chills. On examination, her doctor noticed that her tonsils and **uvula** were red and slightly swollen and that she had cervical **lymphadenopathy**. Suspecting a bacterial infection, her doctor prescribed a course of ampicillin and suggested that she took a couple of days off work. The patient returned a week later with worsening symptoms. She was now very lethargic with a temperature of 38.5°C and a widespread maculopapular rash as shown in Figure 10.1.

On the second visit to her doctor, she had patches of white exudate on the tonsils (Figure 10.2), **petechial** hemorrhages on the soft palate, and generalized lymphadenopathy. The doctor could also palpate an enlarged spleen and noticed that she was slightly tender over the right **hypochondrium**. He suspected that she might have infectious mononucleosis and sent her for hematological and antibody tests.

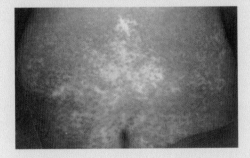

Figure 10.1 Infectious mononucleosis: ampicillin-induced rash. Note the resemblance to measles. *With permission from Elsevier and the original author. This photo was published in 'A Colour Atlas of Infectious Diseases' by Emond, Welsby and Rowland.*

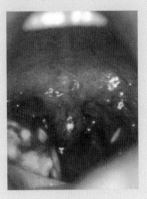

Figure 10.2 Infectious mononucleosis: follicular exudate on the tonsil, which is very swollen; the uvula is red and edematous. *With permission from Elsevier and the original author. This photo was published in 'A Colour Atlas of Infectious Diseases' by Emond, Welsby and Rowland.*

1. WHAT IS THE CAUSATIVE AGENT, HOW DOES IT ENTER THE BODY AND HOW DOES IT SPREAD A) WITHIN THE BODY AND B) FROM PERSON TO PERSON?

CAUSATIVE AGENT

The patient has the clinical symptoms of primary Epstein-Barr virus (EBV) infection – infectious mononucleosis or glandular fever, as it is commonly called due to concomitant swollen lymph nodes (lymphadenopathy). The virus has an outer envelope with an inner icosahedral protein capsid and a linear double-stranded (ds) DNA genome of about 172 kb. The **EBV envelope** has viral glycoproteins on its surface including gp350, gp42, gB, gH, and gL. EBV belongs to the *Herpesviridae*, of which there are nine viruses identified to date as human pathogens (Table 10.1).

After primary infection, herpesviruses remain latent in the body persisting for life, for EBV the site of latency is the B **lymphocyte**. EBV-infected cells can exist in two states. In lytic replication, the resultant virions have a linear DNA genome. In latency, the viral genome is a circular episome and is replicated in dividing cells by cellular enzymes; there is

Table 10.1 Human herpesviruses

Alpha herpesvirus	Herpes simplex virus 1
	Herpes simplex virus 2
	Varicella-zoster virus
Beta herpesvirus	Cytomegalovirus
	Human herpesviruse 6A and 6B
	Human herpesvirus 7
Gamma herpesvirus	Epstein-Barr virus
	Human herpesvirus 8, also known as Kaposi sarcoma-associated herpesvirus

Table 10.2 Patterns of EBV latent gene expression and disease

Latency program	EBV gene product					EBV-associated disease
	EBNA1	EBNA2	EBNA3	LMP1	LMP2	
0	-	-	-	-	-	None
I	+	-	-	-	-	BL, gastric carcinoma
II	+	-	-	+	+	NPC, Hodgkin lymphoma, T-cell lymphoma
III	+	+	+	+	+	Infectious mononucleosis, posttransplant LPD, PCNSL

BL, Burkitt lymphoma; EBNA, Epstein-Barr nuclear antigen; LMP, latent membrane protein; LPD, lymphoproliferative disease; NPC, nasopharyngeal carcinoma; PCNSL, primary central nervous system lymphoma

Table 10.3 EBV latent gene products and their functions

EBV gene product	Function
EBNA1	EBV gene maintenance; EBNA1 is only expressed in dividing cells to ensure proper segregation of the EBV episomal genome to both daughter cells
EBNA2	Activates expression of EBV, and host B cell genes including *c-myc*, & CD21
EBNA3	Represses tumor suppressor gene expression and promotes B lymphocyte transformation into LCLs – see below for LCLs in "Entry and spread within the body"
LMP1	Initiates B-cell activation and proliferation and inhibits apoptosis; mimics CD40 signaling – constitutively activates pathways that mimic the growth and survival signals given to B cells by CD4+ T lymphocytes. Essential for transformation into LCLs.
LMP2	Mimics B-cell receptor signaling.

LCL, lymphoblastoid cell line

limited gene expression and no infectious particle production. Table 10.2 shows the various patterns of latency, distinguished by different latency antigens and Table 10.3 the functions of the latency gene products. The latent state is implicated in several cancers, **neoplasms** (Table 10.2, and see Section 3).

VIRUS REPLICATION

For entry into B cells, gp350 binds to the cellular receptor for the complement component C3d (**CR2/CD21**). Viral gp42 then binds to the major histocompatibility complex (MHC) class II on the cell surface, which serves as co-receptor. Fusion between the virus envelope and the host-cell membrane follows mediated by gH/gL and gB and the virus is endocytosed.

EBV replication in epithelial cells is well established *in vitro* and gp350 may also be important for virus attachment to CD21. Another possibility is that the viral glycoprotein, BMFR2, binds to cellular integrins. In contrast to B cells, epithelial cells lack MHC class II and entry at the plasma membrane is gp42 independent. EBV undoubtedly has clinically significant tropism for epithelial cells *in vivo*, as seen in cancer of the nasopharynx and oral hairy leukoplakia of the tongue in HIV patients, where unchecked lytic replication occurs in the squamous epithelium of the tongue.

EBV virions produced by B cells have reduced gp42 because it binds to MHC class II during budding from the infected cell; the consequence is that the virus infects epithelial cells more efficiently than other B cells. Conversely, virions derived from epithelial cells lacking MHC Class II are better at infecting B lymphocytes as they have high levels of gp42. This reciprocal tropism is proposed to enhance EBV shuttling between B lymphocytes and the oral epithelium.

Natural killer (NK) cells and T cells may occasionally be the targets of EBV during primary infection. CD21 has recently been established as the entry receptor for T cells but the receptor for NK cells is unknown.

ENTRY AND SPREAD WITHIN THE BODY

In primary infection, EBV transmitted in saliva enters the oral cavity, where CD21 is expressed on tonsil epithelium making it a likely site for virus replication rather than non-tonsillar sites. This is logical because tonsils are the site of mucosa-associated lymphoid tissues with consequent circulating B lymphocytes. Although final proof is lacking, it is generally accepted that EBV first replicates in oral epithelial cells and then infects mucosal B cells. Alternatively, the virus may infect B cells in the tonsillar crypts directly.

While some oropharyngeal B cells undergo lytic viral replication infection producing virions, most become latently infected (latency program III); these B lymphoblasts have entered the cell cycle and proliferate continuously in a process termed **transformation** or **immortalization**, which serves to expand rapidly the pool of infected B lymphocytes. These cells can be propagated *in vitro* indefinitely. The ability of EBV to convert peripheral blood B cells into immortalized

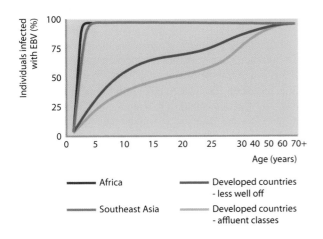

Figure 10.3 Comparison of the ages at which individuals in different populations become infected with Epstein-Barr virus (EBV). *From Epstein MA & Rickinson AB (2010) Virology https://doi. org/10.1002/9780470015902.a0002318.pub2. With permission from John Wiley & Sons.*

lymphoblastoid cell lines is widely used in genomic studies as a means of preserving DNA from volunteer donors for future use.

Virus and infected cells pass via the afferent lymphatics into the cervical lymph nodes where further B cells are infected. These cells reach the bloodstream via the thoracic duct and spread to the spleen and other lymphoid tissues where EBV causes B-lymphocyte proliferation, lymphadenopathy, and **splenomegaly**. The incubation period of infectious mononucleosis (i.e., time from initial infection to development of symptoms) is usually 4–8 weeks.

Even after resolution of symptoms, the virus is not cleared from the body and remains **latent** throughout life with occasional reactivation into a lytic cycle producing virions transmissible to a new host. It is not fully understood how an initial growth transforming infection in latency program III results in selective colonization of the nonproliferating memory B-cell compartment in latency program 0. Thorley-Lawson and colleagues have postulated a model applied to normal B-cell differentiation which provides an attractive explanation. To enter the resting state and become a memory cell, an uninfected B-cell receptor must bind its cognate antigen and receive appropriate signals from helper T cells in the germinal centers of lymphoid tissue. During EBV latent infection III, the viral LMP proteins mimic all these steps, such that the infected B cell could differentiate into a memory cell in the absence of external cues.

PERSON-TO-PERSON SPREAD

EBV is present in saliva throughout life. EBV-infected memory cells can differentiate into plasma cells seeding lytic infection at oropharyngeal mucosal sites. However, it is unclear whether the final source of salivary virus is infected B or epithelial cells. Whatever the mechanism, most, if not all EBV-infected persons shed low levels of virus throughout their life with the potential to infect others.

Spread of EBV is usually through intimate contact between an infected and an uninfected person – often through kissing. Hence, infectious mononucleosis is known as the "kissing disease". In children, the virus is mainly spread by fingers contaminated with saliva and other close contact.

EBV-seronegative individuals can also acquire EBV by blood transfusion or organ transplant.

EPIDEMIOLOGY OF EBV INFECTION AND INFECTIOUS MONONUCLEOSIS

Globally, most adults are infected with EBV (95–98%; Figure 10.3). At least 50% of those who first acquire the virus in the second or third decade develop some symptoms of infectious mononucleosis (see Section 3). The incidence of infectious mononucleosis shows no consistent seasonal peak, and the disease is rare in older adults.

In developing countries, almost all children have acquired the virus by 2–4 years of age, and the infection is usually asymptomatic or so mild that it is undiagnosed. In developed countries with high standards of living and hygiene, the time of infection is delayed for many individuals, more markedly among the affluent than the less well off. Among the very rich, as many as 50% may reach adolescence or young adulthood without having encountered the virus and will undergo delayed primary infection with a high risk that this will be accompanied by the symptoms of infectious mononucleosis.

2. WHAT IS THE HOST RESPONSE TO THE INFECTION AND WHAT IS THE DISEASE PATHOGENESIS?

There is a paucity of information concerning the role that oropharyngeal mucosal immunity plays in host defense against EBV. Mucosal surfaces are protected by a common immune system and about two-thirds of B and T cells in the body are found both within the epithelium and in the submucosa (lamina propria).

MUCOSAL INNATE IMMUNITY

The oral mucosa is flushed by saliva, which protects, hydrates, and lubricates the epithelium. Mucins, high molecular weight glycoproteins serve to aggregate and remove microorganisms and several antimicrobial peptides, such as defensins, have antiviral activity. Rather than being protective, defensins and transmembrane mucins might promote infection by inducing local inflammation. However, oral epithelial cells do not generally secrete high levels of pro-inflammatory cytokines and infection is resolved with minimal inflammatory tissue destruction.

Secretory immunoglobulin A is the principal immunoglobulin isotype present in external secretions. IgA is a non-inflammatory immunoglobulin that does not activate complement. Its role at barrier epithelia is immune exclusion. Two studies from the early 1990s reported that salivary IgA antibody responses to gp350 were detected in a minority (12–19%) of healthy virus carriers and in a higher proportion (49%) of nasopharyngeal carcinoma patients. *In vitro* IgA specific for gp350 promoted EBV infection of the otherwise refractory epithelial cells with virus entering the cell through secretory component-mediated IgA transport. However, EBV coated in IgA no longer infected B lymphocytes. Such an immune-induced shift in EBV tissue tropism might possibly be involved in the pathogenesis of cancer of the nasopharynx, a malignancy of epithelial origin (see Section 3, Non-B-cell malignancies of epithelial origin).

The mouth has keratinized squamous epithelium on the palate, dorsum of the tongue, and attached gingivae as well as non-keratinized squamous epithelium elsewhere. These epithelial cells desquamate continuously taking with them any adherent microorganisms. Importantly, the oropharynx has a ring of mucosa-associated lymphoid tissues, known as Waldeyer's ring, comprising the pharyngeal, tubal, palatine, and lingual tonsils as well as small collections of lymphatic tissue distributed throughout the pharyngeal mucosa.

EBV must penetrate the oral mucosal epithelium to infect B cells. M cells are present in the epithelium overlying Waldeyer's ring; these serve to capture and transcytose microparticles, such as viruses, and might deliver EBV to B cells. Alternatively, EBV may enter via sites of microscopic injury, be picked up by intraepithelial dendritic cells that extend their dendrites onto the mucosal surface or be endocytosed by epithelial cells triggered by interaction between viral surface glycoproteins and lectin-like molecules on the epithelial cell surface. Once EBV reaches the submucosa, it is subject to the innate and lectin complement pathways. However, gamma herpesviruses encode homologs of regulators of complement activation that help them evade the complement system.

Upon entering the mucosal epithelium, EBV will be exposed to intraepithelial lymphocytes (IELs), among which are NK cells. These cells target and kill infected cells with decreased surface expression of MHC class I due to EBV infection. Notably NK cell frequency is higher in early childhood than in adulthood; this may explain why robust CD8+ T-cell responses in adults and young adults are required to control EBV in infectious mononucleosis (see later, Adaptive Immunity) whereas primary infection is usually asymptomatic in young children.

Other key IELs are invariant NK T (iNKT) cells and γδ T cells. In humans, iNKT cells express an invariant T-cell receptor that recognizes glycolipids presented by the non-classical MHC class I molecule CD1d. A small percentage of these cells bear CD8 and can directly lyse EBV-infected cells as well as producing interferon γ to enhance Th1 responses. Notably, patients with X-linked lymphoproliferative disease, (see Section 3, XLPD) lack iNKT cells, which may explain their susceptibility to EBV. The T-cell receptors of γδ T cells may also exert important anti-EBV responses; they recognize antigen directly without a requirement for antigen processing enabling them to function in the recognition of damaged or stressed host cells.

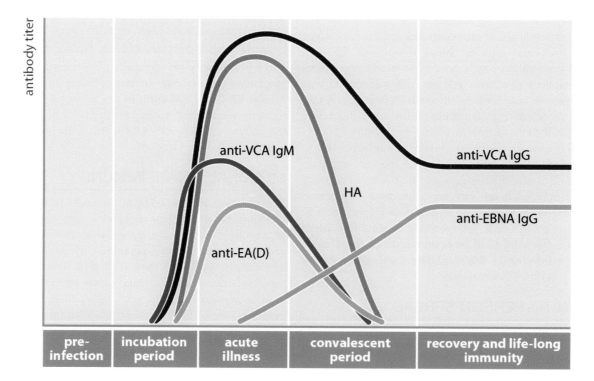

Figure 10.4 Basic pattern of serum antibodies to Epstein-Barr virus-associated antigens before, during, and after primary infection. HA, heterophile antibody; EA(D), diffuse form of early antigen; VCA, virus capsid antigen; EBNA, Epstein-Barr nuclear antigen. Incubation period 4–8 weeks. *Adapted from 'Principles and Practice of Clinical Virology' by Arie J. Zuckerman et al. (2004). With permission from John Wiley and Sons.*

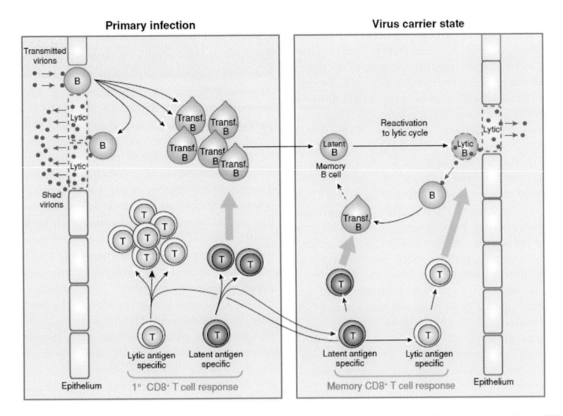

Figure 10.5 Diagrammatic representation of virus–cell interactions and of virus-induced CD8+ T-cell responses in primary EBV infection, as seen in infectious mononucleosis patients, and in healthy virus carriers. Red arrows denote transmission of virus; black arrows denote movement of cells (where the arrow is dotted, this reflects uncertainty as to the level of cell movement via this route); broad shaded arrows denote effector T-cell function. Transf. B = transformed B. *From Hislop AD, Taylor GS, Sauce D & Rickinson AB (2007) Annual Review of Immunology 25:587-617. With permission from Annual Reviews.*

ADAPTIVE IMMUNITY – ANTIBODY RESPONSE

Humoral responses to EBV lytic and latent antigens are seen in infectious mononucleosis. These antigens include the following:

- *Early antigens*   EA, expressed early in lytic replication

- *Late antigens*   Structural proteins – virus capsid antigens (VCA) and envelope, gp350

- *Latency antigens*   EB nuclear antigens (EBNAs), latent membrane proteins (LMPs)

Primary EBV infection also induces heterophile antibodies, which are not EBV-specific and characterized by broad reactivity with antigens of other animal species (see Section 4, How is this disease diagnosed); they are the result of EBV-induced polyclonal B-cell activation with differentiation into plasma cells. Figure 10.4 shows the typical sequence of appearance of EBV-specific and heterophile antibodies in a primary response. **IgM** to VCA is the first to appear followed by **IgG**. Antibodies to EA are highest during the acute phase of

infection but are often detectable at varying levels thereafter. Antibodies to EBNAs arise approximately 4 weeks after onset of symptoms and persist for life as do IgG antibodies to VCA.

EBV-neutralizing antibodies appear late in infectious mononucleosis and persist at stable titers for life. These antibodies block entry of EBV into epithelial cells and/or B cells. These include anti-gp350, anti-gH/gL, and anti-gp42, but this latter cannot neutralize epithelial cell infection.

ADAPTIVE IMMUNITY – T-CELL RESPONSES

Figure 10.5 shows the virus–cell interactions and the virus-induced CD8+ T-cell response in primary EBV infection as seen in infectious mononucleosis patients and healthy virus carriers. T cells initially respond to the EBV-infected cells by proliferation, firstly **CD4+** T cells and then CD8+ T cells which outnumber the B cells by a factor of about 50 to 1. These T cells appear in the circulation as "atypical lymphocytes", a diagnostic feature (see Section 4). During the early response to the virus, up to 50% of the CD8+ T cells in the circulation can be specific for EBV antigens. Infectious mononucleosis is an immunopathologic disease with the immune response itself being responsible for the clinical symptoms and pathogenesis.

The symptoms result from the intense immunologic activity of CD8+ T cells specifically recognizing EBV-encoded peptides presented by infected B cells, accompanied by cytokine release and a strong inflammatory response. At first, CD8+ T cells specific for lytic antigens predominate, but with convalescence, a shift occurs toward T cells that recognize latent antigens, particularly EBNA3, expressed by transformed B cells.

In infectious mononucleosis, the cytolytic CD8+ T cells control the infection by killing most of the virus-infected cells (transformed B cells) but high-level shedding of virus in the throat of infectious mononucleosis patients continues. This might be because of the poor recruitment of the CD8+ T cells to the tonsillar tissue. A few EBV-infected B cells enter a resting state as members of a long-lived memory cell pool, with about 1–50 cells per million circulating B cells in latency program 0. Following recovery from the primary infection, cytotoxic memory CD8+ T cells persist in the blood at a frequency of 0.2–2% of cytotoxic CD8+ T cells, which increases with age. Some individuals aged over 60 years can have up to 40% of their CD8+ T cells specific for EBV antigens.

The importance of T-cell immunity toward EBV is highlighted by the high frequency of EBV-associated malignancies in patients with congenital immunodeficiency syndromes, in patients with HIV/AIDS, or who receive T-cell immunosuppression posttransplant (see Section 3).

VIRUS IMMUNE EVASION

EBV has many strategies that help it evade the immune system and avoid attack by CD8+ T cells. These include the following:

1. A few resting memory B cells containing the latent virus evade the immune response by down-regulation of all EBV antigens (latency program 0). However, in any dividing memory cell, EBV must express EBNA1 to ensure that its genome is replicated. To prevent targeting of this key protein, EBNA1 contains a sequence of expanded glycine-alanine repeats, not necessary for its function in genome maintenance, capable of inhibiting its proteasomal processing. So that EBNA1 peptides cannot be presented on class I MHC molecules and can elude immune surveillance.

2. The envelope protein gp42 is actively shed as a soluble **truncated** molecule during lytic infection and binds directly to MHC class II/peptide complexes and inhibits activation of the CD4+ T cells required for generation of CD8+ cytolytic T cells.

3. BCRF1 is secreted during the EBV lytic cycle. This protein is a viral homolog of the cytokine IL-10, which is an immune modulator. Inhibition of Th1- cell function is the result promoting a shift toward T-helper 2 (Th2) differentiated CD4+ effectors; these cells can provide B-cell help but do not promote the CD8+ responses needed to kill EBV-infected cells. BCRF1 also inhibits the activity of NK cells.

3. WHAT IS THE TYPICAL CLINICAL PRESENTATION AND WHAT COMPLICATIONS CAN OCCUR?

Around 98% of patients presenting with infectious mononucleosis have a triad of symptoms – **pharyngitis**, **pyrexia**, and enlargement of cervical lymph nodes and tonsils. **Myalgia** is common as is fatigue, which may persist for months after the initial presentation. Splenomegaly is frequent and occurs in up to 100% in patients, usually resolving 4 weeks after symptom onset; splenic rupture is a rare but potentially life-threatening complication. Other complications that are less common include **encephalitis**, Guillain-Barré syndrome, upper airway obstruction (due to massive lymphoid hyperplasia of the tonsil) – **tracheostomy** may be required to relieve this, abdominal pain, and liver involvement including **hepatomegaly** and **jaundice**. Older patients are less likely to have sore throat and lymphadenopathy but more likely to have liver involvement. Platelet abnormalities are relatively common with patients having mild **thrombocytopenia**. Autoantibodies induced by EBV infection are believed to be the cause of the thrombocytopenia and, in some cases, this can develop into the more **severe idiopathic thrombocytic purpura** with hemorrhagic consequences. **Aplastic anemia (pancytopenia)** has also been described following EBV infection with a similar autoimmune etiology proposed.

Rashes occur in about 5% of patients and may be macular, petechial, scarlatiniform, **urticarial**, or **erythema multiforme**. Also, 90–100% of patients with infectious mononucleosis who receive ampicillin develop a **pruritic** maculopapular rash usually 7–10 days after administration. This was seen in the patient described here (Figure 10.1). It is not caused by an allergy to penicillin but rather a transient hypersensitivity reaction since transient ampicillin-binding antibodies have been detected in patients with infectious mononucleosis.

OTHER CLINICAL CONSEQUENCES OF EBV INFECTION

As already indicated, recovery is the norm after infectious mononucleosis. However, in addition to the complications already mentioned above, there are several serious clinical consequences that can result from EBV infection.

X-linked lymphoproliferative disease (XLPD) is a rare familial condition affecting young boys characterized by extreme susceptibility to EBV. Primary infection results in a particularly severe form of infectious mononucleosis with a large expansion of infected B cells together with T cells infiltrating the organs of the body, especially the liver. Patients with this condition have several abnormalities of the immune system including defects in NK-cell activity, a lack of iNKT cells, overactivity of **Th1** cells, and poor Th2 responses leading to overexpansion of CD8+ T cells and overproduction of Th1 inflammatory **cytokines** and hence macrophage activation and **hemophagocytosis**. The primary defect has been mapped to a small cytoplasmic protein, SAP, involved in NK- and T-lymphocyte signaling. In 60% of cases, the disease is fatal. Patients with XLPD generally die of acute liver **necrosis** or multi-organ failure. Survivors are at high risk of developing B-cell **lymphoma**.

Chronic active EBV (CAEBV) infection is another rare complication of primary EBV infection with a genetic predisposition. There is histologic evidence of major organ involvement such as interstitial pneumonia, hemophagocytosis, uveitis, lymphadenitis, or persistent hepatitis. It is extremely rare in the US and UK but more frequent in Asia and South America. Diagnostic tests show markedly elevated antibodies to EBV lytic antigens (VCA IgG and EA IgG) and raised EBV DNA levels in the blood. The prognosis for these patients is poor, with most dying of progressive pancytopenia and hypogammaglobulinemia or NK/T cell nasal lymphoma within a few years. The pathogenesis of CAEBV is not well understood but is probably the result of an immune defect that permits the proliferation of EBV-infected T or NK cells.

NEOPLASTIC DISEASES

Several malignancies are associated with one of the three general patterns of latent EBV gene expression (Table 10.2). In lymphoproliferative disease (LPD) many of the EBV transforming genes (latency program III) are expressed but immune surveillance is avoided due to immunocompromise; there are few if any host cell mutations. On the other hand, Burkitt lymphoma (BL), nasopharyngeal carcinoma (NPC), and gastric carcinoma arise in the immunocompetent, latency programs I or II mean fewer EBV genes are expressed and other host cell mutations contribute to pathogenesis.

B-CELL MALIGNANCIES

BL – latency program I. This disease is **endemic** in Central Africa and New Guinea with an annual incidence of 6–7 cases per 100 000 and a peak incidence at 6 or 7 years of age. BL is most frequently found in the "lymphoma belt", a region extending from West to East Africa. The region is characterized by high temperature and humidity, ideal for the malaria parasite, probably the reason why malaria was suspected early to be associated with BL. In Uganda, the association of BL with EBV is very strong (97%) but is weaker outside the "lymphoma belt"; (85% in Algeria; only 10–15% in France and the US).

Persons in regions with endemic disease with high titers of EBV IgG VCA antibodies are at risk of BL and the EBV genome is found in lymphoma cells. In addition, BLs contain a chromosomal translocation involving the *c-myc* gene, a potent oncogene, and an immunoglobulin heavy or light chain locus resulting in continuous cell proliferation due to the unregulated expression of *c-myc*.

EBV LPD – latency program III. Patients at risk of LPD and lymphoma are those with congenital, acquired or iatrogenic immunodeficiencies, for example, XPLD, HIV, post-organ transplant respectively. They have elevated EBV DNA levels in blood and saliva. EBV causes 90% of LPD in immunosuppressed transplant patients – posttransplant lymphoproliferative disease (PTLD). HIV patients are at risk of primary central nervous system lymphoma.

Other B-cell malignancies. EBV has also been associated with some cases of Hodgkin lymphoma in apparently immunocompetent patients.

NON-B-CELL MALIGNANCIES OF EPITHELIAL ORIGIN

NPC – latency program II, a malignancy of epithelial cells – relatively rare (less than 1 per 100 000 in most populations) except in southern China, where there is an annual incidence of more than 20 cases per 100 000. There is also a high incidence in isolated northern populations such as Inuit and Greenlanders, while the incidence is moderate in North Africa, Israel, Kuwait, the Sudan, and parts of Kenya and Uganda. The rate of incidence generally increases from age 20 to around 50 and men are twice as likely to develop NPC than women. Evidence indicates that genetic and environmental factors have a role in tumor development. In the US, Chinese Americans comprise most NPC patients, together with workers exposed to fumes, smoke, and chemicals, implying a role for chemical carcinogenesis. There is also an association between eating highly salted foods and the incidence of NPC.

Gastric carcinoma – 10% are of mucosal epithelial origin and associated with EBV.

OTHER EBV MALIGNANCIES

These include some T- and NK-cell lymphomas.

> ### 4. HOW IS THE DISEASE DIAGNOSED, AND WHAT IS THE DIFFERENTIAL DIAGNOSIS?

Diagnosis of infectious mononucleosis is made on presentation with pharyngitis, pyrexia, and cervical lymphadenopathy and is confirmed by laboratory tests.

HEMATOLOGIC TESTS

The white blood cell count is moderately raised by the second week and a blood film will contain "atypical lymphocytes" up to 10–20% of the blood mononuclear cells, hence mononucleosis (Figure 10.6). Atypical lymphocytes are characterized by the presence of bean-shaped or lobulated nuclei and may persist in the blood for several months. They are cytotoxic CD8+ T cells, produced to control the EBV-induced B-cell proliferation. Atypical lymphocytes are not pathognomonic for infectious

(A)

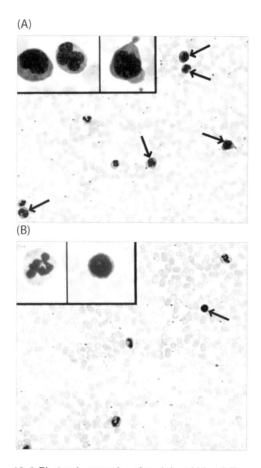

(B)

Figure 10.6 Photomicrographs of peripheral blood films stained with Giemsa (×40; insets ×100). (A) From a patient with infectious mononucleosis; numerous large atypical mononuclear cells are present (arrows and insets) in addition to the normal cells. (B) From a normal individual showing four neutrophils (detail in left inset) and one lymphocyte (arrow and right inset) among a mass of red cells. *From Encyclopaedia of Life Sciences, Infectious Mononucleosis, by MA Epstein, University of Oxford. Doi: 10.1038/npg.els0002318. Permission by John Wiley and Sons. All rights reserved.*

mononucleosis and can be observed with other infections (see differential diagnosis below).

ANTIBODY TESTS

Heterophile antibodies (see Section 2, Adaptive immunity – antibody) are directed against glycolipid antigens on sheep, horse, or ox erythrocytes, agglutination of which forms the basis of the Paul-Bunnell or Monospot tests. Heterophile

antibodies are often absent in children, especially under 4 years.

Detection of EBV-specific antibodies is rarely necessary for the diagnosis of infectious mono-nucleosis because 90% of cases are heterophile antibody positive. The basic pattern of serum antibodies in primary EBV infection is shown in Figure 10.4. The most useful antibody assay is that which detects IgM against the Epstein-Barr VCA. Anti-VCA of the IgM class develops early in the illness and declines rapidly over the following 3 months. **Enzyme immunoassays (ELISAs)** can be used to quantitate IgM anti-VCA, but detection can also be by indirect **immunofluorescence assay** using EBV-infected lymphoblastoid cell lines as the antigenic substrate. Patients also seroconvert to IgG anti-VCA antibodies in primary infection and these serum antibodies persist for life; testing for IgG anti-VCA antibodies can therefore be used to determine whether an individual has been infected with EBV. Antibodies against EBNA are absent early in infectious mononucleosis but persist indefinitely thereafter. They may therefore be helpful in differentiating EBV reactivation from primary infection.

Patients with NPC have elevated levels of serum IgA directed against EBV VCA and EA, which may occur several years before the onset of disease. A screening program for these antibodies has been instituted in southern China to allow early detection and treatment of the cancer.

DIFFERENTIAL DIAGNOSIS

Patients with streptococcal pharyngitis or one of several other infections can present with sore throat, fatigue, and lymphadenopathy (Table 10.4). In the case presented here, the doctor attempted treatment with ampicillin on the supposition of a streptococcal infection. Unfortunately, the patient responded adversely to this antibiotic with a rash, due to an idiosyncratic antibody response peculiar to infectious mononucleosis (see Section 2).

Primary CMV infection, toxoplasmosis and primary HIV infection share many clinical and laboratory features of infectious mononucleosis, including splenomegaly, hepatomegaly, **lymphocytosis**, atypical lymphocytosis, and even (but rarely) false-positive results in the heterophile antibody test. Distinguishing clinically between infectious mononucleosis caused by EBV infection and an infectious mononucleosis-like syndrome caused by toxoplasmosis or

Table 10.4 Infectious mononucleosis: differential diagnosis

Diagnosis	Key distinguishing features
Streptococcal pharyngitis	Absence of splenomegaly or hepatomegaly; fatigue less prominent
Other viral pharyngitis, (e.g. adenoviruses, patients with enteroviruses)	Patients are less likely to have lymphadenopathy, tonsillar exudates, or fever than in streptococcal pharyngitis or infectious mononucleosis
Cytomegalovirus (CMV)	Sore throat less severe, lymphadenopathy may be minimal or absent. Confirm by specific antibody tests for primary CMV infection
Toxoplasmosis	Sore throat less severe; confirm by specific antibody tests
Primary HIV infection	Mucocutaneous lesions, rash, nausea, vomiting, diarrhea and weight loss

CMV may not be possible, or perhaps not useful, since the management of these syndromes is the same. However, diagnostic testing is needed in pregnant women because toxoplasmosis and CMV infections may cause congenital infection resulting in severe damage to the unborn child. If primary HIV infection is suspected, diagnostic tests should also be performed.

The differential diagnosis also includes **leukemia** and lymphoma although these are less likely.

5. HOW IS THE DISEASE MANAGED AND PREVENTED?

MANAGEMENT

There is usually no treatment for infectious mononucleosis other than supportive care. **Nonsteroidal anti-inflammatory drugs (NSAIDs)** or paracetamol (acetaminophen) are given for fever and myalgias; aspirin should not be given to children because it may cause **Reye's syndrome**.

Patients are recommended to rest and return to usual activities based on their energy levels. Although the exact period of increased risk is unknown, patients are advised to avoid contact sports or other activities that may lead to splenic rupture for several weeks.

Corticosteroids may be considered in patients with significant pharyngeal **edema** that might cause or threaten respiratory compromise.

PREVENTION

EBV is a major global health problem. Although it is associated with about 200 000 cancers annually worldwide, a vaccine is not available. It would be useful to protect adolescents and young adults against the debilitating consequences of infectious mononucleosis. In addition, vaccination might reduce the frequency of EBV-positive Hodgkin lymphoma, which is increased in the first 5 years after infectious mononucleosis. An important target for vaccine is the prevention of BL and NPC in at-risk populations. A vaccine is also needed to prevent LPD in seronegative individuals with XLPD or those who receive a transplant from an EBV seropositive donor.

The principal target of EBV-neutralizing antibodies is the major virus surface glycoprotein gp350 and several vaccine candidates based on this molecule have been developed. In 2007, phase I and II studies of a purified gp350 vaccine showed protection from symptoms of infectious mononucleosis but did not prevent asymptomatic infection with EBV, consistent with the possibility of at least reducing cases of infectious mononucleosis. More recently, presentation of gp350 on nanoparticles enhanced the levels of neutralizing antibody in monkeys compared with soluble gp350 and focused the immune response on the CD21 receptor binding site of gp350.

An alternative or complementary strategy is to induce potent T-cell responses to control primary infection, reduce the viral load and thus potentially reduce the risk of EBV-associated malignancy. This is based on the observation that elevated EBV viral loads in blood predict development of EBV-associated LPD in transplant recipients and are present at the onset of NPC. A phase I trial has been completed in Australia with a single EBNA3 EBV epitope. In this trial, two placebo recipients became EBV-infected, and one had symptomatic mononucleosis; four vaccine recipients acquired EBV infection but none were symptomatic. To generate a broad-based cytotoxic T-cell response, a vaccine containing multiple EBV epitopes will be necessary.

SUMMARY

1. WHAT IS THE CAUSATIVE AGENT, HOW DOES IT ENTER THE BODY, AND HOW DOES IT SPREAD A) WITHIN THE BODY AND B) FROM PERSON TO PERSON?

- Epstein-Barr virus (EBV) is a gamma herpesvirus with an outer envelope surrounding an icosahedral capsid enclosing the linear ds DNA genome.
- EBV-infected cells can exist in two states. In lytic replication, the resultant virions have a linear DNA genome. In latency, the viral genome is a circular episome, there is limited gene expression and no infectious particle production.
- There are different patterns of EBV gene expression, latency programs 0, I, II, III associated with various neoplasms.
- For entry into B cells, gp350 on the viral envelope binds to a complement receptor CD21 (CR2) on the surface of B cells. Another envelope glycoprotein, gp42, is responsible for the fusion between the virus envelope and the host-cell membrane.
- EBV can also infect epithelial cells but entry into the cell is gp42 independent.

- During primary infection, the virus probably infects pharyngeal squamous epithelial cells and undergoes a lytic cycle producing many more virions that directly infect B cells in transit through the tonsils.
- EBV causes proliferation of the B cells (B-cell transformation). The ability of EBV to convert peripheral blood B cells into immortalized lymphoblastoid cell lines (LCLs) is widely used in genomic studies.
- Virus and infected cells pass via the afferent lymphatics into the cervical lymph nodes and from thence via the thoracic duct to the spleen and other lymphoid tissues where EBV causes lymphocyte proliferation, lymphadenopathy, and splenomegaly.
- Even after resolution of symptoms, the virus is not cleared from the body and remains latent for life in memory B cells with occasional reactivation into a lytic cycle producing virions transmissible to a new host. *Continued...*

...continued

- EBV virions are shed into the saliva at high level for several months and persist at low level for life. Spread of the infection is usually through intimate contact with an uninfected person often through kissing, and in children probably mainly by fingers contaminated with saliva and other close contact.

- Globally, most adults are infected with EBV (95–98%). In developing countries, most children become infected with the virus early in life. In developed countries, at least 50% of adolescents and young adults undergo primary infection with EBV and develop some symptoms of infectious mononucleosis.

2. WHAT IS THE HOST RESPONSE TO THE INFECTION AND WHAT IS THE DISEASE PATHOGENESIS?

- At the start of the infection, EBV must penetrate the oral mucosal epithelium to infect B cells. This occurs at Waldeyer's ring, comprising pharyngeal, tubal, palatine and lingual tonsils. M cells are present in the overlying epithelium serving to capture and transcytose microparticles such as viruses and might deliver EBV to B cells. Alternatively, the virus might be picked up by intraepithelial dendritic cells.

- After entry, EBV is exposed to intraepithelial lymphocytes including natural killer (NK) cells, invariant NK T (iNKT) cells and γδ T cells. These cells are key components of the innate immune response against EBV.

- EBV antibodies produced at different times as viral antigens are recognized by the immune system of the host. These include early antigens, EA; late antigens such as virus capsid antigens (VCA); and latency antigens including EBV nuclear antigens (EBNAs) and latent membrane proteins (LMPs).

- IgM antibody to VCA is the first to appear followed by IgG. Antibodies to the latency antigens, for example EBNA, are seen later.

- Heterophile antibodies (not EBV-specific) also arise due to polyclonal B-cell activation (B-cell transformation) with consequent differentiation into plasma cells. These IgM antibodies are directed against horse and sheep glycolipid antigens on red blood cells.

- Cytotoxic CD8+ T cells are the main cells that control EBV infection. During primary infection up to 50% of the CD8+ T cells in the circulation can be specific for EBV antigens.

- The symptoms of infectious mononucleosis are attributable to the intense immunologic response to polyclonally activated B cells mounted by CD8+ T cells.

- Cytotoxic T cells can control the primary infection but a few B cells containing latent EBV genome enter a resting state as members of a long-lived memory cell pool.

- Strategies that help EBV to evade the immune system include switching off the expression of viral antigens in resting B memory cells; shedding of gp42, which binds to MHC class II/peptide complexes and inhibits CD4+ T-cell activity; an EBV late protein, BCRF1, is an immunomodulator that inhibits T helper 1 cell activity and hence cytolytic CD8+ T cells.

3. WHAT IS THE TYPICAL CLINICAL PRESENTATION AND WHAT COMPLICATIONS CAN OCCUR?

- Globally, most individuals infected with EBV do not show clinical symptoms of infectious mononucleosis. Nearly all patients with the disease have a triad of symptoms – pharyngitis, pyrexia, and cervical lymph node and tonsillar enlargement. Hepatocellular enzymes are mildly increased in most cases.

- Complications include splenic rupture and upper airway obstruction. Encephalitis and Guillain-Barré syndrome occur rarely.

- Older adults are less likely to have sore throat and lymphadenopathy but more likely to have hepatitis and jaundice.

- Patients with X-linked lymphoproliferative disease (XLPD), a rare familial condition affecting young boys, have extreme sensitivity to EBV. In 60% of cases, primary infection is fatal due to macrophage activation, hemophagocytosis, and destruction of all lymphoid tissue. Survivors are at risk of B-cell lymphoma.

- Chronic active EBV (CAEBV) is rare in the US and UK but more frequent in Asia and South America. The prognosis is poor with most dying of progressive pancytopenia and hypogammaglobulinemia or NK/T cell lymphoma within a few years.

- Several malignancies are associated with EBV latent infection.

- Burkitt lymphoma (BL) is a malignant tumor associated with EBV. BL has limited EBV gene expression and is characterized by a chromosomal translocation involving the *c-myc* oncogene.

- BL is endemic to central parts of Africa and New Guinea, with an annual incidence of 6–7 cases per 100000 and a peak incidence at 6 or 7 years of age. In African countries such as Uganda, the association of BL with EBV is very strong (97%) but is weaker elsewhere in the world (85% in Algeria; only 10–15% in France and the US).

- In the absence of a T-cell immune response in XPLD or HIV infection or post-organ transplant, EBV infection can result in life-threatening lymphoproliferative disease (LPD). There is uncontrolled B-cell proliferation as most, if not all, EBV latency genes are expressed.

- Non-B-cell malignancies associated with EBV include nasopharyngeal carcinoma (NPC) and gastric carcinoma, both of epithelial origin. The virus is also associated with some cases of Hodgkin lymphoma as well as with T- and NK-cell lymphomas.

- NPC is relatively rare (less than 1 per 100000 in most populations) except in southern China, where there is an annual incidence of more than 20 cases per 100000. The rate of incidence generally increases from age 20 to around 50 and men are twice as likely to develop NPC as women. Environmental factors have a role in tumor development.

4. HOW IS THE DISEASE DIAGNOSED, AND WHAT IS THE DIFFERENTIAL DIAGNOSIS?

- Diagnosis of infectious mononucleosis requires persistent pharyngitis with pyrexia and cervical lymphadenopathy confirmed by laboratory tests. *continued...*

...continued

- The white blood count is moderately raised, and a blood film contains "atypical lymphocytes" up to 10–20% of the blood mononuclear cells. These are cytotoxic CD8+ T cells.
- Tests for "heterophile antibodies" such as the Paul-Bunnell or Monospot tests are positive.
- EBV-specific IgM anti-VCA can be detected by indirect immunofluorescence or ELISA.
- Differential diagnosis includes streptococcal pharyngitis, toxoplasmosis, and primary CMV or HIV infection.

5. HOW IS THE DISEASE MANAGED AND PREVENTED?

- Usually, no treatment is given other than good supportive care.
- Non-steroidal anti-inflammatory drugs (NSAIDs) or paracetamol (acetaminophen) may be given to relieve fever and sore throat.

- Patients are recommended to rest and return to usual activities based on their energy levels. Sports activities should be avoided because of the risk of splenic rupture.
- Corticosteroids may be considered in patients with significant pharyngeal edema that might cause respiratory compromise.
- Vaccines targeting the membrane gp350 have already been trialed (phase I and phase I/II) and shown to be safe protecting against symptoms of infectious mononucleosis but not asymptomatic EBV infection.
- A proposed alternative approach is to induce potent T-cell responses to control primary infection, reduce the viral load and thus potentially reduce the risk of EBV-induced malignancy. A phase I trial has been completed in Australia with a single EBNA3 EBV epitope. To generate a broad-based cytotoxic T-cell response, a vaccine containing multiple EBV epitopes will be necessary.

FURTHER READING

Gewurz BE, Longnecker RM, Cohen JI. Epstein-Barr virus. In: Howley PM and Knipe DM (eds). Fields Virology, 7th edition. Wolters Kluwer, Philadelphia, 2022.

Goering RV, Dockrell HM, Zuckerman M, Chiodini PL. Medical Microbiology and Immunology, 6th edition. Elsevier, Oxford, 2019.

Johannsen EC, Kaye KM. Epstein-Barr Virus (Infectious Mononucleosis, Epstein-Barr Virus–Associated Malignant Diseases, and Other Diseases). In: Bennett JE, Dolin R, Blaser MJ, (eds). Mandell, Douglas and Bennett's Principles and Practice of Infectious Diseases, 8th edition. Elsevier Saunders, Philadelphia, 2015.

Rickinson AB, Epstein MA. Epstein-Barr virus. In: Firth J, Conlon C, Cox T, editors. Oxford Textbook of Medicine, 6th edition. Oxford University Press, Oxford, 2020.

REFERENCES

Alari-Pahissa E, Ataya M, Moraitis I, et al. NK Cells Eliminate Epstein-Barr Virus Bound to B Cells Through a Specific Antibody-Mediated Uptake. PLoS Pathog, 17: e1009868, 2021.

Balfour HH Jr. Epstein-Barr Virus Vaccine for the Prevention of Infectious Mononucleosis: And What Else? J Infect Dis, 196: 1724–1726, 2007.

Bharadwaj M, Moss DJ. Epstein-Barr Virus Vaccine: A Cytotoxic T-Cell-Based Approach. Expert Rev Vaccines, 1: 467–476, 2002.

Borza CM, Hutt-Fletcher. Alternate Replication in B Cells and Epithelial Cell Switches Tropism of Epstein-Barr Virus. Nat Med, 8: 594–599, 2002.

Chung BK, Tsai K, Allan LL, et al. Innate Immune Control of EBV-Infected B Cells by Invariant Natural Killer T Cells. Blood, 122: 2600–2608, 2013.

Elliott SL, Suhrbier A, Miles JJ, et al. Phase I Trial of a CD8+ T-Cell Peptide Epitope-Based Vaccine for Infectious Mononucleosis. J Virol, 82: 1448–1457, 2008.

Hislop AD, Taylor GS, Sauce D, Rickinson AB. Cellular Responses to Viral Infection in Humans: Lessons from Epstein-Barr Virus. Ann Rev Immunol, 25: 587–617, 2007.

Jangra S, Yuen K-S, Botelho MG, Jin D-J. Epstein–Barr Virus and Innate Immunity: Friends or Foes? Microorganisms, 7: 183, 2019.

Kanekiyo M, Bu W, Joyce MG, et al. Rational Design of an Epstein-Barr Virus Vaccine Targeting the Receptor-Binding Site. Cell, 162: 1090–1100, 2015.

Moutschen M, Leonard P, Sokai EM, et al. Phase I/II Studies to Evaluate Safety and Immunogenicity of a Recombinant gp350 Epstein-Barr Virus Vaccine in Healthy Adults. Vaccine, 25: 4697–4705, 2007.

Münz C. Epstein–Barr Virus-Specific Immune Control by Innate Lymphocytes. Front Immunol, 8: 1658, 2017.

Sixbey JW, Yao QY. Immunoglobulin A-Induced Shift of Epstein-Barr Virus Tissue Tropism. Science, 255: 1578–1580, 1992.

Thorley-Lawson DA, Gross A. Persistence of the Epstein-Barr Virus and the Origins of Associated Lymphomas. N Engl J Med, 350: 1328–1337, 2004.

Yao QY, Rowe M, Morgan AJ, et al. Salivary and Serum IgA Antibodies to the Epstein-Barr Virus Glycoprotein gp340: Incidence and Potential for Virus Neutralization. Intl J Cancer, 48: 45–50, 1991.

WEBSITES

Centers for Disease Control and Prevention (CDC), National Center of Infectious Diseases, 2008: https://www.cdc.gov/epstein-barr/index.html

Students can test their knowledge of this case study by visiting the Instructor and Student Resources: [www.routledge.com/cw/lydyard] where several multiple choice questions can be found.

Escherichia coli

11

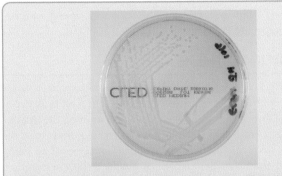

Figure 11.1 *E. coli* **cultured from the urine.** This shows the growth of lactose-fermenting (yellow colonies) *E. coli* on a medium containing lactose and a pH indicator. In routine practice, urine is cultured on cystine lactose electrolyte-deficient (CLED) medium. Cystine is a growth factor for some organisms; lactose is incorporated to differentiate utilization of lactose from non-utilization (e.g. *Pseudomonas*); and the electrolyte deficiency inhibits the swarming of *Proteus*, which may obscure bacterial colonies.

A 50-year-old worker at a school cafeteria went to her primary care physician complaining of tiredness, shaking chills, a pain in her loin, and a burning sensation on passing urine, which she was doing more frequently than normal.

On examination, her doctor noted that she seemed a bit pale and that she had some suprapubic tenderness. He tested her urine with a dipstick and found she had a positive result for nitrite, **pus** cells, and protein.

The doctor took a blood and urine sample for confirmation and sent them to the local hospital laboratory. A diagnosis of **pyelonephritis** was made and she was started on antibiotics. The following day, the laboratory results were available. The full blood count showed a **neutrophilia** and a pure culture of *Escherichia coli* was grown from the urine (Figure 11.1).

1. WHAT IS THE CAUSATIVE AGENT, HOW DOES IT ENTER THE BODY AND HOW DOES IT SPREAD A) WITHIN THE BODY AND B) FROM PERSON TO PERSON?

CAUSATIVE AGENT

Uropathogenic *E. coli* (UPEC) is the major cause of urinary tract infections (UTIs) in anatomically normal, unobstructed urinary tracts. *Escherichia coli* is a motile non-sporing gram-negative bacillus approximately 0.6–1 μm wide and 2–3 μm long (Figure 11.2). It has a typical gram-negative cell wall with an outer hydrophobic membrane containing **lipopolysaccharide** (**LPS**). It has peritrichous flagella and possesses **fimbriae**, which are important in adhesion (Figure 11.3). Eighty percent of all UTIs are caused by *E. coli* in the community and about 60% in hospital.

The LPS is composed of lipid A, an inner core of polysaccharide linked to the lipid A by ketodeoxyoctonate (KDO) and an outer variable polysaccharide (Figure 11.4). The general arrangement of LPS is similar in most gram-negative bacteria, with the highest variability appearing in the outer "O" antigen component. The inner lipid A component comprises different fatty acids (such as hexanoic (6:0), dodecanoic (12:0),

tetradecanoic (14:0), hexadecanoic (16:0), and octadecanoic (18:0), with their monoenoic equivalents) linked by ester amide bonds to KDO via N-acetylglucosamine. This part of LPS is the **endotoxin** – which is phosphorylated – and can have dramatic effects on the clotting and **kallikrein** cascades, complement activation, and **cytokine** production. The core region comprises 6- and 7-carbon sugars (glucose, galactose, heptose) linked to the outer variable region comprising 6-carbon sugars (e.g. glucose, mannose, tyvelose, rhamnose).

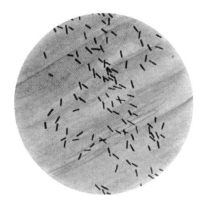

Figure 11.2 *E. coli* **organisms.** This is a Gram stain of *E. coli* showing the red (gram-negative) appearance. *Courtesy of the Centers for Disease Control, Atlanta, Georgia. Image is found in the Public Health Image Library #3211.The photo was taken in 1979.*

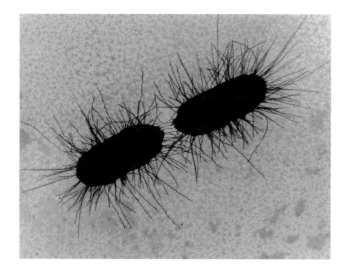

Figure 11.3 The hair-like fimbriae on _E. coli_. This electron-micrograph shows the hair-like fimbriae on _E. coli_, which are important for adhesion. Fimbriae were first demonstrated by Prof. Duguid at the University of Dundee in the 1960s. _Courtesy of the photographer Dennis Kunkel. Copyright Dennis Kunkel Microscopy, Inc._

The polysaccharide sequences of the variable region may be branched and repeated to give long chains projecting from the surface of the bacterium.

Not all _E. coli_ strains cause UTI and, in fact, _E. coli_ (mainly of the K12 strain) is part of the normal flora of the intestine (i.e. a **commensal**) that usually does not cause any problems. However, other diseases caused by _E. coli_ are: gastrointestinal (diarrheagenic _E. coli_ with different pathogenic mechanisms e.g. VeroToxin _E. coli_ O157.H7 Table 11.1); septicemia, pneumonia, meningitis (mainly neonatal cases with _E. coli_ K1 capsule). Whole genome sequencing (WGS) of extra-intestinal pathogenic _E. coli_ reveals that sequential UTI isolates are usually genetically different whereas sequential bloodstream isolates are generally very similar. Furthermore, there was no clustering according to site of isolation. Additionally, several genes were associated with bacteremia such as papA and G (adhesion); traJ (invasion); sat, tosA (toxins). A further genome study of enterotoxigenic _E. coli_ (ETEC) was undertaken and revealed 21 global ETEC lineages and showed that colonization factors and toxins were commonly found in these isolates challenging the suggestion that such factors are usually found on plasmids. This study also showed that no specific lineage correlated with any specific disease.

Other agents that cause UTI are other members of the Enterobacteriaceae such as _Enterobacter_ and _Klebsiella_; _Enterococci_; _Staphylococcus saprophyticus_ in young women; _Proteus_ spp.; and _Pseudomonas_ spp.

ENTRY INTO THE BODY

The GI tract in man is colonized by _E. coli_ within hours to a few days after birth. The organism is ingested in food or water or obtained directly from other individuals handling the infant.

Urinary tract infection with _E. coli_ is acquired from the person's own microflora (Endogenous) as opposed to being acquired from another person, animal or the environment (Exogenous). Enterotoxigenic _E. coli_ are acquired from contaminated food.

SPREAD WITHIN THE BODY

In females, if the _E. coli_ organisms from the feces have the appropriate virulence factors (see below), they can attach to the epithelium of the vagina in the **introitus**, which then becomes colonized. The organisms can enter the urinary tract via the ureter and may eventually be introduced into the bladder where **cystitis** can develop. Once in the bladder, and in the presence of vesicourethral reflux, the organism may

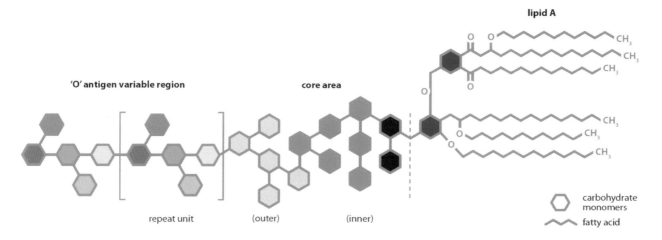

Figure 11.4 The structure of _E. coli_ lipopolysaccharide (LPS). This illustrates the structure of the LPS found in the outer membrane of gram-negative bacteria. It consists of a lipid A component consisting of β-hydroxy fatty acids attached to carbohydrate, a core carbohydrate component consisting of monomers of different carbohydrates, e.g. heptoses and ketodeoxyoctonic acid. This inner region of the LPS is virtually identical in all gram-negative bacteria, although the fatty acids may vary in different genera of bacteria. The outer part of the LPS consists of repeating units consisting of carbohydrates, e.g. mannose, glucose, galactose, rhamnose, etc., as illustrated by the different shading in the figure. This part is highly variable even within the same species of bacteria and is called the variable region or the "O" antigen. Variation between different species also exists in the degree of phosphorylation of LPS. The differences in the fatty acids and level of phosphorylation affect the endotoxin activity of the LPS.

Table 11.1 The different pathovars of *E. coli*

Strain	Designation	Disease
E. coli K12	K12	Commensal
Enteropathogenic *E. coli*	EPEC	Gastroenteritis
Enterotoxigenic *E. coli*	ETEC	Gastroenteritis
Enteroinvasive *E. coli*	EIEC	Gastroenteritis
Enterohemorrhagic *E. coli*	EHEC (O157)	Gastroenteritis, hemolytic uremic syndrome
Enteroaggregative *E. coli*	EAEC	Gastroenteritis
Diffusely adherent *E. coli*	DAEC	Gastroenteritis
E. coli K1	K1	Neonatal meningitis
Uropathogenic *E. coli*	UPEC	Pyelonephritis

spread up the ureter to the renal pelvis and **pyelonephritis** may ensue as it did in this patient.

Some cases of pyelonephritis are hematogenous, the organism localizing in the kidneys from the blood, for example, *Staphylococcus aureus*, although this route is uncommon with *E. coli*, as the ascending route of infection predominates.

PERSON-TO-PERSON SPREAD

Person-to-person spread is not relevant for *E. coli* causing UTI, although enteropathogenic *E. coli* that causes **gastroenteritis** can be transmitted by **feco–oral** spread.

EPIDEMIOLOGY OF UTI

Because of the anatomic differences in the length of the urethra between the male and female, the female generally is far more likely to have a UTI compared with the male, except in neonates where it occurs more commonly in boys. In neonates, the prevalence is about 1–2%. In schoolchildren, the prevalence is 4.5% in girls and 0.5% in boys and where it occurs in the latter it frequently signifies renal tract congenital abnormalities. In adulthood, up to 40% of females will have a UTI at some time in their lives. Sexual intercourse and pregnancy are risk factors for developing a UTI. In men, the prevalence is less than 0.1%. A risk factor in males is an enlarged prostate and the prevalence of UTI in elderly men is up to 10%. In post-menopausal women, the lack of estrogen is a risk factor, as the microbial flora changes from one mainly of lactobacilli to one with gram-negative bacteria. In both males and females, a significant risk factor in hospitals is catheterization. In fact, it is known that catheter-associated infection is the most frequent hospital-acquired infection.

2. WHAT IS THE HOST RESPONSE TO THE INFECTION AND WHAT IS THE DISEASE PATHOGENESIS?

Some virulence characteristics of *E. coli* are shown in Table 11.2 and are often carried on plasmids or on **pathogenicity islands (PAIs)**. These PAIs are DNA sequences recognized as

having a GC content that differs from the average GC content of the organism and occur due to horizontal transfer from other organisms.

UPEC have specific virulence characteristics that allow them to cause infection in the renal tract. These virulence characteristics include (a) specific fimbriae (e.g. P-fimbriae) that bind to the P-blood group antigen present on uroepithelial cells; (b) specific capsular antigens, for example, K1 associated with serum resistance; (c) the production of **hemolysin**; (d) cytotoxic necrotizing factor; and (e) secreted autotransported toxin (SAT) (f) siderophore receptors and the TonB iron uptake mechanism.

The P-fimbriae are composed of the main fimbrial protein (PapA) and the adhesin (PapG). PapG adhesin has several allelic variants, with allele II variant binding to globoside and causing most cases of pyelonephritis and allele III variant binds Forssmann antigen found mainly in children and women causing cystitis or asymptomatic bacteriuria. The GI variant binds globotriaosylceramide. P-fimbriae binding is not inhibited by mannose, but UPEC can bind to mannose receptors on the epithelial cells of the urinary tract by type I fimbriae, which binds mannosyl conjugates, and uroplakins 1a, 1b and which are present in most *E. coli* strains and are composed of the Fim-A structural protein and the Fim-H adhesin. S-fimbriae bind to sialyllactose residues found on the renal tubules and glomeruli. Expression of Dr fimbriae, which bind to **decay accelerating factor** (a natural complement inhibitory protein found on epithelial cells) and collagen IV is associated with cell invasion and related to cystitis in children and pyelonephritis in pregnant women as well as gastroenteritis. Presence of this adhesin also re-attaches dislodged epithelial cells back to the epithelial surface by *E. coli* attached to the cell. Other adhesins are secreted components involved in adhesion and biofilm formation. Curli (a type of fimbriae) are secreted amyloid, which with phosphoethanolamine cellulose, also secreted by *E. coli* strains, provide firm adhesion present in high stress flow conditions.

1. These adhesins with different binding specificities have also been identified and expression of certain of these adhesins is thought to determine whether the *E. coli* moves

Table 11.2 Virulence characteristics of *E. coli*

Virulence characteristic	Strain	Binds	Effect/disease
Adhesins			
Type I fimbriae	Most strains	Mannose	Pyelonephritis/gastroenteritis
P-fimbriae	UPEC	Galα1–4gal	Pyelonephritis
S-fimbriae	UPEC	Sialyl2 – 3lactose	Pyelonephritis
Dr-fimbriae	UPEC	DAF	Cystitis/pyelonephritis
Nonfimbrial adhesins	Most strains	Various ligands	Pyelonephritis/gastroenteritis
CFA I/II	ETEC		Gastroenteritis
AAF type III	EAggEC		Gastroenteritis
K-antigen K12	Commensals		Antiphagocytic and serum resistance
K1	UPEC		Pyelonephritis/neonatal meningitis
Toxins			
Hemolysin	Most	–	Cytotoxic/antiphagocytic
Siderophores	Most	–	Sequester iron
LT (heat-labile toxin)	ETEC	–	Fluid loss/traveler's diarrhea
ST (heat-stable toxin)	ETEC	–	Fluid loss/traveler's diarrhea
Verotoxin, EspP	ETEC	–	Dysentery/hemolytic uremic syndrome
SAT	UPEC	–	Cytotoxic/pyelonephritis
PET/PIC	EAggEC	–	Gastroenteritis
EspC	EPEC	–	Gastroenteritis

DAF, decay accelerating factor; CFA, colonization factor; SAT, secreted autotransported toxin; PET, plasmid-encoded toxin; PIC, protein involved in intestinal colonization; Esp, *E. coli* secreted protein

further up the tract to infect the kidney. Individuals who are of the nonsecretor genotype (do not secrete blood group antigens into, e.g. saliva, urine, mucus) are more prone to binding of *E. coli* and thus to UTI.

2. UPEC protect themselves from the host defenses with the K-capsule, which is antiphagocytic and also provides resistance to complement (serum resistance).
3. UPEC also produces a number of toxins, which are also secreted by other *E. coli* pathovars. The hemolysin is a pore-forming toxin and is cytotoxic.
4. Cytotoxic necrotizing factor affects intracellular signaling by modifying the Rho GTP-binding proteins, actin polymerization and is cytotoxic.
5. UPEC also secrete an autotransporter toxin (SAT): a member of the SPATE (serine protease autotransporters of Enterobacteriaceae) group of toxins, which is also cytotoxic.

Binding of *E. coli* to the uroepithelial cells in the bladder induces widening of the junctions between the squamous epithelium, exposing the underlying basal cells to which the organism can readily bind, and infection also increases desquamation of the superficial uroepithelium. Adhesion of bacteria also induces an acute inflammatory response stimulating the synthesis of **Th1** cytokines by the uroepithelium, **interleukin** (**IL**)-1 and IL-6 with IL-8, which recruits granulocytes and then macrophages and the patient develops a temperature. Production of these cytokines is probably related to the binding of the LPS to **Toll-like receptors** (especially TLR4) on the uroepithelium. In the renal medulla, the high ammonium concentration, osmolarity, and low pH all contribute to **immune-paresis** favoring the organism. The host secretes **defensins** from the uroepithelium and **Tamm Horsfall protein**, which binds *E. coli*, aiding its removal. Binding of the organism also results in exfoliation contributing to the *E. coli* that are found in the urine. The host develops a normal acquired immune response to the presence of the organism in the upper urinary tract, with the production of immunoglobulins of all isotypes. Infection of the bladder has a much-reduced antibody response. Cell-mediated immunity to infection appears to be of little importance in the urinary tract with the exception of cytokine secretion.

The high numbers of granulocytes that accumulate in some parts of the renal tract cause damage to the host in a bystander effect through release of **oxygen free radicals** and proteolytic enzymes. The *E. coli* invade the uroepithelium forming intracellular bacterial colonies and inducing **necrosis** of the cells leading to further inflammation.

The role of the adaptive immune response in pyelonephritis is poorly understood, although **sIgA** antibodies are produced and may inhibit the binding of the organism.

3. WHAT IS THE TYPICAL CLINICAL PRESENTATION AND WHAT COMPLICATIONS CAN OCCUR?

The clinical presentation of UTI depends on the age of the person. In neonates, the symptoms are nonspecific with vomiting, fever, and a "floppy" infant. In older children and adults, localized symptoms occur. Cystitis presents with frequency, dysuria, suprapubic pain, and fever. The urine may contain blood. In cases of pyelonephritis, the patient will present with fever, rigors, loin pain, frequency, and dysuria and **hematuria**. Elderly patients may present with a typical picture or with fever, incontinence, dementia or signs suggestive of a chest infection. A "colicky" pain radiating from the loin to the groin is suggestive of renal stones – which may occur in the absence of infection, although renal calculi are a risk for infection and often associated with Proteus, which has an enzyme (urease) that can hydrolyze urea.

Complications include renal scarring, **septicemia**, papillary necrosis which, if bilateral, can lead to renal failure, parenchymal **abscess**, and perinephric abscess. There is no clear relationship between pyelonephritis and the sequential development of **chronic interstitial nephritis** and **hypertension**.

4. HOW IS THE DISEASE DIAGNOSED, AND WHAT IS THE DIFFERENTIAL DIAGNOSIS?

The organism can be isolated by culturing a mid-stream specimen of urine (see Figure 11.1) and, in 30% of cases, the organism will also be found in the blood culture. Work by Kass showed that if the number of organisms in the urine was greater than 10^5 bacteria per ml then this correlated well with clinical disease and this figure is considered as a "significant" **bacteriuria**. However, this study was in asymptomatic healthy women and it is recognized that UTIs can occur with fewer organisms in the urine. The presence of a single species of bacteria in the urine is also taken as evidence of significance but again UTIs can occur when more than one species is present.

Other organisms can present with signs and symptoms of a UTI but not grow on the medium routinely used yet have a high white cell count in the urine. This is called a sterile pyuria. Two organisms that present like this are *Mycobacterium tuberculosis* and *Actinotignum* spp. For the former, the urine should be grown on Lowenstein-Jennson medium and for the latter it should be grown on blood agar under micro-aerobic conditions.

A number of automated methods for assessing bacteriuria are available, such as turbidometric, bioluminescence, electrical impedance, flow cytometry, and radiometric tests.

Microscopy is generally not useful in the diagnosis of infection, although the presence of **white cell casts** is suggestive but not diagnostic of pyelonephritis.

The differential diagnosis of pyelonephritis depends on the context and age of the patients. Dermatologic conditions such as **shingles** (before the appearance of the rash) and musculoskeletal injury may give rise to loin pain but without urinary symptoms or a fever in the case of musculoskeletal injury. Renal vein **thrombosis** can give rise to severe pain and fever; renal abscess, papillary necrosis, and **urolithiasis** will give rise to pain and fever. The presence of a "sterile" **pyuria** can indicate renal tuberculosis, urolithiasis or neoplasm.

5. HOW IS THE DISEASE MANAGED AND PREVENTED?

There is no strong reason for treatment of asymptomatic bacteriuria in either non-pregnant women or the elderly. On the other hand, in pregnant women and children, particularly in the latter if there are renal congenital abnormalities allowing **vesico-ureteric reflux**, then treatment is required even in the absence of symptoms. If not treated, renal scarring or even renal failure may occur.

The choice of antimicrobial agent to treat UTI is often empirical but is usually based on the recognition that the commonest infectious agent is *E. coli* and a knowledge of the local antibiotic resistance patterns. Cystitis can be treated with 3 days of appropriate antibiotics. Trimethoprim, nitrofurantoin or amoxycillin are often used.

In cases of pyelonephritis, as there is parenchymal disease, a 2-week course of antibiotics is required, for example, cefuroxime or a third-generation cephalosporin, an aminoglycoside or a fluoroquinolone depending on the resistance pattern.

Because of the spread of extended-spectrum β-lactamases (ESBLs), particularly in bacteria such as *Enterobacter*, no β-lactam antibiotic is effective (except the carbapenems) and a different antibiotic class should be used, for example, a fluoroquinolone.

Frequent relapses of UTIs may require prophylaxis, usually with nitrofurantoin or trimethoprim taken at night, so high concentrations of the antibiotic are present in the bladder overnight.

SUMMARY

1. WHAT IS THE CAUSATIVE AGENT, HOW DOES IT ENTER THE BODY, AND HOW DOES IT SPREAD A) WITHIN THE BODY AND B) FROM PERSON TO PERSON?

- The commonest infecting agent for urinary tract infections is *E. coli*.
- Other agents causing UTI are enterococci, *Proteus*, *Staph. saprophyticus*, *Pseudomonas*, and other enteric bacteria.
- *E. coli* has a typical gram-negative cell wall structure with an outer lipopolysaccharide (LPS) membrane.
- LPS consists of an inner lipid A component, a common core region of polysaccharides, and an outer variable region of polysaccharides.
- The infection is endogenous, the source of the bacteria being the fecal flora.
- Organisms gain entry to the bladder from the perineum and in the presence of vesico-ureteric reflux may infect the renal parenchyma.
- UTI is more common in females because of the short urethra.

2. WHAT IS THE HOST RESPONSE TO THE INFECTION AND WHAT IS THE DISEASE PATHOGENESIS?

- UPEC have several adhesins that bind to uroepithelial cells.
- UPEC produce a number of toxins that increase desquamation of uroepithelial cells and are also cytotoxic.
- UPEC are protected from the innate host defense system by the K1 capsular antigen, which is antiphagocytic.
- In the renal tract, the immune defenses are relatively inactive due to the osmolarity and the concentration of ammonium ions.
- Binding of UPEC to uroepithelial cells stimulates the production of IL-8 and the recruitment of granulocytes to the renal tract.
- The role of the adaptive immune system in protection is uncertain although IgA and IgG are produced.

3. WHAT IS THE TYPICAL CLINICAL PRESENTATION AND WHAT COMPLICATIONS CAN OCCUR?

- Infection may be asymptomatic.
- In pregnant women and children, asymptomatic infection can have serious consequences.
- Cystitis presents with frequency of micturition, dysuria, and suprapubic pain.
- Pyelonephritis presents with loin pain, fever, rigors, frequency of micturition, and dysuria.
- Neonates have a nonspecific presentation with fever, vomiting, and failure to thrive.
- Elderly patients may have fever, incontinence, and dementia.
- Complications are septicemia, renal scarring, renal abscess, and renal failure.

4. HOW IS THE DISEASE DIAGNOSED, AND WHAT IS THE DIFFERENTIAL DIAGNOSIS?

- Microscopy is generally unhelpful although a sterile pyuria may indicate renal tuberculosis.
- Laboratory diagnosis is by culture.
- A significant bacteriuria is a single species in numbers greater than 10^5 organisms per ml of urine.
- Differential diagnosis includes musculoskeletal injury, renal vein thrombosis, urolithiasis.

5. HOW IS THE DISEASE MANAGED AND PREVENTED?

- Asymptomatic bacteriuria in pregnant women or children should be treated.
- Antibiotic use depends upon the local resistance pattern.
- Three days treatment with an appropriate antibiotic is usually sufficient for uncomplicated cystitis.
- Two weeks of treatment with the same antibiotic is required for pyelonephritis.

FURTHER READING

Murphy K, Weaver C. Janeway's Immunobiology, 9th edition. Garland Science, New York/London, 2016.

Rogers MC, Peterson ND. E. coli Infections: Causes, Treatment & Prevention. Nova Science, 2011.

Samie, A. Escherichia coli – Recent Advances on Physiology, Pathogenesis and Biotechnological Applications. In Tech Open, 2017.

Zaslau S. Bluprints Urology. Blackwell Publishing, Oxford, 2004.

REFERENCES

Barnhart MM, Chapman MR. Curli Biogenesis and Function. Annu Rev Microbiol, 60: 131–147, 2006.

Browne P, Ki M, Foxman B. Acute Pyelonephritis Among Adults: Cost of Illness and Considerations for the Economic Evaluation of Therapy. Pharmacoeconomics, 23: 1123–1142, 2005.

Katchman EA, Milo G, Paul M, et al. Three Day Versus Longer Duration of Antibiotic Treatment for Cystitis in Women: Systematic Review and Metaanalysis. Am J Med, 118: 1196–1207, 2005.

Kau AL, Hunstad DA, Hultgren SJ. Interaction of Uropathogenic Escherichia coli with Host Uroepithelium. Curr Opin Microbiol, 8: 54–59, 2005.

Larcombe J. Urinary Tract Infection in Children. Clin Evid, 14: 429–440, 2005.

MacDonald RA, Levitin H, Mallory GK, Kass EH. Relation Between Pyelonephritis and Bacterial Counts in the Urine. N Engl J Med, 256: 915–922, 1957.

Neumann I, Fernanda Rojas M, Moore P. Pyelonephritis in Non-Pregnant Women. Clin Evid, 14: 2352–2357, 2005.

Sheffield JS, Cunningham FG. Urinary Tract Infection in Women. Obstet Gynecol, 106: 1085–1092, 2005.

WEBSITES

American Urological Association: http://www.auanet.org/

Centers for Disease Control and Prevention, E. coli (Escherichia coli): https://www.cdc.gov/ecoli/index.html

EcoCyc, E. coli Database: https://ecocyc.org

European Association of Urology: http://www.uroweb.org/

Medscape, Chronic Pyelonephritis, 2019: http://www.emedicine.com/med/topic2841.htm

Todar's Online Textbook of Bacteriology, Pathogenic *E. coli* (page 1), © Kenneth Todar, 2008: http://textbookofbacteriology.net/e.coli.html

Students can test their knowledge of this case study by visiting the Instructor and Student Resources: [www.routledge.com/cw/lydyard] where several multiple choice questions can be found.

Giardia intestinalis

12

A 24-year-old man went on a 3-month backpacking trip across India. He drank bottled water and reportedly ate well-cooked food in hotels and restaurants. While in India, his stools were looser than normal. In the week before his return he developed frequent watery, non-bloody diarrhea. This settled enough for him to fly home. He immediately went to his doctor and a stool culture grew *Campylobacter*. His bowels improved over 10 days without treatment but 2 weeks after his return he developed more diarrhea, with loss of appetite, bloating, and flatulence. For the first time, his stools failed to flush away completely in the toilet and were particularly offensive in smell. He began to lose weight. His doctor requested three stool specimens for culture and also microscopy for ova, cysts, and parasites. One out of three specimens contained *Giardia* cysts. He was treated with a course of metronidazole and his symptoms improved.

1. WHAT IS THE CAUSATIVE AGENT, HOW DOES IT ENTER THE BODY, AND HOW DOES IT SPREAD A) WITHIN THE BODY AND B) FROM PERSON TO PERSON?

CAUSATIVE AGENT

Giardia is a protozoan consisting of six species that infect mammals, amphibians, and birds. The only species that infects humans is *Giarda intestinalis* (alt. *lamblia* and *duodenalis*). It has two stages to its life cycle (Figure 12.1): (a) a trophozoite (feeding and pathology-causing stage) that is flagellated (with four pairs of flagellae), pear-shaped, with two nuclei, a ventral "sucking" disk, and median bodies. It also has a rigid cytoskeleton composed of microtubules and microribbons. It measures 9–21 µm long by 5–15 µm wide; (b) a cyst, with a highly resistant wall that enables it to remain viable outside the body of the host for long periods. The cyst is smooth-walled and oval in shape, measuring 8–12 µm long by 7–10 µm wide. The genome of the parasite has five chromosomes of different sizes and variation of the number of genome equivalents (ploidy) occurs during the life cycle: trophozoite >cyst>excyzoite>trophozoite. The genome ploidy is important in gene regulation and differentiation. *Giardia* is the oldest known extant eukaryote, having prokaryotic characteristics: no mitochondria and metabolism similar to prokaryotes.

G. *intestinalis* is a multispecies complex with at least eight recognized assemblages or genotypes (A–H) based on a number of clinical samples. The two genotypes A and B are the major human pathogens. Molecular analysis of 2800 samples indicates that genotype B accounts for around 58% giardiasis cases with genotype A found in around 37% of cases worldwide. There are, however, differences in percentages of these in different countries. For example in Spain, there were 27.4% of A and 72% of B recorded in Madrid. A and B genotypes were equally represented in Rioja. In Spain, children also appear to be more commonly infected with the B phenotype than adults.

ENTRY INTO THE BODY

Cysts are ingested from contaminated water or food and having passed through the stomach, begin to open up at one end (excystation), releasing trophozoites into the intestine lumen, which then divide within 12 hours (Figure 12.1). They settle in the small intestine (predominantly in the mid-jejunum). A minimum of 10–25 cysts are necessary to produce an infection. The trophozoites attach to the intestinal wall through their ventral "sucking" disk and feed on nutrients (Figure 12.2). They increase in number by binary fission and colonize large areas of epithelial surface causing diarrhea and damage to the epithelium (see Section 2). At regular intervals and following detachment and movement down toward the colon (and probably exposure to biliary secretions), some of the trophozoites become encysted (encystation), with each trophozoite forming a single cyst. Both trophozoites and cysts pass out of the body in the feces.

SPREAD WITHIN THE BODY

Penetration of the epithelial surface by the trophozoites is very rare, as is migration of the trophozoites to systemic sites. Invasion of the gallbladder, pancreas, and urinary tract have been reported but the trophozoites normally remain in the intestine/colon and do not cross the mucosal barrier.

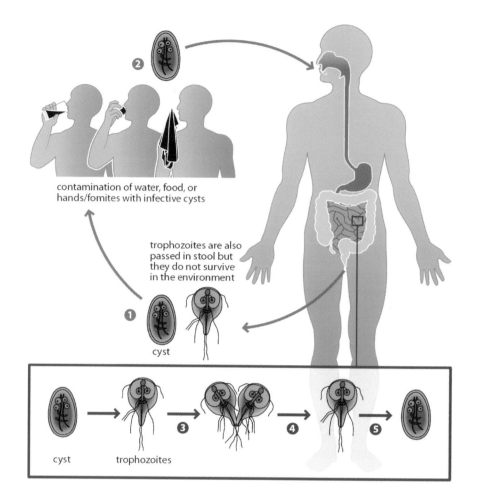

Figure 12.1 The life cycle of *G. intestinalis*. Both cysts and trophozoites are found in feces (1). The cysts are hardy and can survive 2–3 months or more in cold water. Cysts in contaminated water, food, or by the fecal-oral route (hands or fomites) cause infection (2). In the small intestine, the cysts give rise to trophozoites (each cyst producing two trophozoites) (3). The trophozoites multiply by binary fission and remain in the lumen of the small bowel where they can be free in the mucus or attached to the epithelial cells by their ventral sucking disk (4). Trophozoites encyst on transit toward the colon. The cyst is the stage found most commonly in non-diarrheal feces (5). The cysts are infectious when passed in the stool or shortly afterward and, if ingested by another person, the cycle begins again. *Courtesy of the Centers for Disease Control, Atlanta, Georgia. Image is found in the Public Health Image Library #3394. Additional photo credit is given to Alexander J. da Silva, PhD, and Melanie Moser who created the image in 2002.*

Within the figure the labels read: "contamination of water, food, or hands/fomites with infective cysts"; "trophozoites are also passed in stool but they do not survive in the environment"; "cyst"; "cyst"; "trophozoites".

PERSON-TO-PERSON SPREAD

Cysts can remain dormant for up to 3 months in cold water. Spread is through ingestion of contaminated food and also via the **fecal–oral** route (hands and **fomites**), although water is probably the main source. Contamination of public-drinking supplies has led to giardiasis epidemics. When children become infected, up to 25% of their family members also become infected. Individuals can shed cysts in their feces and remain symptom-free but are an important source of person-to-person transmission (see Section 3). Sexual transmission of *Giardia* has been described in men who have sex with men. *Giardia* is found in a wide variety of different animal species and has been regarded as a **zoonosis**, although there is little evidence for animals being a significant source of human giardiasis.

EPIDEMIOLOGY

Giardiasis is one of the most common water-borne diseases that infect humans and the most common enteric protozoal infection worldwide, with 280 million cases a year and up to 33% of individuals infected. Prevalence rates vary from 4% to 42%. It affects nearly 2% of adults and 8% of children in developed countries. Infection is linked to poor hygiene and sanitation and is more prevalent in warm climates.

The prevalence within the US is estimated to be roughly 1.2 million, with the majority of cases not identified due to the carrier being asymptomatic. In 2012, the CDC reported 15 223 cases. The most affected are children 0 to 4 years of age, with the largest percentage of cases being in the northwest US. Infection is most common in late summer and early fall due to outdoor water activities.

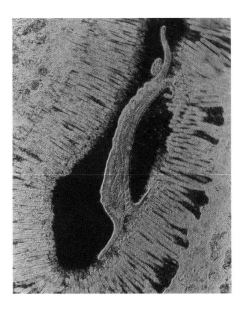

Figure 12.2 *G. intestinalis* **attached to microvilli in the small intestine.** This colored TEM (transmission electron micrograph) shows a *G. intestinalis* trophozoite attached by means of its ventral sucking disks to microvilli in the human small intestine. *From CNRI / Science Photo Library, with permission.*

Water-borne outbreaks appear to be the most common source of infection. In Canada, in 2016, there were 3818 reported cases of giardiasis with a similar seasonal variation to that seen in the US. In the Western world, giardiasis is more likely to be diagnosed as a cause of diarrhea that occurs or persists after travel to a developing country. This is due to its relatively long incubation period and persistent symptoms. Thus, the organism is a cause of "traveler's diarrhea", also called "backpacker's diarrhea" and "beavers' fever" (since it was originally believed to be transmitted from beavers to man).

It is not unusual for outbreaks of *Giardia* to occur on cruise ships.

2. WHAT IS THE HOST RESPONSE TO THE INFECTION AND WHAT IS THE DISEASE PATHOGENESIS?

In a study of prison "volunteers" in the 1950s given the same infectious dose, 50% of subjects developed asymptomatic infections, 35% self-limiting symptomatic infections, and 15% troublesome persistent diarrhea. Given that they received the same dose and strain this illustrates the impact of host susceptibility and resistance. However, host defenses against *G. intestinalis* are not well characterized but they are believed to involve both non-immunologic mucosal processes and immune mechanisms.

HOST DEFENSES

Non-Immunologic Defenses

Intestinal epithelial cells are shed and replaced every 3–5 days and therefore the trophozoites need to constantly detach and reattach to new epithelial surfaces. The mucus produced by the goblet cells impedes access of the trophozoites to the epithelium. It has been proved that certain commensal bacterial species enable mice to be resistant against *Giardia* colonization suggesting that differences in microbiotic composition between individuals and species could explain the variability in pathology and susceptibility to infection (see later).

IMMUNE RESPONSES

There appears to be little or no mucosal inflammation in human *Giardia* infection, which indicates that local defense must be occurring without systemic recruitment. Most of the data on immune responses to *Giardia* come from experimental animal models.

INNATE IMMUNITY

Antimicrobial peptides such as **defensins** and **cathelicidins** secreted by intestinal epithelial cells have anti-giardial activity *in vitro* and may have activity *in vivo*. Nitric oxide (NO) produced by gut epithelial cells inhibits growth, encystation, and excystation *in vitro*, but has no effect on viability. However, *Giardia* has developed strategies to evade this host defense mechanism. Trophozoites down-regulate expression of iNOS in intestinal epithelial cells. This probably leads to the reduced expression of iNOS reported in pediatric patients infected with *Giardia*. Although monocytes/macrophages and polymorphs can kill trophozoites *in vitro* by oxidative mechanisms, very few are found in the human intestinal lumen during infection. Mast cells are currently recognized as effector cells of the immune response against several parasites. Interestingly, in tissues from the small intestine that have been infected with *G. intestinalis*, the most strongly induced transcripts are mast-cell proteases. Mast cells and NO could act together to induce peristalsis. The maturation and activation of dendritic cells (DCs) can be induced through *Giardia* lysates, excretory secretory products, and other specific proteins. This is seen as an increase in pro-inflammatory cytokines, for example IL-6, TNFα and IL-12 and of surface molecules such as CD80, CD86, and MHC class II. *Giardia* may activate the maturation and migration of DCs to the site of infection with their subsequent release of immunomodulatory cytokines. The production of IL6 by DCs seems to be of major importance in the clearance of *Giardia*.

ADAPTIVE IMMUNITY

T-Cell Responses

IL-6 produced by DC is important for T-cell differentiation and may determine the type of T-cell response that ensues. In both humans and animals, *Giardia* infection induces a strong protective adaptive immune response, in which the main player is CD4+ T lymphocytes. This includes Th-1, Th-2, and Th-17 cells. Recent findings in a murine model of *Giardia* has shown local elevations in the ratio between Th-17 and Treg lymphocytes within the small intestinal lamina propria and Peyer's patches related to their increased resistance to infection. Interestingly, duodenal mucosal lymphocyte alterations are maintained for many months after the onset of infection in human patients. Th-17 cells producing IL-17 have been found in the blood of humans infected with *Giardia*. The evidence is clear that CD4+ T cells are important protection against *Giardia* infection and specific depletion of **CD4**+ T cells results in the development of chronic giardiasis. CD8+ T cells appear to play no role in host protection against *Giardia*.

Antibody Responses

Although *Giardia* trophozoites do not usually cross the mucosal barrier, some antigenic *Giardia* products antigens must penetrate the barrier. This might be through the M cells of Peyer's patches or the DCs with intraepithelial processes into the gut lumen. Antibodies to *Giardia* are found in both mucosal secretions and serum and it has been established that antibodies, particularly of the IgA isotype, contribute to the maintenance of protective immunity against giardiasis. Anti-*Giardia* secretory IgA can be detected in human saliva and breast milk. Antibodies of up to 16 immunogenic proteins in infected patients have been identified by serum IgG. These include variant-specific surface proteins (VSP: see below), cytoskeletal proteins unique to *Giardia* (α- and β-tubulin, α- and β-giardin), and enzymes (e.g. arginine deaminase, orthinine carbamoyl transferase, and enolase) are found (see below). These enzymes have been found to be induced following contact of the parasite with the intestinal epithelial cells. **IgG** antibodies to *Giardia* have been shown to kill *Giardia* trophozoites *in vitro* through complement. Although it is unlikely that this mechanism could occur in the intestinal lumen, it might be one explanation as to why *Giardia* does not invade.

GIARDIA-SPECIFIC ANTIGENS AND ANTIGENIC VARIATION

Each trophozoite expresses only one of many highly immunogenic variant-specific surface proteins (VSP) on *Giardia*. The trophozoites being able to switch the VSPs in their surface coats. The mechanism of this variation is believed to be through interference RNA (iRNA) and mRNA. Disruption of this pathway generates trophozoites that simultaneously express numerous VSP. It is likely that the variation is driven by antibody. This **antigenic variation** (at least in mice) is

thought to be a mechanism whereby the trophozoites can avoid the immune system. An alternative, but not mutually exclusive, biological explanation for antigenic variation is their adaptation to different intestinal environments. There is some evidence for this possibility in that different VSPs have different susceptibility to proteases.

Interestingly, the repertoire of VSP antigens is much smaller than that seen in trypanosomes (see Case 37) and the mechanisms leading to their "switching" are different.

PATHOGENESIS

The mechanisms by which giardiasis causes diarrhea and malabsorption are incompletely understood. Several have been postulated and include damage to the endothelial brush border, enterotoxins, immunologic reactions, changes in gut motility, and fluid hypersecretion via increased adenylate cyclase activity. Direct attachment of trophozoites to the epithelium has been demonstrated to cause increased epithelial permeability. *Giardia*-induced loss of intestinal brush border surface area, villus flattening, inhibition of disaccharidase activities, and eventual overgrowth of enteric bacterial flora appear to be involved in the pathophysiology of giardiasis but have yet to be causatively linked to the disease's clinical manifestations. There is little evidence of any exotoxins producing epithelial damage. Regarding the damage to the brush border, which provides a large surface area for absorption, biopsies from only 3% of patients with infection showed villus shortening and there was little inflammation. In experimental infection of 10 human volunteers with *Giardia* type B genotype, only five individuals developed symptoms and only two of these showed any change to the brush border. Thus, microvillus shortening and inflammation are not directly correlated with the symptoms and indeed clearance of the organism from the intestinal tract. From *in vitro* studies, there is some evidence for *Giardia* inducing a change in the cytoskeleton of human duodenal cells with increased **apoptosis** and disruption of tight junctions in monolayers of intestinal cells. Cysteine proteases secreted by *G. intestinalis* may disrupt intestinal epithelial cell junctional complexes and degrade chemokines. Although disruption of tight junctions has not been confirmed by clinical observation, there is evidence for a correlation of infection with impairment of both absorption and digestive functions. In fact, there are varying degrees of malabsorption of sugars (e.g. xylose, disaccharides), fats, and fat-soluble vitamins (e.g. vitamins A and E) but these might contribute to substantial weight loss. Although CD8+ T cells do not contribute to protection against *Giardia*, there is some suggestion that they are involved in enterocytic damage.

Giardia infections can produce symptoms that persist long after infection although, again, the mechanisms for this are unclear.

Recent research has emphasized the importance of the intestinal microbiota in *Giardia* pathogenicity. Conventional, germ-free, or germ-free mice that were reconstituted with duodenal microbes from patients with symptomatic giardiasis, were infected with *G. lamblia* trophozoites. The infected non-reconstituted conventional mice showed the intestinal pathology among the three groups, with the reconstituted germ-free mice showing moderate pathology. The infected germ-free mice, however, did not develop intestinal pathology compared with the other groups, suggesting a requirement for intestinal microbes to stimulate pathology.

Colonization by *Giardia* in mice causes a dysbiosis with an increase of the *Proteobacteria* and a decrease in the *Firmicutes*. Segmented filamentous bacteria (SFB), found in many animals including humans, are present in the small intestine near the terminal ileum and are responsible for stimulating IL-17 production. Resistance to *Giardia* in mice occurs with a high intestinal density of SFBs and susceptibility with a low density. In humans, SFBs are scarce in neonates, who are prone to infection, and are more common in adults. A reduction in *Clostridia* has also been noted in infected animals and the presence of the organism is linked to effective Treg responses.

3. WHAT IS THE TYPICAL CLINICAL PRESENTATION AND WHAT COMPLICATIONS CAN OCCUR?

Most persons infected with *G. intestinalis*, on a global basis, are either asymptomatic or minimally symptomatic. The clinical effects of *Giardia* infection range from asymptomatic carrier status to severe malabsorption (see Section 2). Factors contributing to the variations in presentation include the **virulence** of particular *Giardia* strains (see Section 2), their genotype (A or B), the numbers of cysts ingested, the age of the host, and the state of the immune system (see later). Asymptomatic carriers often have a large number of cysts in their stools.

If symptoms are present, they occur about 1–3 weeks after ingestion of the parasite. These include:

- watery offensive-smelling diarrhea with abdominal cramps;
- severe flatulence;
- nausea with or without vomiting;
- fatigue;
- and possibly fever.

A slower onset may occur with development of yellowish loose, soft and foul-smelling stools – often floating due to the high lipid content. Stools may be watery or even constipation can occur. Initial symptoms usually last 3–4 days or can become chronic leading to recurrent symptoms, severe malabsorption and debilitation may occur. Other symptoms include anorexia, malaise, and weight loss.

Children with malabsorption associated with *Giardia* infection often show failure to thrive and **protein-losing enteropathy** can be a complication leading to stunted growth of children, commonly seen in Africa. Reduced uptake of lipids

across the gut epithelium causes deficiency in lipid-soluble vitamins, which is an additional problem for children.

Poor nutrition can also contribute to an increased risk of a person having symptoms with the infection. More serious infections, which can lead to death, are seen in people with a weakened immune system, such as patients with HIV/AIDS, cancer, transplant patients, and the elderly. *Helicobacter pylori* may predispose to *Giardia* due to **hypochlorhydria**. Patients may develop lactose intolerance.

Giardia infection has also been associated with development of post-infectious complications including irritable bowel syndrome and chronic fatigue syndrome. Changes in the intestinal microbiota profile in children with *Giardia* might contribute to some of these complications.

4. HOW IS THE DISEASE DIAGNOSED, AND WHAT IS THE DIFFERENTIAL DIAGNOSIS?

Clinical diagnosis is often difficult because the same symptoms can occur with a number of intestinal parasites. Giardiasis is therefore diagnosed by the identification of cysts (Figure 12.3) or trophozoites (Figure 12.4) in the feces, and this is still regarded as the gold standard method. Due

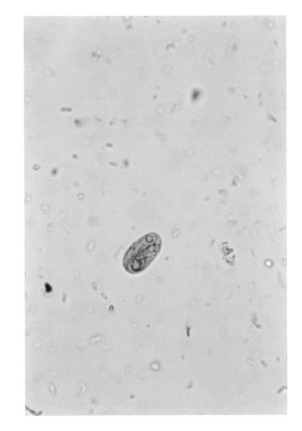

Figure 12.3 *G. intestinalis* **cyst in a wet mount stained with iodine.**
Courtesy of the Centers for Disease Control, Atlanta, Georgia. Image is found in the Public Health Image Library #3741. Additional photo credit is given to Dr Mae Melvin who took the photo in 1977.

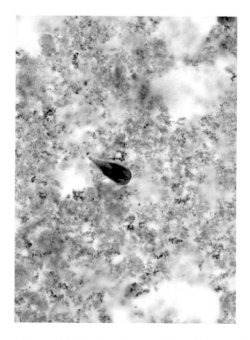

Figure 12.4 *G. lamblia* **trophozoite stained with trichrome.** *Courtesy of the Centers for Disease Control, Atlanta, Georgia. Image is found in the Public Health Image Library #7833. Additional photo credit is given to DPDX / Melanie Moser who created the original image.*

to the intermittent and low levels of cysts, the fecal samples are usually concentrated before direct examination of wet mounts under the microscope. A number of methods have been used including formalin-ether and sucrose gradients for concentration. Samples can be stained with iodine, methylene blue or trichrome.

Alternate methods for detection of the parasite in the stool samples include antigen detection tests by **ELISA** and by a direct fluorescence assay (DFA) (Figure 12.5). Commercial kits for both of these are available.

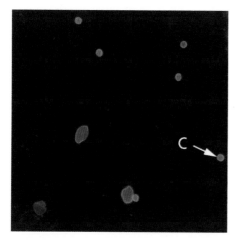

Figure 12.5 Identification of cysts of *G. intestinalis* by fluorescent-labeled *Giardia* antibodies. A formalin-fixed preparation stained with commercially available fluorescent antibodies to *Giardia* and visualized under a fluorescence microscope. Cysts of *Giardia* are seen as large green ovoid objects (labeled C). Oocysts of *Corynebacterium parvum* are also seen in this preparation. *Courtesy of the Centers for Disease Control, Atlanta, Georgia. Image is found in the Laboratory Identification of Parasites of Public Health Concern.*

A "string" test (entero-test) can also be performed. This involves swallowing a weighted gelatine capsule on a piece of string. After the gelatine dissolves in the stomach, the weight carries the string into the duodenum. The string is left for 4–6 hours or overnight while the patient is fasting and then examined for bilious staining. This indicates successful passage into the duodenum and mucus from the string can be examined for trophozoites after fixation and staining.

The **polymerase chain reaction** (**PCR**) detection of *Giardia* DNA is often restricted to laboratory use and mostly for subtyping of *G. intestinalis* (see Section 1).

A duodenal biopsy can be taken and this may be the most sensitive test. This is often taken in cases of unexplained diarrhea.

DIFFERENTIAL DIAGNOSIS

Other causes of gastroenteritis need to be considered including amebiasis, bacterial overgrowth syndromes, Crohn ileitis, *Cryptosporidium* enteritis, irritable bowel syndrome, celiac sprue, tropical sprue, strongyloidiasis, viral gastroenteritis, and lactose intolerance.

5. HOW IS THE DISEASE MANAGED AND PREVENTED?

MANAGEMENT

With mild infection, giardiasis can resolve in 6 weeks or so. However, there are several drugs used in the treatment of giardiasis. They include:

1. Nitroimidazole derivatives (metronidazole (Flagyl®), tinidazole, secnidazole, ornidazole,): most frequently used first-line drugs, especially metronidazole.

2. Benzimidazoles (albendazole and mebendazole): these widely used anti-helminthic drugs are used to treat giardiasis but with variable efficacy.

3. Nitrofuran derivatives (furazolidone): these have been reported to have high efficacy in first-line therapy.

4. Acridine compounds (mepacrine and quinacrine): mechanism of action unknown but has been shown to be efficient in therapy.

5. Aminoglycoside: the oral aminoglycoside paromomycin is the drug of choice for pregnant women as it is poorly absorbed and has no systemic effects.

6. Nitazoxanide: in controlled studies efficacies of between 44% and 91% have been reported.

Different countries may have a preference for the use of different types of drugs. In addition, from a worldwide perspective, albendazole (a benzimidazole compound) is used, which has a much broader range of action than metronidazole and the other agents listed. It kills *Giardia* very well, but also

Entamoeba, Ascaris, Enterobius, and hookworms, and can do this in one single dose. In developing countries, a single-dose albendazole is being given to schoolchildren and has been associated with improved school attendance and educational attainment. They feel better for being cleared of protozoa and helminths.

TREATMENT OF REFRACTORY GIARDIASIS

In general, efficacy of treatment with nitroimidazoles is 90%. However, nitroimidazole failure (especially with metronidazole) has been reported in up to 50% giardiasis cases in both travelers and in high endemic countries. A number of studies have indicated the use of a different class of drug, combination therapy or repeated courses with increased dose/duration of the same drug can be used with variable efficacy.

Pregnant Patients

Treatment of pregnant patients with *Giardia* is difficult because of the potential adverse effects of anti-*Giardia* agents on the fetus. If possible, drug treatment should be avoided during the first trimester. Mildly symptomatic women should have their treatment delayed until after delivery. If left untreated, however, adequate nutrition and hydration maintenance is important. Paromomycin is now the only anti-*Giardia* drug considered completely safe during early pregnancy (see different drugs used to treat *Giardia* above).

PREVENTION

* Good hygiene is very important.
* Contaminated water should be avoided: untreated water should not be consumed. Outbreaks of giardiasis in developed countries are often traced back to breakdown in filtration systems of drinking water supplies.
* Individuals traveling to warm climates where Giardia is found should take extra care with drinking water and consumption of raw food – boil drinking water, and so forth.
* There is no known chemoprophylaxis for humans.
* There is no vaccine for humans yet although there is an effective killed vaccine for dogs (Giardiavax®).

SUMMARY

1. WHAT IS THE CAUSATIVE AGENT, HOW DOES IT ENTER THE BODY, AND HOW DOES IT SPREAD A) WITHIN THE BODY AND B) FROM PERSON TO PERSON?

* *Giardia* is a protozoan flagellate. It has two stages – a trophozoite and a cyst with a highly resistant wall.
* *Giardia* consists of six species. Only one species infects humans and this is variously referred to as *G. intestinalis, lamblia* or *duodenalis*. There are at least eight genotypes or assemblages (A–H). Only A and B are human pathogens, with B being the most frequent globally. Other genotypes are seen in other mammals and birds.
* The main infectious stage is the cyst; cysts are ingested in contaminated water or food. They lose their cell wall in the duodenum and emerge as trophozoites, which attach to the intestinal wall through their ventral "sucking" disk and feed. They colonize large areas of epithelial surface. They rarely invade the epithelium and spread systemically. They become encysted again and both trophozoites and cysts pass out of the body in stools.
* Contamination of public drinking supplies has led to giardiasis epidemics. When children become infected, up to 25% of their family members also become infected. Individuals can shed cysts in their feces and remain symptom-free. Sexual transmission of *Giardia* has been described in homosexual males.
* *Giardia* is found in a wide variety of different animal species and has been regarded as a zoonosis, although there is little evidence for animals being a significant source of human giardiasis.
* Giardiasis is one of the most common causes of diarrhea worldwide. There are around 280 million cases a year. It is more commonly found in children where, in developing countries, it is estimated that up to 20% are infected.
* *Giardia* is widespread in the US, with a prevalence of an estimated 1.2 million with the majority not identified due to the carrier being asymptomatic. Infection is most common in late summer and early fall due to outdoor water activities. Similar seasonal variation is seen in Canada.
* In the Western world, giardiasis is a cause of diarrhea that occurs or persists after travel to a developing country – "traveler's diarrhea", "backpacker's diarrhea", and "beavers' fever".

2. WHAT IS THE HOST RESPONSE TO THE INFECTION AND WHAT IS THE DISEASE PATHOGENESIS?

* The host defenses are clearly effective since individuals infected with *Giardia* are often asymptomatic and some are able to clear the organism without treatment.
* Host defenses are not well characterized but include both non-immunologic and immunologic mechanisms.
* Mucus prevents immediate access of trophozoites. The intestinal microbiota may also play a role in preventing attachment/inhibiting proliferation.
* Antimicrobial peptides such as defensins and cathelicidins secreted by intestinal epithelial cells have anti-giardial activity *in vitro* and may have activity *in vivo*. Monocytes/macrophages and polymorphs can kill trophozoites *in vitro* by oxidative mechanisms but very few are found in the human intestinal lumen during infection IL-6 produced by dendritic cells appears to be of major importance in the clearance of *Giardia*.

Continued...

Case Studies in Infectious Disease

Continued...

- Most information on immune responses to *Giardia* infections comes from animal models. CD4+ T cells including Th-1, Th-2, and Th-17 appear to play a role in *Giardia* infections in mice but their exact protective role in humans is unclear. CD8+ T cells do not play a role.
- Antibody responses to human *Giardia* do have some protective role. Specific IgA antibodies in human saliva and breast milk can protect children against infection in early life. Immunodominant antigens include VSPs, cytoskeletal structures, giardin, and enzymes. Serum antibodies of IgG class to 16 immunogenic proteins of *Giardia* have been seen in humans.
- *Giardia* shows antigenic variation with each trophozoite expressing one VSP expressed and switching of these probably providing escape from the immune system.
- The mechanisms by which giardiasis causes diarrhea and malabsorption are unclear but have been postulated that include damage to the endothelial brush border, enterotoxins, immunologic reactions, changes in gut motility, and fluid hypersecretion via increased adenylate cyclase activity. Malabsorption of sugars (e.g. xylose, disaccharides), fats, and fat-soluble vitamins (e.g. vitamins A and E) might contribute to substantial weight loss.

3. WHAT IS THE TYPICAL CLINICAL PRESENTATION AND WHAT COMPLICATIONS CAN OCCUR?

- The clinical effects of *Giardia* infection range from asymptomatic carrier status to severe malabsorption.
- Factors contributing to the variations in presentation include the virulence of particular *Giardia* strains, their genotype (A or B), the numbers of cysts ingested, the age of the host, and the state of the immune system. Carriers often have a large number of cysts in their stools.
- If symptoms are present, they occur about 1–3 weeks after ingestion of the parasite. These include: watery diarrhea with abdominal cramps, severe flatulence, nausea with or without vomiting, fatigue, and possibly fever.
- Infection can become chronic leading to recurrent symptoms, severe malabsorption, and debilitation. Other symptoms include anorexia, malaise, and weight loss. Children with malabsorption syndrome often show failure to thrive. Patients may develop lactose intolerance.

- Patients with a weakened immune system such as patients with HIV/AIDS, cancer, transplant patients or the elderly can develop more severe infections.
- Infection has been associated with development of post-infectious complications including irritable bowel syndrome and chronic fatigue syndrome.

4. HOW IS THE DISEASE DIAGNOSED, AND WHAT IS THE DIFFERENTIAL DIAGNOSIS?

- Giardiasis is diagnosed by the identification of cysts or trophozoites in the feces which is still the "gold standard". Direct mounts for microscopy as well as concentration procedures may be used. Samples can be stained with iodine or trichrome.
- Commercial kits are available for antigen detection tests by ELISA and by immunofluorescence. PCR testing for *Giardia* DNA can also be used but is usually restricted to laboratory use.
- A "string" test (entero-test) can also be performed.
- Differential diagnosis for other causes of gastroenteritis includes amebiasis, bacterial overgrowth syndromes, Crohn ileitis, *Cryptosporidium* enteritis, irritable bowel syndrome, celiac sprue, and tropical sprue.

5. HOW IS THE DISEASE MANAGED AND PREVENTED?

- Drugs used: nitroimidazole derivatives (e.g. metronidazole, tinidazole, ornidazole, and secnidazole); benzidamazoles (albendazole and mebendazole) nitrofuran derivatives (e.g. furazolidone), acridine compounds (e.g. mepacrine and quinacrine) aminoglycoside (especially for pregnant women) and nitazoxanide.
- Metronidazole is the most common antibiotic treatment for giardiasis but more and more refractory cases are being described.
- Aminoglycoside is the drug of choice for pregnant patients since it avoids the potential adverse effects of the other anti-*Giardia* agents on the fetus.
- Prevention should include: good hygiene, avoidance of contaminated food and water, extra care during traveling to warm climates, boiling water, etc.
- No known chemoprophylaxis and no human vaccine as yet. There is an effective vaccine for dogs.

FURTHER READING

Goering R, Dockrell HM, Zuckerman M, Chiodini PL. Mims' Medical Microbiology and Immunology, 6th edition. Elsevier, Philadelphia, 2018.

Murphy K, Weaver C. Janeway's Immunobiology, 9th edition. Garland Science, New York/London, 2016.

REFERENCES

Adam RD. Giardia duodenalis: Biology and Pathogenesis. Clin Microbiol Rev, 34: e00024–19, 2021.

Buret AG, Caccio SM, Favennec L, Svard S. Update on Giardia: Highlights from the Seventh International Giardia and Cryptosporidium Conference. Parasite, 27: 49, 2020.

Hooshyar H, Rostamkhani P, Arbabi M, Delavari M. Giardia lamblia Infection: Review of Current Diagnosis Strategies. Gasteroenterol Hepatol Bed Bench, 12: 3–12, 2019.

Lopez-Romero G, Quintero J, Astiazaran-Garcia H, Velazquez C. Host Defences Against Giardia lamblia. Parasit Immunol, 37: 394–406, 2015.

Maertens B, Gagnaire A, Paerewijck O, et al. Regulatory Role of the Intestinal Microbiota in the Immune Response Against Giardia. Sci Rep, 11: 10601, 2021.

Morch K, Hanvik K. Giardia Treatment: an Update with a Focus on Refractory Disease. Curr Opin Infect Dis, 33: 355–364, 2020.

Touz MC, Feliziani C, Ropolo AS. Membrane-Associated Proteins in Giardia lamblia. Genes, 9: 404, 2018.

Wang Y, Gonzalez-Moreno O, Roellig DM, et al. Epidemiological Distribution of Genotypes of Giardia duodenalis in Humans in Spain. Parasit Vectors, 12: 432–442, 2019.

WEBSITES

Centers for Disease Control and Prevention, Parasites – Giardia: https://www.cdc.gov/parasites/giardia/index.html

European Centre for Disease Prevention and Control, Giardiasis: https://www.ecdc.europa.eu/en/giardiasis/facts

WebMD, Giardiasis, 2020: https://www.webmd.com/digestive-disorders/giardiasis-overview

Students can test their knowledge of this case study by visiting the Instructor and Student Resources: [www.routledge.com/cw/lydyard] where several multiple choice questions can be found.

Helicobacter pylori

13

A 50-year-old advertising executive consulted his primary healthcare provider because of tiredness, lethargy, and an abdominal pain centered around the lower end of his sternum, which woke him in the early hours of the morning. The pain was relieved by food and antacids. His uncle had died of stomach cancer and he was worried that he had the same illness.

On examination, his doctor noted that he seemed a bit pale and that he had a **tachycardia**. His blood pressure was low. He was slightly tender in his upper abdomen but there was no guarding or rebound tenderness.

The doctor took blood and feces samples and organized for an upper gastrointestinal (GI) **endoscopy**. The full blood count showed a hypochromic normocytic **anemia** with a hemoglobin of 8.9 consistent with iron-deficiency anemia. The gastroscopy showed a 3 cm ulcer in the pre-pyloric region of the stomach (Figure 13.1). The fecal antigen test for *Helicobacter pylori* was positive. The patient was started on routine treatment for a duodenal ulcer.

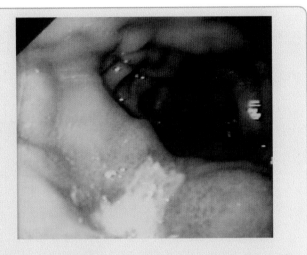

Figure 13.1 Gastroscopy showing a duodenal ulcer in the prepyloric region of the stomach. *Courtesy of Dino Varia, First Medical Clinical University of Bologna, Italy.*

1. WHAT IS THE CAUSATIVE AGENT, HOW DOES IT ENTER THE BODY AND HOW DOES IT SPREAD A) WITHIN THE BODY AND B) FROM PERSON TO PERSON?

CAUSATIVE AGENT

Helicobacter pylori is a nonspore-forming, motile gram-negative bacterium with a helical shape measuring 2.5–4.5 × 0.5–1.0 μm. It has one to five unipolar sheathed flagellae. The organism has a high rate of polymorphism. The genome has at least five regions that it may have acquired from other organisms by lateral transfer of DNA. These are called "**pathogenicity islands**" (PAIs) and carry **virulence** genes. PAIs are found in many other bacteria. In *H. pylori* the cag-PAI (Figure 13.2) comprises about 30 genes that are responsible for the production of a **type IV secretion system** (T4SS) (Figure 13.3), which is used to transfer the CagA protein and other bacterial factors into host cells (see below and Section 2). Although the organism has a typical gram-negative cell wall, containing **lipopolysaccharide** (**LPS**) it is much less of an **endotoxin** compared with *Escherichia coli*. *H. pylori* grows in an atmosphere of 5–15% oxygen, 5–12% carbon dioxide,

and 70–90% nitrogen (i.e. it is **micro-aerobic**) taking up to 5 days to grow on primary isolation and producing small (1–2 mm diameter) colonies on horse blood agar (5% horse blood in Columbia agar base). The colonies are domed, glistening, entire, gray or water-clear, and are sufficiently characteristic to suggest the presence of the organism (Figure 13.4). *H. pylori* can metabolize glucose, but its main carbon and energy source is from catabolism of amino acids. The organism has a urease, which is found both on the surface of the bacterium and in the cytoplasm and which is important for regulating the periplasmic pH. *H. pylori* can survive an acid environment for a short time but is not an **acidophile**. Many other *Helicobacter* species have been isolated from a wide range of animals. Some of these *Helicobacter* species can cause **gastroenteritis** in humans (e.g. *Helicobacter cinedae*) and some cause stomach ulcers in the animal host. Many are associated with the lower intestinal tract and particularly the hepato-biliary system where, in animals, the organisms are the cause of **hepatoma**.

ENTRY INTO THE BODY

The route of infection is not clear but is either feco–oral or oro–oral. In some countries (e.g. Peru), there is some evidence that infection may be acquired from sewage contamination of water supplies or vegetables. Following entry into the body

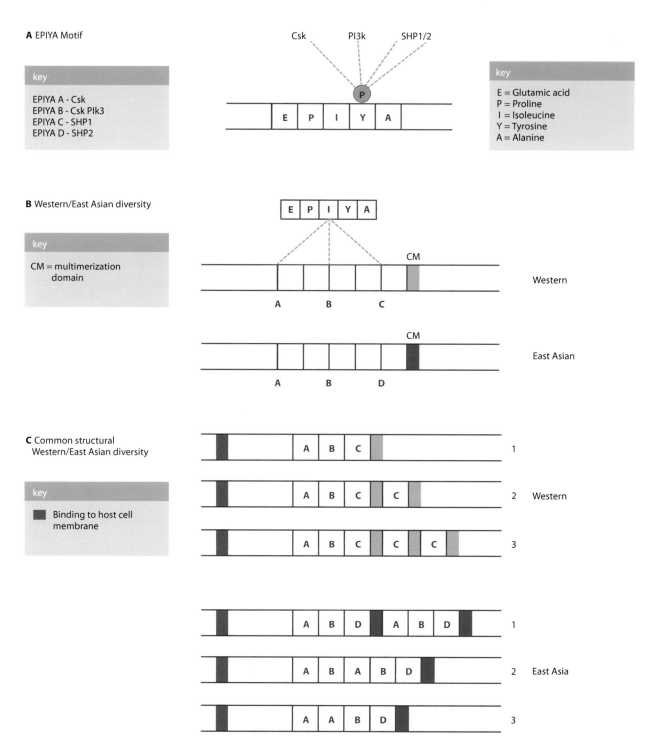

A EPIYA Motif

key

EPIYA A - Csk
EPIYA B - Csk PIk3
EPIYA C - SHP1
EPIYA D - SHP2

Csk   PI3k   SHP1/2

key

E = Glutamic acid
P = Proline
I = Isoleucine
Y = Tyrosine
A = Alanine

B Western/East Asian diversity

key

CM = multimerization
domain

CM

Western

A   B   C

CM

East Asian

A   B   D

C Common structural
Western/East Asian diversity

key

Binding to host cell
membrane

A B C   1

A B C C   2   Western

A B C C C   3

A B D A B D   1

A B A B D   2   East Asia

A A B D   3

Figure 13.2 **The cytotoxin-associated gene (CagA) pathogenicity island (cag-PAI).** This shows the general structure of the cag-PAI. Strains with a cag-PAI are called Type I strains and those lacking a cag-PAI Type II strains. Type I strains are more likely to be linked to severe disease than Type II strains. The cag-PAI comprises about 30 genes involved in the synthesis of the Type IV secretion system and a gene for the Cag A protein. The PAI may be complete, it may be separated into two sections (CagI and CagII) by an insertion element, as indicated in the upper part of the diagram, part of the PAI may be missing or the strain may lack a PAI. Variation occurs in the 3′ end of the PAI and several sequence variations have been identified. The sequences belong to two groups, one found principally in strains isolated from Western countries (WSS) Type A, B, C and one from Asian (Eastern) countries (EASS) Type A, B, D. *Adapted with permission from a map created by the Helicobacter Foundation and originally published at the following web address: http://www.helico.com/h_epidemiology.html.*

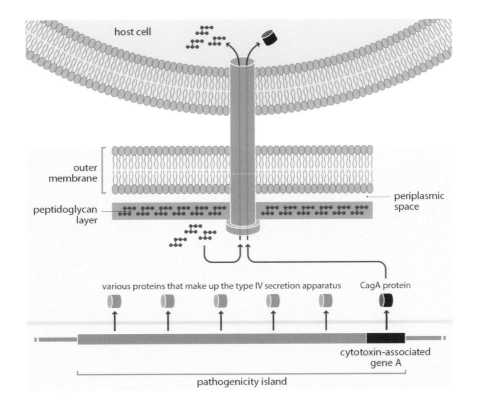

Figure 13.3 Type IV secretion system. The Type IV secretion system is a complex structure that acts as a microsyringe and is used to transfer material such as the CagA protein and part of the peptidoglycan of the cell wall of *H. pylori* into the host cell. The CagA protein is an oncoprotein and affects cellular signaling events, It is also important in evasion of the immune system by H. pylori. The T4SS is the largest of any bacterial species and up-regulates Toll-like receptor 4 (TLR4) on the host cell. Cag L, a component of the secretion system, binds to integrin receptor α5β1 on the gastric epithelial cell but also has a number of non-structural functions. Cag L stimulates epidermal growth factor receptor (EGFR) 4 and 5, stimulates cell spreading and increases gastrin secretion. -*Helicobacter pylori* also has other Type IV secretion systems: ComB transports free DNA, Tfs3 secretes CtkA (cell-translocating kinase A: activation of NF-kB) and along with Tfs4 is involved with conjugative DNA transfer. Various other proteins result in activation of NF-kB and a pro-inflammatory response. *Adapted with permission from an image created by The Nobel Committee for Physiology or Medicine and originally published at the following web address: http://nobelprize.org/nobel_prizes/medicine/laureates/2005/press.html.*

Figure 13.4 Colonies of *H. pylori* on blood agar.

via the oral route *H. pylori* localizes in the stomach. Here, it is found mainly in the antrum but can also be found in all parts of the stomach and duodenum. In the micro-environment of the stomach, the *H. pylori* locates to different regions (Figure 13.5). Migration between the regions occurs and appears to be more frequent between the corpus and the fundus compared to the corpus and the antrum, which probably reflects the different physiologic environments of the antrum compared to the corpus/fundus, which is the main site of oxyntic cells secreting acid. On initial entry into the stomach, it can only survive for a short time before it is killed by the acid. However, the presence of the enzyme urease on its surface (by hydrolyzing urea and producing ammonium ion) protects it for sufficient time enabling it to penetrate the mucus layer. The spiral shape and motility of *Helicobacter*, and the production of phospholipases and the ammonium ion (which affects the tertiary structure of the mucus making it thin and watery) allows the penetration of the mucus layer very quickly. *H. pylori* can only adhere to gastric epithelial tissue, which is found in the stomach. Many **adhesins** on the surface of the organism include Sialic acid binding adhesin (SabA) and OipA but the principle adhesin is Blood Group Antigen Binding Adhesin (BabA), which binds to the Lewis b blood group antigen expressed on gastric epithelial cells. BabA is not only important as an adhesin but also anchors the bacterial secretion system to the host cell surface for efficient injection of bacterial factors into the epithelial cell cytosol. Once attached through BabA, the T4SS system host cell signaling is triggered to induce transcription of genes that enhance inflammation and leads to development of intestinal

117

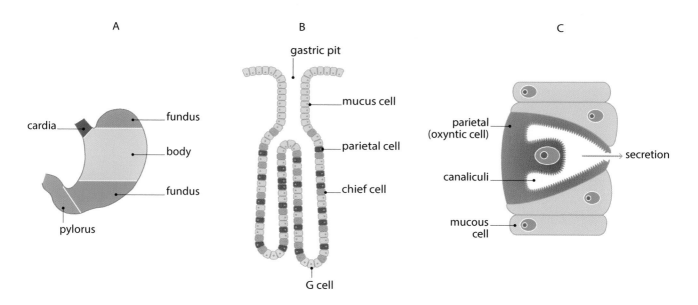

Figure 13.5 Anatomic areas of the stomach with main cell types. The main areas of the stomach which bind *H. pylori* are shown in green and red. (A) Gastric ulcers typically develop on the lesser curve of the body and duodenal ulcers in the antrum and pylorus. (B) illustrates a gastric pit with associated cell types, (C) illustrates a parietal cell which secretes hydrochloric acid.

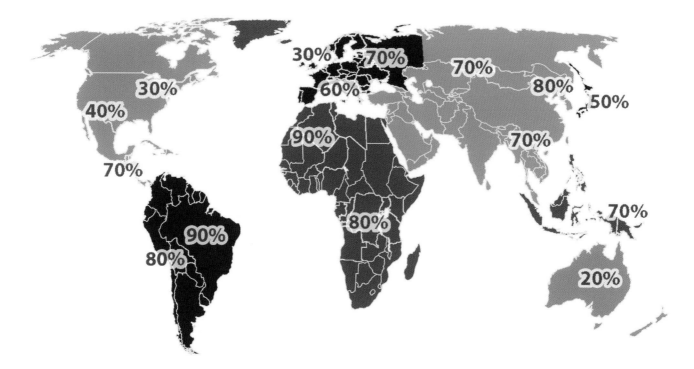

Figure 13.6 Global prevalence of *H. pylori* infection. The prevalence of *H. pylori* infection correlates with local socio-economic status. *H. pylori* has been linked with the human population as far back as 100–150 000 years and different racial groups are colonized by different genotypes of the organism to such an extent that human migration patterns can be discerned from the strain of *Helicobacter* colonized by the person.

A   B   C   D

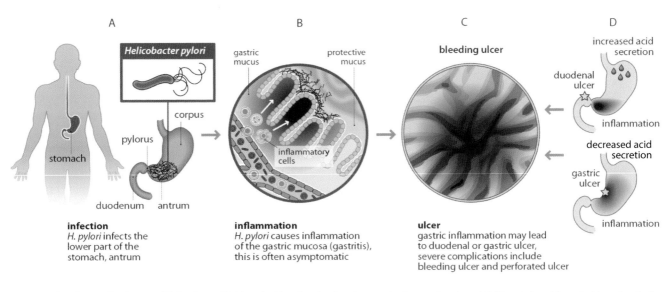

infection
H. pylori infects the
lower part of the
stomach, antrum

inflammation
H. pylori causes inflammation
of the gastric mucosa (gastritis),
this is often asymptomatic

ulcer
gastric inflammation may lead
to duodenal or gastric ulcer,
severe complications include
bleeding ulcer and perforated ulcer

Figure 13.7 General scheme of infection with *H. pylori* leading to development of a peptic ulcer. (A) *H. pylori* and its antral location in the stomach; (B) *H. pylori* on the surface of the epithelium with inflammatory cells in the tissue (C) duodenal ulcer (D) the location of a duodenal and gastric ulcer. The former can be in the first part of the duodenum or the pre-pyloric region, as shown in (C) and the latter is usually on the lesser curvature of the corpus of the stomach.

metaplasia and associated precancerous transformation (see Section 2, Pathogenesis).

SPREAD WITHIN THE BODY

The organism only grows within the stomach either only in the antrum or throughout the whole of the stomach lining. *H. pylori* can invade the gastric epithelium where they become sequestered and survive in phagosomes (see Section 2, host response). They can also be found in the vacuoles of yeasts present in the stomach. It is thought that this may be a source that could allow relapse of infection.

PERSON-TO-PERSON SPREAD

Spread is via the **feco–oral** route or **oro–oral** route mainly, and it is believed to be from mother to child. Adults can become infected with *H. pylori* but the route is not clear and may involve the environment, food or water.

EPIDEMIOLOGY

Globally, the prevalence is high in developing countries with the infection being acquired at a young age. Here, the incidence is 3–10% of the population each year compared with 0.5% in developed countries. However, the incidence of *H. pylori* infection, gastric cancer, and ulcer disease are declining. Worldwide, more than 1 billion people are estimated to be infected with *H. pylori*. The global prevalence of *H. pylori* is shown in Figure 13.6. Sero-epidemiologic studies have identified several risk factors. Thus, infection is higher in Social Class IV and V compared with I and II. Infection in industrialized countries (Western Europe, North America) is about 5–10% in the first decade rising to 60% in the sixth decade of life. In non-industrialized countries (Africa, South America, Middle and Far East) infection in the first decade is

about 60–70% with little increase with age. Overcrowding and a poor public hygiene infrastructure are factors in the spread of *Helicobacter*. Spread within families has been documented by molecular typing techniques. In industrialized countries, the low rate in childhood and high rate in the elderly can be explained by improving social standards of housing (with less overcrowding), potable water supplies, and sewage disposal. The high rate in the elderly can be explained as a cohort effect, reflecting the housing and public health standards when they were children. Although some other species (see Causative agent, above) can infect humans, there is no evidence that *H. pylori* can be acquired from animals (i.e. it is not a **zoonosis**).

2. WHAT IS THE HOST RESPONSE TO THE INFECTION AND WHAT IS THE DISEASE PATHOGENESIS?

The overall sequence of infection of the host with *H. pylori* is shown in Figure 13.7, illustrating its macroscopic and microscopic location and the location of ulcer formation. On initial infection of the gastric tissue, the host responds strongly to the presence of *H. pylori* by both the innate and acquired immune systems.

INNATE IMMUNITY

The mucus is the first barrier to *H. pylori* but having successfully dealt with that, it binds to the gastric epithelium using several adhesins in its outer membrane (see Section 1, entry into body). Although *H. pylori* does not normally penetrate the gastric epithelium it can invade epithelial cells and even enter the submucosa. The organisms can remain in vesicles in the epithelial cells when conditions are not optimum for the organism. The gastric epithelial cells express TLR2 and

TLR4 that interact with microbial-associated molecular patterns (MAMPS) on *Helicobacter* resulting in production of defensins and other antimicrobial peptides. In addition, the epithelial cells produce a variety of cytokines (including IL-8) and chemokines that recruit monocytes, macrophages, and neutrophils to the submucosa resulting in an acute inflammatory response. These cells also have various pattern-recognition receptors (PRRs) including TLRs that, through microbial products (including LPS) of *H. pylori*, result in their activation producing pro-inflammatory cytokines to further amplify the acute inflammatory response. Macrophages and neutrophils release reactive oxygen species (ROS) which damage the epithelial surface. This acute reaction is followed by a more chronic inflammatory reaction.

ADAPTIVE IMMUNITY

Dendritic cells in the gastric mucosa are believed to be activated by *H. pylori* through their PRRs. They up-regulate their CD86, increase expression MHC class II, and release pro-inflammatory cytokines such as IL-12. This is consistent with a role of inducing Th0 cells to Th-1. Th-17 also plays a role by secreting IL-8 and other pro-inflammatory cytokines. Gastric inflammation, mediated by the adaptive immune system in *H. pylori* infection, is believed to be caused primarily by Th-1 cells through the production of IFNγ. Dendritic cells carrying processed antigen from *H. pylori* are likely to traffic to local draining lymph nodes where specific IgG and IgA antibodies are presumably made. It has been speculated that antibodies arising from molecular mimicry (autoantibodies) could contribute to epithelial cell damage during *H. pylori* infections. In addition, experimental models of *H. pylori* in mice have indicated that, instead of having a protective effect, specific antibodies facilitate bacterial colonization and counteract resistance against infection. Thus, although a good cellular and antibody immune response is mounted, the organism is able to evade the immune defenses. The inflammation mediated by *H. pylori* is shown in Figure 13.8.

MECHANISMS OF EVASTION OF THE IMMUNE SYSTEM BY *H. PYLORI*

H. pylori has a number of mechanisms to avoid the immune system:

1. *Evasion of recognition by PRRs through a number of mechanisms*, e.g. Toll-like receptors, (TLRs), RIG-1-like receptors (RIRs), and C lectin-type receptors (CLRs).
2. *LPS*: of low immunogenicity.
3. *CagA and other material transferred by the 4TSS*: suppresses phagocytosis, decreases production of antimicrobial peptides by the epithelial cells, induces tolerogenic dendritic cells suppressing T-cell induction blocking T-cell effector responses.
4. *Vacuolating toxin A*: suppresses phagocytosis, induces tolerogenic dendritic cells blocking T-cell effector responses.
5. *Gamma-glutamyl transpeptidase*: induces tolerogenic dendritic cells blocking T-cell effector responses.
6. *Cholesterol-α-glucosyl transferase*: suppresses phagocytosis.
7. *Catalase superoxide dismutase*: suppresses production of ROS and NO.
8. *Arginase*: suppresses production of ROS and NO, blocks effector T-cell function.

PATHOGENESIS

There are five main ways in which tissue damage can occur (Figure 13.9).

Colonization by *H. pylori* leads to excess acid in the stomach (and peptic ulcers) and both a **hypergastrinemia** and **hyperpepsinogenemia**, unless gastric **atrophy** occurs. This destroys the acid-producing cells and leads to hypoacidity which is linked to gastric cancer. Hyperacidity is caused by inhibition of somatostatin by *H. pylori*. Somatostatin is a negative feedback for acid secretion by oxyntic cells.

Vacuolating cytotoxin is a multimeric pore-forming protein consisting of a 55kD-binding protein and a 33kDa protein which forms a pore in membranes and induces vacuolation in cells by fusion of dysfunctional phagosomes to form megasomes. It has multiple cellular effects: disrupting the mitochondria inducing apoptosis, affecting autophagy, disrupting the tight junction between cells, blocking T-cell proliferation and thus the immune response, and disrupting acid production by oxyntic cells.

After injection into the gastric cell by the T4 secretion apparatus, CagA is phosphorylated at a tyrosine residue and the activated CagA binds and activates SHP-2 and Csk. The activation of SHP-2 stimulates a change in the shape of the cell to a "hummingbird" phenotype. Activation of Csk which

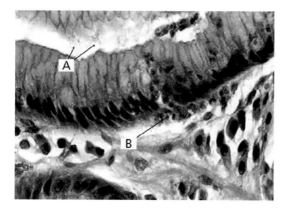

Figure 13.8 *Helicobacter pylori* – mediated inflammation in the stomach. Hematoxylin and eosin (H&E) stain of the gastric mucosa showing in the mucus and on the epithelial surface (A). *Helicobacter pylori* stains poorly with H&E and is better seen with a silver or Giemsa stain. Infiltrating granulocytes and mononuclear cells can be seen in the epithelial layer (lamina propria) indicating that the infection is becoming chronic (B). *With permission from Dr Lorraine Racusen, Johns Hopkins Medical School, Johns Hopkins University, Baltimore, MD.*

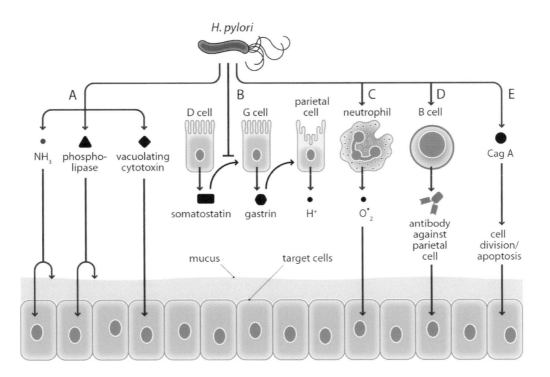

Figure 13.9 Mechanisms of pathogenesis of *H. pylori*. There are five main ways in which tissue damage can occur. These are: (A) local damage caused by a vacuolating cytotoxin (VacA), the ammonium ion as a result of the urease activity and the production of phospholipase that contribute to the formation of a poor-quality mucus barrier; (B) alteration of gastric physiology with enhanced acid production; (C) bystander damage caused by activation of granulocytes; (D) autoimmunity; (E) alteration of the balance of cell division and apoptosis induced by CagA injected into the epithelial cell through T4SS.

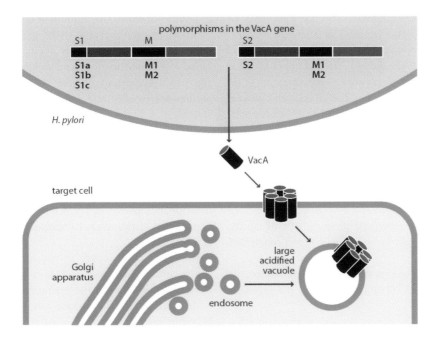

Figure 13.10 The vacuolating cytotoxin. The toxin is activated by the acid of the stomach, and the monomers of the toxin oligomerize in the host cytoplasmic membrane, enter the cytoplasm and affect the normal endocytic cycle, producing large, acidified vacuoles that lead to cell death. Various polymorphisms are found in the gene. Its leader sequence varies with the strain: the S1 leader sequence is found in type I strains and the S2 in type II strains, which produces less cytotoxin. Polymorphisms within the S1 leader sequence, S1a, S1b, and S1c, are found in different geographic regions of the world, are associated with different racial groupings, and can be used as a surrogate marker for human migration patterns. S1a is found mainly in North America and Europe; S1b is found mainly in Iberia and South America, and S1c mainly in the Far East. In Mexico, a1b/m1 is common although S1d , a novel subtype of vacA has been found in children with recurrent abdominal pain. An intermediate region (i) between s and m has two subtypes i1, i2. A further region between i and m called d with one type called d1 (no deletion) and d2 (with an 81bp deletion) The role of i1/i2 in carcinogenesis is unclear because of conflicting results.

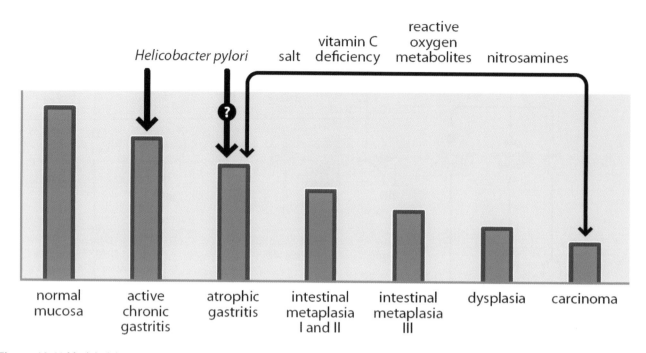

Figure 13.11 Model of the sequential steps in the development of gastric cancer mediated by *H. pylori*. This shows the multiple stages to development of gastric cancer with *H. pylori* being the initial insult but other factors such as salt intake, and a lack of vitamin C, also playing a role. Evidence from short-term studies suggests that eradication of *H. pylori* can lead to reversal of the histologic changes of atrophy, intestinal metaplasia, methylation changes, and epithelial-mesenchymal transition leading to a reduction in gastric cancer development. Long-term studies are not yet available.

then activates the Src tyrosine kinases to form a Csk-CagA complex acts as a negative feedback for the initial activation of CagA. The CagA-SHP-2 complex induce apoptosis of infected cells. Non-phosphorylated CagA interacts with PAR1 kinase (regulating cell polarity and tight junctions) and induces a "scattering" cell phenotype. The net effect of these interactions leads to a complex array of cellular alterations that induce stem-cell transformation of epithelial cells and **gastric** carcinogenesis. The route to duodenal ulcer (high acid, antral gastritis, low cancer risk) or gastric ulcer (low acid, pan gastritis, high cancer risk) may depend on the interaction between host and organism polymorphisms. The development of atrophic gastritis (atrophy of the mucosal epithelium) is a risk factor for the development of gastric cancer and the evolution to cancer proceeds via stages of metaplasia and dysplasia. Other factors are also important in the eventual development of cancer such as the amount of dietary antioxidants and salt consumed (Figure 13.11).

Some of the extra GI effects of *H. pylori* (e.g. idiopathic thrombocytopenic purpura (ITP) are caused by the presence of autoantibodies.

3. WHAT IS THE TYPICAL CLINICAL PRESENTATION AND WHAT COMPLICATIONS CAN OCCUR?

Most individuals colonized by *H. pylori* will remain symptom-free. Antral gastritis is the most common manifestation in children. About 20% of adults will go on to develop peptic

ulcer disease and about 1% gastric cancer. *H. pylori* is the principal cause of peptic ulcer disease (gastric or duodenal ulcer) in adults but this is uncommon in children.

Ulcer disease presents with epigastric pain, heartburn or **dyspepsia** or may be totally asymptomatic. Anemia (due to blood loss from the ulcer) and weight loss may also occur and signs and symptoms of perforation (acute abdominal pain, abdominal rigidity and **guarding**, rebound tenderness and shock).

H. pylori is a good example of a "slow" infection, because infection occurs in childhood but related diseases occur in adulthood. Its role in gastric cancer formation is described as a "hit as and run" model since even if the original insult (*H. pylori* infection) is removed, the process leading to cancer is unstoppable. However, short-term studies have demonstrated histopathologic recovery to normal histology after *Helicobacter* eradication.

The role of *H. pylori* in non-ulcer dyspepsia (NUD) is controversial but it may be that a subset of individuals who have NUD do benefit from eradication of the organism.

In addition to GI diseases, it has been suggested that *H. pylori* may be related to a wide range of extra-GI disease such as coronary heart disease, **stroke**, migraine, **rosaceae**, Behcet's disease, and gallbladder disease. Some evidence exists for most of these but again there is contrary evidence.

Confirmed extra-gastric conditions caused by *H. pylori* are iron deficiency anemia, ITP, and vitamin B_{12} deficiency. Also, *H. pylori* may be protective for reflux esophagitis, Barrett's esophagus, and esophageal carcinoma and in extra-gastric conditions: asthma.

The most serious consequence of *H. pylori* infection is the development of cancer.

H. pylori is a Class I carcinogen and the cause of the majority of cases of gastric **adenocarcinoma** (except that of the cardia of the stomach) and mucosal associated lymphoid tissue (MALT) **lymphoma**. Clinical presentation of carcinoma is very nonspecific and is usually associated with gastric ulcer rather than duodenal ulcer. It presents with abdominal pain or a mass with so-called "alarm symptoms" – weight loss and anemia. Gastric cancer originating at the **cardia** (gastro-esophageal junction) does not appear to be related to colonization by *H. pylori*.

4. HOW IS THE DISEASE DIAGNOSED, AND WHAT IS THE DIFFERENTIAL DIAGNOSIS?

There are numerous ways in which *H. pylori* can be diagnosed. The first general method involves an **endoscopy** and biopsy. The biopsy can then be cultured under micro-aerobic conditions for *H. pylori*, which takes 5 days; histology can demonstrate the characteristically shaped organism on the surface of the epithelial cells (Giemsa or Genta stain) and the inflammatory cell type can be identified (hematoxylin and eosin); the urease of the bacterium can be detected using a rapid urease test (which involves putting one of the biopsies into a urea solution with a pH indicator); and finally *H. pylori* can be detected using the **polymerase chain reaction** (**PCR**) with appropriate primers. With the development of whole genome sequencing (WGS), rapid detection of antibiotic resistance and virulence markers become a possible option. An additional further development in molecular diagnostics is detection of miRNA. These non-coding 18–25 nucleotide RNAs are found in body fluids/tissues and are involved in regulatory systems. Deregulation can lead to panels of biomarkers to identify gastric cancer or *H. pylori*. One miRNA (miR-223) is higher with gastric cancer and *H. pylori*.

Routinely, in the primary care setting, diagnosis is by noninvasive tests. These are:

1. *the stool antigen test* – these are immunoassay tests that use either polyclonal or **monoclonal antibodies** to detect the *Helicobacter* antigen in the feces;
2. *the urea breath test* – this test is performed by giving the patient a drink containing labeled urea (^{13}C) and 20 minutes later collecting the breath and measuring the amount of labeled CO_2 using an isotope ratio mass spectroscopy The basis of the breath test is that the *Helicobacter* urease, in the stomach, hydrolyzes the labeled urea to labeled CO_2 which is exhaled in the breath.

Either of these tests is recommended for use in the Maastricht Guidelines (a consensus document from gastroenterologists in Europe) in the cost-effective "Test & Treat" policy – a person complaining of upper GI symptoms will be tested for *Helicobacter* and, if positive, will be treated.

Serology is of no help in diagnosis of active disease but is useful for epidemiology studies.

DIFFERENTIAL DIAGNOSIS

This includes: stomach ulcer disease caused by nonsteroidal anti-inflammatory (NSAID) drug usage, particularly as the global prevalence of *H. pylori* is decreasing; acute gastritis; atrophic gastritis; gastric cancer; gastrinoma; gastroesophageal reflux disease, non-Hodgkin lymphoma; peptic ulcer disease and stress induced gastritis.

5. HOW IS THE DISEASE MANAGED AND PREVENTED?

MANAGEMENT

The main problem in treatment is the increasing and widespread multi-antibiotic resistance of the strains, leading to failure of the usual first-line regimen: a proton pump inhibitor (PPI) and two antibiotics (amoxicillin, clarithromycin, metronidazole). Otherwise, it should be bismuth quadruple (PPI + bismuth + 2 antibiotics, e.g. metronidazole, tetracycline) The choice depends on the local resistance pattern of the strains. Several other treatment regimens have evolved to cope with the resistance problem: the use of different antibiotics such as tetracycline, furazolidone, fluoroquinolones; quadruple therapy with addition of bismuth citrate (now suggested as first line); sequential therapy with a short course (10 days) of a PPI + amoxicillin followed by the addition of clarithromycin/metronidazole (final 5 days); hybrid therapy with a PPI + amoxicillin (10 days) followed by the addition of clarithromycin/metronidazole (final 7 days); and sensitivity testing of the isolate. Additionally, vonoprazan (a potassium channel blocker) can be substituted for the PPI and studies so far indicate higher eradication rates compared to a PPI. Additionally, a new regime of vonoprazan + amoxicillin is being tried. If treatment fails, a biopsy should be taken and sensitivity of the isolate determined.

In summary, first-line treatment should be bismuth quadruple or concomitant therapy but in areas of low prevalence of clarithromycin resistance a two-week clarithromycin-containing triple therapy can be used in macrolide naïve patients. Second-line treatments include levofloxacin-containing triple therapy and bismuth quadruple therapy. Sequential therapy has reduced success owing to resistance and is going out of favor. Probiotic can be used along with the antibiotics to reduce antibiotic-related side-effects.

Other approaches being assessed are the addition of probiotics to the antibiotic regimen, administering the antibiotic cocktail during endoscopy directly to the stomach lining, using plant derived phytochemicals: antibiofilm medication derived from *Acorus calamus* and inflammatory 7,8 dihydroxycoumarins from *Changbai daphne*. Finally, laboratory evidence suggests that *H. pylori* is sensitive to photodynamic therapy.

PREVENTION

The prevalence of *H. pylori* globally is decreasing possibly due to increasing standards of hygiene. Currently, there is no effective vaccine. The main technical challenges for vaccine development are the ability of the organism to avoid the host response with no obvious protein(s), the inhibition of which could lead to the eradication of the organism. This difficulty is amplified by the extensive polymorphisms and sequence variation present in *H. pylori*. The other barrier to an *H. pylori* vaccine is a reluctance of many pharmaceutical companies to invest in vaccine development owing to a lack of appreciation that, despite the falling prevalence of the organism, it is still a common infection worldwide with serious outcomes (ulcer and cancer) for many patients. Two alternative vaccine approaches are being tried. One is a vaccine against HtrA (a protein which affects cell junctions allowing *H. pylori* to bind to the basolateral surfaces of gastric epithelium where the oncogenic CagA protein is injected into gastric cells. The other approach is a vaccine against g-glutamyltranspeptidase, a protein that inhibits T-cell responses against *H. pylori*.

SUMMARY

1. WHAT IS THE CAUSATIVE AGENT, HOW DOES IT ENTER THE BODY, AND HOW DOES IT SPREAD A) WITHIN THE BODY AND B) FROM PERSON TO PERSON?

- The cell wall is a typical gram-negative structure.
- The lipopolysaccharide has considerably less endotoxin activity compared with other gram-negative bacteria.
- *H. pylori* occurs in over 50% of the global population but prevalence globally is decreasing.
- Colonization occurs in childhood.
- Transmission is feco–oral or oro–oral. In some locations, transmission may be from water supplies.
- Colonization is related to local social conditions and the public health infrastructure.
- The organism colonizes the gastric tissue either in the stomach or the duodenum.

2. WHAT IS THE HOST RESPONSE TO THE INFECTION AND WHAT IS THE DISEASE PATHOGENESIS?

- There is a strong innate immune response with infiltration by macrophages and neutrophils (acute inflammation).
- *Helicobacter* has several ways of avoiding the immune response including inhibition of phagocytosis, reduction in ROS and NO and inhibition of T-cell effector function.
- There is a strong acquired immune response with antibody production but this is generally ineffective. Th-1 cells are mainly responsible for inflammation.
- Certain virulence markers, e.g. CagA, VacA, are associated with more severe disease.
- Direct damage is brought about by the secretion of enzymes that destroy the mucus barrier and vacuolating cytotoxin that kills the surface epithelial cells.
- Gastric regulation of acid production is disturbed by the inhibition of somatostatin caused by the LPS of *Helicobacter*.
- Autoantibodies are induced by *Helicobacter* that kill the acid-secreting parietal cells and cause some extra-intestinal manifestations.

- Gastric cell dynamics are affected by interference with normal cell signaling events caused by introduction of the CagA protein of *Helicobacter*.
- Bystander damage is caused by release of free radicals from the granulocytes.

3. WHAT IS THE TYPICAL CLINICAL PRESENTATION AND WHAT COMPLICATIONS CAN OCCUR?

- *Helicobacter* is a "slow" infection, with colonization occurring in childhood and disease occurring years later.
- The vast majority of persons colonized by *H. pylori* remain asymptomatic.
- *H. pylori* is the main cause of peptic ulcer disease and gastric cancers.
- *H. pylori* may be associated with some extra-gastrointestinal diseases.

4. HOW IS THE DISEASE DIAGNOSED, AND WHAT IS THE DIFFERENTIAL DIAGNOSIS?

- Diagnosis is by invasive (endoscopy) or noninvasive tests.
- Invasive tests are culture, histology, rapid urease test, PCR.
- Noninvasive tests are serology, antigen detection, and the urea breath test.
- A cost-effective strategy is "Test & Treat".
- Recommended tests are the urea breath test and the fecal antigen tests.
- Whole genome sequencing (WGS) and assays for micro inhibitory RNAs are emerging tests.

5. HOW IS THE DISEASE MANAGED AND PREVENTED?

- The first-line treatment is a PPI plus clarithromycin and amoxycillin or metronidazole and amoxycillin for 7–10 days in areas with low levels of resistance.
- There is a global increase in resistant and multi-resistant isolates making treatment difficult.
- Rescue regimens should be guided by the sensitivity of the isolate to antibiotics.

FURTHER READING

Malik TF, Gnanapandithan K, Singh K. Peptic Ulcer Disease. StatPearls, 2021.

Mims C, Dockrell HM, Goering RV, et al. Medical Microbiology, 3rd edition. Mosby, London, 2004.

Olga P Nyssen, Bordin D, Tepes B, et al. European Registry on Helicobacter pylori Management (Hp-EuReg): Patterns and Trends in First-Line Empirical Eradication Prescription and Outcomes of 5 Years and 21 533 Patients. Gut, 70: 40–54, 2021.

Parikh NS, Ahlawat R. Helicobacter pylori. StatPearls, 2021.

REFERENCES

Alipour M. Molecular Mechanisms of Helicobacter pylori-Induced Gastric Cancer. J Gastrointest Cancer, 52: 23–30, 2021.

Bedwell J, Holton J, Vaira D, et al. *In Vitro* Killing Helicobacter pylori with Photodynamic Therapy. Lancet, 335: 1287, 1990.

Chisty A. Update on Indigestion. Med Clin North Am, 105: 19–30, 2021.

Eslick GD. Helicobacter Infection Causes Gastric Cancer? A Review of the Epidemiological, Meta-Analytic and Experimental Evidence. World J Gastroenterol, 12: 2991–2999, 2006.

Fischer W, Tegtmeyer N, Stingl K, Backert S. Four Chromosomal Type IV Secretion Systems in Helicobacter pylori: Composition, Structure and Function. Front Microbiol, 11: 1592, 2020.

Ford AC, Delaney BC, Forman D, Moayyedi P. Eradication Therapy for Peptic Ulcer Disease in Helicobacter pylori Positive Patients. Cochrane Database Syst Rev, 4: CD003840, 2016.

Karkhaha A, Ebrahimpour S, Rostamtabar M, et al. Helicobacter pylori Evasion Strategies of the Host Innate and Adaptive Immune Responses to Survive and Develop Gastrointestinal Diseases. Microbiol Res, 218: 49–57, 2019.

Kusters JG, Van Vliet AH, Kuipers EJ. Pathogenesis of Helicobacter pylori Infection. Clin Microbiol Rev, 19: 449–490, 2006.

Moyat M, Velin D. Immune Responses to Helicobacter pylori Infection. World J Gastroenterol, 20: 5583–5593, 2014.

Munoz-Ramirez ZY, Pascoe B, Mendez-Tenorio A, et al. A 500-year Tale of Co-Evolution, Adaption and Virulence: Helicobacter pylori in the Americas. ISME J, 15: 78–92, 2021.

O'Morain C. Role of Helicobacter pylori in Functional Dyspepsia. World J Gastroenterol, 12: 2677–2680, 2006.

Robinson K, Atherton JC. The Spectrum of Helicobacter-Mediated Diseases. Ann Rev Pathol, 16: 123–144, 2021.

Saracino IM, Pavoni M, Sacconanno L, et al. Antimicrobial Efficacy of Five Probiotic Strains against Helicobacter pylori. Antibiotics, 9: 244, 2020.

Zagari RM, Frazzoni L, Marasco G, et al. Treatment of Helicobacter pylori Infection: A Clinical Practice Update. Minerva Med, 112: 281–287, 2021.

WEBSITES

European Helicobacter and Microbiota Study Group: www.helicobacter.org

Nature Portfolio, Helicobacter pylori: https://www.nature.com/subjects/helicobacter-pylori

National Cancer Institute, Helicobacter pylori and Cancer, 2013: https://www.cancer.gov/about-cancer/causes-prevention/risk/infectious-agents/h-pylori-fact-sheet

The Helicobacter Foundation, 2006: www.helico.com

United European Gastroenterology Federation: www.uegf.org

WebMD, What Is H. pylori? 2020: https://www.webmd.com/digestive-disorders/h-pylori-helicobacter-pylori

Students can test their knowledge of this case study by visiting the Instructor and Student Resources: [www.routledge.com/cw/lydyard] where several multiple choice questions can be found.

Hepatitis B virus

14

Figure 14.1 Jaundice as demonstrated by a yellow discoloration of the sclera of the eye. *From the Centers for Disease Control & Prevention, Atlanta, Georgia. Image is found in the Public Health Image Library #2860. Additional photographic credit is given to Thomas F. Sellers and Emory University who took the photo in 1963.*

A 28-year-old stockbroker has been feeling generally unwell for the last 4 days, and off his food. Although usually a smoker (10–15 cigarettes per day), he hasn't been able to face "lighting up" since his illness began. Since yesterday, he has complained of a vague ache below his ribs on the right side. He noticed that his urine had become very dark, and his friends told him today that his eyes looked yellow (Figure 14.1). He has no relevant past medical history.

On examination by his local primary care physician, he was **pyrexial** (38.5°C) and clinically jaundiced. The only other sign of note was some right upper quadrant abdominal tenderness, but no **guarding**.

A clotted blood sample was sent to the laboratory, and 3 hours later, the laboratory called to convey the preliminary results, suggesting that the patient was suffering from acute hepatitis B virus infection.

1. WHAT IS THE CAUSATIVE AGENT, HOW DOES IT ENTER THE BODY AND HOW DOES IT SPREAD A) WITHIN THE BODY AND B) FROM PERSON TO PERSON?

CAUSATIVE AGENT

Hepatitis B virus (HBV) belongs to the virus family *Hepadnaviridae* – that is hepa (referring to the liver), DNA (referring to the nature of the viral genome), viruses – each member of which infects the liver of its particular host species. HBV is the member of the family that infects humans.

The *Hepadnaviridae* have an unusual partially double-stranded circular DNA genome (Figure 14.2). The genome is small – around 3200 bases – and therefore encodes only a small number of viral proteins, or antigens. There are four open reading frames (ORFs) identifiable within the genome, as follows.

- ORF C encodes the core antigen (HBcAg), which forms a protective **nucleocapsid** around the genome.
- ORF S encodes the surface antigen (HBsAg). This is composed of the large, middle, and small surface proteins, the different sizes arising from use of different start codons within the gene (large contains the pre-S1, pre-S2, and S regions; middle contains the pre-S2 and S regions; small contains just the S region – see Figure 14.2A), which are embedded in a lipid bilayer derived from internal membranes of the infected hepatocyte, surrounding the nucleocapsid.
- ORF P encodes the polymerase enzyme, which has both DNA- and RNA-dependent polymerase activities.
- ORF X encodes the X antigen, which is known to act as a transactivator of transcription and is likely involved in carcinogenesis.

The hepatitis B viral structure is shown schematically in Figure 14.2B.

The replication cycle of HBV is unique among human viral pathogens. The virus attaches to the sodium taurocholate co-transporting polypeptide (NTCP, a bile acid transporter) expressed on the surface of hepatocytes which has been identified as the primary receptor for the virus. After entry into a susceptible cell and uncoating of the viral genome, the first step is synthesis of the missing part of the positive DNA strand, to form a complete double-stranded (ds) DNA molecule. Within the cell nucleus, the dsDNA is converted into covalently closed circular (ccc) DNA, which is an extremely stable molecule, behaving like a mini chromosome. A number of RNA species are transcribed from the DNA, encoding the various viral proteins referred to above, but importantly, there is also a 3.5 kb RNA copy of the whole genome (pregenomic RNA). This RNA is packaged within newly synthesized core protein, together with the viral polymerase, to form immature new virus particles. The final step in viral maturation is then the reverse transcription of this pregenomic RNA into a DNA copy, followed by synthesis of an incomplete complementary DNA strand to yield partial dsDNA.

ENTRY AND SPREAD WITHIN THE BODY

HBV enters a new host via the genital tract or following direct inoculation of virus into the bloodstream (see below). Once within the blood, virus travels to the liver, where it infects hepatocytes. Once within a liver cell, the virus replicates, and new virus particles are released directly into the bloodstream. From here, they gain access to every bodily compartment, including the genital tract.

PERSON-TO-PERSON SPREAD

HBV infection is spread by three routes.

1. *Mother to baby, or vertical transmission.* Babies acquire infection at the time of birth, through exposure to infected maternal blood and/or genital tract secretions. On a global scale, this is by far the most important route of infection. Over 90% of babies of carrier mothers become infected.

2. *Sexual transmission.* In an infected individual, both seminal fluid (male) and the female genital tract will contain virus. Unprotected sexual intercourse may therefore result in transmission of infection from one partner to another. This is especially the case with men who have sex with men, where the act of intercourse may also involve exposure to blood through mucosal tears.

3. *Direct blood-to-blood transfer of virus (parenteral transmission).* This is a somewhat artificial, but nevertheless important, route of transmission, as humans are not normally exposed to blood from each other (except during childbirth). The most obvious way in which this can arise is via blood transfusion – if the blood donor is infected with HBV, then the blood will transmit infection to the recipient. More subtle ways of achieving this include the re-use/sharing of contaminated needles/syringes when injecting medicinal or recreational drugs; exposure to contaminated needles, for example, via tattooing, ear or body piercing or acupuncture; needlestick injuries as suffered by healthcare workers, i.e. accidental stabbing of a needle derived from an infected patient into the healthcare worker's own finger; or by sharing contaminated razors

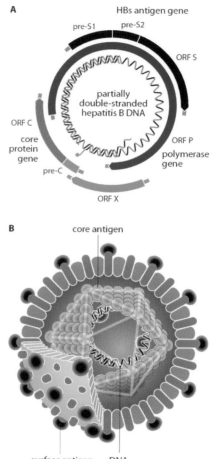

Figure 14.2 (A) Diagram of hepatitis B virus (HBV) genome showing the partially double-stranded DNA genome, and the four open reading frames (S, encodes hepatitis B surface antigen; C, encodes hepatitis B core antigen; P, encodes DNA polymerase enzyme; X, encodes X antigen) from which mRNA is synthesized. (B) Schematic diagram of the structure of an HBV particle. The partially double-stranded DNA genome is enclosed within the core antigen/protein which is, in turn, surrounded by an envelope consisting of a lipid bilayer derived from internal membranes of the hepatocyte into which are embedded the large, middle, and small surface proteins, which together constitute the surface antigen.

or toothbrushes. Horizontal transmission between young children also occurs through inapparent blood contamination of cuts and scratches.

EPIDEMIOLOGY

The World Health Organization (WHO) estimates that there are more than 290 million people chronically infected with HBV (see below). Rates vary across the globe (Figure 14.3). The WHO classifies countries into three groups according to their prevalence of chronic HBV infection – "high" means that over 8% of a country's population are infected, "intermediate" equates to 2–8%, while "low" is <2%. High prevalence countries include China, Japan, SE Asia, and much of sub-Saharan Africa and South America – carriage rates in

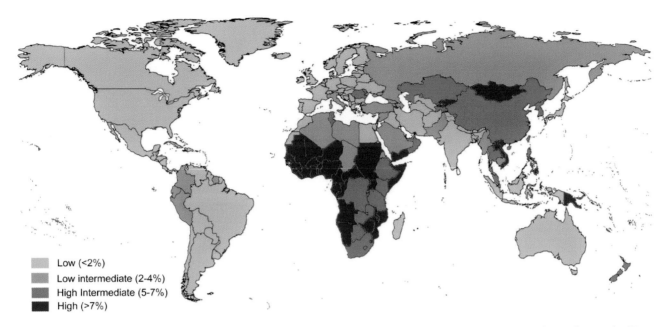

Figure 14.3 Global map of hepatitis B virus (HBV) carriage rates. *Adapted from a map created by the United States Centers for Disease Control and Prevention.*

these countries may exceed 20%. Northern Europe and the US (except for parts of Alaska) are low prevalence areas – the carriage rate in the UK is 0.1–0.5%.

Armed with the above information on the routes of transmission and epidemiology of HBV infection, it is possible to draw up a list of individuals who have a higher risk of being infected (Table 14.1). An understanding of risk factors is helpful when constructing policies both for screening for HBV infection and for selective vaccination (see Section 5 below).

2. WHAT IS THE HOST RESPONSE TO THE INFECTION AND WHAT IS THE DISEASE PATHOGENESIS?

INNATE IMMUNITY

In the early stages of infection within the liver, innate immune responses, particularly **interferon** (**IFN**) induction, and

natural killer (**NK**)-cell activity are thought to be important in determining the eventual outcome of infection. Innate immune activation functions through the recognition of pathogen-associated molecular patterns (PAMPs) by pattern recognition receptors (PRRs). Locally released IFN binds to surface receptors on neighboring cells, triggering the activation of various IFN response genes within those cells, the net effect of which will be to render the cells relatively resistant to virus infection. Viral interference with these PRR responses may lead to failure of viral elimination and hence chronicity of infection. NK-cell dysfunction is also thought to contribute to viral persistence.

ADAPTIVE IMMUNITY

The virus encodes a number of distinct antigens. In addition to HBsAg and HBcAg, there is a third important antigen, namely the "e" antigen, or HBeAg. This is a soluble protein released from infected cells. It is, in fact, derived from the same gene

Table 14.1 Individuals at high risk of being HBV infected

Babies, children of HBV-infected mothers

Sexual partners of HBV-infected individuals

Household contacts of HBV-infected individuals

Members of ethnic groups with high rates of HBV carriage

People with multiple heterosexual or homosexual partners, including sex workers

People who inject drugs who share needles/paraphernalia

Healthcare workers (exposed to blood from infected patients)

Prisoners (high rates of injecting drug use)

Patients who receive regular transfusion of blood or blood products (e.g. hemophiliacs)

Patients with chronic renal failure (because of risks from hemodialysis)

Patients with chronic liver disease (where additional HBV infection may be life-threatening)

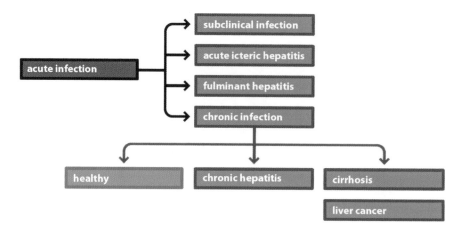

Figure 14.4 Outcomes following hepatitis B virus (HBV) infection of an adult. Around 55% of acute HBV infections are asymptomatic; 1% of infections result in fulminant hepatitis, i.e. acute liver failure; 5–10% of adults fail to clear infection and become chronically infected. The potential outcomes of such chronic infection are also shown.

as the core antigen. In an immunocompetent individual, the adaptive immune response should lead to the generation of antibodies to all viral antigens – those routinely measured in the laboratory are to the surface, core, and e antigens, i.e. anti-HBs, anti-HBc, and anti-HBe, respectively. The production of anti-HBs antibodies is particularly important in enabling the host to overcome the infection and eliminate the virus, as these antibodies are potentially neutralizing, and can therefore prevent newly released virus particles from infecting susceptible hepatocytes. Cellular immune responses are also generated, which lead to virus-specific cytotoxic **CD8+** T lymphocytes (CTLs) within the liver being able to kill infected hepatocytes through recognition of viral antigens present on the infected cell surface in association with HLA class I molecules. In chronic HBV infection, the CD8+ T cells are dysfunctional, and express markers of exhaustion such as programmed cell death protein (PD-1) and cytotoxic T-lymphocyte associated protein 4 (CTLA-4), presumably due to chronic antigenic stimulation.

PATHOGENESIS

Most damage to hepatocytes in HBV infection is thought to arise through the host CTL response killing infected cells in this way – HBV replication within hepatocytes of itself is not cytolytic and does not result in death of the infected cell.

3. WHAT IS THE TYPICAL CLINICAL PRESENTATION AND WHAT COMPLICATIONS CAN OCCUR?

There are a number of potential outcomes of HBV infection (Figure 14.4). Around 55% of acute HBV infections result in no detectable disease – a phenomenon known as asymptomatic **seroconversion**, or **subclinical hepatitis**.

ACUTE INFECTION

If there is sufficient damage to the liver, then the patient will present with an acute hepatitis. The clinical features can be split into those that occur before the patient becomes jaundiced (the pre-icteric phase – **icterus** is the same as **jaundice**), and those that arise once the patient is jaundiced (icteric phase). The pre-icteric phase is fairly nonspecific – the patient may complain of lethargy, loss of appetite, anorexia, nausea, alcohol and cigarette intolerance, and fever. The only clue to the fact the disease process is happening in the liver is that the patient may also mention right upper quadrant abdominal pain, which arises as the inflamed liver swells and stretches its innervated capsule. One of the many functions of the liver is to process pigments derived from hemoglobin in the blood, resulting in excretion of these into the bile, and thence in the feces. As the liver becomes damaged, this process may start to fail, resulting in these pigments accumulating in the bloodstream. This gives rise to the clinical sign of jaundice (due to an excess of circulating **bilirubin**) – a yellow discoloration of the skin, most easily seen by looking at the sclera of the eyes (see Figure 14.1). The circulating pigments are filtered in the kidney and excreted into the urine, which becomes very dark, while the absence of their excretion into the bowel means that the feces become very pale.

Although acute hepatitis is not a trivial illness, and patients may take many weeks to recover full health, most patients will survive. However, there may be such overwhelming liver damage that the patient goes into acute liver failure – a condition known as fulminant hepatitis. This occurs in about 1% of all acute HBV infections and carries a high mortality (i.e. >70%), the only really effective form of therapy being a liver transplant.

CHRONIC INFECTION

A proportion of patients acutely infected with HBV will fail to eliminate virus from their liver, and therefore become chronically infected. In chronic infection, virus continues to replicate in infected hepatocytes, and is continually released from these cells into the bloodstream, but without interfering with the normal lifespan of the cells. Such chronic carriers serve as the source of infection for other individuals, via the routes of transmission discussed above. The chances of an acute HBV infection becoming chronic are dependent on a number of factors, most importantly the age and immune status of the patient. Babies infected from their carrier mothers almost always become chronic carriers themselves, reflecting the immaturity of the immune system at birth. About 10% of infected children and 5% of infected adults become chronically infected – in these individuals, it is believed that a failure of the IFN response at the time of acute infection results in chronicity. Immunodeficient patients (e.g. those with HIV infection) are also highly likely to become chronic carriers once infected, as the absence of an effective CTL arm of the immune response mitigates against clearance of HBV-infected hepatocytes.

Chronic carriers may not be symptomatic – the liver has a large functional reserve, and liver function may be preserved even although there are virally infected hepatocytes present. However, such individuals are at risk of long-term chronic inflammatory hepatitis, as the infected cells are killed by the host CTL response. The usual response to death of host cells within the liver is the laying down of fibrous tissue. Over time, the continual death of liver cells, and their replacement by fibrous tissue, results in complete loss of normal liver architecture and function – a condition known as **cirrhosis** of the liver, which has a number of potentially life-threatening complications. The average time from acute infection to cirrhosis is around 20 years. It is estimated that about 20% of chronic HBV carriers will develop cirrhosis and its complications within their lifetimes.

One further serious complication of chronic HBV infection is the development of **hepatocellular carcinoma (HCC)**, that is, a malignant proliferation of hepatocytes. There are a number of molecular mechanisms underlying this, including a possible role for the X protein of HBV, which may interfere with normal cell division regulatory mechanisms. Additionally, HBV DNA may integrate into host cell chromosomes. Dependent on precisely where this integration takes place (it varies between individuals), the control and function of cell cycle regulatory and other genes such as oncogenes may be disturbed, leading to cell division. Chronically infected individuals are about 300 times more likely to develop HCC than individuals who are not infected. Thus, in parts of the world where chronic infection is common, primary liver-cell tumors are one of the commonest forms of malignant disease – globally, this is the third commonest malignant cause of death. On average, HCC appears about 5 years later than cirrhosis.

Note that the propensity for HBV infection of babies to become chronic, plus the fact that the life-threatening complications of chronic carriage do not usually become manifest for at least 20 years, provides a rational explanation of the epidemiology of HBV infection. Infected baby girls become chronically infected but reach child-bearing age without yet having suffered the serious consequences of disease. Thus, they in turn infect their babies. Passage of virus from generation to generation in this way results in very high rates of carriage within a population.

4. HOW IS THE DISEASE DIAGNOSED, AND WHAT IS THE DIFFERENTIAL DIAGNOSIS?

Laboratory diagnosis of HBV infection depends on the detection of various markers of infection (Table 14.2). Potentially, the laboratory can detect three viral antigens (HBsAg, HBcAg, and HBeAg) and antibodies to these three antigens (anti-HBs, anti-HBc, and anti-HBe, respectively), and the availability of sensitive genome-detection techniques also means that HBV DNA levels can be measured.

The first test to be performed is to detect HBsAg in a blood sample. This protein is excreted in vast quantities by infected

Table 14.2 Labratory markers and their interpretation

Status	HBsAg	IgM anti-HBc	IgG anti-HBc	HBeAg	Anti-HBe	Anti-HBs
Acute HBV infection	+	+	+	+	-	-
Cleared HBV infection	-	-	+	-	+	+
Chronic HBV infection, high risk	+	-	+	+	-	-
Chronic HBV infection, low risk	+	-	+	-	+	-
Responder to HBV vaccine	-	-	-	-	-	+

liver cells and is easy to detect. The presence of HBsAg in a blood sample means only one thing – on the day the blood sample was taken, the patient was infected with HBV. Of itself, this marker cannot distinguish between acute or chronic infection, as it will be present in both.

All patients infected with HBV will make an antibody response to the core antigen (anti-HBc). This marker can be used to distinguish between acute (recent) and chronic infection, as **IgM** anti-HBc will be present in the former, IgM antibodies being a marker of recent infection.

Patients diagnosed with acute infection (HBsAg-positive, IgM anti-HBc-positive) must be followed to see if they eliminate virus over the following 6 months. This will be manifest by the disappearance of HBsAg from the peripheral blood, followed shortly by the appearance of anti-HBs antibodies.

If HBsAg persists for longer than 6 months, then the infection has become chronic, by definition. However, not all chronic carriers are equally infectious or equally at risk of chronic liver disease. The HBeAg/anti-HBe system distinguishes between two types of chronic infection. HBeAg-positive individuals are extremely infectious – HBV DNA levels are usually in excess of 10^8 IU/ml blood (eAg is derived from the core gene and is essentially a surrogate marker of virus replication). Over time, some HBeAg-positive carriers will lose HBeAg from their blood, as virus replication slows down, and shortly afterward, it is possible to detect anti-HBe. An anti-HBe-positive patient is much less infectious, with HBV DNA levels between 10 and 10^3 IU/ml. Note that there are mutants of HBV that do not obey the above generalizations, particularly those that have stop mutations within the coding region of e antigen and are therefore incapable of synthesizing this antigen. These viruses may nevertheless be fully replication competent, and a patient infected with such a mutant may have high levels of viral DNA in the absence of HBeAg. Further discussion of these viral mutants is beyond the scope of this text.

Anti-HBs arise not only in patients who clear infection, but also in individuals who are vaccinated (see below) – measurement of anti-HBs responses in vaccines is an important marker of protection.

HBV DNA is a true marker of viral replication and infectivity. HBV DNA monitoring has become an essential part of the management of patients undergoing antiviral therapy.

DIFFERENTIAL DIAGNOSIS

There are many causes of acute hepatitis, and therefore the differential diagnosis of an acutely jaundiced patient is a long one. Acute alcoholic hepatitis can present in an identical fashion. There are also a number of other infectious agents that can damage the liver. In particular, there are a series of viruses that have a tropism for the liver, which are known by letters of the alphabet. Note that these are quite distinct viruses and not related to each other – the only thing they have in common is that they infect the liver.

Other Hepatitis Viruses

The important features of these viral pathogens are as follows:

Hepatitis A virus (HAV) – this is a **picornavirus** and therefore has a positive single-stranded RNA genome. An infected individual excretes virus in the feces, and therefore the transmission route is fecal–oral through contaminated food or water. The possible outcomes of HAV infection include asymptomatic seroconversion (the majority), acute hepatitis (clinically indistinguishable from acute HBV infection), and fulminant hepatitis (less common than with HBV). Note that there are no chronic sequelae of HAV infection – as patients recover from infection, virus is eliminated from the body and the patient is then immune from further infection. Diagnosis of acute infection is by detection of IgM antibodies against HAV.

Hepatitis E virus – also has an RNA genome (with unusual genomic organization – HEV is classified into the hepevirus family) and bears many clinical and epidemiologic similarities to HAV. Spread is via the **fecal–oral route**, and huge outbreaks involving thousands of individuals have been reported from fecal contamination of water supplies in various countries. Mortality from acute HEV is greater than with acute HAV, especially in pregnant women (>10%). Chronic carriage of HEV does not arise in immunocompetent hosts but may occur if the patient is immunodeficient. Diagnosis of acute infection is by detection of IgM antibodies against HEV, and/or detection of HEV RNA in a blood or feces sample.

Hepatitis C virus – this virus is discussed in detail in a separate case.

Hepatitis D virus – this is an incomplete virus. It has a small RNA genome, encoding its capsid protein, referred to as the delta antigen. However, it requires an outer protein coat and it uses HBV as a helper virus to provide it with HBsAg to enable cellular egress and entry. Thus, HDV infection can only occur in patients already infected with HBV (superinfection), or at the same time as acute HBV infection (co-infection). The risk groups for HDV infection are therefore essentially the same as for HBV infection. Co-infection increases the risk of acute fulminant hepatitis, while patients who become chronic carriers of both HBV and HDV are at increased risk of the development of serious liver disease. Diagnosis is by the detection of delta antigen and antibodies to delta antigen, and chronicity of infection by detection of HDV RNA.

5. HOW IS THE DISEASE MANAGED AND PREVENTED?

MANAGEMENT

There is no specific treatment for acute hepatitis B infection – advice is mainly for bed rest and avoidance of alcohol (a potent liver toxin). In cases of fulminant hepatitis, liver transplantation may be the only resort.

Patients with chronic HBV infection should be advised that they are potentially infectious to others, particularly their sexual partners and close household contacts. Under no circumstances should they donate blood, or share razors or toothbrushes. Blood spillages (e.g. from cuts) should be appropriately dealt with.

The management of patients with chronic hepatitis B has changed dramatically over the last 15 years, as the convoluted replication cycle of the virus has been elucidated. In particular, the recognition that there is a reverse transcription (RT) step in this process has led to the use of RT inhibitors in chronic hepatitis B, the relevant drugs being developed initially as RT inhibitors for the treatment of patients with HIV infection. Options for therapy (see Table 14.3) therefore now include the following:

1. *Interferon-alpha (IFN-α)* – this acts as an immunomodulatory drug in this context, not as an antiviral. It induces the expression of class I HLA molecules on the hepatocyte surface, and therefore facilitates recognition of virally infected cells by circulating virus-specific CTLs, which kill the infected hepatocytes and thereby reduce the burden of infected cells within the liver. This mechanism will only be effective in patients with an intact immune system whose T cells are capable of cytotoxic activity against HBV-infected hepatocytes – thus, it is unlikely to work in immunodeficient patients, or in patients infected with HBV at birth, in whom immunologic tolerance is induced such that their immune system does not recognize the virus as foreign. IFN-α is given as a 6–12-month course of thrice-weekly injections. More recently, preparations of IFN-α linked to polyethylene glycol have become available (pegylated- or PEG-IFN). This has a longer half-life than standard IFN, and therefore only requires once-weekly injection. The longer maintenance of plasma levels over time with PEG-IFN is also associated with a better therapeutic activity. IFNs may induce a plethora of unwanted side effects. Muscle aches, fatigue, and headache are common, but most patients become tolerant to these effects. Bone marrow suppression (low white cell count) and psychiatric manifestations such as depression and even suicide are more worrying.

2. *HBV DNA polymerase inhibitors* – these are collectively referred to as nucleos(t)ide analogs. Currently licensed ones in the UK include lamivudine (3TC, 3-thia-cytidine), tenofovir (as this molecule has a phosphate group attached, it is a nucleotide analog), and entecavir. These are effective against the polymerase activity of HBV, and therefore directly suppress viral replication. Note that their use does not eliminate virally infected hepatocytes, and therefore viral relapse after cessation of therapy is common. As with HIV, there is also a risk that mutations may arise within the viral polymerase (Pol) gene, which confer resistance to these drugs. Around 75% of patients treated with

lamivudine for more than 3 years will have such drug-resistant mutants, but resistance rates to tenofovir and entcavir are substantially lower.

Management of patients with chronic HBV infection is complex and should be performed by physicians with appropriate training and expertise (hepatologists). Difficulties include when and in whom to initiate therapy, with what therapeutic modality (IFN-α or nucleoside analog-based), and when is it safe to stop therapy. Suppression of viral replication, as evidenced by reduction of peripheral blood HBV DNA, is a valuable therapeutic endpoint, as it is associated with the lowering of risk of progression of disease to the life-threatening complications of cirrhosis and HCC development. However, current research is aimed at the development of drugs acting in different ways to not only reduce viral replication but also eliminate virus from infected hepatocytes and thereby achieve cure. These novel approaches include entry inhibitors, small RNA molecules that interfere with viral-derived mRNA, drugs which interfere with capsid formation or viral exit from cells, and immunomodulatory drugs such as Toll-like receptor (TLR)-7 agonists designed to encourage the host immune system to eliminate infected cells.

PREVENTION

HBV infection can be prevented by means of vaccination, using the HBsAg as the immunizing antigen. Original vaccines consisted of HBsAg purified from human plasma, and plasma-derived vaccines may still be used in some parts of the world. However, in Europe and the US, the current vaccines are subunit vaccines prepared by recombinant DNA technology. The gene encoding the HBsAg has been cut out of the virus and inserted into a yeast. As the yeast replicates in culture, large amounts of HBsAg are synthesized within the cytoplasm of the cell. Cell lysis followed by purification yields HBsAg, which is administered by intramuscular injection (usually as a course of three injections given at 0, 1, and 6 months). As the only part of the virus present in the vaccine is

Table 14.3 List of agents available for treatment of patients with chronic HBV infection

Interferons
Standard interferon-alpha
Pegylated interferon-alpha
Reverse transcriptase inhibitors
Lamivudine
Adefovir
Entecavir
Telbivudine
Tenofovir

HBsAg, the response to the vaccine can be easily measured by quantifying the anti-HBs in the vaccinee 6–8 weeks after the last dose. Protection against HBV infection is proportional to the amount of anti-HBs produced. Around 5–10% of adults do not mount an anti-HBs response to this vaccine and therefore remain susceptible to infection. A further 5–10% will make a rather weak response, which may offer only limited protection.

Having developed an effective vaccine, undoubtedly a major medical advance, the question arises as to who should be vaccinated. The WHO unequivocally recommends that the vaccine be included as one of the routine immunizations in childhood, and almost all countries have now adopted this approach, although the UK only switched from a selective policy targeted at identifiable risk groups (see Table 14.3), to a universal policy as recently as 2017.

Vaccines for Other Hepatitis Viruses

Note that there is no effective antiviral agent against HAV, but there is a vaccine – heat-killed whole virus. This is currently offered to individuals at risk of infection, for example, travelers to countries where HAV infection is **endemic**. There is currently no effective antiviral agent or licensed vaccine against HEV, although there are experimental vaccines undergoing clinical trials that have shown initial promise. Treatment and prevention of HCV infection is discussed in a separate case (see Case 14).

There is currently no licensed effective antiviral treatment for HDV but, as with HBV, there are promising new drugs in the pipeline such as bulevirtide, which blocks HBsAg from binding to the NTCP receptor. However, successful vaccination against HBV also prevents infection with HDV.

SUMMARY

1. WHAT IS THE CAUSATIVE AGENT, HOW DOES IT ENTER THE BODY, AND HOW DOES IT SPREAD A) WITHIN THE BODY AND B) FROM PERSON TO PERSON?

- Hepatitis B virus is a hepadnavirus.
- Entry is via the genital tract, or through direct inoculation of the virus into the bloodstream.
- Once virus has entered hepatocytes, viral replication results in release of new virus particles directly into the bloodstream, and from there into every body compartment.
- The three routes of transmission are mother to baby (infected birth canal, exposure to infected maternal blood); sexual; and parenteral, for example, via contaminated needles.

2. WHAT IS THE HOST RESPONSE TO THE INFECTION AND WHAT IS THE DISEASE PATHOGENESIS?

- IFN production within the liver renders neighboring noninfected cells relatively resistant to infection.
- Exposure to viral antigens leads to the development of both humoral and cellular responses.
- Cytotoxic CD8+ T lymphocytes are crucial in eliminating virally infected hepatocytes.
- Antibody against the surface protein (anti-HBs) is vital to enable neutralization of released virus particles.
- An inadequate immune response results in failure to eliminate the virus from the liver. The patient then becomes chronically infected.
- HBV is not a cytolytic virus. Chronic liver damage arises from the host inflammatory response, particularly CD8+ T-cell destruction of hepatocytes.

3. WHAT IS THE TYPICAL CLINICAL PRESENTATION AND WHAT COMPLICATIONS CAN OCCUR?

- Acute infection may be asymptomatic, that is there is insufficient damage to the liver to cause disease.
- Patients may present with acute hepatitis – initially (pre-icteric) the features are nonspecific – fever, malaise, lethargy, anorexia, right upper quadrant abdominal pain.
- Liver damage results in failure to excrete pigments derived from hemoglobin, resulting in dark urine, pale stools, and clinical jaundice.
- Liver damage may be so overwhelming that the patient goes into liver failure – fulminant hepatitis.
- Recovery from acute hepatitis does not always equate with clearance of infection: 90% of neonates, 10% of children, and 5% of adults will become chronically infected (carriers).
- Chronically infected individuals may be healthy or may suffer varying degrees of chronic inflammatory liver disease, leading eventually (i.e. after 20+ years) to cirrhosis of the liver.
- Carriers are also at a 300-fold increased risk of developing hepatocellular carcinoma.

4. HOW IS THE DISEASE DIAGNOSED, AND WHAT IS THE DIFFERENTIAL DIAGNOSIS?

- There are a number of laboratory markers of HBV infection.
- The presence of HBsAg in a serum sample indicates that HBV infection is present but does not distinguish between acute and chronic infection.
- The presence of IgM anti-HBc indicates recent infection.
- Chronic infection is defined as the persistence of HBsAg for more than 6 months.

Continued...

Continued...

- Chronic carriers can be split into those who are highly infectious or those much less infectious on the basis of the HBeAg/anti-HBe system.
- There are mutants of HBV in which HBeAg testing is misleading, with high levels of viral DNA in the absence of HBeAg.
- Measurement of HBV DNA levels is useful in monitoring the effectiveness of antiviral therapy.
- There are many causes of acute hepatitis. Of particular relevance, a number of distinct viruses have a hepatic tropism.
- HAV is a picornavirus, is spread by the fecal–oral route, causes asymptomatic, acute, and fulminant hepatitis, but does not result in chronic disease.
- HEV is a hepevirus, spread via the fecal–oral route, and has a higher mortality than HAV in the acute stage. Chronic infection only arises in immunodeficient individuals.
- HCV is a flavivirus, spread via blood–borne routes, and usually results in chronic infection, with consequent risks of cirrhosis and hepatocellular carcinoma.
- HDV is an incomplete virus that requires the presence of HBV in order to replicate. This virus often results in chronic infection with an accelerated progression of liver disease.

5. HOW IS THE DISEASE MANAGED AND PREVENTED?

- Management of acute infection is symptomatic only. Fulminant hepatitis may necessitate liver transplantation.
- Carriers of HBV should be educated about their condition and advised as to how to prevent transmission.
- Chronic HBV infection may be treated with IFN-α, which acts as an immunomodulatory agent, or with HBV polymerase inhibitors such as lamivudine, tenofovir, and entecavir.
- Long-term therapy may result in the emergence of resistant viral mutants.
- Prevention is by means of a subunit vaccine consisting of purified HBsAg, given as a course of three intramuscular injections.
- The WHO recommends universal vaccination in childhood.
- Vaccine-induced protection is related to the titer of anti-HBs produced. Not all vaccinees generate adequate anti-HBs responses.
- HAV can be prevented by vaccination using a killed whole virus vaccine, currently offered only to high-risk groups.

FURTHER READING

Barer M, Irving W, Swann A, Perera N. Medical Microbiology, 19th edition. Elsevier, Philadelphia, 2019.

Humphreys H, Irving WL, Atkins BL, Woodhouse AF. Oxford Case Histories in Infectious Diseases and Microbiology, 3rd edition. Oxford University Press, Oxford, 2020.

Murphy K, Weaver C. Janeway's Immunobiology, 9th edition. Garland Science, New York/London, 2016.

Richman DD, Whitley RJ, Hayden FG. Clinical Virology, 4th edition. ASM Press, 2016.

REFERENCES

Castaneda D, Gonzalez AJ, Alomari M, et al. From Hepatitis A to E: A Critical Review of Viral Hepatitis. World J Gastroenterol, 27: 1691–1715, 2021.

Lin S, Zhang Y-J. Advances in Hepatitis E Virus Biology and Pathogenesis. Viruses, 13: 267, 2021.

Nguyen MH, Wong G, Gane E, et al. Hepatitis B Virus: Advances in Prevention, Diagnosis, and Therapy. Clin Microbiol Rev, 33: e00046–19, 2020.

Philips CA, Ahamed R, Abduljaleel JK, et al. Critical Updates on Chronic Hepatitis B Virus Infection in 2021. Cureus, 13: e19152, 2021.

Roberts H, Ly KN, Yin S, et al. Prevalence of HBV Infection, Vaccine-Induced Immunity, and Susceptibility Among At-Risk Populations: US Households, 2013-2018. Hepatology, 74: 2353–2365, 2021.

WEBSITES

All the Virology on the WWW, Specific Virus Information, 1995: http://www.virology.net/garryfavweb12.html#Hepad

Centers for Disease Control and Prevention, Viral Hepatitis, 2021: https://www.cdc.gov/hepatitis/index.htm

GOV.UK, Guidance: Hepatitis B: the green book, chapter 18, 2013: https://www.gov.uk/government/publications/hepatitis-b-the-green-book-chapter-18

Virology Online, Hepatitis B: http://virology-online.com/viruses/HepatitisB.htm

World Health Organization, Hepatitis B, 2022: https://www.who.int/news-room/fact-sheets/detail/hepatitis-b

Students can test their knowledge of this case study by visiting the Instructor and Student Resources: [www.routledge.com/cw/lydyard] where several multiple choice questions can be found.

Hepatitis C virus

15

A 25-year-old man attends a needle exchange service to access a supply of clean needles and syringes in order to support his drug injecting habit. As he is provided with a pack of equipment, he is advised to be tested for evidence of any blood–borne virus infection that he might have acquired during his injecting career. He accepts this advice, and a fingerprick blood sample is taken and sent to the laboratory. On his next visit to the needle exchange, one month later, he is given the results by a specialist hepatitis nurse attending the drug treatment centre (Figure 15.1).

The nurse persuades him to have a venous blood sample taken, to confirm these initial results, quantify the HCV viral load, determine the HCV genotype and check on his liver function. She also performs a transient elastography test, which indicates a liver stiffness of 6.3 kPa. Subsequent results indicate he has chronic HCV infection, genotype 3, with a raised alanine aminotransferase. The specialist nurse arranges for him to receive a total of 12 weeks oral antiviral therapy. Three months after he has finished his course of treatment, a repeat blood test shows the absence of HCV RNA.

Greendale hospital Microbiology/Virology

Hospital number: K 1234567   Name: John Smith   DoB: 01/01/1997

Sample number: 21X528134   Date: 01/01/2022

Referral Source: Wellbeing clinic, Long Road, NG1 1AB

Serology results:
HIV antigen/antibody   Negative
Hepatitis B surface antigen   Negative
Hepatitis C antibody   Positive

Further tests:
HCV RNA   Detected
HCV RNA Load

Comment: Indicates current infection with hepatitis C virus. Please refer patient to the hepatitis clinic for further assessment and management

Figure 15.1 Laboratory results.

1. WHAT IS THE CAUSATIVE AGENT, HOW DOES IT ENTER THE BODY AND HOW DOES IT SPREAD A) WITHIN THE BODY AND B) FROM PERSON TO PERSON?

CAUSATIVE AGENT

Hepatitis C virus (HCV) is an enveloped virus containing positive single-stranded RNA of around 9.2 kb. When first characterized in 1989, HCV was classified as a flavivirus, i.e. a very distant relative of yellow fever virus, on the basis of its genome architecture and similarities in the structure of the polymerase enzymes encoded by the two viruses. However, viruses from many different species having much closer similarity to HCV have since been discovered. Currently, these are classified into the genera hepaciviruses and pegiviruses. Many differences between these viruses and classical flaviviruses have now been described, and it is possible these genera will, in time, be reclassified into their own family.

The HCV genome was first identified by molecular biologists in California in 1989. Shortly afterward, a Japanese group also sequenced a full-length HCV, but this showed around 15% difference in nucleotide sequence to the Californian report. In fact, HCV is extremely genetically diverse – 10-fold more

so than HIV. This has led to the classification of hepatitis C viruses into eight distinct genotypes, numbered 1–8, with >30% difference in nucleotide sequence between genotypes. Within genotypes, subtypes, with more than 15% sequence difference, are labeled with letters of the alphabet – currently nearly 100 different subtypes have been identified. Even within a patient, the virus exists as a quasispecies, i.e. a series of closely related, but not identical sequences.

The genome structure of HCV is shown in Figure 15.2. The 5′ non-coding region (NCR) contains a binding site for micro-RNA 122 (miRNA-122, the most abundant miRNA in hepatocytes) which is essential for virus replication. It also contains an internal ribosomal entry site (IRES) so the positive sense RNA can be translated into proteins in a cap-independent manner. Three structural (i.e. present in mature viral particles) viral proteins are encoded at the 5′ end of the genome – a core protein which surrounds and protects the RNA genome, and two envelope glycoproteins which are inserted into the lipid envelope. The 3′ end of the genome encodes a number of non-structural (NS) proteins including two protease enzymes (NS2, and an NS3/4a complex), and an RNA polymerase (NS5b) enzyme. NS4b forms a membranous web within an infected cell which acts as a scaffold for virus replication, while NS5b interacts with a number of host proteins and is involved in viral replication and egress from the cell.

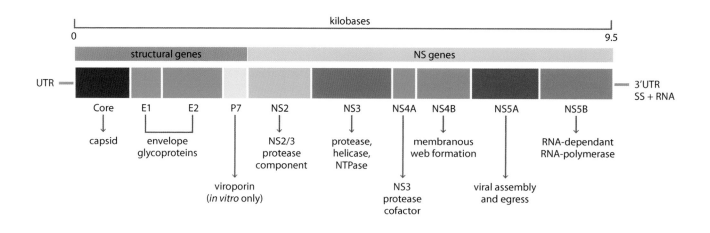

Figure 15.2 **The genome structure and encoded proteins of HCV.** UTR = untranslated region; E = envelope; NS = non-structural.

The replication cycle of HCV starts with binding of virus particles to cell surface molecules such as heparan sulphate and the low-density lipoprotein receptor. Further interaction with a number of other receptor molecules including CD81, scavenger receptor B1, claudin-1 and occludin results in virus entry into hepatocytes via endocytosis and fusion of viral and clathrin-coated endosomal membranes. In addition to translation of the genome into viral proteins, full-length anti-sense copies of the genome are synthesized, which act as a template for the synthesis of multiple positive-strand RNA copies. Replication takes place within a membrane-associated replication complex within a double-membrane vesicle (DMV). New viral particles bud through the endoplasmic reticulum and egress from the cell as a lipoprotein-associated "lipoviral particle" (LVP) via the Golgi secretory pathway.

ENTRY AND SPREAD WITHIN IN THE BODY

Acquisition of HCV infection requires inoculation of infected blood. Once virus particles have entered the bloodstream, they gain access to the liver. Hepatocytes express all of the viral receptors listed above, hence the observed hepatotropism of this virus. There is little evidence that HCV is able to replicate in any other cell type. New virus particles formed within hepatocytes are released into the bloodstream, rendering that person a potential infection source to any contacts exposed to their blood. While it is impossible to prove that HCV *cannot* be transmitted sexually, in marked contrast to other blood–borne viruses (hepatitis B and HIV) there is no increased rate of HCV infection in groups such as commercial sex workers, individuals with multiple sexual partners, men who have sex with men, who traditionally are at increased risk of a sexually transmitted infection. Mother-to-baby (vertical) transmission of HCV is also unusual – only around 1% of babies of HCV-infected mothers acquire infection. Again, this is in marked contrast to HBV (>90%) and HIV (20–25%).

PERSON-TO-PERSON SPREAD

Virus is present within the bloodstream of patients with chronic HCV infection. Infection therefore requires exposure to infected blood. This can occur through a variety of means. Most obviously, transfusion of blood or blood products derived from an HCV-infected donor will result in transmission. In the past, this led to a significant number of infections, particularly in certain groups of patients such as hemophiliacs who require factor VIII replacement. However, introduction of blood-donor screening since 1989 has reduced such transmission to very small numbers worldwide.

More subtle blood-to-blood exposure can occur through sharing of contaminated needles and syringes by people who inject drugs, contaminated body- or ear-piercing needles, and acupuncture needles. Healthcare workers are at increased risk of HCV infection through accidental needlestick injuries from HCV-infected patients. Any re-usable hypodermic needles which are not properly sterilized between use can potentially transmit HCV infection.

EPIDEMIOLOGY

The WHO estimate there are in excess of 70 million individuals with chronic HCV infection globally. These are unevenly spread around the world, and the underlying drivers of HCV epidemics are different in different countries. Thus, in the UK (0.5% of the population), the epidemic has been driven by recreational injecting drug use – the majority of HCV-infected individuals give a history of having injected drugs, maybe decades ago. In contrast, in Egypt, with a prevalence approaching 20%, the epidemic was driven by bilharzia eradication campaigns in the 1960s to 1980s which involved the injection of an anti-Schistosoma drug with inadequately sterilized re-usable needles and syringes. This "medically-driven" means of transmission has also been a major route of HCV transmission in a number of other countries.

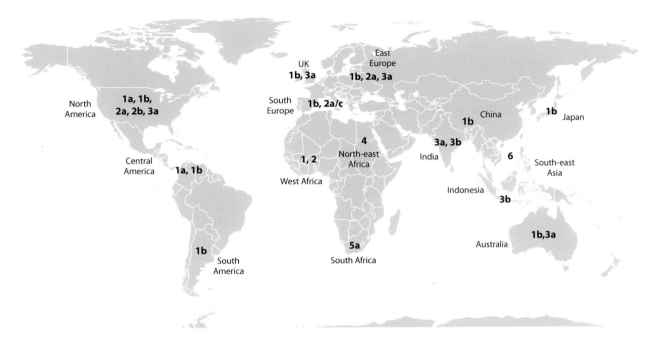

Figure 15.3 Distribution of the major genotypes of hepatitis C virus around the world. *Adapted from a figure courtesy of Dr Alexander Tarr.*

The virus has clearly evolved into different genotypes within different host populations around the world. This is reflected in different genotype distributions in different countries – see Figure 15.3.

2. WHAT IS THE HOST RESPONSE TO THE INFECTION AND WHAT IS THE DISEASE PATHOGENESIS?

INNATE IMMUNITY

The presence of HCV in hepatocytes activates a number of pathogen recognition receptors, including Toll-like receptor 3, protein kinase R (PKR), retinoic acid-inducible gene 1 (RIG-I) and melanoma differentiation-associated protein 5 (MDA5), leading to activation of transcription factors such as interferon regulatory factor 3 (IRF3) and nuclear factor kappa B (NF-κB) which, in turn, enhance expression of interferon genes. In chronic infection, this results in a state of continuous innate immune activation, with increased expression of pro-inflammatory gene expression which will encourage the development of tissue fibrosis. Activation of innate immune cells such as NK cells and liver-resident macrophages (Kupfer cells) also contribute to this process.

HCV has evolved a number of ways of subverting this response – intracellular viral replication takes place within a double membrane (comprised of the NS4b viral protein) bound vesicle thereby "hiding" away within the cell; the NS3/NS4a protease enzyme cleaves mitochondrial antiviral signaling (MAVS) protein, a key protein within the RIG-I signaling pathway; HCV also interferes with the downstream signaling actions of interferons via STAT1, STAT2, IRF9, and JAK-STAT pathways.

Host genetic studies have identified a number of single nucleotide polymorphisms (SNPs) within the interferon lambda (*IFNL*) locus which are associated with the likelihood of spontaneous clearance of infection, and also the likelihood of clearance of chronic infection following treatment with interferon-α. These SNPs alter the activity of interferon-λ4, although the precise mechanisms linking the functionality of IFN-λ4, and viral clearance have yet to be elucidated.

ADAPTIVE IMMUNITY

HCV infection stimulates production of antibodies, including those directed against the two surface envelope glycoproteins (E1 and E2). Most antibodies are directed at a region of the E2 protein named hypervariable region 1 (HVR1). The name reflects the fact that mutations in this part of the coding gene are common and lead to amino-acid changes in the protein, this being one way in which HCV evades the adaptive humoral response. Only a minority of anti-E2 antibodies are able to neutralize viral binding to target hepatocytes. Generation of broadly neutralizing antibodies (Nabs), defined as antibodies that can neutralize cell-culture infection with viruses of more than one genotype during acute infection, is associated with viral clearance.

CD4+ and CD8+ T-cell responses are also generated during infection. In patients with chronic infection, these T-cell responses are weak or undetectable, whereas in patients who have achieved spontaneous clearance, T-cell responses are stronger and more broadly targeted (i.e. recognize a number of epitopes in different viral proteins).

PATHOGENESIS

HCV infection of hepatocytes is not cytolytic – cell death arises from host immune responses, both innate and adaptive. Alcohol is a potent hepatotoxin and acts synergistically with HCV to cause liver damage – patients should therefore be advised to abstain, at least until their HCV infection is cleared. Age does not appear to influence spontaneous clearance of infection, but older age at infection is associated with more progressive liver disease. All the life-threatening complications of chronic infection can arise with any genotype. There is some evidence that disease may progress more rapidly with genotype 3 infection, which is also associated with more fatty liver disease than any of the other genotypes.

3. WHAT IS THE TYPICAL CLINICAL PRESENTATION AND WHAT COMPLICATIONS CAN OCCUR?

ACUTE INFECTION

The majority of acute HCV infections are asymptomatic – it is estimated that only about 15% will present with an acute hepatic illness. Initial symptoms of acute hepatitis are non-specific – lethargy, fever, nausea, loss of appetite, with the only localizing symptom being right upper quadrant abdominal pain as the inflamed liver swells inside its innervated capsule. Damage to liver cells may impact on liver function, with resultant generation of jaundice, dark urine and pale stools (see Case 13). While theoretically possible, acute liver damage sufficient to cause liver failure – fulminant hepatitis – is rare.

However, 50–75% of patients will fail to clear infection at the acute stage and will therefore progress to chronic infection (defined as persistence of HCV RNA for more than 6 months). Female sex, and the presence of a vigorous immune response

sufficient to induce symptomatic disease (jaundice) are factors associated with an increased chance of viral clearance. In contrast to HBV, age does not appear to be a determinant of clearance/chronicity.

CHRONIC INFECTION

During chronic infection, there is continual virus replication and release of new virus particles into the bloodstream. Death of hepatocytes may arise through the detrimental effects of virus replication going on in the cell, or through immune recognition of infected cells by NK and cytotoxic T cells. The repair mechanism in the liver is to replace dead hepatocytes by fibrous tissue. Over many years, this process results in the generation of cirrhosis (see Case 13) with all its attendant life-threatening complications. Also similar to HBV, patients with chronic HCV infection are at hugely increased risk of developing primary liver cell tumors (hepatocellular carcinoma, HCC). For HCV, this appears to be a consequence of the liver becoming cirrhotic – HCV does not possess an equivalent protein to the HBX protein, and also has no DNA intermediate which might integrate into the host cell chromosomes. It is estimated that between them, chronic HBV and HCV infections account for over 75% of the global burden of HCC, which is the third commonest cause of death from malignant tumors.

COMPLICATIONS

In addition to chronic hepatitis, cirrhosis, and HCC, chronic HCV infection is also associated with a number of extra-hepatic complications (Figure 15.4), most notably mixed cryoglobulinemia. A cryoglobulin is a protein which precipitates out at cold temperatures. Circulating cryoglobulins in HCV infection include immune complexes of HCV and anti-HCV. Precipitation in blood-vessel walls at cold temperatures stimulates a vasculitis which typically affects

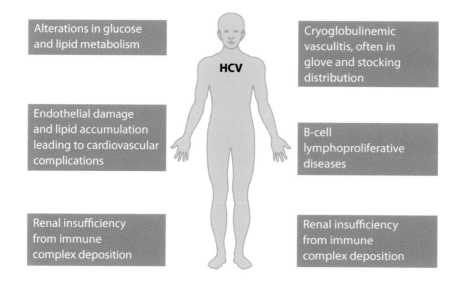

Figure 15.4 Possible extrahepatic manifestations of chronic hepatitis C virus infection.

Table 15.1 Risk factors for HCV infection for patients in the UK

Current or past history of recreational drug use.

People in prison (more likely to have history of injecting drug use).

Recipients of blood (before Sept 1991) or blood product (before 1986) transfusion.

UK recipients of organ or tissue transplants before 1992.

People who have lived or had medical treatment in countries where HCV infection is common.

Babies and children of HCV-infected mothers.

Recipients of tattoos or piercing where equipment may not have been properly sterilized.

Sexual partners, family members and close household contacts of people with HCV.

hands and feet – a so-called "glove and stocking" distribution, due to the lower temperatures of those limb extremities. Deposition in the kidneys can result in glomerulonephritis. There is also evidence that chronic HCV infection may be causally associated with some types of lymphoma. Genotype 3 infection specifically alters lipid metabolism within the liver and may predispose to insulin resistance.

4. HOW IS THE DISEASE DIAGNOSED, AND WHAT IS THE DIFFERENTIAL DIAGNOSIS?

Any patient with risk factors for HCV infection (Table 15.1) should be tested for antibodies to HCV (anti-HCV). This is a straightforward assay usually done in an enzyme-linked immunosorbent assay format. Anti-HCV positivity means that the patient has, at some stage, been infected with HCV, but gives no indication as to when this might have occurred, or, more importantly, whether the patient succeeded in clearing the infection or became chronically infected. A subsequent test for HCV RNA (i.e. a genome detection test, usually the **polymerase chain reaction [PCR]** assay) distinguishes those individuals who have cleared the infection (HCV RNA-negative) from those who are chronic carriers (HCV RNA-positive). The latter should then be further tested to determine which particular genotype and subtype the patient is infected with as this will, in turn, determine which is the optimal form of antiviral therapy. Genotyping is done by sequencing all or part of the viral genome and comparing the sequence with standard databases to identify which geno/sub-type the virus is most closely related to.

DIFFERENTIAL DIAGNOSIS

There is a long list of differential diagnoses of acute jaundice, including acute alcoholic hepatitis, and infection with one of many other hepatitis viruses – see Case 13 for diagnosis of acute HBV, HAV, HEV, and HDV.

Similarly, there are many causes of chronic liver damage and cirrhosis of the liver. Common etiologies of cirrhosis include chronic HBV or HCV infection, alcoholic liver disease, non-alcoholic fatty liver disease, hemochromatosis, Wilson's disease, and auto-immune hepatitis. It is beyond the scope of this text to go into detail about how each of these conditions is diagnosed.

5. HOW IS THE DISEASE MANAGED AND PREVENTED?

MANAGEMENT

All patients with HCV infection should be informed about the potential synergistic liver-damaging effects of alcohol and advised to abstain if possible.

The treatment of chronic HCV infection has undergone a revolution in the past 10 years. Initially, therapy consisted of a combination of interferon and ribavirin given for 6 or 12 months (depending on genotype). This was a difficult therapy for patients to comply with, and had considerable side effects and a cure rate overall of only around 50%. However, the development of directly acting antiviral agents (DAAs) targeted at the products of the viral NS genes NS3 (a protease enzyme), NS5a (a protein essential for several steps in the

Table 15.2 Combination therapies for the treatment of HCV infection

Drug name	NS3 inhibitor	NS5a inhibitor	NS5b inhibitor
Harvoni		Ledipasvir	Sofosbuvir
Zepatier	Grazoprevir	Elbasvir	
Maviret	Glecaprevir	Pibrentasvir	
Epclusa		Velpatasvir	Sofosbuvir
Vosevi	Voxilaprevir	Velpatasvir	Sofosbuvir

viral replication cycle), and NS5b (the viral RNA polymerase enzyme) has allowed the abandonment of interferon-based therapy. The NS3/4a inhibitors have the suffix -*previr*, the NS5a inhibitors have the suffix -*asvir*, and the NS5b inhibitors are -*buvir*. Some of these NS proteins are quite variable between different genotypes of the virus, and therefore some DAAs are genotype-specific, while others are pan-genotypic. Combination therapy, chosen on the basis of which geno/subtype the patient is infected with, with two or more classes of DAA administered as tablets once a day for 8–12 weeks has been shown to achieve cure of chronic HCV infection in over 95% of patients, and is associated with very few adverse events. Commonly used combinations of DAAs marketed in single tablets are shown in Table 15.2.

PREVENTION

It has proven extremely difficult to develop an effective vaccine for HCV. The major challenge is the huge amount of genetic diversity such that vaccine candidates may protect experimental animals against infection with virus of the same genotype, but not against other genotypes. There has been only one large-scale phase 3 trial of a candidate vaccine, which had no demonstrable effect on the rate of infection in people who inject drugs in Baltimore.

The story of HCV, from identification in 1989 to the development of drugs with almost 100% cure rates 25 years later, is a triumph of modern science and medicine – and has duly been recognized as such by the award of the 2020 Nobel prize for Medicine or Physiology to three of the major players, Harvey Alter, Charles Rice, and Michael Houghton. The outstanding success of DAA therapy has contributed to the notion that it may be possible to eradicate HCV altogether. While the WHO has declared a target of 2030 for the elimination of HCV as a public health hazard, and many countries have therefore adopted elimination strategies based around increased diagnosis and better linkage to treatment for infected individuals, it is unlikely that eradication will be achieved in the absence of a prophylactic vaccine.

SUMMARY

1. WHAT IS THE CAUSATIVE AGENT, HOW DOES IT ENTER THE BODY, AND HOW DOES IT SPREAD A) WITHIN THE BODY AND B) FROM PERSON TO PERSON?

- Hepatitis C virus belongs to the Flaviviridae family, with a positive ssRNA genome.
- Virus replicates in the liver and is released into the bloodstream.
- Transmission is via contaminated blood, blood products or needles and syringes.
- Sexual transmission is unusual, as is mother-to-baby (around 1%).

2. WHAT IS THE HOST RESPONSE TO THE INFECTION AND WHAT IS THE DISEASE PATHOGENESIS?

- Recognition of pathogen-associated molecular patterns by pathogen-recognition receptors triggers potent innate immune responses, dominated by production of interferon and interferon-responsive gene products.
- Clearance of virus at the acute stage is associated with certain SNPs in the IFN-λ4 gene.
- Humoral and cellular adaptive immune responses are also critical in determining the outcome of acute infection. Viral clearance is associated with generation of broadly neutralizing antibodies and broad and potent T-cell responses.
- Liver cell death is mediated by host inflammatory responses.
- Hepatocyte loss leads to replacement with fibrous tissue, leading eventually to cirrhosis.

3. WHAT IS THE TYPICAL CLINICAL PRESENTATION AND WHAT COMPLICATIONS CAN OCCUR?

- Most acute infections are asymptomatic; 15% of cases present as an acute jaundice.

- The majority of infections become chronic, with continuous viral replication, stimulating immune-mediated liver cell death.
- Loss of hepatocytes leads to replacement of the normal liver architecture by fibrous tissue, leading to cirrhosis.
- Chronic infection with cirrhosis carries a significant risk of development of hepatocellular carcinoma.
- Chronic HCV infection is also associated with mixed essential cryoglobulinemia, glomerulonephritis, and development of some types of lymphoma.

4. HOW IS THE DISEASE DIAGNOSED, AND WHAT IS THE DIFFERENTIAL DIAGNOSIS?

- The primary diagnostic test is for anti-HCV by ELISA.
- There are many causes of acute jaundice including infection with other hepatitis viruses and acute alcoholic hepatitis.
- There are many causes of cirrhosis of the liver, including chronic HBV infection, alcoholic liver disease, and non-alcoholic fatty liver disease.
- Anti-HCV positive samples should be further tested for HCV RNA by a genome amplification assay.

5. HOW IS THE DISEASE MANAGED AND PREVENTED?

- Treatment of HCV infection consists of combination therapy with two or more directly acting antiviral agents targeting key viral proteins (NS3, NS5a or NS5b). Such therapy is successful in over 95% of patients.
- There is currently no vaccine available for prevention of HCV infection.

FURTHER READING

Barer M, Irving W, Swann A, Perera N. Medical Microbiology, 19th edition. Elsevier, Philadelphia, 2019.

Humphreys H, Irving WL, Atkins BL, Woodhouse AF. Oxford Case Histories in Infectious Diseases and Microbiology, 3rd edition. Oxford University Press, Oxford, 2020.

Murphy K, Weaver C. Janeway's Immunobiology, 9th edition. Garland Science, New York/London, 2016.

Richman DD, Whitley RJ, Hayden FG. Clinical Virology, 4th edition. ASM Press, New York, 2016.

REFERENCES

Castaneda D, Gonzalez AJ, Alomari M, et al. From Hepatitis A to E: A Critical Review of Viral Hepatitis. World J Gastroenterol, 27: 1691–1715, 2021.

Dennis BB, Naji L, Jajarmi Y, et al. New Hope for Hepatitis C Virus: Summary of Global Epidemiologic Changes and Novel Innovations Over 20 Years. World J Gastroenterol, 27: 4818–4830, 2021.

European Association for the Study of the Liver. EASL Recommendations on Treatment of Hepatitis C: Final Update of the Series. J Hepatol, 73: 1170–1218, 2020.

Mazzaro C, Quartuccio L, Adinolfi LE, et al. A Review on Extrahepatic Manifestations of Chronic Hepatitis C Virus Infection and the Impact of Direct-Acting Antiviral Therapy. Viruses, 13: 2249, 2021.

Schwerk J, Negash A, Savan R, Gale M. Innate Immunity in Hepatitis C Virus Infection. Cold Spring Harb Perspect Med, 11: a036988, 2021.

WEBSITES

All the Virology on the WWW Website, Specific Virus Information: http://www.virology.net/garryfavweb12.html#Hepad

Centers for Disease Control and Prevention, Viral Hepatitis, 2021: https://www.cdc.gov/hepatitis/index.htm

Virology Online, Hepatitis B: http://virology-online.com/viruses/HepatitisB.htm

World Health Organisation, Hepatitis C, 2022: https://www.who.int/news-room/fact-sheets/detail/hepatitis-c

Students can test their knowledge of this case study by visiting the Instructor and Student Resources: [www.routledge.com/cw/lydyard] where several multiple choice questions can be found.

Herpes simplex virus

16

A 13-year-old boy was taken to the hospital's emergency department by ambulance. His mother had called out the emergency services as he had had a generalized **fit** that morning. Since he stopped fitting, he had been very drowsy. His mother reported that he was well until 24 hours previously, when he started acting strangely – including wandering around the house not knowing where he was. He vomited once the previous evening, but otherwise has not complained of any specific problems.

On examination, he was drowsy but responsive. He was unable to give a coherent history and had difficulty in understanding where he was. He was **febrile** (38.5°C) but had no other abnormal physical signs.

An intravenous (IV) line was set up, and he was started on empirical antibiotic therapy, together with IV aciclovir. A **magnetic resonance imaging (MRI) scan** of his brain was organized for later in the day, which was reported by the duty radiologist as

showing "an area of low attenuation in the right temporal lobe extending into the frontal lobe gray matter. There was mass effect with displacement of the right middle cerebral artery, appearances most compatible with herpes simplex **encephalitis**".

A lumbar puncture was performed that evening. The cerebrospinal fluid (CSF) was noted to be slightly bloodstained. The microbiology technician reported the presence of 500×10^6 red blood cells l^{-1}, 57×10^6 white cells l^{-1}, predominantly lymphocytes, a normal CSF sugar, and a CSF protein level of $4.8\,g\,l^{-1}$, just above the upper limit of normal. No organisms were seen on a gram-stained film.

A provisional diagnosis of herpes simplex encephalitis was made. This was confirmed 2 days later when a report was phoned through from the virology reference laboratory indicating that the CSF was positive for the presence of HSV DNA as tested by **polymerase chain reaction (PCR)**.

1. WHAT IS THE CAUSATIVE AGENT, HOW DOES IT ENTER THE BODY AND HOW DOES IT SPREAD A) WITHIN THE BODY AND B) FROM PERSON TO PERSON?

CAUSATIVE AGENT

Herpes simplex virus (HSV) type 1 is a herpesvirus. The *Herpesviridae* family of viruses is characterized by having a double-stranded DNA genome of 125–240kb in size (i.e. enough genetic material to encode for 150–200 viral proteins), surrounded by an icosahedral **capsid**, and a **lipid envelope**. HSV genomes are around 150–155kb and encode around 100 viral proteins. The space between the capsid and the envelope is referred to as the **tegument**. This contains a number of virally encoded proteins, some of which are thought to play a role in viral transport within nerves (see later). Embedded within the envelope are several virally encoded glycoproteins, which are important for binding of the virus to target cells and subsequent cell entry. Thus far, eight herpesviruses have been identified as pathogens of humans – see Case 9.

The genome sequences of HSV-1 and HSV-2 share considerable homology and the biological properties of these two viruses are indeed similar. Disease arising from infection with each virus is clinically indistinguishable. However, HSV-1 infections tend to occur "above the waist", that is oral,

while HSV-2 infections are classically "below the waist", that is genital. Herpes simplex encephalitis (HSE) is almost always due to HSV-1 infection.

ENTRY AND SPREAD WITHIN THE BODY

HSV infects at mucosal surfaces (oropharynx and nasopharynx, conjunctivae, genital tract). Intact skin is impervious to HSV, the difference being that skin is covered by a thick layer of keratin, while mucosa has nonkeratinized epithelium.

Glycoprotein B in the envelope appears to be one of the molecules involved in attachment of HSV to mucosal epithelial cells. The interaction of HSV with epithelial cells is **cytolytic**, with cell death occurring 24–48 hours after infection. Virus is released from infected cells by budding, and one infected cell may give rise to thousands of progeny virus particles. These new virus particles infect neighboring cells, leading potentially to extensive local spread at the site of infection.

LATENCY OF HSV

At some stage during this process, virus enters nerve terminals and travels in a retrograde direction up the nerve axon to reach the nerve-cell body, the site of **latency** of HSV (Figure 16.1). Following HSV infection in the mouth, the trigeminal ganglia are typically infected, with occasional extension to cervical ganglia; genital infection, in contrast, results in infection of the lumbosacral nerve root ganglia. Within a latently infected

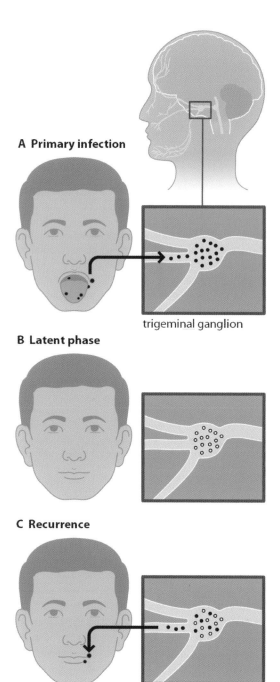

A Primary infection

trigeminal ganglion

B Latent phase

C Recurrence

Figure 16.1 Schematic diagram illustrating HSV latency. (A) During primary infection of the mouth, there is active viral replication within the oropharyngeal mucosa (red dots). Some of this virus enters nerve terminals and travels up the nerve axon toward the nerve cell body – in this case, the trigeminal ganglion – which is the site of latency for this virus. (B) After recovery from the acute infection, the patient enters the latent phase. There is no viral replication within the mouth or trigeminal ganglion (absence of red dots), but the viral genome lies latent within the latter (open red circles). (C) Following reactivation, latent virus (open red circles) within the trigeminal ganglion may start to replicate (red dots), and newly formed viral particles travel back down the nerve axon to reach the periphery (i.e. the mouth), where further virus replication takes place, although this is limited by the effects of the host immune system. The clinical manifestation of this recurrence, i.e. a cold sore, is therefore much more localized than in primary infection. *Adapted with permission from the Wellcome Trust.*

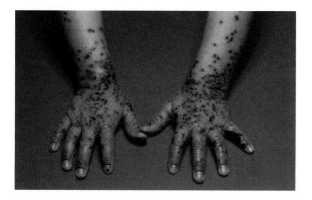

Figure 16.2 Eczema herpeticum. Virus is able to spread across nonintact skin in patients with eczema, giving rise to eczema herpeticum. *From St Bartholomew's Hospital, London / Science Photo Library, with permission.*

cell, viral genome is present as an **episome** within the nucleus, but there is no detectable production of virus proteins, and no production of new virus particles. Molecular mechanisms underlying entry into, maintenance of, and subsequent reactivation from latency are poorly understood, although well-recognized triggers for reactivation include local trauma, immunosuppression, ultraviolet light, and "stress". While virus is in a latent state, there is no evident damage to the infected cell. Reactivation from latency occurs when the viral replication cycle resumes, resulting in new virus particles that travel down the axon to reach the periphery, where they may give rise to clinical disease. The first exposure of an individual to HSV infection is referred to as the *primary* infection, while a reactivated infection is referred to as a *recurrent*, or *secondary*, infection. The clinical manifestations of these two forms of herpesvirus infection are often quite distinct as, in the former case, the infection is occurring in an immunologically naïve individual while, in the latter case, virus is replicating in a host whose immune system has seen the virus before.

In patients with oropharyngeal or genital HSV infection, an additional risk is spread to other mucosal sites via direct self-inoculation of virus, for example, infection of the fingernail bed (known as a herpetic whitlow, or **paronychia**); virally contaminated fingers can spread infection to the eyes, with conjunctival infection and the risk of **keratitis**. Virus may be scratched into the skin, resulting in a local crop of vesicles. If skin is not intact, for example, in patients with a chronic dermatitis such as **eczema**, virus may spread across the skin and cause very extensive lesions – so-called eczema herpeticum (this is usually associated with oral infection, see Figure 16.2). This is potentially life threatening, as virus may gain access to the bloodstream through the skin abrasions, and thereby seed internal organs– such **viremia** is thus potentially life threatening.

Spread of virus into the brain, an event which necessarily precedes the onset of HSE, is discussed as part of the disease pathogenesis raised under Question 2 below.

PERSON-TO-PERSON SPREAD

Herpetic lesions present externally, for example, in the **oropharynx** or genital tract, and contain very large amounts of virus. In addition to the development of symptomatic recurrent disease, reactivation of virus may occur in the absence of clinical disease. Such asymptomatic recurrences result in transient (e.g. 24 hours) shedding of virus from the mouth or genital tract. Infection can then be transmitted to others through direct contact with infected mucosal sites, for example, through kissing or unprotected sexual intercourse. Oral herpes infection can also be transmitted in this way to the genital area via oro-genital sex.

One very important route of person-to-person spread of genital HSV is from mother to baby. This may arise when a baby is born through an infected birth canal due to maternal primary or even reactivated infection and may have devastating consequences for the baby (see clinical features below).

EPIDEMIOLOGY

Oral and genital HSV infections occur worldwide. Oral infection is usually with HSV-1, and genital infection with HSV-2, but these boundaries have become blurred with increasing safe-sex practices leading to oro-genital transmission of HSV-1 (and vice versa, i.e. oro-genital transmission of HSV-2). In the UK, up to half of all new genital herpes infections may be due to HSV-1. However, as genital HSV-2 infection is much more likely to recur than is genital HSV-1, HSV-2 remains the principal cause of recurrent genital herpes.

Large-scale seroprevalence surveys looking for antibodies specific to HSV-2, mostly conducted within the US, at a time before genital HSV-1 infection was common, suggest that up to 20% of the adult population is HSV-2 **seropositive**. Seroprevalence increases with age and number of lifetime sexual partners and is higher in black as compared with white populations. However, only a small proportion of individuals with antibodies to either HSV-1 or HSV-2 give a history of primary or recurrent oral or genital herpes, demonstrating the largely asymptomatic nature of the infection.

2. WHAT IS THE HOST RESPONSE TO THE INFECTION AND WHAT IS THE DISEASE PATHOGENESIS?

IMMUNE RESPONSE

Innate Immunity

Innate immunity is key to prevention or control of a primary infection. The importance of intact skin has been referred to above. Initial infection at mucosal sites will lead to **interferon** (**IFN**) production, as HSV-associated pattern-associated molecular patterns (PAMPs) are recognized by pattern recognition receptors (PRRs). Interferons bind to their cognate receptors leading to a signaling cascade which ultimately generates induction of interferon-stimulated genes, the products of which induce an antiviral state in the infected cells and their neighbors. Toll-like receptors (TLRs) are a large family of PRRs. Glycoprotein B of HSV binds to TLR-2 leading to expression of several pro-inflammatory cytokines. TLR-3 recognizes double-stranded (ds) RNA and has an important role in controlling HSV in neuronal cells – patients with TLR-3 deficiencies or mutations are more susceptible to developing HSE. TLR-9 is expressed in microglia, astrocytes, and antigen-presenting cells. It recognizes dsDNA and mediates early and rapid production of type 1 interferons. Mice deficient in this TLR are unable to mount an effective immune response to HSV and therefore die when infected.

Another important cytosolic DNA sensor is cyclic GMP-AMP Synthase Stimulator of Interferon Genes (cGAS-STING). cGAS-STING signaling also leads to expression of type 1 IFN genes. HSV has evolved multiple ways of avoiding cGAS-STING responses – the HSV protein UL41 degrades cGAS mRNA; HSV protein VP22 interacts directly with cGAS and inhibits its enzymatic activity. HSV protein UL46 interferes with the downstream signaling pathway following cGAS-STING activation.

HSV has also evolved ways of subverting activation of complement and the downstream effectors resulting from this process, suggesting that such processes are an important defense mechanism. Thus, HSV glycoprotein C (gpC) inhibits the binding of C5 and properdin to C3b, thereby blocking the alternative pathway of complement activation. gpC also accelerates the decay of the alternative pathway C3 convertase.

Adaptive Immunity

Carriage of the virus by immature dendritic cells to draining lymph nodes leads to the development of an adaptive immune response. Both **CD4+** T cells and cytotoxic **CD8+** T cells are produced in response to infection and play a role in elimination of the virus. Memory T lymphocytes appear to be effective in response to local episodes of recurrent infection and home efficiently to sites of infection.

In experimental mouse models of genital herpes infections, **IgG** antibodies appear to be protective but **IgA** is less so.

Once virus has become latent in nerve-cell bodies, the adaptive immune response, particularly the T-cell arm, is vital in preventing, or at least limiting, disease caused by reactivation. Thus, patients with cellular immunodeficiency, such as HIV-infected individuals, or transplant recipients, are much more prone to symptomatic recurrent disease.

Infection with HSV-1 does not induce protective immunity against HSV-2 infection, and vice versa.

HSV infection is common and may give rise to disease at a number of distinct sites within the body – see Table 16.1. HSE is fortunately rare – estimated incidence is around 1–3 per million population per year.

HSV-1 is the cause of encephalitis in more than 90% of cases. More than two-thirds of cases of HSE are the result of reactivation of latent virus in previously infected individuals,

Table 16.1 Diseases caused by herpes simplex infection

Orolabial infection

    Primary – gingivostomatitis

    Recurrent – cold sore

Genital infection

Keratitis (infection of the cornea)

Herpetic whitlow – infection of the nail bed

Eczema herpeticum

Herpes encephalitis

Herpes gladiatorum – scratching of virus into the skin, e.g. in wrestlers

with the remainder due to primary infection in children and young adults. Neonatal HSE is distinct in that HSV-2 is the more usual pathogen because of maternal genital infection resulting in intrapartum transmission.

Patients suffering from HSE undoubtedly have virus present within the brain substance itself, but it is unclear when and how virus reaches this site. The precise anatomic pathways whereby virus enters the brain substance are not known with certainty. Virus may travel via either or both the trigeminal and olfactory nerve tracts. In patients with reactivated disease, there is also controversy concerning the site of the latent virus that reactivates. Virus may reactivate peripherally (within the trigeminal ganglion or olfactory bulb) and enter the central nervous system (CNS) by retrograde transport or may reactivate centrally within latently infected brain tissue. HSV DNA has been demonstrated by sensitive genome-amplification assays within post-mortem brain tissue in up to one third of asymptomatic HSV **seropositive** adults. However, the possibility of reactivation of virus from this site remains a speculative hypothesis at the moment.

Despite the above uncertainties, what is not in doubt is that virus is indeed present and replicating within the brain substance in a patient with HSE. HSV infection in this situation

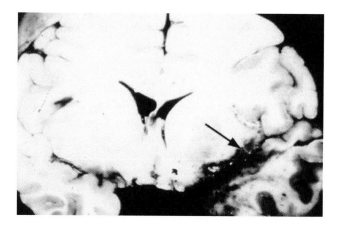

Figure 16.3 Herpes simplex encephalitis: gross pathology. The arrow indicates a large area of hemorrhagic necrosis in the left frontotemporal region of the brain.

is cytolytic. The presence of replicating virus generates an acute inflammatory host response, with an infiltration of mononuclear cells, the characteristic immune response to a virus infection. Cell death arises from both virus infection and the host immune response, resulting in hemorrhagic **necrosis**, the pathologic hallmark of HSE (see Figure 16.3). Involved necrotic brain tissue becomes liquefied, and the disease spreads outward as new virus particles are released. Microscopic examination of brain tissue reveals damage extending much further than that visible macroscopically.

3. WHAT IS THE TYPICAL CLINICAL PRESENTATION AND WHAT COMPLICATIONS CAN OCCUR?

HERPES SIMPLEX ENCEPHALITIS

HSE can arise in any part of the brain, and the neurologic manifestations are dependent on the precise anatomic site. Large case series reported from the National Institutes of Health in the US show that around 75% of patients demonstrate altered behavior or personality change – reflecting the commonest underlying anatomic site of disease in the frontotemporal regions. Also, 80% of patients complain of headache, 90% have fever, and almost all have evidence of decreased level of consciousness. Other presenting manifestations include **seizures** (67%), vomiting (46%), **hemiparesis** (33%), and memory loss (25%).

The pathologic process underlying the disease is such that the only relevant "complication" of HSE is continuing and worsening brain damage, leading to death. Untreated, the case mortality of HSE is around 70%, and only a small fraction (10%) of survivors are able to return to normal function after recovery from the acute illness. These figures have changed with the advent of antiviral therapy, although mortality may still be as high as 20–30%.

GENITAL HSV INFECTION

As explained above, most primary genital (or oral) infections with HSV do not give rise to clinical disease. The assumption must be that expression of disease is a result of the race between replication and spread of the virus on the one hand, and the ability of the innate immune response to control this on the other.

In the minority of patients where disease does become clinically evident, primary genital herpes is not a trivial disease. Lesions are extensive and bilateral, extending from the labia to the cervix. There may be spread onto adjacent skin. There will be painful inguinal **adenopathy**, and a systemic response, (e.g. fever) reflecting the release of **cytokines** from the intense inflammatory reaction taking place. In females, passing urine may be exquisitely painful if there are lesions near the urethral meatus. It may take 2–3 weeks before the patient fully recovers.

In marked contrast, recurrent disease, even when symptomatic, is usually mild, as the reactivation is taking place in an individual whose immune system has seen the virus before. Thus, the lesions will be much more localized, mostly unilateral, with no spread onto the adjacent skin; no, or only unilateral, inguinal adenopathy; and no, or only mild, systemic symptoms. The natural history of recurrent lesions is resolution within 5–7 days. Many recurrences are asymptomatic – and it must be individuals who are not aware of asymptomatic reactivation of disease who are the source of spread to their sexual partners.

There are a number of possible complications. In primary disease, there may be meningeal irritation, such that the patient presents with a clinical diagnosis of **meningitis**. This may rarely happen even with recurrent disease – so-called **Mollaret's meningitis**. HSV meningitis is more common in women. Irritation of the sacral nerve roots (**radiculomyelopathy**) may present with aching pain in the sacral **dermatomes** associated with **parasthesiae** or **dysasthesiae** in the lower limbs. Accidental inoculation elsewhere in the body has been mentioned above, for example, giving rise to herpetic keratitis (infection of the cornea). One disastrous complication of genital herpes in a female is spread to her baby, resulting in neonatal herpes. This usually arises in women suffering a primary attack of genital herpes in late pregnancy, of which she may be unaware. However, the birth canal is rich in virus, and there has not yet been time for the mother to generate and pass protective antibodies transplacentally to the fetus. The majority of neonates infected in this way acquire internally disseminated infection, including herpes **encephalitis**, pneumonitis or hepatitis, which have a high mortality and morbidity even with appropriate therapy. Only about half of these neonates have herpetic lesions evident on their skin or mucous membranes, making the diagnosis very difficult. It is only the babies whose infection is limited to the skin and mucous membrane who make a complete recovery – only about 10–15% of all neonatally infected babies. Fortunately, neonatal HSV infection is not common – up to 17.5 in 100 000 live births in the UK, but perhaps 30 in 100 000 in the US.

In immunocompromised patients, for example with HIV infection, or those on cytotoxic therapy, HSV recurrences are not only more frequent, but may also be somewhat atypical, with extensive local lesions, and the potential for systemic and life-threatening spread.

Genital ulceration of any cause, and certainly including that due to HSV infection, increases the risk of spread of HIV infection, most likely through multiple mechanisms, for example, damage to the skin barrier, attraction of susceptible T cells and macrophages to the site, and up-regulation of HIV replication. HSV-2 infection is associated with a threefold increased risk of sexually acquired HIV infection. Recent data suggest that one way to reduce the spread of HIV infection may be to treat individuals with anti-herpes drugs to prevent the occurrence of genital ulceration.

4. HOW IS THE DISEASE DIAGNOSED, AND WHAT IS THE DIFFERENTIAL DIAGNOSIS?

HERPES SIMPLEX ENCEPHALITIS

A diagnosis of HSE should be considered in any patient presenting with altered consciousness, fever, and/or focal neurologic signs, including seizures. Confirmation of the diagnosis may require both virologic and neuroimaging investigations and may not be straightforward.

Examination of CSF obtained via a lumbar puncture is a routine investigation when assessing a patient with possible encephalitis. However, the findings in HSE are not specific. The CSF typically demonstrates elevated protein and **pleocytosis**, with mononuclear cells of between 10 and 500 cells/mm^3 and red cells predominating, but even a completely normal CSF does not exclude the possibility of HSE. The most sensitive and specific test on CSF is detection of HSV DNA using a genome-amplification technique such as PCR, but this may be negative, especially if the sample is taken less than 72 hours from onset of symptoms. This approach has more or less replaced the need for brain biopsy, which used to be regarded as the definitive test, and may still be necessary in cases of diagnostic difficulty. Note that despite the presence of HSV DNA in CSF, successful isolation of HSV in cell culture is very unusual.

Testing of serum samples for the presence and titer (amount) of anti-HSV antibodies, whether taken in the acute stage, at a later stage, or both, is not particularly helpful in the diagnosis of HSE. Most patients with HSE already have serum antibodies at the time of onset of illness, as the disease represents a reactivation of virus rather than a primary infection. Even a rise in antibody titer taken a few days apart does not confirm HSE, as reactivation of virus, with increased antigen levels boosting an antibody response, may be a consequence of fever or general illness. However, measurement of antiviral antibody synthesis within the CSF may be helpful, provided that serial samples demonstrate a rise in this antibody titer, although such a rise may take several days if not weeks to occur.

The cardinal feature of HSE revealed by neuroimaging is the focal nature of the disease process and its predilection for frontotemporal lobe involvement – as opposed to other causes of encephalitis, which may affect the brain more globally (Figure 16.4). Thus, whichever imaging technique is used, the demonstration of a focal lesion or lesions (disease may spread bilaterally, especially if therapy is delayed) should be interpreted as indicating HSE until proven otherwise. Techniques include (proceeding from the more historical to the most modern) **electroencephalogram (EEG)**, **radionuclide brain scanning**, **CT scan** or **MRI**. MRI scans can often show evidence of HSE before the damage is evident on a CT scan.

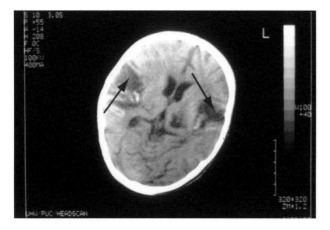

Figure 16.4 Neuroimaging in herpes simplex encephalitis: CT scan of a patient with herpes simplex encephalitis. The arrows point to bilateral areas of low attenuation in the frontotemporal regions. *Originally published in Humphreys and Irving, Figure 42.1, page 191. With permission from Oxford University Press.*

DIFFERENTIAL DIAGNOSIS

The differential diagnosis of HSE is that of any encephalitic illness, and there is therefore a potentially long and complex list of diseases that can mimic HSE. The more common of these are listed in Table 16.2. Some of these, for example Japanese encephalitis virus, West Nile virus, and rabies, are specific to certain geographic locations. Measles virus encephalitis has been virtually eliminated in those countries with adequate immunization rates. CMV and EBV encephalitis is restricted to patients with underlying immunodeficiency.

OROLABIAL OR GENITAL HERPES

The clinical picture of a patient with extensive ulceration in the mouth or genital area is characteristic enough to make a diagnosis, although it is still imperative to confirm this by sending an appropriate sample to the laboratory, as a diagnosis of herpes has important implications both for the individual patient and for his/her sexual partners. In addition, typing of the virus has prognostic significance, as type-2 genital infections are more likely to recur than are type 1 infections. Note that, in any patient with one sexually transmitted infection (STI), it is also appropriate to test for the presence of other STIs.

Recurrent disease, where there may only be one or two ulcers, is much more difficult, and laboratory confirmation should always be sought, as the differential diagnosis of genital ulceration is not straightforward.

Samples sent to the laboratory should be taken by abrading the base of an ulcer or vesicle and breaking the swab off into viral transport medium (isotonic fluid containing antibiotic to prevent bacterial overgrowth). Most laboratories will make the diagnosis by demonstrating the presence of HSV DNA by a genome-amplification assay such as PCR. By use of appropriate primers it is possible to distinguish between HSV-1 and HSV-2 using this approach. Alternatively, immunofluorescence of the abraded cells with **monoclonal antibodies** against HSV-1 and HSV-2 can be used, or virus can be isolated in cell culture. There is even sufficient virus within vesicle fluid to be visualized by electron microscopy, although very few laboratories will perform that these days.

Table 16.2 Differential diagnosis of HSE

Other viral infections that can cause encephalitis	Cytomegalovirus
	Enteroviruses
	Epstein-Barr virus
	Influenza viruses
	Japanese encephalitis virus
	Measles virus
	Mumps virus
	Nipah virus
	Rabies virus
	West Nile virus
	Varicella-zoster virus
Space-occupying lesions	Abscess/subdural empyema
	Bacterial infections (e.g. *Listeria*, *Mycobacterium tuberculosis*, *Mycoplasma*)
	Other infectious agents (e.g. *Rickettsia*, *Toxoplasma gondii*, *Cryptococcus neoformans*)
	Intracerebral tumor
	Subdural hematoma
Systemic diseases	Vascular disease
	Systemic lupus erythematosus
	Toxic encephalopathy

The differential diagnosis is anything that can cause oral/genital ulceration. In the mouth, the most common condition to mimic HSV infection is hand, foot and mouth disease caused by an enterovirus infection. In the genital area, ulceration may arise from physical trauma including excoriation of marked acute candidal vulvitis, chemical burns from disinfectants, **pyogenic infection**, fixed drug eruption, systemic disease (e.g. **Behcet's syndrome**, **erythema multiforme**), other infections, for example the **chancre** of primary syphilis, **chancroid**, **granuloma** inguinale, lymphogranuloma venereum (see *Chlamydia trachomatis*), and dermatologic conditions (e.g. **lichen sclerosis et atrophicus**).

5. HOW IS THE DISEASE MANAGED AND PREVENTED?

MANAGEMENT

The treatment of HSV disease is with aciclovir (acycloguanosine, see Figure 16.5) and its derivatives. Thymidine kinase, a virally encoded enzyme, is able to mono-phosphorylate aciclovir, which is then di- and tri-phosphorylated by host-cell enzymes. Aciclovir triphosphate then competes with GTP for incorporation into the growing viral DNA chain. Once aciclovir is incorporated, the absence of a complete deoxyribose ring means that there are no hydroxyl groups available to participate in formation of phosphodiester linkages with any incoming bases, and therefore further synthesis of the DNA chain is terminated. The antiviral selectivity of aciclovir arises because:

- the drug can only be phosphorylated in virally infected cells as host cells do not possess the necessary thymidine kinase enzyme;
- not only is the drug only activated in infected cells, but because phosphorylation of the molecule effectively reduces the concentration of free aciclovir within the cell, more free drug diffuses into the cell from the extracellular space, resulting in concentration of the drug specifically in virally infected cells;
- in addition to acting as a DNA chain terminator, aciclovir triphosphate binds to, and strongly inhibits, viral DNA polymerase with a much greater affinity than it does cellular DNA polymerases.

Valaciclovir is a valine ester of aciclovir that is hydrolyzed after absorption into valine and aciclovir. However, it is much better absorbed when administered orally than is aciclovir. Penciclovir is a similar molecule to aciclovir, and works in a similar fashion, that is, it requires initial phosphorylation which can only be achieved by viral thymidine kinase, and the triphosphate derivative acts as a potent viral DNA polymerase inhibitor. It is marketed as famciclovir, a complex ester of penciclovir that is better absorbed orally.

Viral resistance to aciclovir may arise through a number of mechanisms. Viral variants lacking a thymidine kinase enzyme (TK⁻) exist in nature and are selected for when aciclovir (or a similar agent) is used. However, the TK gene is an important **virulence** factor for HSV, and TK⁻ variants are not pathogenic. Mutations may arise in the *TK* gene (TK mutants) such that the enzyme no longer phosphorylates aciclovir, but again, many of these mutants have decreased virulence and are not a clinical problem. Mutations may also arise in the viral DNA polymerase gene such that the enzyme no longer binds aciclovir triphosphate. Such DNA pol mutants are fully virulent and are a clinical problem but have, thus far, only been described in heavily immunocompromised individuals who take prolonged courses of aciclovir because of frequent and severe recurrences.

Untreated HSE has a high mortality and survivors are left with major brain damage. Aciclovir should be administered IV in high dosage, $10 \, mg/kg^{-1}$ every 8 hours and continued for 2–3 weeks. The longer the delay in starting antiviral therapy, the more residual damage the patient will be left with, so it is vital that therapy be given as soon as possible. Thus, if a diagnosis of HSE is thought to be a clinical possibility in any given patient, aciclovir therapy must be started immediately, long before the necessary diagnostic tests have been conducted to confirm or refute the diagnosis. In clinical practice, this means that many more patients receive IV aciclovir therapy than turn out actually to have HSE, but given the safety profile of the drug, and the catastrophic consequences of not treating the disease promptly, this is acceptable.

Disease outcome varies from death to full recovery. Key poor prognostic indicators include older age, lower **Glasgow Coma Scale score** at presentation (i.e. decreasing levels of consciousness) and increasing length of time between onset of illness and start of antiviral therapy.

In patients with symptomatic primary oral or genital HSV infection, antiviral therapy should be initiated as soon as possible. This results in rapid cessation of viral replication and resolution of lesions several days faster than in the absence of therapy. Management of recurrent genital herpes is, however, more controversial. This is, as explained above, a much less severe clinical condition, with a natural history of evolution of lesions to clearance of the order of 6–7 days. Oral therapy

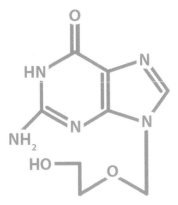

Figure 16.5 Structure of aciclovir. Aciclovir is acycloguanosine – note the absence of a complete deoxyribose ring.

with aciclovir (or derivatives), even if initiated by the patient as soon as he/she is aware that a recurrence is imminent, results in shortening of this disease period by around 24 hours. Thus, in general, there is not a great deal of benefit to be gained by routine therapy of recurrent HSV infection. However, there are exceptions to this. Some unfortunate individuals suffer from rather atypical genital herpes, with frequent attacks (e.g. once a month), which may last for up to 2 weeks. The immunologic reasons for this are poorly understood, but such disease can be managed by use of continuous prophylactic aciclovir. This may continue for several months or even years.

PREVENTION

There is currently no prophylactic vaccine available that protects against infection with HSV.

SUMMARY

1. WHAT IS THE CAUSATIVE AGENT, HOW DOES IT ENTER THE BODY, AND HOW DOES IT SPREAD A) WITHIN THE BODY AND B) FROM PERSON TO PERSON?

- Herpes simplex viruses 1 and 2 are herpesviruses.
- Herpesviruses carry a double-stranded DNA genome in an icosahedral capsid surrounded by a lipid envelope.
- HSV enters at mucosal surfaces (oropharynx and nasopharynx, conjunctivae, genital tract) and replicates within epithelial cells.
- Spread is local, through release of new virus particles from dying cells. Intact skin is an effective barrier to HSV infection.
- Virus is also able to enter nerve cells and travel retrograde to the nerve-cell body, the site of latency of HSV.
- Virus may be reactivated from latency, leading to the presence of virus at the same mucosal sites. Such shedding of virus most commonly is asymptomatic.
- Person-to-person spread arises through direct contact with virus shed from mucosal surfaces gaining access to recipient mucosal surfaces.

2. WHAT IS THE HOST RESPONSE TO THE INFECTION AND WHAT IS THE DISEASE PATHOGENESIS?

- Innate immunity is important in early responses to HSV. Glycoprotein B of HSV binds to TLR-2 leading to expression of several pro-inflammatory cytokines. TLR-3 recognizes double-stranded (ds) RNA and has an important role in controlling HSV in neuronal cells. Mice deficient in TLR 9 are unable to mount an effective immune response to HSV and therefore die when infected. Other PRRs are important.
- An antibody response develops, which is first detectable about 7 days after onset of infection, initially IgM only, followed by IgG.
- T-cell responses are key to preventing symptomatic recurrent disease, as patients with impaired T-cell-mediated immunity are at risk of frequent and severe recurrences.
- Routes of entry into the brain substance are not clear but include via the trigeminal nerve and the olfactory nerve.
- Timing of viral entry into the brain is also controversial – this may occur at the time of primary infection, or at the time of peripheral reactivation of virus in the trigeminal ganglion or olfactory bulb.
- Virus infects brain cells in a cytolytic fashion, resulting in cell death.

- Virus particles released from dying cells spread to adjacent brain tissue.
- The host acute inflammatory response is characterized by an influx of mononuclear cells.
- The pathologic process is one of hemorrhagic necrosis with liquefaction of affected brain tissue.

3. WHAT IS THE TYPICAL CLINICAL PRESENTATION AND WHAT COMPLICATIONS CAN OCCUR?

- Clinical presentation of herpes simplex encephalitis (HSE) is nonspecific, common symptoms and signs including altered behavior, headache, fever, seizures, and altered consciousness.
- Less common manifestations include vomiting, paralysis (hemiparesis), and memory loss.
- Untreated, the disease is progressive, with a high mortality (70%). Survivors may be left with considerable brain damage.
- The majority of primary oral or genital HSV infections are asymptomatic.
- When disease occurs, it is often extensive, bilateral, spread onto adjacent skin, with a systemic response evidenced by regional **lymphadenopathy** and a **febrile** response, lasting for 2–3 weeks.
- Recurrent disease is milder, with fewer crops of lesions, unilateral, no spread onto adjacent skin; no, or only mild, systemic response; and lasting 5–6 days.
- Complications of genital HSV include meningitis (usually women), lumbosacral meningeal irritation presenting as radiculomyelopathy, self-inoculation to other parts of the body, for example, herpetic paronychia, conjunctival infection.
- Genital HSV in pregnancy (usually primary) may give rise to neonatal herpes, which has a high morbidity and mortality.

4. HOW IS THE DISEASE DIAGNOSED, AND WHAT IS THE DIFFERENTIAL DIAGNOSIS?

- HSE should be considered in any patient with altered levels of consciousness, fever, and/or acute onset of focal neurology, including seizures.
- CSF findings are not specific. The commonest CSF picture is an increase in red cells, around 100 white cells μl^{-1}, mostly mononuclear, increased protein, and normal sugar. However, CSF may be completely normal.
- The detection of HSV DNA in CSF by PCR is the most sensitive and specific diagnostic test, but this may be negative, especially early in the course of the disease.

Continued...

...continued

- Neuroimaging, whether by EEG, or radionuclide, CT, or MRI scan, is helpful, the cardinal feature being the presence of focal, rather than diffuse, disease.
- There is a long list of differential diagnoses, including other viral (many), bacterial, fungal or parasitic infections; intracerebral space-occupying lesions such as tumor; systemic disease, for example, vasculitis.
- Oral or genital HSV is diagnosed by demonstration of HSV DNA in vesicle fluid or in a swab of an ulcer base by genome amplification (usually PCR).
- Differential diagnosis includes physical trauma, chemical burns, pyogenic or other infections, fixed drug eruptions, systemic blistering diseases, dermatologic conditions.

5. HOW IS THE DISEASE MANAGED AND PREVENTED?

- HSV is sensitive to aciclovir (and derivatives).
- Treatment is recommended for symptomatic primary disease.
- Prophylactic therapy may be offered to patients with frequent, atypically aggressive recurrent disease.
- Treatment of HSE comprises high-dose aciclovir therapy given intravenously for a minimum of 2 weeks is essential, administered as soon as the diagnosis is considered.
- There is no prophylactic vaccine against HSV infection.

FURTHER READING

Barer M, Irving W, Swann A, Perera N. Medical Microbiology, 19th edition. Elsevier, Philadelphia, 2019.

Goering RV, Dockrell H, Zuckerman M, Chiodini PL. Mims' Medical Microbiology and Immunology, 6th edition. Elsevier, Philadelphia, 2018.

Humphreys H, Irving WL, Atkins BL, Woodhouse AF. Oxford Case Histories in Infectious Diseases and Microbiology, 3rd edition. Oxford University Press, Oxford, 2020.

Murphy K, Weaver C. Janeway's Immunobiology, 9th edition. Garland Science, New York/London, 2016.

Richman DD, Whitley RJ, Hayden FG. Clinical Virology, 4th edition. ASM Press, Philadelphia, 2016.

REFERENCES

Banerjee A, Kulkarni S, Mukherjee A. Herpes Simplex Virus: The Guest That Takes Over Your Home. Front Microbiol, 11: 733, 2020.

Kim HC, Lee HK. Vaccines Against Genital Herpes: Where Are We? Vaccines, 8: 420, 2020.

Samies NL, James SH, Kimberlin DW. Neonatal Herpes Simplex Virus Disease: Updates and Continued Challenges. Clin Perinatol, 48: 263–274, 2021.

Verzosa AL, McGeever LA, Bhark S-J, et al. Herpes Simplex Virus 1 Infection of Neuronal and Non-Neuronal Cells Elicits Specific Innate Immune Responses and Immune Evasion Mechanisms. Front Immunol, 12: 644664, 2021.

Whitley R, Baines J. Clinical Management of Herpes Simplex Virus Infections: Past, Present and Future. F1000Res, 7: F1000 Faculty Rev–1726, 2018.

WEBSITES

All the Virology on the WWW Website, Specific Virus Information, 1995: http://www.virology.net/garryfavweb12.html#Herpe

NHS, Genital herpes, 2020: https://www.nhs.uk/conditions/genital-herpes/

Patient UK: http://www.patient.co.uk/showdoc/40000500/

Wikipedia, Herpes simplex virus, 2022: https://en.wikipedia.org/wiki/Herpes_simplex_virus

World Health Organization, Herpes simplex virus, 2022: https://www.who.int/news-room/fact-sheets/detail/herpes-simplex-virus

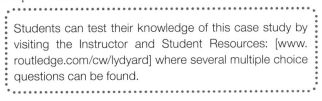

Students can test their knowledge of this case study by visiting the Instructor and Student Resources: [www.routledge.com/cw/lydyard] where several multiple choice questions can be found.

Histoplasma capsulatum

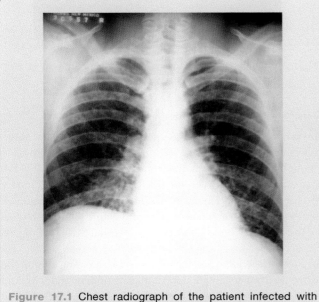

17

Figure 17.1 Chest radiograph of the patient infected with *H. capsulatum*, revealing bilateral nodular infiltrates. *Courtesy of the Centers for Disease Control & Prevention, Atlanta, Georgia. Image is found in the Public Health Image Library #3954. Additional photographic credit is given to M Renz who took the photo in 1963.*

A 60-year-old resident of Louisville (Ohio) had suffered from **rheumatoid arthritis** for 9 years and was currently being treated with 10 mg of methotrexate weekly and 8 mg of methylprednisolone daily followed by monthly injections of 3 mg kg^{-1} infliximab **monoclonal antibody**. Ten weeks after the start of infliximab, he felt severely ill and was hospitalized with the symptoms of **dyspnea** and cough, quickly followed by respiratory failure, requiring mechanical ventilation. A chest radiograph revealed bilateral nodular infiltrates (Figure 17.1). **Bronchoalveolar lavage** fluid contained yeast forms resembling *Histoplasma capsulatum*. Laboratory tests showed normal blood cell counts, but positive *Histoplasma* urine antigen (10.3 U, normal levels <1 U). The findings were confirmed by yeast cell culture and complement fixation titers 1:2048 to the mycelial M antigen and 1:256 to the yeast Y antigen (normal levels < 1:8). The diagnosis of histoplasmosis was further confirmed by immunodiffusion and the patient was given antifungal drugs amphotericin B lipid complex 5 mg kg^{-1} per day for 11 days, followed by itraconazole 200 mg per day for 2 months. Therapy resulted in improvement of the respiratory function, although the patient required ventilation support throughout the treatment.

1. WHAT IS THE CAUSATIVE AGENT, HOW DOES IT ENTER THE BODY AND HOW DOES IT SPREAD A) WITHIN THE BODY AND B) FROM PERSON TO PERSON?

CAUSATIVE AGENT

Histoplasma capsulatum is a dimorphic fungus which causes a systemic endemic **mycosis** called histoplasmosis (sometimes called Darling's disease by the name of a pathologist, Samuel Darling who discovered the disease in 1905). The genus *Histoplasma* belongs to *Ajellomycetaceae* family (order *Onygenales*) and contains one species, *Histoplasma capsulatum*. There are three varieties: *H. capsulatum* var. *capsulatum*, which causes the common histoplasmosis, *H. capsulatum* var. *duboisii*, a cause of African histoplasmosis (histoplasmosis *duboisii*), and *H. capsulatum* var. *farciminosum*, which causes **lymphangitis** in horses.

H. capsulatum is a thermally dimorphic ascomycete, which means that it can survive at two different temperatures. At ambient temperatures below 30°C *H. capsulatum* remains in a saprophytic mycelial mold form, but at mammalian body temperature (37°C) it grows as a parasitic yeast. The two varieties of *H. capsulatum* (*capsulatum* and *duboisii*) that infect humans are similar in **saprophytic** mold form but differ in their parasitic tissue morphology (see below).

The saprophytic mycelial growth of *H. capsulatum* requires an acidic damp soil environment with high organic content. This is provided by bird droppings, particularly those of chickens and starlings, or excrement of bats. The fungus has been found in poultry house litter, caves, areas harboring bats, and in bird roosts. Birds cannot be infected by *Histoplasma* due to the high body temperature of 40°C but can carry *H. capsulatum* on their feathers. In contrast, bats can become infected, and they transmit the fungus through droppings. Contaminated soil is the common natural habitat for *Histoplasma* and it remains potentially infectious for years.

At 25°C on Sabouraud dextrose agar (SDA), or brain heart infusion agar (BHIA) supplemented with 5–10% sheep blood, *H. capsulatum* grows slowly into granular suede-like to cottony colonies, initially white and then brown with a pale yellow-brown or yellow-orange reverse (Figure 17.2). The colonies can also be glabrous or verrucose, sometimes with a red pigmented strain. *H. capsulatum* produces hyphae-like short, hyaline, undifferentiated conidiophores, which arise at right angles to the parent hyphae. Macroconidia appear as large (8–14 μm in diameter), thick-walled, round, unicellular, hyaline, and tuberculate with finger-like projections on the surface (Figure 17.3A). Microconidia (microaleurioconidia) are small (2–4 μm in diameter), unicellular, hyaline, round with a smooth or rough wall, and borne on short branches or directly on the sides of the hyphae.

However, at 37°C the fungal morphology of *H. capsulatum* is totally different: numerous small round to oval, budding yeast-like cells, 2–4 μm in size can be observed under the microscope (Figure 17.3B). Colonies are creamy, smooth, moist, white, and yeast-like (Figure 17.4). This change in morphology under temperature-controlled regulation is used as a diagnostic test for *Histoplasma* (see Section 4).

H. capsulatum var. *duboisii* usually grows as a large yeast (7–15 μm in diameter) within the cytoplasm of **histiocytes** and multinucleate giant cells, but sometimes can appear as small yeast cells similar to those of var. *capsulatum*. The yeasts may form rudimentary pseudohyphae consisting of four or five cells and aggregates, which can be observed within giant cells and extracellularly following **necrosis** of the host tissue.

As many other fungi, *H. capsulatum* have chitin as the deepest layer of the cell wall and a substantial amount of β-(1,3)-glucan above that layer.

ENTRY INTO THE BODY

Conidia and mycelial fragments of *H. capsulatum* from contaminated disrupted soil can be air-borne, are found in aerosols, and can be inhaled. Once inhaled, they settle

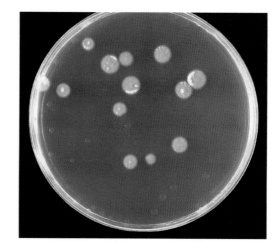

Figure 17.2 *Histoplasma capsulatum* colonies growing at 25°C. *Courtesy of the Centers for Disease Control & Prevention, Atlanta, Georgia. Image is found in the Public Health Image Library #3191. Additional photographic credit is given to Dr William Kaplan who took the photo in 1969.*

in bronchioles and alveolar spaces where they encounter phagocytic macrophages. Binding of the microconidia is thought to be through the complement receptor 3 (CR3), pattern recognition receptor (PRR) Dectin-1, which binds to β-(1,3)-glucan, and integrins **CD11/CD18**. **Opsonization** via antibodies and complement is therefore important for uptake of the fungus. Conversion from the mycelial to the pathogenic yeast phase is critical for infectivity of *H. capsulatum* and occurs intracellularly inside the macrophages. The stimuli that drive the conversion are not completely understood, although the temperature change appears to be a major factor. Upon conversion, the intracellular budding yeasts enlarge and reach approximately 3 μm in diameter.

It must be stated that cultures of *H. capsulatum* represent a severe biohazard to laboratory staff and must be handled with extreme caution and under appropriate conditions in a safety cabinet.

A
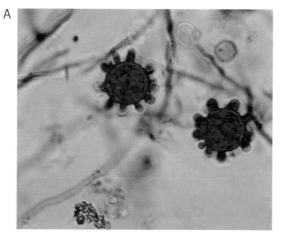

B
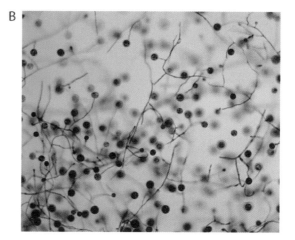

Figure 17.3 Two forms of *Histoplasma capsulatum* demonstrating features of a thermal dimorph. In nature at about 25°C, it grows as a mycelial filamentous form with macroconidia and smaller microconidia (A). At body temperature of 37°C it grows as a yeast (B). *Courtesy of the Centers for Disease Control & Prevention, Atlanta, Georgia. Image is found in the Public Health Image Library (A) #4023 and (B) #4022. Additional photographic credit is given to Dr Libero Ajello who took the photo in 1968.*

Figure 17.4 Culture of *H. capsulatum* on BHIA with 10% sheep blood, incubated at 37°C. *Courtesy of Professor Michael R McGinnis of DoctorFungus.org. www.doctorfungus.org.*

SPREAD WITHIN THE BODY

Following the initial infection, *H. capsulatum* var, *capsulatum* may spread – carried via pulmonary macrophages in draining lymphatics and the bloodstream – to many organs that contain mononuclear phagocytes. These include the liver and spleen and regional lymph nodes.

Infection with *H. capsulatum* var. *duboisii* rarely involves the lungs but mostly causes cutaneous histoplasmosis. It can also infect the liver, lymphatic system, and subcutaneous and bony tissues. The infection presents as nodular and ulcerative cutaneous and osteolytic bone lesions, disseminated or localized.

PERSON-TO-PERSON SPREAD

This organism is generally not spread from person to person.

EPIDEMIOLOGY

H. capsulatum is found in temperate climates, predominantly in river valleys between latitudes 45° north and 30° south in North and Central America. It is **endemic** to the Ohio, Missouri, and Mississippi River valleys in the US (Figure 17.5). In these regions of the US, approximately 90% of residents have been exposed to the fungus, and quite likely on a continuous basis. Between 1938 and 2013, over 100 outbreaks with 3000 cases were reported in 26 states and in Puerto Rico. *H. capsulatum* has five to seven chromosomes, and this lays the foundation for eight clades of *H. capsulatum*, which are distributed in different parts of the world. There are two North American clades, two South American clades, one Australian, one Indonesian, one African, and one Eurasian clade. The genetic differences often define the symptoms. For example, unlike South American clades, those in North America do not cause primary skin disease. The African clade includes all of the *H. capsulatum* var. *duboisii*. It is endemic in Central and West Africa and in the island of Madagascar.

Approximately 250000 individuals are infected annually with *H. capsulatum* in the US and 500000 worldwide. Of these, 50000–200000 develop symptoms of histoplasmosis and 1500–4000 require hospitalization. Infection is more prevalent in immunocompromised individuals (see Section 2) and therefore *Histoplasma* is regarded as an opportunistic pathogen.

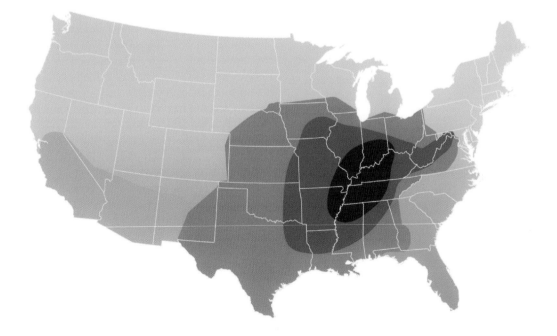

Figure 17.5 Distribution of histoplasmosis in the US, with dark red areas indicating the highest incidence of the disease. *Courtesy of Professor Michael R McGinnis of DoctorFungus.org. www.doctorfungus.org.*

2. WHAT IS THE HOST RESPONSE TO THE INFECTION AND WHAT IS THE DISEASE PATHOGENESIS?

INNATE IMMUNITY

As well as being the site of the infection, phagocytes, especially macrophages, are the main cells responsible for removing the fungus (Figure 17.6). After ingestion, *H. capsulatum* yeasts reside in phagocytic **phagosomes** and replicate there approximately every 15–18 hours. One mechanism used by the fungus to reduce intracellular killing is by inhibiting phagolysosomal fusion and thus preventing exposure to lysosomal hydrolytic enzymes. The fungus also increases the phagosomal pH to 6.5, which potentially enhances the availability of iron it needs for growth. Decrease in free iron via constitutive and inducible iron and zinc sequestration represents an important host antimicrobial defense. Since β-(1,3)-glucan elicits the strongest immune response, in *H. capsulatum* it is often masked by α-(1,3)-glucan as a part of the pathogen's strategy to avoid immune response.

Human macrophages respond to ingestion of *H. capsulatum* by a vigorous oxidative burst mediated by NADPH oxidase, and the release of nitric oxide (NO) and other nitrogen intermediates, yet the fungus is still able to survive this response and even replicate in the presence of **reactive oxygen species (ROS)**. It eventually lyses the alveolar macrophage and the released yeasts are ingested by other resident macrophages and polymorphonuclear neutrophils (PMNs) newly recruited to the site of infection. In immunocompromised individuals, the infection/lytic cycle may be repeated several times.

Interaction of Dectin-1 with CR3 triggers the cytokine response to *H. capsulatum* yeasts via the activation of Syk tyrosine kinase. Intracellular growth of the fungus is inhibited by **interleukin (IL)**-3, granulocyte-macrophage colony stimulating factor (**GM-CSF**), and macrophage-CSF (M-CSF). GM-CSF limits intracellular *H. capsulatum* growth by producing zinc-sequestering metallothioneines.

PMNs from immunocompetent hosts can inhibit the growth of *H. capsulatum* through release of **defensins**. With the development of immunity mediated by neutrophils and macrophages, yeast growth ceases within 1–2 weeks after exposure.

ANTIBODY-MEDIATED IMMUNITY

Humoral immunity has little or no role in host defense against the fungus. Although exposure to *H. capsulatum* does induce an antibody response, and the **IgG** fraction of the sera from infected individuals contains complement-fixing and precipitating antibodies, these paradoxically are associated with progressive disease. Passive transfer of immune serum has not mediated protection from the fungus and B-lymphocyte-deficient mice were not susceptible to infection. No correlation was found between the titers of antibodies specific to *H. capsulatum* and immunity to the infection.

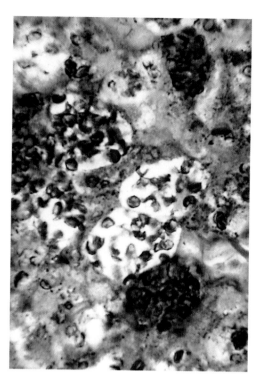

Figure 17.6 Two macrophages with engulfed small yeast-like fungus cells of *Histoplasma capsulatum* var. *capsulatum* (hematoxylin & eosin stain). *Courtesy of the Centers for Disease Control & Prevention, Atlanta, Georgia. Image is found in the Public Health Image Library #4223. Additional photographic credit is given to Dr Libero Ajello who took the photo in 1972.*

CELL-MEDIATED IMMUNITY

Of all the human mycoses, histoplasmosis illustrates best the importance of the T-mediated immune system in limiting the extent of *H. capsulatum* infection. Susceptibility to dissemination of *H. capsulatum* is associated with T-cellular immunodeficiency. In immunocompetent individuals, delayed-type hypersensitivity to histoplasma is observed 3–6 weeks after exposure to the fungus and, in up to 85–90% of cases, a positive response to a skin antigen test for *Histoplasma* can be detected (see Section 4).

It is thought that *H. capsulatum*, like other foreign organisms, is taken up by immature **dendritic cells** and processed with antigens then presented to T cells.

Specific T cells contribute to protective immunity to *H. capsulatum* mainly via production of **cytokines** that activate phagocytes, with **interferon-γ (IFN-γ)** and **tumor necrosis factor-α (TNF-α)** being particularly important. It has recently been shown that invariant NKT (iNKT) cells, a relatively rare population of T cells, defined by expression of an invariant TCR α-chain, are also instrumental in the production of IFN-γ in response to *H. capsulatum* and IL-12. Deficiency in IFN-γ and TNF-α causes a decrease in NO production by phagocytic cells, which is essential for controlling *H. capsulatum* infection. IL-12 regulates the induction of IFN-γ, which is a critical **Th-1** cytokine in primary host resistance. During the acute phase of the infection IL-12, TNF-α, and IFN-γ are released and enhance the influx of the myeloid cells into the lungs followed by T and B cells. GM-CSF is also important for the production of TNF-α, IFN-γ, and NO, and is responsible for down-regulation of **Th-2** cytokines IL-4 and IL-10. High levels of Th-2 cytokines weaken the efficiency of protective immunity by inducing inflammatory responses and chronic infection.

For secondary immune responses, TNF-α is critical in both pulmonary and disseminated infections. Its deficiency leads to the dramatic elevation of IL-4 and IL-10 in the lungs. In most immunocompetent individuals, activation of a protective T-lymphocyte-mediated immunity results in containment of the infection. However, even in normal individuals, but particularly in immunodeficient hosts, the organism may not be completely eradicated and may persist as a latent infection. This can erupt as active disease at a later time when and if the host–pathogen balance is disrupted or host immune responses decline with age or through viral infection.

PATHOGENESIS

Chronic infection will lead to the inflammatory responses that can last over weeks to months and result in the development in the affected organs of calcified fibrinous granulomatous lesions with areas of caseous necrosis. *H. capsulatum* may remain latent in healed **granulomas** and recur resulting in impairment of cell-mediated immunity. The function of the granulomas is to contain fungal growth. In case of perturbed cellular immune response, the patient develops progressive disseminated histoplasmosis. Formation of the granulomas requires the generation of IFN-γ, TNF-α, and IL-17. Since X-linked hyper-immunoglobulin M immunodeficiency is associated with defective IL-17 generation, the majority of these individuals develop disseminated histoplasmosis (see Section 3).

3. WHAT IS THE TYPICAL CLINICAL PRESENTATION AND WHAT COMPLICATIONS CAN OCCUR?

In about 80–90% of infected immunocompetent individuals, histoplasmosis is asymptomatic, subclinical (showing self-limiting influenza-like symptoms) or benign. Clinical symptoms of histoplasmosis develop mostly in immunocompromised individuals. In addition, the inoculum size often defines symptomatic disease.

ACUTE PULMONARY HISTOPLASMOSIS

The onset of acute pulmonary histoplasmosis (APH) in symptomatic cases develops 7–21 days after exposure to the fungus. Common symptoms include fever, headache, malaise, **myalgia**, abdominal pain, and chills. Individuals exposed to a large inoculum of fungus may develop severe dyspnea due to diffuse pulmonary involvement. Diffuse or localized **pneumonitis** may be severe enough to require ventilation support (see the case). Weight loss, night sweats,

and fatigue may persist for weeks after the acute symptoms resolve. A small number of patients may show rheumatologic manifestations such as **erythema multiforme**, **arthritis**, and **erythema nodosum**, which can be of a diagnostic value. Sometimes pulmonary **auscultation** may detect **rales** or wheezes. Severe **hypoxemia** and associated acute respiratory distress syndrome only develop in patients inhaling a high inoculum. In a small group of patients (about 6%), **pericarditis** may be present. More often, the acute pericarditis is due to the granulomatous inflammatory response mounted in mediastinal lymph nodes adjacent to the pericardium. In very rare cases, it is a direct *H. capsulatum* infection that leads to granulomatous inflammation within the pericardial sac. Approximately 10% of patients have asymptomatic pleural **effusions**. Hepatosplenomegaly is rare.

Lymphadenopathy reflected by enlarged hilar and mediastinal lymph nodes is present in 5–10% of patients. In some cases, it can cause local obstructions such as superior vena cava (SVC) syndrome, which results from the compression of the SVC. Obstruction of venous drainage may lead to cerebral symptoms: headache, visual distortion, **tinnitus**, and altered consciousness. Compression of the pulmonary airway and circulation, or of trachea or bronchi presents as cough, hemoptysis, dyspnea, and/or chest pain. Compression of the esophagus with subsequent **dysphagia** is rare.

Rarely, in endemic regions during the healing from APH, the enlarged mediastinal nodes can cause retraction of the airways, leading to post-obstructive pneumonia and bronchiectasis.

CHRONIC PULMONARY HISTOPLASMOSIS

Chronic pulmonary histoplasmosis (CPH) can present as cavitary or noncavitary disease. Cavities occur approximately in 40% of patients. However, if histoplasmosis is pre-conditioned by tuberculosis, the cavities develop in approximately 90% of cases and may lead to necrosis.

Generally, CPH occurs mostly in patients with underlying pulmonary chronic lung disease such as emphysema and may mimic tuberculosis. Histoplasmosis may actually coexist with tuberculosis, actinomycosis, other mycoses, and **sarcoidosis**. In individuals with underlying pulmonary disease chronic cavitary pulmonary histoplasmosis (CCPH) is the condition most commonly observed, with a frequency of 1 per 100000 persons per year in endemic areas. CPH does not have as strong association with a chronic obstructive pulmonary disease (COPD) as previously believed, although 20% of patients present with COPD.

The main symptoms include cough, weight loss, fever, and malaise. In case of **cavitations**, additional symptoms of hemoptysis, sputum production, and increasing dyspnea develop. Untreated cases may result in progressive pulmonary fibrosis and subsequently in respiratory and cardiac failure and recurrent infections.

Chronic fungal infection leads to the development of well-organized solitary pulmonary nodules with a circumferential rim of calcification, which allows their identification on chest X-ray (see Figure 17.1). Fungi within these nodules are usually dead. Older lesions show well-developed granuloma and have a central area of **caseation** resembling tuberculosis. These centers are usually occupied by the fungi.

PROGRESSIVE DISSEMINATED HISTOPLASMOSIS

Progressive disseminated histoplasmosis (PDH) occurs in 1:2000 cases of infected immunocompetent adults and in 4–27% of infected immunosuppressed individuals with impaired cellular immunity, children, or the elderly. All infections with histoplasmosis become disseminated as the macrophages are constantly trafficking between the lungs and other organs. Histoplasmosis becomes progressive when uncontrolled growth of the organism occurs in multiple organ systems. Symptoms vary depending on duration of illness. There is an acute and subacute form of PDH.

The *acute form* is associated with fever, worsening cough, weight loss, malaise, and dyspnea. Dissemination sites include the central nervous system (CNS; 5–20% of patients) and hematopoietic systems, liver, spleen, and eye. In the bone marrow **anemia**, **leukopenia**, and **thrombocytopenia** are detected. Adrenal glands undergo enlargement, and **Addison's disease** may develop as well. Genitourinary tract symptoms include **hydronephrosis**, bladder ulcers, penile ulcers, and **prostatitis**. Oral ulcers and small bowel micro and macro ulcers develop in the gastrointestinal (GI) tract. Skin symptoms include **papular** to nodular **rash**, while ocular problems include **uveitis** and choroiditis. Among CNS-related conditions chronic **meningitis** and **cerebritis** should be noted. The acute form, if untreated, results in death within weeks. About 10% of individuals develop fatal hyperacute syndrome.

The *subacute form* presents with a wide spectrum of symptoms based on dissemination in the affected organs. GI involvement leads to diarrhea and abdominal pain, while cardiac dissemination results in valvular disease, cardiac insufficiency, dyspnea, peripheral edema, angina, and fever. Pericarditis may develop, often with pleural effusions. CNS involvement is reflected by headaches, visual and gait disturbances, confusion, **seizures**, altered consciousness, and neck stiffness or pain. If untreated, the subacute form is fatal within 2–24 months. Approximately 5–10% of patients, treated or not, develop adrenal insufficiency.

In 50–60% of patients with the chronic form of disseminated histoplasmosis constitutional symptoms such as mouth and gum pain are present due to mucosal ulcers. Extensive ulceration including the oropharyngeal area, buccal mucosa, tongue, gingiva, and larynx is characteristic of chronic PDH (Figure 17.7). Rare, isolated lesions can be found in individuals who are immunocompetent.

The patients susceptible to the development of disseminated and potentially fatal histoplasmosis are mostly immunocompromised individuals, particularly HIV-infected patients in the endemic areas (see Section 1), children less than 2 years old, elderly persons, and people exposed to a very large inoculum.

PRESUMED OCCULAR

This syndrome is only found in 1–10% of individuals who live in endemic areas. Ocular histoplasmosis damages the retina of the eyes and leaves scar tissue leading to retinal leakage and a loss of vision similar to macular degeneration. Atrophic scars with lymphocytic cell infiltration (histo spots) may be seen posterior to the equator of the eye. The condition is bilateral in approximately 10% of patients. In the case of macular scarring, retinal hemorrhage, detachment or edema may be present.

4. HOW IS THE DISEASE DIAGNOSED, AND WHAT IS THE DIFFERENTIAL DIAGNOSIS?

A social and occupational history is of great help for the initial evaluation. This includes travel to or residence in an endemic area, interaction with birds or bats, either recent or past. Any possible immunodeficiency in patients due to a medical condition or drug treatment must be taken into account (see the case).

IMAGING

A chest X-ray may be taken to help in the diagnosis of the kind of histoplasmosis. It usually does not show any irregularities in APH. However, diffuse pulmonary involvement caused by exposure to a large inoculum may present with a reticular nodular or **miliary** pattern. Pleural effusions are found in fewer than 10% of uncomplicated cases. Cavitations are very rare. Histoplasmosis can be detected as lesions and residual nodules, 1–4 cm in diameter.

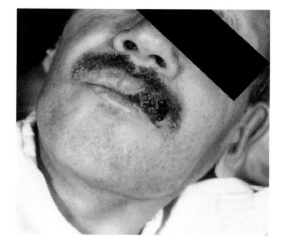

Figure 17.7 Mouth lesions due to *Histoplasma capsulatum* dissemination into oral cavity. *Courtesy of the Centers for Disease Control & Prevention, Atlanta, Georgia. Image is found in the Public Health Image Library #6840. Additional photographic credit is given to Susan Lindsley, VD, who took the photo in 1973.*

In CPH hilar lymphadenopathy is rare. The cavitations are located in the upper lobes of the lungs. Often **emphysematous** changes are present as well as fibrotic scarring, particularly in long-standing cases.

Around 50% of patients with acute PDH have hilar lymphadenopathy with diffuse nodular infiltrates. At the onset of the disease, no chest radiography changes are detected in 33% of patients but, as the disease progresses, pulmonary involvement becomes gradually more apparent. In contrast, chronic PDH does not present with chest radiographic changes.

Head **CT scanning** helps to detect cerebral histoplasmosis before performing a lumbar puncture. Abdominal CT can confirm subacute PDH, which leads to adrenal infection in 80% of patients and presents with bilateral adrenal enlargement on the scan.

RESPIRATORY FUNCTIONAL TESTS

Pulmonary functional tests help characterize the scale of pulmonary involvement. They include evaluation of the restrictive defect, detection of a small airway obstruction, diffusion impairment, and hypoxemia. The functional tests are also used for monitoring the progression of pulmonary disease in patients with CPH.

LABORATORY TESTS

Specimens obtained for analysis include: blood, sputum, bronchial lavage, and cerebrospinal fluid (CSF), by lumbar puncture, if CNS involvement is suspected. Tissue biopsies may be taken of pulmonary lesions and lymph nodes by **bronchoscopy** or thoracoscopy. Biopsies can be also taken from oropharyngeal ulcers.

Serology

Complement-fixing (CF) antibody tests for the presence of M and H precipitin bands with the antigen extract of the Histoplasma mycelial form called histoplasmin. The titers of CF antibodies to yeast and mycelial-phase antigens (Y and M, see the case) are considered positive at dilutions greater than 1:8. A titer of 1:32 or more suggests active histoplasmosis infection. CF antibody to Y antigen is detectable in primary infection, while CF antibody to M antigen appears later. An M band develops with acute infection, persists for months to years and is also present in chronic forms. H band appears after M band and may disappear early. Thus, the presence of both M and H bands suggests active histoplasmosis. Positive results are usually found in 5–15% of cases of acute pulmonary infection 3 weeks after exposure and in 75–95% of cases at 6 weeks. The test usually turns negative in the course of months with resolution of infection. However, in chronic infection, the CF test may remain positive for a long time in 70–90% of patients with CPH or chronic PDH.

Cross-reactivity with *Blastomyces dermatitidis* and *Coccidioides immitis* must be ruled out for differential diagnosis.

Antigen Detection in Urine and Other Fluids

Antigen detection by **ELISA** is useful in individuals who are immunocompromised when antibody production may be impaired. This method is more sensitive for urine samples compared with serum, plasma, CSF, and bronchoalveolar lavage fluid. Detection of the capsular antigen of *H. capsulatum* in urine or serum is a very powerful tool for the diagnosis of the disseminated forms of histoplasmosis. In cases of acute PDH, detection rates are 50% with serum assay and 90% with urine assay. Lower detection rates are observed in APH or CPH.

Histology and Cytology

In rare cases, a tissue biopsy may reveal the presence of large yeast cells with a false capsule. Special stains may reveal budding yeast in areas of necrosis from histoplasmosis and calcified lymph nodes. Direct microscopic evidence of histoplasma infection showing characteristic yeast-like cells from any specimen taken from the patient is considered significant diagnostic proof.

Cell Culture

Isolation of *H. capsulatum* in culture remains the gold standard method for diagnosis of histoplasmosis. However, the fungus takes up to 4 weeks to grow *in vitro*. Specimens are inoculated onto SDA or BHIA supplemented with 5–10% sheep blood and observed for morphology of the growing colonies (see Section 1). A positive culture of diagnostic value has to show conversion of the mold form to the yeast phase by growth at 37°C. Sputum culture results are usually positive in 60% of patients with CPH and in approximately 10–15% of patients with APH. Blood cultures are positive for histoplasma in 50–90% of patients with acute PDH. Cultures of CSF are positive in 30–60% of patients with histoplasmal meningitis.

Skin Tests

Positive histoplasmin skin tests were demonstrated by almost 80% of the people living in areas endemic for *H. capsulatum* and therefore their diagnostic value is very low. During the acute pulmonary infection, a skin test may be positive in more than 90% of cases; while during the chronic pulmonary infections, in 70–90% of cases, and during PDH, in 3–55% of cases.

Polymerase Chain Reaction

Polymerase chain reaction (PCR) tests have been developed for the detection of *H. capsulatum* but are not used in clinical practice. The usual target is *H. capsulatum* ribsomal DNA (rDNA), but it is similar to 18-S rDNA of other fungi. Hence, a nested PCR has been developed with multiple primer sets. This includes both *H. capsulatum* rDNA and a 100-kDa-like protein unique to Histoplasma (Hcp100); or N-acetylated α-linked acidic dipeptidase (NAALAD) and the internal transcribed spacer (ITS) region. This assay can be modified to accommodate fluorescent markers and has the potential to be used in a low budget setting. Fluorescent *in situ* hybridization (FISH) targeting Histoplasma rRNA may be clinically useful.

DIFFERENTIAL DIAGNOSIS

A differential diagnosis for *H. capsulatum* with *Chrysosporium* and *Sepedonium* species that also inhabit soil and plant material should be carried out.

The conidia of *Chrysosporium parvum* morphologically resemble those of *H. capsulatum*. *Chrysosporium* species are also dimorphic, but this is a false dimorphism since they do not convert from mold to yeast at 37°C. Macroconidia of *Sepedonium* species also resemble those of *Histoplasma*.

However, Histoplasma galactomannans (GM) are similar to antigens of other dimorphic fungi such *Blastomyces dermatitidis*, *Paracoccidioides brasiliensis*, *Talaromyces marneffei*, and *Sporothrix* species, and cross-reactivity of these fungi with the antigen test is common. Aspergillus galactomannan testing also cross-reacts with Histoplasma GM.

In the case of *H. capsulatum* var. *duboisii*, differential diagnosis must be made from *B. dermatitidis* since they both occur in Africa.

Importantly, in 2017, a separate genus *Emergomyces*, formerly classified under the genus *Emmonsia* was placed in the *Ajellomycetaceae* family. Emergomyces species is also a dimorphic fungus mostly found among immunocompromised individuals in Asia, Europe, Africa, and North America. The number of emergomycosis cases might be underestimated considering the high incidence of HIV/AIDS. Differential diagnosis remains challenging, with histoplasmosis due to clinical and histopathologic overlap between the two entities. The gold standard for identification would be sequencing the ITS region of ribosomal DNA but this requires substantial resources.

Some patients with acute histoplasmosis may have high serum levels of angiotensin-converting enzyme. This may cause a diagnostic confusion with sarcoidosis, particularly if the patient with histoplasmosis also has hilar **adenopathy**.

5. HOW IS THE DISEASE MANAGED AND PREVENTED?

As discussed in Section 3, in immunocompetent individuals infection with *Histoplasma* species in most cases is self-limiting and does not require therapy. Therapeutic intervention is recommended in cases of prolonged infection, systemic infection or if patients are immunocompromised.

MANAGEMENT

APH: Mild symptoms in immunocompetent individuals need monitoring only. In patients with prolonged symptoms or extensive pulmonary involvement, therapy is recommended.

CPH: No treatment is required for asymptomatic immunocompetent individuals without serious underlying disease. Mild interstitial pneumonitis and/or thin-walled (<3 mm) cavities with serial chest radiographs require monitoring for 2–4 months. Medical treatment is recommended if the lesions are persistent or if patients develop cavities with thick walls (>3 mm). Treatment is indicated for all immunocompromised patients.

PDH: All patients with the symptoms of meningitis require medical therapy. In the case of severe CNS infection, intravenous (IV) antifungal therapy should be supplemented with **intrathecal** or intraventricular injections.

Cutaneous and rheumatologic lesions are usually self-limiting and medical intervention is recommended only for prolonged course of the disease episodes or in immunocompromised patients.

Maculopathy in ocular histoplasmosis is treated with steroids.

Surgery may be required in some cases for resection of pulmonary cavitation lesions, repair of infected heart valves, and **aneurysms**.

Antifungal Drugs

Severe cases of acute histoplasmosis, and all cases of symptomatic chronic and disseminated disease, require treatment with antifungal drugs, if the symptoms persist for more than 4 weeks. The recommended regimen is a 3-month course of itraconazole. In patients with CPH, noncavitary disease, a 6-month therapy is recommended instead, whereas for a cavitary disease the treatment might be prolonged for one year. Patients with PDH require induction therapy with amphotericin-B for 2–4 weeks followed by one year of itraconazole. There are some reports of patients being treated with posaconazole instead of both, itraconazole and amphotericin B. Amphotericin B (AmB) is used to treat APH, CPH, all forms of PDH, meningitis, and endovascular histoplasmosis. There is evidence that the liposomal AmB has better outcomes, particularly in patients with AIDS.

In adults, as well as in children, $0.7-1\,\mathrm{mg\,kg^{-1}}$ per day of AmB or up to a total dose of $35\,\mathrm{mg\,kg^{-1}}$ is used IV. A minimum total dose should be no less than 2 g. In the case of underlying severe immunodeficiency (such as AIDS) when a life-long antifungal maintenance therapy is required to treat acute PDH, the induction dose of $0.7-1\,\mathrm{mg\,kg^{-1}}$ per day AmB to a total of $20-25\,\mathrm{mg\,kg^{-1}}$ is used, followed by a maintenance therapy of 50 mg once per week.

Itraconazole is a fungistatic drug that suppresses fungal cell growth. It is administered 200 mg three times daily for 3 days, then 200 mg twice daily in adults and 5–10 mg/kg (maximum 400 mg daily) in children. However, several studies have indicated that fluconazole is characterized by higher response rates and lower relapse rates.

Ketoconazole can be used as an alternative to AmB in the treatment of CPH or chronic and subacute progressive disseminated histoplasmosis. However, since these drugs do not cross the blood–brain barrier, they are not effective in the treatment of fungal meningitis.

Anti-Inflammatory Drugs

Apart from the fungicidal and fungistatic drugs, nonsteroidal anti-inflammatory drugs (NSAIDs) are used with analgesic, anti-inflammatory, and antipyretic action. They are particularly recommended for patients with pericarditis. Among the most used is ibuprofen given orally at 400 mg every 4–6 hours, 600 mg every 6 hours, or 800 mg every 8 hours, with a maximal dose of 3.2 g per day while symptoms persist. The pediatric dose is 20–70 mg kg^{-1} per day with a low induction dose and maximal dose of 2.4 g per day. Usual NSAIDs contraindications apply.

Among other anti-inflammatory drugs, corticosteroids, mostly prednisone, can be used to decrease hypersensitivity to *Histoplasma*. High-dose steroids are used in patients with extensive maculopathy.

Since IFN-γ augments antifungal activity of PMNs and macrophages it is considered for immunotherapy of *H. capsulatum*.

PROGNOSIS

Prognosis for a complete recovery is good for APH. The relapse rate in CPH is 20%, while in treated acute PDH it is 50%. If lifelong antifungal maintenance is administered, the relapse rate drops to 10–20%. The course of chronic PDH can last for years with long asymptomatic periods.

Fatal outcome of acute PDH and subacute PDH is imminent without treatment.

Full recovery from histoplasmal meningitis with therapy is 50%.

PREVENTION

Prevention includes chemical disinfection and respiratory barrier protection in high-risk areas or during high-risk activities. Since the outbreaks of histoplasmosis are associated with disruption of the soil in endemic areas, decontamination of the infected soil with a 3% formalin solution is highly recommended.

Individuals residing in or traveling to endemic areas, particularly those with a history of immunodeficiency (AIDS, lymphoma, immunosuppressive treatment), must be educated/briefed about exposure risks. The elderly and children are also at risk.

In some areas, special precautions are taken by placing warning signs around particularly contaminated soil.

VACCINES

Although it has been shown that there is no immunity to *Histoplasma* by prior infection and no effective vaccine is currently available, recent studies have indicated that vaccination with highly immunogenic recombinant rHSP60 protein or its F3 fragment protects mice from lethal and sublethal histoplasmosis by stimulating **CD4**+ T lymphocytes. Other approaches to development of a vaccine are associated with targeting the histone H2B-like protein on the surface of *H. capsulatum* (see Section 2) and studies based on glucan nanoparticles extracted from *S. cerevisiae*.

SUMMARY

1. WHAT IS THE CAUSATIVE AGENT, HOW DOES IT ENTER THE BODY, AND HOW DOES IT SPREAD A) WITHIN THE BODY AND B) FROM PERSON TO PERSON?

- The genus *Histoplasma* contains one species, *Histoplasma capsulatum*, with three varieties: *H. capsulatum* var. *capsulatum*, which causes the common histoplasmosis, *H. capsulatum* var. *duboisii*, which causes African histoplasmosis, and *H. capsulatum* var. *farciminosum*, which causes lymphangitis in horses.

- *H. capsulatum* is thermally dimorphic and can survive at two different temperatures. Below 30°C, *H. capsulatum* remains in a saprophytic mycelial mold form, but at an average mammalian body temperature 37°C, it grows as a parasitic yeast.

- For the saprophytic mycelial growth, *H. capsulatum* requires an acidic damp soil with high organic content provided mostly by bird droppings or excrement of bats. Bats can become infected, and they transmit the fungus through droppings.

- In a saprophytic form at 25°C, *H. capsulatum* colonies grow slowly into granular suede-like to cottony colonies brown with a pale yellow-brown or yellow-orange reverse. Macroconidia are

large and tuberculate, microconidia are small and round. At 37°C, *H. capsulatum* grows as creamy, smooth, white and yeast-like colonies.

- In the lungs, inhaled mycelial fragments and microconidia of *H. capsulatum* are ingested by resident phagocytes. Conversion from the mycelial to the pathogenic yeast phase occurs inside the macrophages, mostly due to the temperature change.

- Binding of the microconidia is thought to be through the complement receptor 3 (CR3), pattern recognition receptor (PRR) Dectin-1, which binds to β-(1,3)-glucan, and integrins CD11/CD18. The initial pulmonary infection with *H. capsulatum* var. *capsulatum* disseminates to many organs that contain mononuclear phagocytes and produces extrapulmonary manifestations in the liver and spleen, or in regional lymph nodes. Infection with *H. capsulatum* var. *duboisii* mostly involves cutaneous, subcutaneous, liver, lung, lymphatic, and bony tissues. *H. capsulatum* is endemic to the Ohio, Missouri, and Mississippi River valleys in the US. Approximately 250 000 individuals are infected annually with *H. capsulatum* in the US and 500 000 worldwide. Of these, 50 000–200 000 develop symptoms of histoplasmosis and 1500–4000 require hospitalization.

Continued...

...continued

2. WHAT IS THE HOST RESPONSE TO THE INFECTION AND WHAT IS THE DISEASE PATHOGENESIS?

- After ingestion, *H. capsulatum* yeasts reside in phagosomes and replicate approximately every 15–18 hours. The yeasts inhibit phagolysosomal fusion, prevent exposure to the lysosomal hydrolytic enzymes, block accumulation of vacuolar ATPase, and increase phagosomal pH to 6.5.
- In response to ingestion of *H. capsulatum*, macrophages enhance oxidative burst and the release of nitrogen intermediates. The yeasts are ingested by newly recruited PMNs, which release fungistatic defensins.
- T cells contribute to protective immunity via production of cytokines that activate phagocytes, particularly IFN-γ and TNF-α. GM-CSF is important for the production of TNF-α, IFN-γ, and NO, and down-regulation of Th-2 cytokines IL-4 and IL-10, which are involved in pathogenesis.
- In most immunocompetent individuals, protective T-mediated immunity results in containment of the infection. However, even in normal individuals, and particularly in immunodeficient hosts, the organism may persist as a chronic infection and may lead to inflammatory responses and fibrinous granulomatous lesions with necrosis.
- Humoral immunity has little role in host defense against *H. capsulatum*, although exposure to it induces antibody response.

3. WHAT IS THE TYPICAL CLINICAL PRESENTATION AND WHAT COMPLICATIONS CAN OCCUR?

- In about 80–90% of infected immunocompetent individuals, histoplasmosis is asymptomatic and subclinical with self-limiting influenza-like symptoms or benign.
- The symptoms of severe acute pulmonary syndrome are nonspecific: fever, chills, myalgias, cough, and chest pain. The syndrome can be mild (lasting 1–5 days) or severe (lasting 10–21 days).
- In a small group of the infected individuals (5–10%), histoplasmosis presents as chronic progressive lung disease, chronic cutaneous or systemic disease or an acute fatal systemic disease. The patients are mostly immunocompromised, children less than 2 years old, the elderly, and people exposed to a very large inoculum.
- Chronic pulmonary histoplasmosis (CPH) occurs in patients with underlying pulmonary chronic lung disease (emphysema) and may mimic tuberculosis. The symptoms include cough, weight loss, fever, and malaise. In case of cavitations, hemoptysis, sputum production, and increasing dyspnea develop and lead to necrosis.
- The acute form of progressive disseminated histoplasmosis (PDH) is associated with fever, worsening cough, weight loss, malaise, and dyspnea. The symptoms related to the organ of dissemination include: anemia, leukopenia and thrombocytopenia (bone marrow); hydronephrosis, bladder and penile ulcers, prostatitis (genitourinary tract); oral ulcers, small bowel ulcers (gastrointestinal tract); papular to nodular rash (skin); uveitis and choroiditis (eyes); chronic meningitis and cerebritis (CNS). The subacute form of PDH presents with diarrhea and abdominal pain, valvular disease, cardiac insufficiency, dyspnea, peripheral edema, angina, fever, pericarditis, CNS involvement, and adrenal insufficiency. If untreated, the subacute form is fatal within 2–24 months.
- In 50–60% of patients with the chronic form of PDH, painful granulomatous mucosal ulcers appear as nodular ulcerative or vegetative lesions localized on the oral mucosa, tongue, palate, or lips.
- Presumed ocular histoplasmosis syndrome is only found in 1–10% of individuals who live in endemic areas.

4. HOW IS THE DISEASE DIAGNOSED, AND WHAT IS THE DIFFERENTIAL DIAGNOSIS?

- A social and occupational history is important for the initial evaluation, such as travel to or residence in an endemic area, interaction with birds or bats, possible immunodeficiency.
- CPH can be detected through pulmonary auscultation as rales, wheezes with a history of underlying pneumonitis, consolidation, or cavitation.
- In acute progressive disseminated histoplasmosis, the leading symptoms are hepatosplenomegaly and lymphadenopathy. CNS-related symptoms include a mass lesion, encephalopathy, and meningitis. Subacute PDH presents as abdominal mass or intestinal ulcers and lesions. CNS dissemination may lead to mass lesions or meningismus, muscle weakness, ataxia, altered consciousness, or focal deficits. Endocarditis, murmurs, peripheral edema, petechiae indicate cardiac dissemination.
- Extensive ulceration in the oropharyngeal area, buccal mucosa, tongue, gingiva, and larynx is characteristic of chronic PDH.
- In presumed ocular histoplasmosis syndrome, atrophic scars with lymphocytic cell infiltration (histo spots) may be seen posterior to the equator of the eye. The condition is bilateral in approximately 10% of patients.
- Around 50% of patients with acute PDH have hilar lymphadenopathy with diffuse nodular infiltrates. As the disease progresses, pulmonary involvement becomes more apparent on radiography.
- Abdominal CT can confirm subacute PDH in 80% of patients, presenting with bilateral adrenal enlargement on the scan.
- Direct microscopy evidence demonstrating characteristic yeast-like cells from any specimen taken from the patient is considered significant diagnostic proof.
- Using sputum and blood samples, diagnosis is made on the basis of conversion of the mold form to the yeast phase by growth at 37°C. The titer of complement-fixing antibody is considered positive at dilutions greater than 1:8. Positive results are usually found in 5–15% of cases of acute pulmonary infection 3 weeks after exposure and in 75–95% of cases at 6 weeks. In chronic infection, the test may remain positive for a long time in 70–90% of patients.

continued...

...continued

- Differential diagnosis needs to be carried out on *H. capsulatum* with *Chrysosporium* and *Sepedonium* species. Since Histoplasma GMs are similar to antigens of other dimorphic fungi such as *Blastomyces dermatitidis*, *Paracoccidioides brasiliensis*, *Talaromyces marneffei*, Aspergillus and Sporothrix species, cross-reactivity of these fungi with the antigen test is common.

- In 2017, a separate genus *Emergomyces*, formerly classified under the genus *Emmonsia* was placed in *Ajellomycetaceae* family. Emergomyces species is also a dimorphic fungus mostly found among immunocompromised individuals in Asia, Europe, Africa, and North America. The number of emergomycosis cases might be underestimated considering the high incidence of HIV/AIDS. Differential diagnosis remains challenging.

5. HOW IS THE DISEASE MANAGED AND PREVENTED?

- In most cases, infection with Histoplasma species of **immunocompetent individuals** is **self-limiting** and does not require therapy, which is recommended in cases of prolonged infection, systemic infection or immunocompromised patients.

- All patients with progressive disseminated histoplasmosis and the symptoms of meningitis require medical therapy.

- Surgical resection of pulmonary cavitary lesions is required when repeated relapses or progressive disease occurs despite repeated intensive medical therapy.

- Severe cases of acute histoplasmosis, and all cases of chronic and disseminated disease, require treatment with antifungal drugs. The recommended regimen is a 3-month course of itraconazole. In patients with CPH, noncavitary disease, a 6-month therapy is recommended instead, whereas for a cavitary disease, the treatment might be prolonged for one year. Patients with PDH, require induction therapy with amphotericin-B for 2–4 weeks followed by one year of itraconazole.

- Amphotericin B (AmB) is used to treat acute pulmonary histoplasmosis, chronic pulmonary histoplasmosis, all forms of progressive disseminated histoplasmosis, meningitis, and endovascular histoplasmosis. In adults, as well as in children, 0.7–1 mg kg^{-1} per day of AmB or up to a total dose of 35 mg kg^{-1} is used IV. A minimum total dose should be no less than 2 g. In the case of underlying severe immunodeficiency (such as AIDS) when a life-long antifungal maintenance therapy is required to treat acute progressive disseminated histoplasmosis, the induction dose of 0.7–1 mg kg^{-1} per day AmB to a total of 20–25 mg kg^{-1} is used, followed by a maintenance therapy of 50 mg once per week.

- Itraconazole is administered 200 mg three times daily for 3 days, then 200 mg twice daily in adults and 5–10 mg/kg (maximum 400 mg daily) in children.

- Ketoconazole can be used as an alternative to AmB in the treatment of chronic pulmonary histoplasmosis or chronic and subacute progressive disseminated histoplasmosis.

- Nonsteroidal anti-inflammatory drugs (NSAIDs) such as ibuprofen are used with analgesic, anti-inflammatory, and antipyretic action, particularly for patients with pericarditis. Corticosteroids, mostly prednisone, can be used to decrease hypersensitivity to *Histoplasma*. High-dose steroids are used in patients with extensive maculopathy.

- Prognosis for a complete recovery is good for acute pulmonary histoplasmosis. The relapse rate is 20% in chronic pulmonary histoplasmosis and 50% in treated acute progressive disseminated histoplasmosis, which drops to 10–20% in case of a life-long antifungal maintenance.

- Prevention includes chemical disinfection and respiratory barrier protection in high-risk areas or during high-risk activities. Individuals residing in or traveling to endemic areas, particularly those with a history of immunodeficiency, the elderly, and children, must be educated/briefed about exposure risks.

FURTHER READING

Bongomin F (eds). Histoplasma and Histoplasmosis. IntechOpen, 2020.

Mandell GL, Bennett JE, Dolin R. Histoplasmosis. In: Principles and Practice of Infectious Diseases, 9th edition. Elsevier, Philadelphia, 2019.

Murray JF, Nadel JA. Histoplasmosis. In: Textbook of Respiratory Medicine, 7th edition. Elsevier, Philadelphia, 2021.

REFERENCES

Darling ST. The Morphology of the Parasite (Histoplasma capsulatum) and the Lesions of Histoplasmosis, a Fatal Disease of Tropical America. J Exp Med, 11: 515–530, 1909.

Da Silva RM, da Silva Neto JR, Santos CS, et al. Fluorescent *in situ* Hybridization of Pre-Incubated Blood Culture Material for the Rapid Diagnosis of Histoplasmosis. Med Mycol, 53: 160–164, 2015.

Deepe GS Jr. Immune Response to Early and Late Histoplasma capsulatum Infections. Curr Opin Microbiol, 3: 359–362, 2000.

Kauffman CA. Histoplasmosis: A Clinical and Laboratory Update. Clin Microbiol Rev, 20: 115–132, 2007.

Kirkland N, Fierer J. Innate Immune Receptors and Defense Against Primary Pathogenic Fungi. Vaccines, 8: 303, 2020.

Kurowski R, Ostapchuk M. Overview of Histoplasmosis. Am Fam Physician, 66: 2247–2252, 2002.

Linder KA, Kauffman CA. Current and New Perspectives in the Diagnosis of Blastomycosis and Histoplasmosis. J Fungi, 7: 12, 2021.

Samaddar A, Sharma A. Emergomycosis, an Emerging Systemic Mycosis in Immunocompromised Patients: Current Trends and Future Prospects. Front Med, 8: 670731, 2021.

Segal BH, Kwon-Chung J, Walsh TJ, et al. Immunotherapy for Fungal Infections. Clin Infect Dis, 42: 507–515, 2006.

Shen Q, Rappleye CA. Living Within the Macrophage: Dimorphic Fungal Pathogen Intracellular Metabolism. Front Cell Infect Microbiol, 10: 592259, 2020.

Wheat JL. Current Diagnosis of Histoplasmosis. Trends Microbiol, 11: 488–494, 2003.

Wheat LJ. Antigen Detection, Serology, and Molecular Diagnosis of Invasive Mycoses in the Immunocompromised Host. Transpl Infect Dis, 8: 128–139, 2006.

Wheat LJ, Conces D, Allen SD, et al. Pulmonary Histoplasmosis Syndromes: Recognition, Diagnosis, and Management. Semin Respir Crit Care Med, 25: 129–144, 2004.

Wheat LJ, Musial CE, Jenny-Avital E. Diagnosis and Management of Central Nervous System Histoplasmosis. Clin Infect Dis, 40: 844–852, 2005.

Woods JP. Knocking on the Right Door and Making a Comfortable Home: Histoplasma capsulatum Intracellular Pathogenesis. Curr Opin Microbiol, 6: 327–331, 2003.

WEBSITES

Centers for the Disease Control and Prevention, Fungal Diseases, Histoplasmosis, 2020: https://www.cdc.gov/fungal/diseases/histoplasmosis/index.html

Mayo Clinic, Histoplasmosis, 2022: https://www.mayoclinic.org/diseases-conditions/histoplasmosis/symptoms-causes/syc-20373495

MedicineNet, Histoplasmosis (Cave Disease), 2020: https://www.medicinenet.com/histoplasmosis_facts/article.htm

WebMD, Definition of Histoplamosis, 2021: https://www.webmd.com/a-to-z-guides/histoplasmosis

Wikipedia, The Free Encyclopedia: www.wikipedia.org

Students can test their knowledge of this case study by visiting the Instructor and Student Resources: [www.routledge.com/cw/lydyard] where several multiple choice questions can be found.

Human immunodeficiency virus

18

A 37-year-old accountant began to feel extremely unwell on returning from a 2-week holiday in South America. He had been feeling "under the weather" for 8 weeks before his holiday and had had several male sexual partners during that time. He did not believe in the use of condoms. Over the previous 4 days, he had developed a dry cough, had noticed increasing shortness of breath, and had begun to feel feverish. A painless lump had also developed in his mouth, and on his upper gum.

Because he felt so unwell, he went to the emergency department at his local hospital where they admitted him to one of the wards. On examination of his chest, there were few physical signs, but a chest X-ray showed widespread shadowing (Figure 18.1). Sputum swabs were taken for bacteriology but were found to be negative. A bronchial lavage was therefore performed and the stained cell pellet showed the presence of the organism *Pneumocystis jiroveci* (Figure 18.2). This is an example of an opportunistic infection that is only observed in people with immunodeficiency.

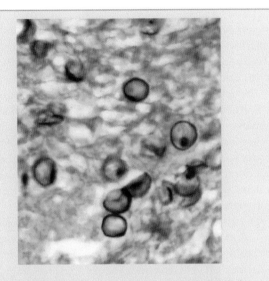

Figure 18.2 Cysts of *Pneumocystis jiroveci*. Histopathology of lung showing characteristic cysts with cup forms and dot-like cyst-wall thickenings. These cysts were detected in the bronchial lavage of the patient (Methenamine silver stain). *Courtesy of the Centers for Disease Control & Prevention, Atlanta, Georgia. Image is found in the Public Health Image Library #960. Additional photographic credit is given to Dr Edwin P. Ewing who took the photo in 1984.*

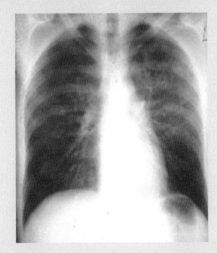

Figure 18.1 Chest X-ray showing bilateral shadowing from both hila to the periphery. *Image originally published in "Immunology 7th edition" by David K Mail, Jonathan Brostroff, Ivan Maurice Roitt, and David Roth, 2006. Reprinted with permission from Elsevier.*

On examination of his mouth, a purplish nodular swelling was visible (Figure 18.3) and this was biopsied during his hospital stay. This lesion histologically showed disordered angiogenesis, formation of new blood vessels characteristic for the so-called **Kaposi sarcoma**, suggesting a severe immunodeficiency, most probably due to HIV infection leading to acquired immunodeficiency syndrome (AIDS). This was confirmed by HIV **seropositivity**. The patient was immediately treated with co-trimoxazole and **highly active antiretroviral therapy** (**HAART**).

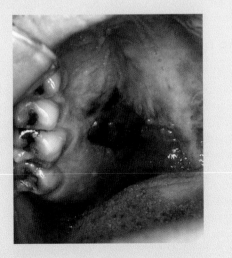

Figure 18.3 Kaposi sarcoma seen as a purple-red growth in the patient's oral cavity. *Courtesy of the Centers for Disease Control & Prevention, Atlanta, Georgia. Image is found in the Public Health Image Library #6070. Additional photographic credit is given to Sol Silverman, Jr, DDS, University of California, San Francisco who took the image in 1987.*

1. WHAT IS THE CAUSATIVE AGENT, HOW DOES IT ENTER THE BODY AND HOW DOES IT SPREAD A) WITHIN THE BODY AND B) FROM PERSON TO PERSON?

CAUSATIVE AGENT

The human immunodeficiency virus (HIV) is a retrovirus belonging to the lentivirus genus. Two species of HIV are recognized, HIV-1 and HIV-2, which differ in origin and nucleotide sequence. HIV-2 is similar to HIV-1 in many properties including replication cycle and transmission. It is much less widespread than HIV-1 being mostly limited to West Africa and the resultant disease is less aggressive. Discussion of this case will be limited to HIV-1.

The HIV-1 genome is composed of approximately 10 kb of RNA, which includes the *gag* gene for capsid and matrix proteins, *pol* gene for three enzymes essential for virus replication, and *env* gene for envelope glycoprotein. The products of *rev* and *tat* genes regulate the synthesis and processing of viral RNA and others known as accessory proteins (nef, vif, vpr, vpu) combat cellular intrinsic defense molecules.

The HIV-1 **virion** (infectious virus particle) is shown in Figure 18.4. Innermost are two copies of the single-stranded (ss) positive sense RNA genome closely associated with several molecules of viral enzymes, namely reverse transcriptase, integrase, and protease. These are enclosed within a conical core made of p24, the **capsid** protein, which is, in turn, surrounded by the matrix protein and a **lipid envelope**. This contains the envelope glycoproteins, gp41 (transmembrane protein) and gp120 (surface protein), the most extensively glycosylated viral protein known. This "sugar coating" or glycan shield impedes access of neutralizing antibodies to target epitopes (see Section 2).

VIRUS REPLICATION

HIV primarily infects **CD4+** T lymphocytes, where it replicates and produces more viruses by budding (Figure 18.5). First, the surface envelope glycoprotein, gp120, binds to CD4 triggering conformational changes allowing viral engagement with a co-receptor, a cellular **chemokine** receptor, usually CCR5 or, less commonly, CXCR4. Interaction with the co-receptor induces changes in the transmembrane envelope glycoprotein, gp41, such that a hydrophobic peptide inserts into the target cell membrane. Fusion with the virus envelope follows and the viral capsid is released into the cytoplasm.

The viral RNA remains within the capsid which protects against recognition by the cell's **Toll-like receptors** (**TLRs**) and prevents initiation of innate immunity. The viral reverse transcriptase converts the viral genome into double-stranded (ds) DNA and the viral capsid passes into the nucleus through a nuclear pore. In the nucleus, the capsid disassembles and the

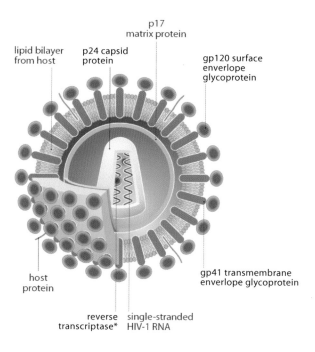

Figure 18.4 The structure of HIV-1 virion. The main structural proteins are shown. *To note as well as reverse transcriptase, protease and integrase are also found within the capsid. *Image originally published in "Immunology 7th edition" by David K Mail, Jonathan Brostroff, Ivan Maurice Roitt, and David Roth, 2006. Reprint With permission from Elsevier.*

viral integrase forms the provirus by inserting the HIV DNA into the host chromosome.

Host-cell transcription of integrated viral DNA produces multiple RNA transcripts, which are transported to the cytoplasm. Some are destined to become encapsidated as progeny viral genomes. Others are translated to form envelope glycoprotein (gp160), a Pol polyprotein and a Gag polyprotein. The host-cell protease, furin, cleaves gp160 into the two *env* products, gp120 and gp41. Viral assembly starts when viral polyproteins and genome associate at areas of the cell's plasma membrane containing the envelope glycoproteins. An immature form of the virus is released from the cell by budding. In the final step, the HIV protease cleaves the Gag and Pol polyproteins into the remaining components, p17, p24, and viral enzymes, facilitating assembly of the viral nucleocapsid and generation of the mature infectious virus (as shown in Figure 18.4).

Virus replication is initiated and completed in activated cells. After viral integration, a small proportion of cells return to the resting state and are latently infected indefinitely.

ENTRY INTO THE BODY

HIV usually enters the body via mucosal surfaces of the vagina and rectum, and rarely of the mouth. The virus passes across the epithelium before entering cells via the cellular receptor CD4 and CCR5. Persons who lack a functional CCR5 receptor are resistant to infection by sexual transmission. HIV strains

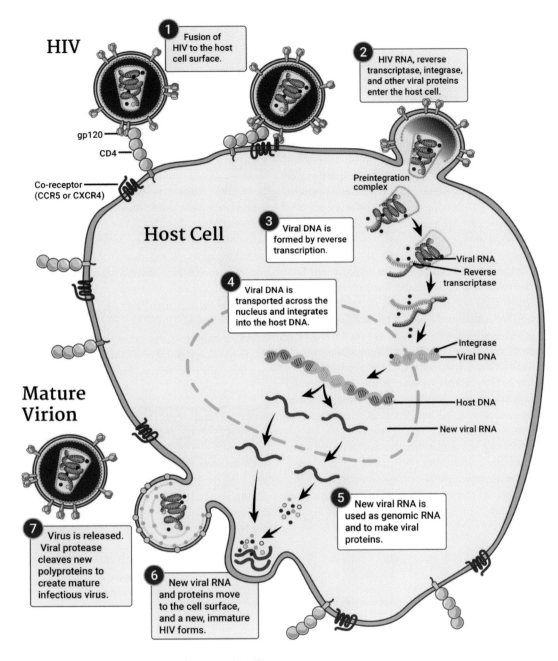

Figure 18.5 Steps in the HIV replication cycle. *Courtesy of NIAID.*

that use CXCR4 to facilitate entry are found late in infection and are not readily transmitted from person to person.

Memory CD4+ T cells are the primary target of HIV because of their high expression of CCR5. **Dendritic cells (DCs)**, and other cells of the macrophage/monocyte lineage, express CD4 and CCR5. Only a subset of roughly 1% of dendritic cells are infected and this infection is less productive than in T cells.

SPREAD WITHIN THE BODY

After entry into the body, virions, infected T cells and dendritic cells travel to the draining lymph node, a rich source of susceptible activated CD4+ T cells. A consequent rapid increase in HIV replication ensues, the virus or infected CD4+ T cells exit the lymph node and travel via the thoracic duct to enter the bloodstream.

The resulting **viremia** enables spread to other mucosal surfaces, which are protected by a common immune system. About two-thirds of B and T cells in the body are found within the epithelium and in the submucosa. The gut-associated lymphoid tissue (GALT) contains about 40% of these cells and CD4+/CCR5 T cells are abundant. This tissue is a major site of HIV reproduction and, in this early stage of primary infection, more than half of the body's memory CD4+ T cells are destroyed. There follows widespread infection of lymphoid organs. Lungs and brain are also affected presumably through infection of **alveolar macrophages** and **microglial cells**, respectively. The body is left with a very restricted repertoire of CD4+ T cells, with which to fight infection, including a stable reservoir of resting long-lived memory T cells latently infected with HIV.

PERSON-PERSON-SPREAD

Transmission of HIV occurs through direct contact with infected body fluids, principally semen, also vaginal and cervical secretions, amniotic fluid, breast milk, and blood and blood products. The main route is through unprotected sexual intercourse, either between men who have sex with men or heterosexuals. HIV is not particularly infectious; many more **virions** are necessary for infection than with other viral pathogens such as hepatitis B virus.

The risk of transmission is higher for the passive (receptive) partner than for the insertive; after a single exposure to HIV by receptive vaginal sex, the risk is about 0.08% but is twice that after receptive anal sex. The vagina is protected by non-keratinized stratified squamous epithelium designed to deal with frictional forces whereas the rectum is not, having only one single layer of columnar epithelium. Underlying sexually transmitted diseases are a risk factor for HIV acquisition. For example, genital herpes disrupts epithelial integrity allowing access to the **lamina propria (submucosa)** or **subepidermal tissues** and increases risk by an order of magnitude. In contrast, circumcision reduces the risk of infection.

Transmission of the virus from mother to infant (vertical transmission) occurs during late pregnancy, and during delivery and breastfeeding. The use of contaminated needles and syringes by people who inject drugs is another important mode of transmission.

The risk of transfusion from infected blood or its components is currently extremely low due to blood screening. In the past, many hemophiliacs were infected with HIV by clotting factor concentrates made from human plasma. Nowadays, these clotting factor concentrates are largely replaced by recombinant products but, if plasma-derived, they are screened or virally inactivated and do not pose a risk of contracting HIV.

EPIDEMIOLOGY OF HIV

Since the discovery of AIDS in 1981, over 36 million individuals have died from the disease. Figure 18.6 shows the worldwide distribution of HIV; more than half of those infected are women. Roughly two thirds of persons with HIV including most of the world's HIV-positive children live in sub-Saharan Africa, where transmission is predominantly heterosexual. In North America, Western and Central Europe, transmission was initially among men who have sex with men but the incidence in other populations has increased with injecting drug use and heterosexual contact now accounting for a significant number of cases (see References for a case of primary HIV infection in a woman).

2. WHAT IS THE HOST RESPONSE TO THE INFECTION AND WHAT IS THE DISEASE PATHOGENESIS?

Much of the data on the interaction between virus and host during mucosal transmission has been derived from animal studies modeled on simian immunodeficiency virus in nonhuman primates. The interaction between HIV and the

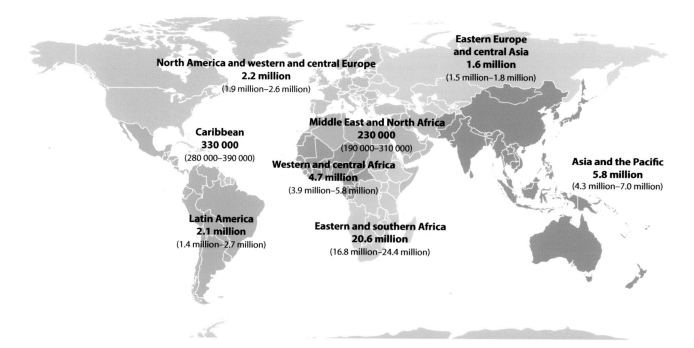

Figure 18.6 The number of children and adults estimated to be living with HIV infection in 2022. *From UNAIDS report 2021. www.unaids.org.*

immune system is complex and not fully understood but there is an immune response to the virus that unfortunately rapidly attenuates. There are three phases to the infection, namely acute, chronic/clinical latency, and AIDS (end stage), with the immune response playing a role in all three.

ACUTE PHASE – HOST RESPONSE AT MUCOSAL SURFACES

Most HIV infections occur at the mucosal epithelia of the vagina and rectum. The vagina and ectocervix are covered by stratified squamous epithelium whereas the endocervix has columnar epithelium as does the rectum. The layers of tightly packed epithelial cells of the vagina are likely to be a more significant barrier to cross as compared to the single cell layer of columnar epithelium.

Mucosal secretions are hypotonic; spermatozoa remain intact but other HIV-infected cells would be rapidly lysed releasing the virus. Mucins, high molecular weight glycoproteins, and defensins are important components. Mucosae are bathed by secreted mucins, which hydrate and lubricate the epithelium and aggregate and flush away microorganisms. Defensins are highly cationic antimicrobial peptides with antiviral activity including destabilization of the HIV envelope. However, rather than being protective, defensins and transmembrane mucins may promote HIV infection by inducing local tissue inflammation and attracting memory T and immature dendritic cells.

Mucosal epithelium and lamina propria are populated with several types of cells including CD4+ lymphocytes, especially T-helper 17 (Th-17), and dendritic cells. Dendritic cells display

pathogen recognition receptors (**PRRs**) including signaling/endocytic receptors such as C-type lectins (**DC-SIGN** and **langerin**), that bind to high mannose glycans. In stratified squamous epithelium, dendritic cells are known as Langerhans cells and display langerin.

Dendrites from intraepithelial immature dendritic cells extend beyond the mucosal epithelium to capture luminal HIV by binding gp120 to DC-SIGN or langerin. These cells transfer HIV on their surface or in endosomes to CD4+ T cells in the epithelium or in the lamina propria, which contains many memory CD4+ T cells. Mature dendritic cells migrate with their cargo of HIV to the draining lymph node to present peptides via the major histocompatibility complex (MHC) I and II to CD8+ and CD4+ T cells, respectively. Presentation of antigenic peptides, whether or not derived from HIV, to the T-cell receptor (immunologic synapse) allows the whole virus in the dendritic cell to encounter CD4 and CCR5 on the CD4+ T cell (Figure 18.7). This process, which is known as trans-infection, can also infect neighboring CD4+ T cells by release of virus from the dendritic cell.

HIV entry via the male genital tract is not well understood. Keratinized, stratified squamous epithelium covers the penile shaft and outer surface of the prepuce (foreskin). This provides a protective barrier against HIV infection. The inner mucosal surface of the foreskin is not keratinized and is rich in Langerhans cells (dendritic cells) (Figure 18.8), making it particularly susceptible to the virus. During sexual intercourse, the foreskin is pulled back down the shaft of the penis, and the whole inner surface of the foreskin is exposed to vaginal secretions, providing a large area where HIV transmission could take place.

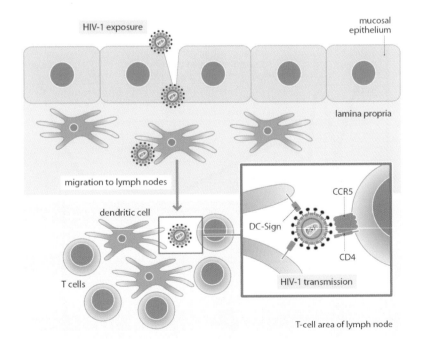

Figure 18.7 Infection of CD4+ T cells through dendritic cells in mucosal surfaces: model of infection through DC-Sign. Dendritic cells (DCs) in the lamina propria bind HIV via their surface DC-Sign. The DCs then migrate to the draining lymph nodes and "present" intact HIV to CD4+ T cells.

Apart from encountering dendritic cells, any cell-free HIV that reaches the lamina propria can infect neighboring CD4+ T cells and macrophages. HIV-infected cells have virally induced decreased MHC I expression and are susceptible to killing by natural-killer (NK) cells, part of the innate immune response. Free virus also activates the lectin complement pathway by virtue of mannose in its glycan coat. HIV is not destroyed by complement-mediated lysis because the virus acquired regulators of complement activation (RCA) during the budding process. RCA tightly control the complement system to prevent spontaneous destruction of host cells and, unfortunately, protects HIV from lysis. Cells of the innate and adaptive immune system bearing complement receptors

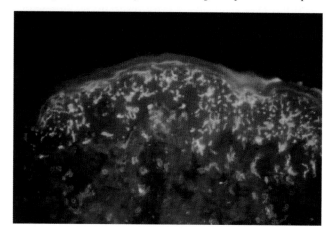

Figure 18.8 The prepuce of the penis showing the distribution of CD4+ dendritic cells in the epidermal/dermal layer (stained green) with the CD4+ T cells labeled orange (green and yellow) in the submucosal tissues. The susceptibility of the prepuce to HIV infection is demonstrated by the protective effect of circumcision. *Courtesy of Professor Leonard Poulter, Professor Emeritus, University College London.*

may take up HIV coated in complement fragments resulting in cell activation and inflammation. Paradoxically, the very nature of the innate immune response is to recruit activated CD4+ T cells fueling progression of the infection while, at the same time, increasing cellular immunity and type-1 interferon responses.

ACUTE PHASE – SYSTEMIC RESPONSE

Around 7–14 days after HIV's first entry into the body and travel to the draining lymph node (see Section 1, entry into and spread within the body), viremia ensues (Figure 18.9). The higher the level of viremia, the greater the spread of virus to other lymphoid organs and tissues. At this stage, there is an increase in HIV-specific **CD8**+ cytotoxic T cells directed against peptides from p24, which destroy virus infected cells and reduce the viral load 10- to 100-fold over a few months. Seroconversion to HIV follows the cytotoxic T-cell response and is marked by the appearance of anti-p24 antibodies but these rapidly evolve to target other viral proteins including the envelope proteins and persist for life at high level.

The levels of circulating CD4+ T cells decrease significantly but increase slightly as the immune response begins to exert some control over the infection. By the end of this phase, the level of viremia settles to a "set point", which presumably represents a balance between production of new virus and control by the immune system. However, it is not possible to clear the virus completely, virus replication continues, and a reservoir of latently infected cells "harboring" the virus remains including T cells and macrophages.

Neutralizing antibodies do not appear until the viremia has significantly subsided, questioning their role in decreasing the viral load. Such is the infidelity of reverse transcriptase that viral genomes recovered from any individual diverge

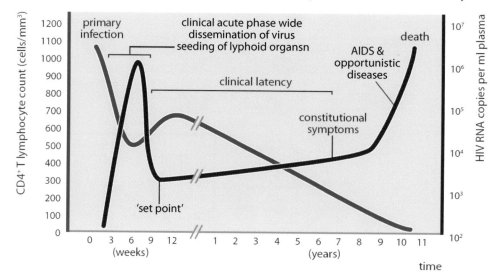

Figure 18.9 The typical course of HIV infection. Following primary infection, CD4+ T cells (blue line) decrease until about 6 weeks and then increase slightly. During the acute phase, which is often symptomatic (see Section 3, clinical presentation), virus levels (red line) increase in the circulation. By the onset of clinical latency/chronic phase, the level of the virus in the circulation settles to a "set point" which predicts the rate of progression to disease. During this period, there is a gradual increase in viremia and a gradual decrease in CD4+ T cells to a level that the patient develops AIDS. *Image originally published in "Immunology 7th edition" by David K Mail, Jonathan Brostroff, Ivan Maurice Roitt, and David Roth, 2006. Reprint with permission from Elsevier.*

steadily with time from the original infecting strain giving rise to countless distinct mutants, collectively known as a **quasispecies**. Variants of HIV are continually being selected for replicative fitness in the face of the evolving immune responses. This variation is most marked in the variable regions of the envelope protein, gp120, the target of virus neutralizing antibody. Unsurprisingly, neutralizing antibodies drive the generation of escape mutants in the gp120 hypervariable region, thus making it even harder for the immune system to control infection.

CHRONIC/CLINICAL LATENCY PHASE

The length of clinical latency, a general lack of symptoms, depends on the "set" point, which varies from individual to individual, the higher the level, the shorter the duration before the onset of AIDS. During this period, CD4+ T cell levels gradually decrease and viremia gradually increases (Figure 18.9). As the immune impairment progresses, there are early signs of symptomatic infection – constitutional symptoms (see Section 3, clinical presentation).

AIDS

Progression of the disease to full-blown AIDS can occur from a few months up to and beyond 10 years. Eventually, when the CD4+ T cell count drops to around 200 cells/mm³ or less, the patient presents with an AIDS-defining illness (see Section 3, clinical presentation).

PATHOGENESIS

Pathogenesis of the disease continues to be elucidated, nevertheless, the key event leading to immunodeficiency is the death of CD4+ T cells.

Several mechanisms have been proposed for the death of CD4+ T cells, the most important of which are:

* Death of CD4+ T cells due directly to HIV infection.
* **Apoptosis** of CD4+ T cells mediated by cytotoxic CD8+ T cells recognizing the infected CD4+ T cell presenting HIV peptides via HLA class I molecules.

The CD4+ T cell is the "conductor" of the immunologic orchestra which, together with dendritic cells, controls the development and function of most immunologic cell types. Thus, ablation of the majority of CD4+ T cells results in major deficiencies in all arms of the adaptive immune response and impairs NK-cell killing of virus infected cells. (Figure 18.10). Depletion and dysregulation occur mostly at mucosal sites, especially in the GALT. T-helper 17 (Th-17) cells mediate immunity against bacterial and fungal pathogens at mucosal surfaces; their loss from the gut increases intestinal permeability and translocation of microbial products from the gut lumen with consequent hyper-immune activation and irreversible damage to the mucosal barrier.

As described above, the dendritic cell is an important cell involved in the trans-infection of CD4+ T cells with HIV. It is presently unclear how much conventional infection of a small

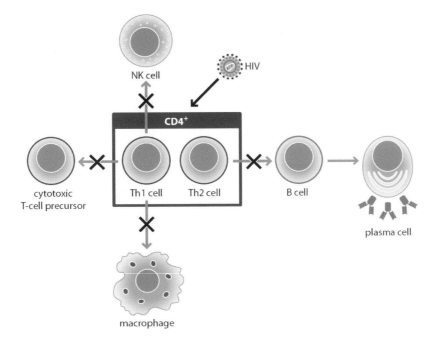

Figure 18.10 HIV compromises the pivotal role of CD4 T cells in immune responses. Th-1 and Th-2 and other Th-cell subsets can be infected with HIV. Th-1 cells help to induce NK-cell function in the innate response against viruses and are important in the development of cytotoxic effector CD8+ T cells from their cytotoxic T-cell precursors and helping macrophages to kill intracellular parasites. Th-1 and Th-2 cells both help B cells to differentiate into antibody-producing plasma cells. In addition, follicular helper T cells (Tfh) – a subset of Th-2 cells, are infected early after HIV infection. Note: At mucosal sites, Th-17 cells are of major importance where they produce pro-inflammatory cytokines activating a variety of different effector cell types. Th-17 numbers are significantly reduced in HIV-infected individuals especially in the GALT.

subset of dendritic cells contributes to disease. It is, however, certain that antigen presentation by dendritic cells is severely handicapped because of HIV infection. Infected dendritic cells might also affect the early differentiation of T helper 0 (Th0) cells into either **Th-1**, **Th-2** or **Th-17**, disrupting all immune responses.

Pathologic changes in primary and secondary lymphoid organs and tissues also contribute to immunodeficiency.

- *Changes to the primary lymphoid organs.* Thymus (in young children): thymocytes become infected with a decrease in the output of CD4+ T cells. Late effects include changes to stromal cell function and loss of tissue architecture. The bone marrow depression seen late in the disease is probably due to effects on stromal cells, but the mechanism remains unclear.

- *Changes to lymph nodes* include polyclonal activation of CD8+ T cells and B cells, as well as a reduction in the number of CD4+ cells. Patients infected with HIV have increased plasma levels of **lipopolysaccharides** (**LPS**) and microbial components due to breakdown of the mucosal barrier of the GI tract. One explanation of the continuous ongoing "inflammation-like" response in lymph nodes is from T-cell activation through these microbial products interacting with TLRs on resident immune cells. This inflammation leads to **fibrosis** and, later in the disease, breakdown in the structure of the lymph nodes including germinal centers.

3. WHAT IS THE TYPICAL CLINICAL PRESENTATION AND WHAT COMPLICATIONS CAN OCCUR?

Figure 18.9 (see Section 2) shows the typical course of HIV infection. Primary infection may be asymptomatic, but from 25% to 65% of patients have some symptoms (acute HIV syndrome or HIV seroconversion illness) that usually begin about 2–8 weeks after exposure – ranging from an infectious mononucleosis-like illness with fever, **maculopapular rash**, sore throat, night sweats, malaise, **lymphadenopathy**, diarrhea, to mouth and genital ulcers to neurologic disease (**aseptic meningitis, encephalitis**). The symptoms are usually mild and subside spontaneously.

This acute phase is followed by a period of clinical latency, which lasts for several years. About one-third of those infected develop **persistent generalized lymphadenopathy**, enlarged lymph nodes in at least two areas of the body for at least 3 months.

PROGRESSION TO AIDS

In recent years, the term AIDS has been used less often, and it is more common to refer to "advanced HIV disease". In the early symptomatic stage, constitutional symptoms develop including fever, weight loss, headache, fatigue, diarrhea, myalgia, arthralgia.

A person living with HIV is said to have advanced disease when the number of CD4+ cells in the blood is <200 cells/mm^3 and certain serious **opportunistic infections** (such as pneumocystis pneumonia) or malignancies (such as Kaposi sarcoma) develop. AIDS-defining illnesses (Table 18.1) include neoplasms and a variety of bacterial, viral, parasitic,

Table 18.1 AIDS-defining illnesses

Neoplasms
Cervical cancer.
Lymphomas – Burkitt's or immunoblastic.
Primary CNS lymphoma.
Kaposi sarcoma.
Bacterial infections
Mycobacterium tuberculosis, pulmonary or extrapulmonary.
Mycobacterium avium complex or *Mycobacterium kansasii*, disseminated or extrapulmonary.
Mycobacterium, other species or unidentified species, disseminated or extrapulmonary.
Pneumonia, recurrent (two or more episodes in 12 months).
Salmonella septicemia, recurrent.
Viral infections
Cytomegalovirus retinitis.
Cytomegalovirus, other (except liver, spleen, glands).
Herpes simplex, ulcer(s) more than 1 month/bronchitis/pneumonitis.
Progressive multifocal leukoencephalopathy (opportunistic JC virus infection).
Wasting syndrome due to HIV.
HIV-associated encephalopathy (dementia).
Parasitic infections
Cerebral toxoplasmosis.
Cryptosporidiosis diarrhea, more than 1 month.
Isosporiasis, more than 1 month.
Atypical disseminated leishmaniasis.
Reactivation of American trypanosomiasis (meningoencephalitis or myocarditis).
Fungal infections
Pneumocystis pneumonia (PCP).
Candidiasis, esophageal.
Candidiasis, bronchial/tracheal/pulmonary.
Cryptococcosis, extrapulmonary.
Histoplasmosis, disseminated/extrapulmonary.
Coccidioidomycosis, disseminated/extrapulmonary.
Penicilliosis, disseminated.

and fungal infections. Most HIV-positive individuals are likely to die of opportunistic infections rather than HIV directly.

NEOPLASMS

The patient in this case study had Kaposi sarcoma, which is seen mainly in men who have sex with men. It frequently presents on the skin but is also found in the oral cavity. Lesions are usually violaceous in color due to formation of new blood vessels. It is invasive and slowly progresses to the visceral form. Human herpesvirus 8 (HHV-8), also known as Kaposi sarcoma-associated herpesvirus (KSHV), is invariably found in tumor tissue; this is an opportunistic infection – the result of immune compromise. The virus infects endothelial cells causing angiogenesis.

Viruses are also implicated in the other types of malignancy found in advanced disease, namely cervical cancer (human papillomaviruses) and some B-cell lymphomas, namely primary central nervous system (CNS) lymphomas (Epstein-Barr virus - EBV).

4. HOW IS THE DISEASE DIAGNOSED, AND WHAT IS THE DIFFERENTIAL DIAGNOSIS?

The diagnosis of HIV infection is based on detecting specific viral antibodies, antigens, and viral nucleic acid; real-time **polymerase chain reaction** (PCR) is used for detection of viral RNA including viral resistance testing (for the latter see Section 5, management), and proviral DNA. Antibodies are the diagnostic tool from which comes the vernacular phrase "testing HIV-positive". Rapid testing kits for HIV antibodies can now provide results in 20 minutes and are used to test plasma, serum, or saliva.

Screening for HIV-1 should be done with an antigen/antibody combination immunoassay that detects HIV-1 antibodies and HIV-1 p24 antigen. Antibody tests alone are not useful during the earliest phase of infection when antibodies are undetectable. The so-called "diagnostic window" is reduced by including testing for p24 antigen as viremia precedes antibody development. Current tests are highly specific but false positives occasionally occur and the diagnosis must always be confirmed using further tests. These should differentiate HIV-1 from HIV-2 antibodies and be supplemented with nucleic acid tests to differentiate the two viral species should antibody be absent early in primary infection. It is also essential to obtain a second serum or plasma sample from the same individual, to confirm that the laboratory result belongs to the correct patient.

In infants born to HIV-positive mothers, maternal HIV-specific IgG will have crossed the placenta persisting for up to 15 months. Direct virus detection is therefore necessary and tests on whole blood for cellular proviral DNA are required for diagnosis.

Levels of CD4+ T cells and viral load are measured for disease staging. Viral load determines the rate of destruction of the immune system and the levels of CD4+ T cells indicate the degree of immunodeficiency and thus the likelihood of developing opportunistic infections. Falling CD4 counts and climbing HIV RNA levels in the face of apparently adequate antiviral therapy suggest the emergence of antiviral drug resistance and should prompt viral resistance testing.

Flow cytometry is used for counting CD4+ T cells and real-time quantitative PCR kits are available to measure the level of HIV RNA copies in blood (plasma or serum). These tests require an advanced laboratory infrastructure and resources. The availability and standardization of tests for diagnosis and monitoring remain a problem in countries where the national health budget is limited. Information on clinical and laboratory strategies to monitor HIV-infected individuals in resource-limited countries can be obtained from the WHO website.

DIFFERENTIAL DIAGNOSIS

The patient in this case study presented with **pneumonia**. The differential diagnosis of pneumonia in an HIV-infected person is wide including *pneumocystis* pneumonia, **miliary** tuberculosis, and **pyogenic** bacterial pneumonia, for example pneumococcal.

5. HOW IS THE DISEASE MANAGED AND PREVENTED?

MANAGEMENT

The patient in our case study was commenced on antiretroviral therapy (ART) and co-trimoxazole to treat his *Pneumocystis jiroveci* infection. Treatment of Kaposi sarcoma is usually to "watch and wait", as it may regress on ART.

A test for antiretroviral-drug resistance would be performed before embarking on a particular regimen. Resistance testing in the clinic is based on HIV-gene amplification by PCR and analysis of the amplicons by direct sequencing allowing, in principle, "tailor-made" drug therapy for each patient, albeit at a considerable cost. The patient would be at risk of "immune reconstitution inflammatory syndrome" (IRIS), which sometimes happens when antiretrovirals are started in someone with AIDS. The immune system begins to recover, but then responds to a previously acquired opportunistic infection with an overwhelming inflammatory response. Infections commonly associated with IRIS include tuberculosis and cryptococcal meningitis.

A nucleoside analog zidovudine, also known as azidothymidine (AZT), was the first drug used against HIV. It is a reverse transcriptase inhibitor (RTI) but, since then, many more drugs have been developed; there are four points in the HIV-replication cycle which are targeted. (See Section 1, Figure 18.5, points 1,3, 4 and 7).

The list of drugs in Table 18.2 is not exhaustive as there are at least 25 anti-HIV drugs available.

Table 18.2 Examples of anti-HIV drugs

Class of antiretroviral drug	Example
Virus entry inhibitors	
Attachment inhibitor	Fostemsavir (FTR)
Chemokine co-receptor antagonist	Maraviroc (MVC)
Fusion inhibitor	Enfuvirtide (T-20)
Nucleoside and nucleotide RT inhibitors (NRTIs)	
Nucleosides	Abacavir (ABC)
	Emtricitabine (FTC)
	Lamivudine (3TC)
	Zidovudine (AZT)
Nucleotide	Tenofovir (TDF)
Nonnucleoside RT inhibitors (NNRTIs)	Efavirenz (EFV)
	Rilpivirine (RPV)
Integrase inhibitors	Cabotegravir (CAB)
	Dolutegravir (DTG)
	Elvitegravir (EVG)
Protease inhibitors	Atazanavir (ATZ)*
	Darunavir (DRV)*
	Ritonavir (RTV)

RT, reverse transcriptase

*Concentration increased by boosting with low-dose ritonavir which, in addition to its antiviral action, inhibits the activity of enzymes in the liver that break down protease inhibitors.

Attachment Inhibitor

Fostemsavir's active metabolite, temsavir, binds to gp120 to stabilize the protein in a conformation that inhibits binding to CD4 preventing virus attachment and replication.

CCR5 Inhibitor (Chemokine Co-Receptor Antagonist)

Maraviroc blocks the binding of HIV to its co-receptor, which is essential for the virus to enter cells. Maraviroc is not effective in the late stages of HIV infection when viruses with CXCR4 tropism predominate.

Fusion Inhibitors

Enfuvirtide, a peptide, binds to the gp41 subunit of the viral envelope glycoprotein and prevents the conformational change required for fusion of viral envelope and cell membrane.

Reverse Transcriptase Inhibitors (RTIs)

There are two classes:

- The nucleoside and nucleotide analogs, together referred to as NRTIs, are activated by cellular phosphorylation and bind competitively to the reverse transcriptase preventing conversion of viral ssRNA into dsDNA.

- The nonnucleoside RTIs (NNRTIs) bind to reverse transcriptase away from the active site leading to its conformational change rendering the enzyme inactive (allosteric inhibition).

Integrase Inhibitors

These block the HIV integrase and prevents the provirus forming in the host cell's chromosome. Resistance mutations against this drug class are uncommon.

Protease Inhibitors

These are synthetic peptide analogs designed to bind to the active site of the viral protease, inhibiting its function and thereby preventing post-translational cleavage of polyproteins and virus particle maturation to its infectious form.

Drug Resistance and Combination Antiretroviral Treatment

In untreated individuals, there is a high rate of viral replication (at least 10^{10} new virions produced per day). This, combined with a high rate of mutation (about one mutation per virion), means that the emergence of drug-resistant strains poses a major problem for ART. When AZT monotherapy was first introduced, incomplete viral suppression allowed continued replication as drug-resistant mutants emerged from the huge diversity of HIV strains in chronically infected individuals.

Highly active antiretroviral therapy (HAART), a combination of antiretroviral drugs, is designed to suppress viral replication to undetectable levels (viral load <50 copies/ml) and prevent the emergence of drug-resistant HIV; for example, two NRTIs as a backbone regimen plus one of the following as a third drug: an integrase inhibitor, an NNRTI or a boosted protease inhibitor. The other three classes of drugs are reserved for when the virus becomes refractory to therapy.

ART has led to a dramatic improvement of the quality of life of persons living with HIV such that their lifespan is only reduced by a decade compared with that of the general population. In 2005, globally less than 10% of people living with HIV were able to access treatment. Encouragingly, in 2020, about 70% were in this position after a remarkable WHO-led effort to overcome financial and logistical barriers.

PREVENTION

Treating HIV is one of the most effective ways to prevent transmission. Whereas access to HAART was previously limited to people with low CD4+ T-cell count, WHO policy is now that people with HIV start therapy at diagnosis. This point is being promoted with the slogan treatment as prevention – undetectable equals untransmissible (U=U).

National awareness at the governmental level of a country's burden of infection is essential in encouraging acceptance of the problem by that country's people. Intervention can be sought in the following ways.

1. **Sexual health**: education, practice of safer sex (condoms), HIV testing and counseling. Prompt treatment of other sexually transmitted diseases and use of vaginal microbicides.

2. **Drug abuse**: education, needle and syringe exchange, outreach, and support.

3. **Blood products**: HIV screening, donor selection, viral inactivation procedures.

4. **Antenatal screening**: identification of HIV-infected mothers during pregnancy allowing antiviral therapy, intervention at the time of delivery, and replacing breast milk with formula milk. ART has virtually eliminated mother to baby transmission of HIV.

All the above interventions require access to accurate, timely, cost-effective, and confidential HIV testing.

Post-Exposure Prophylaxis and Pre-Exposure Prophylaxis

ART has been used successfully as post-exposure prophylaxis (PEP) for many years in occupational and non-occupational settings after exposure to body fluids that contain HIV. It has also been used after sexual exposure, but the difficulty is timely access to therapy.

Recently, several clinical trials have confirmed that pre-exposure prophylaxis (PREP) is 90% effective in uninfected individuals with frequent ongoing sexual exposure to HIV; ART can be given daily, or in a long-acting form, for example cabotegravir. It is hoped that an antiretroviral drug released from a vaginally inserted ring might also be effective against HIV.

Vaccines

HIV presents a formidable challenge for vaccine development. A vaccine is needed to prevent the initial establishment of HIV infection by producing an immune response capable of totally clearing the virus before it infects cells and, in the process, forms a reservoir of latent virus.

Much of the initial effort focused on producing vaccines to generate neutralizing antibodies; however, those that have reached the stage of phase 3 clinical trials have all failed to prevent infection. A major problem in producing such vaccines is HIV's continuous **antigenic drift** due to high mutation rates. In addition, the viral envelope protein, gp120, is heavily glycosylated preventing virus neutralization by antibodies.

More recently, the aim has been to develop candidate HIV vaccines that induce specific CD8+ T cells but these too have been unsuccessful. One of the obstacles to this approach is that latently infected CD4+ T cells do not produce HIV proteins and are invisible to CD8+ cytotoxic T cells. The idea of these "nonperfect" vaccines would be to reduce the initial burst of viremia and the frequency of latently infected cells instead of preventing infection. This would at least reduce transmission of the virus and progression of disease.

Vaccines using new technology such as mRNA platforms are currently on clinical trials.

IS AN HIV CURE POSSIBLE?

Cure is particularly challenging because of the long-lived reservoir of latently infected cells in the body. HAART can control virus replication but cannot remove the provirus. It is estimated that it would take more than 60 years of therapy to eliminate the reservoir.

Cure from HIV has been reported in two patients known as the "Berlin patient" and the "London patient", respectively. This was achieved by an allogeneic bone marrow transplant from a donor homozygous for a lack of CCR5, an essential co-receptor for HIV. Such a transplant is a dangerous procedure with a significant morbidity and mortality, and it would be inappropriate to carry out this procedure just to eliminate latent HIV. The cure might be worse than the disease! Both patients had the transplant to treat another disease, such as leukemia.

The above cases have identified CCR5 as a good target for gene editing to protect cells from infection. The CCR5 gene in either CD4+ T cells or hematopoietic stem cells could be edited *ex vivo* and the modified cells returned to the patient. Gene editing could also be used to target the provirus for excision from the host chromosome. The main obstacle is efficient targeting of the latent viral reservoirs. There is a long way to go before this technology can be transferred from bench to bedside.

SUMMARY

1. WHAT IS THE CAUSATIVE AGENT, HOW DOES IT ENTER THE BODY, AND HOW DOES IT SPREAD A) WITHIN THE BODY AND B) FROM PERSON TO PERSON?

- AIDS is caused by the human immunodeficiency virus (HIV), a retrovirus belonging to the lentivirus group. The virus contains two single strands (ss) of positive sense RNA surrounded by an envelope. There are two viruses, HIV-1 and HIV-2. HIV-2 is much less widespread, and disease is less aggressive. What follows is limited to HIV-1.

- HIV-1 genes include *gag* coding for capsid (p24) and matrix proteins, *pol* coding for reverse transcriptase, integrase, and protease, and *env* coding for the envelope proteins, gp120 and gp41. The products of *rev* and *tat* genes regulate the synthesis and processing of viral mRNA. Other gene products combat cellular intrinsic defense mechanisms.

- HIV infects CD4+ helper T cells. The viral envelope proteins allow attachment and entry at the cell surface via CD4, the viral receptor, and a chemokine co-receptor (usually CCR5 or less commonly CXCR4). A DNA copy of the viral genome is integrated into the host-cell chromosome by the viral integrase forming the provirus, which is transcribed into viral mRNA. Replication is completed by viral budding from the cell. Final assembly of the mature infectious virus requires the viral protease.

- Virus replication is initiated and completed in activated cells. After viral integration, a small proportion of cells return to the resting state prior to viral gene expression and become latently infected indefinitely.

- Upon entry into the body, HIV infects CD4+ memory T cells. Virions and infected cells pass to the draining lymph node, a rich source of susceptible lymphocytes. Following further replication, the virus travels via the bloodstream to 1) primary lymphoid organs, thymus and bone marrow; 2) other mucosal surfaces – especially the gut-associated lymphoid tissue (GALT); 3) other secondary lymphoid tissues – including other lymph nodes; 4) the lungs and the brain. In this early stage of primary infection, more than half of the body's memory T helper cells are lost, many in the GALT.

- HIV transmits through 1) unprotected sexual intercourse; vertical transmission – mother to infant (infection occurs *in utero*, at birth and from breast-feeding); 3) contaminated blood (transfusions or needle sharing) or contaminated blood products (e.g., factor VIII).

- In 2020, 37.7 million individuals, including 19.3 women and 1.7 million children, were estimated to be living with HIV; about two thirds are in sub-Saharan Africa, where transmission is mainly heterosexual. In North America, Western and Central Europe, transmission was initially between men who have sex with men, but the incidence has since risen in persons with heterosexual contact and injecting drug users.

2. WHAT IS THE HOST RESPONSE TO THE INFECTION AND WHAT IS THE DISEASE PATHOGENESIS?

- Much data on the interaction between virus and host during mucosal transmission has been derived from studies of simian immunodeficiency virus in non-human primates.

- There are three phases to the infection, acute, chronic/clinical latency, and AIDS. The immune response plays a role in all.

- At the start of the acute phase, HIV enters the mucosal epithelia of the vagina and rectum. CD4+ T cells in the genital mucosa are memory cells that express CCR5, targets for HIV. Dendritic cells bind HIV via surface DC-Sign. When these cells present HIV peptides or any other antigenic peptides to the T-cell receptor, CD4+ cells also encounter the intact virus and become infected.

- The virus is carried by the above infected cells from the lamina propria to draining lymph nodes where virus replication in CD4+ T cells causes a viremia about 7–14 days after infection. At this stage, there is an increase in HIV-specific CD8+ cytotoxic T cells, which destroy infected cells and reduce the viral load 10–100-fold over a few months.

- At the start of the chronic/clinical latency phase, the level of viremia settles to a particular point called the "set point" or point of equilibrium. This varies from individual to individual, the higher the amount the shorter the time before onset of AIDS.

- The levels of circulating CD4+ T cells decrease significantly but increase slightly as the immune response begins to have some control over the infection. The immune system is unable to clear the virus completely and a reservoir of latently infected CD4+ T cells and cells of the macrophage/monocyte lineage "harboring" the virus remains.

- During the remainder of the chronic phase: 1) the CD4+ T-cell levels gradually decrease further; 2) the viremia gradually increases. There is so-called clinical latency, that is, a general lack of symptoms, despite progression of the immune impairment.

- When the CD4+ T-cell count drops to 200 cells/mm^3 or less the patient develops an AIDS-defining illness.

- The key pathogenic event leading to immunodeficiency is the death of CD4+ T cells. The mechanisms for this include direct apoptosis mediated by virus infection and CD8+ T cytotoxic cells recognizing an HIV peptide with MHC class I on the surface of infected CD4+ T cells.

- Depletion and dysregulation occur mostly at mucosal sites, especially in the gastrointestinal tract. T-helper 17 (Th-17) cells mediate immunity against pathogens at mucosal surfaces; their loss from the gut increases intestinal permeability and translocation of microbial products from the gut lumen with consequent hyper-immune activation and irreversible damage to the mucosal barrier.

Continued...

...continued

- Infection of thymus cells (in young children) decreases the output of CD4+ T cells. Late effects include changes to stromal-cell function and loss of tissue architecture. The bone marrow depression observed late in the disease is probably due to effects on stromal cells, but the mechanism remains unclear.
- There is an "inflammation-like" response in lymph nodes, which become sites of activated CD8+ T cells and B cells and eventually results in breakdown of follicular structure and fibrosis.

3. WHAT IS THE TYPICAL CLINICAL PRESENTATION AND WHAT COMPLICATIONS CAN OCCUR?

- In primary HIV infection, 25–65% of patients have symptoms (acute HIV syndrome or HIV seroconversion illness) beginning about 2–8 weeks after exposure – ranging from an infectious mononucleosis-like illness with fever, maculopapular rash, sore throat, night sweats, malaise, lymphadenopathy, diarrhea, to mouth and genital ulcers to neurologic disease (aseptic meningitis, neuropathy, myelopathy, encephalitis). The symptoms are usually mild and resolve spontaneously.
- The acute phase is followed by a period of clinical latency, which lasts for several years, although about one-third of patients develop *persistent generalized lymphadenopathy*.
- In advanced HIV disease, patients present with AIDS-defining illnesses including a variety of opportunistic bacterial, viral, and fungal infections, and neoplasms such as Kaposi sarcoma and B-cell lymphomas.
- Kaposi sarcoma (seen in this patient) is a disease affecting HIV-infected individuals, in which human herpesvirus 8 (HHV-8) acts as an opportunistic infection.

4. HOW IS THE DISEASE DIAGNOSED, AND WHAT IS THE DIFFERENTIAL DIAGNOSIS?

- The diagnosis of HIV infection is based on detecting specific antibodies, antigens, or both and should distinguish between HIV-1 and HIV-2. Rapid testing kits for HIV antibodies can now provide results in 20 minutes and are used to test plasma, serum, whole blood, or saliva.
- Real-time PCR is used for detection of viral RNA including viral resistance testing, and proviral DNA.
- Because of occasional false positives detected in antibody assays, the diagnosis must always be confirmed using further tests.
- Antibody tests are not useful during the earliest phase of infection when antibodies are undetectable and testing for p24 antigen or viral RNA is needed.
- In the newborn child, maternal IgG will have crossed the placenta and tests on whole blood for cellular proviral DNA are required for diagnosis.
- Advanced laboratory infrastructure and resources are required for disease staging. CD4+ T-cell counts are measured by flow cytometry. Several kits are available to determine plasma viral load as HIV-1 RNA copies/ml.

- The availability and standardization of tests for diagnosis and monitoring remain a problem in countries where the national health budget is limited.

5. HOW IS THE DISEASE MANAGED AND PREVENTED?

- The classes of antiretroviral drugs are virus entry, nucleoside/nucleoside reverse transcriptase inhibitors (NRTI) and non-nucleoside reverse transcriptase inhibitors (NNRTIs), integrase and protease inhibitors.
- Attachment, fusion and CCR5 inhibitors block the entry of the virus into the target cell. These drugs are reserved for situations when the virus becomes refractory to therapy.
- Some examples of RT inhibitors are the following: NRTI: nucleoside analog, e.g. zidovudine; nucleotide analog, e.g. tenofovir, and NNRTI, e.g. efavirenz. These drugs inhibit viral replication after cell entry but before integration of the provirus into the host chromosomes.
- Integrase inhibitors, e.g. cabotegravir, prevent the formation of HIV provirus.
- Protease inhibitors, e.g. atazanavir, prevent post-translational cleavage of polyproteins and virus particle maturation and therefore new virions are not produced.
- The huge diversity of HIV in chronically infected individuals means that the emergence of drug-resistant viral strains within individuals poses a major problem in antiviral therapy. Resistance testing in the clinic is based on HIV-gene amplification by PCR and analysis of the amplicons by direct sequencing.
- Highly active antiretroviral therapy (HAART) is designed to suppress viral replication to undetectable levels, <50 copies/ml, and prevent the emergence of drug-resistant HIV. A combination of antiviral drugs is used, for example, two NRTIs together with an NNRTI or protease inhibitor, or integrase inhibitor.
- Post-exposure prophylaxis with antiretroviral therapy (ART) is used in occupational and non-occupational settings after exposure to body fluids containing HIV. It is also used after sexual exposure, but the difficulty is timely access to therapy.
- Pre-exposure prophylaxis (PREP) with ART is 90% effective in individuals with frequent ongoing sexual exposure to HIV.
- Transmission of HIV can be reduced through the following: education in sexual health and drug abuse; screening of blood and plasma products for HIV; antenatal screening; and replacing breast milk with formula milk.
- The biggest problem in producing HIV vaccines is continuous antigenic drift due to point mutations and high mutation rates. Recent efforts have focused on development of vaccines enhancing the early HIV-specific CD8+ T-cell response to reduce the initial burst of viremia and the frequency of latently infected cells, aimed at increasing the time to progression and reducing transmission of the virus.

continued...

...continued

- HAART can control virus replication but cannot remove the long-lived reservoir of latently infected cells. If the virus load remains undetectable on therapy stopping infection of new cells, eventually elimination of HIV would be achieved. The problem is that the time estimated to achieve this would be more than 60 years.
- Cure from HIV has been reported in two patients, who underwent an allogeneic bone marrow transplant from a donor homozygous for a lack of CCR5, which is an essential co-receptor for HIV.

Both patients had the transplant to treat another disease, such as leukemia.

- Based on the above, CCR5 has been identified as a good target for gene editing to protect cells from infection. Gene editing could also be used to excise and eliminate the provirus from the host chromosome. One of the main obstacles to this approach is efficient targeting of the latent viral reservoirs in humans. There is a long way to go before this technology can be transferred from bench to bedside.

FURTHER READING

Delves PJ, Martin SJ, Burton DR, Roitt IM. Roitt's Essential Immunology, 12th edition. Wiley-Blackwell, Chichester, 2019.

Goering RV, Dockrell H, Zuckerman M, Chiodini PL. Medical Microbiology and Immunology, 6th edition. Elsevier, Philadelphia, 2019.

Reitz MS, Gallo RC. Human Immunodeficiency Viruses. In: Bennett JE, Dolin R, Blaser MJ, editors. Mandell, Douglas and Bennett's Principles and Practice of Infectious Diseases, 8th edition. Elsevier/Saunders, Philadelphia, 2015.

REFERENCES

Altfeld M, Gale, M Jr. Innate Immunity Against HIV-1 Infection, Nat Immunol, 16: 554–562, 2015.

Deeks SG. Antiretroviral Treatment of HIV Infected Adults. BMJ, 332: 1489–1493, 2006.

Deeks SG, Archin N, Cannon P, et al. Research Priorities for an HIV Cure: International AIDS Society Global Scientific Strategy 2021. Nat Med, 27: 2085–2098, 2021.

Goldstein RH, Mehan WA, Hutchinson B, Robbins GK. Case 24-2021: A 63-Year-Old Woman with Fever, Sore Throat, and Confusion. N Eng J Med, 385: 641–648, 2021.

Haase, AT. Targeting Early Infection to Prevent HIV-1 Mucosal Transmission. Nature, 464: 217–223, 2010.

Hutter G, Nowak D, Mossner, M, et al. Long-term Control of HIV by CCR5 Delta32/Delta32 Stem-Cell Transplantation. N Engl J Med, 360: 692–698, 2009.

Johnston MI, Fauci AS. An HIV Vaccine – Evolving Concepts. N Engl J Med, 356: 2073–2081, 2007.

Keele BF, Estes JD. Barriers to Mucosal Transmission of Immunodeficiency Viruses. Blood, 118: 839–846, 2011.

Martín-Moreno A, Muñoz-Fernández, MA. Dendritic Cells, the Double Agent in the War Against HIV-1. Front Immunol, 10: 1306, 2019.

Morris L. mRNA Vaccines Offer Hope for HIV. Nat Med, 27: 2082–2084, 2021.

Simon V, Ho DD, Karim QA. HIV/AIDS Epidemiology, Pathogenesis, Prevention, and Treatment. Lancet, 368: 489–504, 2006.

WEBSITES

Be in the Know, Averting HIV and AIDS, International AIDS charity: www.avert.org

Centers for Disease Control and Prevention, HIV: http://www.cdc.gov/hiv/

GOV.UK (UK Health Security Agency), Search HIV: https://www.gov.uk/search/all?keywords=HIV&order=relevance

UNAIDS, Home: www.unaids.org

World Health Organization, HIV/AIDS: http://www.who.int/hiv/en/

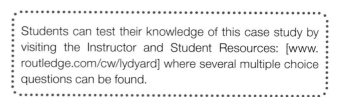

Students can test their knowledge of this case study by visiting the Instructor and Student Resources: [www.routledge.com/cw/lydyard] where several multiple choice questions can be found.

19

Influenza virus

A 59-year-old woman went to see her doctor, as she had been unwell for the previous 3 days. She initially noticed a nonproductive cough, and then she became abruptly worse with a marked fever, headache, and shivering. Since then, she had developed muscle aches all over her body, especially in the legs, and her eyes had become watery and painful to move. She was a nonsmoker, previously fit and well, and on no regular medication. On examination, she was **febrile** (38.2°C), and had difficulty in breathing through her nose, but there were no other abnormal physical signs.

A throat swab was taken, broken off into viral transport medium, and sent to the laboratory. The sample was included in the next respiratory virus panel genome-amplification run, giving an amplification curve for influenza A virus, confirming a diagnosis of acute influenza virus infection.

1. WHAT IS THE CAUSATIVE AGENT, HOW DOES IT ENTER THE BODY AND HOW DOES IT SPREAD A) WITHIN THE BODY AND B) FROM PERSON TO PERSON?

CAUSATIVE AGENT

Influenza A virus belongs to the *Orthomyxoviridae* virus family (myxo = affinity for mucin). The viral genome consists of eight segments of negative single-stranded RNA (i.e. RNA that cannot be translated directly on the ribosome, but has to be first copied into its complementary, positive, strand), which collectively encode 17 viral proteins (**Figure 19.1**). Each RNA segment is closely associated with the nucleoprotein, to form a helical ribonucleoprotein (RNP), or **nucleocapsid**. The RNPs are, in turn, surrounded by a matrix protein and then a **lipid envelope**, which contains two viral glycoproteins, **hemagglutinin** (H or HA) and neuraminidase (N or NA), and also small amounts of the nonglycosylated M2 ion channel protein. Influenza viruses are grouped into types on the basis of the nature of the RNP, which occurs in one of three antigenic forms, hence types A, B, or C influenza viruses. Influenza type A viruses are widespread in nature, infecting many avian species, but also humans, pigs, horses, and occasionally other species such as cats. Influenza B virus is an exclusively human pathogen, while influenza C viruses are not serious pathogens in humans. Influenza type-A viruses are further subdivided into subtypes depending on the nature of their two external glycoproteins. Thus far, 18 distinct hemagglutinins and 11 different neuraminidases have been identified, where each HA or NA molecule differs by at least 20% of its amino-acid sequence from all other HA and NA molecules. When referring to an influenza A virus isolate, it is therefore necessary to specify precisely which subtype it is, for example, influenza A/H1N1 or influenza A/H7N7.

ENTRY AND SPREAD WITHIN THE BODY

Influenza virus enters via the nasal or oral mucosa. In humans and other mammalian species, the virus is **pneumotropic** (in avian species, the virus infects a variety of tissues and is primarily spread through the **fecal–oral** route), that is it preferentially binds to, and infects, respiratory epithelial cells, all the way from the **oropharynx** and nasopharynx right down to the alveolar walls. Influenza virus attaches to target cells via an interaction between the viral ligand, hemagglutinin, and a cellular receptor, comprising sialic acid (SA) residues, a component of the carbohydrate within glycoproteins, on the surface of respiratory epithelial cells throughout the respiratory tract. Human influenza viruses preferentially bind to SA linked to galactose via an $\alpha(2\text{-}6)$-linkage, whereas avian viruses preferentially bind to $\alpha(2\text{-}3)$-linked SA. The virus enters the host cell via endocytotic **vesicles** and uncoats, and free viral RNA is imported to the nucleus by interacting with the cellular importin-α/β for replication. The virus then replicates and new virions are released by the infected cells by budding at the plasma membrane of the host cell. With infections of the lower respiratory tract, direct infection of **pneumocytes** and macrophages can occur. Given the systemic nature of the illness caused by influenza virus infection (see below), it is perhaps surprising that the virus itself does not usually spread beyond the respiratory tract.

PERSON-TO-PERSON SPREAD

Transmission of influenza viruses from person to person is believed to be via large droplets ($\geq 5\,\mu m$ diameter), which are generated from an infected respiratory tract during coughing, sneezing, or even talking. The droplets are deposited on the nasal or oral mucosa of a new susceptible host leading to infection. Additionally, virus can be transmitted via touching a surface contaminated with virus (fomite) and then transferring virus to the mouth, nose or possibly the eyes. The

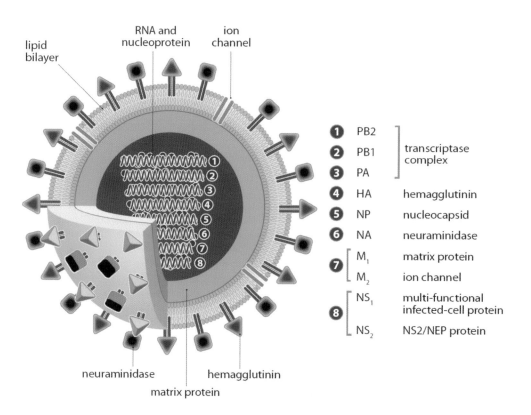

Figure 19.1 **Schematic diagram of an influenza virus.** The helical nucleocapsid contains eight segments of ssRNA each coated with nucleoprotein. This is surrounded by a layer of M1 (membrane or matrix) protein which, in turn, is surrounded by a lipid envelope into which are inserted two viral glycoproteins (hemagglutinin and neuraminidase) and a small amount of the M2 ion channel protein.

incubation period prior to symptom development is around 48 hours. Patients become infectious around 24 hours before symptoms begin, and this usually lasts for a further 5–7 days.

EPIDEMIOLOGY

The epidemiology of influenza has several unusual characteristics (Figure 19.2). Annual outbreaks of infection are highly seasonal, arising each winter in temperate climates,

with a considerable percentage (e.g. 10%) of the population acquiring infection, with concomitant increases in hospital admissions and influenza-related deaths. The size of these outbreaks varies from year to year. In the UK, an outbreak is referred to as an epidemic only when the consultation rate for "influenza-like illness" recorded through the Royal College of General Practitioners surveillance scheme exceeds 400 per 100 000 population. However, superimposed on this regular annual cyclical pattern, unpredictable global epidemics

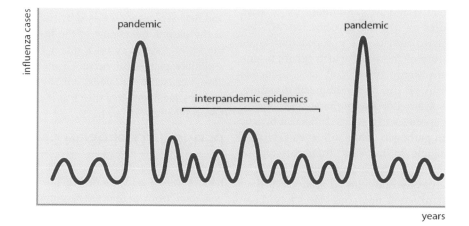

Figure 19.2 **Epidemiology of influenza.** This diagram shows the number of cases of influenza occurring over time. Each peak corresponds to a winter season, illustrating the annual epidemics. Superimposed on that, at irregular intervals averaging about once every 30–40 years, there is a massive peak corresponding to an influenza pandemic. *From Humphreys and Irving, Problem-Orientated Clinical Microbiology and Infection, Figure 17.5, page 72. With permission from Oxford University Press.*

Table 19.1 Pandemic influenza viruses of the 20th century

Year	Virus	Name
1918–19	H1N1	Spanish flu
1956–57	H2N2	Asian flu
1968	H3N2	Hong Kong flu

occur, on a scale much greater than the annual outbreaks, sweeping across the world with huge numbers of infections, and considerable morbidity and mortality. These latter phenomena are referred to as pandemics, and experience in the 20th century, plus careful reading of historic records suggest that these have occurred about every 30 years or so. Influenza epidemics and pandemics arise from the processes of antigenic drift and antigenic shift, respectively.

Antigenic drift results in the emergence of new strains each year. It arises from random spontaneous mutation occurring within the influenza virus genome as it replicates. Virus causing an outbreak in a particular year will have up to 1% genome sequence difference from virus that caused the previous year's outbreak. Although this will occur across the whole viral genome, it is the mutations within the genes encoding the HA and NA surface glycoproteins that are important in this context. The HA protein contains five highly immunogenic regions (Figure 19.3) to which the antibody response to infection is directed. Mutations within these **epitopes** may therefore allow virus to escape the inhibitory effects of antibodies that would otherwise bind to these regions and prevent virus–cell interactions. The important amino-acid differences that accumulate year-on-year within this protein are clustered precisely within these five epitopes. Thus, antigenic drift is an excellent example of Darwinian evolution – mutations occur randomly within the genome, but only those that confer a selective advantage to the virus emerge in the epidemic strain, the selective pressure being the population immune response generated by the previous year's epidemic. Drift occurs in both influenza A and B viruses.

Antigenic shift, which generates the new pandemic strains, is an altogether different process. The viruses causing the influenza pandemics of the 20th century are shown in Table 19.1. Each pandemic arose from the emergence of a new influenza A subtype into the human population. As the new pandemic strain appeared, so the old circulating strain disappeared – thus, in 1956–1957, H2N2 completely replaced H1N1, only to be replaced itself by H3N2 virus in 1968 (an unusual exception to this, arising in 1976, is discussed later). There are two possible underlying mechanisms that can give rise to new pandemic strains, as described below.

1. *Direct transfer of an avian influenza A virus into humans.* This process is undoubtedly happening at the moment, with an increasing number of human infections with the avian H5N1 virus (responsible for large avian epidemics, particularly among chickens) being reported

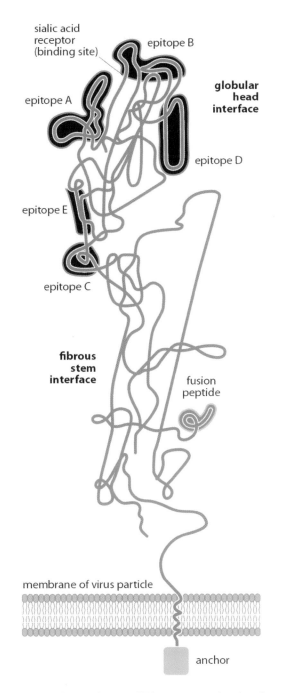

Figure 19.3 Hemagglutinin (HA) structure showing five key epitopes. This shows the protein chain of a single subunit of influenza A hemagglutinin trimer. Epitopes A, B, C, D, and E, are positions where antibody molecules have been shown to bind to HA. The three-dimensional structure was determined by X-ray diffraction of the crystalline protein.

worldwide. However, virus that crosses a species barrier in this way is often not well adjusted for replication in its new host. Currently, avian H5N1 virus is not efficiently transmitted between humans, either because it is not well adapted to grow to high titer in human cells, or because the optimal conformation of SA residues for this virus are only present deep in the lung tissue. Person-to-person spread is therefore very inefficient, as infected individuals are not releasing large amounts of virus in their respiratory secretions. H5N1 virus has thus not emerged (yet) as a new human pandemic virus. There is a worry, however that, as it replicates within human cells, this virus may acquire mutations that could result in adaptation to efficient replication within human cells, at which point person-to-person spread will become more likely, and a true pandemic might eventuate. There is some evidence that the H1N1 virus that caused the 1918–1919 pandemic was entirely avian in origin, and that it had been causing sporadic infections within humans for several years before its emergence as a pandemic virus in 1918. The presumption is that during those preceding years, the virus acquired the necessary mutations to allow adaptation to increased replication within human cells.

2. *Genetic reassortment of human and avian viruses within a co-infected host* (Figure 19.4). Influenza viruses have a segmented genome. Thus, if a cell is infected with two different influenza viruses, it is possible that reassortment (mixing) of these gene segments can occur, such that progeny virus can contain gene segments derived from either one of the "parent" viruses. In Figure 19.4, the emergent virus has six RNA segments derived from the human parent virus, plus segments 4 and 6 from the avian parent. Segments 4 and 6 encode the HA and NA proteins, respectively. Thus, the progeny virus will be one that is well adapted for growth in human cells (all its internal proteins are derived from the human parent) but has two entirely new proteins on its surface (each

HA and NA protein differing by at least 20% amino-acid sequence from all other HA and NA proteins). Such a virus would cause devastating infection across the whole human population, as no-one would have any immunity against these new surface proteins. The H2N2 and H3N2 pandemic viruses from 1956 and 1968 do indeed contain genes derived from both human and avian viruses. It is believed that the reassortment process that generated these viruses took place within pigs (hence referred to as the "mixing vessel"), which are uniquely susceptible to infection with both human and avian viruses. However, it has become at least a theoretical possibility that humans themselves could act as the mixing vessel, for example, if a human was co-infected simultaneously with an avian A/H5N1 and a human A/H1N1 or A/H3N2 virus. The chances of the latter happening will clearly be increased the more humans become infected with avian A/H5N1 virus.

It is worth emphasizing the differences between antigenic drift and shift. The former occurs in both influenza A and B viruses and is a result of random genetic mutation followed by Darwinian selection, resulting in up to 1% differences in amino-acid sequences focused within key epitopes within the surface HA and NA molecules. The latter only occurs in influenza A viruses (presumably because influenza B viruses do not have an animal reservoir) and is a result of either direct cross-species transfer or of genetic reassortment, resulting in the generation of viruses with surface HA and NA proteins that differ by over 20% in amino-acid sequence from those in previously circulating strains.

In 1976, Russian flu emerged, caused by an influenza A/H1N1 virus. This is not presented in Table 19.1 as a pandemic, because the mechanism of emergence of this virus is not believed to have been a natural occurrence. Instead, there is evidence that this virus emerged from a laboratory – its gene sequences were remarkably similar to those of the last previous isolates of H1N1 virus in 1957. Thus, its reappearance in the human population was most likely due to human error.

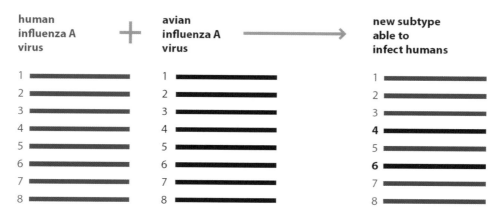

Figure 19.4 Genetic reassortment. Each RNA segment (numbered 1–8) is represented by a horizontal line. The human virus is blue, the avian virus is red. When co-infecting the same cell, emergent viruses may possess RNA segments from either "parent" virus.

Interestingly, this virus did not displace the A/H3N2 virus and ever since 1976, both A/H1N1 and A/H3N2 influenza viruses have co-circulated among humans, together with influenza B viruses. Any one of the three circulating viruses can predominate in a particular year.

The most recent influenza pandemic occurred in 2009. This pandemic virus (known as H1N1 2009 pdm) was unusual in that it was still classified as an H1N1 virus. It had acquired gene segments from avian viruses (segments 1 and 3), human H3N2 virus (segment 2), Eurasian swine virus (segments 6 and 7) and classic swine influenza virus (circulates in N. America, segments 4 and 5). Thus, the two surface glycoproteins (encoded by segments 4 and 6) were derived from swine viruses and were sufficiently different from "human" H1 and N1 proteins so as to avoid existing population immunity to the previously circulating H1N1 virus, and thereby confer pandemic potential. In comparison to the 1918–1919 pandemic, the pandemic caused by H1N1 2009 pdm was relatively mild and more focused on younger age groups, presumably due to at least some measure of immunity in older adults.

2. WHAT IS THE HOST RESPONSE TO THE INFECTION AND WHAT IS THE DISEASE PATHOGENESIS?

Damage to the respiratory epithelial surface occurs due to the **cytolytic** interaction of the virus and the host cell, that is the infected host cells undergo acute cell death. In effect, the virus strips off the inner lining of the respiratory tract, and in so doing, removes two important innate immune defense mechanisms – mucus-secreting cells, and the **mucociliary escalator**. The production of mucus by cells within the respiratory epithelium allows entrapment of inhaled particulate matter (e.g. bacteria). The mucociliary escalator then transports any inhaled particulate matter toward the pharynx, to be coughed out in sputum or swallowed. Removal of these defenses results in potential exposure of the lower respiratory tract to inhaled bacteria.

The above processes result in impairment of lung function to a greater or lesser extent, so patients present with **rhinorrhea**, sore throat, cough, and shortness of breath. However, there is also an important systemic element to the disease influenza (see below). This arises because influenza viruses are potent inducers of **cytokines** such as **interferon-α** (**IFN-α**) and **interleukin** (**IL**)-**6**, and it is these cytokines, not the virus, that circulate in the bloodstream and give rise to the systemic manifestations of fever, headache, muscle aches and pains, and severe malaise. Administration of IFN, for example, as treatment for chronic hepatitis B virus infection, reproduces this symptomatology. Note that the above applies strictly to infection with the currently circulating human influenza viruses – there is emerging evidence that infection of humans with avian influenza A/H5N1 may result in **viremia** and spread to other organs beyond the lungs.

The innate immune response to infection is triggered by recognition of virus-derived pathogen associated molecular patterns (PAMPs) by host pathogen recognition receptors (PRRs), such as retinoic acid-inducible gene-I protein (RIG-I), Toll-like receptors (TLRs), and nucleotide-binding and oligomerization domain (NOD)-like receptors. This leads to activation of specific transcription factors such as nuclear factor kappa B (NF-κB), interferon regulatory factor (IRF)3 and IRF7 and their translocation into the nucleus where they initiate the transcription of genes encoding interferons (IFNs) and other pro-inflammatory cytokines. Production of IFNs then results in stimulation of several genes known as interferon stimulated genes (ISGs). A large number of ISGs have been shown to have specific roles in defense against influenza virus infection – some of which are listed in Table 19.2.

In addition to this innate immune response to infection, adaptive humoral and cellular immune responses are also stimulated. Antibodies (including IgA class antibodies present at mucosal surfaces) to the viral surface proteins, particularly hemagglutinin, may be neutralizing, that is they can prevent the interaction of the HA protein with cellular SA residues and thereby prevent infection. They may also help mediate

Table 19.2 Interferon-stimulated genes with activity against influenza virus

Interferon-stimulated genes	Role
Mx family, MxA, MxB	MxA retains incoming viral genome within cytoplasm
Interferon-induced transmembrane proteins (IFITMs)	Impair fusion of virus and host-cell membranes
Tripartite motif (TRIM) proteins (>80 in humans)	TRIM22 blocks viral genome encapsidation TRIM32 inhibits influenza PB1 RNA polymerase
Zinc finger antiviral protein (ZAP)	Reduces viral RNA expression and translation
Oligoadenylate synthase and ribonuclease-L (OAS-RNase L)	Destroys viral RNA thereby halting protein synthesis
Protein kinase R (PKR)	Binds viral dsDNA, suppresses viral protein synthesis
Viperin (a.k.a. RSAD-2)	Alters formation of lipid raft specific microdomains and inhibits viral budding
Tetherin	Limits export of viral progenies and degrades them

antibody-dependent cellular cytotoxicity. However, antigenic drift results in the generation of strains of virus that can escape this protective immunity. T-cell responses to influenza virus are mostly directed against antigens derived from the internal viral proteins, for example, the nucleoprotein, matrix protein and the RNA polymerase subunit proteins. These proteins are much more conserved within influenza types than the surface proteins, so T-cell immunity may offer some protection each year to emerging drifted viruses.

As would be expected, the virus has evolved multiple mechanisms to assist in evading innate and adaptive immune responses. Influenza HA can enhance ubiquitination and degradation of interferon receptors. The NA protein is thought to play a role in blocking recognition of HA by natural killer (NK)-cell receptors, thus reducing NK-cell clearance of infected cells. Non-structural protein 1 (NS1) is the most important antagonist of interferon-induced responses, for example by inhibiting RIG-I ubiquitination and thereby blocking the RIG-I activation pathway; up-regulating the inactivation of PKR; blocking phosphorylase activity which is essential in activation of the NF-κB mediated induction of ISGs; and reducing phosphorylation of STATs 1-3, which also reduces expression of ISGs.

3. WHAT IS THE TYPICAL CLINICAL PRESENTATION AND WHAT COMPLICATIONS CAN OCCUR?

There are two distinct components to the illness that arise following infection with influenza virus – a respiratory tract component, plus a marked systemic illness characterized by fever, headache, and **myalgia**. Infection does not necessarily result in clinical disease – this will be dependent on the pre-existing state of the patient's lung function, the infecting dose of virus, the presence of preexisting immunity and the extent to which that immunity is able to cross-react with a new viral strain. However, symptomatic influenza virus infection is not a trivial illness. There is considerable morbidity, and it may take several days before patients are well enough to return to their normal daily activities.

The commonest life-threatening complication of influenza virus infection is **pneumonia**, of which there are two pathologic types.

Primary influenzal pneumonia. The virus itself infects right down to the alveoli. There is a mononuclear cell infiltrate into the alveolar walls, and the airspaces become filled with fibrinous inflammatory exudates. This can occur in previously healthy individuals of any age.

Secondary bacterial pneumonia. In recent years, this has been considerably more common than viral pneumonia. Bacteria gain access to the lower respiratory tract for reasons explained above. There is a polymorphonuclear cell infiltrate into the alveoli. This complication is more common in the elderly and in those with preexisting lung disease, of whatever etiology, for example, chronic bronchitis.

There is some evidence that influenza infection can also result in a **myocarditis** – certainly patients with preexisting cardiovascular disease are at increased risk of mortality should they acquire infection. An **encephalitis** (inflammation of the brain substance) is also well recognized. This is not due to the virus itself gaining access to the brain – as explained above, virus is restricted to the lungs. Thus, this is believed to be an immune-mediated phenomenon, or so-called post-infectious encephalitis. In certain individuals, the immune response generated to the influenza virus infection can cross-react with antigens present within the brain, resulting in an encephalitis.

Human infection with avian influenza A H5N1 carries a very high mortality (>50%), and yet, at least thus far, is not easily transmitted person-to-person. This virus induces an explosive acute inflammatory reaction within the lungs – referred to as a **cytokine storm**. The high mortality thus arises from the inappropriate hyperactivity of the host immune response to infection, as opposed to the cytolytic properties of the virus itself.

4. HOW IS THE DISEASE DIAGNOSED, AND WHAT IS THE DIFFERENTIAL DIAGNOSIS?

Infections with a number of different agents (mostly viruses) can result in presentation with an "influenza-like illness". Infection with respiratory syncytial virus, especially in the elderly, is the most common mimic of influenza virus infection. Other possibilities include human metapneumovirus, adenoviruses, and *Mycoplasma pneumoniae*. Clinical "end-of-the-bed" diagnosis is therefore neither sensitive nor specific enough for practical purposes – with the advent of antiviral drugs that are absolutely specific for influenza viruses, accurate diagnosis is necessary to ensure that these drugs are used appropriately and effectively. As the efficacy of these drugs is dependent on initiation of their use as soon as possible after infection, there is a need for rapidity as well as accuracy.

Genome-amplification technology has revolutionized the rapid diagnosis of viral respiratory tract infection. Most laboratories will perform multiplex assays which contain primers capable of amplifying a whole range of respiratory viruses, along with corresponding probes to detect the amplification happening in real time. As most respiratory viruses are RNA viruses, there is an initial reverse transcriptase step. Real-time polymerase chain reaction (PCR) assays have major advantages in terms of their incredible sensitivity as, theoretically, they result in several logs of amplification of the targeted nucleic-acid sequences, and they are also rapid – a positive result can be generated within 2 hours.

Alternative diagnostic approaches include isolation of virus in cell culture, antigen detection by indirect immunofluorescence (the principle of which is illustrated in Figure 19.5) and demonstration of a four-fold or greater rise in anti-influenza virus antibody titers in peripheral blood samples taken 7 days or more apart. Influenza viruses grow

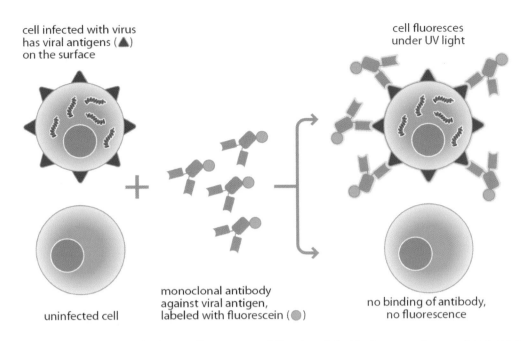

cell infected with virus
has viral antigens (▲)
on the surface

cell fluoresces
under UV light

uninfected cell

monoclonal antibody
against viral antigen,
labeled with fluorescein (●)

no binding of antibody,
no fluorescence

Figure 19.5 Detection of influenza virus by immunofluorescence. A throat swab (or ideally, a nasopharyngeal aspirate) is spotted onto a glass slide, and a fluorescently tagged monoclonal antibody is added. The antibody binds to cells infected with virus, but not to uninfected cells. After an incubation period, any unbound antibody is washed off, and the cells are examined under ultraviolet light. The presence of brightly fluorescent cells is a positive result indicating that the patient is infected with the virus to which the monoclonal antibody was raised. The whole process takes about 2 hours. *From Humphreys and Irving, Problem-Orientated Clinical Microbiology and Infection, Figure 21.2, page 80. With permission from Oxford University Press.*

well in cell culture but may take several days before their presence is revealed by the development of a cytopathic effect.

5. HOW IS THE DISEASE MANAGED AND PREVENTED?

MANAGEMENT

Many cases of influenza virus infection require symptomatic relief only, for example, with anti-pyretics and analgesics such as paracetamol, or other proprietary over-the-counter or home-grown remedies. For the more seriously ill patients, there are a number of anti-influenza agents shown to be effective, acting at different stages of the viral replication cycle.

Amantadine and rimantadine block influenza virus uncoating by binding to the viral matrix M2 protein and thereby blocking ion channels, whose function is essential for the pH-mediated dissolution of the viral capsid (see Section 1). While these agents undoubtedly have efficacy in clinical trials, they have mostly been abandoned in clinical use as there are a number of drawbacks to their use. First, they only work against influenza A viruses; Second, these drugs are also dopamine agonists and therefore have marked central nervous system (CNS) stimulatory activity – in fact, amantadine was originally developed for the treatment of Parkinson's disease. Thus, they are very poorly tolerated in the elderly, the precise group of patients who are most likely to require antiviral therapy, giving rise to hallucinations, insomnia, and agitation. Third, resistance to amantadine emerges within a few days of onset of therapy, due to point mutations in the M2 protein.

Finally, many of the avian influenza viruses, including H5N1 strains, are inherently resistant to amantadine.

The current mainstay of influenza therapy are the neuraminidase inhibitors (Figure 19.6). These are "designer drugs". The influenza neuraminidase was purified, crystallized, and its three-dimensional structure was elucidated. Small molecules were then designed to bind to the active site of the enzyme. One advantage of this class of drugs is that they have activity against all known influenza neuraminidase subtypes. Both of the currently licensed members of this family, zanamavir and oseltamivir, are effective in inhibiting production of infectious viral particles, and are effective in randomized clinical trials. Currently, their use in the UK is reserved for the treatment of seriously ill patients admitted to hospital, or the prophylaxis of patients exposed to influenza who are at increased risk of serious disease, although in the US they may be prescribed by primary care physicians. The importance of these drugs is illustrated by the decision of several governments to stockpile millions of doses as part of their influenza pandemic preparedness plans. Unsurprisingly, however, as these drugs are more widely used, there are increasing reports of viral variants emerging with point mutations within the neuraminidase gene that confer drug resistance. Worryingly, oseltamivir-resistant H1N1 viruses and avian H5N1 strains have now been documented to occur in infected humans.

More recently, baloxavir marboxil has been approved for use in the US and Japan. Baloxavir is an inhibitor of the virus cap-dependent endonuclease activity used in "cap-snatching" by the viral polymerase complex, an essential step in the viral

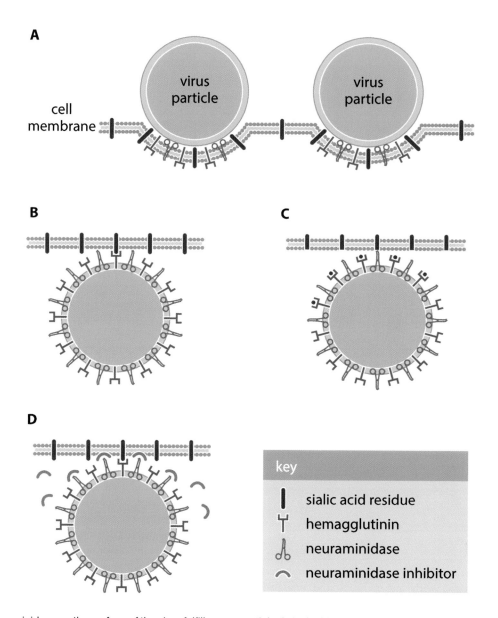

Figure 19.6 Neuraminidase on the surface of the virus fulfills an essential role in the life cycle of the virus. In (A), virus particles are budding through the cell membrane, at the points in the membrane where hemagglutinin (in blue) and neuraminidase (in black) are inserted. The cell membrane contains sialic acid residues (in red). In (B), the virus particle has budded through the cell membrane but the hemagglutinin on the surface of the virus still binds to the sialic acid residues on the surface of the cell. In (C), the neuraminidase (scissors) has cleaved the sialic acid residues thereby allowing the virus particle to move away from the cell and infect new cells. In (D), the neuraminidase is inhibited by the drug (purple semicircles) and therefore the virus particle remains attached to the cell.

replication cycle. Single-point mutations in the polymerase gene may confer resistance.

PREVENTION

Vaccines

Prevention of influenza virus infection is possible through the use of active immunization. There are two different types of vaccine currently in use. Inactivated vaccines are mostly prepared from viruses grown in embryonic hens' eggs, although problems with egg adaptations in the virus resulting in antigenic changes, and with some patients who have severe egg allergy, have led to the development of vaccine derived from virus grown in mammalian cells, and this cell-based vaccine production is likely to increase in the future. Inactivated vaccines contain either whole virus which has been chemically inactivated, or just purified surface HA and NA proteins – the latter are less reactogenic. Live attenuated vaccines are a relatively recent development, containing cold-adapted viruses, and administered by intranasal inoculation. These novel vaccines induce a more robust protection from influenza infection, possibly even from antigenic drift variants. In the UK, live attenuated vaccines are recommended for use in children, because of evidence of superior efficacy in this age group and ease of administration. The propensity for influenza viruses to undergo antigenic drift and shift creates major problems for the vaccine manufacturers, as essentially the viruses are moving targets. In practice, the World Health Organization (WHO) monitors circulating influenza viruses through the collaborative activities of a large number of reference laboratories around the world. Each year, the WHO (based on monitoring current strains in its reference laboratories) announces which particular A/H1N1, A/H3N2, and B viral strains should be used for vaccine being manufactured for the following influenza season. The protection offered by these vaccines depends to a large extent on the degree of antigenic match between the vaccine strains and the strains actually circulating during the season – some years this is better than others! Finally, the success of mRNA-based vaccines in protecting individuals from severe SARS-CoV-2 infection will undoubtedly lead to the development of similar vaccines against other infectious agents. mRNA-based influenza vaccines are well advanced in production and clinical trials.

The availability of a vaccine then begs the question of who should be vaccinated. Most countries adopt a selective policy, that is the recommendation is to vaccinate those subgroups within the population who will fare badly should they acquire infection. Table 19.3 lists those groups in the UK designated in the Department of Health guidelines. This list is reviewed annually – in 2010, pregnancy was added as a risk category, in 2012, the decision was taken to extend immunization to include children aged 2–16 years, and in 2014, morbid obesity was also added to the list. In contrast, in the US, the Advisory Committee on Immunization Practices now recommends influenza vaccination for all persons aged >6 months who do not have contraindications. As the target viruses undergo antigenic drift, immunization is recommended on an annual basis.

Prophylaxis with anti-influenza drugs is also effective at preventing infection, although clearly the protection mediated by this approach only lasts as long as the drugs are administered. However, this will be important if/when the next pandemic emerges. It will take some months before an effective vaccine against the pandemic strain of virus is developed and manufactured in enough doses to offer realistic protection on a population basis. Thus, initial prevention measures once a new pandemic strain is identified may well include the use of prophylaxis with the neuraminidase inhibitors.

Table 19.3 Target groups for influenza vaccination

UK – Department of Health recommendations
Healthcare workers
Patients aged 6 months or older with underlying:
chronic respiratory disease (including asthma)
chronic heart disease
diabetes requiring insulin or oral hypoglycemic drugs
chronic renal disease
immunosuppression
chronic liver disease
Pregnant women
Morbid obesity (BMI > 40)
Individuals over the age of 65
Children aged 2–16
Health and social care staff directly involved in patient care
People who live in nursing homes and other long-term care facilities

SUMMARY

1. WHAT IS THE CAUSATIVE AGENT, HOW DOES IT ENTER THE BODY, AND HOW DOES IT SPREAD A) WITHIN THE BODY AND B) FROM PERSON TO PERSON?

- Influenza A, B, and C viruses, carry a segmented ($n = 8$ for A and B) negative single-stranded RNA genome, are enveloped, and belong to the family *Orthomyxoviridae*.
- Typing into A, B, C is according to the nature of the internal proteins.
- Type A is subtyped according to the nature of the surface H and N proteins.

- Entry is via inhalation, or droplet inoculation onto oropharyngeal mucous membranes.
- Viral tropism is for respiratory epithelial cells, with no evidence of spread beyond the lungs (except perhaps for avian H5N1 infection of humans).
- The viral ligand hemagglutinin binds to cell surface sialic acid receptors.
- On entry, the virus uncoats and replication begins.
- Epidemiology is characterized by pandemics arising from antigenic shift, and interpandemic epidemics, arising from antigenic drift.

Continued...

continued...

- Antigenic drift is due to Darwinian selection of variants with mutations in key neutralizing epitopes within the surface glycoproteins in the face of immune selection pressure.
- Antigenic shift is due to the emergence of a new influenza A subtype.
- Possible mechanisms include genetic reassortment of human and avian influenza viruses within a mixing vessel, either pigs or humans, or direct trans-species transfer of avian viruses to humans, with subsequent adaptive mutations.
- The last pandemic, due to A/H1N1 pdm virus, was in 2009.
- Currently, A/H1N1, A/H3N2, and B viruses co-circulate, giving rise to annual interpandemic epidemics.

2. WHAT IS THE HOST RESPONSE TO THE INFECTION AND WHAT IS THE DISEASE PATHOGENESIS?

- Viral infection results in **lysis** or **apoptosis** of the cell.
- Influenza virus infection therefore effectively strips off the inner lining of the respiratory tract, including mucus-secreting and ciliated epithelial cells.
- This predisposes to inhalation of particulate matter, including bacteria, into the lower respiratory tract.
- Innate immune responses, triggered by recognition of viral associated pathogen associated molecular patterns (PAMPs) by host pathogen recognition receptors (PRRs), include potent induction of interferon and other cytokines.
- Adaptive immune responses include neutralizing antibody production and T-cell responses.

3. WHAT IS THE TYPICAL CLINICAL PRESENTATION AND WHAT COMPLICATIONS CAN OCCUR?

- There are two components to the clinical manifestations of influenza virus infection.
- Respiratory tract symptomatology (e.g. rhinorrhea, cough) arises from local cellular damage and inflammation.
- Systemic manifestations (fever, headache, pronounced myalgia) arise from the effects of circulating cytokines.
- Life-threatening complications include primary influenzal pneumonia, secondary bacterial pneumonia, myocarditis, and post-infectious encephalitis.
- Avian H5N1 infection of humans has a high mortality, due to an intense acute inflammatory reaction within the lungs (a cytokine storm).

4. HOW IS THE DISEASE DIAGNOSED, AND WHAT IS THE DIFFERENTIAL DIAGNOSIS?

- Diagnosis is by demonstration of influenza virus in a respiratory sample – for example, nasopharyngeal aspirate, throat swab.

- Genome amplification, e.g. by reverse transcriptase real-time polymerase chain reaction assay is a rapid, specific, and very sensitive approach.
- Other diagnostic approaches include antigen detection by indirect immunofluorescence, isolation of virus in cell culture (takes several days), or by demonstration of a rise in specific antibody titers (requires a blood sample taken several days after onset of illness).
- Many other infections may present with an "influenza-like illness" – the most common is with respiratory syncytial virus.

5. HOW IS THE DISEASE MANAGED AND PREVENTED?

- The majority of patients infected with influenza virus infection can be managed symptomatically, for example, with appropriate analgesia and antipyretics.
- Amantadine has activity against some (not all) influenza A viruses. It works by binding to the M2 protein and blocking an ion channel necessary for uncoating of the virus.
- It has central nervous system stimulatory side effects and is not well tolerated, particularly in the elderly.
- Resistance emerges rapidly due to mutations in the M2 protein.
- The neuraminidase inhibitors (zanamivir, oseltamivir) have activity against all known influenza virus neuraminidase enzymes. In the UK, their use is focused on seriously ill hospitalized patients, or as prophylaxis in high-risk exposed patients.
- Resistance to the neuraminidase inhibitors has been reported, due to point mutations within the NA gene.
- Baloxavir marboxil, an inhibitor of the viral RNA polymerase complex, has recently been licensed for clinical use.
- Vaccination is with either inactivated whole virus, a subunit derivative containing only purified hemagglutinin and neuraminidase, or live attenuated vaccines.
- The vaccines are trivalent (i.e. contain antigens from all three co-circulating viruses).
- Vaccine composition is adjusted annually to take account of antigenic drift.
- Vaccine targeting may differ in different countries. In the UK, recommendations are to vaccinate high-risk subgroups within the general population, those over the age of 65 and children aged 2–16.
- Targeted individuals require annual vaccination.
- An alternative to vaccination is prophylactic use of neuraminidase inhibitors.

FURTHER READING

Barer M, Irving W, Swann A, Perera N. Medical Microbiology, 19th edition. Elsevier, Philadelphia, 2019.

Humphreys H, Irving WL, Atkins BL, Woodhouse AF. Oxford Case Histories in Infectious Diseases and Microbiology, 3rd Edition. Oxford University Press, Oxford, 2020.

Murphy K, Weaver C. Janeway's Immunobiology, 9th edition. Garland Science, New York/London, 2016.

Richman DD, Whitley RJ, Hayden FG. Clinical Virology, 4th edition. ASM Press, New York, 2016.

REFERENCES

Chen X, Liu S, Goraya MU, et al. Host Immune Response to Influenza: A Virus Infection. Front Immunol, 9: 320, 2018.

Harrington WN, Kackos CM, Webby RJ. The Evolution and Future of Influenza Pandemic Preparedness. Exp Mol Med, 53: 737–749, 2021.

Lampejo T. Influenza and Antiviral Resistance: An Overview. Eur J Clin Microbiol Infect Dis, 39: 1201–1208, 2020.

Nuwarda RF, Alharbi AA, Kayser V. An Overview of Influenza Viruses and Vaccines. Vaccines, 9: 1032, 2021.

Trifonov V, Khiabanian H, Rabadan R. Geographic Dependence, Surveillance and Origins of the 2009 Influenza (H1N1) Virus. New Engl J Med, 3612: 115–119, 2009.

WEBSITES

All the Virology on the WWW Website, Specific Virus Information, 1995: http://www.virology.net/garryfavweb13.html#Ortho

Centers for Disease Control and Prevention, Influenza (Flu), 2022: http://www.cdc.gov/flu/

GOV.UK, Influenza: the green book, chapter 19, 2013: https://www.gov.uk/government/publications/influenza-the-green-book-chapter-19

Virology online, Influenza Viruses: http://virology-online.com/viruses/Influenza.htm

Students can test their knowledge of this case study by visiting the Instructor and Student Resources: [www.routledge.com/cw/lydyard] where several multiple choice questions can be found.

Leishmania spp.

20

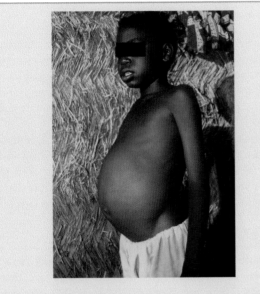

Figure 20.1 A child with visceral leishmaniasis. As in the patient described in the case history, the liver and spleen are enlarged, causing distension of the abdomen. *Courtesy of World Health Organization, Special Programme for Research and Training in Tropical Diseases, http://www.who.int/tdr/index.html. Image #9706290. Additional photographic credit given to Andy Crump who took the photo in Sudan, 1997.*

A 72-year-old man retired to the south of Spain but returned to the UK for the summer months. He began to develop fever, malaise, loss of appetite, and weight loss. He was admitted to hospital and had temperatures reaching 39°C. Both his liver and spleen were palpable. No lymph nodes could be felt. Blood tests showed a **pancytopenia**. Routine investigations for an infection were negative and he did not improve with broad-spectrum antibiotics. His condition deteriorated and the size of the liver and spleen increased (Figure 20.1). A bone marrow examination did not show any sign of hematologic malignancy. No organisms were seen on staining. His history was explored again. Four months before his illness he had been on a camping break in Spain to a coastal area. He recalled seeing many thin dogs in the vicinity. Part of his bone marrow sample was sent to a reference laboratory for *Leishmania* **polymerase chain reaction** (**PCR**). This returned positive. He was successfully treated with a course of liposomal amphotericin B and over the ensuing 3 months his liver and spleen became impalpable and his blood tests returned to normal. His diagnosis was visceral leishmaniasis probably due to *Leishmania infantum*.

1. WHAT IS THE CAUSATIVE AGENT, HOW DOES IT ENTER THE BODY AND HOW DOES IT SPREAD A) WITHIN THE BODY AND B) FROM PERSON TO PERSON?

CAUSATIVE AGENT

Leishmania are protozoan parasites. They have an intracellular form called an amastigote (Figure 20.2) and an extracellular, flagellated form called a promastigote (Figure 20.3).

There is variety in the clinical diseases caused, geographic distribution, and animal reservoirs. The genus *Leishmania* is divided into groups, complexes and species. Classification was classically determined by **isoenzyme typing**; molecular methods (using DNA sequencing) are now more common. Table 20.1 lists species and the diseases they cause.

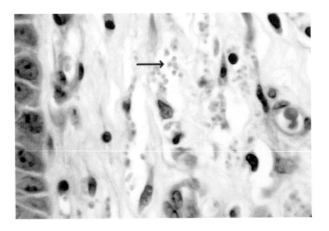

Figure 20.2 A skin biopsy showing *Leishmania* amastigotes (arrowed). *Courtesy of the Centers for Disease Control & Prevention, Atlanta, Georgia. Image is found in the Public Health Image Library #331. Additional photographic credit is given to Dr Martin D. Hicklin who created the image in 1964.*

ENTRY AND SPREAD WITHIN THE BODY

People are infected after the bite of a sandfly laden with *Leishmania* promastigotes. Under the skin, the promastigotes are rapidly phagocytosed by macrophages. For cutaneous disease, lesions are confined to the locality of the sandfly bite. For *L. braziliensis* and *L. panamensis* cutaneous spread can occur and later this can involve mucous membranes of the mouth or nose. *L. donovani* and *L. infantum* are capable of deeper spread within macrophages to the rest of the **mononuclear phagocytic system**, mainly present in organs such as the liver, spleen, and bone marrow. They are responsible for visceral leishmaniasis. In India, visceral leishmaniasis is called **kala-azar**. Relapse of infection after an interval may be manifest as a widespread cutaneous form of disease, called post kala-azar dermal leishmaniasis (PKDL). This occurs in India and East Africa. The life cycle of *Leishmania* is shown in Figure 20.4.

PERSON-TO-PERSON SPREAD

In areas with visceral leishmaniasis, sandflies can ingest protozoa when they feed from the skin. Numbers of *Leishmania* in the skin are even higher in PKDL. However, leishmaniasis is largely a zoonosis. Different animal reservoirs occur in different regions. They include rodents, gerbils, hyraxes, sloths, and the domestic dog.

The sandfly vector is a *Phlebotomus* species in the Old World and *Lutzomyia* species in the New World. Sandflies are small, less than 5 mm in size, and bite at dusk or during the night (Figure 20.5). They are not capable of flying great heights above the ground and usually bite individuals sleeping close to the ground. In the case described above, the patient was probably infected through sandflies when he was lying near the ground on his camp bed. He normally lived in a flat. The sandflies will have carried infection from the local dog population.

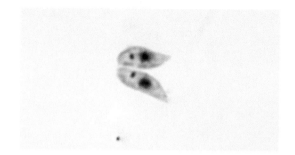

Figure 20.3 Elongated *Leishmania* promastigotes. *Courtesy of the Centers for Disease Control & Prevention, Atlanta, Georgia. Image is found in the Public Health Image Library #544.*

Female sandflies bite and take blood from their target host. Any amastigotes ingested from the skin change into promastigotes. These pass into the sandfly midgut, proliferate, cause damage to the digestive valve system, and are regurgitated to the biting mouthparts and then onto the skin of the next host to be bitten.

Another form of transmission for visceral leishmaniasis has been described among intravenous (IV) drug users in Southern Europe. Infection can be passed on with shared needles and equipment. In one study, about half of discarded needles in Madrid were positive by PCR for *Leishmania*.

EPIDEMIOLOGY

Notification of cases of leishmaniasis is not universal. Numbers of people afflicted by the disease are therefore estimates. There are about 50 000 to 90 000 new cases of visceral leishmaniasis annually, and 95% of these are seen in Brazil, China, Ethiopia, India, Iraq, Kenya, Nepal, Somalia, South Sudan and Sudan. The annual incidence of cutaneous leishmaniasis is 0.6 to 1 million new cases. Ninety-five percent of cutaneous leishmaniasis occurs in the Americas,

Table 20.1 Species of Leishmania and the diseases they cause

Complex	Species	Disease
	L. tropica	Cutaneous leishmaniasis
	L. major	Cutaneous leishmaniasis
	L. aethiopica	Cutaneous leishmaniasis
L. mexicani complex	L. mexicani	Cutaneous leishmaniasis
	L. amazonensis	Cutaneous leishmaniasis
	L. venezuelensis	Cutaneous leishmaniasis
	Subgenus Viannia	
	L. [V.] guyanensis	Cutaneous leishmaniasis
	L. [V.] peruviana	Cutaneous leishmaniasis
	L. [V.] panamensis	Muco/cutaneous leishmaniasis
	L. [V.] braziliensis	Muco/cutaneous leishmaniasis
L. donovani complex	L. donovani	Visceral leishmaniasis
	L. infantum	
	(also known as L. chagasi in New World)	

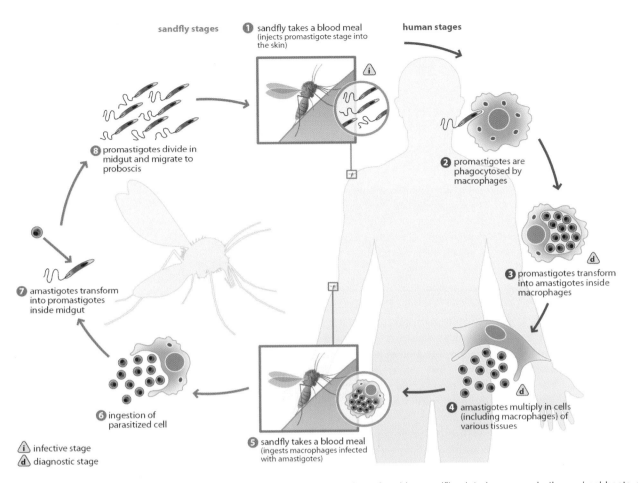

sandfly stages

① sandfly takes a blood meal
(injects promastigote stage into
the skin)

human stages

⑧ promastigotes divide in
midgut and migrate to
proboscis

② promastigotes are
phagocytosed by
macrophages

⑦ amastigotes transform
into promastigotes
inside midgut

③ promastigotes transform
into amastigotes inside
macrophages

⑥ ingestion of
parasitized cell

④ amastigotes multiply in cells
(including macrophages) of
various tissues

⑤ sandfly takes a blood meal
(ingests macrophages infected
with amastigotes)

ⓘ infective stage
ⓓ diagnostic stage

Figure 20.4 Life cycle of *Leishmania* spp. *Leishmania* promastigotes are inoculated by sandflies into human and other animal hosts at the time of taking a blood meal (1) Promastigotes are phagocytosed by macrophages (2) Within macrophages, promastigotes transform into amastigotes (3) Amastigotes can multiply in various cell types (4) Macrophages containing amastigotes are ingested by sandflies taking a blood meal (5) and the life cycle continues within the sandfly vector (6–8). *Courtesy of the Centers for Disease Control & Prevention, Atlanta, Georgia. Image is found in the Public Health Image Library #3400. Additional photographic credit is given to Alexander J da Silva, PhD, and Melanie Moser who created the image in 2002.*

Figure 20.5 Phlebotomus sandfly. *Courtesy of the Centers for Disease Control & Prevention, Atlanta, Georgia. Image is found in the Public Health Image Library #6274. Additional photographic credit is given as follows: World Health Organization (WHO), Geneva, Switzerland.*

the Mediterranean basin, the Middle East, and Central Asia. About 90% of mucocutaneous disease occurs in Bolivia, Brazil, Ethiopia, and Peru.

2. WHAT IS THE HOST RESPONSE TO THE INFECTION AND WHAT IS THE DISEASE PATHOGENESIS?

As promastigotes enter the skin they are phagocytosed by macrophages and neutrophils. They change into amastigotes. Classically, any pathogen engulfed by a phagocyte is wrapped within host-cell plasma membrane. This forms a phagosome. Various membrane molecules are imported and exported as cytoplasmic vesicles fuse with or erupt from the phagosome. Eventually, **lysosomes** fuse with the phagosome and discharge their contents. Lysosomal enzymes lyse susceptible pathogens. On engulfment, phagocytes are activated to produce **reactive oxygen species** and reactive nitrogen intermediates. They secrete tumor necrosis factor-α (**TNF-α**), which contributes to their activation state. Macrophages are activated further by T-helper 1 lymphocytes (**Th-1**) through interferon-γ (**IFN-γ**). These are stimulated by antigen-presenting cells, most efficiently by **dendritic cells**. They secrete **interleukin**-12 (**IL-12**). When it takes time to deal with a pathogen, the combination of Th-1 cells and macrophages organize into **granulomas**.

To survive, *Leishmania* need to subvert the above process. Various effects have been described but the mechanisms by which these occur are not altogether clear. Some leishmanial molecules that have been shown experimentally to play a part are lipophosphoglycan, a surface membrane metalloprotease (gp63), cysteine proteases, and a *Leishmania* homolog of activated C kinase receptor (LACK). The outcome is that macrophage activation and the generation of reactive oxygen and nitrogen intermediates are suppressed, *Leishmania* resist lysosomal attack, dendritic function is compromised, and LACK induces a **Th-2** response. Variations in this interplay occur between different *Leishmania* species. Furthermore, this may be affected by simultaneous co-infections such as viruses infecting *Leishmania* parasites, viruses co-inoculated by sandflies and, importantly, by HIV infection in hosts.

Some *Leishmania* species may use neutrophils as a "Trojan horse". The appearance of neutrophils at the site of the sandfly bite is promoted by the pro-inflammatory effect of sandfly saliva. Cell entry is favored by complement-mediated opsonization, provided that there is no complement-mediated lysis of *Leishmania*. Leishmanial lipophosphoglycan promotes the former and inhibits the latter. Neutrophils fail to kill *Leishmania* after **phagocytosis** and undergo **apoptosis**. Apoptotic fragments may contain *Leishmania*. The neutrophils release **MIP-1β**, a **chemokine** that attracts macrophages. When macrophages phagocytose the apoptotic neutrophil fragments, *Leishmania* enter "silently" and continue to multiply. The macrophages release **transforming growth factor-β** (**TGF-β**), which is anti-inflammatory.

In mice experimentally infected with *L. major*, there is a clear polarization of Th-1 and Th-2 responses. Some strains mount a Th-1 response and control infection, unlike others (BALB/c) which mount a Th-2 response and experience fatal disseminated infection. Humans infected with *L. major* experience localized cutaneous disease. Conversely, *L. donovani*, which causes visceral leishmaniasis in humans, can be controlled by BALB/c mice. *L. donovani* does not lead to a polarized Th-1 and Th-2 response between mouse strains. While there are differences between mice and humans in *Leishmania* infection, the experimental experience with mice indicates the important role of host genetics.

In humans, markers of a Th-1 (IFN-γ) and Th-2 (interleukin (IL)-4) response are both present at the same time. IL-4 down-regulates Th-1 responses and so do other cytokines such as IL-10, IL-13, and TGF-β. These cytokines are more prominent in forms of infection that are not self-limiting like visceral leishmaniasis and PKDL. IL-10 seems to play a greater role in susceptibility to these infections than IL-4. Otherwise, a Th-1 response tends to heal other forms of infection. An exuberant inflammatory response mediated by Th-1 cells may paradoxically enable spread of cutaneous forms of disease.

3. WHAT IS THE TYPICAL CLINICAL PRESENTATION AND WHAT COMPLICATIONS CAN OCCUR?

Infection may be asymptomatic. *Leishmania* may reside in the body for years and only cause clinical disease if the host becomes immunocompromised.

Cutaneous leishmaniasis is seen on exposed parts of the body where the sandflies are likely to bite (Figure 20.6). Sandflies are unable to bite through clothing. Thus, lesions may be found on the face, arms, and lower legs. Lesions may be single or multiple. They are usually apparent 2–6 weeks after the bite. Initially there is a red **papule**. This gradually enlarges over a few weeks. The lesion may take on a raised painless, ulcerated form or is **papulo-nodular**. Secondary infection is possible and then lesions are more likely to be painful. Without specific treatment, lesions will usually self-heal, but over prolonged periods. This may take 6 months to a few years. Over this period, lesions may seem to regress and then relapse. All *Leishmania* species are capable of causing cutaneous disease, but the host immune response may alter the clinical picture. A weakened immune response with a high parasite burden causes diffuse cutaneous leishmaniasis (DCL) with multiple, spreading **papular** lesions. There may also be lymphatic spread with localized nodules along the track of lymphatics. A strong immune response with a low parasite burden causes a condition called leishmania recidivans (LR). The immune response effectively clears the initial site of infection. A series of small papules surround this central clearing and these in turn are cleared.

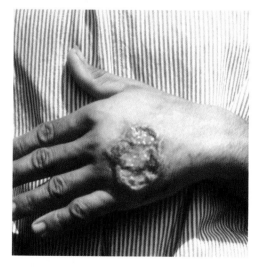

Figure 20.6 Cutaneous leishmaniasis of an ulcerating form. *Courtesy of the Centers for Disease Control & Prevention, Atlanta, Georgia. Image is found in the Public Health Image Library #352. Additional photographic credit is given to Dr DS Martin.*

Mucocutaneous leishmaniasis is associated with *L. braziliensis* and *L. panamensis*. As these species' names suggest, this form of leishmaniasis is restricted to South America. Cutaneous lesions occur first as in purely cutaneous leishmaniasis. These can self-heal, but the parasite does not disappear from the body. After an interval, sometimes of several years, the parasite re-emerges in the mucous membranes of nose or mouth. Local inflammation results in nasal stuffiness. There is then progressive destruction of the anatomy of the nose or mouth and infection can progress backward toward the throat and larynx (Figure 20.7). Eating and drinking become difficult and secondary infections in the upper and lower respiratory tract often occur. These latter effects can prove fatal unless the infection is treated. There can be considerable scarring and disfigurement if treatment is delayed.

Visceral leishmaniasis is associated with *L. donovani* in India and East Africa and with *L. infantum* around the Mediterranean and South America. A cutaneous lesion may not be apparent. After an incubation period of a few months, illness is heralded by fevers. These may continue for about one month before abating. The spleen progressively enlarges first and then the liver (Figure 20.1). Both may become massively enlarged. The enlarged spleen causes **hypersplenism** and consumption of blood cells, but infection within the bone marrow also causes a pancytopenia with **anemia**, **leukopenia**, and **thrombocytopenia**. In dark skins, the anemia plus hormonal effects of chronic infection cause an altered appearance. In India, the graying of the complexion is called kala-azar. Leukopenia predisposes to secondary infections, which themselves may be life-threatening. Thrombocytopenia can predispose to bleeding and there may be life-threatening hemorrhage. On blood tests, there is also a fall in albumin levels. A drop in oncotic pressure can result in **edema**. This may be peripheral in the legs or ascites within the abdomen. There is also a polyclonal stimulation of **IgG** antibodies. The polyclonal stimulation of B lymphocytes can compromise their ability to respond to other infections. Visceral leishmaniasis runs a chronic and progressive course. Patients become wasted. It is invariably fatal unless treated. If treated, some parasites may escape killing and return later to cause post-kala-azar dermal leishmaniasis (PKDL). In India, this interval may be 2–3 years, but shorter intervals have been observed in Sudan. The host now has some immunity from the first spell of visceral leishmaniasis. The parasite is largely confined to the skin, with extensive papulo-nodular lesions starting on the face and peripheries and then spreading to most of the body surface. This may self-cure, only to relapse and remit at a later date.

Leishmaniasis and HIV co-exist in many areas. The immunocompromised nature of HIV has caused more florid manifestations of leishmaniasis. Parasite burdens are higher. Species that may only cause cutaneous disease may become visceral.

4. HOW IS THE DISEASE DIAGNOSED, AND WHAT IS THE DIFFERENTIAL DIAGNOSIS?

In endemic areas, cutaneous and mucocutaneous leishmaniasis may be diagnosed on purely clinical grounds. The clinical picture of fever, **splenomegaly**, and anemia due to visceral leishmaniasis may also be caused by other diseases. Investigations are required to confirm the diagnosis. These could entail direct visualization of the parasite in a tissue sample, culture of samples, detection of antigen, detection of nucleic acid by PCR, or immunodiagnosis. The latter includes **serology** or, historically, a leishmanial skin test. Determining the exact species by culture or PCR can be important in choosing between treatment options.

Deep-tissue samples may be obtained by bone-marrow aspirate, a splenic aspirate, lymph-node aspirate, or sometimes liver biopsy. Splenic sampling runs the risk of serious splenic hemorrhage. Cutaneous lesions may be squeezed firmly with fingers to exclude blood, superficially incised with a scalpel at their edge, and then tissue-fluid expressed and impressed onto a glass slide. On staining of tissue samples, intracellular amastigotes are sought. Their appearance is characteristic with a small kinetoplast body adjacent to the nucleus. This is called a **Donovan body** (Figure 20.8). The sensitivity of tissue sampling varies with the sample – >90% for splenic aspirate, 55–97% for bone marrow, and 60% for lymph

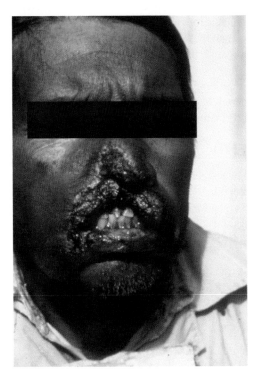

Figure 20.7 Mucocutaneous leishmaniasis in a patient with progressive destruction of tissues around the lips and nose. *Courtesy of World Health Organization, Special Programme for Research and Training in Tropical Diseases, http://www.who.int/tdr/index.html. Image #9106015. Additional photographic credit given to Andy Crump who took the photo in Sudan, 1997.*

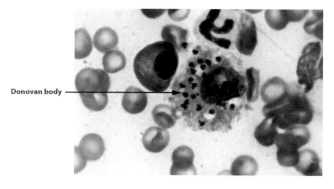

Donovan body

Figure 20.8 Characteristic *Leishmania* amastigote forms (Donovan bodies, arrowed) on an impression smear. Darkly staining small kinetoplast adjacent to nuclei. *Courtesy of the Centers for Disease Control & Prevention, Atlanta, Georgia. Image is found in the Public Health Image Library #30. Additional photographic credit is given to Dr Francis W Chandler who created the image in 1979.*

nodes. If facilities are available, culture can help but now in resource-rich settings PCR is applied with a sensitivity of >95%. Leishmanial antigen can be detected in urine. A latex agglutination technique called KATEX has shown sensitivities of 68–100% for visceral leishmaniasis. After successful treatment, antigen disappears from the urine.

Serology is usually negative in cutaneous leishmaniasis and should be reserved for visceral disease. Serologic tests are unable to distinguish current from past infection. A commonly used test for anti-leishmanial antibody is the direct agglutination test (DAT). Promastigotes are formalin-fixed onto slides and serum is placed on top. Agglutination is observed after 24 hours. DAT has a sensitivity of about 95% and its specificity is about 86%. A dipstick test has been developed with the K39 antigen impregnated on a reagent strip. Blood is added to the strip and a reaction noted after 20 minutes. The K39 dipstick has a sensitivity of about 94% and specificity of about 90%. The K39 antigen is also used in ELISAs.

Historically, a test similar to the tuberculin skin test for tuberculosis was used for leishmaniasis. This was the Montenegro skin test. Leishmanial antigen was implanted in the forearm and the induration after 48–72 hours was measured. Now standardized antigen preparations are no longer available.

DIFFERENTIAL DIAGNOSIS

In endemic areas, cutaneous and mucocutaneous leishmaniasis may have a characteristic appearance. Cutaneous lesions may have to be differentiated from other infected insect bites, tuberculosis, fungal infection, **myiasis**, and skin cancers. Mucosal sites may also be affected by syphilis, histoplasmosis, paracoccidioidomycosis, and leprosy. Visceral leishmaniasis may have to be differentiated from other causes

of fever, splenomegaly, and anemia. The differential diagnosis includes malaria, schistosomiasis, typhoid fever, brucellosis, tuberculosis, rickettsial infection, sarcoidosis, **Still's disease**, and hematologic malignancy.

5. HOW IS THE DISEASE MANAGED AND PREVENTED?

MANAGEMENT

Small, single lesions of cutaneous leishmaniasis in immunocompetent hosts may be left to self-heal in geographic areas with *L. major*. Other cutaneous lesions, mucocutaneous disease, and visceral leishmaniasis require treatment. There are a number of treatment permutations dependent on form of disease, geographic location, species, and availability of agents. Originally, the mainstay of treatment was pentavalent antimony compounds. These include sodium stibogluconate and meglumine antimonate. They are administered by intramuscular (IM) injection on a daily basis for durations up to 28 days. The IM injections can be painful and there can be systemic toxicity. Alternative treatments are very welcome, as about 60% of visceral leishmaniasis infections in Bihar, India are resistant to treatment with pentavalent antimonials. Amphotericin B and the formulation of liposomal amphotericin B represent advances in treatment. However, they require IV administration, are also toxic, and are much more expensive, especially liposomal amphotericin B. An oral, tolerable agent is now available for visceral leishmaniasis. Oral miltefosine for 28 days was shown to be equally effective with amphotericin B for visceral leishmaniasis in India. Depending on geographic location, agents may be used singly or in combination. In certain situations, other options for cutaneous lesions include intra-lesional injections of pentavalent antimonials, topical paromomycin, and oral azoles.

PREVENTION

Vaccines have been trialed for leishmaniasis but have not been encouraging to date. Prevention has therefore focused on sandflies and efforts directed at human and animal reservoirs. Targeting animal reservoirs is difficult, but it is important to identify and treat PKDL patients who are a key human reservoir for visceral leishmaniasis. As sandflies mainly bite at night, sleeping under a bed-net might afford some protection. But the sandflies are small and can get through the mesh of the nets. However, if the nets are impregnated with a pyrethroid insecticide the sandflies are killed. Insecticide-treated bed-nets are also a key component of malaria control programmes. Furthermore, indoor insecticide spraying on walls has been used both for malaria and leishmania control.

SUMMARY

1. WHAT IS THE CAUSATIVE AGENT, HOW DOES IT ENTER THE BODY, AND HOW DOES IT SPREAD A) WITHIN THE BODY AND B) FROM PERSON TO PERSON?

- *Leishmania* are protozoan parasites.
- There is a large number of species.
- The extracellular stage is called the promastigote and the intracellular stage is the amastigote.
- Spread is by sandflies either from animal reservoirs or humans with heavy skin loads of parasite. The latter occurs in a condition called post-kala-azar dermal leishmaniasis.
- The ability to spread within the body is a function of both the species of *Leishmania* and the host immune response.

2. WHAT IS THE HOST RESPONSE TO THE INFECTION AND WHAT IS THE DISEASE PATHOGENESIS?

- *Leishmania* are phagocytosed by neutrophils and macrophages.
- T-helper 1 lymphocytes help macrophages to kill *Leishmania*.
- Some species may be more successful at subverting the immune response and causing disseminated infection.
- Subversion of the immune response involves suppression of macrophage activation and diversion toward a T-helper 2 type of response.
- In mice infected with *L. major*, there is a clear polarization of T-helper 1 and 2 responses, depending on the genetic background of the mice.

3. WHAT IS THE TYPICAL CLINICAL PRESENTATION AND WHAT COMPLICATIONS CAN OCCUR?

- Cutaneous leishmaniasis is caused by a large number of species with infection confined to one locality.
- Lesions enlarge gradually over a few weeks, become papulo-nodular or ulcerate.
- Lesions may self-heal.
- Mucocutaneous leishmaniasis is caused by *L. braziliensis* and *L. panamensis*.
- After an initial cutaneous lesion, there is a later mucosal lesion, which is progressively destructive of nose or mouth.
- Visceral leishmaniasis is due to *L. donovani* or *L. infantum*.
- Infection spreads through the mononuclear phagocytic system with enlargement of spleen and liver, and bone-marrow infiltration.
- Skin complexion changes giving rise to the Indian term, kala-azar.
- After treatment of visceral leishmaniasis relapse may be confined to the skin with post-kala-azar dermal leishmaniasis.

4. HOW IS THE DISEASE DIAGNOSED, AND WHAT IS THE DIFFERENTIAL DIAGNOSIS?

- Diagnosis of cutaneous or mucocutaneous leishmaniasis may be purely clinical.
- Stained-tissue samples may show characteristic Donovan bodies.
- Species-specific PCR has a high sensitivity.
- Assays exist for leishmanial antigen or antibody.

5. HOW IS THE DISEASE MANAGED AND PREVENTED?

- Simple cutaneous lesions may self-heal.
- **Parenteral** pentavalent antimonial drugs have been traditional treatment for all forms of disease.
- Parenteral amphotericin B, either conventional or liposomal, can be used for visceral leishmaniasis.
- Oral miltefosine is easier to administer.
- Combinations of agents may be required for treatment
- Vector control and treatment of the human reservoir of PKDL patients are important preventive measures.

FURTHER READING

Boelaert M, Sundar S. Leishmaniasis. In: Farrar J, Hotez P, Junghanns T, et al, editors. Manson's Tropical Diseases, 23rd edition. Elsevier/Saunders, London, 2014: 631–651.

Murphy K, Weaver C. Janeway's Immunobiology, 9th edition. Garland Science, New York/London, 2016.

REFERENCES

Aronson N, Herwaldt BL, Libman M, et al. Diagnosis and Treatment of Leishmaniasis: Clinical Practice Guidelines by the Infectious Diseases Society of America (IDSA) and the American Society of Tropical Medicine and Hygiene (ASTMH). Am J Trop Med Hyg, 96: 24–45, 2017.

Rossi M, Fasel N. How to Master the Host Immune System? Leishmania Parasites have the Solutions! Int Immunol, 30: 103–111, 2018.

Torres-Guerrero E, Quintanilla-Cedillo MR, Ruiz-Esmenjaud J, Arenas R. Leishmaniasis: A Review. F1000Res, 6: 750, 2017.

WEBSITES

Centers for Disease Control and Prevention, Parasites - Leishmaniasis, 2020: www.cdc.gov/parasites/leishmaniasis

World Health Organization, Leishmaniasis: www.who.int/health-topics/leishmaniasis

Students can test their knowledge of this case study by visiting the Instructor and Student Resources: [www.routledge.com/cw/lydyard] where several multiple choice questions can be found.

Leptospira spp.

21

A Staff Sergeant was repatriated from an exercise in Belize and admitted to hospital in the UK. He was complaining of fever, headache, and **myalgia**. After an initial improvement, he began to deteriorate. He became jaundiced with signs of **pneumonia** and, on examination, had conjunctival inflammation and hepatosplenomegaly. A chest X-ray indicated bi-basal opacities. His blood count showed a **neutrophilia** with a **thrombocytopenia**. Liver function tests showed an elevated conjugated **bilirubin** with mild elevation of transaminases. He was **oliguric** and **uremic**. A diagnosis of Weil's disease (leptospirosis) was made and he was started on benzylpenicillin and renal dialysis.

1. WHAT IS THE CAUSATIVE AGENT, HOW DOES IT ENTER THE BODY AND HOW DOES IT SPREAD A) WITHIN THE BODY AND B) FROM PERSON TO PERSON?

CAUSATIVE AGENT

Leptospira spp. (leptospires) are motile, very thin, tightly coiled spirochetes measuring 0.1 μm by 10–20 μm with a characteristic curve at either end (Figure 21.1) and they have a typical gram-negative cell-wall structure (see Case 11, *E. coli*). They have two axial flagella located between the peptidoglycan and an outer-membrane layer in the periplasmic space (Figure 21.2).

Ellinghausen McCullough Johnson and Harris (EMJH) liquid medium is used to isolate *Leptospira* and consists of bovine serum albumin, Tween 80, glycerol, sodium pyruvate, cyanocobalamin, and various salts (Mg, Fe, Zn, Ca) dissolved in ultrapure water. Growth is slow and may achieve 10^7/ml of organism. In some cases, excessive growth can lead to lysis due to the lipases produced by the organism. The generation time can range from 6 to 16 hours depending on the isolate. The organism is aerobic but requires CO_2 to stimulate growth. Maximal growth is between pH 7.2 and pH 7.6.

Growth temperature depends on whether it is a pathogen (29°C–37°C) or a saprophyte which can grow at 14°C but not at 37°C.

In addition to the growth temperature indicating its identity as a saprophyte or pathogen, growth in EMJH medium containing 8-azaguanine (or copper) allows the growth of saprophytes but inhibits pathogen growth.

Recently in 2019, the taxonomy of the Genus *Leptospira* has undergone a major review. Previously there were 35 recognized species with many named serovars of one pathogenic species *Leptospira interrogans*. The Genus was divided into three divisions: the Saprophytes (e.g. *L. biflexa*), an Intermediate division (*L. fainei, L. inadai*) and the Pathogen division (e.g. *L. interrogans*).

This latest study used whole genome sequencing (WGS) (NextSeq 500 (Illumina), Nextra XT Library preparation and CLC Genomics assembly platform) with additional analysis of Average Nucleotide Identity (ANI) and values of the percentages of conserved proteins (COPD). The study included 90 isolates of *Leptospires* from a variety of locations (e.g. Japan, France, Malaysia, Algeria) and identified 30 new *Leptospira* species giving a total of 64 named species. The taxonomy generated by this study identified two major clades and four subclades P1, P2, S1, S2 (see Table 21.1). A study of core gene sets identified genes and domains linked with each subclade. The two major clades are identified as saprophytes (S) and pathogens (P). Subclades P1 (Pathogens in humans and animals), P2 (those species previously classified as Intermediate), S1 (saprophytes) S2 a new subclade including *L. idonii, L. kobayashii, L. ilyithenesis*, and *L. ognonensis*. Species within this group grew at 14°C but not at 37°C and grew well in the presence of 8-azaguanine, both phenotypic characteristics of saprophytes.

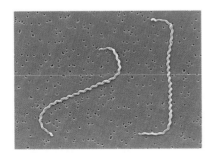

Figure 21.1 Morphology of *Leptospira*. *Reprint permission kindly given by the Centers for Disease Control & Prevention, Atlanta, Georgia. Image is found in the Public Health Image Library #1220. Additional photographic credit is given to Janice Haney Carr who took the image and the CDC NCID and Rob Weyant who provided the image for CDC PHIL website.*

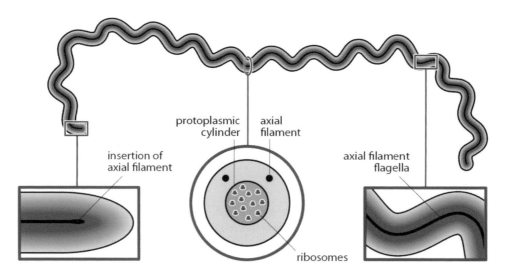

Figure 21.2 The axial internal flagella of *Leptospira*. *Adapted with permission from Dr Samuel Baron and the University of Texas Medical Branch at Galveston, Department of Microbiology and Immunology.*

protoplasmic cylinder

axial filament

insertion of axial filament

axial filament flagella

ribosomes

Table 21.1 Some representative species of *Leptospira*

P1
L. interrogans
L. noguchi
L. alexanderi
L. alstonii
P2
L. licerasiiae
L. hartskeetlii
L. langatensis
L. broomii
S1
L. biflexa
L. levettii
L. perdikensis
L. ellinghauseni
S2
L. idonii
L. ryugenii
L. ilyithenesis
L. kobayashii

ENTRY INTO THE BODY

Infection is acquired by contact with water or soil contaminated by one of the pathogenic *Leptospira* serovars. Entry is via skin lesions, or lesions in the mucosae of the respiratory and digestive tracts or conjunctivae. The organism may also be acquired from contaminated aerosols entering the respiratory tract.

A number of animal species act as a reservoir of infection, the common ones being rodents, but dogs (*L. canicola*) and livestock (*L. pomona*) can also act as reservoirs. Infected animals excrete the bacterium in their urine thus contaminating water sources and soil. Human infection can be acquired directly from contact with the animals or from contact with contaminated water. Globally, *Leptospira* is the commonest **zoonosis**.

SPREAD WITHIN THE BODY

After entry, the organism is spread around the body by the blood, entering all organs and thus giving rise to a wide spectrum of clinical presentations (see later).

EPIDEMIOLOGY

Leptospirosis is a global disease, although it is primarily a disease of tropical and subtropical regions and is relatively uncommon in temperate climates. It is endemic in many countries but outbreaks are also associated with adverse weather. Examples of this include: an outbreak of leptospirosis in Nicaragua following Hurricane Mitch in 1995; an outbreak in Peru and Ecuador following heavy flooding in 1998; a post-cyclone outbreak in Orissa, India in 1999.

The precise number of human cases worldwide is not known but it is estimated that 1.03 million cases and 58 900 deaths occur annually. Incidences range from approximately 0.1–1 per 100 000 per year in temperate climates to 10–100 per 100 000 in the humid tropics; incidence may reach over 100 per 100 000. Reports of cases in the US and Europe are low. In the UK, it varies between 13 and 31 per annum and in continental France is in the order of 300, although in 2005 there were only 212 confirmed cases. In the US, the incidence is between 43 and 93 per annum.

Certain occupations are prone to infection, such as veterinarians, butchers, sewage workers, and farmers. Recreational exposure can occur through adventure holidays such as white-water rafting and other water sports.

2. WHAT IS THE HOST RESPONSE TO THE INFECTION AND WHAT IS THE DISEASE PATHOGENESIS?

The innate immune system constitutes the first line of host defense, playing a crucial role in early recognition and elimination of leptospires.

INNATE IMMUNITY

Pattern-Recognition Receptors, Neutrophils and Macrophages

Microbes are recognized through their microbial-associated molecular patterns (MAMPs) through a set of different pattern-recognition receptors (PRRs). The MAMPS include many different components such as nucleic acids, flagellin, and lipopolysaccharide (LPS). PRRs also recognize endogenous molecules associated with cellular damage (DAMPs) produced through microbial infection, for example. PRRs are expressed on both immune cells and nonimmune cells (e.g. epithelial cells) and include members of the membrane Toll-like receptor (TLR), the cytosolic NOD-like receptor (NLR) families (Nucleotide-binding oligomerization domain-like receptors (including NOD-1 and NOD-2) and CLR (C-type lectin receptors).

MAMP recognition results in a signaling cascade that leads to activation of transcription factors such as NF-kB and IRF3 that are involved in the production of cytokines, chemokines, and antimicrobial peptides. Pro-inflammatory cytokines produced include interleukins (IL-1β, IL-6, IL-12, interferons (IFNs) and tumor necrosis factors (TNFs), as well as chemokines. Both cytokines and chemokines lead to activation and recruitment of phagocytes, such as neutrophils, macrophages and dendritic cells to the infection site.

Inflammation may not only lead to pathogen destruction but also, if unregulated, can result in a "cytokine storm" observed in patients with sepsis (see pathogenesis).

TLR4 normally senses LPS on gram-negative bacteria but for *Leptospira* LPS, TLR2 appears to be the PRR which can potentially stimulate macrophages to release pro-inflammatory cytokines. Human monocytes are activated, *in vitro*, by *Leptospira* to produce pro-inflammatory cytokines as seen by up-regulation of TNFα and IL1β genes. Intracellular uptake of leptospires through NLR triggers reactive oxygen species (ROS) and reactive nitrogen species (RNS). In addition, infection triggers TLR2-mediated production of IL-8 and NLRP3-dependent production of IL-1β.

TLR2 on human macrophages also recognizes the outer-membrane lipoprotein LipL32, the major lipoprotein of leptospires, It is thought that this occurs through dimers of TLR2 and TLR1.

NOD-1 and NOD-2 (NLR family) are intracellular receptors that recognize peptidoglycan fragments of bacterial peptidoglycans (PGs) called muropeptides and are active against *Leptospira* PGs.

There appears to be some controversy around the importance of neutrophils in the host response to *Leptospira*. Data suggest that direct phagocytosis of leptospires is rather poor although early studies on immunity indicated that serum from infected individuals but not "normal serum" results in phagocytosis. This indicates that antibodies were "opsonizing" the organisms for phagocytosis. It appears that neutrophils are mostly effective against saprophytic leptospires rather than pathogenic leptospires. Neutrophils also release myeloperoxidase and ROS that are antimicrobial.

The antimicrobial peptide cathelicidin released by neutrophils has been shown to have anti-leptospiral activity among different *Leptospira* species (*L. interrogans* serovars and *Leptospira biflexa*). In addition, a particular strain of *L. interrogans* triggers the release of DNA extracellular traps (NETs) and kills leptospires through the process of NETosis – extrusion of the neutrophil DNA with bactericidal proteins which results in trapping and/or killing of the pathogens. It has been suggested that these DNA traps are crucial to preventing early leptospiral dissemination.

Complement

The complement system has an important role in protection against foreign microbes. It is made up of many molecules (around 50) and is activated through three pathways, classical (CP), alternative (AP), and lectin (LP). Both LP and AP pathways are involved in the host innate immunity while the CP pathway is triggered by the IgM or IgG antibodies specifically bound to antigens. The lectin pathway is triggered when lectins, including ficolins or mannose-binding lectin, attach to carbohydrate moieties on the microbial surfaces. Activation of all three pathways leads to promotion of phagocytosis (opsonization), recruitment of inflammatory mediators and inflammation and lysis via the MAC attack complex. Recent data on using inhibitors of these pathways and survival in human serum of non-pathogenic or pathogenic serovars of *Leptospira* have suggested that both AP and LP have an important role in eliminating saprophytic leptospires. In all leptospires investigated, the deposition of the lytic MAC attack complex proteins on saprophytic *Leptospira* strains was more pronounced when compared to pathogenic species. The pathogenic species have many evasion mechanisms targeting complement activation – see below.

γδT cells, that recognize a broad range of antigens without the presence of major histocompatibility complex molecules, have been shown to respond against leptospires. Identification of the protein antigens recognized will be important in understanding their role in the leptospiral immune response.

Adaptive immunity

PRR activation is also important for the development of adaptive immunity and results in the expression of costimulatory molecules at the surface of macrophages and dendritic cells that are important for antigen presentation to naive T cells and the subsequent activation of B cells and the production of antibodies.

Experimental animal models of leptospires have provided most of the evidence for antibodies playing a key role in both protection and clearance of infections. Several studies in leptospirosis patients have shown that there is a strong antibody response to leptospire LPS. Both IgM and IgG antibodies are produced against a variety of leptospire-specific antigens. These include LipL32, LipL41, and leptospiral immunoglobulin-like A (LigA) which are present only in pathogenic leptospires and have been shown to be exposed on their cell surface. The major protective role of IgG antibodies (which is increased in infections with pathogenic leptospires) seems to be opsonization. The role of T cells is unclear except for their help with the production of antibodies.

Host-response evasion mechanisms

Most leptospire infections are asymptomatic and/or resolve fairly quickly, probably as the result of an effective immune response not compromised by extensive evasion mechanisms. Other strong pathogenic leptospires that do have extensive evasion mechanisms produce more severe and sometimes fatal illness – see pathogenesis below.

PRRs: NOD-1 and NOD-2 receptors cannot sense *Leptospira* since its outer-membrane protein, LipL21 lipoprotein, binds to the PG and impairs its degradation into muropeptides, which therefore cannot signal through NOD-1 and NOD-2.

Complement evasion: The complement system provides an early innate defense. Pathogenic leptospires use different sophisticated strategies to subvert or inactivate all three pathways (classical, alternative, and lectin) of the complement cascade.

Pathogenic leptospires evade complement attack by binding Factor H (FH) and C4 binding protein (C4BP) of the alternative and classical pathways onto their surfaces. FH, a plasma protein, inhibits the alternative pathway of complement by preventing binding of Factor B to C3b, accelerating decay of the C3-convertase C4BP and acting as a co-factor for the cleavage of C3b by Factor I. LenA and LenB are leptospiral ligands for human FH. C4BP, a plasma glycoprotein, inhibits the classical pathway of complement by interfering with the assembly and decay of the C3-convertase C4bC2a and acts as a co-factor for Factor I in the proteolytic inactivation of C4b. LcpA is a leptospiral outer-membrane protein which interacts with human C4BP. Thus, complement activation is down-regulated preventing opsonization and the formation of the lytic membrane attack complex on its surface.

Phagocyte function: the outer-membrane proteins of *L. interrogans* serovar Copenhageni – LipL21 and LipL45 have been shown to be myeloperoxidase inhibitors.

Pathogenesis

Leptospirosis starts with an acute phase that is self-resolving in most of the cases and is followed potentially by a chronic phase (especially kidney colonization) depending on the virulence of the strain, severity of the acute phase, and the overall immune defense of the host. As shown above, pathogenic leptospires have many evasion mechanisms to overcome the immune system. Around 10% of leptospirosis cases develop into severe forms.

The organism migrates to the interstitium, renal tubules, and tubular lumen of the kidney causing an interstitial nephritis and tubular necrosis. Liver involvement is seen as centrilobular necrosis with proliferation of Kupffer cells. Leptospires also invade skeletal muscle, causing edema, vacuolization of myofibrils, and focal necrosis. In the lungs, leptospires induce intra-alveolar hemorrhages. A consistent pathologic finding in all organs is a vasculitis. Whether organ damage is due directly to secreted toxins of leptospires or is secondary to the vasculitis induced by cell-wall components of the organism such as collagenase or bystander damage is unclear. A study has shown that the peptidoglycan of leptospires can increase the adhesion of the organism to vascular endothelial cells and is, thus, an important virulence factor. The illness is characterized by a bleeding diathesis although its pathophysiology is uncertain. Infection causes a thrombocytopenia, but it is unclear whether this is due to a direct action of the leptospires on thrombopoiesis or secondary to disseminated intravascular coagulation or a specific anti-platelet antibody.

During infection, the triggering of the inflammatory response through the production of pro-inflammatory cytokines is important for the early elimination of pathogens. However, unregulated cytokine production can result in a cytokine storm that might be followed by a state of immunoparalysis, which can lead to sepsis, associated organ failures, and subsequent death of some patients with severe leptospirosis.

3. WHAT IS THE TYPICAL CLINICAL PRESENTATION AND WHAT COMPLICATIONS CAN OCCUR?

Leptopirosis was first recognized in sewage workers in 1883. The incubation period is about 10 days but can range from 5 to 30 days. Because the organism spreads to all organs of the body, the clinical presentation may vary. The infection may be asymptomatic and exposure is only recognized serologically. Symptomatic disease may present as a biphasic illness with an initial nonspecific phase where the patient complains of:

*   a high temperature with rigors;
*   headache;
*   myalgia;
*   retro-orbital pain and photophobia;
*   conjunctivitis;
*   nausea, vomiting, diarrhea;
*   dry cough;
*   pre-tibial rash.

This is the leptospiremic phase of the illness lasting about 7 days. After the temperature falls, more specific system-based symptoms may develop within a few days clinically presenting as:

- aseptic meningitis (headache, stiff neck, photophobia, lymphocytic CSF);
- hepatitis (jaundice);
- renal failure with jaundice and hemorrhagic features (Weil's disease);
- pulmonary infection (cough, hemoptysis);
- systemic inflammatory syndrome or shock.

The commonest presentation is anicteric, for example aseptic meningitis (80–90%) compared to icteric with renal failure (Weil's Disease 10–20%).

Laboratory abnormalities occur associated with specific syndromes, for example increased serum creatinine, thrombocytopenia, leucocytosis, and hyperbilirubinemia.

On examination, the liver and / or spleen may be enlarged. This phase corresponds to the *immune phase* and may last from 1 to 6 weeks. In some cases, the two phases may not be apparent or the illness appears to start with the immune phase. The severity of the disease may vary from a mild illness to more severe disease (Weil's disease) with complications including cardiac dysrhythmias, liver or renal failure, myelitis, and Guillain-Barré syndrome. Weil's disease carries a high mortality.

4. HOW IS THE DISEASE DIAGNOSED, AND WHAT IS THE DIFFERENTIAL DIAGNOSIS?

The laboratory diagnosis is made with:

- dark-ground microscopy;
- culture;
- serology;
- genomics.

Dark-Ground Microscopy

Leptospira spp. can be detected by dark-ground microscopy in blood or cerebrospinal fluid (CSF) during the first week of the illness and in the urine about 10 days thereafter. It has a low sensitivity and specificity and requires a concentration of 10^4 *leptospires*/ml for detection.

Culture

Leptospires can also be cultured from the blood, CSF or urine using EMJH medium incubated at 30°C for 6–8 weeks. Because of the low sensitivity and specificity of dark-ground microscopy, and the long timescales of culture, the routine method of diagnosis is serologic. Culture, however, is important in epidemiologic studies with specimens obtained from soil, water, animals, and humans although, for the latter, it is not an important diagnostic test because of the long timescales.

Serology

The microscopic agglutination test (MAT) is difficult to perform and relies on mixing serum from the patient with live leptospires from the different serogroups and looking for agglutination by dark-ground microscopy. This test also has a low sensitivity due to failure to **seroconvert** in a proportion of patients and cross-reactivity with a number of other infections and illnesses including other spirochetal diseases (Lyme disease, relapsing fever, treponemal disease), *Legionella*, HIV, and autoimmune diseases. Although the MAT is serogroup-specific, cross-reactions between serogroups occur and the individual serovars cannot be determined. A result of > 1:100 or a four-fold rise in titer is considered positive.

An IgM ELISA directed against LipL32 is most often used for diagnosis and detects antibodies 7–10 after infection. The majority of IgM (95%) are directed against this antigen and it is common in pathogenic leptospires. It has a high sensitivity and specificity.

Other formats such as latex agglutination, indirect hemagglutination and lateral flow tests (LeptoTek Lateral Flow) are also commercially available and have the advantage that can be used as point-of-care (POC) assays.

Genomics

Various genomic assays are available:

Polymerase chain reaction (PCR) directed against the *sec Y*-gene or the *LipL32* gene have a high sensitivity and specificity. Other PCR modifications are also available, for example chip-based RT-PCR (Real-time PCR), loop-mediated isothermal amplification (LAMP).

Whole Genome Sequencing (WGS) will probably be the standard genomic method of leptospire detection as it has more specificity of the identity of leptospire species. For example, PCR using 16S RNA is useless, as 16 different leptospire species all have the same 16S RNA profile. However, one gene that may prove useful to detect all leptospires is the *ppk* gene (*polyphosphate kinase*) as it evolves rapidly and gives the identical taxonomy to WGS.

DIFFERENTIAL DIAGNOSIS

Leptospirosis is endemic in areas where the following are found:

- Dengue virus (co-infection occurs in about 8% of cases);
- malaria;
- hantavirus;
- scrub typhus.

Patients presenting with the initial febrile illness may be diagnosed with Dengue fever which is more common than leptospirosis. If presenting with signs of meningitis, then leptospirosis may be misdiagnosed as viral meningitis or if with petechiae then meningococcal meningitis. A CSF white cell count would suggest a viral etiology rather than meningococcal, as the cell response in leptospirosis is lymphocytic. If presenting with jaundice the differential diagnosis includes:

- viral hepatitis;
- malaria;
- schistosomiasis;
- relapsing fever;
- tularemia

and if with jaundice and renal failure then it includes:

- Legionnaires disease;
- hemolytic uremic syndrome.

If the pulmonary syndrome is prominent, it may be confused with hantavirus.

5. HOW IS THE DISEASE MANAGED AND PREVENTED?

MANAGEMENT

Antimicrobials

As previously indicated, the majority of cases of leptospirosis are self-limiting and do not require antimicrobial therapy. If the illness is severe enough to cause clinically recognized symptoms and diagnosis is made, antibiotic therapy needs to be started to reduce the duration of illness and shedding of organisms in the urine.

Mild disease: patients are treated with doxycycline or azithromycin, ampicillin or amoxicillin, clarithromycin, fluoroquinolone such as ciprofloxacin or levofloxacin. These antibiotics also have activity against rickettsial disease, which can often be confused with leptospirosis.

Severe disease: for hospitalized patients, IV penicillin, and oral doxycycline for up to 7 days. Ceftriaxone or cefotaxime can also be given.

A Jarisch-Herxheimer reaction may occur following antimicrobial therapy for leptospirosis and is characterized clinically by fever, rigors, and hypotension.

This reaction is due to the release of endotoxin when large numbers of organisms are killed by the antibiotics leading to increase in immune complexes and the release of cytokines, particularly TNF. This reaction also occurs typically with syphilis and Lyme disease. Supportive measures may be required, such as hemodialysis, in severe disease.

There is also some evidence for the benefit of plasmapheresis in severe leptospirosis.

Additional clinical support

There is insufficient data evidence for the routine use of IV steroids but it has been proposed given the vasculitic nature of severe leptospirosis, especially in the setting of pulmonary involvement. Some data have indicated a benefit for the use of steroids as an adjunct to antibiotic therapy in severe disease but further study is needed.

PREVENTION

Prevention includes avoiding potential sources of infection, prophylaxis for individuals at high risk of exposure, and animal vaccination.

The most important control measures for preventing human leptospirosis include avoiding potential sources of infection such as stagnant water and animal farm water runoff, rodent control, and protection of food from animal contamination.

In a study of antimicrobial prophylaxis with doxycycline for individuals at high risk of exposure, fewer cases of clinical leptospirosis were observed in the antibiotic-treated groups of soldiers on jungle exercises versus the placebo group.

Vaccines

The real challenge is to produce a universal vaccine against all of the >300 serovars of *Leptospira*. After many years of research, this has not yet been achieved. Inactivated whole cells (bacterins) are the only vaccines commercially available, primarily for veterinary use. Vaccines that have been produced are mainly serovar dependent and based on lipopolysaccharide antigens. In addition, inactivated vaccines do not promote long-term protection and require annual boosters. Safety has also been reported as an issue. Vaccination of domestic and farm animals against leptospirosis has been shown to give variable levels of protection with some immunized animals becoming infected and excreting leptospires in their urine.

Many approaches are being made to produce effective vaccines and these have been helped by the WGS of leptospires. These include live vaccines, multiepitope vaccines against some OMP, DNA vaccines, and use of NOD-1 and NOD-2 agonists. The development of mRNA vaccines (recently used in SARS-2 vaccines) coding for multiple-surface proteins is also predicted to be an attractive future approach for creating a successful vaccine.

SUMMARY

1. WHAT IS THE CAUSATIVE AGENT, HOW DOES IT ENTER THE BODY, AND HOW DOES IT SPREAD A) WITHIN THE BODY AND B) FROM PERSON TO PERSON?

- Two main identified species exist: *L. interrogans* and *L. biflexa*.
- There are two clades – Pathogens and Saprophytes.
- Leptospirosis is a zoonosis, the organism is carried by many animals and excreted in the urine.
- Infection is acquired by contact with contaminated water or soil.
- The organism enters through the conjunctiva, mucosa or skin abrasions.
- Leptospires spread to all body locations by the blood.

2. WHAT IS THE HOST RESPONSE TO THE INFECTION AND WHAT IS THE DISEASE PATHOGENESIS?

- TLR2 senses outer-membrane protein LipL32 and NOD-like receptors sense peptidoglycan fragments of leptospires. Stimulation through these receptors results in production of pro-inflammatory cytokines TNFα, IL1β and IL8 leading to recruitment of neutrophils, macrophages and dendritic cells to the site of infection.
- The antimicrobial peptide cathelicidin released by neutrophils is effective against some leptospires. Neutrophils also release myeloperoxidase and ROS.
- Complement is important as a host mechanism against leptospires, especially against the non-pathogenic serovars (saprophytes).
- A strong antibody response is made against leptospire LPS. Both IgM and IgG antibodies are made against a variety of specific antigens including LipL32, LipL41, and LigA. The major role of IgG seems to be in opsonization.
- Leptospires, especially those that are pathogenic, have many ways of evading the immune system including: inhibition of degradation of PG into muropeptides that are not recognized by NOD-1 and NOD-2 receptors; subversion or inactivation of complement pathways.
- Pathogenesis: around 10% of leptospirosis cases develop into severe forms. Pathogenic leptospires cause vasculitis in many organs. Liver, lungs, heart, meninges, kidneys, and muscles are all affected; Thrombocytopenia can occur.

3. WHAT IS THE TYPICAL CLINICAL PRESENTATION AND WHAT COMPLICATIONS CAN OCCUR?

- The incubation period is about 10 days but can be 5–30 days.
- The clinical presentation is classically biphasic.
- The initial illness is nonspecific with fever and myalgia.
- The second phase of the illness may present as meningitis, pneumonia, jaundice or renal failure.
- Complications include cardiac arrhythmia, Guillain-Barré syndrome and renal failure.

4. HOW IS THE DISEASE DIAGNOSED, AND WHAT IS THE DIFFERENTIAL DIAGNOSIS?

- Leptospires can be detected in the blood, CSF or urine by dark-ground microscopy.
- The organism can be cultured from the blood, CSF or urine using EMJH medium incubated at 30°C for 6–8 weeks.
- Serologically, the disease can be diagnosed by the macroscopic agglutination test (MALT) or the IgM ELISA.
- Various genomic assays exist: PCR, RT-PCR (real-time PCR), LAMP, and WGS.
- The differential diagnosis includes Dengue virus, viral meningitis, viral hepatitis, malaria, hantavirus and Legionnaires disease.

5. HOW IS THE DISEASE MANAGED AND PREVENTED?

- Most cases of leptospirosis are self-limiting.
- Patients with mild disease are treated with doxycycline or azithromycin.
- Hospitalized patients with severe cases are given IV penicillin, and oral doxycycline and ceftriaxone or cefotaxime is given for up to 7 days.
- A Jarisch-Herxheimer reaction may occur following antimicrobial therapy and is characterized clinically by fever, rigors, and hypotension.
- Prevention is by identifying risks, antimicrobial prophylaxis for individuals at high risk, and animal immunization (although shown to have variable levels of protection).
- Vaccines pose a major challenge because of the many different serovars. There are many new approaches, helped by WGS, and include live vaccines against some OMP, DNA vaccines, and the use of NOD-1 and 2 agonists and mRNA.

FURTHER READING

Bennett JE, Blaser MJ, Dolin R. Mandell, Douglas, and Bennett's Principles and Practice of Infectious Diseases, 9th edition. Elsevier, Philadelphia, 2020.

Goering R, Dockrell H, Zuckerman M, Chiodini PL (eds). Mims' Medical Microbiology and Immunology, 6th edition. Elsevier, Cambridge, 2018.

REFERENCES

Barazzone GC, Teixeira AF, Azevedo BOP, et al. Revisiting the Development of Vaccines Against Pathogenic Leptospira: Innovative Approaches, Present Challenges, and Future Perspectives. Front Immunol, 12: 760291, 2022.

Bierquel E, Thibeaux R, Girault D, et al. A Systematic Review of Leptospira in Water and Soil Environments. PLoS ONE, 15 e0227055, 2020.

Caimi K, Ruybal P. Leptospira, a Genus in the Stage of Diversity and Genomic Data Expansion. Infect Genet Evol, 81: 104241, 2020.

Chin VK, Basir R, Nordin SA, et al. Pathology and Host Immune Evasion During Human Leptospirosis: A Review. Int Microbiol, 23: 127–136, 2020.

Haake DA, Levett PN. Leptospirosis in Humans. Curr Top Microbiol Immunol, 387: 65–97, 2015.

Jiménez JIS, Marroquin JLH, Richards GA, Amin P. Leptospirosis: Report from the Task Force on Tropical Diseases by the World Federation of Societies of Intensive and Critical Care Medicine. J Crit Care, 43: 361–365, 2018.

Karpagam KB, Ganesh B. Leptospirosis: A Neglected Tropical Zoonotic Infection of Public Health Importance—An Updated Review. Eur J Clin Microbiol Infect Dis, 39: 835–846, 2020.

Levett PN, Branch SL. Evaluation of Two Enzyme Linked Immunosorbent Assay Methods for Detection of Immunoglobulin M Antibodies in Acute Leptospirosis. Am J Trop Med Hyg, 66: 745–748, 2002.

Pinto GV, Senthilkumar K, Rai P, et al. Current Methods for the Diagnosis of Leptospirosis: Issues and Challenges. J Microbiol Methods, 195: 106438, 2022.

Santecchia I, Ferrer MF, Vieira MR, et al. Phagocyte Escape of Leptospira: The Role of TLRs and NLRs. Front Immunol, 11: 571816, 2020.

Vincent AT, Schiettekatte O, Goarant C, et al. Revisiting the Taxonomy and Evolution of Pathogenicity of the Genus Leptospira Through the Prism of Genomics. PLoS Negl Trop Dis, 13: e0007270, 2019.

Wagenaar JFP, Goris MGA, Sakudarno MS, et al. What Role Do Coagulation Disorders Play in the Pathogenesis of Leptospirosis. Trop Med Int Health, 12: 111–122, 2007.

WEBSITES

Centers for Disease Control and Prevention, Leptospirosis, 2019: https://www.cdc.gov/leptospirosis/

European Centre for Disease Prevention and Control, Leptospirosis: https://www.ecdc.europa.eu/en/leptospirosis

GOV.UK, Leptospirosis, 2013: https://www.gov.uk/guidance/leptospirosis

Medscape, Leptospirosis, 2021: https://emedicine.medscape.com/article/220563-overview

The Leptospirosis Information Center, 2008: http://www.leptospirosis.org/

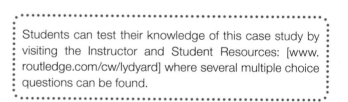

Students can test their knowledge of this case study by visiting the Instructor and Student Resources: [www.routledge.com/cw/lydyard] where several multiple choice questions can be found.

Listeria monocytogenes

22

Figure 22.1 Colonies of *Listeria monocytogenes* on blood agar. *From CC Studio / Science Photo Library, with permission.*

A young woman was admitted to hospital with fever, headache, **myalgia**, and joint pains of 48 hours duration. The previous day, she had attended a lunch where she had eaten ham and a soft cheese. She was 26 weeks pregnant. The admitting physician thought the patient may have listeriosis and a blood culture was taken in addition to a full blood count. Gram-positive cocci were reported on the Gram stain of the blood culture and the patient was started empirically on teicoplanin. The following day, small hemolytic colonies were present on the blood agar (Figure 22.1) and a Gram stain revealed gram-positive rods. Further testing showed the organism was motile with tumbling motility and it was biochemically identified as *Listeria monocytogenes*. A diagnosis of listeriosis was made and the patient's treatment was changed to ampicillin and gentamicin.

1. WHAT IS THE CAUSATIVE AGENT, HOW DOES IT ENTER THE BODY AND HOW DOES IT SPREAD A) WITHIN THE BODY AND B) FROM PERSON TO PERSON?

CAUSATIVE AGENT

Listeria monocytogenes is the cause of both animal and human disease and is widespread in the environment. As such, the organism is a good example of the "One Health" approach to medicine, as agreed at the G7 summit in 2016.

Listeria monocytogenes is a gram-positive nonsporing motile bacillus (Figure 22.2). As of 2020, there were 21 species of *Listeria* identified and of these six species form a phenotypic homogenous group (*Listeria sensu strictu*); the remainder, *Listeria sensu lato*, may be better placed into a new Genus. They are phenotypically different from *Listeria* notably that they cannot grow below 7°C. The clade *Listeria sensu strictu* comprise *L. monocytogenes*, *L. ivanovii*, *L. welshimeri*, *L. innocua*, *L. seeligeri*, and *L. marthii*. However, in 2021, 27 isolates that grouped into five clusters representing five novel *Listeria* species were identified from soil and water samples across the US. Three of the new species, *L. cossartiae*, *L. farberi*, and *L. immobilis* belong to the *sensu stricto* clade and the other two belong to the *sensu lato* clade.

The most common cause of human disease is *L. monocytogenes*, although *L. ivanovii* can rarely cause disease. *Listeria* species grow on blood agar (see Figure 22.1) and *L. monocytogenes* produces a narrow zone of β-hemolysis often only seen beneath the colonies. The organism grows at 37°C but can also grow slowly at 4°C and this can be used as an

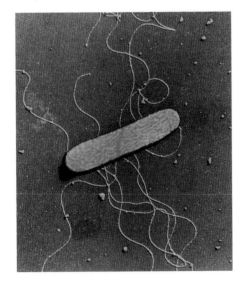

Figure 22.2 Scanning electron microscopic (EM) image of *Listeria monocytogenes* showing flagella. *From AB Dowsett / Science Photo Library, with permission.*

enrichment technique when examining foodstuff. *Listeria* can survive freezing which is relevant to the food-chain and infection. The organism shows a characteristic tumbling (end-over-end) movement at 25°C, which is diagnostic. *L. monocytogenes* has 14 serotypes based on cell wall (O) and flagellar (H) antigens and divided into three divisions I, II, and III. The majority of disease is caused by serotypes 1/2a, 1/2b, and 4b. Several molecular subtyping techniques (**multilocus enzyme electrophoresis**, **pulse field gel electrophoresis**, and **ribotyping**) have been found useful in epidemiologic investigations.

SOURCE OF INFECTION

L. monocytogenes has been isolated from a variety of natural sources including soil, water, animals, and vegetables and is widely distributed in the animal kingdom, with over 40 species of wild and food-source animals including birds, crustaceans, and fish being colonized. The organism can also be carried in the intestine of about 5% of the human population without any symptoms of disease. Infection is acquired by consumption of contaminated food (Figure 22.3) such as fish, salad, pâté, soft cheeses, salami, ham, and coleslaw, where contamination rates as high as 70% may occur. Ingestion of *Listeria* probably occurs frequently and the organism can be carried in the intestine of the human population for short periods without any symptoms of disease. Disease develops mainly in specific groups of individuals including pregnant women, neonates, and immunocompromised patients (see Section 3).

ONE HEALTH

In line with the One Health approach to disease, multi-disciplinary research groups have investigated *Listeria monocytogenes* in horticulture, the Fresh Leafy Produce Supply Chain (FLPSC), animal welfare, and human disease.

An environmental study showed that *L. monocytogenes* was detected in 0.7% of urban soils, 24% of soils on ruminant farms, and 9% in fruit and vegetable farms. Other studies have demonstrated *Listeria* was able to survive for at least 84 days in 71 different soil types, but not in 29 other soil types – the total set representing all the different soil types in France. Short-term survival (<14 days) in the 29 soil types was related to the ratio of cations and long-term survival was related to soil texture and was best in clay soils. Further environmental studies have demonstrated that the two-component regulatory system of *L. monocytogenes*, AgrC-AgrA, are required for successful survival in various soil types and if the gene was deleted, there were large transcriptional profile differences depending on the soil type with limited survival. The transcriptional profile was very limited in sterilized soil types.

Studies on colonization of various vegetables have demonstrated that 10% of each of ten different vegetables were positive for *L. monocytogenes* and its presence in the soil from which the vegetable was taken was found in 5%. *L. monocytogenes* was not detected from four of the vegetables

Figure 22.3 There are a variety of sources of infection with *Listeria*. *From Maximilian Stock Ltd / Science Photo Library, with permission.*

(cabbage, broccoli, cowpea, and palak) nor in the rhizosphere. Both the organism from the vegetable and associated rhizosphere had distinct typing profiles.

In studies using *L. monocytogenes* grown on lettuce leaves, one gene (*lcp*), containing a potential cellulose-binding domain, was significantly up-regulated during growth. Mutants of *lcp* bound less frequently to the leaves compared to the wild type, suggesting cellulose is a key ligand for adhesion. Once bound on the surface of lettuce leaves, co-operation between the bacteria would stimulate biofilm formation, strengthening adhesion to the surface. Subsequent to binding, the organism penetrates the lettuce leaves. After binding to lettuce seeds the organism can be found in every tissue type of the plant: epidermis (4%), cortex (21%), xylem (17%), and phloem (9%).

Studies on the FLPSC have identified where contamination/recontamination may occur from the farm through to the kitchen of the end user and highlighted appropriate remedial action.

Animal models are frequently used as a surrogate for human medicine but, in the case of *Listeria*, the organism also causes disease in animals, such as food-source animals: cattle, goats and sheep, companion animals such as rabbits, cats and dogs and wild animals such as rats, deer, fish and birds. The disease is characterized by septicemia, encephalitis, meningitis, meningo- and rhombencephalitis, abortion and stillbirth, perinatal infections, and gastroenteritis – all identical to diseases also found in humans, clearly suggesting a common pathologic process with similar diagnostic and management approaches.

ENTRY AND SPREAD WITHIN THE BODY

The organism enters the epithelial cells of the intestine using specific adhesion molecules (see Section 2) and then enters the bloodstream. This gives rise to **septicemia** or **meningitis** as the two most common presentations. Localization of the organism in the central nervous system (CNS) is frequently seen. An alternative route of infection to the CNS may be intra-axonal spread directly from the peripheral nerves in the gastrointestinal (GI) tract. The organism can also cross the placenta to give rise to fetal disease, and subsequent abortion.

EPIDEMIOLOGY

This organism is found worldwide but most countries do not keep accurate records.

In the UK, there are about 150 cases per annum. In 2019, there were 142 cases with the incidence varying from 0.14/100 000 (East of England) to 0.39/100 000 (London). Four outbreaks were investigated associated with pre-packed sandwiches; 17.6% were pregnancy associated with one third associated with fetal death. In the US, there are approximately 1600 cases annually with about 260 deaths. Since 2019, there have been five outbreaks associated with hard-boiled eggs, mushrooms, deli meats, and soft cheese. In Europe in 2019, there were 2621 cases of invasive *Listeria* disease over 28 member states, giving a notification rate of 0.46/100 000 population. Since 2015, the rate has remained stable. In 2019, the fatality rate was 17.6%, an increase over the two previous years (13.6%, 15.6%). There were 21 outbreaks, which was higher than the previous year (14). The largest outbreaks were in Spain involving 225 individuals of which three died. In 2018, a pan-European outbreak occurred caused by frozen vegetables produced in a factory in Hungary which involved 7 countries and infected 47 individuals. The largest global outbreak occurred in 2017 in S. Africa affecting over 1000 individuals with 200 deaths.

2. WHAT IS THE HOST RESPONSE TO THE INFECTION AND WHAT IS THE DISEASE PATHOGENESIS?

Listeria monocytogenes is a foodborne intracellular pathogen that attaches to intestinal epithelial cells, penetrates the epithelial barrier, and escapes from the epithelial cells into the lamina propria. It is taken up by phagocytic cells (neutrophils and macrophages) and is distributed around the body causing disease.

Two key virulence factors are Listeria adhesion protein (LAP) and internalin A (InlA). LAP binds to its receptor Hsp60 on the enterocyte, activating NFκB and myosin light chain kinase (MLCK): the former activates an inflammatory response (IL-6, TNFα, IL-8) and the latter induces reconfiguration of the intercellular tight junction (occluding/claudin) and the adherins junction (Ecadherin) by relocating the component proteins. InlA then binds to its receptor (Ecadherin) following relocation of the tight junction proteins. Ecadherin can also be found on the villi of the epithelial cells and goblet cells, which may be an additional route of entry. Once the organism binds to Ecadherin, it is taken up by receptor-mediated endocytosis into a membrane-bound vacuole (phagosome in phagocytic cells). Release from the vacuole into the cytoplasm of the epithelial cell (or phagocytic cell) is assisted by *Listeria* hemolysin (Hyl) and phospholipases. Once in the cytoplasm, the organism replicates and induces polymerization of actin by Act A located at one pole of the cell where the actin precipitates and polymerizes. This polymerized actin forms a growing scaffold that pushes the bacterium through the cytoplasm to the plasma membrane where it makes contact with the plasma membrane of an adjacent cell, and is promptly ingested, and the whole process is repeated (Figure 22.4). This allows the organism to spread through the epithelial layer and eventually entry into the lamina propria. The infected host cells eventually die by **apoptosis** or **necrosis**.

In the CNS, *Listeria* will predominately penetrate **microglia** and to a lesser extent astrocytes and oligodendrocytes. Direct penetration into neurons is rare, although if co-cultured with macrophages, *Listeria* can penetrate neurons following cell-to-cell contact. *L. monocytogenes* may invade the CNS by several routes: hematogenous spread and direct invasion of endothelial cells of the blood–brain barrier; circulating mononuclear cells carrying *Listeria*; or a direct neural route by intra-axonal spread. In some animals, *Listeria* may initially be taken up by tissue macrophages in the mouth and spread by cell-to-cell contact with distal nerve axons and thence by intra-axonal spread into the brain. The role of this route in humans is unclear.

Once bacteria arrive in the brain, **encephalitis** is established by direct cell-to-cell spread. Pathologically in the brain following infection there is necrosis and focal hemorrhages with **meningoencephalitis** (Figure 22.5), micro-abscesses, **vasculitis**, and perivascular lymphocytic infiltration. *L. monocytogenes* is present in the necrotic parenchymal lesions and can be seen in the cytoplasm of macrophages and endothelial cells.

Listeria avoids both the innate and humoral immune system by its intracellular localization. However, antibodies to *Listeria* can be detected along with complement. These can enhance **opsonization**, resulting in killing of *Listeria* by **NK (natural killer) cells**, **dendritic cells**, and macrophages. **Interferon-γ (IFN-γ)** is a key cytokine produced by NK cells, **CD4+ Th-1** cells, and **CD8+** T cells and is important in the immune response to infection with *Listeria*. Escape from the phagosome in dendritic cells and **interleukin (IL)-10** are both important for the development of protective cytotoxic CD8 memory cells. This is presumably via enhancement of Th1 responses.

Once taken up by activated macrophages many *Listeria* will be killed following phagolysosome fusion and subsequent oxygen-dependent and independent routes. However, bacteria localized in nonactivated macrophages and epithelial cells and those escaping from the phagosome before phagolysosome fusion, can survive in **w** produced by overactive CD4+ Th-1 cells, as in tuberculosis (TB).

3. WHAT IS THE TYPICAL CLINICAL PRESENTATION AND WHAT COMPLICATIONS CAN OCCUR?

L. monocytogenes affects mainly pregnant women, neonates, and immunocompromised individuals.

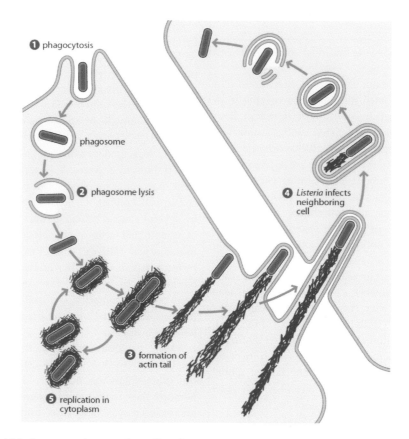

① phagocytosis

phagosome

② phagosome lysis

④ *Listeria* infects neighboring cell

③ formation of actin tail

⑤ replication in cytoplasm

Figure 22.4 Escape of *Listeria monocytogenes* from the phagosome, actin polymerization, and movement into an adjacent cell. 1, uptake; 2, escape from the phagosome by listeriolysin O and phospholipases; 3, actin polymerization; 4, penetration to an adjacent cell; 5, intracytoplasmic replication.

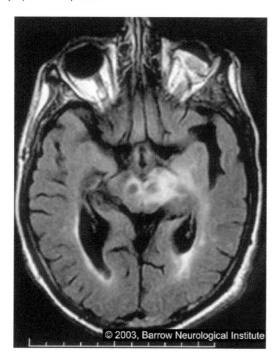

© 2003, Barrow Neurological Institute

Figure 22.5 *Listeria monocytogenes* meningoencephalitis. MRI scan showing enhancement around the left midbrain and thalamus and cerebral peduncle, indicating meningoencephalitis in a case of *Listeria* infection in an immunocompromised patient. *Figure 3 from Barrow Quarterly (2003), 19: 20–24. With permission from © Barrow Neurological Institute.*

In pregnancy, the disease may present as an acute **febrile** flu-like illness and meningitis is an uncommon presentation Most disease occurs in the third trimester. However, about 25% of perinatal infections result in fetal death (leading to spontaneous abortion) and, of those that survive, about 70% will develop neonatal disease.

Infection acquired, *in utero*, may result in granulomatosis infantiseptica, which has a high mortality and is due to widespread granulomatous **abscesses** in many organs.

Neonatal infection presents either as an early onset **bacteremia** or as a late onset (1–2 weeks post-partum) meningitis.

In immunocompromised patients, and those over 60 years of age, listeriosis presents most frequently with meningitis (60%) and bacteremia (30%). Brain abscess, abscesses in other organs or **endocarditis** may complicate bacteremia. *Listeria* infection in the CNS may present with a meningitis that is indistinguishable from other causes of meningitis or as a meningoencephalitis with altered consciousness and **seizures**.

Large outbreaks of *Listeria* gastroenteritis, in otherwise healthy individuals, have occasionally been reported, which present as diarrhea, fever, and myalgia. This presentation has been linked to consumption of high counts of *Listeria* in food. Rarely, otherwise healthy individuals may develop an encephalomyelitis of the brainstem caused by *Listeria*, which

Figure 22.6 BBL CHROMagar showing the characteristic blue-green color of *L. monocytogenes* surrounded by a white opaque halo. *From http://www.bd.com/ © Beckton, Dickinson and Company, with permission.*

presents with a flu-like prodrome and, a few days later, with an acute onset of cerebellar signs, cranial nerve palsies, and respiratory failure.

4. HOW IS THE DISEASE DIAGNOSED, AND WHAT IS THE DIFFERENTIAL DIAGNOSIS?

Diagnosis of listeriosis is confirmed by isolation of *L. monocytogenes* from a normally sterile site such as the blood or the cerebrospinal fluid (CSF). Mis-identification of *Listeria* can occur as they may appear like diphtheroids or gram-positive cocci in specimens.

Listeria grows well on routine diagnostic media and on blood agar. *L. monocytogenes* (as well as *L. seeligeri* and *L. ivanovii*) are beta-hemolytic. *L. monocytogenes* and *L. seeligeri* have a narrow zone of beta-**hemolysis** whereas *L. ivanovii* has a wide double zone of hemolysis. A number of selective media exist for the isolation of *Listeria* from food/feces (Table 22.1). A number of chromogenic media have also been developed to differentiate *L. monocytogenes* from other *Listeria* spp. and other bacteria (Figure 22.6).

Serology is not useful for diagnosis of acute cases, although listeriolysin O antibodies have been used to identify patients with noninvasive illness.

DIFFERENTIAL DIAGNOSIS

As *Listeria* infection presents as a **pyrexia of unknown origin (PUO)**, a bacteremia or a meningitis, any of the microorganisms causing these conditions must be considered in the differential diagnosis. There should be a high index of suspicion for *Listeria* infection if PUO, bacteremia or meningitis occur in the at-risk groups. In particular, the other main causes of neonatal bacteremia and meningitis (group B *Streptococcus* and *E.coli*) should be considered.

5. HOW IS THE DISEASE MANAGED AND PREVENTED?

MANAGEMENT

Disease caused by *Listeria* is usually treated with a combination of ampicillin and gentamicin. Co-trimoxazole can be used in patients who are hypersensitive to β-lactams or in place of gentamicin. In patients who are pregnant and who cannot take folic acid inhibitors, a glycopeptide can be used in place of ampicillin.

Bacteremia is usually treated for 2 weeks and meningitis for 3 weeks, although because of the affinity of *Listeria* for the CNS one can argue that bacteremia should also be treated for 3 weeks. Brain abscess or endocarditis should be treated for 6 weeks.

PREVENTION

At-risk individuals (pregnancy, neonates, immunocompromised) should avoid food in which *Listeria* can grow to high cell numbers (e.g. pâté and deli meats, soft cheese, raw seafood, salad, and coleslaw) and correct food-handling precautions will reduce the chance of infection. There are no vaccines for *Listeria* as yet.

Table 22.1 Selective media for the isolation of *Listeria* from food/feces

Selective medium	Constituents
Listeria selective agar	Tryptose, glucose, sodium chloride, thiamine, acriflavine, nalidixic acid
Oxford *Listeria* agar	Peptone, corn starch, sodium chloride, esculin, lithium chloride, ferric ammonium citrate
Al-Zoreky Sadine *Listeria* agar	Acriflavine, ceftazidime, moxalactam
Listeria PALCAM agar	Peptone, starch, sodium chloride, D-mannitol, ammonium ferric citrate, esculin, glucose, lithium chloride, polymyxin, ceftazidime, phenol red
LPM agar	Casein, peptone, beef extract, sodium chloride, lithium chloride, glycine, phenylethanol, moxalactam
HiChrome *Listeria* agar	Meat extract, yeast extract, peptone, rhamnose, sodium chloride, lithium chloride, chromogenic mixture
Listeria selective supplement I	Acriflavine, 10cycloheximide, nalidixic acid
Listeria selective supplement II	Acriflavine, 11cycloheximide, colistin, cefotetan, fosfomycin

SUMMARY

1. WHAT IS THE CAUSATIVE AGENT, HOW DOES IT ENTER THE BODY, AND HOW DOES IT SPREAD A) WITHIN THE BODY AND B) FROM PERSON TO PERSON?

- There are several species of *Listeria* but *L. monocytogenes* is responsible for the majority of disease.
- *L. monocytogenes* is a motile nonspore-forming gram-positive rod.
- *L. monocytogenes* can grow at 4°C and shows characteristic tumbling motility at 25°C.
- *L. monocytogenes* produces a narrow zone of beta-hemolysis on blood agar.
- *Listeria* spp. are ubiquitous in animals and the environment and therefore found globally.
- Infection is usually acquired from consumption of contaminated food.
- Neonatal infection can be acquired transplacentally.

2. WHAT IS THE HOST RESPONSE TO THE INFECTION AND WHAT IS THE DISEASE PATHOGENESIS?

- *Listeria* is an intracellular pathogen avoiding the host immune system.
- *Listeria* penetrates both phagocytic and nonphagocytic cells.
- *Listeria* escapes from the phagosome with the aid of listeriolysin O.
- *Listeria* precipitates actin when in the cytoplasm and penetrates adjacent cells.
- *Listeria* infection is limited by CD8 cytotoxic T cells.

3. WHAT IS THE TYPICAL CLINICAL PRESENTATION AND WHAT COMPLICATIONS CAN OCCUR?

- *L. monocytogenes* typically infects pregnant women, neonates, and immunocompromised individuals.

- Disease in pregnant women presents as a flu-like illness but can cause spontaneous abortion and granulomatosis infantiseptica.
- In neonates, early post-partum disease presents as a bacteremia and late onset disease as a meningitis.
- Disease in immunocompromised adults presents as a meningitis or bacteremia.
- Disease caused by *L. monocytogenes* in otherwise healthy adults is infrequent but may present as a brainstem meningoencephalitis or as a febrile gastroenteritis.
- Bacteremia may be complicated by abscesses or endocarditis.

4. HOW IS THE DISEASE DIAGNOSED, AND WHAT IS THE DIFFERENTIAL DIAGNOSIS?

- Disease caused by *Listeria* is diagnosed by isolation of the organism from sterile sites such as blood or CSF on routine blood agar.
- Several selective media exist for the isolation of *Listeria* from food or feces.
- Serology is not helpful in diagnosis of acute illness.
- Any of the many causes of bacteremia, endocarditis, and meningitis can mimic the presentation of listeriosis and a high index of suspicion should be held in at-risk groups.

5. HOW IS THE DISEASE MANAGED AND PREVENTED?

- Treatment is with ampicillin and gentamicin.
- Co-trimoxazole is an alternative treatment in case of penicillin hypersensitivity.
- In penicillin hypersensitivity in pregnancy, a glycopeptide should be used.
- At-risk individuals (pregnancy, neonates, immunocompromised) should avoid food in which *Listeria* can grow to high cell numbers.

FURTHER READING

Cimolai N. Laboratory Diagnosis of Bacterial Infections. Marcel Dekker, New York, 2001.

Mandell GL, Bennet JE, Dolin R. Principles & Practice of Infectious Diseases, 6th edition, Vol 2. Churchill Livingstone, Philadelphia, 2005.

Mims C, Dockrell H, Goering RV, et al. Medical Microbiology, 3rd edition. Mosby, London, 2004.

Murphy K, Weaver C. Janeway's Immunobiology, 9th edition. Garland Science, New York/London, 2016.

Wilson M, McNab R, Henderson B. Bacterial Disease Mechanisms: An Introduction to Cellular Microbiology. Cambridge University Press, London, 2002.

REFERENCES

Arslan S, Ozdemir F. Prevalence and Antimicrobial Resistance of Listeria Species and Molecular Characterization of Listeria monocytogenes Isolated from Retail Ready-to-Eat Foods. FEMS Microbiol Lett, 367: fnaa006, 2020.

Bahjat KS, Liu W, Lemmens EE, et al. Cytosolic Entry Controls CD8+ T-Cell Potency During Bacterial Infection. Infect Immun, 74: 6387–6397, 2006.

Bierne H, Cossart P. InlB a Surface Protein of Listeria Monocytogenes that Behaves as an Invasin and a Growth Factor. J Cell Sci, 115: 3357–3367, 2002.

Bierne H, Cossart P. Listeria monocytogenes Surface Proteins: From Genome Predictions to Function. Microbiol Mol Biol Rev, 71: 377–397, 2007.

Carlin CR, Liao J, Weller D, et al. Listeria cossartiae sp. nov., Listeria immobilis sp. nov., Listeria portnoyi sp. nov. and Listeria rustica sp. nov., Isolated from Agricultural Water and Natural Environments. Int J Syst Evol Microbiol, 71: 004795, 2021.

Dhamaa K, Karthik K, Tiwaric R, et al. Listeriosis in Animals, its Public Health Significance (Food-Borne Zoonosis) and Advances in Diagnosis and Control: A Comprehensive Review. Vet Q, 35: 211–235, 2015.

Dharmendra Kumar Soni DK, Singh M, Singh DV, Dubey SK. Virulence and Genotypic Characterization of Listeria monocytogenes Isolated from Vegetable and Soil Samples. BMC Microbiol, 14: 241, 2014.

D'Orazio SE, Troese MJ, Starnbach MN. Cytosolic Localization of Listeria monocytogenes Triggers an Early IFN Gamma Response by CD8 T Cells That Correlates with Innate Resistance to Infection. J Immunol, 177: 7146–7154, 2007.

Foulds KE, Rotte MJ, Seder RA. IL-10 is Required for Optimal CD8 T Cell Memory Following Listeria monocytogenes Infection. J Immunol, 177: 2565–2574, 2006.

Gandhi M, Chikindas ML. Listeria: A Foodborne Pathogen That Knows How to Survive. Int J Food Microbiol, 113: 1–15, 2007.

Hoelzer K, Pouillot R, Dennis S. Animal Models of Listeriosis: A Comparative Review of the Current State of the Art and Lessons Learned. Vet Res, 43: 18, 2012.

Jamshidi A, Zeinali T. Significance and Characteristics of Listeria monocytogenes in Poultry Products. Int J Food Sci, 2019: 7835253, 2019.

Kim KP, Jagadeesan B, Burkholder KM, et al. Adhesion Characteristics of Listeria Adhesion Protein (LAP)-Expressing Escherichia coli to Caco-2 Cells and of Recombinant PAP to Eukaryotic Receptor Hsp60 as Examined in a Surface Plasmon Resonance Sensor. FEMS Microbiol Lett, 256: 324–332, 2006.

Locatelli A, Spor A, Jolivet C, et al. Biotic and Abiotic Soil Properties Influence Survival of Listeria monocytogenes in Soil. PLoS One, 8: e75969, 2013.

Lecuit M. Understanding how Listeria monocytogenes Targets and Crosses Host Barriers. Clin Microbiol Infect, 1: 430–436, 2005.

Nwaiwu O. What Are the Recognized Species of the Genus Listeria? Access Microbiol, 2: acmi000153, 2020.

Olaimat AN, Al-Holy MA, Shahbaz HM, et al. Emergence of Antibiotic Resistance in Listeria monocytogenes Isolated from Food Products: A Comprehensive Review. Compr Rev Food Sci Food Saf, 17: 1277–1292, 2018.

Osek J, Lachtara B, Wieczorek K. Listeria monocytogenes – How This Pathogen Survives in Food-Production Environments? Front Microbiol, 13: 866462, 2022.

Ramaswamy V, Cresence VM, Rejitha JS, et al. Listeria – Review of Epidemiology and Pathogenesis. J Microbiol Immunol Infect, 40: 4–13, 2007.

Shaughnessy LM, Swanson JA. The Role of Activated Macrophages in Clearing Listeria monocytogenes Infection. Front Biosci, 12: 2683–2692, 2007.

Shenoy AG. Persistence and Internalization of Listeria monocytogenes in Romaine Lettuce, Lactuca sativa var. Longifolia. MSC Thesis, Purduh University, USA, 2015.

Vivant AL, Garmyn D, Gal L, et al. Survival of Listeria monocytogenese in Soil Requires AgrA-mediated Regulation. Appl Environ Microbiol, 81:5073–5084, 2015.

Vivant AL, Garmyn D, Piveteau P. Listeria monocytogeneses, a Down-to-Earth Pathogen. Front Cell Infect Microbiol, 3: 87, 2013.

WEBSITES

Centers for Disease Control and Prevention, *Listeria* (Listeriosis), 2022: https://www.cdc.gov/listeria/index.html

Centers for Disease Control and Prevention, *Listeria* (Listeriosis), 2022: https://www.cdc.gove/outbreaks/mono cytogenes-06-22

Food Safety News, UK authorities renew warning in deadly fish Listeria outbreak, 2022: https://www.foodsafetynews.com/2022/08/uk-authorities-renew-warning-in-deadly-fish-listeria-outbreak/

GOV.UK, Listeria: guidance, data and analysis, 2013: https://www.gov.uk/government/collections/

Public Health Scotland, Listeria: https://www.hps.scot.nhs.uk/a-to-z-of-topics/listeria/

Purdue University, Persistence and internalization of Listeria monocytogenes in romaine lettuce, Lactuca sativa var. longifolia, 2015: https://docs.lib.purdue.edu/cgi/viewcontent.cgi?article=1629&context=open_access_theses

World Health Organization, Listeriosis, 2018: https://www.who.int/news-room/fact-sheets/detail/listeriosis

World Health Organization, One Health, 2017: who.int

Students can test their knowledge of this case study by visiting the Instructor and Student Resources: [www.routledge.com/cw/lydyard] where several multiple choice questions can be found.

Mycobacterium leprae

23

A 37-year-old Hispanic woman, a native of Southern Mexico, went to see her primary care physician after complaining of a persistent **rash** throughout her body. The woman had three children, was a nonsmoker, and appeared to be well-nourished.

Her symptoms had started 5 years before with spasms, with needle-like pain, in her arms. She also felt tired and stressed and had been initially diagnosed with depression. Her skin examination indicated **atopic dermatitis** and **urticaria** and she was prescribed ibuprofen, fluoxetine, and hydroxyzine.

Physical examination revealed numerous hypopigmented skin lesions (Figure 23.1), especially those on her arms, nasal bridge area, cheeks, abdomen, back, and her legs. Her eyebrows had started thinning and she had numbness in her forearm.

The patient had a biopsy of skin lesions on the abdomen. Acid-fast bacilli were detected and a diagnosis of lepromatous leprosy (LL) was made. The patient was counseled about leprosy and was prescribed a course of a standard multidrug therapy (MDT). As required by Mexican law, her case was reported to the Public Health Department.

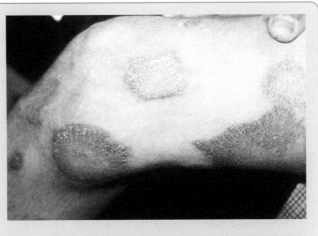

Figure 23.1 Skin lesions of a patient with lepromatous leprosy. This is usually accompanied by the loss of sensation around the affected area. *Courtesy of the US Department of Health and Human Services.*

1. WHAT IS THE CAUSATIVE AGENT, HOW DOES IT ENTER THE BODY AND HOW DOES IT SPREAD A) WITHIN THE BODY AND B) FROM PERSON TO PERSON?

CAUSATIVE AGENT

Leprosy (also called Hansen's disease, HD) is a chronic granulomatous disease of the skin and peripheral nervous system. For many years, *Mycobacterium leprae* was considered as a sole pathogen causing leprosy. However, in 2008, *Mycobacterium lepromatosis* was identified as second causative agent of HD. *M. lepromatosis* has been implicated in a small number of HD cases, and clinical aspects of the disease caused by *M. lepromatosis* are not yet clearly defined. *M. lepromatosis* was initially associated with diffuse LL, but subsequent studies have linked it to other forms of HD. *M. leprae* and *M. lepromatosis* comprise a so-called "*Mycobacterium leprae* complex*". Based on their DNA sequences, both mycobacteria are classified as different species: *M. leprae* and *M. lepromatosis* might have diverged from a common ancestor more than 13 million years ago. However, they are both obligate intracellular organisms and share many important features.

In this chapter, we focus on *Mycobacterium leprae* as the most frequent causative agent of leprosy.

M. leprae is a member of the Mycobacteriaceae family. It is an obligate intracellular gram-positive bacillus that requires the environment of the host macrophage for survival and propagation by binary fission. It shows preferential tropism toward macrophages and Schwann cells that surround the axons of nerve cells. The bacilli resist intracellular degradation by macrophages, possibly by escaping from the **phagosome** into the cytoplasm and preventing fusion of the phagosome with **lysosomes**.

M. leprae is an acid-fast microorganism. This means that it is resistant to decolorization by acids during staining procedures. The bacilli appear as straight or curved rod-shaped organisms, 1–8 μm long and approximately 0.3 μm in diameter. In common with all pathogens belonging to the genus *Mycobacterium*, *M. leprae* can be stained using the Ziehl-Neelsen method, in which the bacteria are stained bright red and stand out clearly against a blue background (Figure 23.2). A Fite modification of this method allows identification of more bacilli using their ability to accumulate large amounts of hyaluronic acid.

M. leprae, unlike *M. tuberculosis*, cannot be cultured in the laboratory. *M. leprae* can, however, be propagated in mouse footpad and in its natural host – the nine-banded armadillo – for experimental purposes.

The genome of *M. leprae* has been sequenced and it includes 1605 genes coding for proteins and 50 genes for stable RNA molecules. The genome of *M. leprae* is smaller than that of *M. tuberculosis* and less than half of the genome contains functional genes but there are many pseudogenes, with intact counterparts in *M. tuberculosis* (Table 23.1). Gene deletion and decay compared with *M. tuberculosis* have eliminated many important metabolic activities. *M. leprae* is therefore even more dependent on host metabolism and this is why it is characterized by a slower growth rate than that of *M. tuberculosis*.

The cell wall of *M. leprae* is similar to that of *M. tuberculosis*, but it includes the species-specific phenolic glycolipid I (PGL-I) widely used for serodiagnosis of LL and disease pathogenesis (see Section 2); and genus-specific lipoarabinomannan (LAM) among other molecules (see Section 4).

The major reservoir of *M. leprae* is in humans, rarely apes, and some species of monkeys: chimpanzees, nine-banded armadillos, mangabey monkeys, and cynomolgus macaque. These, especially armadillos, are characterized by a natural low body temperature of 34°C that favors the survival of *M. leprae*.

ENTRY AND SPREAD WITHIN THE BODY

The current view is that *M. leprae* enters the body via nasal epithelial cells or poorly differentiated keratinocytes, fibroblasts and endothelial cells of the skin through minor injuries. Tropism for these cells is likely to be mediated via mammalian cell entry protein 1A (Mce1A). In the respiratory system, the microorganism is taken up by **alveolar macrophages** where they live and proliferate very slowly. In early infections, subsequent to entry into respiratory mucosa, the bacilli are spread by the circulation, and reach neural tissue where they infect Schwann cells. It is still unclear how *M. leprae* travels through blood but it is likely to be via macrophages and **dendritic cells (DCs)**. Fibroblasts, which can be invaded by *M. leprae,* can also contribute to dissemination of the bacteria.

M. leprae binds to Schwann cells via the cell wall PGL-I (an **adhesin**), which attaches to the G-domain of α2-chain of

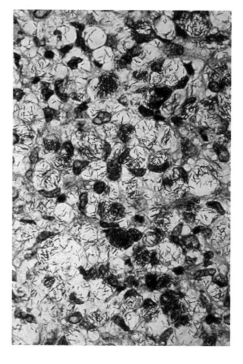

Figure 23.2 Skin lesion fluid from the leprosy patient stained by the Ziehl-Neelsen acid-fast method. The reagent stains *M. leprae* red against the blue background. *Courtesy of the US Department of Health and Human Services.*

laminin 2 isoform in the basal lamina of Schwann cells and is restricted to peripheral nerves. However, the subsequent uptake of *M. leprae* by the Schwann cells is facilitated by its α-dystroglycan – a receptor for laminin on the cell membrane. The bacilli live in the infected Schwann cells and macrophages and are characterized by slow growth (sometimes for years) that is reflected in the long incubation period of the disease. The average division rate of *M. leprae* is very low for microbes, 10–12 days for one division. Optimal temperature for growth of *M. leprae* is lower than core body temperature (27–33°C), thus it prefers to grow in the cooler parts of the body, that is skin and superficial nerves. Several factors enable *M. leprae* to survive in Schwann cells: affecting host-cell metabolism by increasing the uptake of glucose and lipid synthesis leading to the reduction in intracellular oxidative stress. To avoid

Table 23.1 Comparative genomics of *M. leprae* and *M. tuberculosis*

Parameter	*M. leprae* (strain TN)	*M. tuberculosis* (strain H37Rv)
Genome size (bp)	3268203	4411532
Protein genes	1614	3993
tRNA genes	45	45
rRNA genes	3	3
Unknown genes	142	606
Pseudogenes	1133	6
Gene density (bases/gene)	2024	1106
% protein coding	49.5	91.2
Single nucleotide polymorphism frequency	1/24000 bp	1/3000 bp

Adapted from Scollard et al., 2006

host-cell apoptosis, Schwann cell survival factor insulin-l like growth factor-I (IGF-I) is induced by *M. leprae* as well as 2′–5′ oligoadenylate synthetase-like (OASL), which impairs autophagy and antimicrobial peptide expression.

Another possible route of entry for *M. leprae* is through minor injuries in the epidermis which consists of almost 90% keratinocytes. The bacteria attach to keratinocytes probably via laminin- 5, the major form of laminin in the skin, prompting keratinocytes to launch innate immune responses.

PERSON-TO-PERSON SPREAD

Leprosy sufferers, particularly with multibacillary leprosy (predominantly those with LL – see Sections 2 and 3) represent the major source of the infection, and transmission is through aerosols to the respiratory tract. Its dissemination through skin lesions seems to be less important. There is increasing evidence that those infected individuals that do not develop the symptoms of the disease may nevertheless transiently excrete the microbe nasally and spread the infection. As a result, close proximal contacts of the infected individuals (household, neighbors, social contacts) have an increased risk of contracting the disease. The risk varies for different forms of the disease (see Section 3): 8–10 times higher for LL and 2–4 times higher for the tuberculoid form (TT). Factors such as probability of a frequent contact, similar genetic and immunologic background, and similar environment may all play a role. In the southern states of the US, molecular typing has proven that *M. leprae* can be transmitted from armadillos to humans. Immunosuppression is one of the contributing factors. In immunocompromised individuals, leprosy can develop after solid-organ transplantation, chemotherapy, anti-rheumatoid treatment or HIV infection, although the latter cases are rare.

It is a slowly developing disease with an incubation period from 6 months to up to 40 years or even longer! Again, the mean incubation time is different for TT and LL, being 4 years and 10 years, respectively.

EPIDEMIOLOGY

According to the World Health Organization (WHO) in 1997, 2 million people worldwide were infected with *M. leprae*. Introduction in the mid-1980s of MDT significantly reduced the prevalence of leprosy. In 2000, the WHO declared leprosy eliminated as a public health problem on the basis of overall reduction in prevalence to less than 1 case per 10 000 people. Since 2010, however, a fairly stable number of new leprosy cases of around 200 000 are reported each year indicating that MDT upon diagnosis is not sufficient to eliminate the disease. A high frequency (17%) of infection in children should be noted. In adults, the LL form is more common in men than women with an approximate 3:2 ratio. The risk for the acquisition of leprosy actually might be bimodal: first wave at 5–15 years of age and then after 30 years.

Leprosy can affect races all over the world. However, it is most common in warm, wet areas in the tropics and subtropics. Leprosy is endemic in Asia, Africa, the Pacific region, and Latin America. About 75% of the patients have been registered with the disease in South-East Asia, particularly in India where leprosy is epidemic (Figure 23.3). Significant regional clustering has been detected in individual countries and even communities. Interestingly, African blacks have a high incidence of TT, while people with white skin and Chinese people mostly have LL.

In the last decade of the 20th century, the case detection rate remained stable (India), decreased (China) or increased (Bangladesh). However, these data rely on the measurement

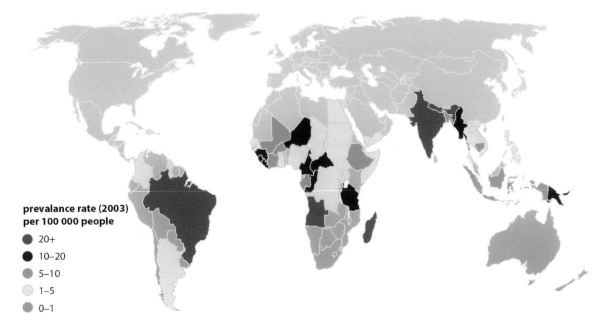

prevalance rate (2003)
per 100 000 people

- 20+
- 10–20
- 5–10
- 1–5
- 0–1

Figure 23.3 Global distribution of leprosy in 2003 according to WHO. South-East Asia, Africa, the Pacific region and South America are the regions the most affected.

of incidence through the number of patients registered for the treatment that is far from optimal. Often long sufferers from leprosy remain unregistered for various reasons (stigma, costs, lack of health services). For example, a door-to-door inquiry in Bangladesh, produced nearly a five-fold increase in cases compared to self-reported findings only. Consequently, the information currently available from endemic countries might often be inadequate and does not anticipate trends in leprosy epidemiology and transmission.

In non-endemic countries, the symptoms of leprosy are often overlooked. In the UK, in more than 80% of cases, leprosy was not suspected at the first visit to a medical doctor since it simply was not expected. This often leads to a median diagnostic delay of 1.8 years.

In the US, 75% of cases were shown to be from immigrants, and for the rest international exposure was the contributing factor, although transmission from infected armadillos remains a possibility.

2. WHAT IS THE HOST RESPONSE TO THE INFECTION AND WHAT IS THE DISEASE PATHOGENESIS?

Host immune responses to *M. leprae* differ for TT and LL types of leprosy, with a prevalence of either T-cell immunity (TT) or antibody production (LL). For the definition of different types of leprosy see Section 3.

TUBERCULOID LEPROSY

A strong T-cell-mediated immunity combined with a weak antibody production leads to the development of a mild clinical form of TT with few nerves involved and low bacterial load. The replication of *M. leprae* in Schwann cells leads to the stimulation of T-cell response and subsequently to chronic inflammation. This is mediated through the bacterial cell wall glycolipid PGL-I. The Schwann cells can express HLA class II molecules and therefore stimulate **CD4+** T cells. **γ-interferon** (**γ-IFN**)-producing CD4+ T cells are found infiltrating the lesions and forming **granulomas** with epithelioid and giant cells.

In addition, interaction of *M. leprae* with **Toll-like receptors** (TLR1 and TLR2) on Schwann cells in patients with TT is believed to induce **apoptosis** thus leading to nerve damage. A chronic inflammatory reaction results in swelling of perineural areas, **ischemia**, **fibrosis**, and death of axons. However, very little circulating antibody against *M. leprae* is present in TT.

Keratinocytes can also function as antigen-presenting cells (APC) by presenting *M. leprae* antigens to CD4+ T cells. In turn, production of γ-IFN by CD4+ T cells can activate keratinocytes in lesions of tuberculoid leprosy patients.

LEPROMATOUS LEPROSY

Antibody-mediated responses with no protective effect are a feature of this form of leprosy. In LL, PGL-I stimulates production of specific **IgM** that is proportional to the bacterial load and this has diagnostic and prognostic importance. These antibodies are not protective and are thought to down-regulate cell-mediated immunity. They form **immune complexes** that are deposited in tissues attracting neutrophils to the deposition sites and causing complement activation. This leads to the development of intense and aggravated inflammation and tissue/organ damage. A systemic inflammatory response mediated by the deposition of extravascular immune complexes is called **erythema nodosum** leprosum (ENL). Women with leprosy may develop post-partum nerve damage. Although pregnancy itself does not seem to aggravate the disease or cause a relapse, LL patients experience ENL reactions throughout pregnancy and lactation, possibly associated with earlier loss of nerve function.

As might be expected by the predominant cellular or humoral response, the clinical form of the disease is affected by the balance of **Th-1/Th-2** cytokines in TT and LL forms of leprosy. Since T-cell immunity is enhanced by the T helper 1 (Th-1) **cytokine** pathway (**interleukin (IL)**-2, IL-15, IL-12, IFN-γ), and humoral immunity is supported by Th-2 cytokines (IL-4, IL-5 and IL-10), the severity of antibody-dominated LL is dependent upon activation of Th-2 cytokines.

DCs are likely to encounter the bacilli in the nasal mucosa or skin abrasions, where they play a key role in the subsequent regulation of inflammation and balance of Th-1/Th-2 cell-mediated immunity – summarized in Figure 23.4.

Epithelial cells can direct local DCs to stimulate IgA switch in B lymphocytes. IgA can confer protection to *M. leprae* infection by efficiently clearing the bacteria in the airways. Elevated levels of IgA directed to PGL-I and LAM were detected in the saliva of untreated leprosy patients, their contacts and household members indicating a potential protective role of these antibodies.

Genetic polymorphism was also found to be associated with the pattern of the disease: HLA DR2 and DR3 alleles are associated with TT, and DQ1 with LL forms of leprosy (see Section 3).

The borderline cases appear immunologically unstable and susceptible to aggressive, inflammatory episodes affecting the peripheral nerves which can lead to leprosy-associated disabilities.

PATHOGENESIS

Hypersensitivity reactions vastly contribute to the pathogenesis of the disease. There are two types of reactions that can develop: type 1 reaction (T1R) also called reversal; or type 2 reaction (T2R) ENL. T1R are delayed hypersensitivity

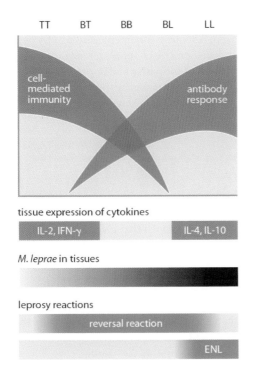

Figure 23.4 Spectrum of clinical and immunologic characteristics of leprosy. Immune responses regulated by Th-1 cytokines (IL-2, IFN-γ) are characteristic of the less severe TT form with the prevalence of T-mediated immune responses, while Th-2 cytokines (IL-4, IL-10) enhance the more aggressive antibody-dependent LL form of leprosy. Intermediate forms (BT, BB, and BL, see Section 3) are characterized by an increasing shift in balance from Th-1 toward Th-2 cytokine-mediated immune responses accompanied by a concomitant appearance of *M. leprae* in the tissues. Intermediate forms can revert to the less severe form of the disease, but LL is irreversible and presents as ENL. BT: borderline turberculoid leprosy; BB: mid-borderline leprosy; BL: borderline lepromatous leprosy. *From Britton WJ & Lockwood DNJ (2004) The Lancet 363:1209-1219. With permission from Elsevier.*

reactions associated with an increase in cell-mediated immune responses in the lesions. These are characteristic for the TT form of leprosy.

T2R develops as a result of antibody responses to *M. leprae* antigens leading to the formation and deposition of immune complexes into tissues and subsequent infiltration by neutrophils and systemic effects such as high fever and edema. Bacterial antigens involved in the immune complex formation are PGL-I, LAM and major membrane protein II (MMP-II). T2R predominantly occurs in BL/LL patients (see Section 3) with high bacillary loads and leads to the nerve damage.

Patients with untreated LL might develop a rare condition, called the *Lucio phenomenon*, mostly present in cases with T2R which manifests with the onset of necrotizing vasculopathy.

Excessive accumulation of collagen as a result of chronic activation of fibroblasts can lead to fibrosis, causing irreversible nerve impairments and neuropathy.

Surprisingly, and different from *M. tuberculosis*, co-infection with *M. leprae* and HIV does not have any appreciable effect on clinical symptoms, although HIV infection impairs T-cell-mediated immune responses. However, in patients with preexisting leprosy and subsequent infection with HIV, a paradoxical clinical deterioration may occur after highly active antiretroviral therapy (HAART). This treatment would flair up a so-called reversal reaction (RR) which manifests as an increase of the inflammatory process in the skin, nerve, or both, leading to the appearance of new lesions.

3. WHAT IS THE TYPICAL CLINICAL PRESENTATION AND WHAT COMPLICATIONS CAN OCCUR?

THE DIFFERENT FORMS OF LEPROSY

The forms of leprosy range from the mildest indeterminate form to the most severe lepromatous type. Severe forms arise because of a less effective immune response to the infection (see Section 2). Most of those infected develop an appropriate immune response and never develop signs of leprosy.

The classification of leprosy is currently based on clinical grounds. Some authors identify the least aggressive indeterminate leprosy (IL) as the very first stage of the disease.

According to the accepted Ridley and Jopling (1966) classification, there are five more advanced stages of leprosy, given here in increasing severity.

- Tuberculoid leprosy (TT), also called type I leprosy.
- Borderline tuberculoid leprosy (BT).
- Mid-borderline leprosy (BB).
- Borderline lepromatous leprosy (BL).
- Lepromatous leprosy (LL), also called type II leprosy.

IL is an early form with no loss of sensation and one or several skin hypopigmentation areas and **erythematous** macules. About 75% of these marks heal spontaneously. In some affected individuals, supposedly with weak immune responses, the disease progresses to other forms.

Both IL and TT forms are characterized by fewer than five skin lesions, with no bacilli being detected on smear tests and are collectively called paucibacillary (PB) leprosy.

In patients with TT, one or several erythematous plaques are usually found on the face and limbs, but not on intertriginous areas or the scalp. Loss of sensation and hair loss (**alopecia**) are also characteristic. Neural involvement includes thickening of peripheral nerves, their tenderness, and subsequent loss of function. The TT form can either resolve itself spontaneously after several years, or develop to borderline BT and, very rarely, can progress to the more aggressive LL form.

LL and, often, BL are characterized by more than five skin lesions and considerable numbers of bacilli in smears and are collectively called multibacillary (MB) leprosy. Patients with LL have small diffuse and symmetric macular lesions. Lepromatous infiltrations are called lepromas or plaques and lead to the development of leonine faces. The loss of sensation and nerve loss progress slowly. Bone and cartilage damage is common as well as alopecia of eyebrows and eyelashes (**madarosis**). Testicular **atrophy** leads to sterility and **gynecomastia**.

Within each type of leprosy, a patient may remain in that stage, improve to a less debilitating form or worsen to a more debilitating form depending on their immune status (see Section 2). LL is the only form that never reverts to a less severe form.

CLINICAL PRESENTATION

The initial symptoms of leprosy are not demonstrative and often go unnoticed. Most often, patients start to lose sensation to heat and cold years before skin lesions occur (see Case details). The next sensation to be affected is light touch, then pain, and finally hard pressure. The most usual sites for numbness are hands and feet, and the first symptoms that patients report are usually burns and ulcers at the numb sites. However, the disease is often recognized through the development of dermal eruptions. At this stage, other symptoms appear such as damage to the anterior chambers of the eyes, chin, earlobes, knees or testes.

SKIN

The most common skin presentations are lesions in the form of **macules** or plaques, more rarely **papules** and nodules. The TT form is characterized by few hypopigmented lesions with lost or reduced sensation. In patients with LL, there are usually many lesions, sometimes confluent, with not so apparent loss of sensation. Alopecia is common in LL patients and eyebrows and eyelashes are affected. The scalp remains intact.

NERVES

Peripheral nerve damage occurs characteristically near the surface of the skin, particularly in the neck, elbow, wrist, neck of the fibula, and the medial malleolus (posterior tibial nerve). The latter is the most affected, followed by ulnar, median, lateral popliteal, and facial nerves. Involvement of small dermal nerves leads to **hypoesthesia** and **anhidrosis** in TT patients and "glove and stocking" loss of sensation in the case of LL. In the final stages of nerve injury segmental demyelination is quite frequent.

EYES

Both nerve damage and bacterial invasion can lead to blindness in up to 3% and potential blindness in almost 11% of MB patients (mostly the LL form). Patients suffer from loss of corneal sensation and dryness in the cornea and conjuctiva due to the reduction in blinking. This increases the risk of microtrauma and ulceration of the cornea. **Photophobia** and **glaucoma** can also develop.

SYSTEMIC

Systemic damage is mostly associated with the LL form of the disease and is a result of bacillary infiltration. Particularly

important is testicular atrophy, followed by **azoospermia** and gynecomastia. Adequate treatment significantly decreases the risk of the development of renal damage and **amyloidosis**. However, **lymphadenopathy** and **hepatomegaly** can develop in LL patients. Joint invasions by the microorganism may lead to aseptic **necrosis** and **osteomyelitis**.

4. HOW IS THE DISEASE DIAGNOSED, AND WHAT IS THE DIFFERENTIAL DIAGNOSIS?

Diagnosis of leprosy is mainly clinical, and a panel of clinical criteria has been developed that gives an average reliability of 91–97% for MB and 76–86% for PB leprosy. The panel consists of three major criteria, as follows:

1. Hypopigmentation or reddish patches of the skin with the loss of sensation. The latter can be tested by touching the skin (tactile test) with a wisp of cotton, while the patient's eyes are closed. Test tubes with hot and cold water may be used to test temperature perception.

2. Detection of acid-fast bacilli in biopsies or in skin smears.

3. Thickening of peripheral nerves. Enlarged nerves, especially in the areas where a superficial nerve may be involved, are examined by palpation. Attention should be paid for wincing during nerve palpation, indicating pain.

A skin biopsy for testing the presence of inflamed nerves is currently the standard criterion for diagnosis. Hematoxylin and eosin sectioning is used. In T2R, the polymorphonuclear leukocytes are characteristic, while in Lucio's phenomena lesions, fibrin thrombi can be found. In addition, biopsies are examined for the presence of acid-fast bacilli and the morphologic index (MI) is calculated that represents the number of viable bacilli per 100 bacilli. Histologic diagnosis is used as the gold standard.

Tissue/skin smear tests using the Ziehl-Neelsen staining method (see Figure 23.2) or Fite modification allow the visualization of bacilli under the microscope and calculation of the bacterial index (BI). However, this method is relatively insensitive and smears from almost 70% of patients are found to be negative.

Two tests can be used to diagnose nerve injury.

1. In the histamine test, a drop of histamine diphosphate is placed on non-affected and affected skin. It normally causes the formation of a wheal on normal skin, but not where the peripheral nerves have been damaged. However, this test is not very clear in dark-skinned patients.

2. In the methacholine sweat test, an intracutaneous injection of methacholine results in sweating in normal skin but no sweating in lepromatous lesions.

Establishment of neural inflammation histologically differentiates leprosy from other granulomatous diseases.

Serologic tests detect antibodies to *M. leprae*. Usually high titers of antibodies indicate untreated LL. Antibodies to PGL-1 are present in 90% of LL patients, but only in 40–50% of TT cases. They are also detected in 1–5% of healthy individuals.

Polymerase chain reaction (**PCR**) assay on biopsy material for *M. leprae* specific genes or repeat sequences has been able to confirm 95% of MB and 55% of PB cases. PCR is more applicable as a detector rather than an identifier of the disease with a sensitivity of more than 90% and specificity of 100%. However, in cases with TT, a sensitivity dropped to 34% and specificity – to 80%.

The **Lepromin test** allows the assessment of delayed hypersensitivity to antigens of *M. leprae* or other mycobacteria that cross-react with *M. leprae* by injection of a standardized extract of inactivated bacteria under the skin. A positive test indicates cell-mediated immunity associated with TT. This test is only of value in the classification of leprosy. It should not be used to establish a diagnosis of leprosy.

Early diagnosis is often prevented by difficulties in reaching some affected rural communities in developing countries. Delays in the diagnosis and treatment lead to an increase in leprosy-related disabilities. Overcoming the stigma of leprosy in certain countries and public education about this disease are important. Effective therapy further reduces the stigma attached.

5. HOW IS THE DISEASE MANAGED AND PREVENTED?

Leprosy control includes early detection of the disease, adequate treatment of patients, effective care for the prevention of disabilities, and patient rehabilitation.

MANAGEMENT

All forms and stages of newly diagnosed leprosy are normally treated with rifampicin, clofazimine, minocycline, ofloxacin, and dapsone as a part of a standard MDT recommended by the WHO. MDT was introduced for leprosy control as a result of a resistance to monotherapy with dapsone (diaminodiphenylsulfone). However, the doses and combinations of these antibiotics differ for PB and MB leprosy.

1. For the one lesion PB leprosy, a single shot of rifampicin 600 mg, ofloxacin 400 mg, and minocycline 100 mg (also known as ROM) is used, which is both acceptable and cost-effective.

2. For PB leprosy with more than one lesion, a 6-month treatment includes: daily self-administered doses of dapsone 100 mg and a monthly supervised administration of rifampicin 600 mg.

3. For MB leprosy, the treatment is prolonged over 24 months and consists of: daily self-administered combinations of dapsone 100 mg and clofazimine 50 mg, together with monthly supervised administration of rifampicin 600 mg and clofazimine 300 mg.

In 2018, the WHO issued new guidelines and introduced uniform MDT for administering three drugs: rifampin, dapsone, and clofazimine, to all patients irrespective of the classified type. The prior regimens are still used with the addition of daily clofazimine for the PB disease. No changes were introduced to the treatment of MB cases. Using the uniform MDT treatment decreases the chances of MB disease misclassified as PB.

Other potential drugs include minocycline, ofloxacin, moxifloxacin, levofloxacin, and clarithromycin.

A study on more than 1900 patients in 2009–2015 demonstrated rising bacterial resistance to some drugs: 3.8% to rifampin, 5.3% to dapsone, and 1.3% to ofloxacin.

Treatment of Inflammatory Responses

Severe neuritis may need immediate treatment with corticosteroids to minimize irreversible nerve damage: prednisone, 40 to 60 mg/day, is recommended. Treatment with elevated doses for extended time is preferable for reducing pain and inflammation as compared to smaller and shorter doses of corticosteroids. In patients with diabetes, methotrexate can be substituted for a steroid regimen.

T1R can be treated with cyclosporine in patients unresponsive to corticosteroids, although toxicity has to be taken into account.

In mild cases of acute T2R, clofazimine can be used in increasing doses of approximately 300 mg daily for one month with a gradual decrease to 100 mg/day in one year.

Thalidomide is a powerful drug for the treatment of T2R and especially ENL, but it should not be used in pregnancy. The initial dose of 300–400 mg daily should be soon lowered to a maintenance concentration of around 100 mg daily, every couple of months. Neuropathy advancement is an indication of an immediate withholding of treatment.

PREVENTION

Vaccines for Leprosy

Because of the dormant nature of *M. leprae*, the long incubation period, effective treatment with MDT, low prevalence of leprosy, and insufficient funding, vaccine development has not been a priority. However, both, BCG and *Mycobacterium indicus pranii* (MIP) have a potential preventive role against leprosy and tuberculosis and are generally well tolerated. The key challenge is the development of a vaccine that prevents *M. leprae*-associated neuropathy.

A major recent advance has been the identification of potent vaccine adjuvants capable of stimulating Th-1 responses by macrophages and DCs. These include formulated TLR ligands (TLRL). The vaccine prototype LepVax (containing LEP-F1, a fusion protein mixed with a cocktail of ML2055, ML2380, and ML2028 antigens, in conjunction with glucopyranosyl lipid adjuvant in stable emulsion GLA-SE) was successfully tested in armadillos. It appeared to be safe and effective, reducing sensory nerve damage and delaying motor nerve damage in animals, and is now progressing to stage 1 clinical trials.

Prevention of Disabilities and Rehabilitation

Twenty-five percent of leprosy patients, particularly with LL and BL forms, have some degree of disability. It increases with age and is more severe in males. The gravity of disability correlates with the duration of the disease. Hands are most frequently affected followed by the feet and eyes.

There are currently around 3 million people with physical deformities and disabilities caused by leprosy.

The socioeconomic impact of leprosy continues to be a burden in endemic countries. Rehabilitation strategies for leprosy should contain physical, psychological, social, and economic aspects. Up to now, most leprosy control programs have been focused on physical rehabilitation only. Socioeconomic rehabilitation has been introduced by some local governments, nongovernmental organizations (NGOs), and community-based rehabilitation (CBR) programs focused on the patient's self-awareness and combating the stigma attached to the disease. These are specifically adapted for the local needs and differ among countries and from area to area.

SUMMARY

1. WHAT IS THE CAUSATIVE AGENT, HOW DOES IT ENTER THE BODY, AND HOW DOES IT SPREAD A) WITHIN THE BODY AND B) FROM PERSON TO PERSON?

- For many years, *Mycobacterium leprae* was considered as a sole pathogen causing leprosy. However, in 2008, *Mycobacterium lepromatosis* was identified as second causative agent of Hansen's disease (HD). *M. lepromatosis* has been implicated in a small number of HD cases, and clinical aspects of the disease caused by *M. lepromatosis* are not yet clearly defined. *M. lepromatosis* was initially associated with diffuse lepromatous leprosy, but subsequent studies have linked it to other forms of HD.

- *M. leprae* and *M. lepromatosis* comprise a so-called "*Mycobacterium leprae* complex".

- *Mycobacterium leprae* is the most frequent causative agent of leprosy. It is an obligate intracellular gram-positive bacillus that requires the environment of the host macrophage for survival and propagation by binary fission.

- It shows preferential tropism toward macrophages and Schwann cells that surround the axons of nerve cells.

- *M. leprae* is a straight or curved rod-shaped organism, 1–8 µm long and approximately 0.3 µm in diameter.

- An intracellular parasite, *M. leprae* cannot be cultured by conventional laboratory methods and can be propagated in mouse footpad or in its natural host – the nine-banded armadillo.

- The genome of *M. leprae*, contains less than 50% of the functional genes of *M. tuberculosis*. These lost genes coding for the metabolic pathways have been replaced in *M. leprae* by pseudogenes or inactivated genes, and it is even more dependent on host metabolism for survival than *M. tuberculosis*.

- The cell wall of *M. leprae* is similar to that of *M. tuberculosis*, but its cell wall includes the species-specific phenolic glycolipid I (PGL-I) used for serodiagnosis of lepromatous leprosy and disease pathogenesis and genus-specific lipoarabinomannan (LAM).

- *M. leprae* enters the body via nasal epithelial cells or poorly differentiated keratinocytes, fibroblasts, and endothelial cells of the skin through minor injuries. Tropism for these cells is likely to be mediated via mammalian cell entry protein 1A (Mce1A). In the respiratory system, the microorganism is taken up by **alveolar macrophages**.

- *M. leprae* binds to the G-domain of α2-chain of laminin 2 in the basal lamina of Schwann cells. Uptake of *M. leprae* by the Schwann cells is facilitated by α-dystroglycan – a receptor for laminin on cell membrane.

- Optimal temperature for growth for *M. leprae* is low (30–35°C), so it prefers to grow in the cooler parts of the body: skin and superficial nerves. Natural hosts of *M. leprae* – armadillos – are characterized by low body temperature.

- Asymptomatic infected individuals may transiently excrete the microbe nasally and spread the infection among close proximal contacts.

- In 1997, two million people worldwide were infected with *M. leprae*, with around 800 000 new cases of leprosy detected per year.

2. WHAT IS THE HOST RESPONSE TO THE INFECTION AND WHAT IS THE DISEASE PATHOGENESIS?

- A strong T-cell-mediated immunity, combined with a weak antibody production, leads to the development of a mild clinical form of tuberculoid leprosy (type I or reversal) with few nerves involved and low bacterial load.

- In the tuberculoid form of leprosy, the most damage is inflicted by chronic inflammation caused by immune responses.

- In lepromatous (type II) leprosy high titers of antibodies in combination with unappreciable cellular immune response result in widespread skin lesions, extensive nerve involvement, and a high bacterial load.

- In LL, systemic inflammatory response to the deposition of extravascular immune complexes leads to erythema nodosum leprosum (ENL) resulting in neutrophil infiltration and complement activation.

- The borderline forms of leprosy are characterized by an increasing reduction of cellular immune responses associated with increasing bacillary load.

- The clinical form of the disease is affected by the balance of Th-1/Th-2 cytokines that favor T-cell- or B-cell-(antibody) mediated immune responses, respectively.

Continued...

...continued

3. WHAT IS THE TYPICAL CLINICAL PRESENTATION AND WHAT COMPLICATIONS CAN OCCUR?

- The forms of leprosy distinguished by the Ridley-Jopling classification are: indeterminate leprosy (IL), tuberculoid leprosy (TT), borderline forms (BT, BB, and BL) and lepromatous leprosy (LL).
- IL and TT are characterized by fewer than five lesions, no bacilli on smear tests, and are called paucibacillary (PB) disease.
- LL and, often, BL are characterized by more than five lesions with bacilli and are called a multibacillary (MB) disease.
- *M. leprae* is not very pathogenic and most infections do not result in leprosy. Some early symptoms such as skin lesions are self-limiting and can heal spontaneously.
- The initial symptoms of leprosy are the loss of sensation of temperature, followed by the loss of the sensation of a light touch, then pain, and finally hard pressure.
- The disease is mostly recognized through the development of dermal eruptions such as lesions in the form of macules or plaques, more rarely papules and nodules.
- The TT form is characterized by few hypopigmented lesions with lost or low sensation. The LL form is characterized by many lesions, sometimes confluent, with less apparent loss of sensation.
- Peripheral nerve damage appears near the surface of the skin, particularly in the neck, elbow, wrist, neck of the fibula, and medial malleolus.
- Both nerve damage and bacterial invasion can lead to blindness in up to 3% and potential blindness in almost 11% of multibacillary patients.
- Systemic damage in the patients with the LL form is a result of bacillary infiltration and comprises testicular atrophy, azoospermia, and gynecomastia.

4. HOW IS THE DISEASE DIAGNOSED, AND WHAT IS THE DIFFERENTIAL DIAGNOSIS?

- The three main diagnostic criteria are: hypopigmentation or reddish patches of the skin with the loss of sensation; acid-fast bacilli in biopsies or in skin smears; thickening of peripheral nerves.
- Skin biopsy for testing of the presence of inflamed nerve is currently the standard criterion for diagnosis.
- Tissue/skin smear test using Ziehl-Neelsen staining method or Fite modification allows bacilli to be visualized under the microscope, but this method is not highly sensitive and smears from almost 70% of all patients are found to be negative.
- Serologic tests are used to detect antibodies to *M. leprae*. High titers of antibodies indicate untreated LL.
- To confirm diagnosis, automated PCR- and RT-PCR-based techniques have been implemented in many reference laboratories in countries with endemic leprosy.
- Early diagnosis is often prevented by difficulties in reaching some affected rural communities in the developing countries.

5. HOW IS THE DISEASE MANAGED AND PREVENTED?

- Leprosy control includes early detection, adequate treatment, comprehensive and effective care for the prevention of disabilities, and rehabilitation.
- Newly diagnosed leprosy is treated with a combination of rifampicin, clofazimine, minocycline, ofloxacin, and dapsone as a part of a standard multidrug therapy (MDT) recommended by the WHO.
- For the one lesion, PB leprosy a single shot of rifampicin 600 mg, ofloxacin 400 mg, and minocycline 100 mg (also known as ROM) is used.
- For PB leprosy with more than one lesion a 6-month treatment includes: a daily self-administered dose of dapsone 100 mg and a monthly supervised administration of rifampicin 600 mg.
- For MB leprosy, the treatment is prolonged to 24 months and consists of: daily self-administered combination of dapsone 100 mg and clofazimine 50 mg together with monthly supervised administration of rifampicin 600 mg and clofazimine 300 mg.
- BCG administration, together with killed *M. leprae*, leads to a shift in immune response from multibacillary to paucibacillary leprosy, which is beneficial for a patient.
- The development of BCG-based and recombinant vaccines is currently under way.
- Rehabilitation strategies for leprosy should contain physical, psychological, social and economic aspects.

FURTHER READING

Britton WJ. Leprosy. In: Cohen J, Powerly WG, editors. Infectious Diseases, 2nd edition. Mosby, London: 1507–1513, 2004.

Cogen AL, Lebas E, De Barros B, et al. Biologics in Leprosy: A Systematic Review and Case Report. Am J Trop Med Hyg, 102: 1131–1136, 2020.

REFERENCES

Barker L. Mycobacterium leprae Interactions with the Host Cell: Recent Advances. Indian J Med Res, 123: 748–759, 2006.

Brosch R, Gordon SV, Eiglmeier K, et al. Comparative Genomics of the Leprosy and Tubercle bacilli. Res Microbiol, 151: 135–142, 2000.

Cambau E, Saunderson P, Matsuoka M, et al. Antimicrobial Resistance in Leprosy: Results of the First Prospective Opensurvey Conducted by a WHO Surveillance Network for the Period 2009-15. Clin Microbiol Infect, 24: 1305–1310, 2018.

Demangel C, Britton WJ. Interaction of Dendritic Cells with Mycobacteria: Where the Action Starts. Immunol Cell Biol, 78: 318–324, 2000.

Deps P. Collin SM. Mycobacterium lepromatosis as a Second Agent of Hansen's Disease. Front Microbiol, 12: 698588, 2021.

Ebenezer GJ, Scollard DM. Treatment and Evaluation Advances in Leprosy Neuropathy. Neurotherapeutics, 12: 698588, 2021.

Eiglmeier K, Simon S, Garnier T, Cole ST. The Integrated Genome Map of Mycobacterium leprae. Lepr Rev, 72: 387–398, 2001.

Gebre S, Saunderson P, Messele T, Byass P. The Effect of HIV Status on the Clinical Picture of Leprosy: A Prospective Study in Ethiopia. Lepr Rev, 71: 338–343, 2000.

Haanpaa M, Lockwood DN, Hietaharju A. Neuropathic Pain in Leprosy. Lepr Rev, 75: 7–18, 2004.

van Hooij A, Geluk A. In Search of Biomarkers for Leprosy by Unraveling Host Immune Response to Mycobacterium leprae. Immunol Rev, 301: 175–192, 2021.

Lockwood DN, Suneetha S. Leprosy: Too Complex a Disease for a Simple Elimination Paradigm. Bull WHO, 83: 230–235, 2005.

Pereira de Silva T, Bittercourt TL, Leite de Oliveira AL, et al. Macrophage Polarization in Leprosy–HIV Co-Infected Patients. Front Immunol, 11: 1493, 2020.

Ridley DS, Jopling WH. Classification of Leprosy According to Immunity. A Five-Group System. Int J Lepr Other Mycobact Dis, 34: 255–273, 1966.

Scollard DM, Adams LB, Gillis TP, et al. The Continuing Challenges of Leprosy. Clin Microbiol Rev, 19: 338–381, 2006.

You EY, Kang TJ, Kim SK, et al. Mutations in Genes Related to Drug Resistance in Mycobacterium leprae Isolates from Leprosy Patients in Korea. J Infect, 50: 6–11, 2005.

WEBSITES

American Leprosy Missions: https://www.leprosy.org/

Lepra Society, Health in Action, Andhra Pradesh, India: http://www.leprasociety.org

World Health Organization, Leprosy (Hansen's disease) 2008: https://www.who.int/health-topics/leprosy

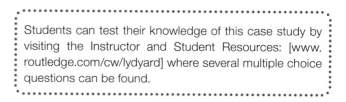

Students can test their knowledge of this case study by visiting the Instructor and Student Resources: [www.routledge.com/cw/lydyard] where several multiple choice questions can be found.

Mycobacterium tuberculosis

24

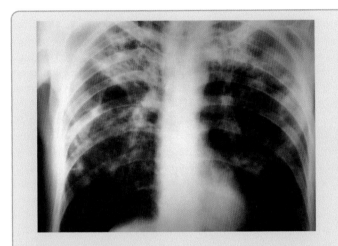

Figure 24.1 Chest X-ray of patient showing typical apical consolidation with possible cavities. There is also some consolidation adjacent to the mediastinum on the right. The bases are relatively spared. The medial aspect of the right hemi-diaphragm is just visible.

A 63-year-old man lived in a hostel for the homeless and sold magazines outside a railway station. He had been finding it difficult to cope with this recently, as he had been feeling weak, had lost weight, and often had a fever at night.

One month previously, he started coughing up blood and feeling breathless, which had really worried him. He was not registered with a primary healthcare provider but a friend told him about a walk-in practice for homeless people. Next day, he went to the practice and was seen by the physician on duty, who found that the patient had a low-grade fever and detected bronchial breathing when he listened to his chest. The doctor sent him for a chest X-ray and asked him to return for the results. When the X-ray result came back, it showed that he had apical shadowing and large **cavitation** consistent with tuberculosis (TB).

The X-ray is shown in Figure 24.1. A sputum sample was taken since the doctor suspected that the patient had TB and the patient was started an anti- tuberculosis therapy.

1. WHAT IS THE CAUSATIVE AGENT, HOW DOES IT ENTER THE BODY AND HOW DOES IT SPREAD A) WITHIN THE BODY AND B) FROM PERSON TO PERSON?

CAUSATIVE AGENT

This patient is infected with *Mycobacterium tuberculosis*. The stain used to identify mycobacteria is called the Ziehl-Neelsen (ZN) stain and characteristically stains mycobacteria red while all other organisms stain green (Figure 24.2). Culture and molecular tests are also used to identify Mtb and its antibiotic resistance (see diagnosis).

In common with Gram-positive bacteria, the cell wall contains a thick layer of peptidoglycan. Overlying this is a layer of arabinogalactan, which is covalently linked to the outer layer composed of mycolic acid, long-chain fatty acids specific for the mycobacterial genus, with other components such as glycophospholipids and trehalose dimycolate (also called cord factor as on staining the organism, it has the appearance of cords). Running vertically through the whole of the cell wall and linked to the cytoplasmic membrane is lipoarabinomannan – LAM (Figure 24.3).

The genus *Mycobacterium* can broadly be divided into rapid and slow growers and non-cultivable species. *Mycobacterium leprae*, the causative agent of leprosy and a close relative of the tubercle bacillus, cannot be grown on artificial media and can only be propagated in armadillos. Rapid-growing mycobacteria produce colonies within 2–3 days, for example *M. smegmatis*, while slow growers take more than 7 days. *M. tuberculosis*, a slow grower, can take 2–4 weeks to produce

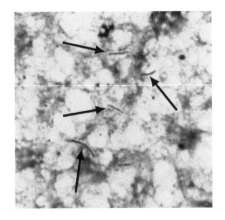

Figure 24.2 Ziehl-Neelsen stain of sputum: note the red bacilli (arrowed) against a green background.

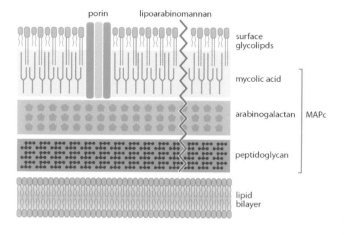

Figure 24.3 Model of the structure of the cell wall of mycobacteria. Note the mycolate-arabinogalactan-peptidoglycan-complex (MAPc). The mycobacterial cell wall is highly hydrophobic. The mycolic acid layer is impervious to many substances necessitating the presence of porin channels to allow entry of hydrophilic compounds. This layer plays a major role in the defence of the cell because few antibodies can penetrate it and it is relatively resistant to desiccation and some disinfectants.

colonies. This means that clinical decisions affecting treatment of TB and leprosy and the diagnosis of these conditions do not now rely primarily on culture. The slow growth also raises problems in determining the actual species causing illness (as the treatment may vary depending on the causative agent) and in determining antibiotic resistance. The usual medium for the isolation and growth of mycobacteria is Lowenstein-Jensen, which contains egg yolk and a dye (malachite green) that inhibits growth of more rapidly growing bacteria (**Figure 24.4**). There are also newer and faster methods of diagnosis used (see Section 4).

ENTRY AND SPREAD WITHIN THE BODY

Healthy individuals can be infected via the respiratory tract, the digestive tract, damaged skin, and mucous membranes. In primary infection, *M. tuberculosis* usually enters the respiratory tract in aerosol droplets, with 1–5 bacilli probably sufficient to cause infection. However, first they have to "run the gauntlet" while passing down the bronchus and bronchioles where mucosal epithelium produces anti-bacterial peptides (see Section 2). Those that make it to the alveoli interact with and enter alveolar macrophages. Entry into alveolar macrophages is mediated through a variety of surface receptors expressed by these phagocytic cells. Although the organism can multiply extracellularly to some extent within the alveolus, the organism is able to survive and multiply within the macrophages (due to mechanisms that prevent killing within the **phagosome** – see below). At this stage, macrophages die by programmed cell death (**apoptosis**) and release the mycobacteria. **Dendritic cells** also take up mycobacteria and become activated, which induces their migration to draining lymph nodes where they prime/activate T cells. Activated T

cells recognizing mycobacterial antigens migrate into the lung and induce formation of small **granulomas** (see Section 2).

The site of infection in the lungs tends to be at the lung apex and close to the pleura. After a maximum of 3 weeks, and usually before cell-mediated immunity develops to any great extent, the microorganisms are released from macrophages and spread via the bloodstream to draining regional lymph nodes such as hilar or mediastinal. In addition, the bacilli spread to every organ in the body principally the lung apices, **meninges**, kidneys, and bones. Macrophages can also carry viable microorganisms around the body and how much of the spread occurs via this mechanism is unclear. Spread via the pulmonary arteries can give rise to **miliary** TB of the lungs which is, however, primarily found in immunocompromised patients and small children. Miliary TB is so named because the many tiny foci that form in the lungs resemble millet seed. Entry of the organism into the body via the mouth can cause laryngeal TB or intestinal TB.

It should be emphasized that primary infection with *M. tuberculosis* leads to **active** disease in only a small number of individuals (5–10%). Thus, most individuals are able to control the initial infection, showing either no symptoms or mild clinical manifestations similar to those seen for a common cold. However, most infected individuals carry the organism in a **latent** state for life under the control of an effective immune system (see below) although Mtb has many ways of avoiding the immune system (see Section 2). In the latent state, they are unable to transmit the infection but some individuals, estimated to be 10% or so in the US, develop active disease (reactivation) many years after primary infection. Reactivation TB often occurs when persons become immunosuppressed through HIV infection, diabetes or other conditions affecting the immune system. It is estimated that 20% of the world population have latent TB infection (LTBI) although the percentage varies from country to country.

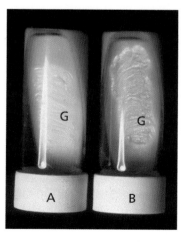

Figure 24.4 Growth of mycobacteria (G) on Lowenstein Jensen medium. The growth of *M. tuberculosis* does not produce a pigment (A) but the growth of another mycobacterial species produces a yellow pigment (B).

PERSON-TO-PERSON SPREAD

In patients with active TB, *M. tuberculosis* bacilli are released from granulomas into the bronchi and are spread through coughing. The aerosols produced contain droplet nuclei and survive for quite long periods of time outside the body. It is estimated that each infected person infects, on average, 20 other individuals. Repeated contact with an infected individual, particularly in a closed environment, produces higher transmission rates than casual contact. Similarly, if the infected person is smear-positive (mycobacteria seen in the sputum by ZN stain), indicating that bacteria are present in large numbers in the airways they are much more contagious, that is 50% of contacts may become infected. Whereas, if the index case is smear-negative (i.e. mycobacteria not seen by ZN stain of sputum) but is culture-positive, then only about 5% of contacts will be infected. The number of times an index case coughs is also directly related to the transmission rate.

EPIDEMIOLOGY

TB has been a human pathogen for thousands of years and, in 2019, was the eighth of the top ten causes of death in low-income countries. It was the seventh in middle-income countries and did not figure in the top ten of high-income countries.

It is estimated that about one quarter of the world's population (2 billion) is infected with *M. tuberculosis* and 5–10% of them will develop disease in their lifetime.

A total of 10 million people fell ill and 1.4 million died because of TB in 2019 worldwide. There were 5.6 million men, 1.2 million women and 1.2 million children. Eight countries accounted for two thirds of the total new TB cases in 2019, with India leading the count, followed by Indonesia, China, the Philippines, Pakistan, Nigeria, Bangladesh, and South Africa. The incidence of TB for the above-mentioned countries

and several other countries for the purpose of comparison is shown in Table 24.1.

At-risk groups for development of active disease in the population include prison inmates, the homeless, alcoholics, intravenous (IV) drug users, and those who are suffering social deprivation.

The good news is that the global incidence of TB fell at about 2% per year and between 2015 and 2019, the cumulative reduction was 9%. However, multidrug-resistant TB (MDR-TB) has increased and remains a public health crisis and a health security threat (see Section 5).

M. TUBERCULOSIS AND CO-MORBIDITIES

HIV

People living with HIV are on average 18 (15–21) times more likely to develop active TB disease than people without HIV. Probably, this is due to the immunosuppressive properties of HIV. The average prevalence of HIV in patients testing positive for TB in 2019 in the top eight high-burden countries was 7.8%. In 2019, around 208 000 people died of HIV-associated TB.

SARS-CoV-2 (COVID-19)

There are little data available at this time, but it is predicted that the pandemic of COVID-19, which was made official in March 2020, will have an impact on both detection of *M. tuberculosis* and clinical outcome of TB patients infected with COVID-19.

Due to lockdowns in many countries, and reluctance of individuals to attend clinics and due, in part, to GPs not having face-to-face meetings but relying on virtual appointments, the detection of the disease is likely to have been reduced. There is already evidence from several high TB-burden countries of large reductions in the monthly number of people with TB being detected and officially reported in 2020, especially in India, Indonesia, the Philippines, Sierra Leone, and South

Table 24.1 Estimated incidence of TB in 2019

Country	Actual cases	Rate/100 000 population
India	2 640 000	193
Indonesia	845 000	312
China	833 000	58
Philippines	599 000	554
Pakistan	570 000	263
Nigeria	440 000	219
Bangladesh	361 000	221
South Africa	360 000	615
Vietnam	170 000	176
Ethiopia	157 000	140
Russian Federation	73 000	50
Congo	20 000	373
UK	5132	7.7
USA	8920	2.7

Data from WHO Global Tuberculosis Report – 2020 Licence: CC BY-NC-SA 3.0 IGO.

Africa. The longer TB cases remain undetected, the greater the mortality risks.

Since both TB and Covid-19 are respiratory diseases, co-infection is likely to produce poorer treatment outcomes. This is also likely to be exacerbated by patients becoming infected with Covid-19 in countries where there is a high frequency of co-infections with TB and HIV.

Other co-morbidities affecting the risk of TB are diabetes, smoking, chronic obstructive pulmonary disease (COPD), and alcohol abuse.

2. WHAT IS THE HOST RESPONSE TO THE INFECTION AND WHAT IS THE DISEASE PATHOGENESIS?

INNATE IMMUNITY

On entry into the respiratory tract, the mycobacteria in small droplets are exposed to the epithelial cells in the trachea, bronchus and bronchioles that through pattern recognition receptors (PRRs), including Toll-like receptors (TLRs), produce a variety of antibacterial substances, for example defensins. Mucus produced by goblet cells contains defensins, together with some antibodies (especially IgA), lysozyme and cytokines. Defensins can disrupt the mycobacterial envelope. The mucociliary escalator can eliminate around 90% of inhaled microbes. If mycobacteria make it to the lower respiratory tract (the alveoli) they are then confronted with alveolar macrophages. These have a variety of receptors including complement receptors, scavenger receptors, Fcγ receptors, and PRRs that allow them to sense the mycobacteria and induce pro-inflammatory cytokines. PRRs include the TLRs, 2, 4, and 9 as well as others. Thus, recognition by the macrophage PRRs of the microbial-associated molecular patterns (MAMPs) on the mycobacteria leads to macrophage activation resulting in the production of the cytokines TNFα and IL6, and the chemokine CXCL8. This initiates an inflammatory response, which induces the recruitment of macrophages/monocytes from the circulation. Other important outcomes of macrophage activation include up-regulation of numerous antimicrobial effectors (NADPH oxidase, which generates reactive oxygen species, and iNOS which generates reactive nitrogen species, NRAMP). TLRs therefore play a critical role in early responses to M. tuberculosis. In fact, TLR polymorphisms have been associated with increased susceptibility to tuberculosis among different populations. The alveolar secretion itself also contains antibodies and molecules of the innate immune system including antibacterial peptides including collectins and complement proteins that all play a role in antibacterial immunity. Normally, having entered macrophages through phagocytosis, bacteria are killed within the phagolysosome. However, mycobacteria can live and divide within macrophages. They do this by inhibiting maturation of their phagosomes to prevent fusion with lysosomes and phagolysosome formation. This is mediated by some of their cell-wall glycolipids such as the cord factor or trehalose dimycolate. Thus, they are not exposed to the bactericidal content of the lysosome. There are many different innate immune evasion mechanisms used by M. tuberculosis (see below). Damage to the phagosomes also leads to escape of the organisms into the cytosol. The host may also respond by inducing a mechanism called "autophagy, which is not only the recycling system of the host cell but also a mechanism to target intracellular microorganisms to lysosomes. Neutrophils are also attracted to activated alveolar macrophages and are thought to provide antimicrobial proteins. NK cells recognize M. tuberculosis organisms and are thought to control their growth through granulysin and perforin, in a contact-dependent manner and also by enhancing macrophage-mediated killing. Gamma/delta T cells (Tγδ) are also activated by mycobacterial components and, like NK cells, directly kill organisms and enhance intracellular killing by macrophages. Other cells are thought to be important in immune responses to M. tuberculosis. These include NKT cells, and invariant NKT cells (iNKT cells). In addition, innate lymphoid cells (ILCs1, 2, and 3) and lymphoid tissue inducer cells (LTi) are also involved in anti-M. tuberculosis immunity. From experimental animal models, ILC3s are believed to be important in early protection against M. tuberculosis.

ADAPTIVE IMMUNITY

An adaptive cell-mediated response does not occur until 2–3 weeks after infection following the migration of lung macrophages and dendritic cells containing whole, or pieces of M. tuberculosis to the draining pulmonary lymph nodes. The mycobacterial components are broken down by proteolytic enzymes to produce peptides that are then presented via class II human leukocyte antigens (HLA) to CD4+ T cells, which, in the presence of interleukin (IL)-12/IL-18 produced by macrophages and dendritic cells, will differentiate to Th-1 cells. These Th-1 cells produce the pro-inflammatory cytokines interferon-γ (IFN-γ) and tumour necrosis factor-α (TNF-α). Together with IFN-γ and IL-1, mainly produced by the macrophages themselves, these cytokines further activate the bactericidal effectors of macrophages such as defensins (e.g. cathelicidin), nitric oxide (NO) production and autophagy induction. Th-17 cells are also induced through the cytokines IL-1β, IL-6, and IL-23 produced by dendritic cells, macrophages, osteoblasts, or brain microglial cells, either in local lesions or in draining lymph nodes. Reduced levels of CD4+ T cells in HIV infection has been linked to the higher mycobacterial load and increased extrapulmonary dissemination in TB infection.

During mycobacterial infection, cytotoxic CD8+ T cells are induced through peptide antigens presented to them by class I HLA. These T cells produce IFN-γ but are also able to kill cells infected with mycobacteria by perforin and granzymes. Granulysin, released by cytotoxic T cells can also kill mycobacteria.

Although controversial, there is accumulating evidence to suggest that antibodies play a role in immune protection. Thus, it has been reported that antibodies to *M. tuberculosis* increase their uptake through opsonization (enhancement) by macrophages and inhibit the intracellular survival of the microbes. From animal models, there is also some evidence for a role for particular subsets of B cells playing a part in immunity to *M. tuberculosis*.

IMMUNE ESCAPE MECHANISMS OF M. TUBERCULOSIS

There are a large number of escape mechanisms used by Mtb allowing the organism to lie latent in the host and these are summarized in Table 24.2.

The immune response to *M. tuberculosis* through activation of pro-inflammatory responses (Th-1 cells and M1 polarized macrophages) is usually more damaging (immunopathology) than the bacterial infection itself. Early on in the infection, when the adaptive system becomes activated, small granulomas are formed in the lungs.

Table 24.2 Summary of the immune escape mechanisms used by *M. tuberculosis*

Inhibition of maturation of phagolysosomes
Inhibition of acidification of phagolysosome
Inhibition of oxygen stress
Inhibition of function of reactive oxygen and reactive nitrogen intermediates
Inhibition of apoptosis (strongly virulent Mtb)
Inhibition of autophagy
Formation of granulomas (see below)

Granuloma Formation

In most cases, the organism will survive and persist within some macrophages due to the numerous immune escape strategies used by Mtb (see above). They do not proliferate but persistence leads to continuous activation of both CD4 and CD8+ T cells. The cytokines released by these T cells and macrophages lead to the development of granulomas termed "tubercles" which gave the disease its name. Granuloma formation is a way of limiting the spread of infection and tissue involvement (Figure 24.5) and is one mechanism of escape from the immune system. The inability to eliminate the organism completely results in the production of a connective tissue layer to "wall off" the organism from the rest of the body. This is the containment phase (**latent** infection with dormant mycobacteria) and is called LTBI (latent TB infection).

Histologically, a granuloma is a collection of activated macrophages called epithelioid cells and a center that frequently shows an area of tissue **necrosis**. In tuberculosis, the necrosis is characteristically "cheesy" and is called caseous necrosis. Sometimes, the macrophages fuse to form giant cells. Lymphocytes, particularly of the CD4 T-cell subset but

also CD8+ T cells, are also present in the granulomas and actively produce cytokines. They are formed by macrophages containing mycobacteria, epithelioid cells, and multinucleate giant cells all surrounded by T cells, which are mainly CD4+. Fibroblasts are also found in the outer layers, together with some dendritic cells and B lymphocytes.

The persistence of mycobacterial antigens in live and dead mycobacteria means that the T cells are continuously activated and the granulomatous response in tuberculosis is considered to be a **type IV hypersensitivity** response. This is a classic form of chronic inflammation through persistence

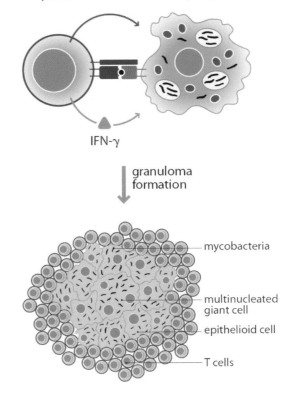

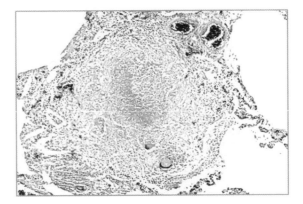

histology of a granuloma

Figure 24.5 Granuloma showing macrophages containing mycobacteria, epithelioid cells, and multinucleate giant cells all surrounded by T cells, which are mainly CD4+.

of the infectious agent. However, M2 macrophages (anti-inflammatory macrophages) were found to predominate in both necrotic and non-necrotic granulomas from tuberculosis patients, while both M1 and M2 polarized macrophages were found in the non-granulomatous lung tissues. M1 macrophages promote granuloma formation and macrophage bactericidal activity *in vitro*, while M2 macrophages inhibit these effects. This suggests that M2 macrophages help to modulate the immune mechanism and are important in down-regulating the chronic inflammation resulting in an equilibrium between the immune system and the organism that keeps live organisms "in check".

At any time during life, the disease can be reactivated, leading to a change in status of these granulomas. Some of the granulomas cavitate (decay into a structureless mass of cellular debris), rupture, and spill thousands of viable, infectious bacilli into the airways (if they are in the lung). This leads to the person becoming infectious, as described above. Reactivation can also take place within granulomas at other sites in the body leading to active disease, for example, in the brain causing **meningitis**.

The host response depends on the immune conditions and dose of microorganisms.

1. Strong T-cell immunity and high dosage: there is greater tissue damage and **caseation**.
2. Strong T-cell immunity and low dosage: granulomas are produced.
3. Weak T-cell immunity (e.g. HIV co-infection): there is a poor granulomatous response, and many microorganisms are produced.

Figure 24.6 shows the possible sequence of events following infection.

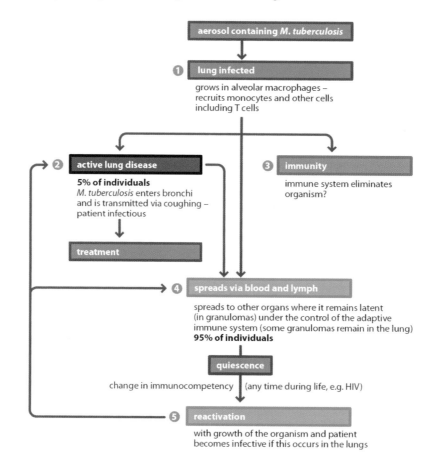

Figure 24.6 Sequence of possible events following tuberculosis infection. (1) In primary infections, *M. tuberculosis* organisms enter the lungs via aerosols and are taken up by alveolar macrophages and dendritic cells. The infection then moves to the lung parenchyma where monocytes and T cells are recruited through cytokines produced by the infected macrophages. The organism grows slowly within the resident and monocyte-derived macrophages to produce primary lesions or Ghon foci. These are small granulomas. Organisms from these foci can produce active disease in 5–35% of cases (2) and spread to other organs. In active disease, the foci become larger and result in significant tissue damage, release of the organism into the bronchi and transmission by aerosols (infectious stage). The patient develops symptoms at this stage and can infect other people. It is possible, but not proven, in some individuals that the organisms are totally eliminated from the body after primary infection (3). In primary infection, some microorganisms spread via the bloodstream and lymphatic system to other organs in the body where the immune system keeps the *M. tuberculosis* sequestered in granulomas (inactive) (4). At a later date, and probably as the result of a decrease in immunocompetence (e.g. HIV co-infection), reactivation of the growth of the *M. tuberculosis* within the granuloma occurs. This results in further spread and growth of the organism (5). Growth and active disease can develop in several organs other than the lung, including the pleura, brain (meningitis), lymph nodes, genitourinary tract, skin, joints, and bones. The rare infection of the spine (tuberculous spondylitis) is called Pott's Disease. If reactivation occurs in the lung, the individual becomes infectious.

3. WHAT IS THE TYPICAL CLINICAL PRESENTATION AND WHAT COMPLICATIONS CAN OCCUR?

The most common clinical presentation is fever, chronic productive cough that may be streaked with blood (hemoptysis), and weight loss. The release of T-cell and macrophage cytokines, particularly TNF-α, leads to fever (by its action on the thermoregulatory system of the hypothalamus) and weight loss. Adjacent granulomas in lymph nodes may fuse to produce a sizeable lump, which can be seen in a chest X-ray in the mediastinum or tissue destruction and cavities produced by dead tissue (cavitation, see Figure 24.1). The tissue destruction in the lungs resulting in cavitation can lead to loss of lung volume and erosion of bronchial arteries (cavitation, see Figure 24.1). This leads to hemoptysis. Spread of the organism through the body can lead to granulomas developing in other organs such as the brain, bone, liver, and so forth; perhaps the most common complication being the "space-occupying" effects of granulomas, for example, in the brain, where it can lead to **seizures**. Thus, TB can present with protean manifestations such as adrenal failure (**Addison's disease**) and fractures if it occurs in bone, for example, vertebral collapse (**Pott's disease**).

Several studies have been carried out that identify genes that associate with protection or predisposition to developing clinical TB, for example, alleles of HLADR (DRB1*1501 and 1502); alleles of NRAMP1 and alleles of IFN-γ. For more detailed information see Moller et al. (2018).

4. HOW IS THE DISEASE DIAGNOSED, AND WHAT IS THE DIFFERENTIAL DIAGNOSIS?

If pulmonary TB is suspected, then a chest X-ray is performed, and mycobacterial culture of sputum samples (Figure 24.4) is still considered the "gold standard" for TB diagnosis. The time to detection using culture for TB can be up to 2 months, significantly delaying the diagnosis of active TB in a large number of patients. However, culture methods that utilize both a liquid and a solid medium are now recommended and should allow detection within as little as 10–14 days. Although liquid systems are more sensitive and may increase the case yield by as much as 10%, they are also more prone to contamination. Although less sensitive, ZN-stained smear and microscopy is rapid and widely applied but it is only 50% sensitive at 5000–10 000 organisms per ml of specimen. However, Auramine which is a fluorescent stain can be used to rapidly scan a large area of a slide, resulting in a slightly higher sensitivity of 60%. Neither of these stains are specific for *M. tuberculosis* infection, they cannot differentiate between viable and dead bacteria, and are unable to detect drug-resistance. SybrGold is a DNA/RNA stain that has 90% sensitivity.

During the last 10 years or so, more rapid detection methods have been developed and endorsed by the WHO (consolidated guidelines on Tuberculosis, 2021). Nucleic acid amplification techniques (NAAT) using the **polymerase chain reaction (PCR)**, are now used, usually in conjunction with culture. NAAT amplify specific sequences of the genome of *M. tuberculosis*, thereby detecting its presence in the specimen and, if used with liquid culture, can speed up the detection of the organism greatly. This method is both highly specific and sensitive and is particularly useful in detecting the presence of multidrug-resistant TB, as drug resistance is correlated with characteristic mutations in specific genes that can be detected by PCR. Using this test, Mtb can be detected in less than 2 hours outside the conventional laboratory and requiring only minimal healthcare skills. NAAT can identify mutations in the *rpoB* gene which codes for resistance to rifamycin, one of the first-line drugs used to treat TB and because this often coincides with isoniazid resistance, can serve as a surrogate marker for MDR (see treatments). A second-generation NAAT platform now enables detection of second-line drug resistance, i.e. fluoroquinolone and aminoglycosides/capreomycin.

In some laboratories, whole genome sequencing (WGS) is used for identifying specific mutations that lead to antibiotic resistance. Some of the common genes with mutations are: for isoniazid – *KatG*, *inhA*, *ahpC*; for rifampicin – *rpoB*; pyrazinamide – *pncA*, *rpsA*; ethambutol – *embB*; aminoglycosides – *rrs*, *tlyA*; fluroquinolones – *GyrA*, *GyrB*; *PAS – thyA*, *folC*.

An alternative diagnostic available outside the US utilizes unprocessed urine to detect the LAM antigen (lipoarabinomannan) an outer mycobacterial cell wall component that is shed into, and cleared by, the kidney (see Figure 24.3). However, the currently available urinary LAM assays have suboptimal sensitivity and specificity and are therefore not suitable as diagnostic tests for TB in all populations. Also, unlike traditional diagnostic methods, urinary LAM assays demonstrate improved sensitivity for the diagnosis of TB among individuals co-infected with HIV.

Blood diagnostic TB tests are available in some laboratories, such as the assay for IFN-γ from T lymphocytes activated by mycobacterial antigen (IGRA-interferon gamma release assay) and an **Elispot test** assaying the T-cell responses to mycobacterial antigens e.g. early secretory antigen target-6, (ESAT 6).

Diagnosis of Latent Tuberculosis

Transcriptosome signatures are able to distinguish between patients with active TB and TBLI and those patients with LTBI who go on to active diseases. A blood diagnostic test can also be used to test for LTBI.

The Tuberculin Skin Test (TST)

A small amount of tuberculin, also known as purified protein derivative, is a combination of proteins from Mtb that is injected into the skin on the lower part of the arm (also called the Mantoux tuberculin skin test). A positive reaction for TB is indicated by a raised hard area or swelling as the result of a delayed-type hypersensitivity reaction after 48 to 72 hours. This indicates that the patient has been exposed to *M. tuberculosis* bacilli previously and could have LTBI but they will also require further testing for active disease.

DIFFERENTIAL DIAGNOSIS

It is important to note that the typical symptoms of weight loss, chronic cough, and fever may also occur in patients with tumors of the lung, for example, adenocarcinoma, squamous cell carcinoma, and oat cell carcinoma. Because *M. tuberculosis* spreads throughout the body and can present with signs and symptoms referrable to many systems, the infection can mimic many other diseases (e.g. brucellosis, lymphoma, glioma, meningitis).

Histologically, there are many other causes of granuloma that are infectious, for example, *Histoplasma* and *Bartonella* or non-infectious, for example berylliosis and silicosis and, in some cases, the antigen(s) is (are) unknown, for example sarcoidosis.

5. HOW IS THE DISEASE MANAGED AND PREVENTED?

MANAGEMENT

Infected patients in hospital should be isolated in a negative-pressure room where the air pressure outside the room is greater than that in the room. Thus, any airflow is into the room. Healthcare workers, such as physiotherapists, should wear close-fitting masks if they are involved in activities likely to induce coughing/expectoration by the patient. Once the patient has been on adequate treatment for 2 weeks, they can leave the isolation room.

The standard anti-TB regimen in the UK is a combination of isoniazid (with pyridoxine), rifampicin, pyrazinamide, and ethambutol for 2 months then isoniazid (with pyridoxine) and rifampicin for a further 4 months. This length of treatment, especially since patients may feel better before the end of treatment, may lead to lack of compliance.

A 4-month treatment for TB in which rifampicin and ethambutol substitute for rifapentine and moxifloxacin has been shown to have an equivalent success rate to the 6-month regimen.

TB is a notifiable disease in the UK. The contacts of patients with active TB should be screened by a chest X-ray or given prophylaxis, if appropriate. Close contacts of children with primary TB should be screened as there is likely to be a source that should be identified. Failure of therapy either due to inappropriate treatment or lack of compliance is important in development of drug-resistant strains. In the community, this can be controlled by giving directly observed therapy (DOT). This requires a healthcare infrastructure and appropriate finances.

TREATMENT OF LATENT TB

More than 80% of people who contract active TB disease in the US each year already have untreated LTBI. Treatment involves 1) isoniazid with rifapentine once weekly for three months, or 2) isoniazid with rifampin daily for three months. In addition, the WHO also recommend isoniazid daily for six or nine months as a primary regimen, whereas the CDC (USA) now lists these as alternative rather than preferred regimens. In contrast, rifampin daily for four months is the preferred regimen for the CDC (USA), but the WHO lists this as an alternative. The WHO guidelines also include two other regimens not included in the CDC guidelines: 1) isoniazid with rifapentine daily for one month, as an alternative regimen; and 2) for HIV-infected individuals in high TB transmission settings, isoniazid daily for 36 months.

MULTIDRUG-RESISTANT TB

Anti-TB medicines have been used for many decades and drug resistance emerges when anti-TB medicines are used inappropriately, through incorrect prescription by healthcare providers, poor quality drugs, and patients stopping treatment prematurely. TB strains resistant to one or more of the anti-TB drugs have been described in every country surveyed. In 2019, there was a global total of 206 030 people with multidrug (MDR) – or rifampicin-resistant (RR) TB (MDR/RR-TB), a 10% increase from 186 883 in 2018. About half of the global burden of MDR-TB is in three countries – India, China and the Russian Federation. There were 1.4 million incident cases of isoniazid-resistant TB, of which 1.1 million were susceptible to rifampicin.

MDR-TB is usually treatable and curable by using second-line drugs. However, second-line treatment options are limited, and require extensive drug treatment (up to 2 years of treatment) that involves expensive toxic drugs. In 2020, the WHO recommended a new shorter (9–11 months) fully-oral regimen for patients with MDB-TB. Research has shown that patients find it easier to complete the regimen, compared with the longer regimens that last up to 20 months.

Patients thought to have drug-resistant TB may be started on the above three drugs plus ethambutol until the actual sensitivity of the isolate is determined, when an appropriate combination of drugs can be given. Examples of second-line drugs include levofloxacin, moxifloxacin, bedaquiline, delamanid, linezolid, and pretomanid.

PREVENTION

Vaccination: BCG (Bacillus Calmette-Guerin) vaccine is an attenuated form of *M. bovis* (which causes TB in cattle) grown for many years on artificial medium. This is believed to provide variable protection against *M. tuberculosis* depending on the geographic location. It is generally 70–80% effective against the most severe forms of TB, such as TB meningitis, but less effective in preventing pulmonary TB. In most European countries, Canada, and the US, vaccination was abolished some years ago, due to their low TB-transmission status and potential interference of vaccination with skin-test diagnosis. In the UK, US, and some other countries, BCG vaccination is now only given to specific high-risk populations. Such populations include: healthcare workers (see European policies for TB prevention in healthcare workers in references), laboratory staff, military personnel, and children in high-risk communities and countries. In high-risk countries, BCG is given to neonates.

A number of new TB vaccines are being developed that include various platforms including whole-cell vaccines, adjuvanted proteins, and recombinant subunit vector vaccines. One candidate vaccine (M72/AS01E) proved to be significantly protective against TB disease in a Phase IIb trial conducted in Kenya, South Africa, and Zambia, in individuals with evidence of LTBI. The point estimate of vaccine efficacy was 50% (90% CI, 12–71), over approximately three years of follow-up.

SUMMARY

1. WHAT IS THE CAUSATIVE AGENT, HOW DOES IT ENTER THE BODY, AND HOW DOES IT SPREAD A) WITHIN THE BODY AND B) FROM PERSON TO PERSON?

- *Mycobacterium tuberculosis* is an acid-fast bacillus, which grows very slowly.
- Because the cell wall is very impervious, few antibiotics can penetrate, and it is relatively resistant to desiccation and some disinfectants.
- The route of transmission is aerial, by inhalation of infected droplet nuclei.
- *M. tuberculosis* organisms grow in alveolar macrophages, can be released into the bronchi, and are spread through coughing.
- Active disease occurs in only 5% of individuals following primary infection.
- Most infected individuals carry the organism in a latent state for life, as the result of the many immune avoidance mechanisms used by *M. tuberculosis*.
- Some infected individuals may develop active disease many years after primary infection (5–10%) if they become immunosuppressed (reactivation).
- About 25% of the world's population (2 billion) is infected with *M. tuberculosis*, with around 10 million individuals developing active disease annually and 1.4 million dying in 2019.
- The main foci of infection are in India, Indonesia, China, the Philippines, Pakistan, Nigeria, Bangladesh, and South Africa.
- People living with HIV are 18 (15–21) times more likely to develop active TB.

2. WHAT IS THE HOST RESPONSE TO THE INFECTION AND WHAT IS THE DISEASE PATHOGENESIS?

- Both innate and adaptive immunity play a role in host defense.
- *M. tuberculosis* has many immune escape mechanisms including: inhibition of maturation and acidification of phagolysosome, inhibition of oxygen stress and function of both reactive oxygen and nitrogen intermediates, inhibition of apoptosis and autophagy. This allows the mycobacteria to live within the macrophage.
- Local tissue destruction is caused by the host response to the presence of mycobacterial antigens.
- Immune cells are activated, releasing cytokines, which cause further damage.

- CD4+ T cells are activated to produce cytokines to help with elimination of the *M. tuberculosis* organisms from the macrophages. Inability to get rid of all the microorganisms results in cytokine production leading to granuloma formation. This host response limits the spread of mycobacteria. A granuloma is a characteristic histologic feature of chronic inflammation. It is a collection of activated and resting macrophages called epithelioid cells surrounding an area of necrosis.
- Reactivation of the microorganisms within granulomas can occur at any time during later life through decreased immunocompetency (e.g. co-infection with HIV).

3. WHAT IS THE TYPICAL CLINICAL PRESENTATION AND WHAT COMPLICATIONS CAN OCCUR?

- Tuberculosis commonly presents with fever, weight loss, chronic cough, and there may be hemoptysis.
- Many sites other than the respiratory tract can be affected and it may present with signs and symptoms referrable to other organ systems.
- *M. tuberculosis* may cause seizures (CNS), meningitis (CNS), fractures (bones – Pott's disease) or Addison's disease (adrenal gland).

4. HOW IS THE DISEASE DIAGNOSED, AND WHAT IS THE DIFFERENTIAL DIAGNOSIS?

- Diagnosis of active tuberculosis is initially a clinical decision based on an X-ray (for pulmonary TB) and treatment is started on that basis.
- The simplest laboratory procedure is the Ziehl-Neelsen (ZN) stain but it is not absolutely specific.
- Specimens are cultured on Lowenstein-Jensen medium (which can take up to two months for diagnosis) and culture remains the gold standard for accurate diagnosis of tuberculosis although the bacterium can now be detected within 10–14 days.
- Newer nucleic acid amplification techniques (NAAT) using the polymerase chain reaction (PCR) are now used to detect Mtb present in the specimen which is also more rapid than culture and useful for detecting multidrug-resistant TB. WGS is used in some laboratories for identifying multidrug resistance genes and transcriptosome signatures can distinguish between TB and TBLI.
- Serologic tests detect the LAM antigen and blood diagnostic tests, e.g. IGRA are now available in some laboratories.

Continued...

...continued

- Tuberculin skin tests are used to test for latent TB.
- Tuberculosis may mimic tumors of the lung, brain, bone, intestine, blood, and other systemic granulomatous infections, for example, brucellosis.

5. HOW IS THE DISEASE MANAGED AND PREVENTED?

- The standard anti-TB regimen is a combination of rifampicin plus isoniazid plus pyrazinamide for 2 months, followed by rifampicin and isoniazid for a further 4 months, so the total treatment time is 6 months.
- Some regimens include ethambutol in the initial phase.
- MDR-TB has increased 10% since 2018 and about half of the global burden of MDR-TB is in India, China and the Russian Federation.

- A fully oral regimen for patients for second-line treatment of MDR-TB has been recommended by the WHO and second-line drugs include levofloxacin, moxifloxacin, bedaquiline, delamanid, linezolid, and pretomanid.
- TB is a notifiable disease in the UK.
- Effectiveness of BCG vaccination is highly variable between geographic areas. It is thought to be 70–80% effective against the most severe forms of TB such as TB meningitis but less effective against pulmonary TB. It is given only to "at-risk" groups including healthcare workers, children in high-risk communities and countries, military personnel and laboratory staff dealing with TB.
- A number of new TB vaccines are being developed that include various platforms including whole-cell vaccines, adjuvanted proteins, and recombinant subunit vector vaccines.

FURTHER READING

Friedman LN, Dedicoat M, Davies PDO, (eds). Clinical Tuberculosis, 6th edition. CRC Press, Boca Raton, 2020.

Goering R, Dockrell HM, Zuckerman M, Chiodini PL, (eds). Mims' Medical Microbiology and Immunology, 6th edition. Academic Press/Elsevier, Cambridge, 2018.

Lydyard PM, Whelan A, Fanger MW. Instant Notes in Immunology, 3rd edition. Taylor and Francis, New York/London, 2011.

Murphy K, Weaver C. Janeway's Immunobiology, 9th edition. Garland Science, New York/London, 2016.

REFERENCES

Dockrell HM, Smith SG. What Have We Learnt About BCG Vaccination in the Last 20 Years? Front Immunol, 8: 1134, 2017.

Dyatlov AV, Apt AS, Linge IA. B Lymphocytes in Anti-Mycobacterial Immune Responses: Pathogenesis or Protection? Tuberculosis, 114: 1–8, 2019.

Eddabra R, Benhassou HA. Rapid Molecular Assays for Detection of Tuberculosis. Pneumonia, 10: 4, 2018.

Faridgohara M, Nikoueinejadb H. New Findings of Toll-Like Receptors Involved in Mycobacterium tuberculosis Infection. Pathog Glob Health, 111: 256, 2017.

Hoang LT, Jain P, Pillay TD, et al. Transcriptomic Signatures for Diagnosing Tuberculosis in Clinical Practice: A Prospective, Multicentre Cohort Study. Lancet Infect Dis, 21: 366–375, 2021.

Joosten SA, Ottenhoff THM, Lewinsohn DM, et al. Harnessing Donor Unrestricted T-Cells for New Vaccines Against Tuberculosis. Vaccine, 37: 3022, 2019.

Lee A, Xie YL, Barry CE, Chen RY. Current and Future Treatments for Tuberculosis. BMJ, 368: m216, 2020.

Marco B, Zotti CM. European Policies on Tuberculosis Prevention in Healthcare Workers: Which Role for BCG? A Systematic Review. Hum Vaccin Immunother, 12: 2753–2764, 2016.

Mayer-Barber KD, Barber DL. Innate and Adaptive Cellular Responses to Mycobacterium tuberculosis Infection. Cold Spring Harb Perspect Med, 5: a018424, 2015.

McQuaid CF, Vassall, A, Cohen T, et al. The Impact of Covid-19 on TB: A Review of the Data. Int J Tuberc Lung Dis, 6: 436, 2021.

Mehrotra P, Jamwal SV, Saquib N, et al. Pathogenicity of Mycobacterium tuberculosis is Expressed by Regulating Metabolic Thresholds of the Host Macrophage. PLoS Pathog, 10: e1004265, 2014.

Miggiano R, Rizzi M, Ferraris DM. Mycobacterium tuberculosis Pathogenesis, Infection Prevention and Treatment. Pathogens, 9: 385, 2020.

Moller M, Kinnear CJ, Orlova M, et al. Genetic Resistance to Mycobacterium tuberculosis Infection and Disease. Front Immunol, 9: 2219, 2018.

Park H-E, Lee W, Shin M-K, Shin SJ. Understanding the Reciprocal Interplay Between Antibiotics and Host Immune System: How Can We Improve the Anti-Mycobacterial Activity of Current Drugs to Better Control Tuberculosis? Front Immunol, 12: 703060, 2021.

Seifert M, Catanzaro D, Catanzaro A, Rodwell TC. Genetic Mutations Associated with Isoniazid Resistance in Mycobacterium tuberculosis: A Systematic Review. PLoS ONE, 10: e0119628, 2015.

Zhai W, Wu F, Zhang Y, et al. The Immune Escape Mechanisms of Mycobacterium tuberculosis. Int J Mol Sci, 20: 340, 2019.

WEBSITES

Centers for Disease Control and Prevention, Tuberculosis (TB), 2022: https://www.cdc.gov/tb/topic/treatment/tbdisease.htm

TB FACTS, Drug-Resistant TB, 2022: https://tbfacts.org/treatment-drug-resistant-tb/

Treatment of Multi-Drug resistant TB: https://tbfacts.org/treatment-drug-resistant-tb/

World Health Organization, Global Tuberculosis Programme, 1997: https://www.who.int/teams/global-tuberculosis-programme/tb-reports

World Health Organization, Tuberculosis: https://www.who.int/news-room/fact-sheets/detail/tuberculosis

Students can test their knowledge of this case study by visiting the Instructor and Student Resources: [www.routledge.com/cw/lydyard] where several multiple choice questions can be found.

Neisseria gonorrhoeae

25

A 15-year-old heterosexual male was brought to the emergency room by his sister. He gave a 24-hour history of **dysuria** and noted some "**pus**-like" drainage in his underwear and the tip of his penis (Figure 25.1). Urine appeared clear and urine culture was negative, although urinalysis was positive for leukocyte esterase and multiple white cells were seen on microscopic examination of urine. He gave a history of being sexually active with five or six partners in the previous 6 months. He claimed that he and his partners had not had any sexually transmitted diseases. His physical examination was significant for a yellow urethral discharge and tenderness at the tip of the penis. A gram stain of the discharge was performed in the emergency room (Figure 25.2). He was given antimicrobial agents and scheduled for a follow-up visit one week later. He was asked to provide the names and addresses of his sexual partners to the Health Department so that they could be examined and treated if necessary. One of the sexual partners, a 15-year-old female who reported having had sexual relations with her boyfriend frequently over the last 6 months, was asymptomatic until 2 days before coming to the hospital when she developed fever, shaking chills, and lower abdominal pain. Her symptoms worsened and she presented with fever of 42°C, generalized abdominal pain, and a swollen right knee, with blood pressure 120/80 and pulse 150/min and regular. The date of her last menstrual period, which was described as normal, was one week before admission. The patient was oriented as to time, person, and place. Physical examination was unremarkable except for tender abdomen and rigidity, and decreased bowel sounds; the right knee was red, hot, tender, and swollen. Pelvic examination showed some white discharge of the cervical os (Figure 25.3) A swab was obtained from her cervix for culture (Figure 25.4). Her right knee was tapped.

Laboratory findings:

- Hemoglobin: 12 g dl^{-1};
- white blood cell (WBC) count: 26000 μl^{-1};
- differential: 70 PMN, 5 bands, 25 lymphocytes;
- urinalysis: specific gravity, 1.010; protein, 2+; sugar, negative; WBC, 8–10 per HPF; no casts.

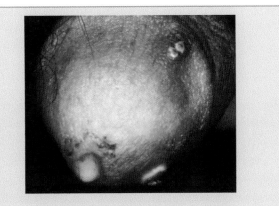

Figure 25.1 Gonorrhea infection in the male. Creamy purulent discharge from the urinary meatus. In many cases, infection is asymptomatic, but may cause painful urination or a purulent discharge, as seen here. In severe cases, it may also cause inflammation of the testicles and prostate gland, and infertility. Treatment is with antibiotics. *Courtesy of the Centers for Disease Control & Prevention, Atlanta, Georgia. Image is found in the Public Health Image Library #4065.*

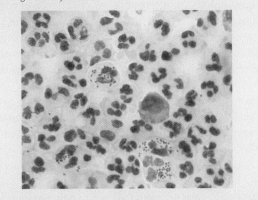

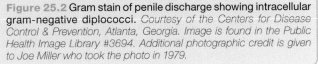

Figure 25.2 Gram stain of penile discharge showing intracellular gram-negative diplococci. *Courtesy of the Centers for Disease Control & Prevention, Atlanta, Georgia. Image is found in the Public Health Image Library #3694. Additional photographic credit is given to Joe Miller who took the photo in 1979.*

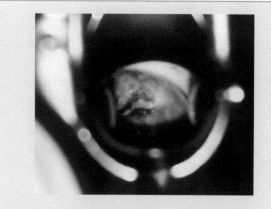

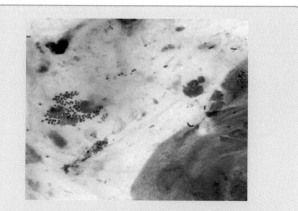

Figure 25.3 Cervicitis with purulent discharge. *Courtesy of the Centers for Disease Control & Prevention, Atlanta, Georgia. Image is found in the Public Health Image Library #4087.*

Figure 25.4 Gram stain of a cervical smear showing extracellular and intracellular gram-negative diplococci. *Courtesy of the Centers for Disease Control & Prevention, Atlanta, Georgia. Image is found in the Public Health Image Library #5516. Additional photographic credit is given to Joe Miller who took the photo in 1975.*

1. WHAT IS THE CAUSATIVE AGENT, HOW DOES IT ENTER THE BODY AND HOW DOES IT SPREAD A) WITHIN THE BODY AND B) FROM PERSON TO PERSON?

CAUSATIVE AGENT

The patient has gonorrhea and his girlfriend has gonorrhea-related **pelvic inflammatory disease (PID)** and disseminated gonococcal infection (DGI) caused by the bacterium *Neisseria gonorrhoeae* (often termed the gonococcus). *Neisseria* species are **capnophilic**, gram-negative cocci. The cocci are often found in pairs where their adjacent sides are flattened giving them a coffee-bean appearance. Their habitat is the mucous membranes of mammals and many species are **commensals** of these surfaces. Two species, however, *N. gonorrhoeae* and *N. meningitidis* (termed the meningococcus), are obligate pathogens of humans. *Neisseria* are oxidase-positive, catalase-positive, and produce acid from a variety of sugars by oxidation.

Neisseria species have a typical gram-negative envelope but the outer leaflet of their outer membrane contains lipooligosaccharide (LOS) instead of **lipopolysaccharide (LPS)**. Gonococcal and meningococcal LOS lack the repeating "O" somatic antigenic side chains present in the *Enterobacteriaceae*, for example. The core polysaccharide of LOS of both the gonococcus and the meningococcus undergoes **antigenic variation** exposing a variety of epitopes on the cell surface. The cell surface of both the gonococcus and the meningococcus (see Case 26) is decorated with pili (type IV pili), which are hair-like projections composed of polymers of the structural protein **pilin** (PilE). The protein PilC is located at the tip of the **pilus** and is the **adhesin** that mediates initial attachment of the bacterium to the surface of mucosal epithelium. The PilC protein binds CD46, a complement

regulatory protein that has co-factor activity that inactivates the complement components C3b and C4b by serum factor I.

The pilin (PilE) protein contains a constant N-terminal domain, a hypervariable C-terminal domain, and several variable regions termed mini-cassettes, which are encoded by genes with varying DNA sequence. The gonococcus has a single complete copy of the pilin gene termed *pilE* but as many as 15 truncated genes with variable DNA sequence. The truncation is at the 5' end, resulting in lack of the sequence encoding the N-terminal constant domain and promoter elements. These truncated genes are termed *pilS* (silent) and form the *pilS* locus. By recombination of *pilS* sequences into the *pilE* gene, the bacterium can express a high number of antigenically distinct pili. In addition to antigenic variation, the pili undergo phase variation. In phase variation, the bacterium has the ability to turn pilus expression on or off at a high frequency.

Among the outer-membrane proteins are a family of opacity-associated proteins (Opa), so named because they give rise to an opaque colony phenotype. Opa proteins are important in the ability of the organism to adhere tightly to epithelia. They also dictate the tissue tropism of the gonococcus and its ability to be taken up by epithelial cells. There are as many as 12 genes encoding Opa proteins and they undergo phase variation such that a neisserial population will contain bacteria expressing none, one or several Opa proteins. There are two hypervariable domains within the extracellular portion of the molecule that give rise to new Opa variants as a result of point mutation and by modular exchange of domains between different Opa proteins. These hypervariable domains are also the sites of interaction with cellular receptors. The Opa proteins of the gonococcus and the meningococcus can be divided into two major groups based on the cellular receptors to which they bind. The OpaHS-type proteins bind to heparan sulfate and extracellular matrix molecules such as fibronectin and vitronectin, while the OpaCEA-type proteins bind to carcinoembryonic antigen-related cell-adhesion molecule (CEACAM). OpaHS and OpaCEA adhesins mediate internalization of gonococci by various human cells. Either directly through engagement with heparan sulfate proteoglycans on the cell surface or indirectly using fibronectin or vitronectin to bridge OpaHS-type proteins and cell-surface integrins, the actin cytoskeleton of the epithelial cell is signaled to rearrange to enable engulfment of the bacterial cell. Engagement of OpaCEA-type proteins with members of the CEACAM receptor family also triggers actin cytoskeleton rearrangement, permitting internalization by neutrophils, macrophages, and various other types of cells.

Gonococcal porin proteins are named protein IA (PorPIA) and protein IB (PorPIB) and they assemble into trimers that form channels that traverse the gonococcal outer membrane. When gonococci are apposed to the epithelial cell membrane, porin PIB appears to be able to insert into the eukaryotic cell membrane and translocate to the cytosol where it participates in the reorganization of the actin cytoskeleton. However, neisserial porins can inhibit neutrophil actin polymerization, degranulation, expression of opsonin receptors, and the respiratory burst.

N. gonorrhoeae and *N. meningitidis* are adept at acquiring iron from their human host from numerous iron sources, which contributes to their ability to replicate on mucosal surfaces, intracellularly, and in the blood.

Both *N. gonorrhoeae* and *N. meningitidis* secrete an IgA1 protease. The enzyme cleaves IgA1 at the hinge region to produce Fab and Fc fragments. While it is logical to assume that IgA1 protease contributes to the **virulence** of the gonococcus by subverting the protective effects of sIgA, it should be realized that half of sIgA is of subclass 2 that is resistant to IgA1 protease. Moreover, it has been demonstrated that experimental urethral infections of male volunteers with an IgA1 protease-negative mutant of *N. gonorrhoeae* matching the parent strain in expression of Opa proteins, LOS, and pilin was indistinguishable from that of the parent strain. A role for IgA1 protease may lie in its ability to cleave **lysosome**-associated membrane protein 1 (h-lamp-1). As their name implies h-lamp-1 and h-lamp-2 are found in the membranes of mature lysosomes but also in the membranes of **phagosomes**/endosomes. Their functions are not fully understood but they are thought to protect the membrane from the action of degradative enzymes within the lysosome and appear to be required for fusion of lysosomes with phagosomes. It has been shown that gonococcal IgA1 protease can cleave the less glycosylated form of h-lamp-1 found in epithelial cell phagosomes/endosomes, which may enable the bacteria to escape into the cytosol of the cell and prolong their intracellular survival.

N. gonorrhoeae has become increasingly resistant to a range of antibiotics that include tetracyclines, macrolides (including azithromycin), sulphonamides, trimethoprim combinations, and quinolones.

ENTRY AND SPREAD WITHIN THE BODY

In uncomplicated gonorrhea, the bacteria adhere to urethral epithelium of males and to the cervical epithelium and urethral epithelium of females. The apical surface of the epithelium is induced to take up the gonococci via engagement of OpaCEA with CEACAMs. Rearrangement of the actin cytoskeleton facilitates internalization and the bacteria pass through the cells via transcytosis and are released at the basolateral surface into the sub-epithelial space where infection is established. From this site, the gonococci may seed the bloodstream and from there, the joints and skin. The up-regulation of CEACAM1 on endothelium brought about by LOS and pro-inflammatory **cytokines** contribute to trafficking of the bacteria into blood. In women, the cervical infection may ascend to the fallopian tubes (**salpingitis**), which can lead to scarring, **ectopic pregnancy**, sterility, and chronic pelvic pain.

PERSON-TO-PERSON SPREAD

The gonococcus is a sexually transmitted pathogen and it is acquired and spread horizontally (person to person) by vaginal, anal or oral sex. Ejaculation need not occur for the gonococcus to be transmitted or acquired. Gonorrhea can also be spread vertically from mother to baby during delivery. Persons who have had gonorrhea and received treatment may become infected again if they have sexual contact with a person with gonorrhea.

EPIDEMIOLOGY

In 2019, a total of 616392 cases of gonorrhea were reported to the CDC (Centers for Disease Control and Prevention) in the US. By law, gonorrhea is a reportable infectious disease in the US, as are chlamydia and syphilis, the other major bacterial sexually transmitted infections (STIs), but only about half of these infections are reported to the CDC. Gonorrhea is the second most commonly reported bacterial STI in the US after chlamydia (caused by *Chlamydia trachomatis*). Very often, individuals are co-infected with both *N. gonorrhoeae* and *Chlamydia trachomatis*. The WHO estimates that there are 30.6 million cases of gonorrhea worldwide annually. The UK recorded 17443 cases in 2018–2019 and Europe recorded more than 89000 cases in 2017.

2. WHAT IS THE HOST RESPONSE TO THE INFECTION AND WHAT IS THE DISEASE PATHOGENESIS?

Innate immunity in the male and female urinary tract is largely mediated by the antimicrobial proteins β-defensins (HBD), cathelicidin, lactoferrin, and lipocalin. **Tamm-Horsfall protein** (**THP**) is an abundant urinary glycoprotein that binds specific mannosylated residues and blocks adhesion of certain bacterial urinary tract pathogens such as *Escherichia* *coli.* The endocervix provides a protective mucous layer rich in HBD-1, HD-5, lysozyme, and lactoferrin for the vagina. Cervical, vaginal, and urethral mucosal IgA antibodies likely contribute to immune protection at these mucosal surfaces. Both the gonococcus and the meningococcus produce an IgA1 protease (see above) that may subvert mucosal IgA antibodies. The vaginal pH varies between 3.5 and 5.0, which restricts the establishment of all but aciduric bacteria. Thus, gonococci must successfully negotiate these defense mechanisms. Gonococci are readily taken up by neutrophils and macrophages in the submucosa. LOS and other components of the outer membrane as well as IgA1 protease are pro-inflammatory and induce release of **tumor necrosis factor-α (TNF-α)**, **interleukin (IL)**-1β, IL-6, and IL-8 from phagocytes, which induces a florid inflammatory reaction.

The complement system serves as the primary mechanism of killing gonococci. This is supported by the fact that individuals with complement deficiencies are at increased risk for disseminated neisserial infections although, interestingly, they have less severe symptomatology than do persons with an intact complement system. Individuals deficient in terminal complement proteins C5, C6, C7, C8, and/or C9 commonly exhibit recurrent, often systemic, neisserial infections. Consistent with the importance of complement, DGIs are associated predominantly with gonococci that express type PI.A porin molecules that bind factor H, which inhibits the **alternative complement pathway**. Similarly, the PI.A porin binds C4-binding protein (C4bp), which results in the inhibition of the classical **complement pathway**. The **membrane attack complex (MAC)** is crucial to the bactericidal action of the complement system and MAC assembly occurs in an aberrant manner on the surface of serum-resistant gonococci.

Gonococcal infection does not appear to result in protective humoral or cellular immunity despite the intense inflammatory reaction engendered by the bacteria. Although local and systemic antibodies can be detected in infected persons, they are at low levels and appear not to protect against re-infection. The reason for this is thought to be the extensive and rapid variation of pili, LOS, and the Opas. In addition, both the male and female genital tracts lack mucosa-associated lymphoid tissue required to induce mucosal immunity. Furthermore, antibodies to the reduction-modifiable protein (Rmp) also termed outer-membrane protein (OMP)3, interfere with the ability of antibodies directed against other surface components of the gonococcus to activate complement properly. It has been shown that gonococcal binding of CEACAM1 on **CD4**+ T cells can down-regulate the activation and proliferation of these cells, which may contribute to the suboptimal adaptive immune response.

3. WHAT IS THE TYPICAL CLINICAL PRESENTATION AND WHAT COMPLICATIONS CAN OCCUR?

In the male, the most common clinical presentation is infection of the genitourinary tract producing **urethritis**. The most common symptoms are urethral discomfort, dysuria, and discharge of varying severity. If the infection ascends to the epididymis, **epididymitis** presents as unilateral pain and swelling localized posteriorly within the scrotum.

In the female, the most common clinical presentation is **endocervicitis**. The most common symptom is a thin, purulent, and unpleasant-smelling vaginal discharge, although many women may be asymptomatic. Women may also have urethritis as well as **cervicitis**, which manifest as dysuria or a slight urethral discharge.

Ascending infection of the endometrium, fallopian tubes, ovaries, and peritoneum manifests as pelvic or lower abdominal pain, which may be in the midline, unilateral, or bilateral. The pain may be accompanied by fever, nausea, and vomiting. Infection of the peritoneum may spread to that covering the liver (peri-hepatitis, Fitz-Hugh-Curtis syndrome) resulting in right upper quadrant pain.

Rectal infection may follow receptive anal intercourse and, in women, by local spread of the gonococcus from the vaginal **introitus**. Often, rectal infection is asymptomatic, but pain, **pruritus** (itch), **tenesmus** (the constant feeling of the need to empty the bowel), discharge, and bloody diarrhea may occur.

Oral sex may result in **pharyngitis**, which usually is asymptomatic. Conjunctivitis may follow accidental inoculation of the eye(s) with fingers. In neonates, vertical transmission may occur from an infected mother to her neonate during vaginal delivery. This results in bilateral conjunctivitis (ophthalmia neonatorum). The symptoms of gonococcal conjunctivitis are pain, redness, and a purulent discharge. Corneal ulceration, perforation, and blindness can occur if treatment is not given promptly. Blindness from neonatal gonococcal infection is a serious problem in developing countries but is uncommon in the US and Europe where neonatal prophylaxis is routine.

In a few percent of cases, gonococci disseminate from the site of mucosal infection via the bloodstream giving rise to systemic disease termed disseminated gonorrheal infection (DGI). Gonococcemia (gonococci in the blood) occurs most frequently in the adolescent and young adult population, with a peak incidence in males aged 20–24 years and females aged 15–19 years.

The clinical manifestations of DGI are biphasic with an early **bacteremic** phase consisting of **tenosynovitis** of the hands, **arthralgias**, and **rash**, followed by a localized phase consisting of localized **septic arthritis** typically of the knee. The rash consists of small **papules** that become pustules on broad **erythematous** bases with necrotic centers (Figure 25.5) and is usually found below the neck and also may involve the palms and the soles but usually spares the face, scalp, and mouth. The gonococci may also seed the bone (**osteomyelitis**), the central nervous system (**meningitis**) and the heart (**endocarditis**). Other serious complications are **adult respiratory distress syndrome** (**ARDS**), and fatal **septic shock**. Pregnant or menstruating women are particularly susceptible to DGI, as are individuals with complement deficiencies, HIV or **systemic lupus erythematosus** (**SLE**).

There is a strong association between the acquisition of HIV-1 and other sexually transmitted diseases (STDs) including gonorrhea. An increased HIV load has been found in semen of men with gonorrhea. The mechanisms by which the gonococcus enhances HIV-1 infection are not fully understood but may include the ability of the bacterium to enhance infection and replication of HIV-1 in **dendritic cells**.

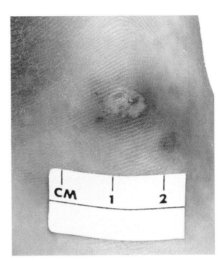

Figure 25.5 Cutaneous gonococcal lesion due to a disseminated *N. gonorrhoeae* infection. *Courtesy of the Centers for Disease Control & Prevention, Atlanta, Georgia. Image is found in the Public Health Image Library #6384. Additional photographic credit is given to Dr Weisner who took the photo in 1972.*

4. HOW IS THE DISEASE DIAGNOSED, AND WHAT IS THE DIFFERENTIAL DIAGNOSIS?

Nucleic acid amplification tests (NAATs) are more specific and sensitive in detecting *Neisseria gonorrhoeae* and *N. meningitidis* than culture. Furthermore, for STDs, NAATs can detect *Chlamydia trachomatis* at the same time. Therefore, they are recommended for detection if available, otherwise culture and determinative physiologic tests can be used. Although pathogenic *Neisseria* are fastidious, they will grow on blood agar as well as chocolate agar (Figure 25.6). However, they are usually isolated on selective media based on chocolate agar because clinical specimens are frequently obtained from mucosal surfaces and growth of the gonococcus may be inhibited by the attendant commensal **microbiota** in the specimen. Mucosal specimens should be collected using Dacron or Rayon swabs rather than alginate or cotton because both of the latter may be inhibitory for gonococci. Modified Thayer-Martin (MTM) medium is commonly used in the US and contains various antimicrobial agents to suppress gram-positive and gram-negative bacteria and fungi to allow the selective recovery of *N. gonorrhoeae* and *N. meningitidis* from mucosal surfaces (Figure 25.7). Plates are incubated at 35–37°C in 3–7% CO_2 in a humid environment. A candle extinction jar is suitable. Identification of the gonococcus and meningococcus is achieved by subjecting oxidase-positive (see *N. meningitidis* case, Figure 25.4) and catalase-positive (30% H_2O_2) gram-negative diplococci (Figure 25.8) to a panel of sugars for carbohydrate utilization tests, specific fluorescent antibodies, chromogenic substrates, or DNA probes. Differential utilization of the sugars glucose, maltose, sucrose, and lactose is a simple and common way to speciate the pathogenic *Neisseria*. *N. gonorrhoeae* utilizes glucose only (Figure 25.9). Because they grow on MTM, the selective medium used to isolate *N. gonorrhoeae* and *N. meningitidis*, *Neisseria lactamica*, and *Moraxella catarrhalis* must be differentiated from these bacteria. *N. lactamica* can be differentiated because it oxidizes lactose in addition to glucose and maltose, whereas *M. catarrhalis* does not oxidize glucose, maltose, sucrose or lactose. Of the four species, *M. catarrhalis* alone produces butyrate esterase and deoxyribonuclease (DNase). In STD clinics, a diagnosis of gonococcal urethritis in adult males is usually made by observing intracellular (neutrophils) gram-negative diplococci on smears of urethral discharge (see Figure 25.2). However, confirmatory tests are required in females and for all extra-genital infections because there can be social and medicolegal issues resulting from the findings.

DIFFERENTIAL DIAGNOSIS

For uncomplicated gonococcal urethritis and cervicitis, the differential diagnosis should include chlamydial genitourinary infections, male and female urinary tract infection, and **vaginitis**.

For ascending infections in the male, the differential diagnosis should include testicular torsion and, in the female, **endometriosis**, **endometritis**, and ectopic pregnancy.

For DGI, meningococcemia should be considered.

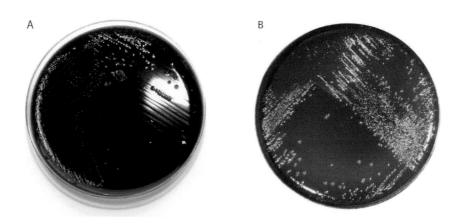

Figure 25.6 A pure culture of *N. gonorrhoeae* growing on (A) blood agar and (B) chocolate agar. *From CC Studio / Science Photo Library, with permission.*

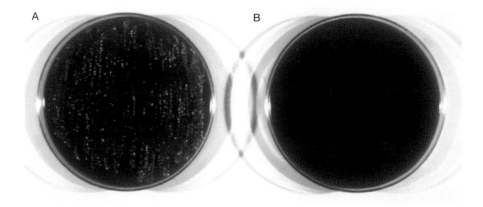

Figure 25.7 A rectal swab specimen plated onto chocolate agar (A) and the selective chocolate agar-based Thayer-Martin (T-M) agar medium (B). Note the overgrowth of the endogenous rectal microbiota on the chocolate agar plate (A), which suppressed the growth of the gonococcus. On the selective T-M agar shown in (B), there is a pure culture of the gonococcus because T-M agar contains antimicrobials that inhibit the growth of organisms other than *N. gonorrhoeae*. *From the Centers for Disease Control & Prevention, Atlanta, Georgia. Image is found in the Public Health Image Library #6505. Additional photographic credit is given to Renelle Woodall who took the photo in 1969.*

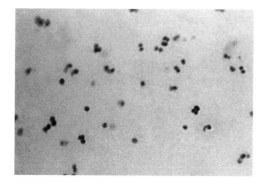

Figure 25.8 Gram stain of *N. gonorrhoeae* showing small, gram-negative, kidney-shaped diplococci. *From the Centers for Disease Control & Prevention, Atlanta, Georgia. Image is found in the Public Health Image Library #3798. Additional photographic credit is given to Joe Miller who took the photo in 1980.*

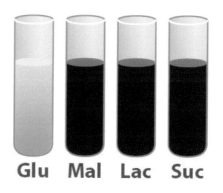

Figure 25.9 *N. gonorrhoeae* produces acid by oxidation of glucose (Glu) but not from maltose (Mal), sucrose (Suc), or lactose (Lac). The acid turns the pH indicator, phenol red, from red to yellow. *Courtesy of Gary E Kaiser, Professor of Microbiology, Community College of Baltimore County, Cantonsville Campus, Baltimore, MD.*

5. HOW IS THE DISEASE MANAGED AND PREVENTED?

MANAGEMENT

The CDC currently recommends a single 500 mg IM dose of ceftriaxone for treatment of uncomplicated urogenital, anorectal, and pharyngeal gonorrhea. For persons weighing ≥150 kg (300 lb), a single 1g IM dose of ceftriaxone should be given. If concurrent chlamydial infection has not been excluded, the addition of doxycycline (100 mg orally twice a day for 7 days) is recommended. The substitution of doxycycline for azithromycin in co-infections with the gonococcus and chlamydia reflects increasing concern about azithromycin resistance in *N. gonorrhoeae*. In contrast, the 2020 European guideline for the diagnosis and treatment of gonorrhoea in adults recommends dual antimicrobial therapy including high-dose ceftriaxone and azithromycin (ceftriaxone 1g plus azithromycin 2g) OR ceftriaxone 1g monotherapy for uncomplicated gonorrhea. In the UK, treatment for uncomplicated ano-genital and pharyngeal infection in adults is a single IM dose of 1g ceftriaxone.

Gonorrhea is a notifiable disease in the US but not in the UK. However, ophthalmia neonatorum, which is caused by *N. gonorrhoeae* or *Chlamydia trachomatis*, is notifiable in the UK.

PREVENTION

The use of latex condoms to prevent gonorrhea and other STIs is recommended for all sexually active individuals unless in a long-term monogamous relationship in which both partners are proven free of STI.

SUMMARY

1. WHAT IS THE CAUSATIVE AGENT, HOW DOES IT ENTER THE BODY, AND HOW DOES IT SPREAD A) WITHIN THE BODY AND B) FROM PERSON TO PERSON?

- Gonorrhea is the second most commonly reported bacterial STI in the US after chlamydia (caused by *Chlamydia trachomatis*). Very often, individuals are co-infected with both bacteria.
- *Neisseria* species are capnophilic, gram-negative cocci. The cocci are often found in pairs (diplococci) where their adjacent sides are flattened giving them a coffee-bean appearance.
- Their habitat is the mucous membranes of mammals and many species are commensals of these surfaces. Two species, however, *N. gonorrhoeae* and *N. meningitidis*, are obligate pathogens of humans.
- *Neisseria* are oxidase-positive, catalase-positive, and produce acid from a variety of sugars by oxidation. *N. gonorrhoeae* utilizes glucose only.
- *Neisseria* species have a typical gram-negative envelope but the outer leaflet of their outer membrane contains lipooligosaccharide (LOS), which lacks repeating 'O' somatic antigenic side chains.
- The core polysaccharide of LOS of the gonococcus undergoes antigenic variation exposing a variety of epitopes on the cell surface.
- The cell surface of the gonococcus is decorated with pili, which are hair-like projections composed of polymers of the structural protein pillin (PilE).
- Pili mediate initial attachment to mucosal epithelia. Pili are antigenically variable and pilus expression can be turned on or off at a high frequency.
- Opa outer-membrane proteins enable the organism to adhere tightly to epithelia and facilitate uptake by various types of host cell.
- *N. gonorrhoeae* and *N. meningitidis* are adept at acquiring iron from their human host.
- Both *N. gonorrhoeae* and *N. meningitidis* secrete an IgA1 protease.

- The gonococcus is a sexually transmitted pathogen and it is acquired and spread horizontally (person to person) by vaginal, anal or oral sex.
- Gonorrhea can also be spread vertically from mother to baby during delivery.
- From the subepithelial space where infection is established, the gonococci may seed the bloodstream and disseminate to various organs.

2. WHAT IS THE HOST RESPONSE TO THE INFECTION AND WHAT IS THE DISEASE PATHOGENESIS?

- In the genitourinary tract, gonococci encounter various innate antimicrobial factors such as β-defensins, cathelicidin, lactoferrin, lipocalin, lysozyme, and Tamm-Horsfall protein.
- Low levels of secretory IgA antibodies are induced that are ineffective at preventing re-infection.
- Por proteins in the outer membrane facilitate uptake by epithelial cells and inhibit phagocytosis by neutrophils.
- The complement system serves as the primary mechanism of killing gonococci.
- Gonococci can subvert the alternative and classical complement pathways.
- Gonococcal infection does not result in protective humoral or cellular immunity because of antigenic variation and other mechanisms of immune evasion.

3. WHAT IS THE TYPICAL CLINICAL PRESENTATION AND WHAT COMPLICATIONS CAN OCCUR?

- In the male, the most common clinical presentation is infection of the genitourinary tract producing urethritis.
- In the female, the most common clinical presentation is endocervicitis. The most common symptom is a thin, purulent, and unpleasant-smelling vaginal discharge, although many women may be asymptomatic.

Continued...

...continued

- Women may experience ascending infection of the endometrium, fallopian tubes, ovaries, and peritoneum termed pelvic inflammatory disease (PID).
- Rectal infection may follow receptive anal intercourse and, in women, by local spread of the gonococcus from the vaginal introitus.
- Oral sex may result in pharyngitis, which usually is asymptomatic. Conjunctivitis may follow accidental inoculation of the eye(s) with fingers.
- In neonates, vertical transmission may occur from an infected mother to her neonate during vaginal delivery. This results in bilateral conjunctivitis (ophthalmia neonatorum), which can lead to blindness.
- Rarely, gonococci disseminate via the bloodstream giving rise to systemic disease termed disseminated gonorrheal infection (DGI).
- Bacteremia results in seeding of the joints and skin resulting in tenosynovitis of the hands, arthralgias, and rash.

4. HOW IS THE DISEASE DIAGNOSED, AND WHAT IS THE DIFFERENTIAL DIAGNOSIS?

- Pathogenic *Neisseria* are fastidious and are usually isolated on selective media based on chocolate agar.
- Modified Thayer-Martin (MTM) medium is commonly used in the US to allow the selective recovery of *N. gonorrhoeae* from mucosal surfaces. Plates are incubated at 35–37°C in 3–7% CO_2 in a humid environment.

- Identification of the gonococcus is achieved by subjecting oxidase-positive and catalase-positive gram-negative diplococci to carbohydrate utilization tests. *N. gonorrhoeae* oxidizes glucose only.
- Specific fluorescent antibodies, chromogenic substrates, or DNA probes can also be used for identification.
- In STD clinics, a diagnosis of gonococcal urethritis in adult males is usually made by observing intracellular (neutrophils) gram-negative diplococci on smears of urethral discharge.
- Confirmatory tests are required in females and for all extra-genital infections because there can be social and medicolegal issues resulting from the findings.

5. HOW IS THE DISEASE MANAGED AND PREVENTED?

- The use of latex condoms to prevent gonorrhea and other sexually transmitted infections (STIs) is recommended for all sexually active individuals unless in a long-term monogamous relationship in which both partners are proven free of STI.
- Many gonococcal strains are resistant to penicillins, tetracyclines, spectinomycin, and fluoroquinolones.
- Currently, only the cephalosporins are recommended for the treatment of gonorrhea.
- Gonorrhea is a notifiable disease in the US but not in the UK. However, ophthalmia neonatorum, which is caused by *Neisseria gonorrhoeae* or *Chlamydia trachomatis*, is notifiable in the UK.

FURTHER READING

Goering R, Dockrell H, Zuckerman M, Chiodini PL. Mims' Medical Microbiology and Immunology, 6th edition. Elsevier, Philadelphia, 2018.

Murray PR, Rosenthal KS, Pfaller MA. Medical Microbiology, 9th edition. Elsevier/Mosby, Philadelphia, 2021.

REFERENCES

Ángel, ST, Elizabeth GC, Patricia RSD, et al. Neisseria gonorrhoeae Infection, Current State of Antimicrobial Susceptibility. Literature Review. J Bacteriol Mycol, 8: 117–122, 2020.

Cyr S. St, Barbee L, Workowski KA, et al. Update to CDC's Treatment Guidelines for Gonococcal Infection, 2020. MMWR Morb Mortal Wkly Rep, 69: 1911–1916, 2020.

Lenz JD, Dillard JP. Pathogenesis of Neisseria gonorrhoeae and the Host Defense in Ascending Infections of Human Fallopian Tube. Front Immunol, 9: 2710, 2018.

Quillin SJ, Seifert HS. Neisseria gonorrhoeae Host Adaptation and Pathogenesis. Nat Rev Microbiol, 16: 226–240, 2018.

Unemo M, Ross JDC, Serwin AB, et al. European Guideline for the Diagnosis and Treatment of Gonorrhea in Adults. Int J STD AIDS, 956462420949126, 2020.

Unemo M, Seifert HS, Hook III EW, et al. Gonorrhoea. Nat Rev Dis Primers, 5: 79, 2019.

WEBSITES

Centers for Disease Control and Prevention, Gonorrhea, 2022: https://www.cdc.gov/std/gonorrhea/default.htm

European Centre for Disease Prevention and Control, Gonorrhoea: https://www.ecdc.europa.eu/en/gonorrhoea

World Health Organization, WHO guidelines for the treatment of Neisseria gonorrhoeae: https://www.who.int/publications/i/item/9789241549691

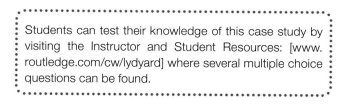

Students can test their knowledge of this case study by visiting the Instructor and Student Resources: [www.routledge.com/cw/lydyard] where several multiple choice questions can be found.

Neisseria meningitidis

A 19-year-old college student was in his usual state of health until the evening before admission, when he went to bed with a headache. He told his room-mate that he felt feverish, and, on the following morning, his room-mate found him in bed, moaning and lethargic. He was taken to the emergency room, where he appeared toxic and drowsy but oriented to time and place. His temperature was 40°C, his heart rate was 126/min, and his blood pressure was 100/60 mm Hg. His neck was supple. He had an impressive, nonblanching purpuric rash, most prominent on the trunk, wrists, and legs (Figure 26.1). His white blood cell count was 26 000 μl⁻¹ with 25% band forms. The platelet count was 80 000 μl⁻¹. Blood cultures were obtained and the patient was started on intravenous (IV) ceftriaxone. Blood cultures revealed gram-negative diplococci (Figure 26.2).

It should be noted that in North America, this patient would have received a lumbar puncture (LP) as an essential component in the diagnosis of meningococcal disease except where explicitly contraindicated. However, this is not the case in the UK where severe sepsis is a contraindication to LP (www.meningitis.org). In the UK, many pediatricians and adult physicians feel that LP should not be performed acutely in patients suspected of having meningococcal disease and, if LP is considered to be necessary, it is done once the patient is stable. In the case considered above, the cerebrospinal fluid (CSF) glucose, protein, and white blood cell count were normal, and CSF bacterial culture was negative.

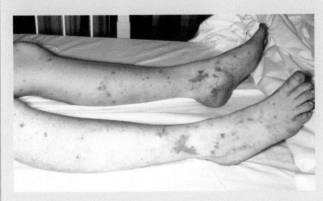

Figure 26.1 Legs of patient showing a purpuric rash typical of meningococcal septicemia. *From John Radcliffe Hospital / Science Photo Library, with permission.*

Figure 26.2 Gram stain of *N. meningitidis*. *From the Centers for Disease Control & Prevention, Atlanta, Georgia. Image is found in the Public Health Image Library #6423. Additional photographic credit is given to Dr Brodsky who took the photo in 1966.*

1. WHAT IS THE CAUSATIVE AGENT, HOW DOES IT ENTER THE BODY AND HOW DOES IT SPREAD A) WITHIN THE BODY AND B) FROM PERSON TO PERSON?

CAUSATIVE AGENT

The patient has **bacteremia** (meningococcemia) caused by *Neisseria meningitidis* (often termed the meningococcus). This bacterium is one of the three principal causes of bacterial **meningitis**, the other two being *Streptococcus pneumoniae* (Sp) and *Haemophilus influenzae* serotype b (Hib). However,

the introduction of **conjugate vaccines** for immunization of infants has reduced invasive disease caused by Sp and Hib significantly.

The pathogenic *Neisseria* species *N. meningitidis* and *N. gonorrhoeae* are **capnophilic**, gram-negative cocci. The cocci are found in pairs where their adjacent sides are flattened giving them a coffee-bean appearance (Figure 26.3). *Neisseria* species are oxidase-positive (Figure 26.4), catalase-positive, and produce acid from sugars by oxidation. Their habitat is the mucous membranes of mammals and many species are **commensals** of these surfaces. However, *N. meningitidis* is an obligate human pathogen that exhibits intermittent carriage in the nasopharynx reaching a maximum of about 10% in those

26

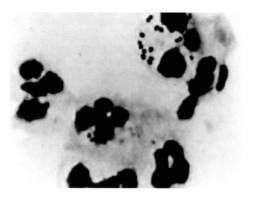

Figure 26.3 Gram stain of cerebrospinal fluid: *N. meningitidis*, intracellular, gram-negative diplococci. *Reprint permission kindly given for image in Laboratory Methods for the Diagnosis of Meningitis caused by Neisseria meningitidis, Streptococcus pneumoniae, and Haemophilus influenzae CDC, Centers for Disease Control and Prevention August 1998, Figure 9a, page 26.*

aged between 20 and 24 years. In addition, the meningococcus colonizes the mucosae of the endocervix, urethra, and anus.

Neisseria species have a typical gram-negative envelope (Figure 26.5) but the outer leaflet of their outer membrane contains lipooligosaccharide (LOS) instead of **lipopolysaccharide (LPS)**. The O antigen is absent. The core polysaccharide of LOS of the meningococcus undergoes **antigenic variation** exposing a variety of epitopes on the cell surface. Phase variation of LOS directs interconversion between invasive and antibody/complement-resistant phenotypes of *N. meningitidis*. LOS variants containing low amounts of sialic acid enter human mucosal cells efficiently whereas those with LOS containing high amounts of sialic acid or expressing capsule do not. On the other hand, these variants with low LOS sialylation are highly susceptible to killing by antibody and complement in the bloodstream whereas variants with highly sialated LOS and expressing capsule are serum-resistant. The cell surface is decorated with **pili**, which are hair-like projections composed of polymers

of the structural protein **pilin** (PilE). Pili mediate initial attachment to nasopharyngeal epithelium and enable the bacterium to resist **phagocytosis**. The PilC protein present at the tip of the pilus is the **adhesin** that binds a receptor(s) on the host cell surface thought to be CD46 which is a complement regulatory protein. CD147 is a receptor for type IV pilus–mediated adhesion to the human brain or peripheral endothelial cells through interaction with PilE and PilV.

The pilin (PilE) protein contains a constant N-terminal domain, a hypervariable C-terminal domain, and several variable regions termed mini-cassettes, which are encoded by genes with varying DNA sequence. The meningococcus has a single complete copy of the pilin gene termed *pilE* but many truncated genes with variable DNA sequence. The truncation is at the 5′ end, resulting in lack of the sequence encoding the N-terminal constant domain and promoter elements. These truncated genes are termed *pilS* (silent) and form the *pilS* locus. By recombination of *pilS* sequences into the *pilE* gene, the bacterium can express a high number of antigenically distinct pili. In addition to antigenic variation, the meningococcus undergoes phase variation. In phase variation, the bacterium has the ability to turn pilus expression on or off at a high frequency.

The major outer-membrane proteins (OMPs) of the meningococcus are subdivided into five classes based on descending molecular weight. Class I, II, and III are porin proteins with class I being PorA, and class II and III, PorB. The class IV protein is the reduction modifiable protein (Rmp) and the class V proteins are the opacity-associated proteins (Opa), so named because they give rise to an opaque colony phenotype. Opa proteins are important in the ability of the organism to adhere tightly to epithelia. They also dictate the tissue tropism of the meningococcus and its ability to be up-taken by epithelial cells. There are three or four genes encoding Opa proteins and they undergo phase variation such that a neisserial population will contain bacteria expressing none, one or several Opa proteins. There are two

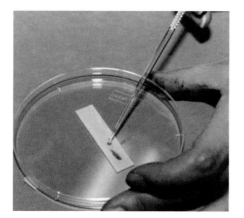

Figure 26.4 Kovac's oxidase test – positive reaction. *Reprint permission kindly given for image in Laboratory Methods for the Diagnosis of Meningitis caused by Neisseria meningitidis, Streptococcus pneumoniae, and Haemophilus influenzae CDC, Centers for Disease Control and Prevention August 1998, Figure 16, page 37.*

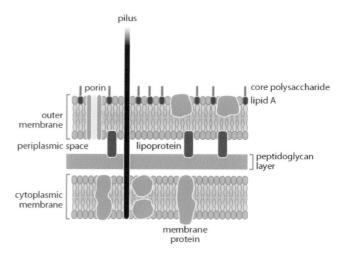

Figure 26.5 Schematic representation of the *N. meningitidis* envelope.

hypervariable domains within the extracellular portion of the molecule that give rise to new Opa variants as a result of point mutation and by modular exchange of domains between different Opa proteins. These hypervariable domains are also the sites of interaction with cellular receptors. The Opa proteins of the meningococcus can be divided into two major groups based on the cellular receptors to which they bind. The OpaHS-type proteins (Opc) bind to heparan sulfate and extracellular matrix molecules such as fibronectin and vitronectin, while the OpaCEA-type proteins (Opa) bind to carcinoembryonic antigen (CEA) and related molecules. OpaHS-type proteins and OpaCEA-type proteins mediate internalization of meningococci by various human cells. Either directly through engagement with heparan sulfate proteoglycans on the epithelial cell surface or indirectly using fibronectin or vitronectin to bridge OpaHS-type proteins and cell-surface integrins, the actin cytoskeleton of the epithelial cell is signaled to rearrange to enable engulfment of the bacterial cell. Engagement of OpaCEA-type proteins with CEA-family cell-adhesion molecules (CEA-CAM) also triggers actin cytoskeleton rearrangement, permitting internalization by neutrophils, macrophages, and various other types of cells.

The porin proteins that form channels that traverse the outer membrane are thought to play several roles in pathogenesis. For example, when meningococci are apposed to the host-cell membrane, PorB appears to be able to translocate into the eukaryotic cell membrane and affect the maturation of **phagosomes**.

An important **virulence** determinant of the meningococcus is the production of a **capsule**. Non-capsulate strains rarely cause invasive disease and are often found colonizing the nasopharynx of asymptomatic carriers. Although there are 13 different types of capsule, only five commonly cause invasive disease, A, B, C, Y, and W-135. However, recently, there have been outbreaks of invasive disease caused by serogroup X in West Africa. The capsule of serogroup A *N. meningitidis* is composed of N-acetyl mannosamine-1-phosphate whereas the capsules of serogroups B, C, Y, and W-135 meningococci are composed completely of poly-sialic acid or sialic acid linked to glucose or galactose. *N. meningitidis* serogroup X synthesizes capsular polymers of $\alpha1{\rightarrow}4$-linked N-acetylglucosamine 1-phosphate. The capsule is a major vaccine antigen (see later). Differences in the chemical structure of the capsular polysaccharide determine the meningococcal serogroups whereas the serotype is defined by differences in the class 2/3 OMPs. Subtypes based on class 1 OMP have also been defined. Meningococci can switch the type of capsule they express and this is probably accomplished by transformation and horizontal DNA exchange with other capsule types of *N. meningitidis*.

N. meningitidis is adept at acquiring iron from numerous iron sources in its human host. This property contributes to its ability to replicate on mucosal surfaces, intracellularly, and in the blood.

ENTRY AND SPREAD WITHIN THE BODY

The nasopharyngeal epithelium is induced to take up the meningococci by receptor-mediated **endocytosis** and the bacteria pass through the cells into the sub-epithelial space where infection is established. From this site, the meningococci may seed the bloodstream where production of capsule is important for survival and from there to the meninges and many other sites including joints and the skin. Type IV pili mediate adhesion of *N. meningitidis* to endothelial cells and induce Rho- and Cdc42-dependent cortical actin polymerization that results in the formation of membrane protrusions that lead to bacterial uptake. An essential step in the formation of membrane protrusions appears to be the phosphorylation of cortactin because it controls the polymerization of cortical actin. LOS plays an essential role in the induction of membrane protrusions because it induces the recruitment and phosphorylation of cortactin.

PERSON-TO-PERSON SPREAD

The meningococcus is spread horizontally (person to person) by respiratory droplets or direct contact. At any given time, about 10% of a population harbors the meningococcus in the nasopharynx. The rate of carriage increases with age to a maximum in late adolescence and early adulthood, when as many as one third of individuals may harbor the meningococcus. The carriage rate is related to the degree of crowding such as is encountered in barracks, prisons, university dormitories (halls of residence), and many other host and environmental factors.

EPIDEMIOLOGY

For the most part, *N. meningitidis* exists in a state of asymptomatic carriage in the human nasopharynx and it has been estimated that as many as 10% of persons are colonized by the organism (see above). Why the meningococcus on rare occasions penetrates the mucosa and invades the blood is not completely understood. Meningococcal meningitis occurs sporadically and unpredictably in small clusters throughout the world. In temperate regions, the number of cases increases in winter and spring. Rates of meningococcal disease have been declining in the US since the late 1990s. In 2019, there were about 371 total cases of meningococcal disease reported. Similarly, there has been a decline in England where the national Public Health England Meningococcal Reference Unit (MRU) confirmed 461 cases of invasive meningococcal disease during 2019 to 2020. In 2017, 3221 confirmed cases of invasive meningococcal disease were reported in 30 EU/EEA member states. Infections are most commonly caused by serogroups B and C. The predominant serogroups in various parts of the world are as follows: Asia and Africa, A, C, W-135, and X; Europe, North and South America, B, C, and Y. The highest incidence of meningococcal disease occurs in sub-Saharan

Africa, in an area that stretches from Senegal in the west to Ethiopia in the east. The population of this area is about 300 million people and it is referred to as the "meningitis belt". There are several factors that contribute to the high incidence of disease in this area. Dust winds and cold nights during the dry season are thought to reduce the local immunity of the pharynx and transmission of *N. meningitidis* is facilitated by overcrowded housing, pilgrimages, and traditional markets. Attack rates as high as 1000 cases per 100 000 population have been reported.

There is considerable interest in determining why outbreaks of meningococcal disease occur. *N. meningitidis* is a highly polymorphic species and meningococci isolated from asymptomatic carriers are highly diverse. Based on the use of **multilocus enzyme electrophoresis** (**MLEE**) or **multilocus sequence typing**, the carriage population has been found to comprise clonal complexes or lineages. Some of these clonal complexes are disproportionately the source of strains isolated from invasive disease such that about 10 of these lineages, termed "hyperinvasive lineages", caused the majority of meningococcal disease during the twentieth century.

2. WHAT IS THE HOST RESPONSE TO THE INFECTION AND WHAT IS THE DISEASE PATHOGENESIS?

The innate secretory immune system in the nasopharynx serves as a particle filter and a sticky microbicidal trap to prevent pathogenic microorganisms from adhering to the nasal mucosa and from reaching the lower respiratory tree. The anatomy of the nasal cavities is designed to create turbulent airflow to maximize the contact of microorganisms with the nasal hairs and the mucous coating overlying the epithelium. Turbulent airflow is created by the turbinate bones located laterally in the nasal cavity. These are covered by ciliated, pseudostratified columnar epithelium supported by a thick, vascular, and erectile glandular **lamina propria**. The inhaled air is warmed and humidified and this may facilitate binding of the polysaccharide capsule of the meningococcus and other capsulate pathogenic bacteria. Microorganisms become trapped in the mucus, which is moved backward toward the throat by cilial action (mucociliary ladder/blanket) where it can be swallowed and enter the stomach acid bath. Mucus is a viscoelastic gel, which is complemented by an armamentarium of antimicrobial factors such as lysozyme, lactoferrin, and **defensins**.

Acquired humoral immunity in the nasopharynx and at other mucosal surfaces is mediated by secretory immunoglobulin A (SIgA). The principal purpose of SIgA is immune exclusion, that is, prevention of pathogen adherence to, and subsequent invasion of the mucosal epithelium. SIgA can also neutralize exotoxins and microbial enzymes. SIgA is anchored in the mucus where it can participate in pathogen clearance mediated by the mucociliary ladder. It can also agglutinate microorganisms in nasopharyngeal secretions

facilitating their clearance. With some viruses, polymeric IgA has been shown to play a role in viral clearance from the lamina propria by binding the virions prior to its attachment to the poly-Ig receptor and transporting them through epithelial cells as the immunoglobulin traffics to the apical surface and ultimately into secretions. In addition, during transport through epithelial cells to the mucosal surface, SIgA is able to intercept and neutralize virions as they make their way in the other direction (from the apical surface to the basal surface of the cell). Whether SIgA can clear meningococci and other pathogenic bacteria from the lamina propria and intercept them within epithelial cells remains to be determined. It is thought that an **IgA1 protease** secreted by *N. meningitidis*, like *N. gonorrhoeae*, plays a role in subverting the host antibody responses. This enzyme cleaves IgA1 at the hinge region to produce Fab and Fc fragments. However, although IgA in plasma is almost all IgA of subclass 1, half of SIgA at mucosal surfaces is SIgA of subclass 2 that is resistant to IgA1 protease.

Once meningococci have transcytosed the mucosal barrier by adherence to epithelium and receptor-mediated endocytosis (described earlier), the **mannose-binding lectin** and the **alternative pathways** of the complement system serve as the primary mechanism of meningococcal killing. Individuals with deficiency in the **membrane attack components** (**C5-9**) of the complement cascade are highly susceptible to invasive disease. Anti-meningococcal bactericidal **IgM** and particularly **IgG** antibodies play the principal role in acquired immune defense. Neonates are highly resistant to meningococcal disease as a result of trans-placental transfer of maternal IgG antibodies, but they are extremely susceptible by 6 months of age. Thereafter, antibodies are induced and maintained as a result of intermittent carriage of strains of *N. meningitidis*, although carriage is rare in children under 5 years, and constant exposure to the commensal *N. lactamica*. In addition, *E. coli* strain K1 and *Bacillus pumilis* share structurally and immunologically identical capsules with *N. meningitidis* serogroup B and serogroup A strains, respectively, and may induce cross-reactive antibodies that protect against the meningococcus. The role of cell-mediated immunity in protection against *N. meningitidis* is not well understood.

PATHOGENESIS

Pathogenesis is a function of the characteristics of the strain of *N. meningitidis*, particularly the propensity to release LOS and the characteristics of host immunity. The feared outcomes of meningococcemia are shock, meningitis, **disseminated intravascular coagulation** (**DIC**) and myocardial dysfunction. Shock and DIC are interrelated processes that are initiated by LOS. During growth and **lysis**, the meningococcus releases LOS in blebs. LOS activates the complement cascade while the lipid A component of LOS binds the CD14/TLR-4/MD-2 receptor complex in the membrane of many cell types, particularly macrophages. Binding induces the release of large amounts of pro-inflammatory **cytokines**, particularly **tumor necrosis factor-α** (**TNF-α**) but also **interleukin-1**

(**IL-1**), IL-6, IL-8, GM-CSF, IL-10, IL-12, and **interferon-γ** (**IFN-γ**), the levels of which are closely correlated with disease severity and risk of death. These cytokines activate vascular endothelium, lymphocytes, and **natural killer** (**NK**) cells and recruit neutrophils, basophils, and T cells to the site of infection. Furthermore, TNF-α, IL-1, and IL-6 induce fever and the production of acute phase proteins. The release of LOS into the bloodstream brings about the release of TNF-α by macrophages in the liver, spleen, and other sites, causing massive vasodilation resulting in a fall in blood pressure and increased vascular permeability leading to loss of albumin followed by loss of fluid and electrolytes. Initially, vasoconstriction of arterial and venous vascular beds attempt to compensate for the decrease in plasma volume but, as fluid continues to be lost, venous return to the heart is impaired and cardiac output falls. Extravascular fluid accumulates in tissues and organs leading in the lungs to pulmonary **edema** and respiratory failure. Intravascular coagulation occurs as a result of marked **thrombocytopenia** and prolonged coagulation. The end point is organ failure.

proceeds, pustules, **bullae**, and hemorrhagic lesions with central **necrosis** may develop. From the blood, meningococci may seed the brain. Fulminant meningococcal sepsis (FMS), comprising shock and DIC, occurs within a matter of hours, which may or may not be accompanied by meningitis. Patients with invasive meningococcal disease fall into four groups: (1) patients with bacteremia without **shock**; (2) patients with bacteremia and shock but without meningitis (FTS); (3) patients with shock and meningitis; and (4) patients with meningitis alone.

Complications of meningococcal meningitis include **seizures**, increased intracranial pressure, cerebral venous and sagittal sinus **thrombosis**, and **hydrocephalus**. Cerebral herniation is rare. Communicating hydrocephalus can lead to gait difficulty, mental status changes, incontinence, and hearing loss. Fulminant meningococcemia may result in severe DIC leading to bleeding into the lungs, urinary tract, and gastrointestinal (GI) tract. Suppurative complications are infrequent but include **septic arthritis**, purulent **pericarditis**, **endophthalmitis**, and **pneumonia**.

3. WHAT IS THE TYPICAL CLINICAL PRESENTATION AND WHAT COMPLICATIONS CAN OCCUR?

The onset of meningococcemia may be insidious and can result in several outcomes. If the bacteria are rapidly cleared from the blood by the mononuclear phagocyte system, the patient manifests a short episode of **febrile**, flu-like symptoms; if not, overt disease develops. Multiplication of meningococci in the blood is associated with chills, fever, and localized or generalized **myalgia**. A hemorrhagic rash develops in about 80% of patients, although half of patients with meningitis do not have a rash. The rash begins as **petechiae** or **maculopapules** that occur initially on the extremities and trunk but may progress to involve any part of the body. As meningococcemia

4. HOW IS THE DISEASE DIAGNOSED, AND WHAT IS THE DIFFERENTIAL DIAGNOSIS?

A gram stain of CSF or material obtained from the characteristic skin lesion is informative and will reveal extracellular and intracellular coffee bean-shaped gram-negative diplococci (Figure 26.3). Pathogenic *Neisseria* are fastidious and are usually cultured using chocolate agar, although they will grow on blood agar (Figures 26.6–26.8). For clinical specimens obtained from mucosal surfaces modified Thayer-Martin (MTM) medium is commonly used in the US. This medium contains various antimicrobial agents to suppress the growth of commensal gram-positive and gram-negative bacteria and fungi to allow the selective recovery of

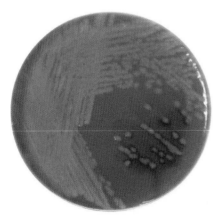

Figure 26.6 *N. meningitidis* growing on chocolate agar. The colonies in the lower right quadrant of the plate have been tested for oxidase activity and are positive (the colonies are surrounded by a purple ring). *Courtesy of the Department of Pathology, University of Pittsburgh School of Medicine.*

Figure 26.7 *N. meningitidis* growing on blood agar. *Reprint permission kindly given for image in Laboratory Methods for the Diagnosis of Meningitis caused by Neisseria meningitidis, Streptococcus pneumoniae, and Haemophilus influenzae CDC, Centers for Disease Control and Prevention August 1998, Figure 6, page 21.*

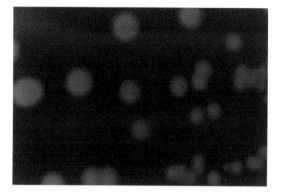

Figure 26.8 Overnight growth of *N. meningitidis* on blood agar plate appears as round, moist, glistening, and convex colonies. *Reprint permission kindly given for image in Laboratory Methods for the Diagnosis of Meningitis caused by Neisseria meningitidis, Streptococcus pneumoniae, and Haemophilus influenzae CDC, Centers for Disease Control and Prevention August 1998, Figure 111, page 31.*

N. meningitidis and *N. gonorrhoeae.* Agar plates are incubated at 35–37°C in 3–7% CO_2 in a humid environment. A candle extinction jar can be used for this purpose. Identification of the meningococcus is achieved by subjecting oxidase-positive and catalase-positive (30% H_2O_2) gram-negative diplococci to a panel of sugars for carbohydrate utilization tests, specific fluorescent antibodies, chromogenic substrates, or DNA probes. A flow chart of a simple scheme to identify *N. meningitidis* is shown in Figure 26.9. Differential utilization of the sugars glucose, maltose, sucrose, and lactose is a simple way to speciate the pathogenic *Neisseria. N. meningitidis* utilizes glucose and maltose (Figure 26.10). Rapid tests are available commercially that detect capsule antigen in CSF and urine by latex agglutination. Serogrouping can be accomplished by agglutination of bacterial cells from a colony with specific antisera. Polymerase chain reaction (PCR) on blood and CSF is the diagnostic method of choice.

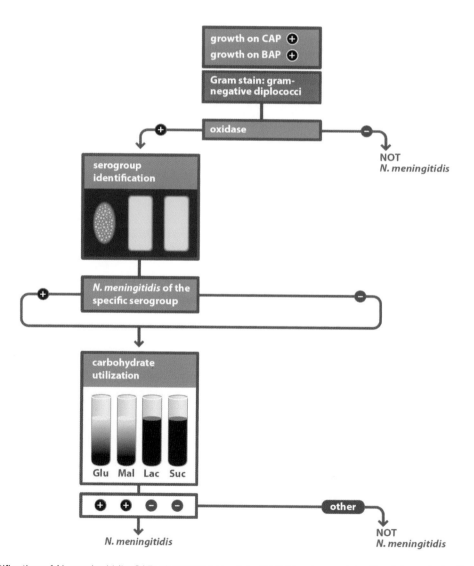

Figure 26.9 Identification of *N. meningitidis.* CAP: chocolate agar plate; BAP: blood agar plate. *Reprint permission kindly given for image in Laboratory Methods for the Diagnosis of Meningitis caused by Neisseria meningitidis, Streptococcus pneumoniae, and Haemophilus influenzae CDC, Centers for Disease Control and Prevention August 1998, Figure 15, Page 36.*

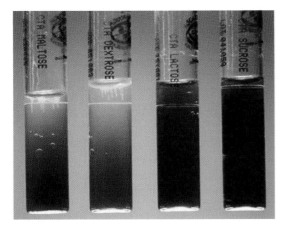

Figure 26.10 Cystine trypticase agar-sugar reactions differentiating *N. meningitidis* from other *Neisseria* species. Acid production (yellow color) shows oxidative utilization of dextrose (glucose) and maltose with no utilization of lactose and sucrose. *Reprint permission kindly given for image in Laboratory Methods for the Diagnosis of Meningitis caused by Neisseria meningitidis, Streptococcus pneumoniae, and Haemophilus influenzae CDC, Centers for Disease Control and Prevention August 1998, Figure 18, page 41.*

DIFFERENTIAL DIAGNOSIS

Bacterial meningitis caused by *Haemophilus influenzae*, *Streptococcus pneumoniae*, *Borrelia burgdorferi*, *Mycobacterium tuberculosis*, *Streptococcus agalactiae*, *Escherichia coli*, *Listeria* species, *Staphylococcus aureus*, and viral meningitis should be considered in the differential diagnosis. In immunocompromised patients such as those with HIV infection, meningitis caused by the fungus *Cryptococcus neoformans*, the protozoan *Toxoplasma gondii*, and the herpes simplex viruses should be considered. In patients presenting with the purpuric rash, **erythema multiforme**, **Henoch-Schönlein purpura**, and hypersensitivity, **vasculitis** could be considered. Other conditions to be considered are **Dengue fever**, viral hemorrhagic fever, enterovirus infection, gonococcal infection, infective **endocarditis**, malaria, Rocky Mountain spotted fever, and thrombotic thrombocytopenic purpura.

5. HOW IS THE DISEASE MANAGED AND PREVENTED?

MANAGEMENT

Because of the rapidity and severity of deterioration in the condition of patients with meningococcal disease, antibiotic therapy should never await the results of laboratory diagnostic procedures. Successful treatment includes early recognition, prompt parenteral antibiotics, aggressive fluid resuscitation in cases of meningococcal sepsis, frequent re-evaluation and, in patients with a poor prognosis, supportive therapy and transfer to an intensive care unit.

The Infectious Disease Society of America currently recommends ceftriaxone or cefotaxime for the treatment of meningococcal meningitis and septicemia. Penicillin G,

ampicillin, chloramphenicol, fluoroquinolone, and aztreonam are alternative therapies.

Individuals with close (kissing) contact during the week before onset of illness are reported to have a 1000-fold increased risk of being infected. Such individuals include household members, day-care contacts, cell-mates, or individuals exposed to infected nasopharyngeal secretions through kissing, mouth-to-mouth resuscitation or other means of contact with oral secretions. For such individuals, the US and several European countries recommend chemoprophylaxis with rifampin, ciprofloxacin or ceftriaxone. Prophylaxis must also be given to the index case to prevent onward spread.

ANTIBIOTIC RESISTANCE

Unlike *N. gonorrhoeae* that readily acquires antibiotic resistance, *N. meningitidis* has remained incredibly sensitive to antibiotics. However, a strain of beta-lactamase-producing *Neisseria meningitidis* that is resistant to both penicillin and ciprofloxacin has been observed in the US and parts of Europe. Chloramphenicol-resistant isolates have been detected in South-East Asia.

PREVENTION

As regards vaccines, currently, a tetravalent meningococcal capsular polysaccharide-protein conjugate vaccine (MCV4) that protects against meningococcal disease caused by serogroups A, C, Y, and W-135 is licensed for use in persons aged 11–55 years. The conjugate vaccine (MCV4) is a T-cell-dependent immunogen that induces protective immunity in infants, confers long-lasting protection, and provides herd immunity by reducing nasopharyngeal carriage and transmission. In the UK serogroup, C conjugate vaccine is given as three doses at 3, 4, and 12 months of age. The Advisory Committee on Immunization Practices (ACIP) of the US Centers for Disease Control and Prevention (CDC) recommends routine immunization with MCV4 of persons aged 11–12 years, at high-school entry if previously unvaccinated, and of college freshmen living in dormitories. Immunization is also recommended for persons with deficiencies in the complement pathway (C3, C5–9) or who are asplenic. Travelers to Mecca, Saudi Arabia during the annual Hajj and Umrah pilgrimage must have proof of immunization. Immunization is recommended for persons traveling to the meningitis belt in Africa during the dry season, which extends from December through June. A serogroup X conjugate vaccine is under development for the prevention of epidemic meningitis in Africa.

In contrast to the MVC4 conjugate capsule vaccines, the serogroup B vaccines consists of one or more non-capsule immunogens.

The reason for this is that the serogroup B capsule is composed of repeating units of α2-8 *N*-acetyl neuraminic acid, which is not only a poor immunogen in humans, but is identical to polysialic acid commonly found on human cell-surface glycoproteins. Thus, there is the possibility that a capsule vaccine might induce autoreactive antibodies that might result in immunopathology. There are two serogroup B vaccines available: Bexsero that contains Factor H Binding Protein (fHbp), *Neisseria* heparin-binding Antigen (NHBA), Neisserial adhesin A (NadA) and outer-membrane vesicles (OMV) from *Neisseria meningitidis* group B strain NZ98/254; Trumenba that contains two recombinant fHbp one from each of the two fHbp sub-families A and B. In the US, the serogroup B vaccine is recommended for the following groups of people age 10 years and older: those who have functional

or anatomic asplenia (including sickle cell disease); those with a persistent complement component deficiency; and those who are exposed during an outbreak caused by serogroup B. For adolescents and young adults who do not fall into these risk groups, the United States Advisory Committee on Immunization Practices (ACIP) recommends that a MenB series may be administered to people 16 through 23 years of age (preferred age 16 through 18 years) on the basis of shared clinical decision-making between the doctor and the patient. To date, the MenB vaccine has been introduced into publicly funded national routine immunization programmes in Ireland, Italy, and the UK.

SUMMARY

1. WHAT IS THE CAUSATIVE AGENT, HOW DOES IT ENTER THE BODY, AND HOW DOES IT SPREAD A) WITHIN THE BODY AND B) FROM PERSON TO PERSON?

- *Neisseria meningitidis* is one of the three principal causes of bacterial meningitis, the others being *Hemophilus influenzae* and *Streptococcus pneumoniae*.
- *Neisseria* species are capnophilic, gram-negative cocci. The cocci are often found in pairs where their adjacent sides are flattened giving them a coffee-bean appearance.
- *N. meningitidis* is oxidase-positive, catalase-positive, and produces acid from glucose and maltose by oxidation.
- *N. meningitidis* is an obligate human pathogen whose principal habitat is the nasopharynx.
- *Neisseria* species have a typical gram-negative envelope but the outer leaflet of their outer membrane contains lipooligosaccharide (LOS) instead of lipopolysaccharide (LPS).
- The core polysaccharide of LOS undergoes antigenic variation exposing a variety of epitopes on the cell surface.
- The cell surface is decorated with pili, which undergo antigenic and phase variation.
- Opacity-associated proteins (Opa) are important in the ability of the organism to adhere tightly to epithelia.
- The most important virulence determinant of the meningococcus is the production of a capsule.
- Meningococci can switch the type of capsule they express.
- *N. meningitidis* is adept at acquiring iron from its human host, which contributes to its ability to replicate on mucosal surfaces, intracellularly, and in the blood.
- *N. meningitidis* secretes an IgA1 protease, which may contribute to virulence.
- Following entry via the nasopharyngeal epithelium, the meningococci may seed the bloodstream and, from there, the meninges and many other sites including joints and the skin.
- The meningococcus is spread horizontally (person to person) by respiratory droplets or direct contact.
- At any given time, about 10% of a population harbors the meningococcus in the nasopharynx.

- The rate of carriage increases with age to a maximum in late adolescence and early adulthood, when as many as one third of individuals may harbor the meningococcus.
- The carriage rate is related to the degree of crowding such as is encountered in barracks, dormitories (halls of residence), prisons, and so forth.
- Meningococcal meningitis occurs sporadically in small clusters throughout the world. In temperate regions, the number of cases increases in winter and spring.
- In the US, the incidence of meningococcal disease is approximately 1 case per 100000 population, with a fatality rate of about 10%.
- Infections are most commonly caused by serogroups B and C.
- The predominant serogroups in various parts of the world are as follows: Asia and Africa, A, C, W135, and X; Europe, North and South America, B, C, and Y.
- The highest incidence of meningococcal disease occurs in sub-Saharan Africa, in an area from Senegal to Ethiopia termed the "meningitis belt", where attack rates as high as 1000 cases per 100000 population have been reported.

2. WHAT IS THE HOST RESPONSE TO THE INFECTION AND WHAT IS THE DISEASE PATHOGENESIS?

- At the mucosal surface, defensins are likely the principal mediator of innate defense and SIgA antibody mediates humoral immunity.
- In the tissues, the innate mannose-binding lectin and the **alternative pathways** of the complement system serve as the primary mechanism of meningococcal killing.
- Anti-meningococcal bactericidal IgM and IgG antibodies play the principal role in acquired immune defense. Antibodies are induced and maintained as a result of intermittent carriage of strains of *N. meningitidis*, and constant exposure to the commensal *N. lactamica*.
- The role of cell-mediated immunity in protection against *N. meningitidis* is not well understood.
- *N. meningitidis* secretes an IgA1 protease, which may subvert the immune response and contribute to virulence.

Continued...

...continued

- Pathogenesis is a function of the characteristics of the strain of *N. meningitidis*, particularly the propensity to release LOS in the form of blebs.
- During growth and lysis, the meningococcus releases LOS, which induces the release of large amounts of pro-inflammatory cytokines, particularly tumor necrosis factor-α (TNF-α).

3. WHAT IS THE TYPICAL CLINICAL PRESENTATION AND WHAT COMPLICATIONS CAN OCCUR?

- The onset of meningococcemia may be insidious and can result in several outcomes ranging from a short episode of febrile, flu-like symptoms to **septicemia** and meningitis.
- A hemorrhagic rash develops in about 80% of bacteremic patients, which begins as petechiae that occur initially on the extremities and trunk but may progress to involve any part of the body.
- As meningococcemia proceeds, pustules, bullae, and hemorrhagic lesions with central necrosis may develop.
- Fulminant meningococcal sepsis (FMS), comprising shock and DIC, occurs within a matter of hours, which may or may not be accompanied by meningitis.

4. HOW IS THE DISEASE DIAGNOSED, AND WHAT IS THE DIFFERENTIAL DIAGNOSIS?

- A gram stain of blood or CSF or material obtained from the characteristic skin lesion is informative and will reveal extracellular and intracellular coffee-bean shaped gram-negative diplococci.
- *Neisseria meningitidis* is fastidious and is usually cultured using chocolate agar although it will grow on blood agar.
- For clinical specimens obtained from mucosal surfaces, modified Thayer-Martin (MTM) medium is commonly used in the US.
- Plates are incubated at 35–37°C in 3–7% CO_2 in a humid environment.
- Identification of the meningococcus is achieved by subjecting oxidase-positive and catalase-positive, gram-negative diplococci to a panel of sugars for carbohydrate utilization tests. The meningococcus oxidizes glucose and maltose.
- Specific fluorescent antibodies, chromogenic substrates, or DNA probes may be used.
- Rapid tests are available commercially that detect capsule antigen in CSF and urine by latex agglutination and PCR.

- Serogrouping can be accomplished by agglutination of bacterial cells from a colony with specific anti-capsule antisera.
- Bacterial meningitis caused by *H. influenzae* and *Streptococcus pneumoniae* should be considered in the differential diagnosis.
- In patients presenting with the purpuric rash, erythema multiforme, and hypersensitivity vasculitis are part of the differential diagnosis.
- Other conditions to be considered are Dengue fever, Ebola virus, enterovirus infection, gonococcal infection, infective endocarditis, malaria, Rocky Mountain spotted fever, and thrombotic thrombocytopenic purpura.

5. HOW IS THE DISEASE MANAGED AND PREVENTED?

- Because of the rapidity and severity of deterioration in the condition of patients with meningococcal disease, antibiotic therapy should never await laboratory diagnostic procedures.
- Successful treatment includes early recognition, prompt parenteral antibiotics, frequent re-evaluation and, in patients with a poor prognosis, supportive therapy and transfer to an intensive care unit.
- Antibiotics for the treatment of meningococcal disease include penicillin G and ceftriaxone, cefotaxime, cefuroxime. In patients hypersensitive to β-lactam antibiotics, chloramphenicol is used.
- Decreased susceptibility of meningococci to penicillin has been observed in several countries.
- Close contacts of the index case should receive chemoprophylaxis with rifampin, ciprofloxacin or ceftriaxone.
- A tetravalent meningococcal capsular polysaccharide-protein conjugate vaccine (MCV4) that protects against serogroups A, C, Y, and W-135 is given to children aged 11–12 years, at high-school entry if previously unvaccinated, and to college freshmen living in dormitories. Immunization is also recommended for persons with deficiencies in the complement pathway (C3, C5–9) or who are asplenic.
- A conjugate vaccine against serogroup B is recommended for certain susceptible groups in the US but is included as part of the routine immunization programs in certain European countries.
- Immunization is recommended for persons traveling to the meningitis belt in Africa during the dry season, which extends from December through June.

FURTHER READING

Goering R, Dockrell H, Zuckerman M, Chiodini PL. Mims' Medical Microbiology and Immunology, 6th edition, Elsevier, Philadelphia, 2018.

Murray PR, Rosenthal KS, Pfaller MA. Medical Microbiology, 9th edition. Elsevier/Mosby, Philadelphia, 2021.

REFERENCES

Batista RS, Gomes AP, Gazineo JLD, et al. Meningococcal Disease, a Clinical and Epidemiological Review. Asian Pac J Trop Med, 10: 1019–1029, 2017.

Booya R, Gentilec A, Nissend M, et al. Recent Changes in the Epidemiology of Neisseria meningitidis Serogroup W

Across the World, Current Vaccination Policy Choices and Possible Future Strategies. Hum Vaccin Immunother, 15: 470–480, 2019.

Ladhani SN, Lucidarme J, Parikh SR, et al. Meningococcal Disease and Sexual Transmission: Urogenital and Anorectal Infections and Invasive Disease due to Neisseria meningitidis. Lancet, 395: 1865–1877, 2020.

Pizza M, Rappuoli R. Neisseria meningitidis: Pathogenesis and Immunity. Curr Opin Microbiol, 23: 68–72, 2015.

Soriani M. Unraveling *Neisseria meningitidis* Pathogenesis: From Functional Genomics to Experimental Models. F1000Res, 6: 1228, 2017.

Vaz LE. Meningococcal Disease. Pediatr Rev, 38: 158–169, 2017.

WEBSITES

Centers for Disease Control and Prevention, Meningococcal Disease, 2022: https://www.cdc.gov/meningococcal/index.html

Centers for Disease Control and Prevention, Morbidity and Mortality Weekly Report (MMWR), 2020: https://www.cdc.gov/mmwr/volumes/69/wr/mm6924a2.htm

European Centre for Disease Prevention and Control, Meningococcal disease: https://www.ecdc.europa.eu/en/meningococcal-disease

Nguyen N, Ashong, D. Neisseria meningitidis, StatPearls, 2020: https://www.ncbi.nlm.nih.gov/books/NBK549849/

StatPearls, Neisseria meningitidis, 2022: https://www.statpearls.com/ArticleLibrary/viewarticle/25640

WebMD, An Overview of Meningococcal meningitis, 2019: https://www.webmd.com/children/meningococcal-meningitis-symptoms-causes-treatments-and-vaccines

World Health Organization, Meningococcal meningitis, 2021: https://www.who.int/news-room/fact-sheets/detail/meningococcal-meningitis

Students can test their knowledge of this case study by visiting the Instructor and Student Resources: [www.routledge.com/cw/lydyard] where several multiple choice questions can be found.

Norovirus

27

A doctor was called to a home for the elderly because two residents had developed vomiting and watery diarrhea within 2 days of each other – an 80-year-old woman followed by an 87-year-old man. On taking a history, the doctor found that both patients had flatulence and stomach cramps during the first 24 hours of illness followed by vomiting, diarrhea, aching joints, and neck pains. The doctor examined the patients; the first case had recovered, and the symptoms were resolving in the second case. She requested a stool sample from both patients to be sent to the local microbiology laboratory to be tested for bacterial and viral pathogens. On the doctor's return to the home the next day, both patients were well and a telephone message from the laboratory reported norovirus detected in both stool samples. The local health authority was notified of a potential outbreak. In response, the community physician advised scrupulous attention to handwashing to prevent further spread of the infection.

However, the next day, another resident became ill with diarrhea and vomiting and, a few days later, a member of staff developed the same symptoms and had to leave work as she was unwell. Five days afterward, a five further staff members telephoned to say that they were sick. A nonresident day-patient also reported vomiting 2 days after her last attendance at the home. An outbreak of norovirus-induced vomiting and diarrhea was declared.

COURSE OF THE OUTBREAK

Setting

The setting was a recently refurbished home for the elderly. A total of 51 full-time and part-time staff cared for 50 resident and 60 nonresident day-patients, who each attended 1 day a week.

Course

The entire outbreak lasted 9 weeks (Figure 27.1). Most of those affected recovered completely within 72 hours. The first case reported was in a resident who had little contact with the outside community. The first day-patient to be affected became ill 2 days after the initial case. None of the staff became unwell until 9 days after the outbreak began. In all, 34/51 (67%) staff, 27/50 (53%) residents, and 9/60 (15%) day-patients were affected.

Investigation of the outbreak

The time course of the outbreak (Figure 27.1) suggested serial transfer from one individual to another. After several visits to the home, Environmental Health Officers reported that, although the kitchens and preparation of food were adequate, hand-washing facilities were poor throughout the home, with very few paper towel dispensers. Because of the design of the building, soiled bedpans

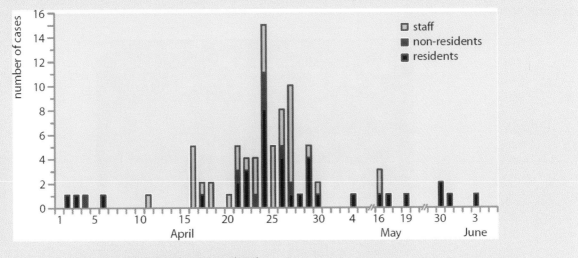

Figure 27.1 Time course of the norovirus outbreak.

Continued...

...continued

had to be carried though the dining area so that they could be emptied into uncovered sluices located in rooms alongside baths. This meant that the surrounding surfaces in both the dining area and around the sluices would be contaminated with virus particles due to air-borne dispersal of fecal material. Thus, fecal contamination of the environment in the dining and washing areas accounted, at least in part, for the continuing person-to-person spread of norovirus infection.

Control measures

Several measures, which caused considerable practical difficulties for both residents and staff, were instituted to restrict the spread of infection. The home was closed to new admissions and the meals-on-wheels service to the local community was suspended. However, no attempt was made to isolate affected residents and visiting continued. The transfer of day-patients from the home to a local hospital was discontinued after a secondary outbreak of **gastroenteritis** involving several hospital patients. Three bathrooms in the home were designated sluice rooms, the baths being no longer used, and the remaining three were used for bathing only. All toilets were separate from the bathrooms and some were designated for the exclusive use of those with gastroenteritis. Staff

who wished to return to work as soon as their symptoms subsided were advised to remain away for an additional 48 hours so that they did not re-introduce the infection into the home. Soiled bedpans were placed in polythene bags and then into a closed bin before being transported through the dining room into the sluice room. Handwashing facilities were improved with the installation of more paper towel dispensers.

Comment

The high attack rate and long duration of the outbreak were related to the vulnerability of the elderly population (average age 84 years) and the difficulties in maintaining high standards of hygiene. Although the home had recently been refurbished and internally redesigned, the planners had not fully considered the facilities necessary for the care of aged and infirm residents, nor had they consulted community physicians or microbiologists at any stage of the planning process. The standards should have been comparable to those of a hospital ward for the care of the elderly. More building work, at considerable cost, was undertaken to redesign parts of the building to separate sluice rooms from rooms with baths.

Note – This description of an outbreak is adapted from a report of a real one that happened over 30 years ago but the principles it illustrates still stand. Case taken from report in Communicable Diseases in 1988 by JJ Gray, KN Ward and IR Clarke.

1. WHAT IS THE CAUSATIVE AGENT, HOW DOES IT ENTER THE BODY AND HOW DOES IT SPREAD A) WITHIN THE BODY AND B) FROM PERSON TO PERSON?

CAUSATIVE AGENT

Noroviruses have a positive sense single-stranded (ss) RNA genome with a polyadenylated tail; this molecule is linked covalently to a viral protein (VPg), which may be important in infectivity and genome expression. The virus

has an icosahedral capsid and is unenveloped, hence relatively heat and acid stable, and ether resistant. There are three genogroups, I, II, and IV, of human noroviruses subdivided into altogether about 30 genotypes.

The noroviruses were formerly known as Norwalk viruses (because of their first identification in Norwalk, USA) or small, round structured viruses (SRSVs) – the latter term arose because of their characteristic electron microscopic appearance – feathery, ragged lacking a distinct surface structure (Figure 27.2A). The human noroviruses belong to the family of caliciviruses – the other group are the sapoviruses, which have a structure distinct from noroviruses (Figure 27.2B).

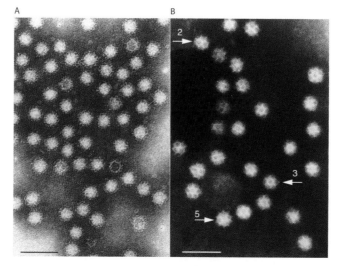

Figure 27.2 Electron microscopic appearance of human enteric caliciviruses. (A) Noroviruses have a characteristic ragged edge and lack the distinctive surface morphology of sapoviruses. (B) The distinctive surface structures on sapoviruses with cup-shaped depressions giving a "Star of David" appearance. Two-, three- and five-fold axes of symmetry indicated. Bar: 100 nm for each panel. *From Topley & Wilson Microbiology and Microbial Infection, 10th edition CD Set, 2005, Figure 42.7, Brian E J Mahey et al. With permission from John Wiley and Sons.*

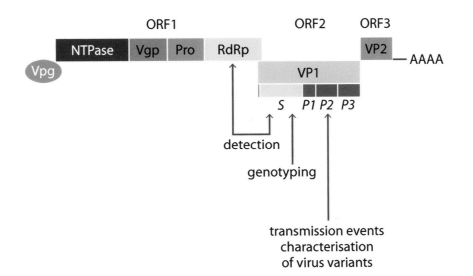

Figure 27.3 Norovirus genome and its ORFs showing virus structural and non-structural proteins and their use for virus detection and characterization (see Section 4).

The genomic structure of noroviruses is shown in Figure 27.3. There are three open reading frames (ORFs):

- the first encodes a non-structural polyprotein containing nucleotide triphosphatase (NTPase)/helicase, viral protein genome-linked (VPg), protease (Pro) and RNA-dependent RNA polymerase (RdRp);

- the second encodes the capsid protein (VP1), which determines the antigenic phenotype and interacts with host-cell receptors. VP1 spontaneously assembles into virus-like particles (VLPs) when expressed in insect cells by a recombinant baculovirus. The X-ray crystallographic structure of these empty capsids shows two domains, a shell (S), and a protruding (P) (Figure 27.4). The P domain is variable and further divided into the P1 and P2 subdomains. The surface exposed P2 subdomain is thought to dictate binding to the cell, while the conserved S domain forms the core of the viral particle;

- the third, a minor structural component of the virion (VP2).

Due to the lack of an efficient cell-culture system for human norovirus propagation, little is known of the virus replication cycle and functions of the viral-encoded proteins in host cells are still poorly understood.

ENTRY AND SPREAD WITHIN THE BODY

After ingestion with food or drink, the virus passes through the stomach and, due to its inherent resistance to acid, arrives undamaged in the small intestine. Infection is thought to be limited to the small intestine but direct evidence for this is very difficult to obtain in the human host (see Section 2). The **incubation period** is 12–48 hours. Virus is found in vomit and feces during the illness and fecal shedding continues for 3 weeks or more after recovery.

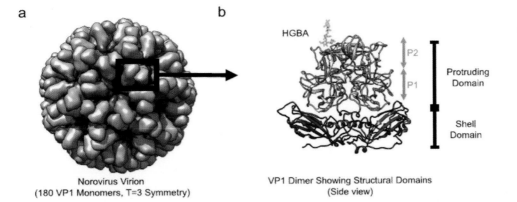

Norovirus Virion
(180 VP1 Monomers, T=3 Symmetry)

VP1 Dimer Showing Structural Domains
(Side view)

Figure 27.4 Structure and variable antigenic sites of the human norovirus major capsid protein, VP1. (A) The norovirus capsid is composed of 90 dimers of VP1 arranged in icosahedral symmetry. (B) The VP1 protein is divided into the conserved shell (S) in blue and the variable protruding (P) domains. The P domain is further divided into the P1 and P2 subdomains. The surface-exposed P2 subdomain is thought to dictate binding to the cellular attachment factors, histo-blood group antigen (HGBA) carbohydrates (highlighted in green), while the S domain forms the core of the viral particle as shown in blue. *From Ford-Siltz LA, et al. (2021) Gut Microbes, 13:1–13. doi: 10.1080/19490976.2021.1900994. Published under CC BY 4.0.*

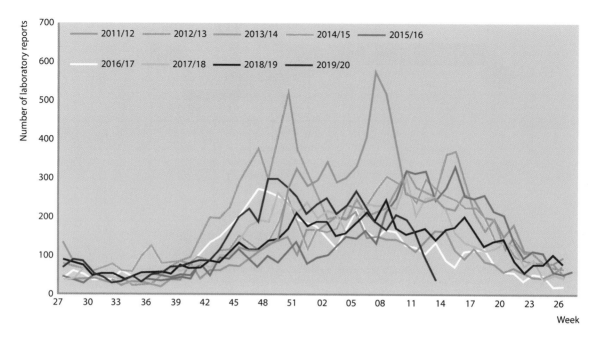

Figure 27.5 Seasonal comparison of laboratory reports of norovirus 2011/12–2019/20 (England and Wales). *From PHE National norovirus and rotavirus Report: https://assets.publishing.service.gov.uk/government/uploads/system/uploads/attachment_data/file/878578/Norovirus_ update_2020_weeks_12_to_13.pdf*

PERSON-TO-PERSON SPREAD

Noroviruses are spread directly from person to person via the **fecal-oral** route but also via ingested aerosolized vomit. Indirect transmission occurs via environmental surfaces (**fomites**) where the virion can remain for at least 2 weeks. Food is contaminated most often by infected food handlers but also by using contaminated water in food preparation. Shellfish are a common source of infection, especially oysters – these are filter feeders and concentrate virus from sewage-polluted water. Swimming in sewage-polluted water also carries the risk of norovirus infection. The infection is highly contagious – the minimal infectious dose is low, being 10–100 **virions** – and the attack rate in family/close contacts is at least 30%.

EPIDEMIOLOGY

Norovirus accounts for more than 90% of viral gastroenteritis outbreaks worldwide; one fifth of children younger than 5 years are hospitalized with noroviral gastroenteritis causing up to 200 000 deaths in the developing world.

Noroviruses cause most of the gastroenteritis cases identified in community-based studies in developed countries being the commonest cause of outbreaks of nonbacterial gastroenteritis in both children and adults. In the UK and US, sporadic infection occurs throughout the year but an epidemic sweeps through the community in the winter with a high attack rate (Figure 27.5). These winter outbreaks are a major problem in hospitals with consequent ward closures.

Figure 27.6 gives an example of the settings and presumptive modes of transmission for outbreaks of norovirus gastroenteritis. Outbreaks are a common problem in hospitals and in residential homes for the elderly. Persistent outbreaks occur on cruise ships and frequently involve multiple routes of transmission including consumption of food or water, directly from person to person, and from contamination of the environment (fomites). Such outbreaks can only be terminated by closing the ship and deep cleaning or removing all possibly

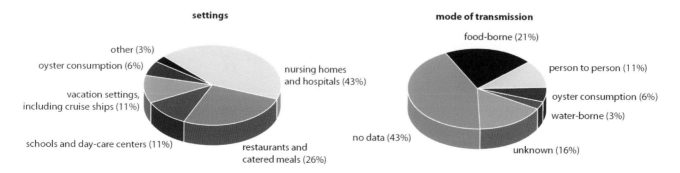

Figure 27.6 Settings and presumptive modes of transmission for 90 outbreaks of nonbacterial gastroenteritis in the US from January 1996 to June 1997. Noroviruses were detected in 86 (96%) of the 90 outbreaks by reverse transcription PCR.

infected surfaces including carpets and curtains. Norovirus was also identified as a cause of outbreaks among American military personnel during the Gulf War.

2. WHAT IS THE HOST RESPONSE TO THE INFECTION AND WHAT IS THE DISEASE PATHOGENESIS?

HOST RESPONSE TO INFECTION

Investigation of the nature of immunity to norovirus infection has been hampered by the lack of *in vitro* cultivation. Limited information came from volunteer studies conducted about four decades ago. Immunity to norovirus re-infection persisted for up to about 14 weeks in volunteers after previously induced norovirus illness. Viral genetic variation cannot solely explain the absence of long-term immunity to noroviruses; these human volunteer studies involved repeated challenge of the same individual with an inoculum of identical virus. Thus, it seems that mucosal immunity to norovirus is inherently short-lived.

In one volunteer study, six of 12 individuals developed gastroenteritis upon exposure to norovirus and the same six individuals became ill again upon rechallenge 27–42 months later. The other six individuals were resistant to the first infection and on rechallenge. It later transpired that the genetic make-up of an individual plays a role in susceptibility to infection; about 20% of the European population is resistant to norovirus infection. An explanation has recently emerged in that noroviruses bind to intestinal mucosal epithelial cell-surface carbohydrates of the ABH histo-blood group antigens. These antigens are not expressed on mucosa in some individuals who are so-called non-secretors. Non-secretor status does not provide total resistance to disease as some noroviruses have an identified receptor different from the blood-group antigens.

The genetic variability of norovirus is an important cause of re-infection. New variants of genogroup II-4 appear every 2–8 years causing an increased number of outbreaks throughout the world and a shift in seasonality (Figure 27.7). Genetic and antigenic variability rapidly evolves due to point mutations, as the viral RNA-dependent RNA polymerase has no proof-reading capacity and also recombination between different strains. For a new genotypic variant to predominate over other cocirculating genotypes requires replicative and other fitness advantages including greater transmissibility with a lower infectious dose, and a mechanism to evade immune surveillance.

Finally, although preexisting serum antibodies do not protect against re-infection, the importance of the adaptive immune response for the rapid resolution of norovirus infection is apparent because immunocompromised persons may be symptomatic and shed virus for months.

PATHOGENESIS

It has been proposed that abnormal gastric motor function is responsible for nausea and vomiting, as a marked delay in gastric emptying was observed in volunteers who became ill after experimental infection, but the precise mechanism is unknown. Another equally plausible explanation would be inflammation of the pyloric junction between the stomach and the intestine.

The pathogenesis of norovirus-induced diarrhea is not fully established but is presumed to be noninflammatory or secretory as evidenced by histopathologic changes in the small intestine showing damage to and blunting of the small intestinal villi (Figure 27.8). Enterocytes appear to be the primary site of norovirus infection. Viral capsid antigen has been detected in these cells when biopsies from immunocompromised patients with chronic norovirus gastroenteritis were examined. Viral antigen has also been detected in macrophages, dendritic cells, and T cells in the **lamina propria** underlying the enterocytes.

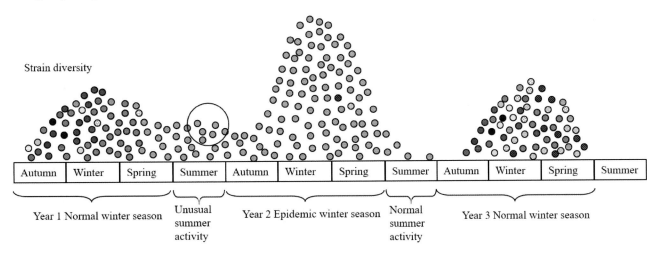

Figure 27.7 Schematic representation of evolution of a new norovirus variant. Year 1: Autumn narrowing diversity of strains as new variant (circled) emerges. Spring new variant causes out-of-season outbreaks. Year 2: Epidemic winter season as population has no immunity to new variant. Year 3: Return to normal season, wide diversity of new strains at the beginning narrowing as season progresses. Population protected in the short term against new variant but susceptible to other strains due to short-term immune protection. *Courtesy of Dr Jim Gray.*

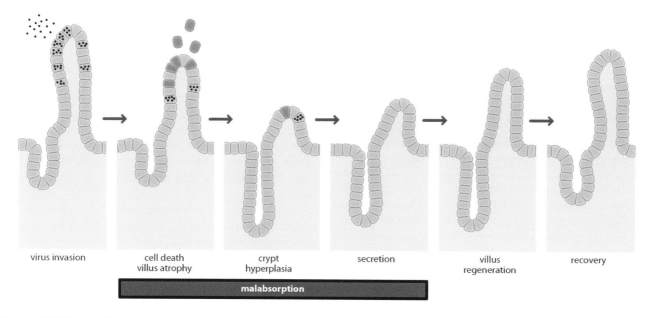

Figure 27.8 Presumed mechanism of norovirus pathogenesis – development of damage to gut mucosa and ensuing diarrhea. Destruction of infected enterocytes results in decreased fluid absorption and loss of water, sugar, and salts.

However, it is not clear whether virus is replicating in these cells or merely the result of phagocytosis of infected epithelial cells.

3. WHAT IS THE TYPICAL CLINICAL PRESENTATION AND WHAT COMPLICATIONS CAN OCCUR?

The incubation period is short – 12 to 48 hours. Because of its seasonal occurrence, the illness is known as "winter vomiting disease". Symptoms begin abruptly with nausea, abdominal cramps, and vomiting which is characteristically projectile. Diarrhea may be absent, mild, or severe but, when present, it is not bloody, lacks mucus, and may be watery. Headache and myalgia are common and low-grade fever occurs in 50% of cases. Although infections in the very young and elderly can be quite severe, the disease is usually self-limiting, and symptoms subside within 24 to 48 hours.

Complications are rare but the effects of dehydration (metabolic alkalosis, **hyponatremia**, **hypokalemia**, and renal failure) may occur. Immunosuppressed individuals are at risk of prolonged diarrhea, perhaps lasting for months with malnutrition and weight loss.

4. HOW IS THE DISEASE DIAGNOSED, AND WHAT IS THE DIFFERENTIAL DIAGNOSIS?

Sporadic cases are rarely investigated and what follows concerns the diagnosis of outbreaks of gastroenteritis. Such diagnosis is based on the characteristic clinical features of the illness, that is vomiting in over 50% of cases and mild diarrhea without blood or mucus lasting for about 48 hours, together with detection of norovirus in feces. A fecal sample, preferably obtained while the patient is symptomatic, should be tested by reverse transcription **polymerase chain reaction (PCR)** to detect the RNA-dependent RNA polymerase and VP1 RNA (Figure 27.3). In an outbreak at least five fecal samples from five different individuals should be tested to ensure detection. Genotyping and identification of variants for epidemiologic purpose requires VP1 gene rescue by PCR.

Before the development of such highly sensitive molecular testing, electron microscopy was widely used but this technique is limited by its insensitivity as it requires large numbers of norovirus particles in a stool sample, that is $>10^6$ per ml for detection. Many decades ago, immunoelectron microscopy showing antibody **seroconversion** was used to prove that norovirus detected in feces was the cause of gastroenteritis (Figure 27.9A and 27.9B) and is still used today in the research setting to concentrate virus.

DIFFERENTIAL DIAGNOSIS

The other possible viral causes are given in Table 27.1 – notably outbreaks of diarrhea are commonly caused by rotavirus in young children and the elderly. Fecal samples should be tested for *Salmonella*, *Shigella*, and *Campylobacter* spp. In addition, *Clostridium difficile* can cause serious outbreaks with high morbidity and even mortality in hospitals.

5. HOW IS THE DISEASE MANAGED AND PREVENTED?

MANAGEMENT

The illness is usually short and requires attention only to fluid and electrolyte replacement. Severe cases of persistent

Table 27.1 Viruses causing gastroenteritis and their epidemiologic features

Family	Virus	Endemic Infection in children	Epidemiologic features Outbreaks of infection in all ages	Food- or water-borne
	Group A rotavirus	++++	Children and elderly in hospitals or residential homes	No
	Group B rotavirus	–	Large outbreaks in China	No
	Group C rotavirus	–	Occur but uncommon	No
Caliciviruses	Noroviruses	+++	Major cause of outbreaks	Yes
	Sapoviruses	+/-	–	Yes
Adenovirus	Adenovirus types 40 and 41	++	–	No
Astrovirus	Human astrovirus	+	Family outbreaks	No

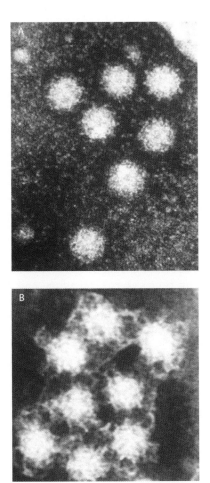

Figure 27.9 **Immunoelectron microscopy to show antibody seroconversion.** Noroviruses (size 27nm) from a patient with gastroenteritis visualized by electron microscopy (A) after incubation with serum taken from the patient in the acute phase of illness and (B) after incubation with serum taken from the patient in the convalescent phase of illness. Specific antibody molecules in the convalescent serum of the patient aggregate and coat the virus particles. *Reprint permission kindly given by Dr Jim Gray of the Health Protection Agency (originator of the image) and by Elsevier.*

reduced risk of spread	increased risk of spread
ward closure	**virus variant/mutant**
cohort nursing adherence to hand washing practices, soap and water	projectile vomiting **low infectoius dose** **high attack rate** staff affected (reduced staffing)
elbow/automatic taps on handbasins	soft furnishings contaminated
single rooms with en suite	**hand contact surfaces contaminated**
surface finishes allowing decontamination	unrestricted visiting/admissions open wards

affected patients

environment

Figure 27.10 Risk of norovirus spread within the hospital environment. *Courtesy of Dr Jim Gray.*

diarrhea in the immunosuppressed may also require **enteral** or **parenteral** nutrition. There is no antiviral drug available for treatment.

OUTBREAK CONTROL

As illustrated in the example given of an outbreak in a residential home, this is a difficult issue and outbreaks can grumble on for long periods and often do not end until most susceptible people have been infected. Because they lack an envelope, noroviruses are very resistant to adverse environmental conditions and even to commonly used disinfectants – this property facilitates their propensity to cause outbreaks (see below).

Control measures (Figure 27.10) rely on:

1. Limiting contact between infected and susceptible patients, by isolation of those affected in single rooms in hospital or nursing homes, and using contact (enteric) precautions – gloves, aprons, and scrupulous handwashing with soap and water as alcohol hand rubs are insufficient. Closing the hospital ward or relevant section of a nursing home to new admissions may be necessary. If the ward is closed, it should not open until more than 48 hours (the longest incubation period) – usually 72 hours.

2. Infected staff should stay off work for at least 48 hours after resolution of symptoms. This is particularly important for those handling food.

Note – although the virus is shed in feces over about 3 weeks, the virus is most communicable in the acute phase of illness when the virus load is highest.

PREVENTION

This relies not only on clean drinking water and efficient sewage disposal but also on good standards of personal and food hygiene plus adequate cleaning arrangements in hospitals and residential homes. Raw shellfish should be cooked before consumption and fruit washed if to be eaten uncooked.

Vaccine: An effective vaccine would give substantial economic and public health benefits. Currently, candidate vaccines are based on VLPs consisting of recombinant capsid proteins that are morphologically and antigenically identical to native virus and react specifically with convalescent human sera. Norovirus VLPs have been shown to be immunogenic when administered orally to human volunteers, inducing serum IgG and mucosal IgA. It seems likely that, as with the influenza vaccine, formulation would need to be changed as new genotypes of human noroviruses emerge.

SUMMARY

1. WHAT IS THE CAUSATIVE AGENT, HOW DOES IT ENTER THE BODY, AND HOW DOES IT SPREAD A) WITHIN THE BODY AND B) FROM PERSON TO PERSON?

- Norovirus – an unenveloped single stranded RNA virus.
- Noroviruses were formerly known as Norwalk or small, round, structured viruses.
- Three norovirus genogroups, GI, GII, and GIV, subdivided into about 30 genotypes infect humans.
- Norovirus infection has a seasonal incidence – "winter vomiting disease".
- New variants of norovirus genotypes regularly appear causing an increased number of outbreaks including summer out of season.
- Noroviruses are spread by the fecal-oral route.
- The main source of virus is another infected person, contaminated food, or water.
- Norovirus outbreaks are common in healthcare settings and cruise ships.

2. WHAT IS THE HOST RESPONSE TO THE INFECTION AND WHAT IS THE DISEASE PATHOGENESIS?

- Immunity is short-lived and re-infection is common. Very little is known regarding the pathogenesis of norovirus infection.
- The cause of the delay in gastric emptying responsible for the high incidence of vomiting episodes is unknown.

- Intestinal enterocytes appear to be the primary sites of viral infection. Viral antigen has been detected in macrophages, dendritic cells and T cells in the lamina propria cells; the significance of this is unclear.

3. WHAT IS THE TYPICAL CLINICAL PRESENTATION AND WHAT COMPLICATIONS CAN OCCUR?

- Self-limiting vomiting and diarrhea.
- Main complication is dehydration.
- Causes persistent diarrhea in the immunocompromised.

4. HOW IS THE DISEASE DIAGNOSED, AND WHAT IS THE DIFFERENTIAL DIAGNOSIS?

- The virus cannot be routinely grown in cell culture, so diagnosis relies on reverse transcription PCR to detect viral nucleic acid in vomit or feces.
- Differential diagnosis includes infection with rotavirus, *Salmonella*, *Shigella*, and *Campylobacter* spp. and, in the hospital setting, *Clostridium difficile*.

5. HOW IS THE DISEASE MANAGED AND PREVENTED?

- There is no specific treatment.
- There is no vaccine.
- Control of outbreaks includes reinforcing good hygiene, closure of hospital wards, and isolation of ill persons until 48 hours after symptoms have resolved.

FURTHER READING

Dolin R, Treanor JJ. Noroviruses and Sapoviruses (caliciviruses). In: Bennett JE, Dolin R, Blaser MJ, editors. Mandell, Douglas and Bennett's Principles and Practice of Infectious Diseases, volume 2, 8th edition. Elsevier/Saunders, Philadelphia, 2122–2127, 2015.

Goering RV, Dockrell HM, Zuckerman M, Chiodini PL. Medical Microbiology and Immunology, 6th edition. Elsevier, Philadelphia, 2019.

Wobus CE, Green KY. Caliciviridae: The Viruses and their Replication. In: Howley PM, Knipe DM, editors. Fields Virology, 7th edition. Wolters Kluwer, Philadelphia, 2021.

REFERENCES

Atmar RL, Ramani S, Estes MK. Human Noroviruses: Recent Advances in a 50-Year History. Curr Opin Infect Dis, 31: 422–432, 2018.

Chong PP, Atmar RL. Norovirus in Healthcare and Implications for the Immunocompromised Host. Curr Opin Infec Dis, 32: 348–355, 2019.

Ford-Siltz LA, Tohma K, Parra GI. Understanding the Relationship Between Norovirus Diversity and Immunity. Gut Microbes, 13: 1–13, 2021.

Glass RI, Parashar UD, Estes MK. Norovirus Gastroenteritis. N Engl J Med, 36: 1776–1785, 2009.

Green KY, Kaufman SS, Nagata BM, et al. Human Norovirus Targets Enteroendocrine Epithelial Cells in the Small Intestine. Nat Commun, 11: 2759, 2020.

Karandikar UC, Crawford SC, Ajami NJ, et al. Detection of Human Norovirus in Intestinal Biopsies from Immunocompromised Transplant Patients. J Gen Virol, 97: 2291–2300, 2016.

Lopman B, Vennema H, Kohli E, et al. Increase in Viral Gastroenteritis Outbreaks in Europe and Epidemic Spread of New Norovirus Variant. Lancet, 363: 682–688, 2004.

Roddie C, Paul J, Benjamin R, et al. Allogeneic Haematopoietic Stem Cell Transplantation and Norovirus Gastroenteritis: A Previously Unrecognised Cause of Morbidity. Clin Infect Dis, 49: 1061–1068, 2009.

WEBSITES

Centers for Disease Control and Prevention, Morbidity and Mortality Weekly Report, 2004: http://www.cdc.gov/mmwr/preview/mmwrhtml/rr5304a1.htm

Guidelines for the management of norovirus outbreaks in acute and community health and social care settings, 2012: https://assets.publishing.service.gov.uk/government/uploads/system/uploads/attachment_data/file/322943/Guidance_for_managing_norovirus_outbreaks_in_healthcare_settings.pdf

Students can test their knowledge of this case study by visiting the Instructor and Student Resources: [www.routledge.com/cw/lydyard] where several multiple choice questions can be found.

Plasmodium spp.

28

A 26-year-old model went to see her doctor about 1 week after returning from a job in the Gambia. She complained of an abrupt onset of bouts of shivering and feeling cold, vomiting, rigors, and profuse sweating accompanied by a headache and nausea. On examination, she was noted to be pale with a temperature of 39.5°C and had tachycardia. She gave a history of having taken anti-malarial tablets before and during her stay in the Gambia but was admitted to hospital with a provisional diagnosis of malaria.

1. WHAT IS THE CAUSATIVE AGENT, HOW DOES IT ENTER THE BODY AND HOW DOES IT SPREAD A) WITHIN THE BODY AND B) FROM PERSON TO PERSON?

CAUSATIVE AGENT

The organism causing malaria is *Plasmodium*, a eukaryotic protozoan that infects the erythrocytes of humans. It has the characteristics of eukaryotes, with a nucleus, mitochondria, endoplasmic reticulum, and so forth. There are approximately 156 named species of *Plasmodium* which infect various vertebrates. Five species of *Plasmodium* are able to infect humans: *P. falciparum*, *P. ovale*, *P. vivax*, *P. malariae* and *P. knowlesi*. *P. falciparum* is the most **virulent** species of malaria but *P. vivax* is the dominant malaria parasite in most countries outside of sub-Saharan Africa. All of these species have similar life cycles in which the organisms undergo both sexual and asexual reproduction in the vector and host and alternate between intracellular and extracellular forms. The female *Anopheles* mosquito is the vector for malaria. The risk of malaria transmission is therefore restricted to those areas where mosquitoes can breed and where the parasite can develop within the mosquito – see Epidemiology below.

ENTRY AND SPREAD WITHIN THE BODY

The transmission stage of *Plasmodium* is the sporozoite, which is injected into the bloodstream of a human when the female *Anopheles* mosquito takes a blood meal (Figure 28.1). The detailed life cycle is shown in Figure 28.2. Following the mosquito bite, at least some of the sporozoites remain in the dermis for some time before crossing the endothelial cells and entering the bloodstream; some pass directly into draining lymph nodes. Only a few dozen sporozoites are transmitted during feeding but there is rapid translocation into the liver to begin the first stage of disease.

Liver Stage (Pre-Erythrocytic Stage)

The blood-borne sporozoites localize in the liver via the sinusoids, where through interactions between their surface circumsporozoite protein (CSP – most abundant protein on the sporozoite surface) and highly sulfated heparan sulfate proteoglycans (HSPGs), they cross the sinusoidal cellular barrier through liver sinusoidal endothelial cells or Kupffer cells (KC). The sporozoites actively enter the hepatocytes (invade – rather than are taken up passively by **endocytosis**), and may traverse several hepatocytes before they switch to a replicative form. There, within a parasitophorous vacuole (to avoid lysosomal degradation), they increase in number and develop into schizonts. This asexual stage takes up to 2 weeks. Rupture of the liver cells releases the schizonts into the bloodstream as merozoites (with about 10–40000 being released from the liver). *P. vivax* and *P. ovale* also produce a resting stage within the liver cell called hypnozoites. In this case, some sporozoites once in the liver do not develop immediately into schizonts, but remain at an uninucleate stage, in a quiescent form named hypnozoite, before resuming

Figure 28.1 *Anopheles funestus* mosquito taking a blood meal from its human host. This mosquito species, together with *Anopheles gambiae*, is one of the two most important malaria vectors in Africa. Note the blood passing through the proboscis. *From the Centers for Disease Control & Prevention, Atlanta, Georgia. Image is found in the Public Health Image Library #7192. Additional photographic credit is given to James Gathany, Dr Frank Collins and the University of Notre Dame. The image was taken in 2005 by James Gathany.*

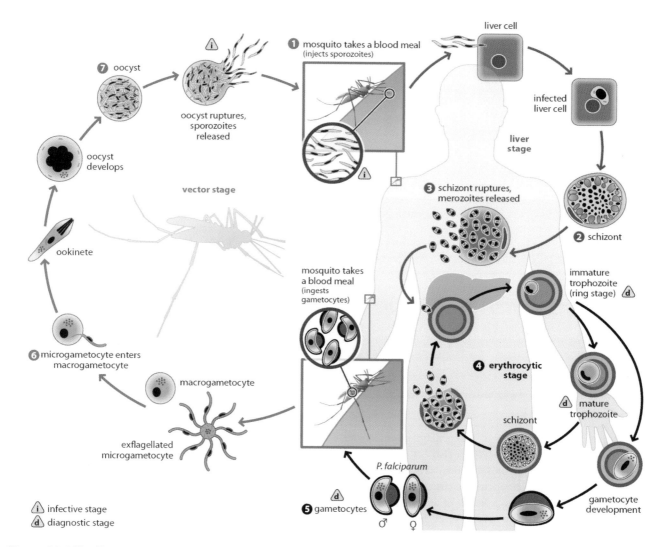

Figure 28.2 The lifecycle of *Plasmodium*. (1) The mosquito injects saliva containing sporozoites as it takes a blood meal and the parasite localizes in the liver (liver stage), where it undergoes a stage of development to produce a schizont which contains developing merozoites, in the infected liver cell (2). *P. vivax* and *P. ovale* also produce a resting stage within the liver cell called hypnozoites, which can persist in the liver and result in relapses months or even years later. The dead liver cell breaks open and the schizont ruptures producing small vacuoles (merosomes) which release their merozoites into the bloodstream. These invade erythrocytes (erythrocytic stage) and undergo developmental stages as trophozoites, which mature and produce schizonts, at which stage, the erythrocyte bursts rupturing the schizonts to release further merozoites (4). Further cycles of asexual development within uninfected erythrocytes occur, releasing more merozoites to infect further erythrocytes. Differentiation of the immature trophozoite into male and female gametocytes occurs in some erythrocytes (5) and these are ingested when a mosquito takes a blood meal. The male (microgametocyte – exflagellated) fertilizes the female macrogametocyte (6) to form a zygote within the intestine of the mosquito (vector stage) and this becomes an ookinete that invades the intestinal wall where it develops into an oocyte (7). The oocyte matures into sporozoites, which are released and migrate to the salivary gland of the mosquito. Here, they will be transmitted to a new human host when the mosquito takes a blood meal and the cycle starts again (1). *From the Centers for Disease Control & Prevention, Atlanta, Georgia. Image is found in the Public Health Image Library #3405. Additional photographic credit is given to Alexander J da Silva, PhD and Melanie Moser. The image was created in 2002.*

hepatic development induced by as yet unknown factors that cause relapses weeks, months or even years after the primary infection. The sporozoite has small vacuoles (micronemes) that release substances onto their surface in a co-ordinated way to ensure successful migration to and invasion of hepatocytes and production of merozoites.

Erythrocyte Stage

Merozoites (the smallest extracellular parasite form) invade and destroy erythrocytes giving rise to symptoms (see Section 3). Invasion is a multistep process that includes firstly binding of the merozoite, reorientation, discharge of secretory organelles (called rhoptries and micronemes), formation of an electron-dense "tight junction" between the merozoite apical end and the erythrocyte membrane. This leads to actinomyosin-powered entry into the erythrocyte with formation of a membrane-bound parasitophorous vacuole and resealing of the erythrocyte cell membrane. Within this vacuole, the parasite feeds on hemoglobin and multiplies. Invasion takes less than 30 seconds. Entry of the merozoites into erythrocytes is achieved through attachment of a number of surface molecules (merozoite surface proteins (MSPs) that are anchored by glycosylphosphatidylinositol (GPI), and a number of erythrocyte binding ligands (EBL). At least four have been identified for *P. falciparum* merozoites. It is generally accepted that glycophorins A, B, C, and D are important "receptors" on erythrocytes and are known for *P. falciparum* Glycophorin A on erythrocytes. Another important structure on erythrocytes is band 3 protein that binds MSP1 complex on *P. falciparum* merozoites. *P. vivax* has a specific reticular binding protein to enable it to attach and invade **reticulocytes** but not mature erythrocytes. In addition, *P. vivax* has surface molecules (**Duffy binding proteins** – DBP-1) that bind to Duffy blood group antigens on the erythrocytes. Although initially thought that DBP-1 was essential for binding of *P. vivax* to erythrocytes, some cases have been seen where infection occurs in individuals that genetically lack DBP-1. The search is on for other receptors for *P. vivax* merozoites on reticulocytes.

Within the erythrocyte, the merozoites undergo further development as a trophozoite (seen as a "ring" stage – see Figure 28.2) and then undergo asexual reproduction to produce schizonts, at which stage, the erythrocyte bursts releasing merosomes containing 16–32 daughter merozoites into the bloodstream.

Each asexual cycle takes 44–48 hours and is followed by cell rupture and re-invasion steps that induce periodic waves of fever in the patient (see Figure 28.3 and Section 3). This erythrocytic cycle may continue for months or years. However, in some erythrocytes, the trophozoites differentiate into male and female gametocytes and a mosquito taking a blood meal will take up some of the gametocyte-containing erythrocytes, heralding the sexual developmental phase in the vector.

Vector Stage

Within the intestine of the insect, male (exflagellated microgametocytes) and female gametes (macrogametocytes)

fuse to become a zygote. These become an ookinete, which then invades the intestinal wall where it develops into an oocyte. The oocyte develops into thousands of sporozoites, which then migrate to the mosquito's salivary gland.

PERSON-TO-PERSON SPREAD

The sporozoites are injected into an individual when an infected female *Anopheles* mosquito feeds and the whole cycle starts again.

NON-MOSQUITO SPREAD

Malaria can also be transmitted through blood transfusion, hypodermic needle sharing or accidents, and from mother to fetus.

EPIDEMIOLOGY

Malaria transmission occurs in five WHO regions. Globally, an estimated 3.4 billion people in 92 countries are at risk of being infected with malaria and developing disease (see map – the atlas project website), and 1.1 billion are at high risk (>1 in 1000 chance of getting malaria in a year). There was a marked reduction in global malaria case incidence and mortality rates between 2000 and 2019. The malaria case incidence rate (cases per 1000 population at risk) fell from 80 in 2000 to 57 in 2019. Total malaria cases declined from 238 million in 2000 to 227 million in 2019. The mortality incidence rate (deaths per 100 000 population at risk) was reduced from 25 in 2000 to 10 in 2019. The total number of deaths fell from 736 000 in 2000 to 409 000 in 2019. Children under 5 years of age are the most vulnerable group affected by malaria and in 2019 they accounted for 67% (274 000) of all malaria deaths worldwide. Africa continues to carry a disproportionately high share of the global malaria burden. In 2019, the region had 94%

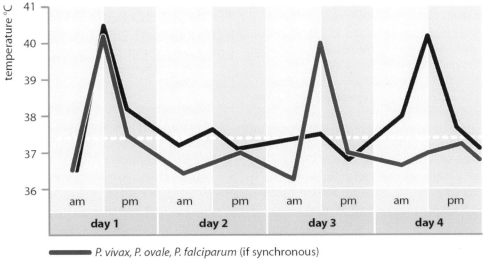

P. vivax, P. ovale, P. falciparum (if synchronous)
P. malariae

Figure 28.3 **Cyclical fever coincident with the release of merozoites.** *Adapted from Goering R, Dockrell H, Zuckerman M et al. (2008) Mim's Medical Microbiology, 4th edition, Figure 27.11. With permission from Elsevier.*

of all malaria cases and deaths. Six countries accounted for approximately half of all malaria deaths worldwide: Nigeria (23%), the Democratic Republic of the Congo (11%), United Republic of Tanzania (5%), Burkina Faso (4%), Mozambique (4%) and Niger (4%). In 2020, there was a slight increase in cases worldwide from 227 (in 2019) to 241 million cases. The estimated number of malaria deaths stood at 627000 in 2020. The WHO African Region carries a disproportionately high share of the global malaria burden. In 2020, the WHO African region was home to 95% of the malaria cases and 96% of malaria deaths.

With increasing international travel, prior to the SARS-2 (COVID-19) pandemic there continued to be a rise in the number of cases of malaria in travelers returning to non-malarious areas from countries where malaria is endemic. In Europe, in 2018, 8349 malaria cases were reported, 8347 (> 99%) of which were confirmed. Among 7338 cases with known importation status, 99.8% were travel related. As in previous years, the overall rate of confirmed malaria cases was higher among men than women (1.6 cases and 0.7 cases per 100000 population, respectively; male-to-female ratio 1.9:1). However, the travel-related cases were reduced during the SARS-CoV-2 pandemic due to limited travel.

Pregnancy is associated with a higher risk of malaria. An estimated 10000 pregnant women and 200000 of their infants died annually in sub-Saharan Africa as a result of malaria infection during pregnancy. HIV-infected pregnant women are at increased risk.

HUMAN GENETIC FACTORS THAT DECREASE THE INFECTION RATES OF PLASMODIUM

As already mentioned, the absence of DBPs in most West Africans prevents most infections by *P. vivax*, since it uses the Duffy blood group antigen-1 as a means of attachment. Sickle cell trait (heterozygous for HbS with HbA) gives an increasing amount of immunologic protection against malaria for young children during their first 10 years of life (a 29% reduction in malaria incidence). A number of mechanisms have been proposed to explain malaria resistance, including sickling of the infected red blood cells, increased splenic phagocytosis, premature hemolysis and parasite death, impaired hemoglobin digestion, weakened cytoadherence, acquired host immunity, translocation of HbS-specific parasite-growth inhibiting microRNAs, and induction of heme-oxygenase-1. Glucose-6-phosphate dehydrogenase (G6PD) deficiency confers resistance to malaria. Although the mechanism is unclear, it has been found that when the parasites are at the ring stage, erythrocytes deficient in the enzyme were phagocytosed 2–3 times more intensely than normal ring-stage erythrocytes. There was, however, no difference when the parasites were at the more mature trophozoite stage.

The Impact of the SARS-CoV-2 (COVID-19) Pandemic on Malaria

During the pandemic, a recent modeling analysis by the WHO predicted a >20% rise in malaria morbidity and >50% mortality in sub-Saharan Africa as a result of 75% reduction in routine malaria control measures. These might significantly increase the longer the pandemic goes on. Other indirect effects of the pandemic, particularly those that affect people's lives and well-being, such as increased malnutrition, poverty, and social instability, may further influence malaria burden. Current guidelines by WHO at the time of writing have included the continuation of all routine malaria control measures while adhering to Covid-19 local personal and physical distancing guidelines established by the authorities.

2. WHAT IS THE HOST RESPONSE TO THE INFECTION AND WHAT IS THE DISEASE PATHOGENESIS?

The understanding of the immune response to *Plasmodium* is far from complete. Some immunity does develop to the parasite with continuous infection (especially to blood-stage parasites) and children who survive early attacks become resistant to severe disease by about 5 years of age. Levels of parasites fall progressively until adulthood, when they are low or absent most of the time. This immunity, however, appears to be lost after spending a year away from exposure, presumably due to lack of repeated antigenic stimulation needed for its maintenance. **IgG** antibodies to blood-stage parasites are transferred across the placenta and limit parasitemia and severe disease in neonates. Thus, although immune responses do develop to different stages of the lifecycle in infected individuals, they are weak and cannot eradicate the parasite.

PRINCIPAL MECHANISMS THAT ARE THOUGHT TO BE RESPONSIBLE FOR IMMUNITY AT DIFFERENT STAGES OF THE LIFECYCLE OF PLASMODIUM

Liver Stage

- Initial entry of the sporozoites into the lymph nodes and bloodstream on the way to the liver: antibodies.
- Evidence for antibodies to sporozoites and CSP to block binding to hepatocytes.
- Activation of complement by the antibodies activates complement fixation, phagocytosis, and lysis by cytotoxic NK and NKT cells through their Fc receptors. It also recognizes parasite neoantigens at the surface of infected hepatocytes and kills through an antibody-dependent cell-mediated mechanism by KC and NK cells.

- Since the liver stage of malaria infection is clinically silent, it has long been thought that parasites inside hepatocytes grow undetected by the innate immune system. However, a number of pattern recognition receptors (PRR) inside hepatocytes are stimulated leading to production of IFN-α. DNA of the parasite probably acts as a (pathogen) microbe-associated molecular pattern (MAMP) in the hepatocyte cytosol. This results in recruitment of macrophages, neutrophils, and lymphocytes.

- Cytotoxic CD8+ T cells that produce interferon-γ are mainly involved in killing of intrahepatic parasites. NK, NKT, and γδ T cells also kill intrahepatic parasites through secretion of type I interferons and IFN-γ.

This stage is "silent" without any clinical symptoms.

Erythrocyte Stage

- Merozoites in the bloodstream are targeted by antibodies to their erythrocyte attachment ligands, for example GPI anchoring MSPs (see Section 1) that can both opsonize and prevent entry.

- GPI is also a major MAMP, recognized by the PRR, Toll-like receptor 2 (TLR 2) on macrophages that induces pro-inflammatory cytokines such as IL-1 and TNFα.

- At the intra-erythrocyte stage, antibodies can mediate cellular killing, block adhesion of infected RBCs to endothelium, and neutralize parasite toxins to prevent the induction of excessive inflammation. Also, complement can lyse infected RBC.

- CD8+ T cells play little role at this stage but CD4+ T cells contribute to activation of macrophages.

- NK cells and γd T cells are involved at this stage. IFN-γ, perforins and granzymes produced by NK cells can also kill *P. falciparum* infected RBCs.

Male and female gametocytes in the bloodstream are also thought to be targeted by antibodies.

Recent studies have indicated a possible role for polymorphisms in the *Fc receptor gamma* (*FcRγ*) genes since they have been associated with either susceptibility or resistance to malaria. *FcRγ* are present on many immune cells and have several functions particularly in relation to antibody-mediated immune mechanisms.

Some innate immunity is also induced in the mosquito to the gametes and sporozoites of *Plasmodium*.

Malaria Immune Escape Mechanisms

Plasmodium has a number of "escape mechanisms" that allow it to avoid the immune response.

The **liver stage**: a) In order to invade hepatocytes, the sporozoites have first to pass through KC and endothelial cells. Binding of CSP to KC surface proteins produces high levels of intracellular cAMP that prevents the formation of reactive oxygen species (ROS). b) Sporozoite contact with KC also down-regulates inflammatory Th-1 cytokines

and up-regulates anti-inflammatory Th-2 cytokines. c) KC apoptosis may be induced and expression of major histocompatibility complex (MHC)-I reduced. d) Once inside the hepatocyte, the parasitophorous vacuole prevents lysosomal degradation. e) Host heme oxygenase-1 (HO-1) also enhances the development of intrahepatic parasites through modulating the host inflammatory response.

The **erythrocyte stage**: a) The intracellular localization in the RBC escapes direct interaction with the immune cells. b) Lack of MHC-I molecule expression on the surface RBCs also avoids recognition by CD8+ T cells. c) Expression of variable antigenic surface proteins on infected RBCs helps the parasite to evade host immune responses. d) The phagocytic functions of macrophages are also hindered by *P. falciparum* malaria pigment or hemozoin. Hemozoin (released from ruptured erythrocytes) inside the macrophages reduces phagocytosis of infected RBC as well as reducing the production of radical oxygen intermediates.

PATHOGENESIS

There are three major pathologic mechanisms that account for the pathology and clinical signs and symptoms in patients with malaria (see below). These are as follows:

- The response of the **mononuclear phagocytic system** to the parasite leads to release of **cytokines** from host cells. Cytokines, such as IL-1 and TNF, act as endogenous pyrogens.

- Cyclical fever is caused by erythrocyte rupture leading to toxins being released into the bloodstream such as hemozoin. Defective production of red blood cells probably induced by cytokines and toxins including hemozoin and, leads to **anemia**. Modification of the erythrocyte membrane makes it less deformable and the cells are removed by the spleen, leading to **splenomegaly**. Additionally, antibody-mediated **lysis** of the erythrocytes occurs as parasite antigens are expressed on the surface of the erythrocytes.

- Obstruction of capillaries is caused by parasitized red blood cells. This is also caused by modification of the erythrocyte membrane as the surface molecules bind to adhesion molecules on the endothelium of the capillaries. This mechanism is particularly important in the pathogenesis of severe *P. falciparum* malaria such as development of cerebral malaria.

3. WHAT IS THE TYPICAL CLINICAL PRESENTATION AND WHAT COMPLICATIONS CAN OCCUR?

In uncomplicated malaria, the first symptoms are fever, headache, chills, vomiting, muscle pains and diarrhea that appear 10–15 days after a person is infected. The pre-erythrocytic stage in the liver is asymptomatic. If not treated promptly with effective medicines, malaria can cause severe illness that is often fatal. Patients may also have splenomegaly due to the sequestering of infected erythrocytes and anemia as the result of removal of erythrocytes.

With time, the fevers take on a periodicity depending on the species of malarial parasite.

Cyclical temperature fluctuations coincide with the rupture of erythrocytes and the release of merozoites into the bloodstream which, during the erythrocytic stage, occurs every 48 hours for *P. falciparum*, *P. ovale*, and *P. vivax*, and every 72 hours for *P. malariae* (see Figure 28.3). The cyclic temperature fluctuations are due to the release of toxins and hemozoin during erythrocyte rupture and GPI on fragments into the circulation. Through PRRs and other receptors, this induces the release of pyrogenic cytokines including TNFα, IL-1β, and IL6 by macrophages that cause rapid elevation of core temperature by acting on the hypothalamus.

COMPLICATIONS

In the most severe form of malaria (*P. falciparum*), there can be cerebral complications – **thrombosis** may occur due to occlusion of the cerebral vessels caused by the increased stickiness of the erythrocytes (see above). Multi-organ damage and renal failure can also result from the infection. The latter is uncommon in children with severe malaria who present with prostration, respiratory distress, severe anemia, and/or cerebral malaria.

Blackwater fever can also occur due to intravascular **hemolysis** leading to **hemoglobinuria** and kidney failure. *P. falciparum* also causes pulmonary **edema** and *P. malariae* causes **nephrotic syndrome**. In patients with splenomegaly, the spleen may rupture due to its large size.

The pattern of severe malaria is not fully understood, although genetic factors, age, and intensity of transmission determine susceptibility to severe anemia and cerebral malaria. Other infectious diseases can interact with malaria and modify susceptibility and/or severity of either disease. HIV infection with malaria increases the risk of both uncomplicated and severe malaria, and *Plasmodium* causes a transient increase in viral load, which might promote transmission of the virus.

Plasmodium and Epstein-Barr virus are concurrent risk factors for **Burkitt's lymphoma** in Africa.

4. HOW IS THE DISEASE DIAGNOSED, AND WHAT IS THE DIFFERENTIAL DIAGNOSIS?

Clinical diagnosis of malaria can be quite difficult because of the overlap of symptoms with other infectious diseases.

There are three main diagnostic tests for malaria:

1. ***Microscopic diagnosis.***

   The gold standard for diagnosis is by a blood film. Thick or thin films are prepared from finger-prick blood. These are stained with Giemsa or Fields stain and viewed under the microscope. The degree of parasitemia is noted and the morphologic appearance of the trophozoite (or gametocyte) can be used to identify the species of malaria (see diagnostic stages in Figure 28.2).

   P. falciparum: ring forms, often resembling stereo headphones, may be seen at the edge of the cell. The red cell is of normal size (Figure 28.4).

   P. vivax: irregular large thick ring. The red cells are enlarged.

   P. ovale: regular dense ring. The red cell is oval.

   P. malariae: dense thick ring or band form. The red cell is normal in size.

   Blood films may also be stained with acridine orange to identify the species.

2. ***The rapid detection test (RDT).***

   Malaria can be diagnosed serologically by the detection of antibodies or antigen, the latter being used for rapid identification of acute cases. A number of commercial kits are available for rapid diagnosis of *P. falciparum*. Although many are highly sensitive, some are not as sensitive as a blood film. They detect *P. falciparum*, histidine-rich protein (HRP), for example, ParaSight F® kits, or lactate dehydrogenase (LDH), for example Kat-Quick® kits.

3. ***Molecular diagnosis (PCR).***

   The detection of parasite nucleic acids using polymerase chain reaction (PCR), although slightly more sensitive than smear microscopy, takes longer and is often not available quickly enough to be of value in establishing the diagnosis of malaria infection. PCR is most useful for confirming the species of malarial parasite after the diagnosis has been established by either smear microscopy or RDT.

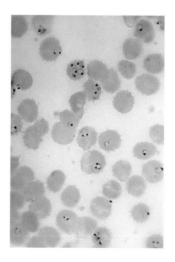

Figure 28.4 Blood film showing the presence of *P. falciparum* rings in human erythrocytes (×1125). Note that some red blood cells contain multiple parasites, which is more common with *P. falciparum* than other *Plasmodium* spp. *From the Centers for Disease Control & Prevention, Atlanta, Georgia. Image is found in the Public Health Image Library #4884. Additional photographic credit is given to Dr Mae Martin who took the photo in 1971.*

DIFFERENTIAL DIAGNOSIS

- Meningitis or encephalitis: Lower respiratory tract infection: COVID-19;
- Influenza and other viral infections such as Epstein-Barr virus or cytomegalovirus;
- Gastroenteritis: Urinary tract infection: Lymphoma: Sepsis: Viral hepatitis: HIV seroconversion: Legionellosis: Leptospirosis.

Malaria may present with similar symptoms to other travel-related infections, including:

- Viral hemorrhagic fevers such as Lassa fever, Crimean-Congo hemorrhagic fever, Marburg, and Ebola – patients suspected of having a viral hemorrhagic fever require strict isolation.
- Enteric fevers such as typhoid or paratyphoid.
- Arboviruses such as Dengue, West Nile virus, and Japanese encephalitis.
- Rickettsial infection such as scrub typhus and relapsing fever.
- Trypanosomiasis.
- Rabies.

5. HOW IS THE DISEASE MANAGED AND PREVENTED?

Early diagnosis and prompt treatment are the basic elements of malaria control and this is crucial to prevent the development of complications and the majority of deaths from malaria. The two main approaches to malaria control are:

- drug treatment of the patients with infection, and
- prevention and control of the vector.

MANAGEMENT

Antimalarial treatment policies vary between countries depending on epidemiology of the disease, transmission, patterns of drug resistance, and political and economic contexts.

Historically, chloroquine has been the first-line drug for treating malaria and is still used for *P. ovale*, *P. vivax*, and *P. malariae* except in Indonesia, Sabah, and Papua New Guinea where there is high-level chloroquine resistance. *P. falciparum* became increasingly resistant to chloroquine and sulfadoxine-pyrimethamine combinations of drugs were used and are still used. However, there is now significant resistance to these drugs. Currently, the first-line drugs against *P. falciparum* and adopted by the WHO, are the artemisinin-based combination therapies (ACTs). Artemisinin, first isolated and developed in China in the 1980s from the plant *Artemisia annua*, has several derivatives including artesunate, artemether, and dihydro-artemisinin. ACTs work by combining a derivative of artemisinin, a fast-acting antimalarial endoperoxide, with a longer-lasting partner drug that continues to reduce the parasite numbers after the artemisinin has dropped below a therapeutic level. ACTs not only act on the asexual blood stages to alleviate symptoms but also on the gametes, therefore reducing the spread of the disease. From April 2019, the WHO recommended **artesunate** as the first-line treatment of **severe** malaria.

Companion drugs for the artemisinin derivatives include lumefantrine, mefloquine, amodiaquine, sulfadoxine/pyrimethamine, piperaquine, and chlorproguanil/dapsone.

One drug that targets the pre-erythrocytic phase of the disease is primaquine and this is still used to kill the hypnozoites of *P. vivax* and *P. ovale* in the liver. The FDA-approved tafenoquine, an anti-plasmodial 8-aminoquinoline derivative is indicated for the radical cure (prevention of relapse) of *P. vivax* malaria in patients aged 16 years or older who are receiving appropriate antimalarial therapy for acute *P. vivax*.

Sequencing of the parasite genome has aided and will continue to aid in the discovery of new drugs to treat malaria.

PREVENTION AND VECTOR CONTROL

Personal prevention measures include the use of a) indoor residual spraying (IRS) of walls in dwelling places using organochlorines, pyrethroides, organophosphates, carbamates or neonicotinoids and b) window screens, repellents (such as DEET) and wearing of light-colored clothes, long pants, and long-sleeved shirts. Insecticide-treated bed nets (ITNs) have been shown to reduce malaria illness, severe disease, and death due to malaria in endemic regions. In several African settings, ITNs have been shown to reduce the death of children under 5 years from all causes by about 20%. Due to the resistance of mosquitos to DDT, only two insecticide classes are approved for net treatment – pyrroles and pyrethroids. These have been shown to pose very low health risks to humans and other mammals but are toxic to insects and kill them. Resistance is, however, already developing against pyrethroides. Nets, to be effective, have to be retreated every 6–12 months and now the WHO recommends that long-lasting insecticidal nets (LLINs) be distributed to and used by all people ("universal coverage") in malarious areas, not only by the most vulnerable groups who are pregnant women and children under 5 years. These nets have been shown to be effective for up to 3 years.

Infection is also prevented by controlling the breeding cycle of the vector, which should significantly reduce the number of cases and rate of parasite infection. Control of the larval stage includes:

- Oils applied to the water surface, suffocating the larvae and pupae. Most currently used oils are rapidly biodegraded.
- Toxins from the bacterium *Bacillus thuringiensis var. israelensis* (Bti) can be applied in the same way as chemical insecticides. They are highly specific, affecting only mosquitoes, black flies, and midges.

- Insect-growth regulators such as methroprene are specific to mosquitoes and can be applied in the same way as chemical insecticides.

These approaches are generally considered to be environmentally friendly methods of mosquito control.

PREVENTION FOR TRAVELERS TO ENDEMIC AREAS

With travel to malaria endemic areas now very common, a number of drugs are given to prevent infection with the parasite, as well as steps taken to avoid mosquito bites. The drugs currently used in prophylaxis vary depending on which area you are travelling to. They include: Atovaquone-proguanil, doxycycline, mefloquine, tafenoquine and primaquine.

The patient described in this case had been taking antimalarial tablets as a preventive measure against getting malaria. Possibilities as to why she still contracted malaria are: a) the patient was taking homeopathic tablets; b) the patient did not take the drugs on a regular basis; or c) the particular malaria species with which the patient was infected was resistant to the prophylactic treatment.

Information as to the appropriate drugs for the area currently to be visited is available at several websites, for example https://www.cdc.gov/malaria/travelers/drugs.html.

VACCINES

Development of highly effective and durable vaccines to prevent infection with *Plasmodium falciparum* and *P. vivax*
have been a key priority. Strategies for producing vaccines to *P. falciparum* have been focused on its life cycle. Vaccines include:

- *whole sporozoite antigens* – using whole attenuated sporozoites,
- *liver-stage vaccines* – using thrombospondin-related adhesion protein linked to a multi-epitope string (ME-TRAP) inserted into a chimpanzee adenovirus vector,
- *blood-stage vaccines* – using immunodominant and non-polymorphic merozoite antigens (e.g. *P. falciparum* reticulocyte-binding protein homolog 5),
- *transmission-blocking vaccines* (targeting the sexual stages to impact the parasite's life cycle in the mosquito vector aiming to prevent sporozoite development and onward transmission) - using ookinete surface protein Pfs25 and the gametocyte antigens Pfs48/45 and Pfs230.

The most extensively tested vaccine candidate for prevention of *P. falciparum* malaria is RTS,S/AS01 which has been under development for several years. This vaccine directs immune responses against the major circumsporozoite protein (PfCSP) covering the surface of the infecting sporozoite. The WHO is now recommending widespread use of the RTS,S/AS01 (RTS,S) malaria vaccine among children in sub-Saharan Africa and in other regions with moderate to high *P. falciparum* malaria transmission. It is recommended that this malaria vaccine be given in a schedule of four doses to children from 5 months of age for the reduction of malaria disease and burden.

SUMMARY

1. WHAT IS THE CAUSATIVE AGENT, HOW DOES IT ENTER THE BODY, AND HOW DOES IT SPREAD A) WITHIN THE BODY AND B) FROM PERSON TO PERSON?

- The organism causing malaria is *Plasmodium*, a protozoan with a complex life cycle.
- Five species of *Plasmodium* are able to infect humans: *P. ovale*; *P. vivax*, and *P. malariae*, *P. knowlesi*, and *P. falciparum* being the most virulent species of malaria.
- The female *Anopheles* mosquito is the vector for malaria.
- The sporozoite is transmitted from the mosquito into the blood during a blood meal and localizes in the liver. Schizonts are produced through asexual reproduction.
- Liver schizonts rupture and release merozoites into the bloodstream, which invade and destroy erythrocytes giving rise to symptoms.
- Within the erythrocyte, the merozoite undergoes another round of schizogony to produce trophozoites.
- Differentiation of the trophozoite into gametocytes occurs in some erythrocytes.
- When another mosquito feeds it takes in the gametocytes, where they fuse to form a zygote, which becomes an oocyte.

- The oocyte matures into sporozoites, which migrate to the salivary gland of the mosquito and are inoculated into another human when the mosquito feeds.
- The malaria case incidence rate, total malaria cases and deaths fell from between 2000 and 2019. In 2020, there was an estimated 241 million malaria cases. The estimated number of malaria deaths stood at 627000 in 2020. The WHO African Region carries a disproportionately high share of the global malaria burden. In 2020, the region was home to 95% of all malaria cases and 96% of all malaria deaths. Children under 5 are still the most vulnerable group.
- Asia, Latin America, the Middle East, and parts of Europe are also affected.
- With increasing international travel, there continues to be a rise in the number of cases of malaria in travelers returning to non-malarious areas from countries where malaria is endemic. However, since the start of the SARS-2 pandemic with reduced global mobility it is likely that travel-associated malaria has decreased.
- During the SAR-2 pandemic, a recent modeling analysis by the WHO predicted a >20% rise in malaria morbidity and >50% mortality in sub-Saharan Africa as a result of 75% reduction in routine malaria control measures. *Continued...*

...continued

- Pregnancy significantly increases the risk of malaria.
- Absence of Duffy blood group antigen in West Africans, sickle cell trait, and glucose 6 phosphate dehydrogenase (G6PD) deficiency are examples of genetic factors leading to reduced infection rates for malaria.

2. WHAT IS THE HOST RESPONSE TO THE INFECTION AND WHAT IS THE DISEASE PATHOGENESIS?

- The parasite uses a number of "escape mechanisms" at both the liver stage and erythrocyte stages to avoid the immune response. These include different life cycle forms and direct immunosuppression by the parasite.
- Some immunity does develop to the parasite with continuous infection but this is weak and does not eliminate the parasite.
- Immune mechanisms that exist are different against the different stages of the life cycle. In the pre-erythrocyte phase: antibodies (and complement) against the sporozoites, phagocytosis, NK and cytotoxic T cells against the hepatocytes. Erythrocyte stage: antibodies against the merozoites and gametes. NK and $\gamma\delta$ T cells. CD8+ T cells play little role but CD4+T cells contribute to macrophage activation.
- Pathogenesis occurs due to: (a) cytokine release leading to spikes of fever at regular intervals, (b) destruction of erythrocytes leading to anemia, and (c) obstruction of capillaries due to binding of erythrocytes through parasite encoded surface molecules to endothelium.

3. WHAT IS THE TYPICAL CLINICAL PRESENTATION AND WHAT COMPLICATIONS CAN OCCUR?

- In uncomplicated malaria, the first symptoms are fever, headache, chills, vomiting, muscle pains, and diarrhea that appear 10–15 days after a person is infected.
- Patients can develop splenomegaly and anemia.
- Cyclical temperature fluctuations coincide with the rupture of erythrocytes and the release of merozoites into the bloodstream which, during the erythrocytic stage, occurs every 48 hours for *P. falciparum*, *P. ovale*, and *P. vivax*, and every 72 hours for *P. malariae*. This gives rise to fevers of differing periodicity.
- The cyclic temperature fluctuations are due to the release of toxins and hemozoin during erythrocyte rupture and GPI on fragments into the circulation. Through PRRs and other receptors, this induces the release of pyrogenic cytokines including TNFα, IL-1β, IL6 by macrophages that cause rapid elevation of core temperature by acting on the hypothalamus.
- Complications include cerebral malaria where thrombosis may occur (with *P. falciparum*, severe anemia and blackwater fever).
- Co-infection with HIV increases the risk of both uncomplicated and severe malaria, and *Plasmodium* causes a transient increase in viral load, which might promote transmission of the virus.
- *Plasmodium* and Epstein-Barr virus are concurrent risk factors for Burkitt's lymphoma in Africa.

4. HOW IS THE DISEASE DIAGNOSED, AND WHAT IS THE DIFFERENTIAL DIAGNOSIS?

- There are three diagnostic tests.
- Microscopic diagnosis: the gold standard is by a blood film; the morphologic appearance of the trophozoite or gametocyte can be used to identify the species.
- The rapid detection test (RDT): malaria can be diagnosed serologically using the detection of antibodies or antigens; there are kits available.
- Molecular diagnosis: PCR used to detect parasite nucleic acids; useful to confirm species after microscopic or RDT diagnosis.
- Clinical diagnosis of malaria can be quite difficult because of the overlap of symptoms with many other infectious diseases, e.g. meningitis, enteric fevers, rickettsial infections, rabies, etc.

5. HOW IS THE DISEASE MANAGED AND PREVENTED?

- The two main approaches to malaria control are drug treatment of the patients with infection and prevention and control of the vector.
- Most drug treatments target the erythrocytic stage of the infection.
- Chloroquine has been the first-line drug for treating malaria and is still in use for *P. ovale*, *P. vivax*, and *P. malariae*. *P. falciparum* has become increasingly resistant to chloroquine.
- Drugs currently used against *P. falciparum* and adopted by WHO are the artemisinin-based combination therapies (ACTs). Companion drugs for the Artemisinin derivatives include lumefantrine, mefloquine, amodiaquine, sulfadoxine/pyrimethamine, piperaquine, and chlorproguanil/dapsone.
- Primaquine targets the pre-erythrocytic stage of the disease and is used to kill the hypnozoites of *P. ovale* and *P. vivax* in liver cells.
- ACTs act not only on the asexual blood stages to alleviate symptoms but also on the gametes, therefore reducing the spread of the disease.
- Sequencing of the parasite genome has aided and will continue to aid in the discovery of new drugs to treat malaria.
- Personal prevention measures include the use of IRS for spraying internal surfaces and window screens, repellents (such as DEET), and wearing of light-colored clothes, long pants, and long-sleeved shirts.
- Insecticide-treated bed nets (ITNs) reduce malaria illness, severe disease, and death due to malaria in endemic regions. In several African settings, ITNs have been shown to reduce the death of children under 5 years from all causes by about 20%. Due to the resistance of mosquitos to DDT, only two insecticide classes are now approved for net treatment – pyrroles and pyrethroids.
- Nets have to be retreated every 6–12 months and now the WHO recommends that long-lasting insecticidal nets (LLINs) be distributed to and used by all people ("universal coverage") in malarious areas.

continued...

...continued

- Mosquito larval control is also used. This is achieved by using oils applied to the water surface, toxins from *Bacillus thuringiensis*, and insect growth regulators such as methoprene.

- Prophylactic drugs for international travelers depend on drug resistance in the endemic areas to be visited and include atovaquone-proguanil, doxycycline, mefloquine, tafenoquine, and primaquine.

- There are a number of strategies for producing vaccines to *P. falciparum* including liver-stage vaccines, blood-stage vaccines, and transmission-blocking vaccines. The WHO have recently recommended the widespread use of a vaccine (RTS,S) against the major circumsporozoite protein (PfCSP) covering the surface of the infecting sporozoite. The vaccine should be used to treat children in sub-Saharan Africa and in other regions with moderate to high *P. falciparum* malaria transmission.

FURTHER READING

Goering R, Dockrell HM, Zuckerman M, Chiodini PL, (eds). Mims' Medical Microbiology and Immunology, 6th edition. Academic Press/Elsevier, Cambridge, 2018.

Male D, Peebles S, Male V. Immunology, 9th edition. Elsevier, Philadelphia, 2020.

Murphy K, Weaver C. Janeway's Immunobiology, 9th edition. Garland Science, New York/London, 2016.

REFERENCES

Amiah MA, Ouattara A, Okou DT, et al. Polymorphisms in Fc Gamma Receptors and Susceptibility to Malaria in an Endemic Population. Front Immunol, 11: 561142, 2020.

Chu CS, White NJ. The Prevention and Treatment of Plasmodium vivax Malaria. PLoS Med, 18: e1003561, 2021.

Cowman AF, Tonkin CJ, Tham W-H, Duraisingh M. The Molecular Basis of Erythrocyte Invasion by Malaria Parasites. Cell Host Microbe, 22: 232–245, 2017.

Draper SJ, Sack BK, King CR, et al. Cell Malaria Vaccines: Recent Advances and New Horizons. Host Microbe, 24: 43–56, 2018.

Gowda DC, Wu X. Parasite Recognition and Signaling Mechanisms in Innate Immune Responses to Malaria. Front Immunol, 9: 3006, 2018.

Jaskiewicz E, Jodłowska M, Kaczmarek R, Zerka A. Erythrocyte Glycophorins as Receptors for Plasmodium merozoites. Parasit Vectors, 12: 317, 2019.

Kurtovic L, Reiling L, Opi DH, Beeson JG. Recent Clinical Trials Inform the Future of Malaria Vaccines. Commun Med, 1: 26, 2021.

Liu Q, Jing W, Kang L, et al. Trends of the Global, Regional and National Incidence of Malaria in 204 Countries from 1990 to 2019 and Implications for Malaria Prevention. J Travel Med, 28: taab046, 2021.

Milner, DA, Jr. Malaria Pathogenesis. Cold Spring Harb Perspect Med, 8: a025569, 2018.

O'Leary, K. A Malaria Vaccine at Last. Nat Med, 27: 2057, 2021.

Talapko J, Škrlec I, Alebic' T, et al. Malaria: The Past and the Present. Microorganisms, 7: 179, 2019.

Thriemer K, Ley B, von Seidlein L. Towards the Elimination of Plasmodium vivax Malaria: Implementing the Radical Cure. PLoS Med, 18: e1003494, 2021.

Tse EG, Korsik M, Todd MH. The Past, Present and Future of Anti-Malarial Medicine. Malar J, 18: 93, 2019.

WEBSITES

BMC, Part of Springer Nature, Malaria Journal: http://www.malariajournal.com/

Centers for Disease Control and Prevention, Parasites – Malaria, 2022: https://www.cdc.gov/parasites/malaria

European Centre for Disease Prevention and Control, Malaria, Annual Epidemiological Report for 2018, 2018: https://www.ecdc.europa.eu/sites/default/files/documents/malaria-annual-epidemiological-report-2018.pdf

Map – The Malaria Atlas Project: https://malariaatlas.org/

National Institute for Health and Care Excellence, Malaria, treatment: https://bnf.nice.org.uk/treatment-summary/malaria-treatment.html

World Health Organization, © Copyright World Health Organization (WHO) 2021. All Rights Reserved: http://www.who.int/mediacentre/factsheets/fs094/en/index.html

World Health Organization, Malaria, 2022: https://www.who.int/news-room/fact-sheets/detail/malaria

Students can test their knowledge of this case study by visiting the Instructor and Student Resources: [www.routledge.com/cw/lydyard] where several multiple choice questions can be found.

29

Respiratory syncytial virus

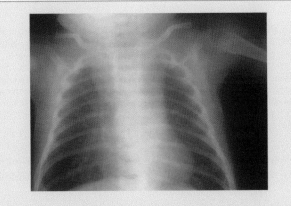

Figure 29.1 Acute bronchiolitis. Early radiographic appearance – flat diaphragm and hyperlucent lung fields indicate hyperinflation. *From Di Nardo M, Perrotta D, Stoppa F et al. (2008) Journal of Medical Case Reports, 212. doi:10.1186/1752-1947-2-212. Published under Creative Commons Attribution License 2.0.*

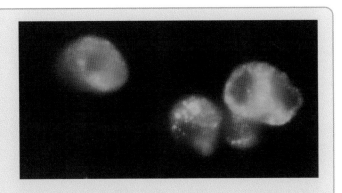

Figure 29.2 Photomicrograph of respiratory epithelial cells taken from a child infected with respiratory syncytial virus (RSV) and stained with a monoclonal antibody specific for RSV. The antibody is labeled with fluorescein so that patchy (speckled) green fluorescence is observed under illumination with ultraviolet light. Note cytoplasmic staining and binucleate cell (result of RSV-induced cell fusion). This method was the mainstay of direct diagnosis for many years but is labor intensive and expertise and is unsuitable for large numbers of samples. *Courtesy of Dr Jeremy A Garson and the Centre for Virology, Department of Infection, Royal Free and University College Medical School, Windeyer Building, London.*

A 2-month-old child presented to a doctor with a 2-day history of a **febrile** upper respiratory tract infection. His clinical condition deteriorated with worsening cough, wheezing and **dyspnea**, necessitating admission to hospital. On clinical examination in the hospital emergency department, the child had obvious evidence of respiratory distress with severe intercostal retraction and **tachypnea** (>70 breaths per minute) together with **tachycardia**. On **auscultation** of the chest, diffuse high-pitched wheezing was audible and there were fine inspiratory crackles. Pulse oximetry gave an oxygen saturation of 92% on air. Radiologic investigation revealed findings typical of **bronchiolitis**, with hyperinflation of both lung fields (**Figure 29.1**). A diagnosis of acute bronchiolitis was made and the child was admitted to hospital for further management. Respiratory syncytial virus (RSV) was detected in nasal secretions (**Figure 29.2**); the child was nursed in isolation on oxygen therapy and made an uneventful recovery.

1. WHAT IS THE CAUSATIVE AGENT, HOW DOES IT ENTER THE BODY AND HOW DOES IT SPREAD A) WITHIN THE BODY AND B) FROM PERSON TO PERSON?

CAUSATIVE AGENT

RSV belongs to the paramyxovirus family in the pneumovirus subfamily. It is closely related to the recently discovered human metapneumovirus. Both are enveloped viruses with a single-stranded negative sense RNA genome with a helical **nucleocapsid**. The envelope contains the virus attachment glycoprotein, G, and the fusion glycoprotein, F.

The viral RNA-dependent RNA polymerase, an integral part of the **virion**, has no proof-reading mechanism. RSV evolves quickly by point mutations that allow for changes in viral virulence and avoidance of the immune response. The G protein is the most variable of the structural proteins, and the resultant antigenic differences form the basis of the two major groups of RSV strains, A and B, which circulate concurrently. At any one time, there may be multiple RSV variants within a population. In contrast, the fusion protein is highly conserved.

ENTRY AND SPREAD WITHIN THE BODY

The RSV G protein is the primary mediator of attachment to the glycocalyx of ciliated airway epithelial cells and type 1 alveolar pneumocytes but the F protein can also facilitate viral attachment although to a lesser extent than G. Virus attachment initiates a conformational change in F protein so that a fusion peptide is revealed and fusion of the closely juxtaposed viral and cell membranes ensues. Subsequently, the viral nucleocapsid is released into the cell's cytoplasm and the viral RNA-dependent RNA polymerase, initiates virus replication with the synthesis of viral messenger RNA.

Virus progeny from the first infected cell then spread to neighboring cells often involving cell-to-cell fusion mediated by the F protein to form syncytia, also known as multinucleate giant cells (Figures 29.3 and 29.4 – hence the name of the virus). Virus shed from the apical surface of infected cells infects more distal mucosal cells via respiratory secretions.

In infants and the elderly, the infection is not limited to the upper respiratory tract and may involve the trachea, bronchi, bronchioles, and alveoli. In all cases, the incubation period is short, being between 2 and 8 days, since the infection is restricted to respiratory mucosa and does not become systemic.

PERSON-TO-PERSON SPREAD

RSV infection is highly contagious, being shed in respiratory secretions for several days. Large droplets arising from coughs and sneezes of an ill person may transmit RSV to others within a radius of about three feet. Longer-distance spread by small-particle (droplet nuclei) aerosols seems to be much less likely. Touching objects (**fomites**) contaminated with infected secretions followed by self-inoculation into the eyes or nose, is an important mode of transmission. Virus in infected secretions is viable for up to 6 hours on nonporous surfaces, up to 45 minutes on cloth, and up to 20 minutes on skin.

EPIDEMIOLOGY

RSV is responsible for about 80% of cases of bronchiolitis and is a leading cause of hospitalizations due to acute lower

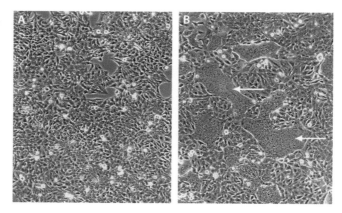

Figure 29.3 Photomicrograph of RSV-infected respiratory epithelial cells as seen in tissue culture. Cultured respiratory epithelial cells on day 1 (A) and day 3 (B) following infection with the virus. After one day, the cells continue to appear healthy. After three days, many of the cells have fused, forming large syncytia (arrows), that is a mass of cytoplasm containing several separate nuclei enclosed in a continuous membrane. Magnification: ×400. *From Domachowske JB & Rosenberg HF (1999) Clinical Microbiology Reviews 12. doi:10.1128/CMR.12.2.298. With permission from the American Society for Microbiology.*

respiratory infection in infants and young children <1 year old. It is associated with significant morbidity worldwide and there are between 66 000 and 200 000 deaths annually, predominantly in low- and middle-income countries. About half of all deaths are in children younger than 5 years. In the UK and US, two-thirds of infants become infected in the first year of life and of these one-third will develop lower respiratory tract symptoms. In the first 6 months of life, bronchiolitis is an important and life-threatening disease; 2–3% of infants up to 1 year old are admitted to hospital each year with bronchiolitis caused by RSV, and many more will be managed in the community. Most children have been infected for the first time by their second birthday; disease burden is highest in the first two years.

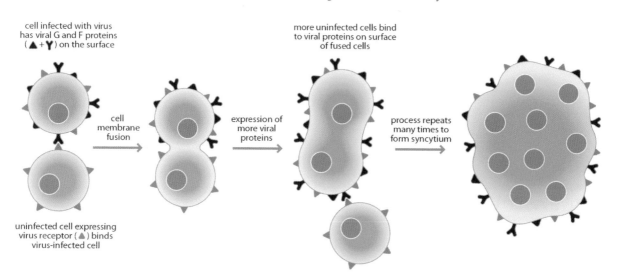

Figure 29.4 Schematic representation of the mechanism of syncytium formation due to RSV infection of respiratory epithelial cells. The virus envelope glycoproteins, G and F, whose roles on virus particles are in receptor binding and membrane fusion, are expressed on the surface of infected cells. Uninfected epithelial cells bearing the cellular virus receptor come into contact with infected cells and are fused together to form a syncytium.

Infection occurs in community-wide annual winter outbreaks in temperate climates (Figure 29.5) in all ages including the elderly, in whom RSV is a leading cause of disease. Exceptionally there was no winter outbreak in 2020–2021 due to social distancing during the Covid-19 pandemic when RSV infections were delayed to mid 2021 (Figure 29.6).

2. WHAT IS THE HOST RESPONSE TO THE INFECTION AND WHAT IS THE DISEASE PATHOGENESIS?

HOST RESPONSE TO INFECTION

Current knowledge of the immune response to RSV is mostly based on *in vitro* studies and animal models. In humans, it is difficult to disentangle the influence of genetics, environment, age, and antigenic experience. What follows is a summary of the salient facts.

At the start of RSV infection, there is a strong inflammatory response with neutrophils predominating. This correlates with disease severity and is mediated by interleukin 8. Other inflammatory cells are recruited to the respiratory tract including macrophages, T cells, natural killer (NK) cells, and eosinophils, these last cells might also be associated with more severe disease. CD8+ T lymphocytes are critical for the control and clearance of infection. CD4+ T helper 2 (Th-2) cells are generally considered to be involved in pathogenesis whereas T helper 1 (Th-1) cells seem protective. Immunocompetent infants infected with RSV shed virus for 1–21 days, children or adults with impaired cellular immunity shed virus for months and may develop fatal pneumonia.

ANTIBODY RESPONSES

RSV primarily infects mucosal respiratory epithelium, and IgA antibodies present in upper and lower respiratory tract are important for disease prevention. Neutralizing antibodies are predominantly directed against the F protein and prevent the fusion of the envelope of the virus with the cell membrane. The RSV humoral response to infection is incomplete and transient, which is typical of mucosal immunity, allowing repeated infections throughout life.

PATHOGENESIS OF BRONCHIOLITIS

Immunopathogenesis has a role to play (see host response to infection, above). There is a peribronchiolar lymphocytic infiltration together with **edema** of the bronchiolar walls. This is followed by **necrosis** of the bronchiolar epithelium and necrotic material blocks the lumen of the small airways. Virus infection also causes increased secretion of mucus, which compounds the problem. The infant's bronchioles are very narrow and hence especially prone to obstruction. On inspiration, negative intrapleural pressure due to obstructed bronchioles is clinically demonstrated by supraclavicular, intercostal, and subcostal recession of the soft tissues. On expiration, positive intrapleural pressure further narrows the bronchiolar lumen and there is resistance to expiration causing

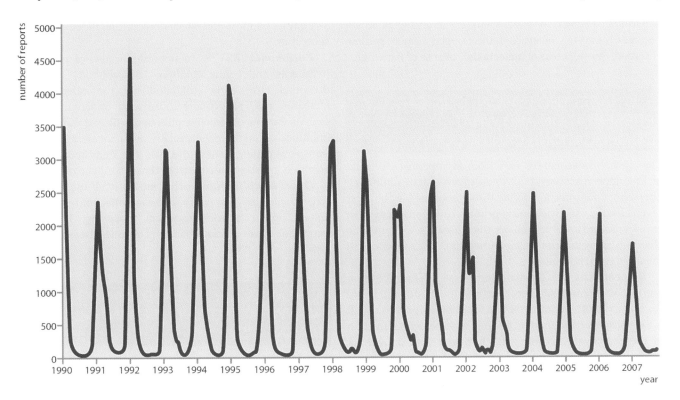

Figure 29.5 Laboratory reports of RSV infections, England and Wales, 1990–2007 (4 weekly). *Adapted with kind permission from the Centre for Infections, Health Protection Agency, England.*

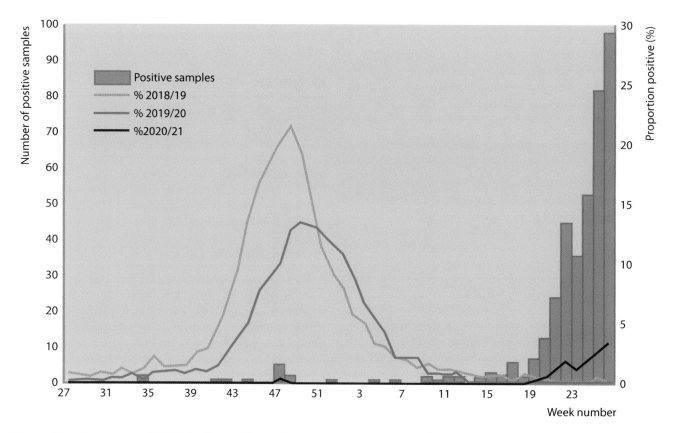

Figure 29.6 Comparison of RSV infections in England in the seasons 2018/9, 2019/20, and 2020/21. Laboratory reports of samples submitted from patients with respiratory infections. Although winter outbreaks are usually invariable as shown in **Figure 29.5**, the Covid-19 pandemic changed this due to lockdown and social distancing, measures which proved very effective in preventing the 2020 RSV winter outbreak. Bar chart shows positive samples in 2020/21. *From UKHSA https://www.gov.uk/government/collections/weekly-national-flu-reports.*

hyperinflated lungs. In areas where the bronchiolar lumen becomes completely obstructed, the trapped air may become absorbed, causing areas of **atelectasis** (collapse of the lung).

3. WHAT IS THE TYPICAL CLINICAL PRESENTATION AND WHAT COMPLICATIONS CAN OCCUR?

The clinical consequences vary from asymptomatic or mild afebrile upper respiratory tract illness to severe and fulminating pneumonia. Re-infection occurs throughout life but the most severe disease usually occurs in primary infection; those most at risk of severe lower respiratory tract disease are infants, the elderly as aging impairs the immune response, and those with impaired cellular immunity.

Bronchiolitis: this is characteristic of the first infection with RSV in infants. There are usually prodromal upper respiratory tract symptoms. As the infection progresses cough is a consistent and prominent finding, together with retraction of the chest wall, dyspnea, and **hypoxia**, with or without **cyanosis**. On auscultation there is a diffuse expiratory wheeze with crackles. Fever is low grade. Hyperinflation is seen on chest X-ray (Figure 29.1). The duration of illness is between 7 and 12 days. Most hospitalized infants improve clinically after

3–4 days, and most previously normal infants are discharged soon after.

Pneumonia: RSV is an important cause of pneumonia in infants and young children. Susceptibility to disease decreases after early childhood but the elderly are at increased risk, particularly those with underlying disease such as chronic obstructive pulmonary disease (COPD). The immunocompromised, especially hematopoietic stem cell transplant patients, are also at risk. The signs and symptoms are often indistinguishable from those of influenza.

Other complications: **otitis media** (RSV infection often accompanied by secondary bacterial infection), respiratory failure, apneic attacks, recurrent wheezing; RSV is an important cause of exacerbation of airway diseases, such as asthma and COPD.

4. HOW IS THE DISEASE DIAGNOSED, AND WHAT IS THE DIFFERENTIAL DIAGNOSIS?

Bronchiolitis is defined by two clinical signs, namely wheezing and hyperinflation of the lungs. Nonspecific supportive findings are normal or slightly raised white cell count with lymphocytes predominating. Chest X-ray shows

Table 29.1 Lower respiratory tract infections in young children, their characteristic clinical features and common causes

Syndrome	Presenting symptoms and signs	Commonest cause
Croup (laryngotracheitis)	Hoarseness, cough, inspiratory stridor with laryngeal obstruction	Parainfluenza viruses (types 1, 2, and 3)
Bronchiolitis	Expiratory wheeze with or without tachypnea, air trapping, and intercostal indrawing	Respiratory syncytial virus, human metapneumovirus
Pneumonia	Crackles on lung auscultation or evidence of pulmonary consolidation on physical examination or chest X-ray	Many different possibilities including respiratory viruses and enteroviruses*

* Whereas croup and bronchiolitis are viral in origin, pneumonia is often caused by bacteria: <1 month old – group B streptococci and *Escherichia coli*; 1–3 months – *Chlamydia trachomatis*; 3 months to 5 years – *Haemophilus influenzae* type b and *Streptococcus pneumoniae*.

hyperinflation, prominent bronchial wall markings, and multiple areas of collapse and consolidation, and there is **hypoxemia**. The assistance of a microbiology laboratory is required to confirm the etiological agent; bronchiolitis due to RSV must be distinguished from that due to other respiratory virus infections, especially the closely related human metapneumovirus but also parainfluenza virus type 3 and influenza, so that appropriate infection control can be instigated.

Samples: a good respiratory sample contains a mixture of respiratory ciliated epithelial cells as well as respiratory secretions. Samples from the upper respiratory tract are sufficient for the diagnosis of RSV infection. These can be a nasopharyngeal swab or combined nose and throat swab – sampling only the throat mucosa is likely to provide mainly squamous epithelial cells, which do not support respiratory virus replication very well.

Specific diagnosis: in the laboratory, the possibilities are **enzyme immunoassay** for viral antigen, or more usually reverse transcription multiplex **polymerase chain reaction (PCR)** for a variety of respiratory viruses (Table 29.1) including RSV. Nowadays this would include SARS-CoV-2, the causative agent of Covid-19, although this is an unlikely cause of bronchiolitis in a young child.

For the most rapid result, so-called "point of care tests" can be used; such tests are based on the detection of RSV antigen by immunochromatography and can be used at the bedside rather than in the laboratory.

Virus isolation in tissue culture requires viable virus and takes from 3 to 7 days, by which time the patient is usually well on the way to recovery. Likewise, diagnosis of RSV using antibody tests is not clinically useful as it takes several weeks for an antibody response to develop.

DIFFERENTIAL DIAGNOSIS

The differential diagnosis of bronchiolitis includes **croup**, **epiglottitis**, pertussis (whooping cough), and pneumonia (Table 29.1), and the most likely noninfectious disease is **asthma**.

5. HOW IS THE DISEASE MANAGED AND PREVENTED?

MANAGEMENT

Bronchiolitis: Patients most at risk of hospital admission are those with congenital heart disease, extreme prematurity, or any preexisting lung disease or immunodeficiency. The patient should be admitted if there is a need for oxygen or tube feeding or impending respiratory failure. Management involves the use of humidified oxygen, maintaining oral nutrition, and respiratory support if required. Nebulized ribavirin (an antiviral drug with *in vitro* activity against RSV) has been advocated for children likely to develop severe disease (infants <2 months, those born prematurely or with chronic cardiorespiratory disease). However, despite promising **clinical trials**, in practice clinical use of ribavirin has given disappointing results and it is now rarely used.

Pneumonia: The use of nebulized ribavirin as treatment for pneumonia in the immunocompromised is controversial and cannot be recommended.

Control of infection: In hospitals, RSV outbreaks are common on pediatric wards. If on a neonatal unit, there is a high risk of mortality. Although initially introduced from the community, subsequent spread is usually **nosocomial**. Hospitalized infants or immunocompromised patients infected with RSV pose a risk of infection to neighboring vulnerable patients. Healthcare staff, parents and siblings become infected and while symptomatic spread the virus on the wards via large droplets and **fomites**.

Rapid identification of infection by the "point of care tests" described above is of paramount importance for timely isolation of infected patients. Scrupulous handwashing together with cohort nursing are also vital.

PREVENTION

Immunoprophylaxis: A humanized **monoclonal antibody** directed against the F protein of RSV (palivizumab) can be used as passive immunoprophylaxis to protect vulnerable infants, in particular preterm infants with bronchopulmonary dysplasia and those with congenital heart disease. However, even for this at-risk population, it is of limited benefit and expensive. For anyone other than an infant, immunoprophylaxis with palivizumab is impractical, as the amount of antibody required is prohibitively expensive. It is likely that long-acting monoclonal antibodies directed specifically against the prefusion conformation of the F protein (see below) will prove more effective.

Vaccine: In the 1960s, a formalin-inactivated RSV vaccine was developed, but the results were unexpected and disastrous. After immunization, subsequent exposure to the virus resulted in worse bronchiolitis and pneumonia than in unimmunized children, with severe morbidity and several deaths. To this day, it is not known exactly why this happened. The vaccine induced a good antibody response, but it was not neutralizing due to formalin-mediated destruction of neutralizing epitopes in the vaccine preparation. Whatever the mechanism, a catastrophic inflammatory response destroyed lung tissue, a process that continued even after viral clearance.

The obvious targets for an RSV vaccine are the G and F proteins. The G protein is important for viral attachment but shows high variability across RSV strains and there is little knowledge of its surface structure. On the other hand, the RSV F protein facilitates viral fusion with host cells, and is highly conserved. The recent understanding of the structure and function of this protein and the stabilization of its prefusion (pre-F) conformation has provided a new target for vaccines as opposed to many attempts using F protein in unmodified form, which tended to flip to the rearranged post-fusion form.

There is still no RSV vaccine currently available. So far, there is no indication that any of the vaccines currently in clinical trials elicit mucosal anti-RSV IgA neutralizing antibodies or long-term IgA B cell memory; two requirements for a vaccine to confer a significant level of protection against RSV infection.

SUMMARY

1. WHAT IS THE CAUSATIVE AGENT, HOW DOES IT ENTER THE BODY, AND HOW DOES IT SPREAD A) WITHIN THE BODY AND B) FROM PERSON TO PERSON?

- The main cause of bronchiolitis is respiratory syncytial virus (RSV) but also human metapneumovirus, parainfluenza and influenza viruses.
- RSV attaches to the mucous membranes of the nose and infects the ciliated epithelial cells, spreading to neighboring cells by the formation of syncytia.
- Shedding of virus from the apical surface of infected cells allows spread via respiratory secretions to more distant mucosal cells in the trachea, bronchi, bronchioles, and alveoli.
- Spread from person to person is via respiratory secretions, either large droplets or contaminated surfaces.

2. WHAT IS THE HOST RESPONSE TO THE INFECTION AND WHAT IS THE DISEASE PATHOGENESIS?

- Immunity to RSV infection is incomplete and re-infections occur in all age groups. Every year or two, the virus will have changed enough, antigenically so the population is effectively immunologically naïve and hence open to re-infection. This has implications for vaccine development since, as for influenza vaccine, the composition of an RSV vaccine might need to be changed each year.
- RSV causes infection and inflammation of the bronchioles, leading to obstruction of expiratory airflow in young children causing wheeze and hyperinflation and sometimes even collapse of the lung.

3. WHAT IS THE TYPICAL CLINICAL PRESENTATION AND WHAT COMPLICATIONS CAN OCCUR?

- Bronchiolitis typically presents with tachypnea, expiratory wheeze, and inspiratory intercostal recession.
- Complications are otitis media, secondary bacterial infection, apnea, hypoxia, and respiratory failure.

4. HOW IS THE DISEASE DIAGNOSED, AND WHAT IS THE DIFFERENTIAL DIAGNOSIS?

- Bronchiolitis is a clinical diagnosis defined by wheezing and hyperinflation of the lungs.
- Identification of the causal agent (usually RSV) requires a respiratory sample, a nasal swab for infected epithelial cells, and a test for the virus, usually reverse transcription multiplex PCR.
- The differential diagnosis includes croup, epiglottitis, pertussis, pneumonia, and asthma.

5. HOW IS THE DISEASE MANAGED AND PREVENTED?

- Admit to hospital if there is a need for oxygen or tube feeding or impending respiratory failure.
- Management involves the use of humidified oxygen, maintenance of oral nutrition, and respiratory support.
- Infection control is necessary to prevent RSV outbreaks on pediatric wards.
- Use of the antiviral drug, ribavirin, is controversial and cannot be recommended.
- A humanized **monoclonal antibody** (palivizumab) directed against the F protein of RSV is used as passive immunoprophylaxis to protect especially vulnerable infants.
- There is no vaccine.

FURTHER READING

Walsh EE, Hall CB. Respiratory syncytial virus (RSV). In: Bennett JE, Dolin R, Blaser MJ, editors. Mandell, Douglas and Bennett's Principles and Practice of Infectious Diseases, Volume 2, 8th edition. Elsevier/Saunders, Philadelphia, 1948–1960, 2015.

REFERENCES

Bush A, Thomson AH. Acute Bronchiolitis. BMJ, 335: 1037–1041, 2007.

Jalal H, Bibby DF, Bennett J, et al. Molecular Investigations of an Outbreak of Parainfluenza 3 and Respiratory Syncytial Viruses in a Haematology Unit. J Clin Microbiol, 45: 1690–1696, 2007.

Mascola JR, Fauci AS. Novel Vaccine Technologies for the 21st Century. Nat Rev Immunol, 20: 87–88, 2020.

Mazur NI, Higgins D, Nunes MC, et al. The Respiratory Syncytial Virus Vaccine Landscape: Lessons from the Graveyard and Promising Candidates. Lancet Inf Dis, 18: e295–e311, 2018.

Mazur NI, Martinonen-Torres F, Baraldi E, et al. Lower Respiratory Tract Infection Caused by Respiratory Syncytial Virus: Current Management and New Therapeutics. Lancet Respir Med, 3: 888–900, 2015.

Meissner HC. Viral Bronchiolitis in Children. N Engl J Med, 374: 62–72, 2016.

Otomaru H, Sornillo JBT, Kamiguchi T, et al. Risk of Transmission and Viral Shedding from the Time of Infection for Respiratory Syncytial Virus in Households. Am J Epidemiol, 190: 2536–2543, 2021.

Russell CD, Unger SA, Walton M, Schwarze J. The Human Immune Response to Respiratory Syncytial Virus Infection. Clin Micro Rev, 30: 481–502, 2017.

Shi T, McAllister DA, O'Brien KL, et al. Global, Regional, and National Disease Burden Estimates of Acute Lower Respiratory Infections due to Respiratory Syncytial Virus in Young Children in 2015: A Systematic Review and Modelling Study. Lancet, 390: 946–958, 2017.

WEBSITES

Centers for Disease Control and Prevention, Respiratory Syncytial Viral Infection (RSV), 2020: https://www.cdc.gov/rsv

UK Health Security Agency (GOV.UK), Guidance: Respiratory Syncytial Virus (RSV): Symptoms, Transmission, Prevention, Treatment, 2021: https://www.gov.uk/government/publications/respiratory-syncytial-virus-rsv-symptoms-transmission-prevention-treatment/respiratory-syncytial-virus-rsv-symptoms-transmission-prevention-treatment

UK Health Security Agency (GOV.UK), Respiratory Syncytial Virus: The Green Book, Chapter 27a, 2015: https://www.gov.uk/government/publications/respiratory-syncytial-virus-the-green-book-chapter-27a

Students can test their knowledge of this case study by visiting the Instructor and Student Resources: [www.routledge.com/cw/lydyard] where several multiple choice questions can be found.

30

Rickettsia spp.

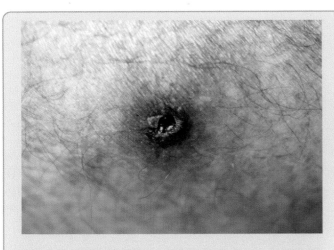

A university professor returned from a botanical expedition to Kenya feeling generally unwell. He was complaining of a headache. He also noticed a large, swollen, black lesion on his thigh that was painful (Figure 30.1). Thinking he had injured himself and that it was now infected, he went to his primary healthcare provider who gave him co-amoxiclav. Over the next few days, the lesion did not respond and he continued to feel unwell with headache and **myalgia**. He presented to a local travel clinic where the doctor identified the lesion on his thigh as a tick bite. He also noticed that the patient had regional **lymphadenopathy**. Making a provisional diagnosis of rickettsiosis the doctor took blood for **serology** and started the patient on an appropriate antibiotic.

Figure 30.1 Tick bite showing the black lesion with surrounding erythema. *From Dr P Marazzi / Science Photo Library, with permission.*

1. WHAT IS THE CAUSATIVE AGENT, HOW DOES IT ENTER THE BODY AND HOW DOES IT SPREAD A) WITHIN THE BODY AND B) FROM PERSON TO PERSON?

CAUSATIVE AGENT

The taxonomy and nomenclature of the *Rickettsiales* in the Alphaproteobacteria division has undergone a major revision based upon 16S rRNA sequences. Currently, the Rickettsiales contains the *Anaplasmataceae* and the *Rickettsiaceae*. The former contains *Anaplasma*, *Ehrlichia*, and *Wolbachia* whereas the latter contains *Rickettsia* and *Orientia* all of which cause human disease. *Coxiella burnetti* is now included in the γ-proteobacteria along with *Legionella* and *Francisella*. Other significant human pathogens in the Alphaproteobacteria are *Brucella* and *Bartonella* belonging to the *Rhizobiales*. An interesting evolutionary issue is the origin of the mitochondria in eukaryotic cells. One view puts the ancestral bacterial origin within the Rickettsiales with the closest genome sequence that of *R. prowazekii*, whereas the other opinion is the organism was a free living alphaproteobacterium that belonged to a now extinct sister group to all Alphaproteobacteria.

Frequently, *Rickettsia* are given species names relating to their original geographic location. There are many species of *Rickettsia* found in nature, only some of which have been linked to illness (Table 30.1) and present clinically as spotted fevers, typhus or scrub typhus. The *Rickettsia* are divided into two main groups, the arthropod-related group containing those associated with human and animal disease and a broader group that infects a wider range of hosts including protozoa, algae, moths, leeches, and beetles. Those found associated with arthropods are separated into four groups: the Ancestral Group (AG) is formed of *R. bellii* and *R. canadensis*, the Transitional Group (TRG) consisting of *R. felis* (cat-flea typhus) and *R. akari* (Rickettsial pox), the Typhus Group (TG) *R. prowazekii*, (Epidemic typhus), *R. typhi* (murine typhus) and the Spotted Fever Group (SFG) which is the largest and consists of, for example, *R. rickettsia* (Rocky Mountain Spotted Fever), *R. conori* (Mediterranean Spotted Fever or Boutonneuse Fever [RMSF]), *R. africae* (African Tick Bite Fever) and *R. sibirica* (Siberian Tick Typhus). An alternative (later) classification scheme suggests a different arrangement of groups based on a wider range of *Rickettsia*-like bacteria with four species closely related to *Rickettsia* such as *Orientia*, and *Pelagibacter* and ten groups of the *Rickettsiae* (Hydra, Torix, Rhizobius, Bellii, Adalia, Onychiurus, Canadensis, Meloidae Spotted Fever, and Transitional).

The *Rickettsia* are small bacteria (0.3–.5 × 0.8–2 μm) that are obligate intracellular pathogens. They have a typical tri-laminar gram-negative cell wall structure and chemistry,

Table 30.1 Species of *Rickettsia* and *Orientia* causing pathogenic human infection and their arthropod vectors

Disease	Tick	Flea	Louse	Mite
Spotted fevers	R. rickettsii	R. felis		R. akari
	R. conori			
	R. japonica			
	R. sibirica			
	R. australis			
	R. slovaca			
	R. africae			
	R. honei			
	R. helvetica			
Typhus		R. typhi	R. prowazekii	
Scrub typhus				O. tsutsugamushi

although they stain poorly with the gram stain but stain well with the Giminez stain. Owing to their obligate intracellular existence there has been an evolutionary trend of genome reduction with reduced or loss of function particularly of amino-acid anabolism and cell-wall constituents, although some genetic markers have increased, particularly small RNA molecules. One of the main cell-wall components is peptidoglycan, which is a microbe-associated molecular pattern (MAMP), and the ability to synthesize this is reduced in a group of obligate intracellular bacteria (*Orientia, Chalmydia,* and *Wolbachia*) leading to a reduced amount of peptidoglycan, although *Rickettsia* still retain a normal peptidoglycan. The *Rickettsia* separated from *Claudobacter* about one and a half to two million years ago with a further event half a million years ago when the organism infected arthropods. The evolutionary pressure for genome reduction was discarding unwanted genomic material and presumably avoidance of stimulating an immune reaction against the organisms. The *Rickettsia* spp. have a small genome size of about 1.3 Mbp. Most of the 40 species of *Rickettsia* and all of the two species of *Orientia* have been sequenced. Comparison of the genomes indicates split genes, pseudogenes and remnants of genes, repeat palindromic sequences, genes with eukaryote motifs and selfish DNA evidencing large gene reductions due to the endosymbiotic lifestyle, horizontal transfer of genes and new gene formation. Plasmids are common in *Rickettsia* and some of them carry a large plasmid p(RF) having about 70 open

reading frames many of which are closely related to genes in *R. bellii* in the AG.

Since their cytosolic habitat is rich in nutrients, amino acids, and nucleotides, they lack enzymes for sugar and amino-acid metabolism, and for lipid and nucleotide synthesis. Their genome therefore codes for several transport proteins, which enable them to utilize host cell products including ATP, although they may also synthesize ATP. They possess a **type IV secretion system** paradoxically with many component duplications, seemingly overcomplicated for its function.

ENTRY AND SPREAD WITHIN THE BODY

The bacterium is injected into the host skin by an arthropod vector (see below). It multiplies locally in the dermal tissue, primarily within the endothelial cells of the vasculature, although *R. akari* and *O. tsutsugamushi* infect monocytes. Subsequently, the organism is spread by the bloodstream to all body organs where it attaches to and reproduces in endothelial cells of the blood vessels.

PERSON-TO-PERSON SPREAD

Some *Rickettsia* spp. are only found in human hosts and are spread from one human host to the next by one of the arthropod vectors. Table 30.2 shows the organisms found in a number of geographic locations with their reservoirs.

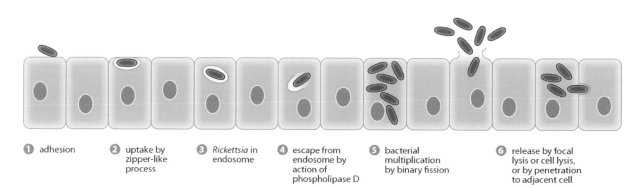

| 1 adhesion | 2 uptake by zipper-like process | 3 *Rickettsia* in endosome | 4 escape from endosome by action of phospholipase D | 5 bacterial multiplication by binary fission | 6 release by focal lysis or cell lysis, or by penetration to adjacent cell |

Figure 30.2 Entry, multiplication, and outcome of Rickettsia inside the endothelial cell.

Pathogenic *Rickettsiae* are spread to humans by different arthropod vectors (Table 30.1). Some *Rickettsia* spp. have a fairly restricted geographic location, while others can be found on all the continents. The distribution of *Rickettsia* spp. is related to that of their arthropod vectors (Figures 30.3–30.6), particularly since ticks, fleas, and lice are ubiquitous. The bacteria are maintained in nature by colonizing/infecting mammalian hosts (dogs, rats, ferrets, deer, etc.), which act as a reservoir for further human infection. In the case of ticks, the bacteria are transmitted **trans-stadially** and thus the vector also acts as a reservoir. Ticks transmit the organism during feeding. The tick may feed over several days and may go unnoticed. Transmission of the *Rickettsia* spp. does not occur immediately and may take a few days. It is thus important to scrutinize one's body if one has been hiking in tick-infested locations and carefully remove any ticks without crushing them.

EPIDEMIOLOGY

Horizontal transmission between ticks co-feeding on a susceptible animal maintains the rickettsia in the tick populations host. Transmission can also occur, although less commonly in immune hosts if they share the same feeding site. In the UK, there are about 20 different tick species. Thus far only one case of tick-borne disease has been acquired in the UK (babesiosis) and two cases of tick-borne encephalitis. Between 2006 and 2018, 16 tick species have been identified in returning travelers of which nine were known to be vectors for rickettsial disease in the country visited. There are many types of flea in the UK the common ones being dog, cat, bird, and human. Lice and mites are also common in the UK and can be vectors of rickettsial disease.

The prevalence of infection with *Rickettsia* spp. is probably underreported. *Rickettsia* spp. are not common in the UK and cases occur in returning travelers. Rickettsia infections in returning travelers in the UK are linked frequently with having visited game parks in sub-Saharan Africa and are usually *R. africae* or *R. conorii*. Travelers visiting Asia often acquire *R. typhi* or *O. tsutsugamushi*. However, *Rickettsia felis* has been detected in 6–12% of the cat fleas in the UK, suggesting that endemic Rickettsial disease is a possibility. Between 1990 and 2002 there were 66 imported cases. The following three species belonging to the SFG have been detected in ticks in the UK: *Rickettsia helvetica*, *R. massiliae* and *R. raoultii* but only one possible case (*R. massiliae*) may have been acquired in the UK. However, *R. helvetica* was detected in 10 of 338 *Ixodes ricinus* ticks collected between 2006 and 2009 from various locations in the UK and it has also been detected in female *Ixodes* collected from UK hedgehogs. Also, a study of 85 *Rhipicephalus sanguineus* ticks demonstrated six were positive for *Rickettsia* and monitoring at country entry points into the UK demonstrated *Rickettsia* are often carried in ticks on imported dogs, although *Rickettsia* have also been found on the ticks of migratory birds.

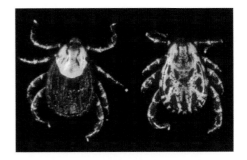

Figure 30.3 A tick – the vector for some rickettsial diseases. *From the Centers for Disease Control & Prevention, Atlanta, Georgia. Image is found in the Public Health Image Library #5977.*

Figure 30.4 A flea – the vector for some rickettsial diseases. *From the Centers for Disease Control & Prevention, Atlanta, Georgia. Image is found in the Public Health Image Library #5636. Additional photographic credit is given to CDC/ DVBID, BZB, Entomology and Ecology Activity, Vector Ecology & Control Laboratory, Fort Collins, CO. and was created in 2004.*

Figure 30.5 A louse – the vector for some rickettsial diseases. *From the Centers for Disease Control & Prevention, Atlanta, Georgia. Image is found in the Public Health Image Library #9217. Additional photographic credit is given to CDC/ Frank Collins, PhD and was created in 2006.*

Figure 30.6 A mite – the vector for some rickettsial diseases. *From the Centers for Disease Control & Prevention, Atlanta, Georgia. Image is found in the Public Health Image Library #5447.*

Table 30.2 The geographic distribution of *Rickettsia* and *Orientia* and their reservoirs

Organism	Location	Reservoir
R. rickettsii	North, Central, and South America	Dogs, deer, rodents, ticks
R. conori	Africa, India, Middle East, Mediterranean, Russia	Dogs, rodents, ticks
R. japonica	Japan, China	Dogs, rodents, ticks
R. sibirica	Russia, China, France, Africa	Dogs, rodents, ticks
R. australis	Australia	Dogs, rodents, ticks
R. slovaca	Europe	Dogs, rodents, ticks
R. africae	sub-Saharan Africa	Dogs, rodents, ticks
R. honei	Flinders Island Australia	Dogs, rodents, ticks
R. felis	America, Brazil, Europe	Dogs, rodents, ticks
R. akari	Eastern USA, Europe, Korea, South Africa	Mice
R. helvetica	Scandinavia	Dogs, rodents, ticks
R. typhi	Worldwide	Rat
R. prowazekii	Worldwide	Human / Flying squirrels in USA
O. tsutsugamushi	Japan, Russia, Australia	Rodent, mites

In Europe, cases of *R. helvetica* have been reported from a number of countries, for example Sweden, Austria, Netherlands, Italy and Slovakia. In the Netherlands, a higher seroprevalence to SFG *Rickettsia* was noted in patients with Lyme disease compared to healthy blood donors suggesting ticks can carry more than one pathogen. Cases of *R. raoultii* have been found in France, Slovakia, and Poland. *R. massiliae* has been detected in Sicily, Spain, and France and has been detected in cats, foxes and hedgehogs.

In the US, *Rickettsial* spotted fever is a notifiable disease. RMSF, caused by the bacterium *R. rickettsii*, is the most severe and most commonly reported SFG rickettsiosis. Tick-borne SFG *Rickettsia* species in the US that are known to cause human illness is not just *R. rickettsii*, but other species have also been identified as causal agents: *R. parkeri* and a newly recognized *Rickettsia* identified as 364D with the proposed nomenclature *Rickettsia philipii*. The number of cases in 2000 was 495 and the most current data, 2017, identified 6248 cases. Incidence of cases depends on the season (when ticks are active) and the State/Country: in Arizona, cases are transmitted by *Rhipicephalus* (dog tick) with a peak April–Oct.

In Australia, a seroprevalence study showed that 5.6% of nearly 1000 individuals were positive for rickettsia, with 24% of these positive for scrub typhus. Despite its importance as a global pathogen, and indeed as an organism that could be used for bioterrorism, it has been little studied to date. Figure 30.7 shows the global distribution of different species of the Rickettsiales.

2. WHAT IS THE HOST RESPONSE TO THE INFECTION AND WHAT IS THE DISEASE PATHOGENESIS?

The pathophysiology of *Rickettsial* infection is currently poorly understood and the details will vary according to the species of *Rickettsia*, particularly the two main groups: the SFG and the TG.

Rickettsia attach to and enter the vascular endothelium by inducing **endocytosis** using a zipper-like process whereby there is sequential interaction between ligand and an **adhesin** (Figure 30.2). Two cell-surface proteins appear to be involved in the adhesion process, OmpA (outer-membrane protein A) and Sca2, which are both members of an autotransporter family of proteins called surface cell antigen (Sca). Other members of this family are Sca1, Sca3, and OmpB. OmpB may also be involved in the adhesion process. A rickettsial phospholipase A2 appears to be involved in the uptake of the rickettsia into the cell. The intracellular signaling events mediating induced endocytosis involve various signaling pathways such as Cdc-42 PI 3 kinase, c-Src, and Arp2/3. Once they have entered the endothelial cell they escape the **endosome** with the aid of bacterial enzymes, for example, phospholipase D, and replicate in the cytoplasm. Like many other intracellular organisms, the rickettsia inhibit **apoptosis** of the host cell to allow their replication to proceed. They spread from an infected cell to adjacent cells by polymerizing actin monomers at one pole of the cell, thereby pushing the organism forward. RickA (a group of proteins found in the SFG of rickettsia but not in the TG) induces nucleation of actin monomers via the Arp2/3 complex, thereby mediating intracellular movement (Figure 30.2). Rickettsia are also released into the bloodstream using the polymerizing actin monomers.

Rickettsia affects the trans-Golgi network (but not the cis-Golgi network) by the production of ankyrin repeat protein 2 which affects protein secretion by the cell including MHC I molecules and thus inhibiting immune recognition.

Infection of vascular endothelium by *Rickettsia* leads to an increase of the membrane-bound CX3CL1 chemokine on the host cell membrane (which is involved with recruiting immune cells to the site of infection). The increased expression

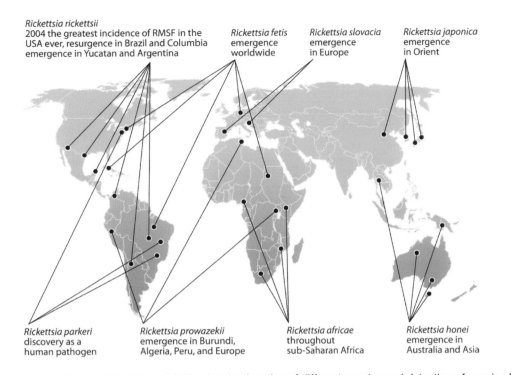

Rickettsia rickettsii
2004 the greatest incidence of RMSF in the USA ever, resurgence in Brazil and Columbia emergence in Yucatan and Argentina

Rickettsia fetis emergence worldwide

Rickettsia slovacia emergence in Europe

Rickettsia japonica emergence in Orient

Rickettsia parkeri discovery as a human pathogen

Rickettsia prowazekii emergence in Burundi, Algeria, Peru, and Europe

Rickettsia africae throughout sub-Saharan Africa

Rickettsia honei emergence in Australia and Asia

Figure 30.7 Distribution of rickettsial diseases globally, showing location of different species and detection of species from locations where they had not previously been detected. *From Walker DH Rickettsiae and Rickettsial Infection: The Current State of Knowledge. Clin Infect Dis, 45: S39–S44, 2007.*

of CX3CL1 is stimulated by miRNA-424. However, following infection by *Rickettsia*, there is a reduction in expression of miRNA-424 inhibiting the immune response to infection. During Rickettsial infection, the Lipid A moiety activates TLR4/MD2 but the clinical effect of Lipid A on outcome is unknown particularly as the structure of the Lipid A from different Rickettsial groups varies. These differences may have clinical relevance in relation to disease pathogenesis or vaccine efficiency.

Damage to the endothelial cells leads to changes in microvascular permeability. Currently, it is believed that the injury is the result of oxidative stress in the endothelial cells mediated by **reactive oxygen species**, which cause lipid peroxidative membrane damage. This is probably the direct effect on the endothelial cells by the infective organism but mostly by cells of the immune system.

Endothelial cell infection induces the production of pro-inflammatory **cytokines** such as **interleukin (IL)**-1α and this causes an up-regulation of E-selectin (*R. rickettsii*) allowing increased adhesion of polymorphs to the vasculature. Following infection with *R. conorii*, **E-selectin, ICAM-1,** and **VCAM-1** are all up-regulated, allowing adhesion of mononuclear cells including macrophages and T cells.

Endothelial damage-mediated changes in microvascular permeability result in **hypovolemia**, hypotension, and pulmonary **edema**. Membrane leakiness induced by nitric oxide, which is a host response by the endothelial cells and macrophages, can also lead to interstitial **pneumonia, myocarditis**, perivascular lesions in the brain and other organs, and in surface peripheral blood vessels leads to a

rash. In skin cells from *R. conorii*-infected patients, the host responds by up-regulation of **tumor necrosis factor-α (TNF-α), interferon-γ (IFN-γ)**, IL-10, RANTES, and inducible nitric oxide synthase (iNOS), and the levels of the cytokines correlate with severity of disease. The **vasculitis** leads to increased local consumption of platelets with a **thrombocytopenia** in a proportion of patients and a pro-coagulation state, although **disseminated intravascular coagulation (DIC)** and **thromboses** are rare.

IMMUNE RESPONSE

The immune response, as well as playing a role in pathogenesis, is eventually able to control the infection. Production of an enzyme **indoleamine 2,3-dioxygenase** by the endothelial cells following the effect of pro-inflammatory cytokines, limits the replication of *Rickettsia* by metabolizing tryptophan, an essential amino acid for rickettsia.

Early in infection it is thought that **natural killer (NK)** cells inhibit growth of *Rickettsia* in endothelial cells through release of IFN-γ. **CD4** T cells and **CD8+** T cells are believed to produce IFN-γ and also RANTES that enhance intracellular killing of *Rickettsia* via nitric oxide production and hydrogen peroxide. CD8+ T cells are able to kill *Rickettsia*-infected endothelial cells via a **perforin**-dependent cytotoxic mechanism, which contributes significantly to recovery from infection. There is also an antibody response generated to some of the *Rickettsia* adhesion molecules, such as OmpA and OmpB, but these antibodies appear in significant amounts only after the infection has resolved. They are therefore thought to be protective against re-infection.

3. WHAT IS THE TYPICAL CLINICAL PRESENTATION AND WHAT COMPLICATIONS CAN OCCUR?

Rickettsial disease presents in one of two ways: (a) as a spotted fever or (b) as **typhus**, depending upon the infecting organism (Tables 30.1 and 30.3). The typical presentation of Rickettsial disease is abrupt onset of fever, headache, myalgia, and a rash.

SPOTTED FEVERS

The incubation period is about 7 days but may be up to 14 days. Patients present with fever, headache, myalgia, and a **maculopapular rash**.

The rash typically begins around the wrists and ankles but can also cover the trunk and may also appear on the palms and soles. In 10% of patients with RMSF or Brazilian Spotted fever caused by *R. rickettsii* the rash may be absent. Following infections with *R. slovaca* and *R. helvetica* typically there is no rash and a rash is uncommon with *R. africae*.

About 25% of patients may show signs of central nervous system (CNS) involvement with encephalitis or **seizures**. Gastrointestinal (GI) symptoms of nausea, vomiting, abdominal pain, and diarrhea may be present.

Despite the low platelets and prolonged coagulation time, bleeding and DIC are uncommon. In infections with *R. conori*, a pro-coagulation state exists and thromboses may occur in 10% of patients.

A tick bite may be visible as a black eschar (tache noire), although this is uncommon in RMSF, or as multiple eschars in infections with *R. africae*. Following infections with *R. slovaca*, the tick bite is commonly in the scalp and may lead to **alopecia**.

Complications include coma, renal failure, **adult respiratory distress syndrome (ARDS)**, **myocarditis**, and death in about 10% of patients. Fulminant infection is more common in patients with glucose 6 phosphate dehydrogenase (G6PD) deficiency.

Laboratory investigations show a **thrombocytopenia** in about 50% of patients, **anemia**, abnormal liver function tests (LFTs), increased urea, **hyponatremia,** and **hypoalbuminemia**. Prolonged coagulation times may be observed and increased levels of creatine kinase. A chest X-ray may reveal pulmonary edema.

RICKETTSIAL POX

This presents with a fever, headache, and a rash. The rash is vesicular and starts as a **papule**, becoming vesicular, and scabs leaving a black eschar. The rash may also appear in the oral cavity as well as the palms and soles. Regional lymphadenopathy is present. The patients may also complain of profuse sweating, rigors, sore throat, and **photophobia**. Laboratory results indicate a **leukopenia** although LFTs, urea, electrolytes, and hemoglobin are usually normal.

TYPHUS

Epidemic Typhus

Epidemic typhus is associated with poor hygiene, homelessness, overcrowding, war, poverty, and natural disasters. The incubation period is about 10 days. The patient presents with fever, headache, myalgia, and a macular rash that is found on the trunk although not usually on the palms and soles. Respiratory and CNS symptoms may also be present. The case fatality rate is 15%. The disease acquired from flying squirrels is less severe.

Laboratory results indicate a leukopenia and thrombocytopenia. Elevated LFTs, creatine kinase, and blood urea are found. A chest X-ray may show interstitial shadowing.

Once infected and after apparently successful treatment, a patient may suffer a recrudescence of typhus (**Brill-Zinsser disease**) with the same signs and symptoms. This is frequently induced by stress or immune suppression.

Murine (Endemic) and Scrub Typhus

For both endemic and scrub typhus, the incubation period is 7–14 days and the patient presents with the typical symptoms of abrupt onset of fever, headache, and myalgia. A

Table 30.3 Common eponyms for some rickettsial diseases

	Species	Disease
Spotted fever	*R. rickettsii*	Rocky Mountain spotted fever (RMSF), Brazilian Spotted fever
	R. conori	Mediterranean spotted fever, Kenya tick typhus, Indian tick typhus, Boutonneuse fever
	R. sibirica	N. Asian tick typhus, Siberian tick typhus
	R. japonica	Japanese tick typhus
	R. australis	Queensland tick typhus
	R. honei	Flinders Island spotted fever
	R. africae	African tick bite fever
	R. akari	Rickettsial pox
Typhus	*R. prowazekii*	Epidemic typhus, Brill-Zinsser disease (recrudescence)
	R. typhi	Murine (or endemic) typhus
	O. tsutsugamushi	Scrub typhus

maculopapular rash on the trunk develops during the course of the illness in the majority of cases. Hepatosplenomegaly may be present in both endemic and scrub typhus. In scrub typhus, the rash is pale and transient and may be missed. GI (nausea, vomiting), respiratory (cough), and CNS symptoms (altered consciousness, seizures) may also occur. Complications similar to RMSF may occur and the case fatality rate is about 5%.

4. HOW IS THE DISEASE DIAGNOSED, AND WHAT IS THE DIFFERENTIAL DIAGNOSIS?

Diagnosis of Rickettsial disease relies on **serology** and **polymerase chain reaction** (PCR) because of the difficulty in culturing these organisms in a routine setting. Rapid detection using loop isothermal PCR, qPCR or unbiased WGS on a clinical sample can be used. Reference and specialist laboratories may offer a culture service where the organism is grown on cell culture using a shell-vial assay or in animals. An accurate diagnosis of the infecting species is frequently not possible using serology because of the many cross-reactions between the *Rickettsia* and other bacteria, particularly *Proteus* spp., which is the basis of the **Weil-Felix test**. Immunohistochemistry can be used on biopsy material from any rash and can help distinguish Rickettsial disease from other causes of a rash. **Immunofluorescence** or **ELISA** for serum antibodies are not useful in making a diagnosis of acute disease. The Weil-Felix agglutination test using *Proteus* OX-2 and OX-19 has a low sensitivity and specificity. More accurate diagnosis can be made using PCR combined with restriction fragment length polymorphism (RFLP) or sequencing to confirm the identity of any amplicon detected. RT-PCR and real-time PCR assays are also available. PCR is most useful using skin biopsies of the rash and is less helpful for blood samples. The primers are based on a number of different genes including those for rOmpA, rOmpB, a 17 kDa genus-specific protein (*htrA*), and citrate synthase (*gltA*).

DIFFERENTIAL DIAGNOSIS

Rickettsial disease presents with a fever and a rash and thus, in the differential diagnosis, any other infection with a similar presentation, including other Rickettsial illnesses, should be considered. Other infectious causes that should be considered are measles, rubella, varicella, arboviral infections, meningococcal and disseminated gonococcal disease, typhoid, secondary syphilis, anthrax, and leptospirosis. Of the noninfectious causes, **idiopathic thrombocytopenic purpura** and **immune complex** disease should be considered.

5. HOW IS THE DISEASE MANAGED AND PREVENTED?

MANAGEMENT

The mainstay of treatment is doxycycline. Other tetracyclines may also be used. An alternative, although less effective treatment, is chloramphenicol, which can be used during pregnancy or in cases of allergy to the tetracyclines. Azithromycin may also be an effective antibiotic in some cases. Quinolones and rifampicin have *in vitro* activity but poor clinical response. β-lactams, aminoglycosides, and sulfonamides are ineffective. In severe cases of RMSF, where IV doxycycline cannot be sourced, tigacycline can be used and is as effective as doxycycline.

Because eschars and skin necrosis can occur with both Covid-19 and Rickettsial disease, in areas where *Rickettsia* is common, it may be prudent to start doxycycline (in case the illness is a Rickettsiosis) until confirmation of the disease is obtained. Furthermore, because flea-borne Rickettsiosis (*R. typhi*, *R. felis*) present with nonspecific symptoms of myalgia, headache and fever and a rash, empirical treatment with doxycycline should be considered.

PREVENTION

Preventive measures include the use of insect repellents and protective clothing. Two vaccines have been developed, although not generally available, because of the potential threat of *Rickettsia* spp. being used as a bioweapon. Studies indicate a live attenuated vaccine is more effective than a killed vaccine.

Targeting the salivary proteins of the main tick vector of *R. prowazekii*, by developing a vaccine against them induces tick death and thus reduces vector transmission.

SUMMARY

1. WHAT IS THE CAUSATIVE AGENT, HOW DOES IT ENTER THE BODY, AND HOW DOES IT SPREAD A) WITHIN THE BODY AND B) FROM PERSON TO PERSON?

- *Rickettsia* and *Orientia* spp. are obligate intracellular gram-negative bacteria.
- They have a small genome size and restricted metabolic activity but the genome codes for several autotransporter proteins.
- The bacteria are spread by the bite of an arthropod vector: tick, flea, louse, mite.
- The bacteria are spread by the bloodstream and infect vascular endothelial cells systemically.
- Adhesins of *Rickettsia* for attachment to endothelial cells belong to an autotransporter family of proteins.
- Induced endocytosis of *Rickettsia* uses a zipperlike action.
- Intracellular signaling events depend upon Arp2/3 complex.
- *Rickettsia* escape from the endosome and replicate in the cytoplasm.
- Infection with Rickettsia up-regulates adhesion molecules on the endothelial cell, inducing adhesion of leukocytes.
- The bacteria are maintained in nature by infecting mammals such as dogs and rodents, which act as a reservoir of infection.
- Some *Rickettsia* spp. are confined to humans.
- The *Rickettsia* and *Orientia* are found on all continents although some have a more restricted geographic distribution.

2. WHAT IS THE HOST RESPONSE TO THE INFECTION AND WHAT IS THE DISEASE PATHOGENESIS?

- Damage to the endothelial cells leads to changes in microvascular permeability through oxidative stress in the endothelial cells mediated by the infection but also by cells of the immune system.
- Membrane leakiness induced by nitric oxide, which is a host response by the endothelial cells and macrophages, can also lead to interstitial pneumonia, myocarditis, or perivascular lesions in the brain.
- The vasculitis leads to increased local consumption of platelets with a thrombocytopenia in a proportion of patients.
- Cellular damage is caused by free radicals.
- Cell damage results in increased vascular permeability and edema.

- Host indoleamine 2,3-dioxygenase limits replication of rickettsia.
- CD8+ T cells are important in recovery from Rickettsial infection.

3. WHAT IS THE TYPICAL CLINICAL PRESENTATION AND WHAT COMPLICATIONS CAN OCCUR?

- Rickettsial disease presents with abrupt onset of fever, headache, myalgia, and rash.
- The incubation period is usually 7–14 days.
- The rash is maculopapular but in rickettsial pox it is vesicular.
- Gastrointestinal, respiratory, and CNS symptoms may occur.
- Evidence of an insect bite in the form of a black eschar may be found.
- Complications include renal failure, ARDS, and death in 5–15% of cases.
- Although damage occurs directly to endothelium, DIC and major vascular incidents are rare.

4. HOW IS THE DISEASE DIAGNOSED, AND WHAT IS THE DIFFERENTIAL DIAGNOSIS?

- Culture of *Rickettsia* spp. is not part of routine diagnosis but is performed in specialist or reference laboratories.
- Detection of serum antibodies is more useful as a retrospective diagnosis.
- Accurate species diagnosis by serology is difficult because of cross-reactions within the genus.
- Immunofluorescence and enzyme immunoassay are frequently used serologic assays.
- The Weil-Felix agglutination reaction has low sensitivity and specificity.
- PCR combined with RFLP or sequencing can give an accurate species identification, particularly on biopsy material from rashes.
- Differential diagnosis includes many illnesses that present with a fever and a rash.

5. HOW IS THE DISEASE MANAGED AND PREVENTED?

- The mainstay of treatment is doxycycline.
- Alternative treatments include chloramphenicol and azithromycin.
- Preventative measures include insect repellents and protective clothing.

FURTHER READING

Cimolai N. Laboratory Diagnosis of Bacterial Infections. Marcel Dekker, New York, 823–860, 2001.

Firth J, Conlon C, Cox T, (eds). Oxford Textbook of Medicine, 6th edition. Oxford University Press, Oxford, 2020.

Goering RV, Dockrell HM, Zuckerman M, Chiodini PL. Mims' Medical Microbiology and Immunology, 6th edition. Elsevier, Philadelphia, 2018.

Heymann DL. Control of Communicable Disease Manual. American Public Health Association, Washington, DC, 459–464, 583–590, 2004.

Mandell GL, Bennet JE, Dolin R. Principles & Practice of Infectious Diseases, 6th edition, Vol 3. Elsevier, Philadelphia, 2284–2310, 2005.

Murphy K, Weaver C. Janeway's Immunobiology, 9th edition. Garland Science, New York/London, 2016.

REFERENCES

Abdad MY, Abdallah RA, Fournier P-E, et al. A Concise Review of the Epidemiology and Diagnostics of Rickettsioses: Rickettsia and Orientia spp. J Clin Microbiol, 56: e01728–17, 2018.

Blanton LS. The Rickettsioses: A Practical Update. Infect Dis Clin North Am, 33: 213–229, 2019.

Carl M, Tibbs CW, Dobson ME, et al. Diagnosis of Acute Typhus Using the Polymerase Chain Reaction. J Infect Dis, 161: 791–793, 1990.

Dignat-George F, Teysseire N, Mutin M, et al. Rickettsia conorii Infection Enhances Vascular Cell Adhesion Molecules-1 and Intercellular Adhesion Molecule 1-Dependent Mononuclear Cell Adherence to Endothelial Cells. J Infect Dis, 175: 1142–1152, 1997.

Eremeeva ME, Dasch GA, Silverman DJ. Evaluation of a PCR Assay for Quantitation of Rickettsia rickettsii and Closely Related Spotted Fever Group Rickettsiae. J Clin Microbiol, 41: 5466–5472, 2003.

Gillespie JJ, Beier MS, Rahman MS, et al. Plasmids and Rickettsial Evolution: Insights from Rickettsia felis. PLoS One, 2: e266, 2007.

Gillingham EL, Cull B, Pietzch ME, et al. The Unexpected Holiday Souvenir: The Public Health Risk to UK Travellers from Ticks Acquired Overseas. Int J Environ Res Public Health, 17: 7957, 2020.

Hechemy KE, Oteo JA, Raoult DA, et al. A Century of Rickettsiology: Emerging, Re-Emerging Rick-Ettsioses, Clinical, Epidemiologic, and Molecular Diagnostic Aspects and Emerging Veterinary Rickettsioses. Ann N Y Acad Sci, 1078: 1–14, 2006.

Jensenius M, Fournier PE, Raoult D. Tick-borne Rickettsioses in International Travellers. Int J Infec Dis, 8: 139–146, 2004.

Johnston VJ, Stockley JM, Dockrell D, et al. Fever in Returned Travellers Presenting in the UK: Recommendations for Investigation and Initial Management. J Infect, 59: 1–18, 2009.

Karpathy SE, Espinosa A, Yoshimizu MH, et al. A Novel TaqMan Assay for Detection of Rickettsia 364D, the Etiologic Agent of Pacific Coast Tick Fever. J Clin Microbiol, 58: e01106–19, 2020.

Kenny MJ, Birtles RJ, Day MJ, Shaw SE. Rickettsia felis in the United Kingdom. Emerg Infect Dis, 9: 1023–1024, 2003.

Li H, Walker DH. rOmpA is a Critical Protein for the Adhesion of Rickettsia rickettsii to Host Cells. Microb Pathog, 24: 289–298, 1998.

Martinez JJ, Cossart P. Early Signalling Events Involved in the Entry of Rickettsia conorii into Mammalian Cells. J Cell Sci, 117: 5097–5106, 2004.

McGinn J, Lamason RL. The Enigmatic Biology of Rickettsiae: Recent Advances, Open Questions and Outlook. Pathog Dis, 79: ftab019, 2021.

Murray GGR, Weinert LA, Rhule EL, et al. The Phylogeny of Rickettsia Using Different Evolutionary Signatures: How Tree-Like Is Bacterial Evolution. Syst Biol, 65: 265–279, 2016.

Olano JP. Rickettsial Infection. Ann N Y Acad Sci, 1063: 187–196, 2005.

Qin XR, Han HJ, Han FJ, et al. Rickettsia japonica and Novel Rickettsia Species in Ticks, China. Emerg Infec Dis, 25: 992–995, 2019.

Raoult D, Fournier PE, Fenollar F, et al. Rickettsia africae, a Tick-Borne Pathogen in Travelers to Sub-Saharan Africa. N Engl J Med, 344: 1504–1510, 2001.

Riley SP, Fish AI, Piero FD, Martinez JJ. Immunity Against the Obligate Intracellular Bacterial Pathogen Rickettsia australis Requires a Functional Complement System. Infect Immun, 86: e00139–18, 2018.

Salje J, Weitzel T, Newton PN, et al. Rickettsial Infections: A Blind Spot in Our View of Neglected Tropical Diseases. PLOS Negl Trop Dis, 15: e0009353, 2021.

Treadwell TA, Holman RC, Clarke MJ, et al. Rocky Mountain Spotted Fever in the United States, 1993–1996. Am J Trop Med Hyg, 63: 21–26, 2006.

Walker DH. Rickettsiae and Rickettsial Infection: The Current State of Knowledge. Clin Infect Dis, 45: S39–S44, 2007.

Weinert LA, Werren JH, Aebi A, et al. Evolution and Diversity of Rickettsia Bacteria. BMC Biol, 7: 6, 2009.

Zurita A, Benkacimi L, El Karkouri K, et al. New Records of Bacteria in Different Species of Fleas from France and Spain. Comp Immun Microb Infec Dis, 76: 101648, 2021.

WEBSITES

BMJ Best Practice, Rickettsial Diseases, 2022: https://bestpractice.bmj.com/topics/en-gb/1604

Centers for Disease Control and Prevention, Division of Viral and Rickettsial Diseases, Atlanta, GA, USA: http://www.cdc.gov/ncidod/dvrd/

Centers for Disease Control and Prevention, Tickborne Diseases of the United States, Rickettsia parkeri Rickettsiosis, 2019: https://www.cdc.gov/ticks/tickbornediseases/rickettsiosis.html

University of South Carolina, Pathology, School of Medicine Columbia, Microbiology and Immunology: http://pathmicro.med.sc.edu/mayer/ricketsia.htm

Students can test their knowledge of this case study by visiting the Instructor and Student Resources: [www.routledge.com/cw/lydyard] where several multiple choice questions can be found.

Salmonella typhi

31

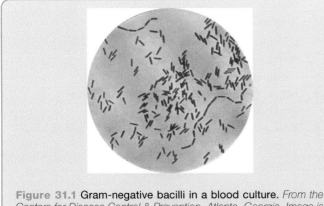

Figure 31.1 **Gram-negative bacilli in a blood culture.** *From the Centers for Disease Control & Prevention, Atlanta, Georgia. Image is found in the Public Health Image Library #2114.*

A 53-year-old woman returned from a visit to Lahore complaining of feeling generally unwell with a fever and a cough. She attended a clinic where she was seen by a doctor who confirmed a temperature of 38°C and he noticed a rash on the upper chest. She was admitted to hospital for investigation which included a thick and thin film for malaria, a full blood count, urea and electrolytes, a chest X-ray, and blood cultures. The malaria investigation was negative, the chest X-ray showed patchy basal consolidation, the full blood count revealed a relative **lymphocytosis** and gram-negative bacilli were seen in the blood culture (Figure 31.1). A provisional diagnosis of enteric fever was made and she was started on appropriate antibiotics. The diagnosis was confirmed by isolation of *Salmonella typhi* from the blood cultures.

1. WHAT IS THE CAUSATIVE AGENT, HOW DOES IT ENTER THE BODY AND HOW DOES IT SPREAD A) WITHIN THE BODY AND B) FROM PERSON TO PERSON?

CAUSATIVE AGENT

The nomenclature of the genus *Salmonella* has undergone a number of revisions leading to two systems of validly published names, the latest version of which (2005) has the disadvantage of not highlighting important human pathogens such as *Salmonella typhi* or *Salmonella enteritidis* by not giving them succinct names. In the current version of the nomenclature, the cause of enteric fever and the organism that is one of the important causes of gastroenteritis would be *S. enterica* subsp. *enterica* serovar Typhi and *S. enterica* subsp. *enterica* serovar Enteritidis, respectively. For pragmatic reasons, this nomenclature will not be used in this text but will be shortened to *Salmonella* Typhi. There are only two species in the genus: *S. enterica* and *S. bongori* of which *Salmonella enterica* is further subdivided into six subspecies: *S. enterica* (1531), *S. salamae*, (505) *S. arizonae*, (99) *S. diarizonae*, (336) *S. houtenae* (73) and *S. indica* (13) with the number of serovars in each group in brackets. New serovars are regularly being identified so these numbers are not constant.

Salmonella organisms are motile nonsporing gram-negative facultative anaerobic rods measuring 2–3×0.4–0.6μm. The genome of *Salmonella* Typhi and *Salmonella* Typhimurium have both been sequenced and contain about 4.8 million base pairs with 4000–5000 coding sequences, of which 98% are homologous between the two serovars. *Salmonella* pathogenicity islands (SPIs) are gene sequences within the *Salmonella* genome that code for virulence characteristics. They usually have a GC content different from the host genome as they are acquired horizontally from other bacteria. The genus *Salmonella* has five SPIs. SPI-1 is involved in penetration of the intestinal epithelium. SPI-2 is involved with intracellular survival. Both SPI-1 and 2 code for a different Type III secretion system. SPI-3 carries virulence functions related to survival within macrophages. SPI-4 codes for a Type I secretion system and probably secretes toxin(s). SPI-5 mediates inflammation and chloride secretion which characterizes non-typhoid *Salmonella* (NTS) infection. SPI-6 codes for a Type VI secretion system and is found in SPI-7 and codes for the Vi antigen.

Frequent genome rearrangements associated with the SPIs lead to outgrowth of strains that are better adapted to different environmental circumstances. Horizontal gene transfer is also an important factor in the evolution of the genus. A large plasmid is carried in *Salmonella* Typhi (pHCM1) that encodes drug resistance. **Virulence** plasmids are also carried in non-typhoidal serovars. There are over 2500 serovars in the genus and they are grouped according to the possession of somatic O antigens, flagellar H antigens, and surface virulence (Vi) antigens in the Kauffman-White scheme. K (capsular) antigens are also present (Figure 31.2). Most *Salmonella* spp. express two types of flagella made up of different proteins (H antigen) and they switch from one phase to the next at characteristic frequencies. Thus, the H antigens occur in two phases: phase I and II. The O antigens are designated by arabic numerals, the phase I antigens by letters a–z, and the phase II antigens by arabic numerals. Thus, a specific serovar may be designated [9,12,Vi/d/–] = *Salmonella* Typhi (*Salmonella* Typhi does not have a phase II antigen) or [1,4,5,12/i/1,2] = *Salmonella* Typhimurium. For epidemiologic purposes, a number of typing methods have been used. These include **phage typing**, plasmid profiling, **ribotyping**, **pulse field gel electrophoresis** (**PFGE**), variable number of tandem repeats (VNTR), enterobacterial repetitive intergenic consensus-**polymerase chain reaction** (ERIC-**PCR**), and whole genome sequencing (WGS).

ENTRY AND SPREAD WITHIN THE BODY

Salmonella are ingested in food or water. The infecting dose is uncertain but has been reported to vary between 10^2 and 10^6 cells. Reduced gastric acidity is a predisposing risk factor and colonization by *Helicobacter pylori* may be an important co-morbidity through its reduction in acid output in cases of pan-gastritis.

Entry into the body is a complex process with sequential expression of various virulence modalities at each stage.

- The initial site is in the stomach where the organism expresses an acid tolerance response.
- Penetration of the mucus layer in the small intestine and adhesion to enterocytes and M cells in Peyer's patches.
- Penetration of cells with bacteria enclosed in a vacuole.
- Disruption of normal endocytic pathway to avoid fusion with the lysosome.
- Penetration of basement membrane and release of bacteria into lamina propria.
- Activation of the innate immune system.
- Uptake by neutrophils, macrophages, and dendritic cells with bacteria in a vacuole.
- Disruption of endocytic pathway again.
- Infected cells enter lymph system then enter bloodstream.
- Relocation to various other organs: liver, spleen, bone marrow.
- Disease onset with clinical symptoms.

Non-Typhoid *Salmonella*

NTS penetrate intestinal epithelial cells but rarely disseminate throughout the body by the bloodstream and are more prone to cause local disease. Co-infection with HIV is a risk factor

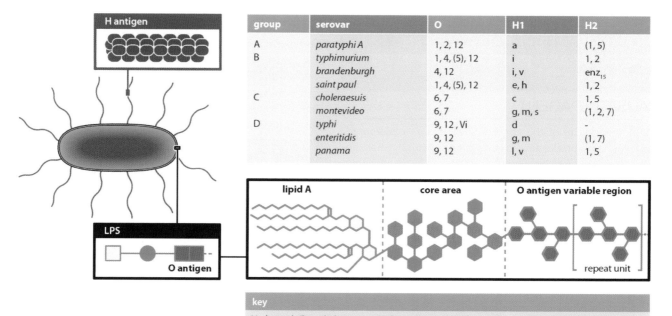

group	serovar	O	H1	H2
A	paratyphi A	1, 2, 12	a	(1, 5)
B	typhimurium	1, 4, (5), 12	i	1, 2
	brandenburgh	4, 12	i, v	enz$_{15}$
	saint paul	1, 4, (5), 12	e, h	1, 2
C	choleraesuis	6, 7	c	1, 5
	montevideo	6, 7	g, m, s	(1, 2, 7)
D	typhi	9, 12 , Vi	d	-
	enteritidis	9, 12	g, m	(1, 7)
	panama	9, 12	l, v	1, 5

key

H haunch (breath) bacteria which produce a thin film on the growth medium similar to the effect of breath on a glass, only motile bacteria do this because they have flagellae, H antigens can occur as two types, H1 and H2

O ohne (without) that is without breath i.e. do not produce a thin film on the growth medium because they do not have flagellae and are nonmotile

Vi virulence

Figure 31.2 The relationship between the O (cell wall) and H (flagella) antigens and the Kaufmann-White serotyping scheme.

for severe non-typhoidal infection and is a significant cause of mortality.

PERSON-TO-PERSON-SPREAD

Salmonella Typhi

Salmonella Typhi and *Salmonella* Paratyphi are strictly human pathogens and are **endemic** in several countries (see below). Endemicity of typhoid fever is associated with a poor social hygiene infrastructure (inadequate sewage disposal, inadequate potable water supplies, and inadequate food hygiene practices) and social upheaval.

Salmonella Typhi is acquired either directly or indirectly from another human or carrier by ingestion. Direct acquisition from a person is uncommon but can occur associated with certain sexual practices. More usually, infection is acquired from fecally contaminated water or food and less commonly in laboratories handling clinical specimens. Outbreaks have been linked to food-handlers who are carriers of the organism – the most notorious being "Typhoid Mary". Mary Mallon was a cook for a New York banker, who infected most of his family with typhoid because she was a healthy carrier excreting *Salmonella* Typhi. Investigations by a civil engineer employed by the banker to identify the source revealed that outbreaks of typhoid had occurred in seven families for whom Mary had been the cook.

An important virulence characteristic is the ability to form biofilms. The main factors in biofilm formation are curli fimbriae and exopolysaccharide cellulose. Biofilm is important in maintaining viability under adverse environmental circumstances making the food chain susceptible as a source of infection.

The organism is endemic in India, Africa, South America, and South-East Asia. In Africa, the incidence is about 900/100000 population. In Indonesia, enteric fever is one of the commonest causes of death in children, with 20000 deaths each year. In nonendemic countries such as the UK and the US, the incidence is about 0.2/100000 population. In the UK, the number of cases in 2000 was 331 (caused by *Salmonella* Typhi, *Salmonella* Paratyphi A and B) and in 2012 the number of cases was 430.

A significant proportion of cases of typhoid in industrialized countries are acquired abroad.

Non-Typhoid *Salmonella*

NTS are found in a variety of animal species. Infection is acquired by ingestion from contaminated food, particularly eggs, poultry, and dairy produce. The sudden increase in *Salmonella* Enteritidis PT4 in the UK was linked to contaminated eggs. Since 2018, a pan-European outbreak of *Salmonella* Enteritidis sequence type 11, associated with breaded chicken, has affected people in Britain, Denmark, Finland, France, Germany, Ireland, the Netherlands, Poland, and Sweden, with 120 infections thus far in the UK. In 2021, five clusters of *Salmonella* Braenderup in the UK are being investigated. The source is currently unknown. Infection may also be acquired from direct contact with infected animals, particularly exotic pets, for example, snakes.

EPIDEMIOLOGY

The number of cases of gastroenteritis caused by the 2500 different serovars of salmonella (e.g. *Salmonella* Enteritidis, *Salmonella* Typhimurium, etc.) has increased over the years. In the US, about 1.4 million cases occur annually with an incidence of 17/100000 population, with *Salmonella* Typhimurium and *Salmonella* Enteritidis being the most common serovars causing infection. In the UK, the incidence of *Salmonella* [Enteriditis, Typhimurium, Typhi, other non-typhoidal salmonellae (NTS)] in 2000 was 15435 [8616, 2688, 331, 3798], which decreased to 8355 [2169, 1902, 430, 3854] in 2012. The most frequent serovar causing disease was *Salmonella* Typhimurium, until 1988 when there was a sudden increase in *Salmonella* Enteritidis, which became the predominant serovar. The predominant strain of *Salmonella* Enteritidis causing disease is phage type 4 (PT4) and in 2000, *Salmonella* Enteritidis accounted for 8616 cases, outnumbering all other *Salmonella* types combined. The age of infection (per 100000 population) in 2000 [<1yr, 1–9yr, 15–64yr, >64] was 104, 91, 54, 34, respectively and in 2012, this had decreased to 79, 58, 30, 21, respectively. Globally, NTS is estimated (WHO) that 500 million persons are infected annually with 3 million deaths and in 2017 this had decreased to 21.7 million cases and 217000 deaths.

2. WHAT IS THE HOST RESPONSE TO THE INFECTION AND WHAT IS THE DISEASE PATHOGENESIS?

Low gastric acid, immunosuppression, and certain polymorphisms predispose to infection, for example Toll-like receptor (TLR) 4, interleukin (IL-) 4, and HLA class II and class III.

INNATE IMMUNITY

Infection with *Salmonella* Typhi induces a monocytic response with little diarrhea, due in part to the immunosuppressive effect of the **Vi antigen**. On the other hand, NTS induces an acute inflammatory reaction with the secretion of **interleukin (IL)-8**, the recruitment of neutrophils, and pronounced diarrhea. The diarrhea is induced in part by translocated proteins and in part by disruption of the epithelial barrier.

In the lamina propria, salmonella enter epithelial cells and macrophages by bacteria-mediated endocytosis or **phagocytosis**, respectively. Once inside the macrophage, the bacterium is protected from the humoral immune system and it survives intracellular killing by the macrophage by up-regulating numerous genes required for intracellular survival and replication, including the Vi antigen, which is principally anti-phagocytic. Changes occur in the cell wall of *Salmonella* making it resistant to bactericidal peptides and reactive oxygen species (ROS) and causing less inflammation. Additionally, products coded by the pathogenicity island SPI-2 are secreted into the macrophage cytoplasm, which inhibit phagolysosome fusion by modulation of the actin cytoskeleton surrounding the vacuole and by incorporating proteins into the membrane of the vacuole.

In response to bacterial invasion, **microbe-associated molecular patterns** (**MAMPs**) are recognized by **Toll-like receptors** (TLR2, lipoprotein; TLR4, **lipopolysaccharide** (**LPS**); TLR5, flagellin) and initiate an inflammatory response in the GI tract. Pro-inflammatory **cytokines** are induced by the *Salmonella*, particularly IL-1, **interferon** (IFN-β and γ), IL-2, IL-6, and the **chemokine** IL-8, which recruits neutrophils. The epithelial monolayer is an important target and mainly secretes the IL-8 in response to the bacterium. The production of IL-8 is a result of the activation of NFκB within the epithelial cells and is dependent on the type III secretion system in the bacterium, which injects bacterial proteins into the cells thereby inducing the cytokine response. Killing by macrophages is also important as defects in macrophage function, for example, lack of IFN-γ, leads to severe disease.

Both Type I and Type II interferons (IFN) are important in infection with *Salmonella* Typhi and NTS. Type II interferon (IFN-γ) promotes phagocytosis and activates macrophages to kill intracellular bacteria. IFN-γ acts in two ways to eradicate *Salmonella* spp. It induces the production of guanylate binding proteins which lyse the vacuole wall protecting the organism, releasing it into the cytoplasm of the cell leading to pyroptotic cell death. The second method by which IFN-γ is involved is activation of the macrophage leading to ROS that kill the *Salmonella* and induce a pro-inflammatory response.

Type I interferons (IFN-β) are protective of *Salmonella* as they have an anti-inflammatory action via IL-10 which diminishes the TLR-inflammosome response and inhibits a pro-inflammatory environment.

ADAPTIVE IMMUNITY

Eradication of the infection and the production of immunity require **CD4+** helper T cells and the production of specific antibodies by B cells. The role of cytotoxic CD8 cells in effective eradication of the organism is less clear.

PATHOGENESIS

NTSs induce a secretory diarrhea with a Th-1 pro-inflammatory response. MAMPs on the organism for example LPS and flagellin bind to TLR4 and 5 respectively, activating the Th-1 response by stimulating the production of IL-8 and recruiting neutrophils to the local area causing inflammation.

With *Salmonella* Typhi, adhesion to epithelial cells occurs via **fimbriae**. *Salmonella* Typhi enter via the M cells and/or **dendritic cells** and enterocytes; *Salmonella* Typhi adheres to the **cystic fibrosis transmembrane conductance receptor** (**CFTM**) on GI epithelial cells. Mutations in this protein may lead to lack of adhesion and thus resistance to infection. Once adherent, the bacteria invade by a process called bacteria-mediated **endocytosis** where the type III secretion system is involved. Adhesion of the bacterium induces cytoskeletal changes in the epithelial cell with membrane ruffling that encloses the organism into a vacuole.

Once inside the intestinal epithelial cells of the host, the organism produces a multi-subunit toxin, B (typhoid toxin TT). The toxin consists of three subunits, one of which is similar to that of cytolethal distending toxins and has DNase activity. The toxin induces single-strand breaks in host DNA and replication stress with accumulation of damaged replication forks. Damage to DNA leads to the DNA damage response (DDR). The regulators of DDR are ataxia telangiectasia mutated kinase (ATM) and ATM and rad3-related kinase (ATR). ATM is involved with resolution of double-stranded breaks (dsDNA) and ATR with replication stress involving single-stranded DNA (ssDNA). Single-stranded DNA binds to the replication protein A complex (RPA) which activates ATR and the cell cycle is stalled in order to repair the replication fork. A link between ATM and ATR is phosphorylation of the H2AX histone which co-ordinates the repair process.

The action of the typhoid toxin is to overload the repair pathway with ssDNA eventually leading to senescence of the cell. This state of senescence is transmitted to adjacent cells by an as yet unidentified factor. Induction of cell senescence allows chronic carriage of the organism making the host a typhoid carrier.

Once endocytosed, a proportion of **vesicles** will fuse with the basolateral membrane and enter the **lamina propria**. During adhesion, *Salmonella* Typhi induces up-regulation of the host CFTM receptor and an increased level of translocation to the lamina propria. Here, the organism enters macrophages and dendritic cells where it survives and replicates. *Salmonella* Typhi surviving and replicating within the monocytic lineage are released and pass to the mesenteric lymph nodes and via the thoracic duct to the general circulation where they localize in the cells of the **mononuclear phagocyte system** (in the liver, spleen, and bone marrow). The bacteria continue to replicate and are shed into the bloodstream with the onset of clinical illness and are circulated in the blood to all body organs and induce organ-specific signs and symptoms. Movement of *Salmonella* Typhi from the GI tract may also occur when dendritic cells and macrophages carrying the intracellular organisms migrate to mesenteric lymph nodes.

Organs particularly affected (through the cells of the mononuclear phagocytic system) are the liver and spleen, bone marrow, gallbladder, and importantly, the Peyer's patches in the intestine. **Kupffer cells** in the liver are a major defense mechanism of the host but surviving *Salmonella* that invade hepatocytes induce **apoptosis** of the cells. At this stage, the patient will show systemic signs of inflammation: fever, **jaundice**, hepatosplenomegaly, **myalgia**, headache, and, in some cases, mental confusion. Later on in the infection, **rose spots** may appear and, if left untreated by the third week, there is an intense monocytic infiltration of Peyer's patches, which may rupture leading to signs of perforation of the small bowel.

3. WHAT IS THE TYPICAL CLINICAL PRESENTATION AND WHAT COMPLICATIONS CAN OCCUR?

SALMONELLA TYPHI

Enteric (or typhoid) fever is a serious infection with an appreciable mortality, if left untreated. It is caused by *Salmonella* Typhi or *Salmonella* Paratyphi. The incubation period is from 5 to 21 days depending on the size of the ingested inoculum and host co-morbidity, for example, immunosuppression. In adults, the disease may present as a nonspecific **febrile** illness (**pyrexia of unknown origin – PUO**) or it may present with fever and GI symptoms. In children, there may be a much more nonspecific presentation. Typical symptoms include fever, headache, abdominal pain, and tenderness, constipation or diarrhea, and delirium. On examination, there may be a relative **bradycardia** (where the pulse rate is less than expected from the temperature of the patient), a pinkish **maculopapular rash** found on the trunk (rose spots – Figure 31.3), and hepatosplenomegaly. Laboratory investigations (see later) demonstrate **anemia, leukopenia**, classically, but not always, with a relative lymphocytosis, abnormal LFTs, and elevated creatine phosphokinase (CPK). Risk factors for infection include immunosuppression and sickle cell anemia. Complications include **cholecystitis**, intestinal perforation and hemorrhage, **osteomyelitis**, and **endocarditis**. *Salmonella* Paratyphi (paratyphoid fever) causes a milder form of the illness compared with *Salmonella* Typhi. Following infection, about 1–5% of patients may become long-term GI carriers of the organism. These individuals are at greater risk of developing hepatobiliary and intestinal carcinoma.

NON-TYPHOID *SALMONELLA*

NTS is one cause of a self-limiting gastroenteritis that is indistinguishable from other bacterial causes. The incubation period is from 12 to 48 hours following ingestion of contaminated food. It presents with fever, abdominal pain, and diarrhea, which may occasionally be bloodstained. Complications include **paralytic ileus** leading to **toxic megacolon**.

4. HOW IS THE DISEASE DIAGNOSED, AND WHAT IS THE DIFFERENTIAL DIAGNOSIS?

DIAGNOSTIC TESTS

Diagnosis of enteric fever is made by isolation of *Salmonella* Typhi or Paratyphi from a blood culture, feces, or bone marrow of a symptomatic patient. The specimen providing the highest diagnostic yield is the bone marrow. Sampling of duodenal secretions using the string test may add to the diagnostic yield. The organism may also be isolated from other specimens such as rose spots, urine, sputum, and cerebrospinal fluid (CSF). NTS are usually isolated from only the feces but occasionally may also be isolated from the blood.

One of a number of selective media may be used for fecal specimens, including xylose-lysine-deoxycholate (XLD), deoxycholate citrate agar (DCA), bismuth sulfite agar, *Salmonella Shigella* (SS) agar, and Hektoen agar. The general principle of these media is that they include sugars and a pH indicator to differentiate *Salmonella* spp. (and *Shigella* spp.) from non-enteric pathogens; a source of sulfur and iron as the *Salmonella* spp. generate hydrogen sulfide, which precipitates the iron to give black colonies; and a selective agent such as bile to which *Salmonella* spp. are resistant (Figure 31.4).

The organism is identified by **MALDI-TOF** or **WGS (whole genome sequencing)**. The serovars are identified by slide agglutination with O and H antibodies and by reference to the Kaufmann-White scheme although this is being replaced by WGS. In the UK, enteric fever is a notifiable disease, as is food poisoning (of any microbiologic cause). Although most industrialized countries have a reporting system for infectious diseases, many countries do not.

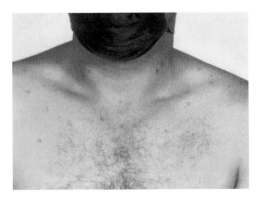

Figure 31.3 Rose spots on the upper trunk in a case of enteric fever. *From the Centers for Disease Control & Prevention, Atlanta, Georgia. Image is found in the Public Health Image Library #2215. Additional photographic credit is given to the Armed Forces Institute of Pathology and Charles N Farmer who created the image in 1964.*

Figure 31.4 Growth of *Salmonella* on DCA medium. *Note the black center in some colonies indicating precipitation of FeS. From the Centers for Disease Control & Prevention, Atlanta, Georgia. Image is found in the Public Health Image Library #6619.*

Serologically, the **Widal test** has been used specifically for *Salmonella* Typhi but is not recommended for diagnosis (in the UK) as it does not have sufficient sensitivity; as many as 50% of patients with culture-confirmed enteric fever are negative in the Widal test. Serologic assays such as a dot **ELISA** test format (Typhidot, TpTest) and a latex agglutination inhibition test (Tubex-detecting IgM to 2009 only) detecting **IgG** and **IgM** in studies are available. In 2016, an comparative assay of TpTest, Tubex, Typhidot was assessed in culture positive cases versus healthy controls in Bangladesh. The sensitivity/specificity was TpTest (96%/96%), Tubex (60%/89%), and Typhidot (59%/80%). PCR and DNA probes are available but not currently in routine use, particularly in resource-poor countries.

DIFFERENTIAL DIAGNOSIS

The differential diagnosis includes a wide range of causes of PUO such as brucellosis, malaria, typhus, and **Dengue fever** as well as causes of fever with intra-abdominal pathology such as **abscess**, amebic dysentery, and abdominal tuberculosis.

5. HOW IS THE DISEASE MANAGED AND PREVENTED?

MANAGEMENT

Salmonella Typhi and Paratyphi

Because of the increasing resistance to antibiotics including chloramphenicol, co-trimoxazole, ampicillin, and ciprofloxacin in India and South-East Asia, the treatment depends on the geographic location of the infecting strain and its susceptibilities. Similar antibiotic resistance also occurs in NTS spp. and, particularly in the UK, the emergence of resistance

to ciprofloxacin was temporally associated with the use of a veterinary fluoroquinolone – enrofloxacin. An outbreak (2016) of a strain in Pakistan was resistant to chloramphenicol, amoxycillin, cotrimoxazole, ciprofloxacin, and ceftriaxone, moving the management of such strains to azithromycin/meropenem. The current recommended treatment in the UK in uncomplicated typhoid fever is either azithromycin or ciprofloxacin depending on the susceptibility. In severe enteric fever, either ceftriaxone, ciprofloxacin or meropenem is used, again depending on the susceptibility.

Enteric fever can be prevented by vaccination. Three main types of vaccine exist including an oral live attenuated strain (-Ty21a), a parenteral vaccine based on the Vi capsular polysaccharide antigen (ViCPS Vi), and a **conjugate vaccine** based on the Vi antigen and recombinant exotoxin A of *Pseudomonas aeruginosa* (Vi-rEPA). An acetone-inactivated parenteral vaccine also exists. The protection afforded by these vaccines ranges from 60% to 90%.

Non-Typhoid *Salmonella* spp.

Generally, gastroenteritis does not require antibiotic treatment as the illness is self-limiting. However, in severe disease, or if there is evidence of systemic spread, then depending on the sensitivities an appropriate antibiotic such as a cephalosporin or ciprofloxacin can be used. An unwanted effect of antibiotics is to increase the carriage rate of the organism.

PREVENTION

Prevention of gastroenteritis relies on control of critical points along the food chain from farm to dining room, which includes infection-free animals, sanitary animal transport, hygienic processes in abattoirs, and correct storage and cooking conditions in the kitchen.

SUMMARY

1. WHAT IS THE CAUSATIVE AGENT, HOW DOES IT ENTER THE BODY, AND HOW DOES IT SPREAD A) WITHIN THE BODY AND B) FROM PERSON TO PERSON?

- *Salmonella* spp. are motile gram-negative rods.
- There are over 2500 serovars; they are grouped in the Kauffman-White scheme based upon the O and H antigens.
- *Salmonella* Typhi is a strictly human pathogen and is acquired from food or water contaminated with organisms from another case or a carrier.
- Non-typhoid *Salmonella* spp. are found in animals and infection is acquired from contaminated food or pets.
- *Salmonella* Typhimurium and *Salmonella* Enteritidis are the two most frequent causes of non-typhoid *Salmonella* spp. gastroenteritis.

- *Salmonella* Typhi penetrates the gastrointestinal epithelium and is disseminated by the bloodstream.
- Non-typhoid *Salmonella* spp. generally remain within the gastrointestinal tract, although some serovars may disseminate in the blood.

2. WHAT IS THE HOST RESPONSE TO THE INFECTION AND WHAT IS THE DISEASE PATHOGENESIS?

- Low gastric acid, immunosuppression, and certain polymorphisms including TLR4 and HLA predispose to infection.
- Survival in macrophages occurs by modulating phagolysosome fusion and the bacterial cell wall to resist inimical bactericides.

Continued...

...continued

- *Salmonella* Typhi induces a monocytic response with few gastrointestinal symptoms.
- Non-typhoid *Salmonella* spp. induce a granulocyte response with pronounced gastrointestinal symptoms.
- *Salmonella* Typhi localizes to cells of the monocyte/macrophage system where it replicates.
- *Salmonella* Typhi is distributed to all body organs by the blood, giving rise to organ-specific disease.

3. WHAT IS THE TYPICAL CLINICAL PRESENTATION AND WHAT COMPLICATIONS CAN OCCUR?

- Enteric fever presents as a PUO and abdominal symptoms.
- The incubation period is 5–21 days.
- A rash (rose spots) may be present.
- There may be a relative bradycardia and a leukopenia.
- Hepatosplenomegaly may be present.
- Complications include cholecystitis, intestinal perforation, endocarditis, and osteomyelitis.
- Non-typhoidal gastroenteritis is self-limiting.
- The incubation period is 12–48 hours.
- It presents with fever, abdominal pain, and diarrhea.

4. HOW IS THE DISEASE DIAGNOSED, AND WHAT IS THE DIFFERENTIAL DIAGNOSIS?

- Enteric fever is diagnosed by isolation and identification of the organism.
- The highest yield of organism is from the bone marrow.
- Identification of *Salmonella* spp. is based on MALDI-TOF, WGS or serologic reactions using the Kaufmann-White scheme.
- In many countries, including the UK, food poisoning and enteric fever are notifiable diseases.

5. HOW IS THE DISEASE MANAGED AND PREVENTED?

- There are increasing levels of antibiotic resistance in *Salmonella*.
- Treatment is with azithromycin/ceftriaxone/meropenem in patients from India and South-East Asia.
- Treatment is with ciprofloxacin in patients from other areas.
- Protection from enteric fever is provided by vaccination.
- Antibiotics are not usually required in cases of gastroenteritis caused by non-typhoid *Salmonella*.
- Prevention of gastroenteritis is by adequate hygiene standards at all points in the food chain.

FURTHER READING

Cimolai N. Laboratory Diagnosis of Bacterial Infections. Marcel Dekker, New York, 2001.

Lydyard P, Lakhani S, Dogan A, et al. Pathology Integrated: An A–Z of Disease and its Pathogenesis. Edward Arnold, London, 2000.

Mandell GL, Bennet JE, Dolin R. Principles & Practice of Infectious Diseases, 6th edition, Vol 2. Elsevier, Philadelphia, 2005.

Mims C, Dockrell HM, Goering RV, et al. Medical Microbiology, 3rd edition. Mosby, Edinburgh, 2004.

Murphy K, Weaver C. Janeway's Immunobiology, 9th edition. Garland Science, New York/London, 2016.

REFERENCES

Baker S, Dougan G. The Genome of Salmonella enterica Serovar Typhi. Clin Infect Dis, 45: S29–S33, 2007.

Bhan MK, Bahl R, Bhatnager S. typhoid and Paratyphoid Fever. Lancet, 366: 749–762, 2005.

Cooke FJ, Wain J. The Emergence of Antibiotic Resistance in Typhoid Fever. Travel Med Infect Dis, 2: 67–74, 2004.

da Silva KE, Tanmoy AM, Pragasam AK, et al. The International and Intercontinental Spread and Expansion of Antimicrobial-Resistant Salmonella typhi: A Genomic Epidemiology Study. Lancet Microbe, 3: e567–77, 2022.

Everest P, Wain J, Roberts M, et al. The Molecular Mechanisms of Severe Typhoid Fever. Trends Microbiol, 9: 316–320, 2001.

Fang FC, Vazques-Torres A. Salmonella Selectively Stops Traffic. Trends Microbiol, 10: 391–392, 2002.

Gentschev I, Spreng S, Sieber H, et al. Vivotif: A Magic Shield for Protection Against Typhoid Fever and Delivery of Heterologous Antigens. Chemotherapy, 53: 177–180, 2007.

Guzman CA, Borsutzky S, Griot-Wenk M, et al. Vaccines Against Typhoid Fever. Vaccine, 24: 3804–3811, 2006.

Haraga A, Ohlson MB, Miller SI. Salmonellae Interplay with Host Cells. Nat Rev Microbiol, 6: 53–66, 2008.

Huang DB, DuPont HL. Problem Pathogens: Extraintestinal Complications of Salmonella enterica Serovar Typhi Infection. Lancet Infect Dis, 5: 341–348, 2005.

Ibler AEM, El Ghazaly M, Naylor KL, et al. Typhoid Toxin Exhausts the RPA Response to DNA Replication Stress Driving Senescence and Salmonella Infection. Nat Commun, 10: 4040, 2019.

Ingram JP, Brodsky IE, Balachandran S. Interferon γ in Salmonella Pathogenesis: New Tricks for an Old Dog. Cytokine, 98: 27–32, 2017.

Kam KM, Luey KY, Chiu AW, et al. Molecular Characterization of Salmonella enterica Serotype Typhi Isolates by Pulsed-Field Gel Electrophoresis in Hong Kong 2000-2004. Foodborne Pathol Dis, 4: 41–49, 2007.

Liu GR, Liu WQ, Johnston RN, et al. Genome Plasticity and Ori-ter Rebalancing in Salmonella typhi. Mol Biol Evol, 23: 365–371, 2006.

Parry CM, Hien TT, Dougan G, et al. Typhoid Fever. N Engl J Med, 347: 1770–1782, 2002.

Pei YM , Jing ET , Edd WH, et al. Review Human Genetic Variation Influences Enteric Fever Progression. Cells, 10: 345, 2021.

Perkins DJ, R Sharon, Tennant M, et al. Salmonella typhimurium Co-opts the Host Type I Interferon System to Restrict Macrophage Innate Immune Transcriptional Responses Selectively. J Immunology, 195: 2461–2471, 2015.

Raffetellu M, Chessa D, Wilson RP, et al. Capsule-Mediated Immune Evasion: A New Hypothesis Explaining Aspects of Typhoid Fever Pathogenesis. Infect Immun, 74: 19–27, 2006.

Roumagnac P, Weill FX, Dolecek C, et al. Evolutionary History of Salmonella typhi. Science, 314: 1301–1304, 2006.

Srikantiah P, Vafokulov S, Luby SP, et al. Epidemiology and Risk Factors for Endemic Typhoid Fever in Uzbekistan. Trop Med Int Health, 12: 838–847, 2007.

Vazques-Torres A, Fang FC. Cellular Routes of Invasion by Enteropathogens. Curr Opin Microbiol, 3: 54–59, 2000.

Watkins LK, Winstead A, Appiah GD, et al. Update on Extensively Drug-Resistant Salmonella Serotype Typhi Infections Among Travelers to or from Pakistan and Report of Ceftriaxone-Resistant Salmonella Serotype Typhi Infections Among Travelers to Iraq – United States, 2018-2019. Morb Mortal Wkly Rep, 69: 618–622, 2020.

Zhang XL, Jeza VT, Pan O. Salmonella typhi: From a Human Pathogen to a Vaccine Vector. Cell Mol Immunol, 5: 91–97, 2008.

WEBSITES

GOV.UK, Typhoid and Paratyphoid: Guidance, Data and Analysis, 2014: http://www.hpa.org.uk/infections/topics_az/typhoid/menu.htm

Medscape, Typhoid Fever, 2022: http://www.emedicine.com/MED/topic2331.htm

WHO, Multi-Country Outbreak of Salmonella typhi murium Linked to Chocolate Products – Europe and the United States of America: https://www.who.int/emergencies/disease-outbreak-news/item/2022-DON369

Students can test their knowledge of this case study by visiting the Instructor and Student Resources: [www.routledge.com/cw/lydyard] where several multiple choice questions can be found.

Schistosoma spp.

32

A 25-year-old male was admitted to hospital vomiting blood. On examination of the abdomen, both the liver and spleen were enlarged. A full blood count revealed a reduced hemoglobin level of 90 g l^{-1} and the white cell differential showed a raised eosinophil count of 1.1×10^9 L^{-1} (normal range $<0.45 \times 10^9$ L^{-1}). Endoscopic examination revealed dilated **esophageal varices** (Figure 32.1). **Cirrhosis** of the liver was suspected, but the patient denied any excessive drinking of alcohol and tests showed that he was negative for hepatitis B and C. A liver biopsy was performed and confirmed an abnormal liver, with extensive granulomatous change and **fibrosis**. On further questioning, it transpired that he had spent much of his childhood living in a small village in Kenya. His parents had been working there on a farm. He would regularly paddle and swim in a nearby lake. He recalled no particular illness while living in Africa. Urine examination was negative but stool microscopy demonstrated the presence of *Schistosoma mansoni* eggs. A rectal biopsy obtained at **sigmoidoscopy** showed **granulomas**, containing *Schistosoma* eggs. Schistosomal **serology** was also positive. He was treated with praziquantel and his stool was monitored for the disappearance of *Schistosoma*

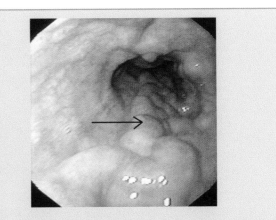

Figure 32.1 Endoscopic view of large esophageal varices (arrowed). *Courtesy of Dr Alan Mills and the Virginia Commonwealth University Department of Pathology.*

eggs. He was referred to a hepatologist for further management of his esophageal varices. Serial ultrasounds were performed to monitor the liver fibrosis.

1. WHAT IS THE CAUSATIVE AGENT, HOW DOES IT ENTER THE BODY AND HOW DOES IT SPREAD A) WITHIN THE BODY AND B) FROM PERSON TO PERSON?

CAUSATIVE AGENT

Human schistosomiasis, also known as bilharzia, is mainly caused by three species of *Schistosoma. S. mansoni* occurs in Africa, parts of the Arabian peninsula, the Caribbean, and South America. *S. haematobium* occurs in Africa and parts of the Arabian peninsula. *S. japonicum* is now found in China, the Philippines, and Indonesia. *S. intercalatum* occurs in pockets in west and central Africa and *S. mekongi* is restricted to small pockets in Cambodia and Laos.

Schistosomes are blood flukes, which are also known as trematodes. Adult worms are less than 2cm long (Figure 32.2). Males have a longitudinal groove in which the female worm resides. They live in the veins around the bladder in the case of *S. haematobium* or in the mesenteric veins of the intestine in the case of *S. mansoni* and *S. japonicum*.

Figure 32.2 Adult schistosome worms. The female (on the left) is slender and fits into a groove on the ventral surface of the male. *From the Centers for Disease Control & Prevention, Atlanta, Georgia. Image is found in the Public Health Image Library #5252.*

ENTRY AND SPREAD WITHIN THE BODY

The lifecycle of schistosomiasis is shown in Figure 32.3. Humans are infected when they enter water containing

Schistosoma cercariae (Figure 32.4). These larvae penetrate the skin. The cercariae transform into a larval stage called the schistosomulum. These enter blood vessels and pass round the venous circulation, through the lungs and then enter the arterial circulation. They reach the liver. The schistosomula mature over about one month to adult worms in the portal vein. Males and females pair and then migrate distally to the veins around the bladder and the intestine. They usually live for 5–7 years. However, there are reports of individuals who have left endemic areas for several years and show signs of active infection many years later. From such reports, it is thought that occasionally adult pairs may survive for up to 30 years. The case history illustrates this prolonged survival.

PERSON-TO-PERSON SPREAD

Schistosomiasis does not spread directly from person to person but the life cycle involves a snail intermediate host. Once adults have matured, eggs are laid after 1–3 months. *S. mansoni* and *S. haematobium* females lay between 20 and 300 eggs per day. The figure can be considerably higher for *S. japonicum*, with an estimated 500–3500 eggs per day. About

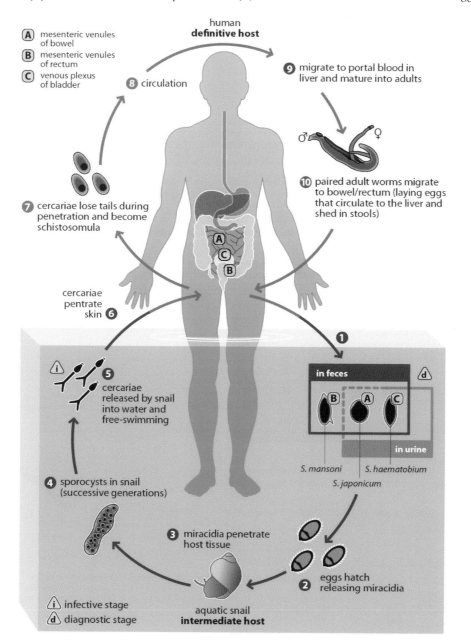

Figure 32.3 Lifecycle of *Schistosoma* spp. Eggs shed in urine or feces (1) from human hosts release miracidia on contact with fresh water (2) and then enter intermediate snail hosts (3). After multiplication in the snail hosts (4) cercariae arise (5), which penetrate human skin on water contact (6). They develop into schistosomula (7), which mature into adult worms and live within blood vessels (10) around the bladder or intestine (A, B, C). Adult male and female pairs shed eggs, about half of which exit from the body and the other half are deposited in tissues, causing pathology. *Adapted from the Centers for Disease Control & Prevention, Atlanta, Georgia. Image is found in the Public Health Image Library #3417. Additional photographic credit is given to Alexander J da Silva, PhD, and Melanie Moser who created the image in 2002.*

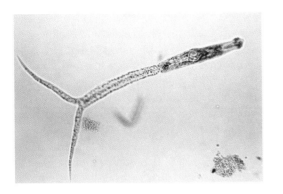

Figure 32.4 A Schistosome cercaria, which emerges from snails and penetrates human skin. *From the Centers for Disease Control & Prevention, Atlanta, Georgia. Image is found in the Public Health Image Library #8556. Additional photographic credit is given to Minnesota Department of Health, RN Barr Library; Librarians Melissa Rethlefsen and Marie Jones, Prof William A Riley and the photo was created in 1942.*

half of these eggs succeed in passing through the wall of the bladder or intestine to emerge in urine or feces. Otherwise, eggs get stuck in the bladder or intestinal wall or embolize in the venous circulation to the liver. In advanced, heavy infections collateral circulation around the liver develops and eggs disseminate into the lungs, brain, and other organs.

Eggs have a spinous process. The shape of the egg and the position of the spine is characteristic for each species of *Schistosoma* (Figure 32.5). *S. haematobium* has a terminal spine. *S. mansoni* and *S. japonicum* have lateral spines, with *S. japonicum* having smaller-sized eggs.

The eggs contain a ciliated larval form called the miracidium, which is released when the eggs come into contact with fresh water in lakes, rivers, and streams. Miracidia go on to infect specific aquatic snails. Within the snails, they multiply into sporocysts from which arise the cercariae that close the life cycle. The species of snails differs between the species of *Schistosoma*. *S. mansoni* is transmitted by *Biomphalaria* snails, *S. haematobium* by *Bulinus*. These are aquatic snails, while *S. japonicum* is transmitted by an amphibious snail called **Oncomelania**.

EPIDEMIOLOGY

Schistosomiasis is endemic in many tropical and subtropical countries and affects more than 240 million people worldwide. Some estimate mortality reaching 200 000 per annum.

2. WHAT IS THE HOST RESPONSE TO THE INFECTION AND WHAT IS THE DISEASE PATHOGENESIS?

A host immune inflammatory response may be directed against the antigens of the schistosomula larvae, the adult worms or the eggs. There is evidence that the host response provides some level of protective immunity to re-infection. Rates of re-infection have been studied in endemic areas when adults or children are treated to clear infection. The rate of re-infection can only partly be explained by behavior and intensity of water contact. Correcting for the extent of water contact, adults who have had a greater lifetime exposure to schistosomiasis are less likely to be re-infected than children.

In vitro studies have shown that schistosomula are more susceptible to immune attack than adult worms. Immunity correlates with levels of **IgE** directed against schistosomulum or adult worm antigens. Key antigens include schistosomal glutathione-*S*-transferase and glyceraldehyde-3-phosphate. The levels of antibodies rise with age. *In vitro*, IgE mediates killing of schistosomula by **opsonization** to eosinophils. The eosinophils degranulate to release toxic molecules such as eosinophil cationic protein. However, immunity is inversely correlated with levels of **IgG$_2$** and IgG$_4$ antibody directed against egg polysaccharide antigens. It is thought that these IgG isotypes also bind to schistosomula and block the binding of effector IgE antibodies.

Larval, adult, and egg antigens are sometimes released in such quantities, that together with significant levels of circulating IgG, an **immune complex** disease develops (**type III hypersensitivity**). This occurs early in infection and is referred to as **Katayama syndrome**. Some observations also implicate **tumor necrosis factor-α (TNF-α)** in severe symptomatology.

Once adults mature, there seems to be a ceiling to the number of adults present in the vasculature. This may be because of the physical space available. Adults become coated with host proteins, including self-HLA molecules, and this "host mimicry" may confuse the immune response, leading the host to believe the worm is "self". Very little, if any, inflammation is seen around adult worms.

A considerable immune and inflammatory response occurs against the eggs that lodge in tissues. Blood **eosinophilia** is

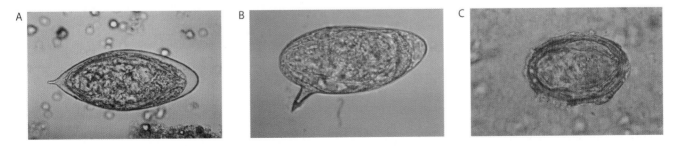

Figure 32.5 Eggs of (A) *S. haematobium* with a terminal spine, (B) *S. mansoni* with a lateral spine, and (C) *S. japonicum*, which are smaller and rounder. *From the Centers for Disease Control & Prevention, Atlanta, Georgia. Images are found in the Public Health Image Library (A) #4843; (B) #4841; (C) #4842.*

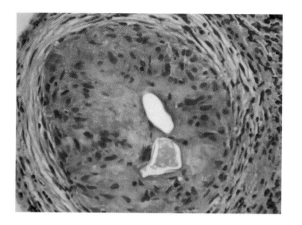

Figure 32.6 Granuloma formed around a schistosome egg.
Courtesy of Dr Yale Rosen and his website Atlas of Granuloma Diseases: http://granuloma.home-stead.com/.

often observed in association with considerable egg deposition. These eggs release soluble antigens that present a strong stimulus to **CD4+** T-helper lymphocytes, which orchestrate an inflammatory response. The human immune response is skewed to a Th-2 response. Granulomas form around the eggs with an influx of lymphocytes and macrophages. Fibroblasts deposit collagen and extracellular matrix proteins. The extent of resulting fibrosis depends on the balance of activities between matrix metalloproteinases (MMPs) and their tissue inhibitors (TIMPs). In turn, these enzymes are affected by the balance of a variety of cytokines. **Transforming growth factor-β (TGF-β)**, **interleukin** (IL)-1, IL-4, IL-13, and TNF-α promote fibrosis, while **interferon (IFN)**-γ is associated with reduced fibrosis. The actions of these **cytokines** in turn are affected by other regulatory cytokines such as IL-10 and IL-12. About 5–10% of patients develop extensive fibrosis (Figure 32.6). Numerous granulomas with surrounding fibrosis constitute a major part of tissue pathology. In time, there can be remodeling of granulomas with some resorption of fibrotic tissue.

3. WHAT IS THE TYPICAL CLINICAL PRESENTATION AND WHAT COMPLICATIONS CAN OCCUR?

Many individuals do not develop symptoms but, when present, clinical features may occur at different stages of infection. Even without symptoms, eosinophilia may be an indicator of infection. When cercariae first penetrate the skin, there can be local irritation, which may appear immediately or within a few days. Raised, red spots may last a few days. Similarly, a swimmers' itch can be caused by trematodes of animal origin, particularly in temperate climatic zones.

A type III hypersensitivity reaction to high-antigen release is referred to as Katayama syndrome, named after a region in Japan. This may occur between 2 and 12 weeks after infection, particularly when egg production increases the antigenic burden. It is more likely in non-immune travelers

unable to curtail the maturation of schistosomula. There is an abrupt onset of fever, malaise, aches, and tiredness. The lung becomes involved, with patchy shadowing on a chest X-ray. This is accompanied by a dry cough. There is also abdominal pain with diarrhea and both liver and spleen can be enlarged. On blood tests, there is a high eosinophil count.

No symptomatology is attributed to the adult worms. Even once eggs are being deposited in tissues, symptoms may be absent or mild. Nonspecific features include malaise and tiredness. When considerable numbers of eggs settle in the intestine, bladder, liver, lung or other tissues, problems arise as described below. These take some time to develop.

With *S. haematobium* granulomatous inflammation of the bladder wall can also result in ulceration (Figure 32.7). Bleeding into urine may be microscopic or macroscopic. Chronic lesions can progress to fibrosis and bladder calcification. This can block the ureters, with back pressure dilating the ureters and compromising kidney function. In the long term, there is a risk that the damaged mucosal epithelium undergoes malignant change, resulting mainly in squamous cell carcinomas rather than transitional cell carcinomas. This is more common in smokers.

With *S. mansoni* and *S. japonicum*, granulomatous inflammation of the intestine is principally found in the rectum and distal bowel. The mucosa may ulcerate. Abdominal pain is accompanied by diarrhea that may contain blood. From the peri-intestinal blood vessels eggs laid by adults embolize through the portal vein to the liver. In heavily infected or genetically predisposed individuals, the granulomatous reaction in the liver results in considerable fibrosis; this periportal fibrosis – described as "clay pipe stem" fibrosis – develops along the portal veins (Figure 32.8). The liver can enlarge and can feel hard and uncomfortable in the right upper abdomen. As blood flow through the liver is impaired by fibrosis, **portal hypertension** develops. This enlarges the spleen, which causes discomfort in the left upper abdomen. Gastro-esophageal varices develop and can bleed, as illustrated in the case history. Eventually, damage to the liver may be so extensive that liver failure results in death.

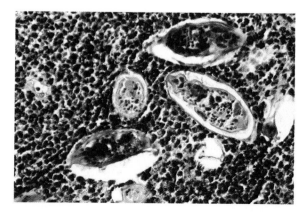

Figure 32.7 Granuloma formed around a schistosome egg.
Courtesy of Dr Yale Rosen and his website Atlas of Granuloma Diseases: http://granuloma.home-stead.com/.

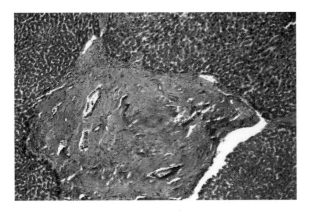

Figure 32.8 Intense eosinophilic inflammation around *S. haematobium* eggs in the bladder wall. *From the Centers for Disease Control & Prevention, Atlanta, Georgia. Image is found in the Public Health Image Library #35. Additional photographic credit is given Dr Edwin P Ewing, Jr who took the photo in 1973.*

In 5–10% of infected individuals, portal hypertension becomes well established and collateral blood vessels open up around the liver. Eggs from the peri-intestinal blood vessels pass toward the liver but then bypass through these dilated blood vessels into the lung. This results in granulomatous inflammation in lung tissue. Eggs may reach any part of the body. *S. japonicum* worms lay more eggs than the other species and, in the Far East, granulomas may develop in the central nervous system (CNS) and be a cause of epilepsy and other neurologic features.

4. HOW IS THE DISEASE DIAGNOSED, AND WHAT IS THE DIFFERENTIAL DIAGNOSIS?

The diagnosis of schistosomiasis can be made by microscopic analysis through finding eggs in urine or stool. Urine collected at the end of micturition provided at mid-day is the best specimen in which to look for *S. haematobium* eggs. This terminal specimen is likely to have a higher concentration of eggs. Egg shedding is also maximal about mid-day. Multiple stool specimens may be required to find eggs, as egg shedding is variable. Stool specimens can be suspended and concentrated. If eggs are not seen in urine or stool, endoscopic examination of bowel or bladder may show inflammatory changes and biopsies may yield granulomas containing eggs. Where available, another direct assay for infection is the detection of schistosomal antigen in urine, stool or blood.

In endemic areas, the presence of **IgM**, IgG and IgE antibodies in serum does not distinguish current from past infection. In returned travelers who have not previously been to a schistosomiasis-endemic area, positive serology may be diagnostic of exposure. Serology does not become positive till about 12 weeks after infection, and occasionally longer.

Eosinophilia is suggestive of ongoing infection in the absence of another explanation. Ultrasonography can be used to detect complications of infection such as changes in the bladder, ureters, kidneys, and liver.

DIFFERENTIAL DIAGNOSIS

The differential diagnoses to be considered depend on the presentation (Table 32.1). In people who have traveled from endemic areas, one may have to consider differential diagnoses for hematuria, diarrhea with or without blood, eosinophilia, and granulomas on biopsy. Diarrhea with blood is also known as **dysentery**. The case history illustrates late-stage presentation when one also has to think of cirrhosis of the liver due, for instance, to alcohol or chronic viral hepatitis B or C. In endemic areas, it is possible to get both schistosomiasis and chronic viral hepatitis.

5. HOW IS THE DISEASE MANAGED AND PREVENTED?

MANAGEMENT

Standard treatment is monotherapy with oral praziquantel in a single dose or two doses 4–6 hours apart. In mass chemotherapy programs, single doses are more practical. Praziquantel acts against the adult worms, causing damage to their surface, increased calcium influx, and paralysis. It does not act on schistosomula and cannot be used very soon after infection, before adults have matured. Treatment should be deferred to 6–12 weeks after the last known fresh-water exposure. Artemisinin derivatives have activity against schistosomula. Praziquantel cures up to 90% of patients. Treatment success is gauged by the disappearance of egg excretion after treatment. Any eosinophilia usually settles by 12 weeks post-treatment, but persistent eosinophilia and detectable egg excretion at that point will merit a repeat course of praziquantel. Concern about reduced efficacy of praziquantel is a concern in some regions but may be overcome by more prolonged courses with higher doses. The cessation of egg production stops new granuloma formation. The host may be able to reverse early fibrotic reactions with some clinical improvement.

As Katayama syndrome is due to a type III hypersensitivity reaction, the mainstay of treatment is steroids as an anti-inflammatory. Steroids also have a role in reducing inflammatory changes in neurologic involvement.

PREVENTION

Vaccines against schistosomiasis are under investigation. While we wait in hope for an effective vaccine, prevention

Table 32.1 Differential diagnoses to be considered with different presentations of schistosomiasis

Hematuria (blood in urine)

    Urinary tract infection

    Bladder or ureteric stones

    Tumor of bladder or prostate in adults

    Glomerulonephritis

    Polycystic kidneys

    Coagulopathy

Dysentery (bloody diarrhea)

    Infections due to

        Campylobacter

        Shigella

        Enterohemorrhagic *Escherichia coli*

        Entamoeba histolytica

    Colonic tumor

    Inflammatory bowel disease, particularly Crohn's disease, which also causes granulomas

Eosinophilia

    Infections due to:

        Strongyloides

        Filaria (e.g. *Wuchereria*, *Onchocerca*)

        Migrating larval stages of intestinal nematodes (e.g. *Ascaris*, hookworms)

    Atopic/allergic reactions

        Asthma

        Eczema

        Drug reactions

    Vasculitis

    Malignancy

Granulomas

    Infections due to:

        Mycobacteria

        Brucella

        Chlamydia granulomatosis

        Francisella tularensis

        Listeria monocytogenes

        Nocardia spp.

        Histoplasma

        Coccidiodes

        Fasciola hepatica

        Paragonimus

Sarcoidosis

Crohn's disease

Vasculitis

Lymphoma

Primary biliary cirrhosis

Chronic granulomatous disease

Berylliosis

depends on health education, control of intermediate host snail populations, reducing contaminated water contact, and mass drug administration to treat infected individuals and reduce egg shedding. Environmental management to reduce intermediate host snail populations is difficult, as is separating humans from water contact. The building of dams for irrigation and hydroelectric purposes has had mainly negative effects on the incidence of infection. The provision of safe water supply and sanitation infrastructure are very important in reducing episodes of contaminated water contact. Several countries have undertaken population-wide mass drug administration with repeated cycles of treatment. This has reaped success in countries such as Brazil, China, and Egypt. The World Health Organization (WHO) has extended this approach to many more countries, especially in the most stricken countries of sub-Saharan Africa.

SUMMARY

1. WHAT IS THE CAUSATIVE AGENT, HOW DOES IT ENTER THE BODY, AND HOW DOES IT SPREAD A) WITHIN THE BODY AND B) FROM PERSON TO PERSON?

- Schistosomiasis in humans is caused by three main species of blood flukes (trematodes) – *S. mansoni*, *S. haematobium*, and *S. japonicum* (two further species are of regional importance only).
- Adult male and female worms lodge together in blood vessels around the bladder (*S. haematobium*) or the intestine (*S. mansoni* and *S. japonicum*). Females lay eggs, about half of which pass into urine (*S. haematobium*) or feces (*S. mansoni* and *S. japonicum*) and others disseminate into various organs.
- Upon fresh-water contact in lakes, rivers, and streams eggs release a larva called a miracidium that infects specific aquatic snails (*Biomphalaria* in the case of *S. mansoni*, *Bulinus* for *S. haematobium*) or the amphibious snail *Oncomelania* (in the case of *S. japonicum*).
- Cercariae arise from snails and transcutaneously infect humans when they are in contact with water. The first larval stage in humans is the schistosomulum.

2. WHAT IS THE HOST RESPONSE TO THE INFECTION AND WHAT IS THE DISEASE PATHOGENESIS?

- Schistosomula are susceptible to immune attack by IgE and eosinophils. The binding of IgE may be blocked by IgG_2 and IgG_4 antibody against egg antigens.
- There is virtually no inflammation around adult worms living in blood vessels.
- Granulomas form around eggs that deposit in tissues. These granulomas are subsequently surrounded by fibrosis. Disease is largely due to granulomatous inflammation and extensive post-granulomatous fibrosis.
- Very occasionally, excessive antigen release results in an immune complex disease.

3. WHAT IS THE TYPICAL CLINICAL PRESENTATION AND WHAT COMPLICATIONS CAN OCCUR?

- An itch and rash may occur when cercariae first penetrate the skin, particularly after primary infection.
- The occasional immune complex disease is called Katayama syndrome and comprises high fever, aches, cough, and later diarrhea.
- Disease of the bladder can cause bleeding into urine (hematuria) and, later, fibrosis can block the flow of urine with ureteral dilatation and renal failure. There is a risk of developing bladder cancer.
- Disease of the intestine can cause diarrhea with (dysentery) or without blood.
- Severe granulomatous change and fibrosis in the liver lead to portal hypertension with the risk of life-threatening bleeding from esophageal varices. The liver and spleen become enlarged.
- When collateral circulation develops, granulomas can appear elsewhere such as in the lung or the central nervous system.

4. HOW IS THE DISEASE DIAGNOSED, AND WHAT IS THE DIFFERENTIAL DIAGNOSIS?

- Diagnosis is based on visualizing characteristic eggs in urine, feces or biopsies.
- Serologic tests are also available and are most useful in returned travelers.
- On routine blood testing, eosinophilia may be found.
- The differential diagnosis includes other causes of hematuria, dysentery, eosinophilia, and granulomas.
- Ultrasonography is useful for identifying structural changes in the renal tract or liver.

5. HOW IS THE DISEASE MANAGED AND PREVENTED?

- Praziquantel is the mainstay of treatment.
- Prevention requires health education, environmental management to control intermediate host snail populations, and decreasing opportunities for water contact (through improved access to clean water and adequate sanitation).
- The provision of a safe water supply and sanitation infrastructure are important for schistosomiasis and many other diseases.
- Some countries undertake population-wide mass drug administration.

FURTHER READING

Bustinduy AL, King CH. Schistosomiasis. In: Farrar J, Hotez P, Junghanns T, et al, editors. Manson's Tropical Diseases, 23rd edition. Elsevier/Saunders, London, 795–819, 2014.

Murphy K, Weaver C. Janeway's Immunobiology, 9th edition. Garland Science, New York/London, 2016.

REFERENCES

Colley DG, Secor WE. Immunology of Human Schistosomiasis. Parasite Immunol, 36: 347–357, 2014.

Gryseels B, Polman K, Clerinx J, Kestens L. Human Schistosomiasis. Lancet, 368: 1106–1118, 2006.

Mutapi F, Maizels R, Fenwick A, et al. Human Schistosomiasis in the Post Mass Drug Administration Era. Lancet Infect Dis, 17: e42–e48, 2017.

Ross AG, Vickers D, Olds GR, et al. Katayama Syndrome. Lancet Infect Dis, 7: 218–224, 2007.

Verjee MA. Schistosomiasis: Still A Cause of Significant Morbidity and Mortality. Res Rep Trop Med, 10: 153–163, 2019.

Weerakoon KGAD, Gobert GN, Cai P, et al. Advances in the Diagnosis of Human Schistosomiasis. Clin Microbiol Rev, 28: 939–967, 2015.

WEBSITES

Centers for Disease Control, Atlanta, GA, USA: https://www.cdc.gov/parasites/schistosomiasis/index.htmlSchistosomiasis

SCI Foundation, Control Initiative: https://schistosomiasiscontrolinitiative.org/

World Health Organization, Schistosomiasis (Bilharzia): https://www.who.int/health-topics/schistosomiasis

Students can test their knowledge of this case study by visiting the Instructor and Student Resources: [www.routledge.com/cw/lydyard] where several multiple choice questions can be found.

Staphylococcus aureus

33

A 44-year-old male presented to the emergency room with complaints of chest pain and was found to have suffered a myocardial infarction. His past medical history was significant for hypertension, noninsulin-dependent diabetes, **hypercholesterolemia**, and a history of heavy smoking (two packs per day for 15 years). A cardiac catheterization on hospital day 3 showed three vessel coronary artery disease, and he underwent triple coronary artery bypass graft surgery on hospital day 5. On hospital day 7, he developed **septic shock** with acute renal and respiratory failure requiring intubation. At that time, he had a fever of 39.3°C, his arterial blood gas revealed a pO_2 of 89 mmHg on 100% O_2, and he had a white blood cell count of 27000/mm³. Two blood cultures were obtained. A chest radiograph showed a left lower lobe infiltrate with pleural **effusion**. A chest tube was placed to drain the effusion. On hospital day 11, pus was noted to be seeping from his sternal wound. His wound was debrided and a rib biopsy was performed. Blood, drainage from his chest tube, tracheal aspirates, pus from his sternal wound, and a rib biopsy were cultured. The bacterium isolated from the clinical specimens was a gram-positive, catalase-positive, coagulase-positive coccus that was resistant to methicillin.

1. WHAT IS THE CAUSATIVE AGENT, HOW DOES IT ENTER THE BODY AND HOW DOES IT SPREAD A) WITHIN THE BODY AND B) FROM PERSON TO PERSON?

CAUSATIVE AGENT

The patient has an infection caused by methicillin (oxacillin)-resistant *Staphylococcus aureus* (MRSA). Bacteria in the genus *Staphylococcus* are gram-positive cocci that grow in grape-like clusters and produce the enzyme catalase (see Section 4). The genus *Staphylococcus* comprises some 35 species, many of which are members of the endogenous **microbiota** of the skin and mucous membranes of the gastrointestinal (GI) and genitourinary tracts of humans. Interestingly, a number of these species have defined habitats on the human body. Three species account for most human disease, namely *S. aureus*, *S. epidermidis*, and *S. saprophyticus*. Of these, *S. aureus* is the most pathogenic. The highly conserved core genome of *S. aureus* that contains essential genes required for cellular metabolism and replication comprises only three-quarters of the total genome. The remaining one-quarter of the genome consists of mobile genetic elements such as pathogenicity islands, bacteriophages, chromosomal cassettes, transposons and plasmids, which are acquired by horizontal transfer. These extrachromosomal elements encode determinants of pathogenicity and antibiotic resistance and are not all present in every strain. They will be discussed below.

S. aureus has a typical gram-positive cell wall that features a thick peptidoglycan layer extensively cross-linked with pentaglycine bridges. The extensive cross-linking makes the cell very resistant to drying so that staphylococci can survive on **fomites** (inanimate objects) for long periods of time. The wall also contains covalently bound teichoic acid and lipoteichoic acid, which are anchored in the wall and cytoplasmic membrane respectively. There are a number of molecules exposed on the cell surface. These are either anchored in the cytoplasmic membrane and traverse the cell wall to the outside or they are anchored in the wall and extend from it (Figure 33.1). Among the important cell surface-exposed molecules of *S. aureus* are various microbial surface components that recognize adhesive matrix molecules (**MSCRAMMs**). These components recognize and bind various extracellular matrix proteins such as fibronectin, collagen, and fibrinogen (clumping factor), and the Fc region of mammalian **IgG** (protein A). In addition to wall-associated clumping factor that binds solid-phase fibrinogen, *S. aureus* secretes coagulase, which binds soluble fibrinogen. Coagulase binds prothrombin in a 1:1 molar ratio to form a complex termed staphylothrombin, which converts fibrinogen to fibrin. Because *S. aureus* is the only staphylococcal species that

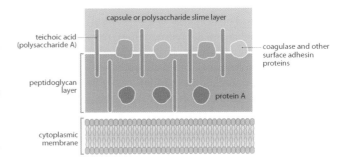

Figure 33.1 Schematic of the envelope of *S. aureus*.

possesses coagulase and protein A their detection serves as a method to identify the bacterium (see Section 4). MSCRAMMs facilitate uptake of *S. aureus* by keratinocytes and endothelial cells. Once taken up by cells, the bacteria survive in vacuoles and free in the cytoplasm where some persist for several days. Bacterial invasion results in cytotoxicity and apoptotic cell death.

Almost all *S. aureus* clinical isolates produce capsular polysaccharides, only isolates that express capsular polysaccharide type 5 (CP5) or 8 (CP8) have been associated with disease.

Almost all strains of *S. aureus* secrete a group of enzymes and cytotoxins that includes four **hemolysins** (alpha, beta, gamma, and delta), nucleases, proteases, lipases, hyaluronidase, and collagenase. Alpha-toxin is a pore-forming toxin that acts on many types of human cells. It is dermonecrotic and neurotoxic. Beta-toxin is a sphingomyelinase C whose role in pathogenesis remains unclear. Delta-toxin is able to damage the membrane of many human cells, possibly by inserting into the cytoplasmic membrane and disordering lipid chain dynamics, although there are suggestions that it may form pores. Two types of bicomponent toxin are made by *S. aureus*, gamma-toxin and Panton-Valentine leukocidin (PVL). While gamma-toxin is produced by almost all strains of *S. aureus*, PVL is produced by only a few percent of strains. The toxins affect neutrophils and macrophages. PVL has pore-forming activity and has been associated with necrotic lesions involving the skin and with severe necrotizing **pneumonia** in community-acquired *S. aureus* infections. Also, *S. aureus* produces a staphylokinase, which is a potent activator of plasminogen, the precursor of the fibrinolytic protease plasmin.

S. aureus produces several families of **exotoxins** that have **superantigen** activity. These include heat-stable **enterotoxins** (A–E, G–I) that cause food poisoning, exfoliative (epidermolytic) toxins A (chromosome encoded) and B (plasmid encoded) (ETA and ETB) that are the cause of staphylococcal scalded skin syndrome (SSSS) occurring predominantly in children, and **toxic shock syndrome** toxin 1 (TSST-1) that causes staphylococcal toxic shock.

ENTRY AND SPREAD WITHIN THE BODY

S. aureus requires a breach in the skin or mucosa such as produced by a catheter, a surgical incision, a burn, a traumatic wound, ulceration or viral skin lesions to facilitate its entry into the tissues. Furthermore, staphylococcal infections are frequently associated with reduced host resistance brought about by **cystic fibrosis**, immunosuppression, diabetes mellitus, viral infections, and drug addiction. *S. aureus* has several enzymes that aid in invasion of the barrier epithelia and serve as spreading factors. Among these are lipases, hyaluronidase, and fibrinolysin.

PERSON-TO-PERSON SPREAD

Because they colonize the surface of the skin staphylococci are readily shed and easily transferred horizontally by direct contact or by contact with fomites such as bed linen, medical instruments, and so forth.

EPIDEMIOLOGY

Infections with *S. aureus* are worldwide and do not show a seasonal prevalence. Transient colonization of the umbilical stump and skin, particularly in the perineal region, is common in neonates. About 15% of children and adults are persistent carriers of *S. aureus* in the anterior nares. There is a high rate of carriage in hospital personnel, IV-drug users, diabetics, and dialysis patients, and MRSA and MSSA (methicillin-sensitive *S. aureus*) colonization is common in the population.

In 2017, an estimated 119 247 *S. aureus* bloodstream infections with 19 832 associated deaths occurred in the US. In England 12 784 *S. aureus* bacteremia cases were reported to Public Health England between 2017 and 2018. In fact, studies suggest that as many as one-third of the US population currently is colonized in the anterior nares by this organism. MRSA is identified as a **nosocomial** (hospital-acquired) pathogen throughout the world. Established risk factors for MRSA infection include recent hospitalization or surgery, residence in a long-term-care facility, dialysis, and indwelling percutaneous medical devices and catheters. However, in the 1990s cases of MRSA were documented in healthy community-dwelling persons without established risk factors for MRSA acquisition. Because they were apparently acquired in the community, these infections are referred to as community-acquired (CA)-MRSA. CA-MRSA infections have been reported in North America, Europe, Australia, and New Zealand. These strains, probably derived from the USA300 clone, contain a unique mobile element that includes the methicillin-resistance gene staphylococcal cassette chromosome mec (SCCmec) IVa, the arginine catabolic mobile element, enterotoxins Seq and Sek, and a prophage containing the Panton-Valentine leucocidin.

2. WHAT IS THE HOST RESPONSE TO THE INFECTION AND WHAT IS THE DISEASE PATHOGENESIS?

S. aureus targets those with a breach of the skin and/or mucosal barriers (burns, traumatic wounds, surgical incisions, ulceration, viral skin lesions) together with those with reduced host resistance (cystic fibrosis, immunosuppression, diabetes mellitus, drug addiction). Intact barrier epithelia, together with innate immune factors such as α-**defensins** and **cathelicidin**, contribute to preventing invasion of *S. aureus*. If *S. aureus* breaches the epithelial barrier, then the bacteria are combated by soluble innate antimicrobial proteins in the intercellular spaces and by neutrophils. α-Defensins and cathelicidin are also secreted by epithelial cells and neutrophils as well as being constituents of phagocyte **lysosomes**. However, staphylokinase and the metalloproteinase, aureolysin, can inactivate these defense factors. *S. aureus* also is resistant to lysozyme, a muramidase found in external secretions and in lysosomal vacuoles. Neutrophil **phagocytosis** is critical in defense against *S. aureus*. *S. aureus* is able to inhibit chemotaxis of neutrophils and monocytes in response to C5a and formylated bacterial peptides by secreting a chemotactic inhibitory protein (CHIP). In addition, nucleases produced by the bacterium are able to degrade the neutrophil extracellular traps (NETs) that are comprised of chromatin and microbicidal granule proteins. Initially, phagocytosis is aided by the innate pathways of the complement cascade (lectin and alternative pathways) and later by opsonic **IgM** and IgG antibodies. Consistent with the importance of opsonophagocytosis in host defense against *S. aureus*, the bacterium has several methods by which to avoid **opsonization** and phagocytosis and death within the phagolysosome. The bacterial **capsule** and protein A impair attachment and internalization. Protein A binds IgG antibodies via the Fc region, which prevents their specific binding that would otherwise lead to opsonization and complement activation resulting in further opsonization by C3b. In addition, plasminogen bound on the surface of *S. aureus* can be cleaved by staphylokinase into plasmin, which can then cleave IgG and C3b. Furthermore, *S. aureus* is able to inhibit the C3 convertase. *S. aureus* resists the bactericidal environment of the phagolysosome by the production of **catalase** and carotenoids. Carotenoids give *S. aureus* colonies their golden color. Both molecules neutralize singlet oxygen and superoxide produced as a result of the respiratory burst. Furthermore, several of the exotoxins produced by *S. aureus* kill neutrophils and monocytes/macrophages as well as other types of cells. The **pus** formed in these infections is a mixture of dead organisms and dead phagocytes and the release of their lysosomal contents contributes to the observed tissue damage. Neutralizing IgG antibodies are important in defense against TSST-1.

PATHOGENESIS

S. aureus is a versatile pathogen that causes a wide range of diseases. Disease may be mediated by invasion, bacterial multiplication leading to formation of **abscesses**, and destruction of a variety of tissues, as well as by a range of exotoxins, some of which are superantigens. Furthermore, *S. aureus* and other staphylococcal species are broadly antibiotic-resistant (see below). Factors contributing to the **virulence** of *S. aureus* are presented in Table 33.1.

Skin infections range from those that are superficial such as impetigo (Figure 33.2) (see also *S. pyogenes*), infections of hair follicles (folliculitis) (Figure 33.3), and boils (furuncles) (Figure 33.4) to deeper infections like carbuncles (Figure 33.5), which occur when furuncles unite and extend deeper into subcutaneous tissues forming sinus tracts. From subcutaneous sites, the bacteria can seed the bloodstream and spread to other tissues.

S. aureus is a frequent cause of **bacteremia** and about half of these instances are nosocomial, associated with surgical wound infection and foreign bodies such as indwelling catheters or sutures. From the blood, *S. aureus* may seed many organs including the heart, lungs, bones, and joints. *S. aureus* **endocarditis** has a high mortality and may throw off septic emboli that may infect other organs. Pneumonia may result if bacteria reach the lungs via the bloodstream or as a result of aspiration. Acute and chronic **osteomyelitis** and **septic arthritis** may result from hematogenous spread or from a skin infection.

Table 33.1 Factors contributing to virulence of *Staphylococcus aureus*

Factor	Action
Cell wall-associated	
Capsule/slime	Adherence; antiphagocytic
Protein A	Prevents opsonization; anticomplementary
Clumping factor	Binds to fibrinogen and platelet–fibrin clots
Secreted products – toxins	
Exfoliative toxins (ETA and ETB)	Serine proteases that disrupt intercellular adhesion in the stratum granulosum; possibly superantigens
α, β, δ, γ, cytotoxins	Pore-forming or detergent-like cytotoxins that are toxic for many types of cell
Panton-Valentine leukocidin	Leukotoxic
Toxic shock syndrome toxin-1	Superantigen
Enterotoxins	Family of nine heat-stable, tasteless proteins that cause food poisoning and are superantigens; proposed biothreat agents
Secreted products – enzymes	
Catalase	Detoxifies H_2O_2 in the phagolysosome of neutrophils
Coagulase	Induces fibrin clots
Hyaluronidase	Aids in tissue spread by hydrolyzing the hyaluronic acid intercellular matrix
Lipases	Hydrolyze skin lipids aiding invasion
Fibrinolysin	Aids spreading by dissolving fibrin clots
Nucleases	Hydrolyzes DNA liquifying pus; degrades neutrophil extracellular traps (NETs)

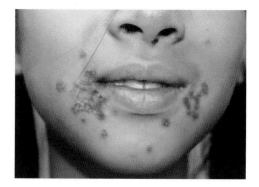

Figure 33.2 Impetigo around the mouth of an 11-year-old girl. This is a highly contagious bacterial skin infection caused by *S. aureus*. *Courtesy of DermAtlas.org © Bernard Cohen.*

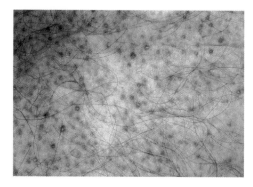

Figure 33.3 Folliculitis caused by *S. aureus*. *From Science Photo Library, with permission.*

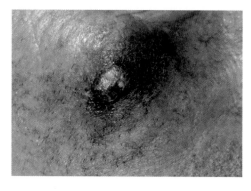

Figure 33.4 Furuncle (boil) on the cheek caused by *S. aureus*. *From Science Photo Library, with permission.*

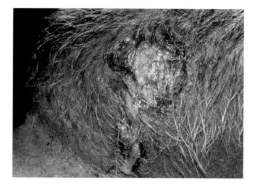

Figure 33.5 Carbuncle on the neck caused by *S. aureus*. *From Science Photo Library, with permission.*

Enterotoxins (A–E, G–I) produced by *S. aureus* are the most common cause of food poisoning, with enterotoxin A causing most cases. Food poisoning results from the production of toxin in the foodstuff and its subsequent ingestion. The most frequently contaminated foods are salted meats, potato salad, custards, and ice cream. Generally, the food is contaminated by *S. aureus* on the hands of food preparers. Holding the food at room temperature allows the staphylococci to proliferate and produce toxin. The enterotoxins are heat-stable and tasteless so that, although heating of the food will kill the staphylococci, the enterotoxins are unaffected. The intoxication has a rapid onset, usually within 4 hours, and the symptoms are nausea, vomiting, diarrhea, and abdominal pain. Symptoms usually resolve within 24 hours. The mechanism(s) of action of the enterotoxins are not understood but may include mast-cell activation and release of inflammatory mediators and neuropeptide **substance P** in the GI tract and elsewhere.

S. aureus secretes several toxins, some of which function as superantigens. These include TSST-1, enterotoxins, and the exfoliative (epidermolytic) toxins. Superantigens bind to both MHC class II molecules on the surface of macrophages and specific Vβ regions of the T-cell receptor on the surface of CD4 T cells. Binding results in the activation of both cell types, leading to excessive production of pro-inflammatory **cytokines** and T-cell proliferation, causing fever, a diffuse macular **rash**, desquamation that includes the palms (Figure 33.6) and the soles, hypotension, and **shock**. Although toxic shock syndrome (TSS) was first reported in 1928, this syndrome came to prominence in 1980 with the introduction of superabsorbent tampons. It was found that *S. aureus* in the vagina could multiply rapidly in these tampons and that tampon fibers acted as an ion exchanger, reducing the level of magnesium ions in the tampon. Under conditions of magnesium limitation, the production of TSST-1 by *S. aureus* is markedly increased. The toxin was then absorbed though the vaginal wall and entered the bloodstream. TSS may also follow localized growth of *S. aureus* in a wound. Interestingly, the vast majority of adults have antibodies that neutralize TSST-1, so only those lacking antibody are at risk.

SSSS is mediated by the exfoliative toxins A and B (ETA, ETB) produced by certain strains of *S. aureus*. The disease is seen mainly in neonates and young children following

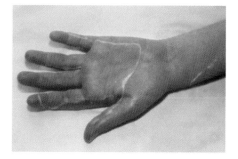

Figure 33.6 Staphylococcal toxic shock syndrome showing desquamation of the left palm. *From the Centers for Disease Control & Prevention, Atlanta, Georgia. Image is found in the Public Health Image Library #5113.*

Figure 33.7 Staphylococcal scalded skin syndrome showing peeling of the skin on the back. *From Mockenhaupt M, Idzko M, Grosber M et al. (2005) Journal of Investigative Dermatology 124: 700–703. With permission from Elsevier.*

an infection of the mouth, nasal cavities, throat acetic or umbilicus. It begins as redness and inflammation around the mouth that extends to the whole body. The superficial layers of skin slip free from the lower layers with a slight rubbing pressure (Nikolsky's sign). Large areas of the skin blister and peel away, leaving wet, red, and painful areas. The blisters or **bullae** do not contain staphylococci or leukocytes. The exotoxins, ETA and ETB, are thought to bind to desmoglein 1 in **desmosomes** in the stratum granulosum, causing it to break down thus releasing the superficial skin layers (Figure 33.7). Healing typically occurs within 1–2 weeks. SSSS can occur in a localized form termed bullous impetigo. The presentation is one of localized bullae (large blisters) without Nikolsky's sign. In contrast to SSSS, the bullae contain *S. aureus* and, therefore, the disease is highly contagious.

3. WHAT IS THE TYPICAL CLINICAL PRESENTATION AND WHAT COMPLICATIONS CAN OCCUR?

Because *S. aureus* is such a versatile pathogen and causes a wide range of disease via invasion and destruction of tissue and by production of toxins, it is difficult to give a "typical" clinical presentation. A retrospective chart review was conducted of patients coded for *S. aureus* sepsis (SAS) requiring admission to a pediatric intensive care unit (PICU) in Auckland, New Zealand over a 10-year period (see References: Miles et al., 2005). Fifty-eight patients were identified with SAS over the 10-year study period. The age distribution of the children was bimodal with peaks at 0–2 and 12–14 years. Almost all of the children had community-acquired SAS; only three had nosocomial SAS. Musculoskeletal symptoms dominated (79%). Pneumonia and/or **empyema** were diagnosed in 78% of children on admission. Two-thirds of the children either presented with multiple site involvement or secondary sites developed during hospitalization that included pneumonia, septic arthritis, osteomyelitis, and soft-tissue involvement (**cellulitis**, **fasciitis**, abscess). A few children had epidural abscesses, vascular **thrombosis**, or endocarditis. Minor limb trauma within the preceding month was reported in half of

the children. Osteomyelitis was also diagnosed in about half of the children, and septic arthritis in slightly less. A pre-existing medical condition was present in eight children and included acute and chronic **myeloid leukemia** and diabetes.

The mortality rate in children with SAS was 8.6% compared with an overall mortality of 6.0% for other patients admitted to the pediatric neonatal intensive care unit. Five children who presented with refractory septic shock died. Complications of SAS occurred in 20 surviving children (38%), including multi-system organ failure, respiratory disease, **paraplegia** from epidural abscesses, and extensive venous thrombosis of the inferior vena cava or femoral veins.

4. HOW IS THE DISEASE DIAGNOSED, AND WHAT IS THE DIFFERENTIAL DIAGNOSIS?

DIAGNOSIS

Microscopy: Depending on the disease and the type of clinical specimen *S. aureus* may be observed by Gram's stain. From clinical specimens, the staphylococci appear as single cells or small clusters of gram-positive cocci (Figure 33.8). However, as several species of staphylococci are **commensals** of the skin and mucous membranes gram staining may not be useful unless the specimen is from a normally sterile body site, but even here as in the case of blood, few bacteria may be present and culture of the specimen is indicated.

Culture: *S. aureus* and other staphylococci grow well on blood agar after 24–48 hours of incubation at 37°C. Staphylococci are **facultative anaerobes**. Colonies of *S. aureus* are large, smooth, glossy domes with an entire edge. If held at room temperature, the colonies will change color from white to a golden yellow (*ergo – S. aureus*, the golden staphylococcus). Almost all strains of *S. aureus* are β-hemolytic due to their secretion of cytotoxins (Figure 33.9). Mannitol-salt agar containing 7.5% sodium chloride and mannitol as the carbon source is a useful selective and differential medium to recover staphylococci from clinical specimens, for example, skin that contain a mixed **microbiota**. Most bacteria

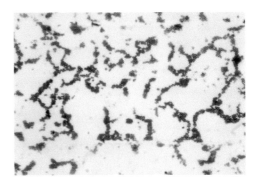

Figure 33.8 Gram stain of *S. aureus*. *From the Centers for Disease Control & Prevention, Atlanta, Georgia. Image is found in the Public Health Image Library #2297. Additional photographic credit is given Dr Richard Facklin who took the photo in 1973.*

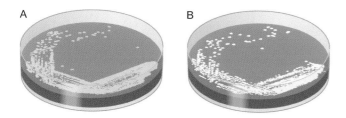

Figure 33.9 *S. aureus* growing on blood agar. (A) *S. aureus* showing typical golden yellow colonies; (B) white colonies of a coagulase-negative *Staphylococcus*. *Courtesy of Professor AM Sefton, Institute of Cell and Molecular Science Centre for Infectious Disease, Barts and the London School of Medicine and Dentistry.*

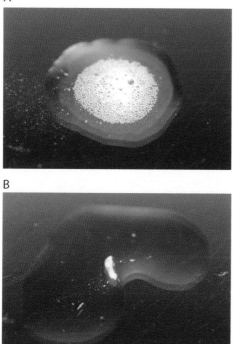

Figure 33.10 *S. aureus* growing on mannitol-salt agar. *With permission from the American Society for Microbiology, MicrobeLibrary Visual Resources Licensing.*

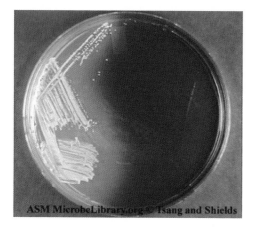

other than staphylococci are inhibited by this concentration of NaCl and *S. aureus* can ferment mannitol whereas most other staphylococcal species cannot. Incorporation of the pH indicator, phenol red, reveals colonies fermenting mannitol because acid produced by the colony changes the color of the agar from pink to yellow (Figure 33.10).

Determinative tests: *S. aureus* are gram-positive cocci growing in grape-like clusters that are positive for the presence of the enzyme catalase. Catalase reduces hydrogen peroxide to water and molecular oxygen and the presence of the enzyme can be readily detected by placing a drop of H_2O_2 on top of staphylococcal cells smeared on a microscope slide. Contact between H_2O_2 and the cells results in the immediate evolution of O_2 that can be seen as bubbles (Figure 33.11). Also, *S. aureus* is coagulase-positive and ferments mannitol (see above). The production of coagulase can be observed by inoculating 0.5 ml of rabbit plasma containing ethylenediamine tetraacetic acid (**EDTA**) with a few colonies of *S. aureus* and incubating for up to 4 hours at 35–37°C. Coagulase will cause the plasma to clot (Figure 33.12). This test can also be performed by emulsifying a colony of *S. aureus* in a drop of rabbit plasma on a microscope slide where the bacterial cells will clump if they produce coagulase. In fact, species of staphylococci other than *S. aureus* are frequently termed "CNS" (coagulase-negative staphylococci). It should be noted that there are two other genera of catalase-positive, gram-positive cocci that are opportunistic pathogens that may be confused with staphylococci, particularly CNS. Members of the genus *Micrococcus* are transients of the skin and mucous membranes while *Stomatococcus mucilaginosus* (weakly catalase-positive) is a commensal of the **oropharynx**. Both have been reported to cause opportunistic infections. *Micrococcus* species may produce bright yellow colonies similar to *S. aureus*, but they are not β-hemolytic, nor do they produce coagulase. *Stomatococcus mucilaginosus* colonies are usually clear to white, mucoid (because of the presence of capsule), and adherent to the surface of the agar. They do not grow in the presence of 7.5% NaCl.

Commercial kits for the rapid detection of clumping factor (cell-bound coagulase) and protein A, based on latex agglutination, are widely available. In these agglutination tests, latex particles coated with fibrinogen and IgG are rapidly agglutinated by *S. aureus* cells. Several nucleic acid

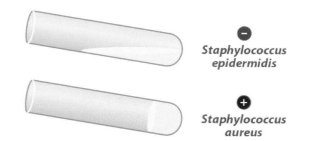

Figure 33.12 Positive and negative coagulase tests of the medically important species of *Staphylococcus* only *S. aureus* produces coagulase. Thus, other staphylococcal species are termed coagulase-negative staphylococci (CNS).

Figure 33.11 Catalase test: (A) positive; (B) negative. *Reprint permission kindly given by Jay Hardy, President of Hardy Diagnostics, www.Hardy.Diagnostics.com.*

Table 33.2 Differential diagnosis of *Staphylococcus aureus* infections

Staphylococcus aureus infection	Differential diagnosis
Impetigo	None
Bullous impetigo	Pemphigus
	Pemphigoid
	Burn
	Stevens-Johnson syndrome
	Dermatitis herpetiformis
Scalded skin syndrome (Ritter disease)	Nonaccidental injury
	Scalding
	Abrasion trauma
	Sunburn
	Erythema multiforme
	Toxic epidermal necrolysis
Bone and joint infections	Bone infarction (in patients with sickle cell disease)
	Toxic synovitis
	Leukemia
Septic arthritis	Trauma
	Deep cellulitis
	Henoch-Schönlein purpura
	Slipped capital femoral epiphysis
	Legg-Calvé-Perthes disease
	Leukemia
	Toxic synovitis
	Metabolic diseases affecting joints (ochronosis*)
Endocarditis – bacteremia	None
Toxic shock syndrome	Staphylococcal scalded skin syndrome
	Meningococcemia
	Rubeola
	Adenoviral infections
	Dengue fever
	Severe allergic drug reactions

Adapted from: Tolan RW Jr, Baorto EP, 2006. *Staphylococcus aureus* infection. Emedicine (from WebMD) http://www.emedicine.com/PED/topic2704.htm
*Ochronosis: bluish-black discoloration of certain tissues, such as the ear cartilage and the ocular tissue, seen with alkaptonuria, a metabolic disorder.

amplification technologies (NAATs) have been implemented or are under development.

DIFFERENTIAL DIAGNOSIS

The differential diagnoses of *S. aureus* infections are shown in Table 33.2.

5. HOW IS THE DISEASE MANAGED AND PREVENTED?

MANAGEMENT

S. aureus and CNS are resistant to many antibiotics that include penicillin, trimethoprim, erythromycin, clindamycin and tetracyclines. Currently, less than 10% of staphylococcal isolates are susceptible to penicillin. Resistance to this antibiotic is mediated by a plasmid-encoded β-lactamase

(penicillinase), which hydrolyzes the β-lactam ring of the molecule. Furthermore, about half of *S. aureus* and CNS isolates are resistant to semi-synthetic penicillins (methicillin, nafcillin, oxacillin) that are resistant to β-lactam ring hydrolysis. Methicillin acts by binding to and competitively inhibiting the transpeptidase used to cross-link the peptide D-alanyl-alanine used in peptidoglycan synthesis. Methicillin is a structural analog of D-alanyl-alanine, and the transpeptidases that are bound by it are termed penicillin-binding proteins (PBPs). Methicillin-resistant *S. aureus* are termed MRSA. Resistance is mediated by acquisition of the gene *mec*A that encodes a PBP for which methicillin and related antibiotics lack affinity. Initially, MRSA were limited to hospitals but, recently, MRSA outbreaks have been reported in the community. Interestingly, the community-associated MRSA strains appear to be clonally related, but unrelated to hospital MRSA. Alarmingly, *S. aureus* and CNS have been isolated with low-level resistance to vancomycin, an antibiotic to which staphylococci had been uniformly sensitive. Such strains of *S. aureus* are termed VISA (vancomycin-intermediate *Staphylococcus aureus*) or GISA (glycopeptide-intermediate *Staphylococcus aureus*). There appear to be two different mechanisms of resistance to glycopeptides. One seems to be related to increased thickness of the cell wall, which prevents penetration of the antibiotic to its target in the cytoplasmic membrane. The other results from the acquisition of the *van*A gene from vancomycin-resistant enterococci (VRE). Glycopeptide antibiotics such as vancomycin act by binding to the D-alanyl-D-alanine terminus of peptidoglycan precursors preventing cell wall synthesis. However, the *van*A gene specifies the synthesis of peptidoglycan precursors terminating in the depsipeptide D-alanyl-D-lactate, to which vancomycin does not bind. Serious MSSA infections are treated with parenteral penicillinase-resistant penicillins or first- or second-generation cephalosporins such as clindamycin. Vancomycin is reserved for staphylococcal strains resistant to penicillinase-resistant penicillins and clindamycin. Mupirocin may be used to treat superficial or localized skin infections caused by *S. aureus* (i.e. impetigo). This antibiotic works by inhibiting isoleucyl-tRNA synthetase, which is required for protein synthesis. However, it should be noted that most community-acquired MRSA strains are already resistant or are acquiring resistance

to mupirocin. Current guidelines for treatment of methicillin-resistant MRSA can be found in the references to this chapter.

In addition to antibiotic therapy, temporary intravascular devices such as catheters should be promptly removed if infection is suspected. Long-term intravascular devices should be removed if infection with *S. aureus* is confirmed. Abscesses must be drained and, usually, if the infection involves a prosthetic joint, it should be removed.

PREVENTION

Prevention of staphylococcal infection relies on the practice of good hygiene. The infectious dose of *S. aureus* required to initiate a wound infection in a healthy person is high unless the wound contains a foreign body of some sort. The Centers for Disease Control and Prevention (CDC) recommend keeping one's hands clean by washing thoroughly with soap and water or using an alcohol-based hand sanitizer; keeping cuts and scrapes clean and covered with a bandage until healed; avoiding contact with other people's wounds or bandages; and avoiding sharing personal items such as towels or razors. In the hospital setting, attempts have been made to eliminate nasal carriage by hospital personnel by the use of antibiotic regimens.

SUMMARY

1. WHAT IS THE CAUSATIVE AGENT, HOW DOES IT ENTER THE BODY, AND HOW DOES IT SPREAD A) WITHIN THE BODY AND B) FROM PERSON TO PERSON?

- Staphylococci are gram-positive cocci that grow in grape-like clusters and produce the enzyme catalase.
- The genus *Staphylococcus* comprises some 35 species, many of which are members of the endogenous microbiota of the skin and mucous membranes of the gastrointestinal and genitourinary tracts of humans.
- *S. aureus* and other staphylococcal species are readily isolated on blood agar incubated aerobically at 37°C for 24–48 hours.
- *S. aureus* is unique among staphylococci because it produces a cell-associated and secreted coagulase, which converts fibrinogen to fibrin and it expresses protein A on its surface, which binds the Fc region of human IgG. These properties are utilized in rapid ID tests to identify *S. aureus*.
- Staphylococci are very tolerant of salt and will grow in the presence of 7.5% NaCl.
- These properties serve as the bases for a selective and differential medium termed mannitol-salt agar that is useful in isolating staphylococci from skin and mucosal surfaces. Mannitol-salt agar differentiates *S. aureus* from coagulase-negative staphylococci (CNS) species that do not ferment mannitol.
- *S. aureus* has a typical gram-positive cell wall that features a thick peptidoglycan layer extensively cross-linked with pentaglycine bridges. The extensive cross-linking makes the cell very resistant to drying so that the bacterium can survive on inanimate objects

for long periods of time. The cell envelope contains membrane-anchored lipoteichoic acids and wall-anchored teichoic acids.
- The anterior nares are the reservoir of *S. aureus*. From this site, it is readily transferred to the hands.
- Because they also colonize the surface of the skin, staphylococci are readily shed and easily transferred horizontally (person to person) by direct contact (bacteria on the hands) or by contact with fomites (inanimate objects) such as bed linen, medical instruments, and so forth.
- There is a high rate of carriage of methicillin-resistant *S. aureus* (MRSA) in hospital personnel, IV drug users, diabetics, and dialysis patients.
- As many as one-third of the US population are colonized in the anterior nares by this organism.
- Transient colonization of the umbilical stump and skin, particularly in the perineal region, is common in neonates.
- *S. aureus* (and other species of staphylococci) produce an extracellular polysaccharide capsule or slime that enables them to form a biofilm on catheters and other prostheses.
- *S. aureus* requires a breach in the skin or mucosa such as produced by a catheter, a surgical incision, a burn, a traumatic wound, ulceration or viral skin lesions to facilitate its entry into the tissues.
- Staphylococcal infections are frequently associated with reduced host resistance brought about by cystic fibrosis, immunosuppression, diabetes mellitus, viral infections, and drug addiction.

Continued...

...continued

- *S. aureus* has several enzymes that aid in invasion of the barrier epithelia and serve as spreading factors. Among these are lipases, hyaluronidase, and fibrinolysin.

2. WHAT IS THE HOST RESPONSE TO THE INFECTION AND WHAT IS THE DISEASE PATHOGENESIS?

- Intact barrier epithelia, together with innate immune factors at these surfaces, are effective in preventing invasion of *S. aureus*.
- If *S. aureus* breaches the epithelial barrier, the bacteria are phagocytosed by neutrophils.
- Phagocytosis is aided by opsonization via the innate pathways of the complement cascade (lectin and alternative pathways) and later by IgM and IgG antibodies.
- *S. aureus* has various mechanisms to resist phagocytosis and intracellular killing.
- Neutralizing IgG antibodies are important in defense against TSST-1.

3. WHAT IS THE TYPICAL CLINICAL PRESENTATION AND WHAT COMPLICATIONS CAN OCCUR?

- *S. aureus* is a versatile pathogen that causes a wide range of disease via invasion and destruction of tissue and by production of toxins.
- Impetigo begins as a small area of **erythema** that progresses into fluid-filled bullae that rupture and heal with the formation of a honey-colored crust. Impetigo is highly contagious.
- Bullous impetigo is a localized form of staphylococcal scalded skin syndrome (SSSS) that presents as bullae without surrounding erythema. The bullae contain staphylococci and rupture, leaving a denuded area. Bullous impetigo is highly contagious.
- Scalded skin syndrome begins with peri-oral erythema that extends over the whole body in 1–2 days. Friction to the skin causes it to be displaced (positive Nikolsky's sign) leading to the formation of fragile bullae that burst, leaving a tender base. This is followed by desquamation of the skin. The patient often is **febrile** and occasionally has a **mucopurulent** eye discharge.
- Folliculitis is a tender pustule that involves the hair follicle.
- Furuncles are small abscesses that involve both the skin and the subcutaneous tissues in areas with hair follicles, such as the neck, axillae, and buttocks.
- Carbuncles are collections of inter-connected furuncles with several sinus tracts opening to the skin surface. Often, they are the source of systemic infection via bacteremic spread of the staphylococci indicated by fever and chills.
- Osteomyelitis often presents with a sudden onset of fever and bone tenderness or limp. There may be little pain or pain can be throbbing and quite severe.

- Septic arthritis presents with decreased range of motion, warmth, erythema, and tenderness of the joint together with constitutional symptoms and fever. However, in infants these signs may be absent.
- Endocarditis presents with high fever, chills, sweats, and malaise. The condition can deteriorate rapidly with reduced cardiac output and septic emboli manifested by **Osler nodes**, subungual hemorrhages, Janeway lesions, and **Roth spots**.
- Toxic shock syndrome (TSS) is a rapidly progressing, life-threatening intoxication. It begins abruptly with fever of 38.9°C or higher, a diffuse macular, **erythematous** rash, and hypotension. Conjunctival and/or vaginal **hyperemia** may be present. Multiple organ systems are affected such as gastrointestinal, musculature, renal, hepatic, hematologic, and nervous. Also, the entire skin including the palms of the hands and the soles of the feet desquamates.
- Staphylococcal pneumonia is a rapidly progressive disease observed primarily at the extremes of age in persons with predisposing systemic conditions such as **chronic obstructive pulmonary disease** (**COPD**), bronchiectasis, cystic fibrosis or influenza infection. There are no unique clinical or radiographic findings but patients may also have prominent gastrointestinal tract symptoms.

4. HOW IS THE DISEASE DIAGNOSED, AND WHAT IS THE DIFFERENTIAL DIAGNOSIS?

- Examination of appropriate clinical specimens by microscopy will reveal gram-positive cocci in grape-like clusters and individual cells.
- Clinical specimens are cultured on blood agar at 37°C for 24–48 hours in air. Colonies are round, glossy, domed with an entire edge, and golden yellow in color. Most *S. aureus* strains are β-hemolytic. Mannitol-salt agar can be used to isolate *S. aureus* from skin or mucosal surfaces. *S. aureus* colonies will be yellow in color and surrounded by a yellow zone indicating fermentation of mannitol.
- Determinative tests include: production of catalase and coagulase; rapid latex agglutination test to detect clumping factor and protein A; mannitol fermentation.
- Bullous impetigo: the differential diagnosis would include pemphigus, pemphigoid, and burns.
- Scalded skin syndrome: the differential diagnosis would include scalding, sunburn, and trauma.
- Toxic shock syndrome: SSSS, meningococcemia, rubeola, adenovirus infections, **Dengue fever**, severe allergic reactions.

continued...

...continued

5. HOW IS THE DISEASE MANAGED AND PREVENTED?

- Prevention of staphylococcal infection relies on the practice of good hygiene. The infectious dose of *S. aureus* required to initiate a wound infection in a healthy person is high unless the wound contains a foreign body of some sort.

- The Centers for Disease Control and Prevention (CDC) recommendations are: keeping one's hands clean by washing thoroughly with soap and water or using an alcohol-based hand sanitizer; keeping cuts and scrapes clean and covered with a bandage until healed; avoiding contact with other people's wounds or bandages; and avoiding sharing personal items such as towels or razors.

- In the hospital setting, attempts have been made to eliminate nasal carriage of *S. aureus* by hospital personnel by the use of antibiotic regimens.

- Intravascular devices such as catheters should be promptly removed if infection is suspected. Long-term intravascular devices should be removed if infection with *S. aureus* is confirmed. Abscesses must be drained and, usually, if the infection involves a prosthetic joint, it should be removed.

- Antibiotic therapy is problematic for staphylococcal infections because of broad antibiotic resistance.

- Currently less than 10% of staphylococcal isolates are susceptible to penicillin. Resistance to this antibiotic is mediated by a plasmid-encoded β-lactamase (penicillinase), which hydrolyzes the β-lactam ring of the molecule.

- About half of *S. aureus* and CNS isolates are resistant to semi-synthetic penicillins (methicillin, nafcillin, oxacillin) that are resistant to β-lactam ring hydrolysis. Methicillin-resistant *S. aureus* are termed MRSA. In MRSA, the gene *mec*A encodes the protein PBP2A (penicillin-binding protein 2A), a transpeptidase involved in the assembly of the cell wall.

- Low-level resistance to vancomycin, an antibiotic to which staphylococci had been uniformly sensitive, has been recognized.

FURTHER READING

Fischetti VA, Novick RP, Ferretti JJ, et al. Gram-Positive Pathogens, 3rd edition. ASM Press, Washington, DC, 2019.

Goering R, Dockrell H, Zuckerman M, Chiodini PL. Mims' Medical Microbiology and Immunology, 6th edition. Elsevier, Philadelphia, 2018.

Murray PR, Rosenthal KS, Pfaller MA. Medical Microbiology, 9th edition. Elsevier/Mosby, Philadelphia, 2021.

REFERENCES

Brown NM, Goodman AL, Horner C, et al. Treatment of Methicillin-Resistant Staphylococcus aureus (MRSA): Updated Guidelines From the UK. JAC Antimicrob Resist, 3: dlaa114, 2021.

Guo Y, Song G, Sun M, et al. Prevalence and Therapies of Antibiotic-Resistance in Staphylococcus aureus. Front Cell Infect Microbiol, 10: 107, 2020.

Jenul C, Horswill AR. Regulation of Staphylococcus aureus Virulence. Microbiol Spectr, 7: 10.1128, 2019.

Miles F, Voss L, Segedin E, et al. Review of Staphylococcus aureus infections requiring admission to a paediatric intensive care unit. Arch Dis Child, 90: 1274–1278, 2005.

Tam K, Torres VJ. Staphylococcus aureus Secreted Toxins and Extracellular Enzymes. Microbiol Spectr, 7: 10.1128, 2019.

Turner NA, Sharma-Kuinkel BA, Maskarinec SA, et al. Methicillin-Resistant Staphylococcus aureus: An Overview of Basic and Clinical Research. Nat Rev Microbiol, 17: 203–218, 2019.

Urish LK, Cassat JE. Staphylococcus aureus Osteomyelitis: Bone, Bugs and Surgery. Infect Immun, 88: 6–22, 2020.

WEBSITES

Johns Hopkins Medicine, Auwaerter PG, Staphylococcus aureus, 2022: https://www.hopkinsguides.com/hopkins/view/Johns_Hopkins_ABX_Guide/540518/all/Staphylococcus_aureus?q=MRSA

News Medical, Mandal A, Staphylococcus aureus treatment, 2019: https://www.news-medical.net/health/Staphylococcus-Aureus-Treatment.aspx

Siddiqui AH, Koirala J, Methicillin Resistant Staphylococcus aureus. StatPearls, 2022: https://www.ncbi.nlm.nih.gov/books/NBK482221/

Taylor TA, Unakal CG, Staphylococcus aureus, StatPearls, 2022: https://www.ncbi.nlm.nih.gov/books/NBK441868/

Students can test their knowledge of this case study by visiting the Instructor and Student Resources: [www.routledge.com/cw/lydyard] where several multiple choice questions can be found.

Streptococcus pneumoniae

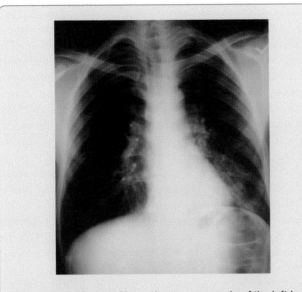

34

Figure 34.1 AP chest X-ray shows pneumonia of the left lower lobe with early consolidation. This disease is caused by the bacterium *Streptococcus pneumoniae*. The alveoli (the air sacs of the lungs, too small to be seen) become blocked with pus, forcing air out and causing the lung to solidify. Symptoms include coughing and chest pain. *From the Centers for Disease Control & Prevention, Atlanta, Georgia. Image is found in the Public Health Image Library #2897. Additional photographic credit is given to Dr Thomas Hooten who took the photo in 1978.*

A 45-year-old male with a known history of alcohol abuse presented at the Emergency Room with a 1-day history of fevers, shaking chills, and productive cough. He stated that he was coughing up blood-tinged sputum that had a rusty appearance. The patient also remarked that he had some left-sided chest pain that was worse when he breathed in or coughed (**pleuritic pain**) and some **dyspnea** (shortness of breath).

He denied any significant past medical history and was not taking any medications. He reported a 20-pack per-year history of smoking and drinking at least one six-pack of beer per day.

On physical examination, he was a well-developed male in moderate distress. His temperature was 40°C, pulse 110 beats per minute, blood pressure 140/80 mmHg, respiratory rate 24, and O$_2$ saturation 90% on room air. His lung examination revealed decreased breath sounds and dullness to percussion in the left lower lung base. There were some **rales** and **egophony** present. A sputum sample and blood were collected and a chest X-ray was ordered. The results of the investigations were as follows:

- Laboratory findings: WBC 18 000, 80% PMNs, 10% bands, 10% lymphocytes.
- Chest X-ray: dense consolidation and air bronchograms in the left lower lobe (Figure 34.1).
- Sputum gram stain: gram-positive, bullet-shaped diplococci and inflammatory cells (Figure 34.2).
- Blood cultures at 24 hours revealed gram-positive diplococci. (Figure 34.3).

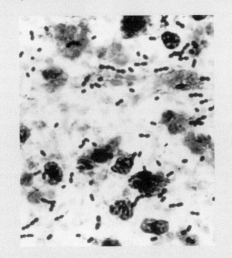

Figure 34.2 Gram stain of *S. pneumoniae* in sputum: note the bullet-shaped diplococci. *From Lebeau / Custom Medical Stock Photo / Science Photo Library, with permission.*

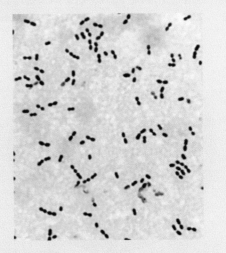

Figure 34.3 Gram stain of *S. pneumoniae* from blood culture broth showing gram-positive bullet-shaped diplococci. *From CNRI / Science Photo Library, with permission.*

1. WHAT IS THE CAUSATIVE AGENT, HOW DOES IT ENTER THE BODY AND HOW DOES IT SPREAD A) WITHIN THE BODY AND B) FROM PERSON TO PERSON?

CAUSATIVE AGENT

The patient is infected with *Streptococcus pneumoniae*, frequently termed the pneumococcus. The World Health Organization (WHO) estimates that *S. pneumoniae* kills close to half a million children under 5 years old worldwide every year. Most of these deaths occur in developing countries. *S. pneumoniae* is the leading cause of pneumonia mortality globally. It accounted for more deaths than all other causes combined in 2016. Most of these deaths occur in countries in Africa and Asia. In the US, it is estimated that more than 150000 hospitalizations from pneumococcal pneumonia occur each year, and about 5–7% of those who are hospitalized from it will die. The death rate is even higher in those age 65 years and older. *S. pneumoniae* accounts for 30–50% (300 million cases annually) of all cases of otitis media worldwide.

S. pneumoniae is a gram-positive, **catalase**-negative diplococcus. The bacterial cells are bullet- or lancet-shaped when observed by Gram's stain (Figure 34.3). The catalase test is used to distinguish streptococci from staphylococci, which are the other medically important genera of gram-positive cocci (see *Staphylococcus aureus*, Case 33, for an image of the catalase test). *S. pneumoniae* grows readily on blood agar incubated at 37°C in the presence of 5% CO_2 for 24–48 hours and forms small-to-medium size colonies with a varying mucoid appearance. The colonies are gray-green in color and are surrounded by a zone of incomplete (greening) **hemolysis** (α-hemolysis) (Figure 34.4). Older colonies of pneumococci collapse in the center due to autolysis and are termed "draughtsman-like" colonies. *S. pneumoniae* is soluble in optochin and bile, unlike other viridans streptococci. In

Figure 34.4 A mucoid strain of *S. pneumoniae* growing on blood agar. The mucoid appearance results from production of capsule. *With permission from the American Society for Microbiology, MicrobeLibrary Visual Resources Licensing.*

2004, a species, *Streptococcus pseudopneumoniae*, closely related to *S. pneumoniae* and *S. mitis*, was described (see *S. mitis* case e8). It is likely that this bacterium was and is often mistaken for *S. pneumoniae*. There is evidence that *S. pneumoniae* evolved from the *S. mitis/S. oralis* complex as all three bacteria are very closely related genetically. *S. pseudopneumoniae* lacks a capsule, is not soluble in bile, and is sensitive to optochin only when incubated in ambient air. The variation in these physiologic determinants would seem to place *S. pseudopneumoniae* between *S. pneumoniae* and *S. mitis*. *S. pseudopneumoniae* appears to be an emerging cause of lower-respiratory tract infections and chronic obstructive pulmonary disease (COPD) and exacerbation of COPD, but little more is known about its clinical importance and epidemiology at this time. The identification of *S. pseudopneumoniae* and *S. mitis* and other alpha-hemolytic streptococci as opportunistic pathogens suggests that more attention should be paid to the non-*S. pneumoniae* viridans streptococci and that they should not continue to be lumped under the umbrella of "Streptococcus viridans" and largely disregarded by clinical microbiologists and the physicians they support.

The most important component of the *S. pneumoniae* outer surface is the polysaccharide **capsule**, since this is the principal **virulence** determinant of the bacterium. The capsule impedes **phagocytosis** primarily by inhibiting deposition of the opsonic complement component, C3b, on the bacterial surface thus impairing the innate and acquired immune response to *S. pneumoniae*. The structure of the capsule is complex and there are some 98 distinct antigenic types. *S. pneumoniae* has a typical gram-positive-type cell wall that features six layers of peptidoglycan with covalently bound teichoic acid (C-polysaccharide) and cell membrane-anchored lipoteichoic acid (Forssman antigen) (Figure 34.5). The two forms of teichoic acid have identical carbohydrate structures and contain choline. Although the vast majority of the teichoic acid is covered by the capsule, anti-phosphocholine antibody can protect against experimental pneumococcal infection and the acute phase reactant, **C-reactive protein**, can interact with choline residues in the cell wall. On the outer surface of the cell wall is a family of 12 choline-binding proteins (CBPs) that are non-covalently bound to the phosphorylcholine moiety of the wall. These include the major autolysin of the pneumococcus, LytA; two other cell-wall hydrolases, LytB and LytC, suggested to be involved in daughter cell separation as well as in bacterial uptake of DNA; PcpA, which is thought to be involved in protein–protein and protein–lipid interactions; PspA that decreases complement deposition on the bacterial surface during sepsis; and CbpA (SpsA), the most abundant of the CBPs, that functions as a cell-surface **adhesin** in the nasopharynx. CbpA has been shown to bind the secretory component of **IgA** and the complement component, C3. The pneumococcus contains a cholesterol-dependent pore-forming cytotoxin termed pneumolysin, which is stored intracellularly in most strains and is released upon cell **lysis** mediated by LytA. This toxin is important in the causation of **meningitis** because it damages ependymal cilia that line

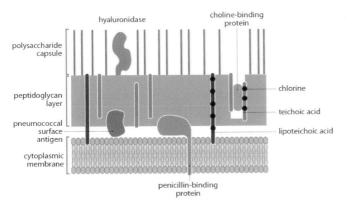

Figure 34.5 Schematic representation of the *S. pneumoniae* envelope. Teichoic acids (TAs) and lipoteichoic acids (LTAs) (both in blue) are carbohydrate phosphate polymers rich in choline (red spheres). TAs are linked to the peptidoglycan via a phosphodiester linkage, whereas LTAs are linked to the cell membrane via a C-terminal fatty acyl group. Choline-binding proteins are linked to cell wall TAs or LTAs via choline-binding domains (CBDs). Pneumococcal surface antigen (PsaA) is located beneath the peptidoglycan layer and is attached to the cell membrane; penicillin-binding proteins (PBPs) are located in the "periplasmic space" (the zone beneath the peptidoglycan and above the cytoplasmic membrane) and interact with the peptidoglycan. Hyaluronate lyase (Hyl) is tethered to the peptidoglycan. New approaches towards the identification of antibiotic and vaccine targets in *Streptococcus pneumoniae. EMBO Reports 3(8):728–734. Taken and adapted from Figure 2 in: Di Guilmi AM & Dessen A. 2002.*

the ventricles of the brain and induces brain cells to undergo **apoptosis**. Also, H_2O_2 produced by the pneumococcus is a potent **hemolysin**. Finally, the bacterium secretes an **IgA1 protease**, which is able to subvert the activity of IgA1 by cleaving the molecule at the hinge region, and a neuraminidase that cleaves terminal sialic acid from glycoconjugates thereby uncovering **epitopes** for pneumococcal adherence.

ENTRY AND SPREAD WITHIN THE BODY

S. pneumoniae enters the body through the oral–nasal route via respiratory droplets and adheres to the oropharyngeal epithelium via the adhesin molecules listed in Table 34.1. Initial binding is thought to involve carbohydrate epitopes in

the respiratory mucosa through an as yet unidentified adhesin. Pneumococci then use CbpA to bind to the poly-Ig (pIg) receptor and phosphocholine to bind to the platelet-activating factor (PAF) receptor, enabling the pneumococcus to be taken up by the epithelium. *S. pneumoniae* transiently colonizes the pharynx and carriage of the bacterium in healthy children has been reported to be as high as 75% depending on detection method and population. Colonization is more common in infants and children than in adults (< 10%). Repeat episodes of carriage are associated with different capsule serotypes and the duration of carriage decreases with each successive colonization episode, in part because of the induction of serotype-specific antibodies. Disease results from the spread of bacteria from the nasopharynx to the sinuses, middle ear, **meninges**, and lungs (as it did in this patient). Bacteremic spread may also occur.

PERSON-TO-PERSON SPREAD

Disease occurs as a result of horizontal spread of the organism by respiratory droplets, particularly in crowded settings such as hospitals, daycare centers, and jails.

EPIDEMIOLOGY

The WHO estimates that *S. pneumoniae* kills close to half a million children under 5 years old worldwide every year. Most of these deaths occur in developing countries. In 2017, the Centers for Disease Control and Prevention (CDC) reported more than 31000 cases and 3500 deaths from invasive pneumococcal disease, more than half of which occurred in adults. In the years 2013–2014, approximately 6000 cases of invasive pneumococcal disease were reported and 192281 hospital admissions for pneumonia in England of which up to 50% may have been caused by *S. pneumoniae*. Despite advances in pneumococcal immunization, the global burden of pneumococcal disease remains high. However, there is evidence that the frequency of community acquired pneumonia caused by *S. pneumoniae* is far lower in North America than in Europe. This may be explained by restricted

Table 34.1 Factors contributing to virulence of *Streptococcus pneumoniae*

Factor	Action
Capsule	Anti-phagocytic
PspA	Anti-phagocytic
Pneumolysin	Cytotoxin
Hydrogen peroxide	Hemolysin
Neuraminidase	Removes terminal sialic acid exposing glycoconjugates for adhesion
Teichoic acid	Adhesin (binds fibronectin), immunomodulatory
CbpA	Adhesin and invasion (binds platelet-activating factor [PAF] receptor and poly-Ig [pIg] receptor)
PsaA	Mn^{2+} binding protein, required for adherence of bacteria to host cells
PcpA	Adhesin (protein–protein and protein–lipid interactions)
IgA1 protease	Cleaves sIgA1 in respiratory secretion

access of children living in many countries to pneumococcal vaccines and the emergence of infection due to non-vaccine serotypes of the pneumococcus.

2. WHAT IS THE HOST RESPONSE TO THE INFECTION AND WHAT IS THE DISEASE PATHOGENESIS?

IMMUNE RESPONSE

Table 34.1 lists the various factors that contribute to the virulence of the organism and some of these help to prevent the organism from being eliminated by the host defenses. The respiratory mucosa with its mucociliary blanket, together with innate immune factors and secretory IgA antibodies in respiratory secretions, are effective barriers to the invasion of *S. pneumoniae* through the epithelium. As for most extracellular, capsulate bacteria, phagocytosis aided by complement and opsonic **IgM** and **IgG** antibodies are the principal modalities of host defense after invasion. Thus, pneumococci are opsonized by activation of the **alternative** and lectin innate **complement pathways** and the **classical** pathway in the presence of anti-capsular antibodies in the plasma and tissue fluid. The pneumococcus targets those with impaired innate and acquired immune systems, for example, persons who lack a spleen, are immunodeficient, or lack early or late complement components. Such individuals are generally the very young and very old who have low levels of anti-capsular antibody and those with impaired bacterial clearance. However, neonates are protected by transplacental transfer of maternal IgG anti-capsular antibodies and infants are protected against pneumococcal disease through antibacterial components of human milk. Impaired bacterial clearance is observed in persons with defective epiglottal and/or cough reflexes, antecedent viral respiratory infections, COPD, chronic cardiovascular disease, alcoholism, renal dysfunction, and diabetes, as well as individuals with certain hereditary or acquired immunodeficiencies.

PATHOGENESIS

The host immune response plays a major role in pathogenesis of the pneumococcus. Resolution of infection occurs when the bacteria are rapidly cleared without neutrophil recruitment. This is accomplished by both humoral and cellular components of the immune system. Lung-tissue injury results from rapid bacterial growth with attendant increase in pro-inflammatory **cytokines** and neutrophil recruitment. Thus, the hallmark of pneumococcal disease is the intense inflammatory infiltrate at the site of infection. This results from the action of pneumolysin and hydrogen peroxide and the pro-inflammatory nature of the bacterial cell-wall components such as teichoic acid and peptidoglycan fragments. These are released from the wall by the autolysins, LytA, B, and C. Pneumococci can be induced to undergo autolysis by exposure to β-lactam

antibiotics (see Section 4), hypertonic media, detergents or extended incubation in the stationary phase of growth. The wall fragments activate the complement cascade releasing the anaphylatoxins, C3a and C5a. In addition, **interleukin (IL)**-1 and tumor necrosis factor-α (TNF-α) are released from activated leukocytes leading to the recruitment of additional inflammatory cells, amplifying tissue damage and causing fever.

3. WHAT IS THE TYPICAL CLINICAL PRESENTATION AND WHAT COMPLICATIONS CAN OCCUR?

Pneumococcal pneumonia typically presents with an abrupt onset of fever to 41°C and shaking chills. These symptoms may follow a viral infection of the respiratory tract. The majority of patients have a productive cough with blood-tinged sputum and **pleurisy**. The mortality from pneumococcal pneumonia is about 5% in the US. Among the over 90 capsule types, capsule type 3 organisms are the most virulent.

S. pneumoniae is also a frequent cause of **sinusitis** and otitis media. From the middle ear or sinuses the bacteria can seed the central nervous system (CNS). CNS infection can also follow **bacteremia**. Meningitis is most commonly observed in children and adults and less frequently in the elderly. Of the bacteria causing meningitis, *S. pneumoniae* infection has the highest mortality and results in the most severe neurologic impairment.

4. HOW IS THE DISEASE DIAGNOSED, AND WHAT IS THE DIFFERENTIAL DIAGNOSIS?

MICROSCOPY

Examination of sputum, blood, CSF, and other specimens stained by Gram's method for gram-positive, bullet-shaped diplococci can provide a rapid preliminary diagnosis. A gram stain of sputum is shown in Figure 34.2.

CULTURE

Specimens should be plated onto blood agar and incubated for 24–48 hours at 37°C in the presence of 5% CO_2. Colonies of various strains vary in the amount of capsule they produce. Highly mucoid, α-hemolytic colonies are shown in Figure 34.4; however, pneumococcal colonies frequently produce little to no capsule (nonmucoid). Examples of colonies of a nonmucoid strain of *S. pneumoniae* are shown in Figure 34.6.

DETERMINATIVE TESTS

S. pneumoniae is lysed by exposure to bile or optochin (ethylhydrocupreine dihydrochloride) while commensal α-hemolytic streptococci in the nasopharynx are resistant to

A

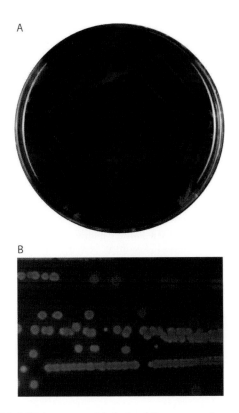

B

Figure 34.6 (A) **A nonmucoid strain of *S. pneumoniae* growing on a blood agar plate.** (B) **Enlargement of** (A). *S. pneumoniae* appear as small grayish colonies with a greenish zone of α-hemolysis surrounding them on the blood agar plate. *Reprint permission kindly given for image in Laboratory Methods for the Diagnosis of Meningitis caused by Neisseria meningitidis, Streptococcus pneumoniae, and Haemophilus influenzae CDC, Centers for Disease Control and Prevention August 1998,* (A) *Figure 7, page 22;* (B) *Figure 12, page 32.*

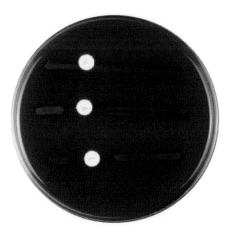

Figure 34.7 **Optochin susceptibility test for identification of *S. pneumoniae*.** α-Hemolytic, gram-positive, catalase-negative cocci are streaked in bands across a blood agar plate and an optochin disc is placed on each streak. After incubation for 24–48 hours in 5% CO_2 in air at 37°C growth around the discs is examined. The strain in the top streak has grown up to the disc and is resistant to optochin and, therefore, is not a pneumococcus. The strains in the center and lower streaks show a zone of inhibition indicating that they are susceptible to optochin and are pneumococci. *Reprint permission kindly given for image in Laboratory Methods for the Diagnosis of Meningitis caused by Neisseria meningitidis, Streptococcus pneumoniae, and Haemophilus influenzae CDC, Centers for Disease Control and Prevention August 1998, Figure 20, page 46.*

both. An optochin disc placed on a blood agar plate inoculated with *S. pneumoniae* will show a wide zone of growth inhibition (Figure 34.7).

SEROLOGY

The capsule type of the *S. pneumoniae* isolate can be determined by the **Quellung test** in which bacteria are mixed into a drop of capsule-specific antiserum and observed under the microscope. Capsule swelling and increased refractility result from the interaction of the antibody with the capsule for which it is specific.

DIFFERENTIAL DIAGNOSIS

The clinical presentation and findings suggest pneumonia of bacterial etiology because of the acute onset and severity of symptoms. Pneumonia caused by viruses has a more gradual onset and is generally less severe. Symptoms of viral pneumonia may include fever, dry cough, headache, and muscle pain. Within 12–16 hours, other symptoms may appear, such as shortness of breath, sore throat, increased cough, and mucus with cough. Bacteria to include in the differential diagnosis are *Haemophilus influenzae*, *Moraxella catarrhalis*, *Klebsiella pneumoniae*, *Staphylococcus aureus*, *Legionella pneumophila*, *Mycoplasma pneumoniae*, and *Chlamydophila pneumoniae*.

5. HOW IS THE DISEASE MANAGED AND PREVENTED?

MANAGEMENT

S. pneumoniae, once highly susceptible to penicillin, has now acquired resistance to this and many other antibiotics. Resistance of the pneumococcus to β-lactams and macrolides is a major concern worldwide with the prevalence of this resistance reaching 20% in Southern European countries. Pneumococci may also be resistant to fluoroquinolones. It is estimated that the worldwide prevalence of multidrug-resistant (MDR) *S. pneumonia*, that is, resistance to three or more antibiotic classes may be as high as 36% in some areas of the world. Penicillin resistance results from the generation of penicillin-binding proteins (PBPs) with decreased affinity for penicillin. PBPs are involved in the assembly of the cell wall. It is likely that these altered PBP genes arose by interspecies recombination in which segments of the PBPs' structural genes were replaced by regions derived from PBP genes of oral streptococci. These altered PBP genes of penicillin-resistant pneumococci can be spread horizontally to sensitive pneumococci by transformation. About one-third of the strains isolated in the US are resistant to penicillin and higher rates of resistance have been observed in other countries. Although there are over 90 *S. pneumoniae* serotypes, over 90% of penicillin-resistant strains are found within seven serotypes (6A, 6B, 9V, 14, 19A, 19F, and 23F). Unfortunately, these are the same serotypes that cause the vast majority of infections in children. Cefotaxime, ceftriaxone, and clindamycin are effective antibiotics for treating pneumonia caused by penicillin-resistant pneumococcal isolates susceptible to these antibiotics. Clindamycin or vancomycin is recommended when a pneumococcal isolate is resistant to cefotaxime or ceftriaxone.

PREVENTION

Two vaccines are available to prevent pneumococcal disease. Both consist of the purified capsular polysaccharide. A 23-valent vaccine consisting of capsular polysaccharide from the 23 serotypes most commonly isolated from infected patients (1, 2, 3, 4, 5, 6B, 7F, 8, 9N, 9V, 10A, 11A, 12F, 14, 15B, 17F, 18C, 19A, 19F, 20, 22F, 23F, and 33F). This vaccine

is only 60% effective. Polysaccharides are T-cell-independent antigens and infants below the age of 2 years respond poorly to them. Therefore, for children younger than 2 years of age, a 13-valent **conjugate vaccine** (capsule types 1, 3, 4, 5, 6A, 6B, 7F, 9V, 14, 18C, 19A, 19F, and 23F) is recommended. In this vaccine, the capsular polysaccharide is conjugated to a protein carrier to render it T-cell-dependent. The protein carrier used is either tetanus toxoid or diphtheria toxoid, themselves vaccine antigens. The 13-valent conjugate vaccine reduces invasive infection in children yet has little effect on the incidence of otitis media and colonization. The long-term efficacy and benefit of this vaccine are, therefore, unknown. In persons 18 or older a 20-valent conjugate vaccine is recommended. This vaccine contains the same capsule types as the 13-valent vaccine plus capsule types 8, 10A, 11A, 12F, 15B, 22F and 33F.

SUMMARY

1. WHAT IS THE CAUSATIVE AGENT, HOW DOES IT ENTER THE BODY, AND HOW DOES IT SPREAD A) WITHIN THE BODY AND B) FROM PERSON TO PERSON?

- *S. pneumoniae* is the most common cause of community-acquired bacterial pneumonia.
- *S. pneumoniae* is a gram-positive, catalase-negative, bullet-shaped diplococcus.
- *S. pneumoniae* grows readily on blood agar incubated at 37°C in the presence of 5% CO_2 for 24–48 hours, forming small, mucoid colonies surrounded by a zone of incomplete hemolysis (α-hemolysis).
- *S. pneumoniae* has a typical gram-positive-type cell wall that features a thick peptidoglycan layer with covalently bound teichoic acid.
- *S. pneumoniae* undergoes autolysis releasing cell-wall fragments that are pro-inflammatory.
- The capsule is the most important virulence factor of *S. pneumoniae* and there are over 90 antigenically distinct types.
- *S. pneumoniae* has many determinants of pathogenesis including a capsule that is antiphagocytic, adhesins, cytotoxins (pneumolysin and H_2O_2), pro-inflammatory cell-wall fragments (teichoic acid, lipoteichoic acid, and peptidoglycan fragments), and IgA1 protease to subvert mucosal immunity.
- *S. pneumoniae* is spread person-to-person via respiratory droplets.
- There is a high rate of carriage in healthy persons, particularly children.
- *S. pneumoniae* adheres to and is taken up by respiratory epithelial cells from where it can seed the bloodstream and central nervous system.

2. WHAT IS THE HOST RESPONSE TO THE INFECTION AND WHAT IS THE DISEASE PATHOGENESIS?

- The barrier epithelium, in concert with innate immune factors at these surfaces, is an effective barrier to the penetration of *S. pneumoniae*.

- Secretory immunoglobulin A (sIgA) antibodies in respiratory mucus are important in acquired specific immunity at the mucosal surface.
- Host phagocytes are a second line of defense against *S. pneumoniae* invasion.
- *S. pneumoniae* are opsonized by activation of the alternative and lectin innate complement pathways and the classical pathway in the presence of anti-capsular antibodies in the plasma and tissue fluid.
- The virulence of *S. pneumoniae* results from its ability to subvert sIgA antibodies, adhere to and will be taken up by host cells, and to avoid opsonization and phagocytosis by means of its capsule.
- The cytotoxins pneumolysin and H_2O_2 lyse host cells.
- *S. pneumoniae* readily undergoes autolysis releasing pneumolysin but also fragments of the cell wall (teichoic acid, lipoteichoic acid, and peptidoglycan) that are strongly pro-inflammatory.

3. WHAT IS THE TYPICAL CLINICAL PRESENTATION AND WHAT COMPLICATIONS CAN OCCUR?

- Pneumonia with a sudden onset comprising shaking chills, fever to 41°C, productive cough with blood-tinged or purulent sputum, and pleurisy.
- *S. pneumoniae* is also a common cause of sinusitis and otitis media.
- Bacteremia and meningitis are serious complications.

4. HOW IS THE DISEASE DIAGNOSED, AND WHAT IS THE DIFFERENTIAL DIAGNOSIS?

- Microscopy: examination of sputum, blood, cerebrospinal fluid or aspirates stained by Gram's method can provide a rapid preliminary diagnosis by revealing gram-positive, bullet-shaped diplococci.
- Culture: clinical specimens are plated onto blood agar to grow the characteristic small-to-medium, mucoid colonies surrounded by a zone of α-hemolysis.
- Determinative tests: susceptible to optochin or bile.

Continued...

...continued

- Antigen detection: capsule typing can be performed using type-specific antisera.

5. HOW IS THE DISEASE MANAGED AND PREVENTED?

- If the strain is susceptible, penicillin is the antibiotic of choice in patients that are not hypersensitive.
- Over the last decade, *S. pneumoniae* has become increasingly resistant to penicillin and other antibiotics.
- Cefotaxime, ceftriaxone, and clindamycin are effective antibiotics for treating pneumonia caused by penicillin-resistant pneumococci susceptible to these antibiotics.

- Clindamycin or vancomycin is recommended for cefotaxime- or ceftriaxone-resistant pneumococci.
- There are currently two vaccines approved for use in the US. The unconjugated vaccine for adults is only 60% effective in the elderly and is ineffective in children. The newly developed 13-valent conjugate vaccine reduces invasive infection in children yet has little effect on the incidence of otitis media and colonization. A 20-valent conjugate vaccine is recommended for persons 18 year or older.
- Pneumococcal disease is complicated by an increasing incidence (currently 30% in the US) of clinical isolates resistant to commonly used antibiotics.

FURTHER READING

Cillóniz C, Garcia-Vidal C, Ceccato A, Torres A. Antimicrobial Resistance Among Streptococcus pneumoniae. In: Fong IW, David Shlaes D, Drlica K, editors. Antimicrobial Resistance in the 21st Century. Springer, Cham/New York, 13–38, 2018.

Fischetti VA, Novick RP, Ferretti JJ, et al. Gram-Positive Pathogens, 3rd edition. ASM Press, Washington, DC, 2019.

Goering R, Dockrell H, Zuckerman M, Chiodini PL. Mims' Medical Microbiology and Immunology, 6th edition. Elsevier, Philadelphia, 2018.

Murray PR, Rosenthal KS, Pfaller MA. Medical Microbiology, 9th edition. Elsevier/Mosby, Philadelphia, 2021.

Torres A, Cillóniz C, Blasi F, et al. Burden of Pneumococcal Community-Acquired Pneumonia in Adults Across Europe: A Literature Review. Respir Med, 137: 6–13, 2018.

Vernet G, Saha S, Satzke C, et al. Laboratory-Based Diagnosis of Pneumococcal Pneumonia: State of the Art and Unmet Needs. Clin Microbiol Infect, 17: 1–13, 2011.

Wahl B, O'Brien KL, Greenbaum A, et al. Burden of Streptococcus pneumoniae and Haemophilus influenzae Type B Disease in Children in the Era of Conjugate Vaccines: Global, Regional, and National Estimates for 2000–15. Lancet Glob Health, 6: e744–e757, 2018.

Weiser JN, Ferreira DM, Paton JC. Streptococcus pneumoniae: Transmission, Colonization and Invasion. Nat Rev Microbiol, 16: 355–367, 2018.

REFERENCES

Brooks LRK, Mias GI. Streptococcus Pneumoniae's Virulence and Host Immunity: Aging, Diagnostics, and Prevention. Front Immunol, 9: 1366, 2018.

Chalmers JD, Campling J, Dicker A, et al. A Systematic Review of the Burden of Vaccine Preventable Pneumococcal Disease in UK Adults. BMC Pulm Med, 16: 77, 2016.

Dupont C, Michon A-L, Normandin M, et al. Streptococcus pseudopneumoniae, an Opportunistic Pathogen in Patients with Cystic Fibrosis. J Cyst Fibros, 19: e28–e31, 2020.

Geno KA, Gilbert GL, Song JY, et al. Pneumococcal Capsules and Their Types: Past, Present, and Future. Clin Microbiol Rev, 28: 871–899, 2015.

Kilian M, Riley DR, Jensen A, et al. Parallel Evolution of Streptococcus pneumoniae and Streptococcus mitis to Pathogenic and Mutualistic Lifestyles. mBio, 5: e01490–14, 2014.

Musher DM, Abers MS, Bartlett JG. Evolving Understanding of the Causes of Pneumonia in Adults, with Special Attention to the Role of Pneumococcus. Clin Infect Dis, 65: 1736–1744, 2017.

Novick S, Shagan M, Blau K, et al. Adhesion and Invasion of Streptococcus pneumoniae to Primary and Secondary Respiratory Epithelial Cells. Mol Med Rep, 6: 65–74, 2016.

WEBSITES

Centers for Disease Control and Prevention, Gierke R, Wodi AP, Kobayashi M, Pneumococcal Disease, 2021: https://www.cdc.gov/vaccines/pubs/pinkbook/pneumo.html

Dion CF, Ashurst JV, Streptococcus pneumoniae, StatPearls, 2022: https://www.ncbi.nlm.nih.gov/books/NBK470537/

European Centre for Disease Prevention and Control, Invasive pneumococcal disease: https://www.ecdc.europa.eu/en/invasive-pneumococcal-disease

Medscape, Sanchez, E, Pneumococcal Infections (Streptococcus pneumoniae), 2020: https://emedicine.medscape.com/article/225811-overview

NHS Inform, Pneumococcal infections, 2022: https://www.nhsinform.scot/illnesses-and-conditions/infections-and-poisoning/pneumococcal-infections

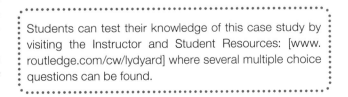

Students can test their knowledge of this case study by visiting the Instructor and Student Resources: [www.routledge.com/cw/lydyard] where several multiple choice questions can be found.

Streptococcus pyogenes

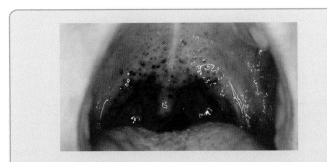

35

Figure 35.1 Pharynx of the patient showing punctate hemorrhages and inflamed tonsils caused by *Streptococcus pyogenes*. *From the Centers for Disease Control & Prevention, Atlanta, Georgia. Image is found in the Public Health Image Library #3185. Additional photographic credit is given to Dr Heinz F Eichenwald who took the photo in 1958.*

A 7-year-old boy was well until yesterday when he developed **dysphagia**, painful anterior lymph nodes, and a fever of 40°C. The patient vomited once and complained of a headache. Physical examination showed an acutely ill patient with a temperature of 39°C and a pulse of 104 beats per minute. Examination of his head, eyes, ear, nose, and throat revealed bilateral tonsillar **hypertrophy** with grayish-white exudates and punctate hemorrhages (Figure 35.1). Bilateral tender submandibular lymph nodes were palpated. The remainder of the physical examination was within normal limits. A throat swab was obtained. Laboratory findings were: hemoglobin, normal; hematocrit, normal; WBC count, 19 000 mm³; differential, 80% PMN, 4% bands, 15% lymphocytes.

1. WHAT IS THE CAUSATIVE AGENT, HOW DOES IT ENTER THE BODY AND HOW DOES IT SPREAD A) WITHIN THE BODY AND B) FROM PERSON TO PERSON?

CAUSATIVE AGENT

The patient has acute **pharyngitis**. The vast majority of cases of acute pharyngitis are caused by viruses, but the principal causative bacterial agent is *Streptococcus pyogenes*, also called the group A streptococcus or GAS. *S. pyogenes* is a gram-positive, catalase-negative coccus that grows in chains. A gram stain of streptococci is shown in Figure 35.2. As for other streptococci, the **catalase** test is used to distinguish them from staphylococci, which are the other medically important genera of gram-positive cocci (see Figure 33.11 in the *Staphylococcus aureus* case [Case 33] for the catalase test). Streptococci grow readily on blood agar incubated at 37°C in the presence of 5% CO₂ for 24–48 hours. GAS form small, gray opalescent colonies and unlike *S. pneumoniae* and *S. mitis*, which show α- (incomplete) **hemolysis**, GAS colonies are surrounded by a large zone of complete hemolysis (β-hemolysis) (Figure 35.3). GAS has a typical gram-positive-type cell wall that features a thick peptidoglycan layer with covalently bound teichoic acid (Figure 35.4). There are a number of molecules exposed on the cell surface. These are either anchored in the cytoplasmic

membrane and traverse the cell wall to the outside or they are anchored in the wall and extend from it. Among the membrane-anchored molecules are M protein and lipoteichoic acid. M protein is the most important **virulence** factor of GAS. M protein is an α-helical coiled-coil fibrillar protein. The amino-acid sequence of the extracellular portion of the molecule is highly variable, giving rise to over 200 serotypes of M protein (see Section 2). The various serotypes are referred to as "M" types. There are two classes of M protein, class I and II. M types with class I M proteins share surface-exposed

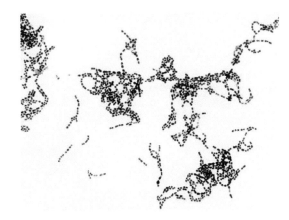

Figure 35.2 Gram stain of *Streptococcus pyogenes*. The streptococci form long chains and appear as purple beads. This form of growth is most obvious when the bacteria are obtained from liquid samples. *From Eye of Science / Science Photo Library, with permission.*

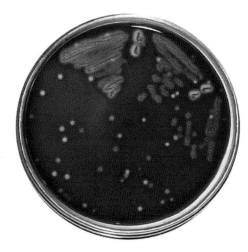

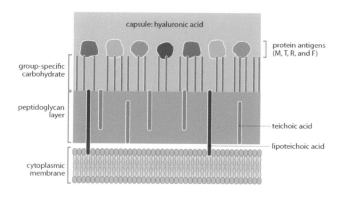

Figure 35.3 *Streptococcus pyogenes* inoculated onto trypticase soy agar containing 5% defibrinated sheep's blood, that is blood agar plate (BAP) that had been "streaked", and "stabbed" with a bacteriologic loop. A clear, colorless region surrounding each colony reflects the fact that the red blood cells in the blood agar medium have been destroyed, or hemolyzed, and indicates that these bacteria are β-hemolytic. Hemolysis is caused by two exotoxins termed streptolysin O and streptolysin S. Streptolysin O is oxygen-labile and so the inoculum is stabbed below the surface of the agar where the oxygen tension is reduced and both hemolysins are active. Stabs are shown at the 12:00 and 2:00 positions. *From the Centers for Disease Control & Prevention, Atlanta, Georgia. Image is found in the Public Health Image Library #8170. Additional photographic credit is given to Richard R Facklam, PhD, who took the photo in 1977.*

Figure 35.4 Schematic of the GAS envelope.

antigenic epitopes and do not produce opacity factor (OF), whereas M types with class II M proteins lack shared epitopes and produce OF. The importance of the division of M types into two classes reflects their differing propensity to cause secondary complications. While both classes cause **suppurative infections** and **glomerulonephritis**, only strains with class I M proteins cause rheumatic fever (see complications later). Another surface appendage, important in typing GAS strains, is the T antigen of which there are about 20 types. T antigens form the backbone of pilus shafts that extend from the cell surface and may be involved in adhesion and invasion. It is interesting that the genes encoding the pili are found on a pathogenicity island (large genomic regions 10–200 kb that are found in pathogenic bacteria but not in non-pathogenic bacteria of the same or closely related species which are acquired by horizontal gene transfer) that also encodes the genes for other extracellular matrix binding adhesins such as fibronectin-binding protein. Lipoteichoic acid is an important pro-inflammatory molecule and contributes to adherence of GAS by binding fibronectin on host cells. This, together with a variety of cell surface molecules including the **adhesins**, F protein (fibronectin-binding protein), glyceraldehyde 3-phosphate dehydrogenase (G3-PD), and enolase (binds plasminogen), are involved in attachment of the bacteria to the epithelium and therefore contribute to pathogenesis (Table 35.1). Some GAS strains produce a capsule composed of hyaluronic acid. The structure of the capsule is identical to that of the mammalian intercellular matrix, thus disguising the organism.

ENTRY AND SPREAD WITHIN THE BODY

GAS enter the body via the **oropharynx** or skin and are very adept at adhering to epithelial cells via the adhesin molecules described above. GAS transiently colonizes the pharynx and skin and carriage of the bacterium is reported to be as high as 20%. GAS is pathogenic when the organism invades and penetrates the epithelial surface or localizes in deeper subcutaneous tissues (see Section 3). The adhesins, particularly M protein and F protein, enable GAS to invade epithelial surfaces.

GAS can seed the lymphatics and bloodstream. What causes the organism to become invasive following a local infection is still not fully understood. However, clotting factors and the level of the cysteine protease (formally known as pyrogenic **exotoxin** SpeB) might be important.

PERSON-TO-PERSON SREAD

GAS is spread person-to-person via respiratory droplets. Crowded environments such as classrooms and daycare centers facilitate spread. In the case of skin infections, GAS is spread via the hands or by **fomites**. Spread via the hands can result in auto-inoculation, that is, spread of the organism to additional parts of the body as well as spread to other persons. This is well illustrated in the case of **pyoderma (impetigo)**, a highly contagious superficial skin infection seen in young children in daycare or kindergarten settings.

EPIDEMIOLOGY

There is a paucity of current data on the epidemiology of *S. pyogenes* (GAS) disease. Unfortunately, reviews published within the last few years cite data obtained in the early 2000s. *S. pyogenes* is reported to cause 15–30% of the pharyngitis cases in children and 5–20% cases in adults out of approximately 11 million pharyngitis cases in the US each year. In Australia, the incidence of culture-positive *S. pyogenes* pharyngitis has been estimated at 13 per 100 person-years whereas the incidence of *S. pyogenes* pharyngitis in less developed countries may be

Table 35.1 Factors contributing to virulence of *Streptococcus pyogenes*

Factor	Action
Capsule	Antiphagocytic
M protein	Antiphagocytic, adhesin, degrades complement C3b,
T protein	Adhesin
Lipoteichoic acid	Adhesin
F protein	Adhesin (ligand fibronectin)
G3-PD	Adhesin
Enolase	Adhesin
Serum opacity factor	Adhesion (ligand fibronectin)
M-like proteins	Bind IgM and IgG, antiphagocytic
Pyrogenic exotoxins	Fever, superantigen, enhances endotoxin activity, shock, scarlet fever
Streptolysin S	Cytotoxin
Streptolysin O	Cytotoxin
Streptokinase	Fibrinolysis
DNase	Hydrolyzes DNA; degrades neutrophil extracellular traps (NETS)
C5a peptidase	Inactivates chemotactic factor C5a
Hyaluronidase	Spreading factor

G3-PD, glyceraldehyde 3-phosphate dehydrogenase.

five to ten times that number. CDC estimates approximately 11000 to 24000 cases of invasive GAS disease occur each year in the US with between 1200 and 1900 deaths. Invasive GAS disease includes cellulitis with blood infection, necrotizing fasciitis, pneumonia, and streptococcal toxic shock syndrome. In developed countries, the incidence of scarlet fever has fallen dramatically over the years with sporadic reports of increased incidence and large-scale outbreaks. A similar decline in **acute rheumatic fever** (**ARF**) is believed to be due to improvements in primary prevention, including access to healthcare and use of antibiotics. In contrast, GAS diseases are highly prevalent in developing countries, and among indigenous populations and low socioeconomic areas in developed countries.

2. WHAT IS THE HOST RESPONSE TO THE INFECTION AND WHAT IS THE DISEASE PATHOGENESIS?

The virulence of GAS results from its ability to adhere to and invade host cells and to avoid **opsonization** and **phagocytosis**. The skin and mucous membranes in concert with innate immune factors at these surfaces are effective barriers to the penetration of GAS. However, GAS are able to invade epithelial cells. It appears the fibronectin-binding proteins and M protein are important co-operative invasins, but it is clear that other surface adhesion molecules listed in Table 35.1 are implicated. Fibronectin may serve as a bridging molecule between the bacterial surface and the α5b1 integrin on the host-cell membrane. Adhesion of GAS to the epithelial cell surface induces cytoskeletal reorganization and cell-membrane ruffling and streptococci appear to be taken up by a zippering mechanism and enter the endosomal pathway.

However, they are able to escape the early **endosome**, perhaps as a result of the action of the pore-forming cytolysin, streptolysin O. Within the cytosol GAS are combated by being enveloped by autophagosomes and killed when autophagosomes fuse with **lysosomes**. Thus, **autophagy** may be a mechanism by which non-phagocytic epithelial cells protect themselves against GAS invasion.

Host phagocytes are a second line of defense against streptococcal invasion. Streptococci are opsonized by activation of the **alternate** and lectin innate **complement pathways** and the **classical** pathway in the presence of anti-M protein antibodies in the plasma and tissue fluid. The hyaluronic **capsule** is poorly immunogenic, antiphagocytic, and serves to mask cell-surface antigens from host immunity. Only M protein extends beyond the capsule. M protein binds factor H, a regulatory protein of the alternative pathway of complement, which degrades the complement component C3b, which is a potent opsonin. Also, by binding fibrinogen, M protein blocks deposition of C3b. In addition, GAS produces a serine protease that inactivates the complement component C5a, which is a powerful chemoattractant for neutrophils and macrophages.

GAS secretes several exotoxins and enzymes that play a role in pathogenesis and are listed in Table 35.1. To date, some 14 pyrogenic exotoxins (superantigens) have been found in *S. pyogenes*. The pyrogenic exotoxins (SpeA, C, H-M) are phage-specified. **Superantigens** bring about the release of **cytokines** (**interleukin (IL)**-1, IL-2, IL-6, **tumor necrosis factor-α** (**TNF-α**), **TNF-β**, **interferon-γ** (**IFN-γ**)) from macrophages and CD4 T cells. These cytokines mediate **shock** and organ failure characteristic of streptococcal **toxic shock syndrome** and give rise to the **rash** associated with scarlet fever. Streptolysins S and O are **hemolysins** responsible for the β-hemolysis of GAS when it is grown on blood agar. However, they also lyse leukocytes, platelets, and likely other host cells. GAS produces three enzymes that facilitate the spread of the organism; hyaluronidase hydrolyzes the intercellular matrix, DNase depolymerizes extracellular DNA reducing the viscosity of **pus** and degrades the neutrophil extracellular traps (NETs), and streptokinase lyses blood clots aiding bloodstream spread. Finally, the immunoglobulin-binding M-like proteins function in blocking phagocytic activity and also degrade complement component C3b.

As mentioned above, it is clear that GAS has several strategies to invade non-phagocytic cells, such as the utilization of the fibronectin-binding protein to bind fibronectin, which is used as a bridging molecule to bind to the α5b1 integrin cellular receptor. However, the invasive phenotype of GAS as seen in the highly virulent sub-clone of serotype M1T1, a major cause of **necrotizing fasciitis**, may be related to the accumulation of plasminogen, fibrinogen, and streptokinase on the cell surface. Down-regulation of a cysteine protease (CP) formally termed pyrogenic exotoxin SpeB appears to favor cell-surface accumulation of these factors. The CP cleaves host proteins and other GAS determinants of pathogenesis among which are plasminogen and streptokinase. So, although the CP is an important determinant of pathogenesis early in the infection cycle, its down-regulation favors invasion of GAS.

GAS expresses several surface-associated immunoglobulin-binding proteins such as M protein, M-related proteins, and M-like proteins. These proteins bind the Fc region of IgA and IgG. GAS also expresses enzymes that can cleave and deglycosylate IgG.

3. WHAT IS THE TYPICAL CLINICAL PRESENTATION AND WHAT COMPLICATIONS CAN OCCUR?

GAS is a versatile pathogen capable of causing a wide spectrum of diseases. It is convenient to divide them into local infections with GAS and its products, invasive infections, and post-group A streptococcal disease.

LOCAL INFECTIONS

Pharyngitis occurs 24–48 hours post-exposure with sudden onset of sore throat, malaise, fever, and headache. The pharynx and tonsils may be **erythematous** with creamy, yellow exudates (see Figure 35.1). There may be cervical **lymphadenopathy**. Complications of streptococcal pharyngitis are scarlet fever, ARF and acute glomerulonephritis (AGN). If the infecting GAS strain is lysogenized by a temperate **bacteriophage** that specifies production of pyrogenic exotoxins, an erythematous rash beginning on the trunk and spreading to the extremities develops 1–2 days following the onset of pharyngitis (Figure 35.5A). The rash spares the palms and soles and the skin around the mouth (circumoral pallor). Initially, the tongue is covered

with a white coating, which is lost to reveal a red, raw surface termed "strawberry tongue" (Figure 35.5B). After about one week, the rash fades and is replaced by desquamation (Figures 35.5C and D). It has been found that about three-quarters of pharyngeal isolates of GAS from children are positive for the pyrogenic exotoxins, speA and speC. Most invasive pediatric GAS strains are identical to acute pharyngitis strains; thus, childhood pharyngitis serves as a major reservoir for strains with invasive potential.

ARF, which can be a nonsuppurative sequela of GAS pharyngitis, is considered below.

Pyoderma (impetigo) is a highly contagious, superficial infection of exposed skin, typically the face, arms, and legs, seen most frequently in young children. It is often a mixed infection of GAS and *Staphylococcus aureus*. GAS enters the subcutaneous tissue via a breach in the integrity of the skin. The course of infection begins as **vesicles**, which progress to pustules that rupture and are replaced by honey-colored crusts (see *Staphylococcus aureus* case, Case 33, Figure 35.2). A complication of impetigo is AGN which will be discussed below.

INVASIVE INFECTIONS

Erysipelas is an acute infection of the skin accompanied by lymphadenopathy, fever, chills, and leukocytosis. The painful, inflamed skin is raised and clearly demarcated from the surrounding healthy skin (Figure 35.6). Although it can occur on any part of the body, the legs are a frequent site of infection due to venous insufficiency and stasis ulcerations.

Cellulitis is an infection similar in nature to erysipelas except that it involves not only the skin but the connective tissues (Figure 35.7).

Necrotizing fasciitis is a deep infection of the connective tissue that spreads along fascial planes and destroys muscle and fat. The bacterium enters through a trivial break in the skin. The course of the infection is rapid, often beginning with severe pain without evidence of injury or wound. Over a matter of hours there is swelling and the appearance of a spreading red or dusky blue skin discoloration, often with fluid-filled **bullae** (Figure 35.8). Flu-like symptoms such as diarrhea, nausea, fever, confusion, dizziness, and weakness are also apparent.

Streptococcal toxic shock syndrome (STSS) often follows necrotizing fasciitis with the disease progressing to shock and organ failure. Patients with STSS are **bacteremic** in contradistinction to patients with staphylococcal toxic shock syndrome. M1 and M3 serotypes are strongly associated with necrotizing fasciitis and STSS. These M types are heavily capsulate and produce SpeA and smeZ.

POST-STREPTOCOCCAL SEQUELAE

ARF is an inflammatory disease of the connective tissue, particularly the heart, joints, blood vessels, subcutaneous tissue, and CNS that occurs about 3 weeks after GAS

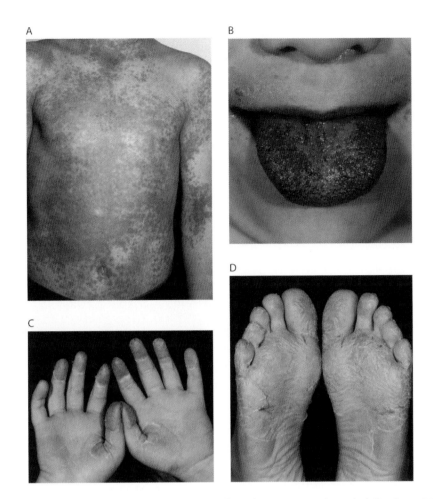

Figure 35.5 (A) Scarlet fever rash on a boy's torso and arms. (B) Strawberry tongue characteristic of scarlet fever. (C) Scarlet fever showing desquamation of palms and fingers. (D) Scarlet fever showing desquamation of the soles of the feet. *From Bio-Photo Associates / Science Photo Library, with permission*

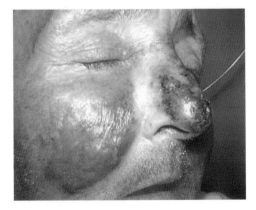

Figure 35.6 Facial erysipelas manifested as severe malar and nasal erythema and swelling. Erysipelas is a dermatologic condition, which involves the inoculation of the skin and subcutaneous tissue with GAS causing edema and bright red erythema of the affected areas. Erysipelas can be differentiated from cellulitis by its characteristically raised advancing edges and sharply demarcated borders, reflecting its more superficial nature. Cellulitis has no lymphatic component and exhibits nondiscrete margins. *From the Centers for Disease Control & Prevention, Atlanta, Georgia. Image is found in the Public Health Image Library #2874. Additional photographic credit is given to Dr Thomas F Sellers and Emory University who took the photo in 1963.*

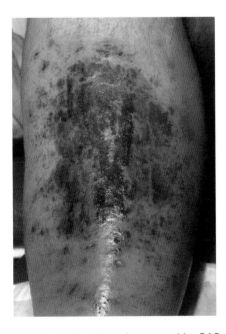

Figure 35.7 Cellulitis of the lower leg caused by GAS. *From Dr P Marazzi / Science Photo Library, with permission.*

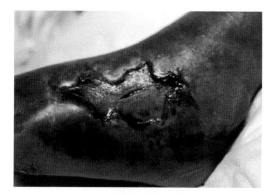

Figure 35.8 **Necrotizing fasciitis.** The course of the infection is rapid, often beginning with severe pain without evidence of injury or wound. Over a matter of hours, there is swelling and the appearance of a spreading red or dusky blue skin discoloration, often with fluid-filled bullae. *From Diseases of the Skin, 2nd edition, found at the following website: http://www.merckmedicus.com/ppdocs/us/hcp/content/white/chapters/white-ch-024-s002.htm. Permission from Elsevier.*

pharyngitis in susceptible subjects. The attack rate is about 3% in untreated pharyngitis. All layers of the heart are involved and there is chronic damage to the valves, particularly the mitral valve. Also, there is a migratory **arthritis**, subcutaneous nodules, a serpiginous, flat, painless rash (erythema marginatum), and chorea (Sydenham's chorea). ARF is an autoimmune disease in which antibodies induced against streptococcal antigens cross-react with human tissues. The basis for this cross-reactivity lies in the coiled-coil nature of M protein and its homology with tropomyosin, myosin, keratin, laminin, desmin, vimentin, and other coiled-coil proteins as well as the hyaluronic acid capsule, N-acetyl-glucosamine (Group A polysaccharide Lancefield antigen), a 60 kDa wall-membrane antigen and a 67 kD Antigen. Also, there may be cross-reactivity with sarcolemmal membranes. ARF is associated with M types 1, 3, 5, 6, and 18.

AGN is a second autoimmune disease that can follow either pharyngeal or skin infection with certain GAS M types. The target organ is the kidney, resulting in inflammation of the glomeruli, **edema**, hypertension, **hematuria**, and proteinuria. The M types causing AGN are termed nephritogenic and the pharyngeal nephritogenic M types are different from the skin nephritogenic M types. In the case of AGN, the autoantigens are glomerular heparin-sulfate proteoglycan and basement membrane type IV collagen and laminin. In addition, **immune complexes** of streptococcal antigens deposit in the glomeruli activating the complement cascade.

Sydenham Chorea and Pediatric Autoimmune Neuropsychiatric Disorder Associated with Streptococcal Infection (PANDAS)

As mentioned above, Sydenham chorea is a post-GAS sequela that occurs in up to 40% of patients with rheumatic fever. Monoclonal antibodies, serum, and CSF antibodies from patients with chorea react with lysoganglioside and target the surface of human basal ganglia. The GAS antigen inducing such antibodies appears to be N-acetyl-β-D-glucosamine.

Other cross-reactive antigens react with tubulin and, importantly, cross-reactive antibodies have been shown to react with and signal the dopamine receptors D1 and D2. The National Institute of Mental Health (https://www.nimh.nih.gov/health/publications/pandas/) defines PANDAS as obsessive-compulsive disorder (OCD), tic disorder, or both that suddenly appear following a streptococcal infection, such as streptococcal pharyngitis throat or scarlet fever, or symptoms of OCD or tic symptoms suddenly become worse following a streptococcal infection. PANDAS remains controversial but there is evidence that similar anti-neuronal autoantibodies and autoantibody mediated neuronal cell signaling are elevated in serum and CSF from PANDAS in Sydenham chorea.

4. HOW IS THE DISEASE DIAGNOSED, AND WHAT IS THE DIFFERENTIAL DIAGNOSIS?

DIAGNOSIS

Microscopy: Examination of infected tissues or body fluids stained by Gram's method can provide a rapid preliminary diagnosis.

Culture: Swabs from the posterior pharynx, undisturbed crusted pustules, tissue, pus, blood, and so forth, are plated onto blood agar, which is incubated for 24–48 hours at 37°C in the presence of 5% CO_2. The first sector of the inoculated plate should be stabbed to promote β-hemolysis (see Figure 35.3).

Determinative tests: GAS are uniformly susceptible to the antibiotic bacitracin. A bacitracin disc placed on a blood agar plate inoculated with GAS will show a wide zone of growth inhibition (Figure 35.9). Lancefield group C and G streptococci may also exhibit a large zone of β-hemolysis; however, GAS can be distinguished from these groups because it produces the enzyme L-pyrrolidonyl arylamide (PYR) that can be detected by a rapid (2 minute) disc test (Figure 35.10).

Antigen detection: Rapid tests are available to directly detect GAS on throat swabs from an individual suspected of having streptococcal pharyngitis. These tests are based on immunologic detection of the Lancefield group A wall carbohydrate antigen. The antigen is extracted from the swab using acid or an enzyme and is detected by antibody immobilized on latex beads (latex agglutination) or on a membrane (Figure 35.11). These tests are highly specific and sensitive if performed correctly. However, a negative rapid test must be confirmed by culture.

Serology: Individuals infected with GAS produce antibodies to many components of the bacterial envelope and proteins secreted by the organism. Immunity is mediated by antibodies to M protein and is M type-specific. Antibody against streptolysin O (the ASO titer) is a useful marker for confirming acute rheumatic fever or acute glomerulonephritis resulting from GAS pharyngitis. Because streptolysin O is inhibited by cholesterol in the skin, patients with AGN following GAS pyoderma do not exhibit an elevated ASO titer. In this case, elevated antibodies against DNase B are a useful indicator.

DIFFERENTIAL DIAGNOSIS

Pharyngitis

- Viral pharyngitis: rhinovirus, adenovirus, herpes simplex virus, influenza virus, parainfluenza virus, coronavirus, enterovirus, respiratory syncytial virus (RSV), cytomegalovirus, HIV.
- Infectious mononucleosis: Epstein-Barr virus (EBV).
- Peritonsillar abscess.
- Pharyngitis caused by groups C and G streptococci.

A

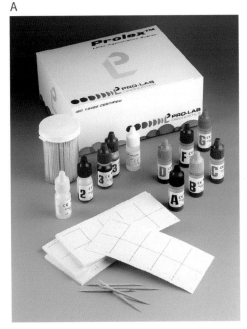

B

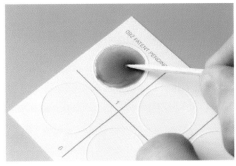

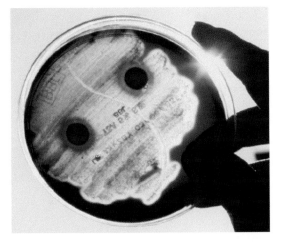

Figure 35.9 Bacitracin sensitivity of *S. pyogenes*. The image shows a blood agar plate inoculated with *S. pyogenes* on which two bacitracin disks have been placed. Growth of the GAS is shown by the clear area of the agar in which hemolysins produced by GAS have lysed the red blood cells. GAS are uniformly sensitive to bacitracin. This is shown by the zone of blood agar surrounding each disk in which the red blood cells are intact. *From Van Bucher / Science Photo Library, with permission.*

Figure 35.11 A latex agglutination kit used in a hospital pathology laboratory for grouping (identifying) hemolytic streptococci. *Reprint permission kindly given by Mark Reed and Pro-Lab Diagnostics, www. pro-lab.com.*

Figure 35.10 PYR test. Hydrolysis of 4-pyrrolidonyl-a-naphthylamide (PYR) is a presumptive test for identification of group A streptococci and *Enterococcus* species. In the disk test shown, colonial growth is applied to the moistened disk containing the PYR substrate. After 2 minutes, dimethylaminocinnamaldehyde reagent is applied to the disk to detect the free α-naphthylamine that is released on hydrolysis of PYR (red color). *Reprint permission kindly given by Jay Hardy, President of Hardy Diagnostics, www.Hardy.Diagnostics.com.*

Pyoderma (Impetigo)

- Atopic dermatitis.
- Bullous pemphigoid.
- Candidiasis, cutaneous.
- Contact dermatitis, allergic and irritant.
- Herpes simplex.
- Insect bites.
- Pemphigus foliaceus.
- Pemphigus vulgaris.
- Scabies.
- Staphylococcal scalded skin syndrome.
- Thermal burns.
- Tinea pedis.

Erysipelas

- None.

Cellulitis

- Angioedema.
- Chemical burns.
- Dermatitis: atopic, contact, and exfoliative.
- Erythema multiforme.
- Gas gangrene.
- Impetigo.

Necrotizing Fasciitis

- Cellulitis.
- Gas gangrene.
- Toxic shock syndrome.

Streptococcal Toxic Shock Syndrome

- Staphylococcal infections (staphylococcal toxic shock syndrome).
- Cellulitis.
- Clostridial gas gangrene.
- Erythema multiforme (Stevens-Johnson syndrome).
- Meningococcemia.

Acute Rheumatic Fever

- Aortic regurgitation.
- Atrial fibrillation.
- Endocarditis.
- Huntington chorea.
- Juvenile rheumatoid arthritis.
- Leukemia.
- Mitral regurgitation.
- Mitral stenosis.
- Myocarditis.

- Pediatrics: Kawasaki disease, scarlet fever.
- Scarlet fever.

Acute Glomerulonephritis

- Post-infectious bacterial etiologies: *S. pneumoniae, Staphylococcus species, Mycobacterium species, Salmonella typhosa, Brucella suis, Treponema pallidum, Corynebacterium bovis,* and *Actinobacillus species.*
- Post-infectious viral etiologies: cytomegalovirus, coxsackie virus, EBV, hepatitis B, rubella, rickettsial scrub typhus, mumps.
- Post-infectious fungal and parasitic etiologies: *Coccidioides immitis* and the following parasites: *Plasmodium malariae, Plasmodium falciparum, Schistosoma mansoni, Toxoplasma gondii,* filariasis, trichinosis, and trypanosomes.
  Systemic causes of glomerular nephritis:
  - Wegener granulomatosis.
  - Hypersensitivity vasculitis.
  - Cryoglobulinemia.
  - Systemic lupus erythematosus.
  - Polyarteritis nodosa.
  - Henoch-Schönlein purpura.
  - Goodpasture syndrome.
  - Renal diseases.
  - Membranoproliferative glomerulonephritis.

5. HOW IS THE DISEASE MANAGED AND PREVENTED?

MANAGEMENT

GAS remains sensitive to penicillin and this antibiotic is the treatment of choice in patients who are not hypersensitive. In this case, erythromycin or a cephalosporin can be substituted. However, it should be noted that during the last few years, erythromycin-resistant *S. pyogenes* has been reported in different parts of the world. Streptococcal pharyngitis is self-limiting, however, antibiotic therapy is indicated because it prevents the development of rheumatic fever but, interestingly, it does not appear to prevent the development of acute glomerulonephritis. Serious soft-tissue infections require drainage and surgical debridement as the first line of therapy.

PREVENTION

In April 2007 (revised 2021), the American Heart Association updated its guidelines for prevention of endocarditis and concluded that there is no convincing evidence linking dental, GI or genitourinary tract procedures with the development of endocarditis. The prophylactic use of antibiotics prior to a dental procedure is now recommended only for those patients with the highest risk of adverse outcome resulting from endocarditis, such as patients with a prosthetic cardiac valve, previous endocarditis, or those with specific forms of congenital heart disease. The guidelines no longer recommend prophylaxis prior to a dental procedure for patients with rheumatic heart disease unless they also have one of the underlying cardiac conditions listed above. Antibiotic prophylaxis solely to prevent bacterial endocarditis is no longer recommended for patients who undergo a GI or genitourinary tract procedure. (See *Streptococcus mitis* case, eCase 8, for prophylactic recommendations for patients with ARF).

SUMMARY

1. WHAT IS THE CAUSATIVE AGENT, HOW DOES IT ENTER THE BODY, AND HOW DOES IT SPREAD A) WITHIN THE BODY AND B) FROM PERSON TO PERSON?

- *S. pyogenes*, also known as the group A streptococcus (GAS), is the most common cause of bacterial pharyngitis.
- The vast majority of cases (about 70%) of pharyngitis are caused by viruses.
- GAS is a gram-positive, catalase-negative coccus that grows in chains.
- GAS grows readily on blood agar incubated at 37°C in the presence of 5% CO_2 for 24–48 hours, forming small, gray opalescent colonies surrounded by a large zone of complete hemolysis (β-hemolysis).
- GAS has a typical gram-positive-type cell wall that features a thick peptidoglycan layer with covalently bound teichoic acid.
- M protein is the most important virulence factor of GAS and there are about 100 types.
- GAS has many determinants of pathogenesis that include lipoteichoic acid (pro-inflammatory), adhesins, superantigens (pyrogenic exotoxins), cytotoxins (streptolysin O and S), and a capsule.
- GAS is spread person-to-person via respiratory droplets, the hands, and fomites.
- GAS adheres to and invades epithelial cells from where it can invade subcutaneous tissues and seed the lymphatics and bloodstream.

2. WHAT IS THE HOST RESPONSE TO THE INFECTION AND WHAT IS THE DISEASE PATHOGENESIS?

- The barrier epithelia, in concert with innate immune factors at these surfaces, are effective barriers to the penetration of GAS.
- Host phagocytes are a second line of defense against GAS invasion.
- GAS is opsonized by activation of the alternate and lectin innate complement pathways and the classical pathway in the presence of anti-M protein antibodies in the plasma and tissue fluid.
- The virulence of GAS results from its ability to adhere to and invade host cells and to avoid opsonization and phagocytosis by means of capsule, M protein, and C5a peptidase.

- The hemolysins, streptolysin S and O, are cytotoxins that can lyse erythrocytes, leukocytes, and platelets and likely other host cells.
- The pyrogenic exotoxins (erythrogenic toxins) are superantigens that result in the release of pro-inflammatory cytokines that mediate shock and organ failure characteristic of streptococcal toxic shock syndrome and give rise to the rash associated with scarlet fever.

3. WHAT IS THE TYPICAL CLINICAL PRESENTATION AND WHAT COMPLICATIONS CAN OCCUR?

- *Pharyngitis* occurs 24–48 hours post-exposure with sudden onset of sore throat, malaise, fever, headache, and erythematous pharynx and tonsils with creamy, yellow exudates and cervical lymphadenopathy.
- *Scarlet fever* is caused by GAS strains lysogenized by a temperate bacteriophage that specifies production of pyrogenic exotoxin resulting in an erythematous rash, circumoral pallor, and strawberry tongue.
- *Pyoderma (impetigo)* is a highly contagious, superficial infection of exposed skin, typically the face, arms, and legs, characterized by vesicles that progress to pustules that rupture and are replaced by honey-colored crusts.
- *Erysipelas* is an acute infection of the skin, most frequently the legs, accompanied by lymphadenopathy, fever, chills, and leukocytosis in which the infected skin is painful, inflamed, raised, and clearly demarcated.
- *Cellulitis* is an infection similar in nature to erysipelas except that it involves not only the skin but the connective tissues.
- *Necrotizing fasciitis* is a deep infection of the connective tissue that spreads along fascial planes and destroys muscle and fat.
- *Streptococcal toxic shock syndrome (STSS)* often follows necrotizing fasciitis; the disease is characterized by shock and organ failure.
- *Acute rheumatic fever (ARF)* is an autoimmune inflammatory disease of the connective tissue, particularly the heart, joints, blood vessels, subcutaneous tissue, and central nervous system that occurs about 3 weeks after GAS pharyngitis in susceptible subjects.
- *Acute glomerulonephritis (AGN)* is an autoimmune disease that can follow either pharyngeal or skin infection causing inflammation of the glomeruli, edema, hypertension, hematuria, and proteinuria.

Continued...

...continued

4. HOW IS THE DISEASE DIAGNOSED, AND WHAT IS THE DIFFERENTIAL DIAGNOSIS?

- *Microscopy*: Examination of infected tissues or body fluids stained by Gram's method can provide a rapid preliminary diagnosis.
- *Culture*: Swabs from the posterior pharynx, undisturbed crusted pustules, tissue, pus, blood, and so forth, are plated onto blood agar to grow the characteristic colonies surrounded by a large zone of β-hemolysis.
- *Determinative tests*: Susceptible to bacitracin and PYR-positive.
- *Antigen detection*: Rapid tests based on immunologic detection of the Lancefield group A wall carbohydrate antigen are available to directly detect GAS on throat swabs; a negative rapid test must be confirmed by culture.
- *Serology*: Anti-streptolysin O titer (the ASO titer) is a useful marker for confirming acute rheumatic fever or acute glomerulonephritis resulting from GAS pharyngitis; anti-DNase B for AGN following skin infection.

5. HOW IS THE DISEASE MANAGED AND PREVENTED?

- Penicillin is the antibiotic of choice in patients who are not hypersensitive.
- Erythromycin or a cephalosporin can be used in patients allergic to penicillin.
- Antibiotic treatment of GAS pharyngitis prevents the development of rheumatic fever.
- Serious soft-tissue infections require drainage and surgical debridement as the first line of therapy.
- Prophylactic antibiotic therapy is required for several years in individuals who have had rheumatic fever.
- Any person with preexisting damage to their heart valves should receive prophylactic antibiotics before any procedure likely to result in transient bacteremia, such as dental treatment. (See *Streptococcus mitis* case, Case 34.)

FURTHER READING

Ferretti JJ, Stevens DL, Fischetti VA (eds). Streptococcus pyogenes: Basic Biology to Clinical Manifestations. The University of Oklahoma Health Sciences Center, Oklahoma, 2016.

Fischetti VA, Novick RP, Ferretti JJ, et al. Gram-Positive Pathogens, 3rd edition. ASM Press, Washington, DC, 2019.

Goering R, Dockrell H, Zuckerman M, Chiodini PL. Mims' Medical Microbiology and Immunology, 6th edition. Elsevier, Philadelphia, 2018.

Murray PR, Rosenthal KS, Pfaller MA. Medical Microbiology, 9th edition. Elsevier/Mosby, Philadelphia, 2021.

REFERENCES

Avire NJ, Whiley H, Ross K. A Review of Streptococcus pyogenes: Public Health Risk Factors, Prevention and Control. Pathogens, 10: 248, 2021.

Carapetis JR, Steer AC, Mulholland EK, Weber M. The Global Burden of Group A Streptococcal Diseases. Lancet Infect Dis, 5: 685–694, 2005.

Castro SA, Dorfmueller HC. A Brief Review on Group A Streptococcus Pathogenesis and Vaccine Development. R Soc Open Sci, 8: 201991, 2021.

Commons RJ, Smeesters PR, Proft T, et al. Streptococcal Superantigens: Categorization and Clinical Associations. Trends in Molecular Medicine, 20: 48–62, 2014.

Cunningham MW. Molecular Mimicry, Autoimmunity and Infection: The Cross-Reactive Antigens of Group A Streptococci and Their Sequelae. Microbiol Spectr, 7: 10.1128/microbiolspec. GPP3-0045-2018, 2019.

Gewitz MH, Baltimore RS, Tani LY, et al. Revision of the Jones Criteria for the Diagnosis of Acute Rheumatic Fever in the Era of Doppler Echocardiography: A Scientific Statement from the American Heart Association. Circulation, 131: 1806–1818, 2015.

Hyman SE. PANDAS: Too Narrow a View of the Neuroimmune Landscape. Am J Psychiatry, 178: 5–7, 2021.

WEBSITES

Centers for Disease Control and Prevention: https://search.cdc.gov/search/index.html?query=Streptococcus+pyogenes&sitelimit=&utf8=%E2%9C%93&affiliate=cdc-main

Kanwal S, Vaitla P, Streptococcus pyogenes, StatPearls, 2022: https://www.ncbi.nlm.nih.gov/books/NBK554528/

Khan ZZ, Salvaggio MR, Group A Streptococcal (GAS) Infections, 2021: https://emedicine.medscape.com/article/228936-overview

Students can test their knowledge of this case study by visiting the Instructor and Student Resources: [www.routledge.com/cw/lydyard] where several multiple choice questions can be found.

Toxoplasma gondii

36

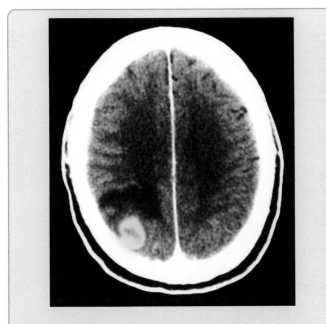

Figure 36.1 A single toxoplasmic encephalitic lesion is seen on the left-hand side of this CT scan of the head. The injection of contrast highlights the lesion, in this case, colored yellow. There is surrounding edema, which gives a black appearance.

A 32-year-old female was admitted to hospital having had two **fits**. She reported a headache over the previous few weeks. Over about 6 months she had been obtaining over-the-counter treatment from the local pharmacy for oral thrush. She had a slight fever on admission, no focal neurologic signs or **papilledema** on **fundoscopy**. On a full blood count, she had a slight **pancytopenia**. A **CT scan** of her head showed one small ring enhancing lesion in the cerebrum (Figure 36.1). After deliberation, it was considered safe and prudent to perform a lumbar puncture. There were 10 lymphocytes (normal <5), a protein of 940 mg L^{-1} (normal range 150–450), but no organisms were seen on microscopy. She was started on lamotrigine to control fits and ceftriaxone and metronidazole to treat possible bacterial brain **abscess**. Later in her admission, the possibility of HIV was considered. After counseling she tested positive for HIV. This changed the differential diagnosis. There had been no improvement in her fever or headache and she was presumptively diagnosed with toxoplasmic **encephalitis**. Treatment was changed to pyrimethamine and sulfadiazine plus folinic acid to help her bone marrow. **Serology** for *Toxoplasma* was **IgG**-positive and **IgM**-negative. Her cerebrospinal fluid (CSF) was positive for *Toxoplasma* by **polymerase chain reaction** (**PCR**). Over the next 3 months, serial CT scans showed resolution of her lesion on treatment.

1. WHAT IS THE CAUSATIVE AGENT, HOW DOES IT ENTER THE BODY AND HOW DOES IT SPREAD A) WITHIN THE BODY AND B) FROM PERSON TO PERSON?

CAUSATIVE AGENT

Toxoplasma gondii is a protozoan intracellular parasite. The definitive host is the cat and other felines. Many warm-blooded animals can serve as an intermediate host including, for example, poultry, rodents, cattle, sheep, and pigs. Humans can become accidentally infected. *T. gondii* is found throughout the world but human infection is more common in some countries than others.

ENTRY AND SPREAD WITHIN THE BODY

T. gondii oocysts (Figure 36.2) are shed in the feces of cats and may survive up to 18 months in the environment.

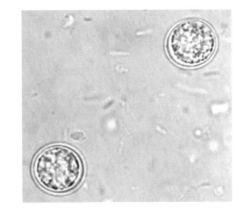

Figure 36.2 *Toxoplasma gondii* oocysts. *From the Centers for Disease Control & Prevention, Atlanta, Georgia. Image is found in the entry for Toxoplasmosis in the DPDx Parasite Image Library at the following website address: http://www.dpd.cdc.gov/dpdx/HTML/Image_Library.htm.*

During this time, they may be ingested by other animals such as rodents (Figure 36.3). Within the intestine of these

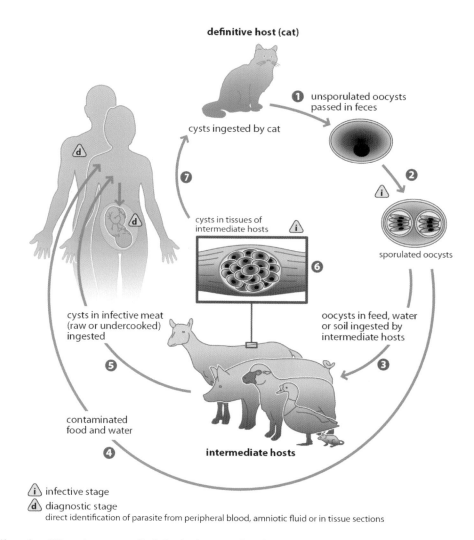

definitive host (cat)

❶ unsporulated oocysts
passed in feces

cysts ingested by cat

❼

⚠ i

❷

⚠ i

sporulated oocysts

cysts in tissues of
intermediate hosts ⚠ i

❻

oocysts in feed, water
or soil ingested by
intermediate hosts

cysts in infective meat
(raw or undercooked)
ingested

❺

❸

contaminated
food and water

❹

intermediate hosts

⚠ i infective stage
⚠ d diagnostic stage
direct identification of parasite from peripheral blood, amniotic fluid or in tissue sections

Figure 36.3 The lifecycle of *Toxoplasma gondii*. Cats shed unsporulated oocysts onto the soil (1). These sporulate (2) and are ingested by the intermediate host when they graze on the ground (3). Humans may ingest oocysts, which contaminate foods or water (4) or eat cysts in infected meat (5). Within the intestine of the intermediate host, the oocysts release sporozoites that invade the intestinal epithelium. They mature into tachyzoites that form tissue cysts at various sites in the body (6). These may then be ingested by cats to continue the lifecycle (7). In humans, tissue cysts develop in the same way as in the intermediate hosts. *From the Centers for Disease Control & Prevention, Atlanta, Georgia. Image is found in the Public Health Image Library #3421. Additional photographic credit is given to Alexander J da Silva, PhD, and Melanie Moser who created the image in 2002.*

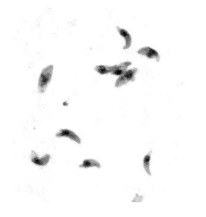

Figure 36.4 *Toxoplasma gondii* tachyzoites. *From the Centers for Disease Control & Prevention, Atlanta, Georgia. Image is found in the entry for Toxoplasmosis in the DPDx Parasite Image Library at the following website address: http://www.dpd.cdc.gov/dpdx/HTML/Image_Library.html.*

intermediate hosts, the oocysts release sporozoites that invade the intestinal epithelium and multiply to form tachyzoites (Figure 36.4). Tachyzoites may spread to local cells or pass via the lymphatics to regional lymph nodes and then to elsewhere in the body. They can infect any cell in the body. However, there is a predilection for lymph nodes, skeletal and cardiac muscle, the brain, and the eye. At each tissue site, tachyzoites enter cells and multiply to form tissue **cysts**, which contain bradyzoites (Figure 36.5). Bradyzoites in tissue cysts multiply more slowly and may pass unnoticed in tissues for prolonged periods. Humans are infected when they ingest bradyzoites in tissue cysts within uncooked or poorly cooked meat from the intermediate hosts or eat food including vegetables contaminated with oocysts from cat feces (Figure 36.3).

Tissue cysts are destroyed by proper cooking or freeze-thawing. The practice of eating undercooked meats is more common in some countries, such as France, where the

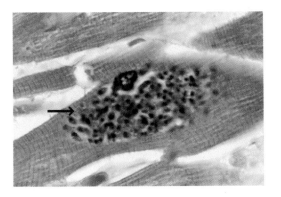

Figure 36.5 Tissue cyst of *Toxoplasma gondii* in a cardiac myocyte. *From the Centers for Disease Control & Prevention, Atlanta, Georgia. Image is found in the Public Health Image Library #966. Additional photographic credit is given to Dr Edwin P Ewing who created the image in 1984.*

prevalence of toxoplasmosis is also higher. Toxoplasmosis observed in vegetarians is presumably through eating vegetables contaminated by oocysts in the soil, which are not completely washed off. Oocysts can also enter the water supply but, if water is properly filtered, oocysts will be excluded. In the developed world, outbreaks of toxoplasmosis have occurred when filtration procedures have been disrupted. In the developing world, the quality of filtration and water supplies cannot be guaranteed.

As infection with *T. gondii* can be silent, its true prevalence is not known. Some indication of prevalence comes from studies on seroprevalence. Highest seroprevalence rates are found in South America and tropical Africa. Moderate rates of 30–50% are found in Southern Europe. Elsewhere, there are lower rates of 10–30%. Genotypes vary by geographic location. In animal models, virulence varies by genotype and, in humans, more severe disease is described in South America and Africa.

PERSON-TO-PERSON SPREAD

Humans can pass infection vertically from pregnant mother to fetus but are not a source of horizontal transmission. If a primary toxoplasmosis infection occurs in a pregnant mother, there is a 30% risk of vertical transmission to the fetus, the risk being less in the first trimester and greatest in the third trimester. Nonimmune cats may become infected by ingesting oocysts lying around in the soil. Alternatively, they could eat animals, for example, rodents, harboring tissue cysts. Bradyzoites then emerge and infect the feline intestinal epithelium. *T. gondii* undergoes sexual reproduction in the intestinal epithelium of the cat. Oocysts shed in cat feces are then a source for human ingestion directly or via an intermediate host.

2. WHAT IS THE HOST RESPONSE TO THE INFECTION AND WHAT IS THE DISEASE PATHOGENESIS?

Targets for the host immune response are tachyzoites or bradyzoites, which are extracellular, or intracellular organisms within phagocytes or tissue cells. An antibody response is mounted and antibodies persist lifelong. On binding to extracellular tachyzoites, the antibodies fix complement and cause target **lysis**.

Within macrophages, *T. gondii* produces a **parasitophorous vacuole**. Somehow lysosomal fusion does not occur with this vacuole and *T. gondii* is spared lysosomal attack. There are other microbicidal and microbistatic mechanisms elaborated by macrophages. *T. gondii* is susceptible to some of these, such as **reactive oxygen species**, reactive nitrogen intermediates, and tryptophan depletion. To enable these mechanisms, macrophages require activation by a type-1 **cytokine** pattern. Interferon-γ (IFN-γ) is central to this pattern. In an experimental mouse model of toxoplasmosis, IFN-γ is produced by **natural killer (NK)** cells, and **CD4+** and **CD8+** T lymphocytes. Other cytokines that can promote the production of IFN-γ are **interleukin (IL)**-1, IL-2, IL-12, and IL-15. Downstream **tumor necrosis factor-α (TNF-α)** released from macrophages enhance their own activation. Some genotypes of *T. gondii* have a rhoptry protein (ROP16) which can down-regulate IL-12 and IFN-γ production.

Macrophages have MHC class II molecules on their surface, which are recognized by CD4+ T-helper cells. Tissue cells only have MHC class I molecules and can only be recognized by CD8+ T cytotoxic cells. If *T. gondii* antigen is processed to appear on the tissue-cell surface in the context of MHC class I molecules, the cells are recognized by *T. gondii* reactive CD8+ T cells and killed. Host-cell survival preserves the intracellular niche for *T. gondii* and it can inhibit caspase 8 mediated cell death through apoptosis.

After primary infection, there is a lag before the immune response is induced. During this lag period, *T. gondii* spreads throughout the body and the immune response has to chase and curtail infection. The eye and the CNS represent relatively immunologically privileged sites. It is harder for immune and inflammatory responses to operate. Lymphocytes do traffic into the CSF and enter the brain substance. However, in the brain, astrocytes and **microglial cells** serve a phagocytic function and are seen to proliferate during acute infection. In the majority of subjects, the immune response brings infection under control without any symptoms. In some cases, the local inflammatory response may cause problems in lymph nodes (**lymphadenopathy**), heart muscle (**myocarditis**), skeletal muscle (**myositis**), and the retina (**chorioretinitis**). Symptomatic disease is more likely if there is an exuberant immune response. IL-10 acts as a counter-regulatory cytokine to damp down the type I cytokine response.

Bradyzoites in chronic infection are intermittently reactivating. Local tissue spread and infection is prevented through immune surveillance. However, in the immunocompromised host, immune surveillance is diminished and bradyzoite reactivation goes unchecked leading to local tissue inflammation.

3. WHAT IS THE TYPICAL CLINICAL PRESENTATION AND WHAT COMPLICATIONS CAN OCCUR?

If pregnant mothers acquire *T. gondii* infection during their pregnancy, there is a risk of transplacental transfer of tachyzoites. The chance of this does not depend on whether the mother is symptomatic or not. If the infection had been acquired before pregnancy there is the possibility that reactivation of tissue cysts will release tachyzoites during pregnancy. However, an immune response will be present and circulating tachyzoites will quickly be destroyed by antibody and complement, thus preventing any fetal infection. The situation may be different in the immunocompromised individual. Severe fetal infection may result in abortion. When trans-placental infection occurs in the third trimester, 80% of neonates are asymptomatic. Earlier infection has a greater fetal impact on development, with the greatest risk of congenital abnormality being between 10 and 24 weeks' gestation. Infection of the brain may cause lesions around the channels for CSF flow. Blockage of these channels causes **hydrocephalus** with swelling of the ventricles. Lesions within the brain substance may subsequently calcify. Disturbance of brain development can stunt its growth resulting in **microcephaly**. Fetal infection at a later stage of infection may not cause any immediate problems. Tissue cysts are established that may reactivate after birth.

Following congenital infection, the commonest manifestation of later reactivation is in the eye with inflammation of the retina called chorioretinitis (Figure 36.6). Reactivation probably occurs elsewhere in skeletal muscle but this goes unnoticed. Chorioretinitis may represent a very small volume of inflammation but becomes very noticeable and troublesome if it affects the visual fields. Understandably, central lesions are more noticeable in causing a patch of visual loss, a **scotoma**. After initial inflammation and **necrosis**, there is scarring and pigmentary change in the retina. More than one episode of chorioretinitis can occur as further tissue cysts reactivate.

Primary infection in children and adults remains asymptomatic in about 80% of subjects. There may be a constitutional illness with fever, malaise, fatigue, and then an organ-specific problem. The most common manifestation is lymphadenopathy. This could be anywhere in the body and

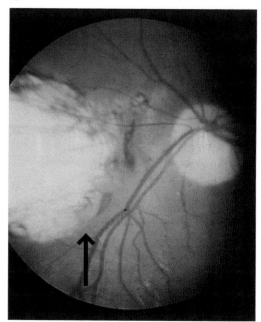

A

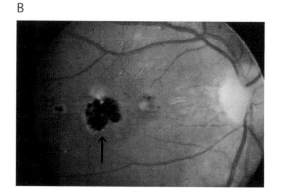

B

Figure 36.6 (A) and (B) Two different stages and appearances of *Toxoplasma* chorioretinitis. *From the Centers for Disease Control & Prevention, Atlanta, Georgia. Image is found in the entry for Toxoplasmosis in the DPDx Parasite Image Library at the following website address: http://www.dpd.cdc.gov/dpdx/ HTML/Image_Library.html.*

be single or multiple. If in the neck, it may be similar to that seen in infectious mononucleosis. Illness can subside without treatment, but the size of lymph nodes can fluctuate for a few months but will invariably settle down.

Immunocompromise that may cause reactivation of toxoplasmosis may arise from hematologic malignancy, transplant immunosuppression, and HIV infection. Prophylaxis with co-trimoxazole, diminishes, but does not preclude reactivation. In HIV, the commonest site for lesions is in the brain. Toxoplasmic encephalitis represents a progressive, focal disease process with central necrosis, surrounded by vascular inflammation. Lesions may be single or multiple and present with headaches, focal fits, and focal neurologic signs depending on the location of lesions. In immunocompromised individuals, many other manifestations are possible including myocarditis, myositis, and **pneumonitis**.

4. HOW IS THE DISEASE DIAGNOSED, AND WHAT IS THE DIFFERENTIAL DIAGNOSIS?

The main diagnostic test used is serology. Histology of an excised lymph node or CT and MRI scans of the head may be highly suggestive of toxoplasmosis but supportive evidence is usually obtained. PCR of amniotic fluid, CSF or samples from the eye has high specificity but variable sensitivity.

Currently, **ELISA** techniques are used to assay for anti-*Toxoplasma* IgG and IgM. Formerly, live or formalin-fixed tachyzoites were used in neutralization, **immunofluorescence**, and agglutination assays. The **Sabin-Feldman dye test** is a neutralization assay of live tachyzoites. It is the standard against which other assays are compared. Serologic tests are useful in excluding *T. gondii* infection if negative on two occasions 3 weeks apart. However, a positive test does not necessarily imply that the current illness is due to toxoplasmosis. IgG antibodies appear 2 weeks after infection, peak at 2 months, and then wane after 2 years, but remain lifelong. The persisting titer of antibody varies between individuals and bears no relationship to disease severity. IgM antibodies in many other infections just appear in the acute phase and are a good indicator of acute infection. In toxoplasmosis, IgM can persist for a few years and is of limited use in defining acute infection.

In high prevalence areas, toxoplasma serologic screening is part of antenatal care to diagnose acute infections in pregnancy. This may be supported by PCR of amniotic fluid to confirm trans-placental transfer of organisms.

Toxoplasmosis has to be considered in the differential diagnosis for congenital abnormalities, lymphadenopathy, chorioretinitis, and ring-enhancing CNS lesions. It is a less common explanation for myocarditis or myositis. "**TORCH**" is a useful mnemonic for key congenital infections that may produce overlapping clinical features. It stands for **T**oxoplasmosis, **O**ther, **R**ubella, **C**ytomegalovirus, and **H**erpes virus. When lymphadenopathy is in the neck, toxoplasmosis may present with similar symptoms to that of infectious mononucleosis. Infectious mononucleosis is most commonly due to Epstein-Barr virus (see Case 9) but can also be due to cytomegalovirus and toxoplasmosis. Differential diagnoses for lymphadenopathy, chorioretinitis, and toxoplasmic encephalitis are shown in Table 36.1.

5. HOW IS THE DISEASE MANAGED AND PREVENTED?

MANAGEMENT

Pyrimethamine and sulfadiazine are effective in killing tachyzoites, but do not clear tissue cysts. Pyrimethamine is a folate antagonist and causes severe bone marrow suppression, unless the bone marrow is spared with folinic acid supplements. Pyrimethamine and sulfadiazine are of proven efficacy in treating toxoplasmic encephalitis in HIV patients. In immunocompetent subjects, disease will self-limit anyway and administering potentially toxic therapy is of no added benefit. One exception may be for large chorioretinitis lesions. Small lesions will probably settle before assessment, diagnosis, and commencement of treatment.

Management of toxoplasmosis in pregnancy is challenging and practice varies from country to country. Mothers who have not previously been infected may be totally asymptomatic if they acquire infection during the pregnancy. If infection is diagnosed on the basis of a **seroconversion**, from being

Table 36.1 Differential diagnoses for various manifestations of toxoplasmosis

Lymphadenopathy	Chorioretinitis	Encephalitis
Pyogenic bacteria	*Infections*:	Focal CNS lesions:
Epstein-Barr virus	Tuberculosis	*Infection*:
Cytomegalovirus (CMV)	Toxocariasis	Brain abscess
Mycobacteria	Histoplasmosis	Tuberculoma
HIV	Syphilis	Cryptococcoma
Syphilis	Cytomegalovirus	Focal CMV encephalitis
Bartonella (cat scratch disease)	Varicella-zoster virus	Nocardia
Kikuchi's syndrome		
Lymphoma	*Connective tissue disease*:	*Neoplasia*:
Sarcoidosis	e.g. rheumatoid arthritis	Primary or secondary brain tumors
		Lymphoma
Various other infections that cause lymphadenopathy may have accompanying diagnostic features, e.g. a measles rash	*Sarcoidosis*	

seronegative early in pregnancy to being seropositive, anti-*Toxoplasma* treatment may commence after the fetus has already been infected and is therefore too late for an impact. Conducting trials on the efficacy of treatment to prevent fetal infection is understandably difficult. Contrasting results have come from various small trials. The drug used in pregnant mothers is spiramycin. This does not cross the placenta and is only used to kill tachyzoites before they can transfer to the fetus. Some of the various trials have shown a benefit with spiramycin. Once the fetus has been infected, perhaps confirmed by amniotic PCR, pyrimethamine and sulfadiazine may be used after the first trimester. In the first trimester, treatment-associated folate antagonism runs a significant risk of fetal malformation.

PREVENTION

Prevention of infection requires care with cooking and hand hygiene. Avoidance of undercooked meats is essential in pregnancy. Adequate cooking of meat to high temperatures eliminates the risk. Hands can become contaminated in the garden or handling a cat litter. Careful handwashing or wearing gloves will reduce risk. Prophylaxis with co-trimoxazole can reduce the risk of disease in those who are already infected and are immunocompromised. The prime indication for co-trimoxazole is actually prevention of *Pneumocystis* pneumonia. If patients develop allergic reactions to co-trimoxazole, they may be switched to other prophylactics for *Pneumocystis* pneumonia, that do not actually prevent *T. gondii* infection.

SUMMARY

1. WHAT IS THE CAUSATIVE AGENT, HOW DOES IT ENTER THE BODY, AND HOW DOES IT SPREAD A) WITHIN THE BODY AND B) FROM PERSON TO PERSON?

- *Toxoplasma gondii* is a protozoan parasite.
- The definitive host is the cat and other felines. From multiplication in the intestinal epithelium, oocysts are shed in the feces.
- Other animals ingest the oocysts. Within their intestine, invasion and multiplication yields tachyzoites, which pass around the body.
- In various tissues, tachyzoites invade cells and produce tissue cysts.
- Humans may be infected by ingesting oocysts from the environment or by eating tissue cysts in undercooked meat.

2. WHAT IS THE HOST RESPONSE TO THE INFECTION AND WHAT IS THE DISEASE PATHOGENESIS?

- Antibody and complement can kill circulating tachyzoites.
- Macrophages can phagocytose tachyzoites and when activated by IFN-γ kill them through reactive oxygen or nitrogen species and tryptophan depletion.
- Tissue cysts are attacked by CD8+ lymphocytes with the help of CD4+ lymphocytes.
- The inflammatory response created by attacking tissue cysts may be low grade and asymptomatic. However, if extensive, with numerous tachyzoites invading cells locally, there can be organ-specific problems such as chorioretinitis, myositis, myocarditis, and encephalitis.

3. WHAT IS THE TYPICAL CLINICAL PRESENTATION AND WHAT COMPLICATIONS CAN OCCUR?

- Congenital infection can result in fetal loss or severe congenital abnormalities involving the brain.

- *In utero* infection may also be silent with later problems as infection reactivates, most commonly in the eye as chorioretinitis.
- Primary infection in children or adults is asymptomatic in the majority but if symptomatic can cause lymphadenopathy.
- In the immunocompromised, infection can reactivate and cause encephalitis. This is most often seen in HIV patients.
- Reactivation may also affect skeletal muscle, the heart, and the lung.

4. HOW IS THE DISEASE DIAGNOSED, AND WHAT IS THE DIFFERENTIAL DIAGNOSIS?

- Antibody to *Toxoplasma* can be measured by a variety of serological tests.
- IgM antibodies can persist for prolonged periods and are not useful for distinguishing acute from chronic infection.
- Antibodies persist lifelong and their presence does not necessarily mean that the current illness is due to toxoplasmosis. The absence of antibody is sometimes more useful in excluding infection.
- PCR for toxoplasmosis has high specificity, but variable sensitivity.
- Histology or the clinical picture can be highly suggestive of toxoplasmosis.

5. HOW IS THE DISEASE MANAGED AND PREVENTED?

- Treatment of toxoplasmosis is usually only required in the immunocompromised and for the infected fetus. Pyrimethamine and sulfadiazine, with folinic acid, are used to kill tachyzoites, although they have no effect on the tissue cysts.
- Immunocompetent individuals will usually have self-limiting disease.
- The management of toxoplasmosis in pregnancy is controversial and practice varies from country to country.
- Avoidance of undercooked meat, thorough cooking, and hand hygiene are important in prevention.

FURTHER READING

Montoya JG, Boothroyd JC, Kovacs JA. Toxoplasma gondii. In: Bennett JE, Dolin R, Blaser MJ, editors. Principles and Practice of Infectious Diseases, 9th edition. Elsevier, Philadelphia, 20675–20877, 2020.

Murphy K, Weaver C. Janeway's Immunobiology, 9th edition. Garland Science, New York/London, 2016.

REFERENCES

Kravetz J. Congenital Toxoplasmosis. BMJ Clin Evid, 0906, 2013.

Robert-Gangneux F, Darde M-L. Epidemiology of and Diagnostic Strategies for Toxoplasmosis. Clin Micro Rev, 2012, 25: 264–296, 2012.

WEBSITES

Centers for Disease Control, Parasites – Toxoplasmosis (Toxoplasma infection), 2018: https://www.cdc.gov/parasites/toxoplasmosis/index.html

Students can test their knowledge of this case study by visiting the Instructor and Student Resources: [www.routledge.com/cw/lydyard] where several multiple choice questions can be found.

37 Trichophyton spp. (Dermatophytes)

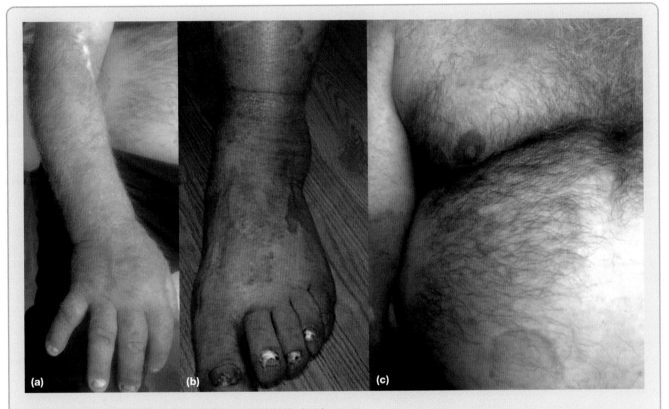

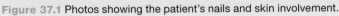

Figure 37.1 Photos showing the patient's nails and skin involvement.

A 45-year-old man noticed a change in color and thickness of his toenails from about age 35. The change started with thickening and crumbling of the nail of the right thumbnails and spread to all nails of both feet, and to all fingers over the past 3 years (Figure 37.1A and B). He did not like going to doctors so decided to see if his nail problem would resolve itself. Over the previous 3 months, red, peripherally growing, sharply defined, scaly plaques with a central clearing and raised edges appeared on the dorsal surfaces of his feet and hands, which spread to the front of his body (Figure 37.1C). The rash was accompanied by prominent itching, which made the patient see his physician. The doctor, on the basis of typical lesions on the nails and skin, suspected fungal infection: onychomycosis with subsequent generalization of the skin (*Tinea pedis*, *Tinea manuum* and *Tinea corporis*). He had no obvious risk factors such as diabetes or being immunocompromised. Samples of the affected nail and skin were taken and sent to the laboratory to check for fungal infection, but he was started on treatment based on clinical diagnosis.

1. WHAT IS THE CAUSATIVE AGENT, HOW DOES IT ENTER THE BODY AND HOW DOES IT SPREAD A) WITHIN THE BODY AND B) FROM PERSON TO PERSON?

CAUSATIVE AGENT

The causative agent in this case is *Trichophyton rubrum*. This is a dermatophyte which, as a group, are characterized by their unique requirement for keratin for their growth. There are up to 40 different species affecting animals and humans. Dermatophytes are fungi that cause infections of the skin, hair and nails and represent the fourth highest cause of global disease, estimated to have an incidence of 20–25% within the healthy population. There are five genera: *Trichophyton, Epidermophyton, Microsporum* with *Arthroderma*, and *Nannizzia* being added more recently. Within these genera, there are different species adapted to particular ecologic niches and hosts.

Dermatophytes colonize 30–70% of the human population without causing clinical disease suggesting that they should be considered as a component of the human microbiota.

The diseases caused by these organisms are called ringworm or tinea followed by the region of the body affected, for example *tinea corporis* (body), *tinea pedis* (foot), *tinea manuum* (hands) *tinea cruris* (groin) and *tinea capitis* (head), etc. Infections can be defined as those infecting the epidermis, hair (scalp and beard), and nails (finger and toe). Infection of the nails is called **onychomycosis**. The dermatophytes consist of septate branching hyphae that grow on appropriate media (dermatophyte test medium) and have characteristic colony morphology and pigmentation.

Dermatophytes can reproduce both asexually and sexually. Asexual reproduction involves the production and dissemination of spores (conidia) produced from a fruiting body (conidiophore) or, in some species, the spores are produced within the hyphae (arthrospores). The conidia and conidiophores have a characteristic species morphology that, along with the macroscopic appearance, are used in laboratory identification. In those species that reproduce sexually, mating can be homothallic (self-fertilizing) or heterothallic (mating between two compatible mating types). Dermatophytes that do not have a recognized sexual stage (anamorphic) are placed in the deuteromycetes and if a sexual stage (telomorphic) has been identified, in the ascomycetes. *T. rubrum* is a deuteromycete.

SKIN INFECTIONS

The initial step in dermatophyte infection is adhesion of the spore to the stratum corneum which is the keratinized outer epithelial layer of the skin. The different dermatophytes use various adhesins to bind to as yet unknown ligands on the keratin and the surface of the host keratinocytes. *Trichophyton* spp., for example, bind to mannose and galactose units and extend fibrils into the keratinized surface of the cell. The fungi secrete a proteolytic enzyme keratinase, the metabolic product of which it uses for catabolism and growth, spreading outward in the stratum corneum producing the typical circular appearance of tinea. These proteolytic enzymes contribute to the virulence of dermatophytes. Although it is not common for dermatophytes to invade the dermis, there are an increasing number of reports of dermatophytes invading deep layers (dermis and hypodermis) in diabetic and immunocompromised patients, as well as in individuals with immunodeficiency. In addition, dermatophytes can spread to the lymph nodes, brain, and bloodstream.

PERSON/ANIMAL-TO-PERSON SPREAD

A patient can become infected from three sources: another human (Anthropophilic); an animal (Zoophilic) or the environment (Geophilic), for example soil. In all cases, close contact is involved although the fungi can be acquired from fomites.

Anthropophilic infections occur in overcrowded areas and areas of social deprivation. Infection may also be transmitted indirectly by using an infected person's comb/razor or contamination from infected skin scales or hairs on towels. Infections can spread from one part of the body to another. Apart from the home environment, other routes of transmission may be swimming baths or exercise gyms, in which case, the dermatophyte is acquired from water or exercise machines. Examples of anthropophilic dermatophytes include *Microsporum audouinii* (tinea capitis, tinea corporis), *Trichophyton rubrum* (tinea corporis, tinea pedis) and *Epidermophyton floccosum* (tinea corporis), *Trichophyton equinus* (tinea corporis, horses), and *Microsporum canis* (tinea corporis, cats and dogs). Again, transmission is acquired by close contact with the animal particularly, for example, horse riding (*T. equinus*) where not only can the dermatophyte be transmitted directly from the horse but also indirectly from the blankets and saddle.

A typical geophilic dermatophyte is *Microsporum gypseum* (tinea capitis) which is acquired from contact with contaminated soil.

EPIDEMIOLOGY

The number of reported dermatophytosis cases has increased over the last decade as a result of immunosuppressive regimes for treating diseases such as cancer, and autoimmune diseases. It is estimated that currently up to 25% of the population worldwide is infected with dermatophytes. However, there is a wide inter-country and in-country diversity of the different types of disease and also of the species of dermatophyte. Globally, the most common dermatophytes are *T. rubrum, T. mentagrophytes*, and *T. interdigitale* affecting females more than males, although *Tinea cruris* is more common in males and common in children. Black races are more likely to have *Tinea capitis* and white races more likely to have onychomycosis.

In Stockholm, Sweden, onychomycosis (14% of the dermatophyte infections) was the most common disease and *T. rubrum* the most common organism. The most common organisms causing tinea capitis (1.5%) were *T. violaceum* and *T. soudanense*.

In Lithuania, *T. rubrum* was the most common pathogen with *T. unguium* the most common disease. Over the period of study (2001–2010) *T. rubrum* decreased as a cause of disease from 55.4% to 11% whereas non-dermatophyte infections increased from 3.4% to 35.9%.

In the UK, between 1980–2005, *M. canis, T. mentagrophytes* var *mentagrophytes, T. verrucosum,* and *E. floccosum* all decreased by 90%. However, *T. tonsurans* and *T. violaceum* increased by 1000%. Also, *T. rubrum* and *T. mentagrophytes* var *interdigitale* increased by 10% to 90% over 1980 figures.

In Addis Ababa, Ethiopia, tinea capitis (48%) was the most common dermatophytosis and *T. violaceum* was the most common dermatophyte (38.4%) and the major cause of tinea capitis (89.7%). Overall, in Africa, tinea capitis was the most common dermatophytosis in all African children and tinea corporis in adults. The most frequent dermatophyte was *T. violaceum* in North and East Africa and *T. soudanense* with *T. audouinii* in West and Central Africa.

In Brazil, *T. rubrum* (59.6%) and *T. interdigitale* (34%) were the most common dermatophytes with tinea unguium (48.5%) and tinea pedis (33.1%) the most common diseases. *M. canis* was the most common dermatophyte causing tinea capitis.

In India, the most common dermatophytosis in males was tinea corporis and tinea unguium in females. *T. metagrophytes* was the most common cause of hair and nail problems and *T. rubrum* the most common cause of skin problems.

2. WHAT IS THE HOST RESPONSE TO THE INFECTION AND WHAT IS THE DISEASE PATHOGENESIS?

Despite significant advances in research into human immunity, there is not a great deal known about the host's immune response against dermatophytes.

The skin is not only a physical barrier but is also part of the innate and adaptive immune systems. Most of the defenses of the epidermis are localized within the stratum corneum.

As already mentioned, dermatophytes colonize 30–70% of the human population without causing clinical disease suggesting that the immune responses in most cases are sufficient to prevent disease. Dermatophyte virulence factors that contribute to their survival and persistence within the host are dependent on the type of dermatophyte (anthropophilic or zoophilic), site of infection (skin or systemic), stage of infection (acute or chronic), and the type of host response.

INNATE IMMUNITY

Dermatophyte colonization starts with adhesion of the spores, and spread of the hyphae, through the keratin, invasion and cell damage in the stratum corneum.

Invading dermatophytes come into contact first with keratinocytes that are the most abundant cells of the stratum corneum. Structures on the spores and germinating fungi such as glycans, glycolipids, and glycoproteins are pathogen-associated or microbe-associated molecular patterns (MAMPs). These MAMPs are recognized by pattern-recognition receptors (PRRs) on the keratinocytes, the macrophages and Langerhans (dendritic) cells. Cells with these PRRs are the first line of defense against all pathogens including bacteria, viruses, and fungi. They comprise four main families that include C-type lectin receptors (CLR), Toll-like receptors (TLR), nucleotide-binding and oligomerization domain (NOD) Leucine Rich Repeats (LRR)-containing receptors (NLR), and retinoic acid inducible gene (RIG)-like receptors (RLR). In the case of dermatophyte infections, Dectin 1 that recognises mannans and Dectin 2 that recognize β-glucans, both C-type lectins, appear to be particularly important as is TLR2.

Initial activation of keratinocytes by dermatophyte MAMPs leads to production of pro-inflammatory cytokines such as IL-1β, IL-6, and IL-17 and chemokines (IL-8, MCP-1), resulting in recruitment of neutrophils into the stratum corneum where they have potent cytotoxicity against hyphae from a number of dermatophyte species. They also phagocytose conidia and produce cytokines including IL-1β, IL6, IL8, IL17, and TNF and form **extracellular nets**. Different dermatophytes stimulate secretion of a different spectrum of cytokines presumably dependent on which PRRs are ligated. The inflammatory response is usually stronger if the dermatophyte is a zoophilic species as opposed to an anthropophilic species.

Activation of the cells via these receptors also leads to the production of antimicrobial peptides that include cathelicidins and defensins. Some defensins are constitutively expressed by epithelial cells and some following ligation with MAMPs. Both cathelicidins and defensins have been shown to inhibit the growth of a number of dermatophyte species and contribute to their demise.

Interestingly, some dermatophyte species (including *T. rubrum*) can avoid the innate immune system by causing a reduction in expression of PRRs.

In addition to neutrophils, macrophages and dendritic cells play an important role in responses to dermatophytes. IL-17 up-regulates production of antifungal peptides particularly β-defensins by phagocytic cells. Macrophages have also been shown to phagocytose a number of different dermatophyte species and produce cytokines and reactive oxygen species. Interestingly, in patients with chronic dermatophytosis, these cells showed decreased phagocytosis, and reduced secretion of reactive oxygen and cytokines TNF-α and IL1β.

Dendritic cells can both kill dermatophytes and present their antigens to T cells. This has been shown particularly for *T. rubrum*, the causative agent in the case presented here. The paradigm being that Langerhans cells (LC) in the skin carry antigen to draining lymph nodes thus acting as a bridge to adaptive immunity (see adaptive immunity).

Cells playing a role in the dermis include macrophages and various subsets of DCs, some of which are migratory.

It is not completely clear why some dermatophyte infections progress to lower levels of the skin (dermis) and beyond (systemic). This is seen more often when patients are immunodeficient or immunocompromised. Severe invasive disease with dermatophytes are associated with mutations in CARD9 (caspase recruitment domain containing protein 9) which is a key adaptor protein in signaling pathways of some PRRs such as Dectin 2 (see above). Killing by neutrophils and their production of IL-17 is severely impaired with this mutation.

ADAPTIVE IMMUNITY

Specific T cells recognizing processed dermatophyte antigens displayed by migrated LCs in the draining lymph nodes of the skin are activated. Th-1 and Th-17 cells appear to be the most important T cells in the adaptive immune response to dermatophytes. IFN-γ-producing Th-1 lymphocytes can elicit a delayed-type hypersensitivity response and, probably through the enhancing effect of IFN-γ on neutrophils and macrophages, help with lesion resolution. IL-7 produced by Th-17 cells also acts on innate cells to help remove the infecting dermatophyte.

The presence of antigens from the dermatophyte can, in some instances, inhibit clonal expansion of B and T cells as shown in an *in vitro* system. Additionally, these antigens can inhibit expression of activation markers such as ICAM-1 on keratinocytes.

Antibodies are produced to dermatophyte infections but are thought to play a minor role, if any, to host defense. Rag2-/- mice (lacking T and B cells) have long-lasting infection but are ultimately able to clear *T. benhamiae* from the skin. Thus, it would appear from animal models that the adaptive immune system contributes a minor part of immunity to dermatophyte infections but is more important in cases of deep and chronic infections. This is seen in HIV patients with dermatophytosis.

PATHOGENESIS

The pathogenesis of dermatophyte infections involves several factors:

1. Determinants of pathogenesis include the secretion of a variety of enzymes including keratinase, protease, phospholipase, lipase and elastase keratinase, the metabolic product of which, the organism uses for catabolism and growth. Other determinants of pathogenesis include non-enzymes and these differ between specific dermatophytes. For example, *T. rubrum* produces a mycotoxin called xanthomegnin.

2. The type of dermatophyte, anthropophilic, zoophilic or geophilic. Zoophilic and geophilic infections produce more inflammation and usually resolve more quickly.

3. The site of infection (skin or systemic) and stage of infection (acute or chronic).

4. The host immune response: individuals with immunodeficiency and immunosuppression can have deeper infections. For example, mutations of genes such as *CARD9* involved in PRR signaling lead to deeper and more chronic infections.

These aspects have been discussed in more detail in the preceding text.

The fungi invade after adhering to stratum corneum cells and producing keratinases or proteases such as subtilisins. Factors which encourage fungal invasion include increased environmental humidity and CO_2 content, both of which may occur in a tropical environment and in the presence of occlusion. Less is known about those factors which determine human susceptibility. Generally, it is thought that most individuals are susceptible to infection. The presence of medium chain length fatty acids in sebaceous secretions may, however, prevent hair shaft invasion by dermatophytes in post-pubertal children. There is evidence that susceptibility to *Tinea imbricata* may be mediated via an autosomal recessive gene. In addition, patients with persistent treatment-unresponsive dermatophytosis affecting the palms and soles are significantly more likely to be atopic than others. Resistance is largely mediated by nonspecific factors such as an increase in epidermal turnover, epidermally derived peptides or by activation of T cell-mediated immunity. Patients with HIV infection, for instance, although not apparently showing an increased incidence of infection, may have clinically atypical and extensive lesions.

3. WHAT IS THE TYPICAL CLINICAL PRESENTATION AND WHAT COMPLICATIONS CAN OCCUR?

Clinical Presentation

As described earlier, there are several major clinical subtypes: dermatophyte infections of the epidermis, hair (scalp and beard) and nails (fingers and toes). These infections present differently depending on the subtype.

Cutaneous Infection

A typical clinical presentation of a skin dermatophyte infection is an erythematous, scaly, pruritic lesion which expands radially with central clearing and a raised border. Infection caused by zoophilic or geophilic dermatophytes produces a more severe erythematous response.

There are three distinct clinical presentations:

- The classical presentation is black dot tinea capitis: in this type, there is an infection with a fracture of the hair.

- Kerion (painful nodules with pus) is another presentation that involves inflammation and may progress to scarring alopecia.

- Favus is the boggy inflammatory type and typically presents with deep-seated oozing nodules, abscesses, crusting, or scutula.

Onychomycosis

Most often, this occurs in adults but less so in children. Clinical manifestations commonly include nail discoloration, **subungual** hyperkeratosis, onycholysis, nail plate splitting, and nail plate destruction.

Potential complications include pain, transmission of fungal infection to other body sites, and, in immunocompromised patients, bacterial cellulitis.

4. HOW IS THE DISEASE DIAGNOSED, AND WHAT IS THE DIFFERENTIAL DIAGNOSIS?

Since a dermatophyte infection can affect several parts of the body at the same time (e.g., tinea pedis and tinea cruris or tinea pedis, tinea unguium, etc.), it is important to perform a complete examination of the skin, hair, and nails.

DIAGNOSIS

Clinical Examination

The infection has typical clinical manifestations which, in most cases, allows the diagnosis to be made clinically.

Direct examination of the skin and hair: Wood's light examination of the skin uses ultraviolet light to detect areas of green fluorescence that may be caused by certain types of dermatophytes.

Laboratory Tests

If a dermatophyte infection is suspected on clinical presentation, confirmation can be made by:

Direct microscopy examination: skin scrapings, nail scrapings, hair or scales are placed on a microscope slide and a clearing agent added (potassium hydroxide or sodium hydroxide solution or Amman's chloral-lactophenol). Examination under the microscope will reveal hyphae and spores. The sensitivity of direct microscopy examination may be greatly enhanced by the use of stains or fluorochromes such as Congo red, Calcofluor white or Lactophenol cotton blue.

Culture: scrapings, hair or scales can also be cultured on dermatophyte agar for several weeks and the culture examined for characteristic microscopic morphology of the conidia and conidiophores. The medium is supplemented with antibiotics and cycloheximide, a reagent specific for the selection of this group of fungi.

Finally, a skin biopsy followed by histology may be performed if the diagnosis is still uncertain or in the case of persisting lesions that have not resolved with prior treatments.

Polymerase chain reaction (**PCR**): Although both direct microscopy and culture can identify some dermatophyte organisms, molecular diagnostic methods have been developed that provide good sensitivity and specificity as well as fast results. These methods include the PCR and real-time PCR techniques. Studies using PCR reported a rate of correct diagnoses of 74–100%. Kits are commercially available.

If a dermatophyte infection is initially misdiagnosed for a similar skin condition and mistakenly treated with topical corticosteroids, the characteristic appearance of the tinea can be masked (i.e. tinea incognito). This can make the diagnostic process more difficult and may delay necessary treatment.

DIFFERENTIAL DIAGNOSIS

Leprosy, lichen planus, psoriasis, sarcoidosis, cutaneous lupus, eczema, Lyme disease, contact dermatitis, seborrheic dermatitis, intertrigo, erythrasma, mycosis fungoides, alopecia areata.

5. HOW IS THE DISEASE MANAGED AND PREVENTED?

MANAGEMENT

Infected patients should be encouraged to wear loose-fitting cotton or synthetic clothing and change clothing more frequently. In addition, they should be discouraged from sharing garments and towels. Socks, caps, and undergarments should be regularly washed and ironed. Absorbent powders and deodorants (to reduce perspiration) should also be advised.

Treatments

Mild cutaneous skin infections of tinea corporis or tinea cruris can usually be resolved using over-the-counter topical creams where available. Allylamines such as terbinafine is commonly used, or azoles such as clotrimazole and ketoconazole. Treatment duration is 2 and 4 weeks. The azoles inhibit lanosterol 14-alpha demethylase which makes the fungal cell membrane leaky and the allylamines inhibit squalene epoxide resulting in accumulation of squalene which kills the fungus.

Chronic or persistent infection of tinea capitis requires systemic treatment. This may require up to 6 months of treatment. Terbinafine is commonly used. Itraconazole and fluconazole have also been shown to be effective. Griseofulvin, thought to inhibit fungal cell mitosis and nucleic acid synthesis, is rarely used because of toxicity and then only for tinea capitis in children in some countries.

Treatment of onychomycosis is often topical ciclopirox (a broad-spectrum anti-fungal with some anti-inflammatory properties), together with oral antifungal agents, including terbinafine (currently drug of choice), itraconazole and fluconazole. Again, oral treatments often require up to 6 months of therapy.

PREVENTION

Prevention of infection is aided by good personal hygienic practices and avoidance of contact of potentially infectious sources.

To avoid dermatophytosis, physicians advise individuals to:

- Keep the skin clean and dry – especially in skin folds.
- Wear shoes that allow air to circulate freely around your feet.
- Wear loose-fitting clothes made of cotton or a breathable material that's designed to keep moisture away from your skin.
- Wash socks, clothes, and bed linen regularly to remove any fungi.

- Clip your fingernails and toenails short and keep them clean.
- Don't walk barefoot in communal areas such as showers, saunas, and swimming pools.
- Don't share towels, hairbrushes or combs.
- If you have diabetes, keep your blood sugar under control.
- If someone in your family has scalp ringworm, wash bedding, hats, combs, and hair accessories in bleach diluted with water.
- If you're an athlete involved in close contact sports, shower immediately after your practice session or match, and keep all of your sports gear and uniform clean. Don't share sports gear (helmet, etc.) with other players.

SUMMARY

1. WHAT IS THE CAUSATIVE AGENT, HOW DOES IT ENTER THE BODY, AND HOW DOES IT SPREAD A) WITHIN THE BODY AND B) FROM PERSON TO PERSON?

- Dermatophytes are fungi that infect skin, scalp, and nails and uniquely require keratin for their growth.
- The three major genera affecting humans are *Trichophyton*, *Epidermophyton*, and *Microsporum*, with *Arthroderma*, and *Nannizzia* recently added.
- Fungal spores attach to the stratum corneum. Dermatophytes use various adhesins to bind to ligands on the surface of the host cell. *Trichophyton rubrum*, for example, binds to mannose and galactose units and extends fibrils into the keratinized surface of the cell. Dermatocytes secrete a proteolytic enzyme keratinase, the metabolic product of which it uses for catabolism and growth. Although it is not common for dermatophytes to invade the dermis, there are an increasing number of reported cases. Some dermatophytes can even spread systemically.
- Infection of humans can occur via three sources: another human (Anthropophilic); an animal (Zoophilic) or the environment (Geophilic), e.g., soil. In all cases, close contact is involved although infections can also be acquired from fomites.

2. WHAT IS THE HOST RESPONSE TO THE INFECTION AND WHAT IS THE DISEASE PATHOGENESIS?

- Dermatophytes colonize 30-70% of the human population without causing clinical disease.
- First contact of germinating spores and then hyphae is with keratinocytes that are the most abundant cells in stratus corneum.
- Glycans, glycolipids, and glycoproteins act as MAMPs and are recognized by a variety of PRRs. Stimulation through these receptors leads to production of cytokines such as IL1β, IL-6,

and IL-17 and antimicrobial peptides (such as defensins and cathelicidins that have anti-fungal effects).

- Keratinocytes stimulated through PRRs produce these cytokines and also chemokines IL8 and MCP-1 that result in recruitment of neutrophils that have potent cytotoxicity against hyphae. Neutrophils also phagocytose conidia and produce TNF-α.
- Although the infection can be contained in the epidermis, infections can penetrate the dermis and beyond. Resident macrophages and Langerhans cells (LC) in the dermis also produce cytokines and can kill dermatophytes, probably through reactive oxygen species. LCs also migrate to the draining lymph nodes where T-cell immunity is stimulated (Th-1 and Th-17 cells).
- Th-1 (through IFN-γ production) and Th-17 cells are thought to contribute to immune defense but data from animal models and mutations such as CARD9 indicate that host defense is mainly mediated by the innate immune system.

3. WHAT IS THE TYPICAL CLINICAL PRESENTATION AND WHAT COMPLICATIONS CAN OCCUR?

- Patients can present with infections of the skin, hair (scalp and beard) or nails (finger and toe).
- The disease is named ringworm or tinea followed by location on the body, e.g. tinea corporis (body), tinea pedis (foot), tinea cruris (or jock itch – groin) and tinea capitis (head). Infection of the nails is called onychomycosis.
- A typical skin dermatophyte infection is an erythematous, scaly, pruritic lesion which expands radially through the epidermis with a central clear region and a raised border.
- There are three distinct clinical presentations of tinea capitis: A black dot in the skin with fracture of hair follicles; painful nodules with pus which may progress to scarring alopecia; favus with deep-seated oozing nodules, abscesses and crusting.
- Zoophilic and geophilic dermatophytes can cause severe inflammatory disease.

Continued...

...continued

- Clinical manifestations of onychomycosis include nail discoloration, subungual hyperkeratosis, onycholysis, nail plate splitting, and nail plate destruction.
- Potential complications include pain, transmission of fungal infection to other body sites, and, in immunocompromised patients, bacterial cellulitis.

4. HOW IS THE DISEASE DIAGNOSED, AND WHAT IS THE DIFFERENTIAL DIAGNOSIS?

- Because the infection has typical clinical manifestations, in most cases, this allows the diagnosis to be made clinically.
- Direct examination of skin and hair using Wood's light.
- Microscopic examination of skin scrapings, nail scrapings, hair or scales using dyes or fluorochromes.
- Culture on dermatophyte agar with antibiotics to prevent growth of bacteria; microscopic appearance of conidia and conidiophores of the culture.
- Specific PCR can be used to identify specific genera and species. Commercial kits are available.
- Should a dermatophyte infection be initially misdiagnosed for a similar skin condition and mistakenly treated with topical corticosteroids, the characteristic appearance of the tinea can be masked, (i.e. tinea incognito). This can make the diagnostic process more difficult and may delay necessary treatment.

- A skin biopsy can also be taken if there is a difficult diagnosis.
- Differential diagnosis includes leprosy, lichen planus, psoriasis, sarcoidosis, cutaneous lupus, eczema, and lyme disease.

5. HOW IS THE DISEASE MANAGED AND PREVENTED?

- Infected patients should be encouraged to wear loose-fitting clothes and change more frequently: should avoid sharing garments or towels.
- Topical treatment for mild cutaneous infections with an azole (e.g. clorimazole) or an allylamine, e.g. terbinafine for 2-4 weeks.
- Chronic or persistent infections require systemic treatment with an azole, or allylamine or griseofulvin an antifungal agent. Treatment may take up to 6 months.
- Onychomycosis treatment is often topical ciclopirox (a broad-spectrum anti-fungal with some anti-inflammatory properties), together with oral antifungal agents including terbinafine (currently drug of choice), itraconazole and fluconazole. Again, 6 months of oral treatments are often required.
- Prevention includes good personal hygiene, keeping the skin dry, wearing loose clothing, not sharing towels, hairbrushes or combs, etc.

ACKNOWLEDGEMENT

The authors of this case would like to thank Professsor Lali Mekokishvili for providing the patient photos and for reading the manuscript.

FURTHER READING

Gnat S, Łagowski D, Nowakiewicz A. Major Challenges and Perspectives in the Diagnostics and Treatment of Dermatophyte Infections. J Appl Microbiol, 129: 212–232, 2020.

Wang R, Huang C, Zhang Y, Li R. Invasive Dermatophyte Infection: A Systematic Review. Mycoses, 64: 340–348, 2021.

REFERENCES

Blutfield MS, Lohre JM, Pawich DA, Vlahovic TC. The Immunologic Response to Trichophyton rubrum in Lower Extremity Fungal Infections. J Fungi, 1: 130–137, 2015.

Burstein VL, Beccacece I, Guasconi L, et al. Skin Immunity to Dermatophytes: From Experimental Infection Models to Human Disease. Front Immunol, 11: 605644, 2020.

Cestrino GA, Veasey JV, Benard G, Sousa MGT. Host Immune Responses in Dermatophytes Infection. Mycoses, 64: 477–483, 2021.

Firat YH, Simanski M, Rademacher F, et al. Infection of Keratinocytes with Trichophitum rubrum Induces Epidermal Growth Factor – Dependent RNase & and Human Beta-Defensin 3 Expression. PLoS One, 9: e93941, 2014.

Lee WJ, Kim SL, Jang YH, et al. Increasing Prevalence of Trichophyton rubrum Identified Through an Analysis of 115,846 Cases Over the Last 37 Years. Korean Med Sci, 30: 639–643, 2015.

Martinez-Rossi NM, Peres NTA, Bitencourt TA, et al. State-of-the-Art Dermatophyte Infections: Epidemiology Aspects, Pathophysiology, and Resistance Mechanisms. J Fungi, 7: 629, 2021.

Michaels BD, Del Rosso JQ. Tinea capitis in Infants: Recognition, Evaluation, and Management Suggestions External Icon. J Clin Aesthet Dermatol, 5: 49–59, 2012.

Piraccini BM, Alessandrini A. Onychomycosis: A Review. J Fungi, 1: 30–43, 2015.

Sharma A, Chandra S, Sharma M. Difference in Keratinase Activity of Dermatophytes at Different Environmental Conditions is an Attribute of Adaptation to Parasitism. Mycoses, 2011, 55: 410–415, 2011.

WEBSITES

Centers for Disease Control and Prevention, Fungal Diseases, 2021: https://www.cdc.gov/fungal/index.html

Mayo Clinic, Ringworm (body), 2022: https://www.mayoclinic.org/search/search-results?q=ringworm

Medscape, Tinea Capitis Medication, 2020: https://www.medscape.com/answers/1091351-36320/which-medications-in-the-drug-class-antifungal-agents-are-used-in-the-treatment-of-tinea-capitis

Patient, Dermatophytosis, Tinea Infections, 2021: https://patient.info/doctor/dermatophytosis-tinea-infections

Students can test their knowledge of this case study by visiting the Instructor and Student Resources: [www.routledge.com/cw/lydyard] where several multiple choice questions can be found.

Trypanosoma spp.

38

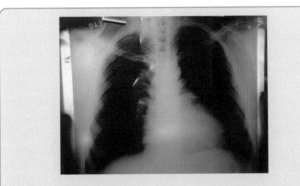

Figure 38.1 A chest X-ray showing enlargement of the heart due to Chagas disease. *From the World Health Organization, Special Programme for Research and Training in Tropical Diseases, http://www.who.int/tdr/index.html. Image #9905349. Additional photographic credit is given to Andy Crump who took the photograph in 1999 in Argentina.*

A 32-year-old female from Brazil presented to her local hospital with a sudden onset of left leg, arm, and facial weakness. She was able to speak and reported being in good health in the past except that she became short of breath running after her 4-year-old son. She grew up as a child in Brazil but came to the UK in her early twenties. A CT scan of her brain showed an acute **stroke**. The ECG tracing of her heart was abnormal with broadened QRS complexes. A chest X-ray showed enlargement of the heart (Figure 38.1). An **ECHO scan** of the heart showed a dilated left ventricle with an apical **aneurysm**, in which there was a small thrombus. An **ELISA** test for *Trypanosoma cruzi* antibodies was performed because of her Brazilian origin and proved positive. She was diagnosed with Chagas disease with cardiomyopathy and an embolic stroke. She was treated with intensive rehabilitation, anticoagulated, and commenced on ACE inhibitors. For her infection, she was treated with benznidazole.

1. WHAT IS THE CAUSATIVE AGENT, HOW DOES IT ENTER THE BODY AND HOW DOES IT SPREAD A) WITHIN THE BODY AND B) FROM PERSON TO PERSON?

CAUSATIVE AGENT

Trypanosomes are flagellated protozoan parasites. In Africa, the species of *Trypanosoma brucei* has two subspecies that infect humans (Figure 38.2). In Central and West Africa, this is *T. brucei gambiense* and in East and Southern Africa it is *T. brucei rhodesiense*. The species in South America is *Trypanosoma cruzi* (Figure 38.3). It is estimated that in the course of evolution *T. brucei* and *T. cruzi* diverged from each other about 100 million years ago. The life cycle and clinical features arising from these two species differ.

ENTRY AND SPREAD WITHIN THE BODY

When *T. brucei* is inoculated into a new human host by tsetse flies local multiplication occurs under the skin. There is spread to lymph nodes and then entry into the bloodstream. Multiplication occurs in blood. Within a few weeks, *T. brucei rhodesiense* passes from the bloodstream, probably through the choroid plexus, into the central nervous system (CNS). This CNS invasion takes several months with *T. brucei gambiense*.

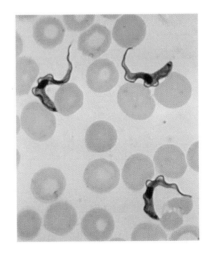

Figure 38.2 *Trypanosoma brucei* spp. in a blood film. The trypanosomes are slender with an undulating membrane leading to the flagellum. *From the Centers for Disease Control & Prevention, Atlanta, Georgia. Image is found in the Public Health Image Library #613. Additional photographic credit is given to Dr Myron G Schultz who took the photo in 1970.*

T. cruzi is transmitted by triatomine bugs (Figure 38.4). Local inflammation occurs at the site of inoculation. Again, there is spread to local lymph nodes and then entry into the bloodstream. Various tissues may be invaded but key targets are the heart and the gastrointestinal (GI) tract.

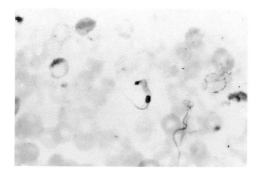

Figure 38.3 *Trypanosoma cruzi* in a blood film. It often appears as a "C" shape. *From the Centers for Disease Control & Prevention, Atlanta, Georgia. Image is found in the Public Health Image Library #543.*

Figure 38.4 *Triatoma infestans*, which transmits *Trypanosoma cruzi*. *From the Centers for Disease Control & Prevention, Atlanta, Georgia. Image is found in the Public Health Image Library #613. Additional photographic credit notes that the images was donated by the World Health Organization, Geneva, Switzerland, in 1976.*

Figure 38.5 Tsetse fly, *Glossina morsitans*, the vector for African trypanosomiasis. *From Martin Dohrn / Science Photo Library, with permission.*

Multiplication occurs in the tissues. The intracellular life form is the amastigote, which lacks a flagellum. Replication of the amastigote produces a pseudocyst within a cell. The extracellular life form is the trypomastigote. *T. cruzi* emerges from infected cells and passes onto other cells.

PERSON-TO-PERSON SPREAD

Human African trypanosomiasis (HAT) is spread by tsetse flies belonging to the genus *Glossina* (Figure 38.5). The geographic distribution of HAT is determined by the ecologic requirements of tsetse flies. This is patchy in countries between the sub-Saharan region and the Kalahari and Namib deserts. Infection of humans can be person to person but also a **zoonosis** with tsetse flies transmitting trypanosomes from a reservoir of ungulates (see below). *Glossina morsitans* usually transmits *T. brucei rhodesiense* and *G. palpalis* usually transmits *T. brucei gambiense*. Trypanosomes ingested from an animal host pass through the mid-gut of the tsetse fly, undergo developmental changes, and reach the salivary glands. Here, they are referred to as metacyclic trypomastigotes. When tsetse flies bite humans, their saliva passes into the bites bearing these trypomastigotes. This is called **salivarian transmission**. The lifecycle of HAT is shown in Figure 38.6.

In South America, various animals can serve as a reservoir, examples include the armadillo and the opossum. Triatomine bugs ingest trypanosomes when biting infected animals. The trypanosomes remain within the intestine of the bug. If they next feed on humans, they bite the skin and defecate at the same time. Humans scratching in the vicinity of the bite rub feces bearing metacyclic trypomastigotes into the open wound. This is called **stercorarian transmission**, the Latin root *sterco* referring to feces. The life cycle of South-American trypanosomiasis is shown in Figure 38.7.

Regrettably, another rare form of transmission is through blood transfusion. In resource-poor settings, blood screening may not be feasible. However, transfusion-related transmission has been described in the US from blood donated by South-American immigrants. Understandably, antibody screening of blood is not routine in nonendemic countries.

EPIDEMIOLOGY

The WHO estimates that 6–7 million people are infected with *T. cruzi*, this representing a considerable decline over previous decades. There has been a dramatic decline in HAT. In 2004, the WHO estimated a prevalence of 0.5 million with 4800 deaths per annum. By contrast, in 2019, the WHO reported that fewer than 1000 cases of HAT were detected, and 98% of these were due to *T. brucei gambiense*. *T. brucei gambiense* primarily affects humans, while *T. brucei rhodesiense* is primarily within ungulates, including cattle and wild game, with occasional human cases.

HAT is endemic in 36 countries and South-American trypanosomiasis in 21 countries.

2. WHAT IS THE HOST RESPONSE TO THE INFECTION AND WHAT IS THE DISEASE PATHOGENESIS?

There are innate and adaptive host responses to trypanosomal infection. Humans may be exposed to nonpathogenic species of trypanosomes. However, in normal human serum, apolipoprotein L-1 (APOL1) binds to the trypanosome surface and is endocytosed. Within the

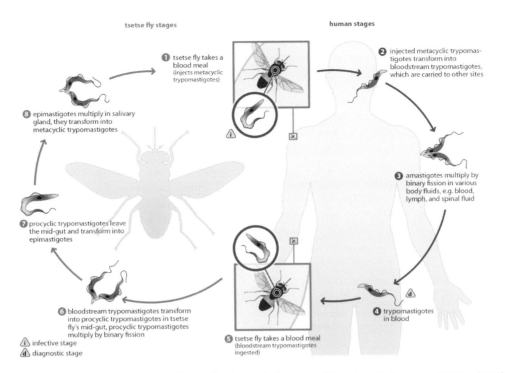

Figure 38.6 Life cycle of African trypanosomiasis. Tsetse flies inoculate humans with metacyclic trypomastigotes when they take a blood meal (1). Trypomastigotes multiply in tissue fluid, then in lymph nodes, and then enter the bloodstream (2). In the bloodstream, they continue multiplying with antigenic variation to avoid the host response. Eventually, they enter the central nervous system (3). Circulating parasite may be engulfed by tsetse flies when they take a blood meal (4). There is then development within the mid-gut of the tsetse fly before migration to the salivary glands (6–8). *From the Centers for Disease Control & Prevention, Atlanta, Georgia. Image is found in the Public Health Image Library #613. Additional photographic credit notes that the images was donated by the World Health Organization, Geneva, Switzerland, in 1976.*

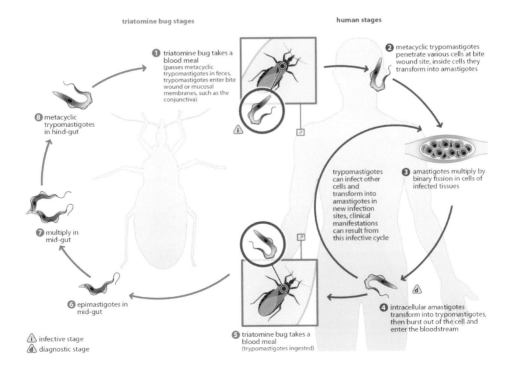

Figure 38.7 Life cycle of South-American trypanosomiasis. Humans are infected by stercorarian transmission from triatomine bugs (1). At the site of infection, metacyclic trypomastigotes enter cells (2), transform into amastigotes and multiply (3). Trypomastigotes arise from infected cells, enter the bloodstream and infect other cells in the body (4). Any circulating parasite may be engulfed by triatomine bugs when they take a blood meal (5). Further development occurs in the mid-gut of the bugs (6-8). *Adapted with kind permission from the Centers for Disease Control & Prevention, Atlanta, Georgia. Image is found in the Public Health Image Library #3384. Additional photographic credit is given to Alexander J da Silva, PhD, and Melanie Moser who created the image in 2002.*

trypanosomal cytoplasm, APOL1 forms pores in **lysosomes**. Release of lysosomal contents causes trypanosomal killing. Species of trypanosomes that are usually nonpathogenic may only cause infection in humans with APOL1 deficiency. The pathogenic species *T. brucei rhodesiense* possesses a serum resistance-associated protein (SRA), which strongly binds to APOL1, inhibiting its toxic action. The other pathogenic species, *T. brucei gambiense*, does not possess SRA, but has reduced uptake of APOL1, increased degradation in the cytoplasm, and increased membrane stiffening.

T. brucei remains extracellular in the bloodstream and is exposed to the host immune response. Specific antibodies appear against the surface glycoprotein and lyse the trypanosomes through activation of complement. However, in *T. brucei rhodesiense*, there are an estimated 1000 different variants of the surface glycoprotein. *T. brucei* switches the gene from one variant to another (**antigenic variation**). Each new antibody response is met with a gene switch and the escaping, new variants of trypanosomes multiply in successive waves (Figure 38.8).

The variant specific glycoproteins (VSG) stimulate B lymphocytes to produce **IgM** in a T-cell independent manner. Eventually, host serum has an excess of polyclonal IgM antibodies. The polyclonal activation of B cells compromises their ability to respond to other pathogens (another escape mechanism, like the antigenic variation to fool the host immune system). Other trypanosomal antigens pass through antigen-presenting cells and stimulate **CD4+** lymphocytes to mount a T-cell-dependent antibody response. Trypanosomes directly release factors, such as Trypanosome-derived

lymphocyte triggering factor (TLTF) and trypanosomal macrophage activating factor (TMAF), that stimulate **CD8+** T lymphocytes and macrophages. Through this cellular activation, various **cytokines** and mediators appear. **Tumor necrosis factor-α (TNF-α)** contributes to the weight loss of chronic infection. It seems to inhibit trypanosomal growth, but conversely **interferon-γ (IFN-γ)** seems to help trypanosomal proliferation.

Dysregulated antibody production leads to the appearance of autoantibodies. **Immune complexes** damage vascular endothelium and on binding to the surface of red blood cells cause **hemolysis**. Chronic infection suppresses bone marrow function, probably through cytokine effects. This, added to hemolysis, causes **anemia**. Equally, platelet numbers may fall and disturbance of clotting may lead to hemorrhage or conversely **thrombosis**.

Inflammation occurs in tissues containing *T. brucei*. This ranges from the skin at the site of local inoculation, to lymph nodes, to organs seeded from the bloodstream such as the heart, the kidneys, and eventually the CNS. Within the CNS, there is progressive inflammation of the **leptomeninges**. The inflammatory response, immune complexes, local cytokines, and **prostaglandins** all contribute to CNS pathology, which eventually involves disruption of the **choroid plexus** and the blood–brain barrier.

While *T. brucei* is extracellular, *T. cruzi* is principally intracellular. The pseudocysts within tissues do not excite an inflammatory response. This only occurs once they rupture from cells and release antigens. *T. cruzi* is coated with a mucin-type glycoprotein that is anchored to the surface

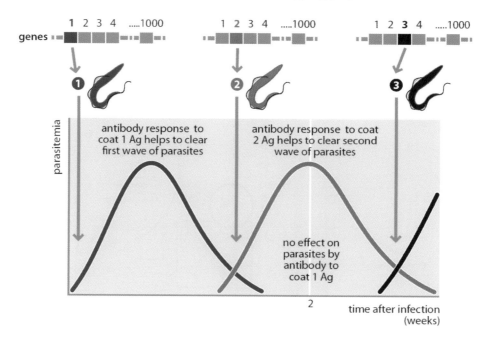

Figure 38.8 Antigenic variation of *T. brucei*. An antigen-specific response is made against the surface coat antigen, which gradually removes the parasites from the bloodstream (coat 1). During this time, some of the parasites escape by switching their surface coat gene and the antibody response has to start from scratch. Antibodies to the second coat antigen (coat 2) are made, which remove these new parasites from the circulation. Some of the parasite will then escape again by switching their gene usage to coat 3 and so on. In *T. brucei rhodesiense*, it is estimated that there are 1000 different variants of the glycoprotein that make up the surface coat. (For simplicity, individual genes are depicted in a linear sequence whereas they are actually scattered about in the genome of the organism.)

by a glycosylphosphatidylinositol (GPI). There are several hundred surface mucin genes, but antigenic variation in *T. cruzi* has not been established. GPI is well recognized as a potent macrophage activator and is thereby pro-inflammatory. Cell-mediated immune responses occur in infected tissue. In mice, CD8+ lymphocytes are essential for survival during the early stages of experimental infection. There is a polyclonal activation of B cells and it is conceivable that autoantibodies arise through dysregulation. Various *T. cruzi* antigens are also cross-reactive with host antigens found in vascular endothelium, muscle interstitium, and cardiac and nervous tissue. Autoimmune pathology is possible as passive transfer of serum or lymphocytes can cause pathology in recipients. However, it is not clear whether chronic pathology is caused by long-term infection-stimulated tissue inflammation or an autoimmune process or both.

3. WHAT IS THE TYPICAL CLINICAL PRESENTATION AND WHAT COMPLICATIONS CAN OCCUR?

In HAT, local inflammation at the site of tsetse fly inoculation is sometimes apparent and referred to as a trypanosomal **chancre**. This may occur in half of *T. brucei rhodesiense* infections but is rare with *T. brucei gambiense*. It appears after 1–2 weeks, may last for 2–3 weeks, and may reach a diameter of 3–5 cm. There may be regional **lymphadenopathy**.

Symptoms occur once trypanosomes multiply and circulate in the bloodstream. These commence about 1–3 weeks after infection with *T. brucei rhodesiense*, but weeks to months with *T. brucei gambiense*. There are fevers, headache, malaise, and loss of appetite. The fevers follow a cyclical pattern as the VSGs change. More generalized lymphadenopathy, **hepatomegaly**, and **splenomegaly** appear. There may be skin **rashes**.

Invasion of the CNS represents the second stage of infection, but there is overlap of first- and second-stage features. There is progressive mental deterioration culminating in coma and death, with a variety of intervening CNS manifestations. There is disturbance of motor function, co-ordination, behavior, and sleep. HAT is called sleeping sickness. Patients may sleep in the daytime and be awake at night. *T. brucei rhodesiense* infection progresses more rapidly, with death within 9 months, while *T. brucei gambiense* may take a few years.

In South-American trypanosomiasis, inflammation at the site of inoculation is called a **chagoma**. The disease was originally described by Carlos Chagas in Brazil in 1907 and the disease is also called Chagas disease. Sometimes, the inflammation occurs around the eyelid and the local swelling is called **Romana's sign** (Figure 38.9). Regional lymphadenopathy is followed by fever, headache, malaise, hepatomegaly, splenomegaly, and rash. In the acute phase, inflammation in seeded organs results in **myocarditis**, diarrhea and vomiting or **meningoencephalitis**.

The clinical features of the acute phase last 1–2 months and features may not return despite continued infection. Up to one

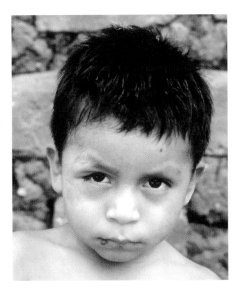

Figure 38.9 **Romana's sign in a boy.** There is swelling by the right eyebrow. The boy will have been infected in that region by a triatomine bug. *From the Centers for Disease Control & Prevention, Atlanta, Georgia. Image is found in the Public Health Image Library #2617. Additional photographic credit notes that the photo was taken by Dr Mae Melvin in 1962.*

third of patients suffer chronic complications after one or two decades. These principally affect the heart and the GI tract.

Cardiac muscle weakens and thins. Impaired cardiac function results in heart failure. Aneurysms can develop, typically at the apex of the left ventricle. Clots formed within the aneurysm can embolize. Cardiac conduction is altered with ECG abnormalities. **Dysrhythmias** or complete heart block may cause sudden death.

In Brazil, more than elsewhere, GI involvement results in dilatation and loss of peristalsis. This results in **megaesophagus** and **megacolon**, with accompanying **dysphagia** and constipation (Figure 38.10). The inability to eat causes **cachexia** and death.

4. HOW IS THE DISEASE DIAGNOSED, AND WHAT IS THE DIFFERENTIAL DIAGNOSIS?

Unfortunately, many individuals suffer from trypanosomiasis distant from diagnostic facilities. Clinical features are nonspecific. In Chagas disease, Romana's sign and local lymphadenopathy are suggestive.

For *T. brucei gambiense* the simplest test is the **C**ard **A**gglutination **T**est for **T**rypanosomiasis (**CATT**). A drop of heparinized whole blood is placed on a card containing freeze-dried trypanosomes and the presence of agglutination by serum antibodies is observed over 5 minutes. This test has high sensitivity but false positive results can occur. Direct visualization of *T. brucei rhodesiense* and *T. brucei gambiense* may be observed in lymph node aspirates, blood films, and cerebrospinal fluid (CSF). If present in reasonable numbers,

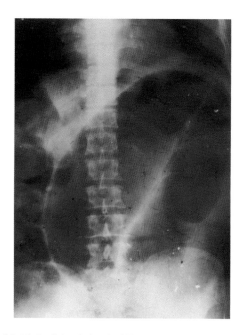

Figure 38.10 A plain abdominal X-ray. There is marked dilatation of the colon, megacolon, which is easily apparent on the X-ray. *From the World Health Organization, Special Programme for Research and Training in Tropical Diseases, http://www.who.int/tdr/index.html. Image #9105027. Additional photographic notes indicate the image was taken in Brazil in 1990.*

the motile trypanosomes are easily observed. However, the level of parasitemia may be low in later stages of infection making parasitologic diagnosis difficult. Examination of the CSF is essential for staging infection and includes microscopy for both parasites and white cell counts.

In Chagas disease, trypanosomes may be seen in a blood film only during the acute phase. The main diagnostic test is **serology**, which may employ an ELISA. Once there are organ complications, ECGs or ECHOs of the heart reveal structural abnormalities and imaging of the esophagus and colon reveal dilatation.

DIFFERENTIAL DIAGNOSIS

Mega syndromes do not have alternative explanations of note. The differential diagnosis for the cardiac complications includes other cardiomyopathies. Otherwise, the systemic **febrile** phase of early Chagas disease and HAT has a long differential diagnosis. In Africa, the CNS phase may have to be distinguished from meningitis, **encephalitis**, cerebral malaria, CNS tuberculosis, HIV-related neurologic disease, and cerebral tumors.

5. HOW IS THE DISEASE MANAGED AND PREVENTED?

MANAGEMENT

The treatments available for HAT are shown in Table 38.1. Treatment choices can depend on species, patient age, weight,

stage of disease, CSF white cell count, and the balance between toxicity and efficacy. Melarsoprol is an arsenical drug and causes a reactive encephalopathy in 20% on treatment and mortality in 2–12%. For *T. brucei gambiense*, it has been superseded by an oral agent, fexinidazole, and the combination of nifurtimox plus eflornithine.

Drugs for Chagas disease include benznidazole (first line) or nifurtimox. Again, they are both toxic and it is debated whether they should be used in the asymptomatic phase of infection before complications arise. Potentially older individuals will die with, rather than from, *T. cruzi* infection. Toxicity seems to be less in children than in adults. There are now studies of benznidazole in children showing that a 60-day course is reasonably well tolerated with disappearance of *T. cruzi* antibodies in almost 60%.

PREVENTION

The control of trypanosomiasis is hampered by its rural occurrence. When possible, it is important to actively diagnose cases and offer treatment for or before complications. Otherwise, vector control is the key strategy employed. Insecticide sprays are used to kill tsetse flies. Insecticide-impregnated traps are positioned in key locations (Figure 38.11). The traps are made of blue or black cloth as these colors attract the tsetse flies. Tsetse populations have been reduced in some localities by the release of sterilized male flies. While tsetse flies are principally outdoors, the triatomine bugs of South America live within cracks in walls or in thatched roofs. Improved housing is important in reducing bug populations. Insecticide spraying has also been very successful in eliminating transmission in many areas. In South America, universal blood screening is also important.

Figure 38.11 Tsetse fly traps in rural Africa. *From the World Health Organization, Special Programme for Research and Training in Tropical Diseases, http://www.who.int/tdr/index.html. Image #9604658. Additional photographic credit is given to Andy Crump who took the photograph in 1996 in Uganda.*

Table 38.1 Drugs used to treat trypanosomiasis

Pathogen	Drugs and comments
T. brucei gambiense	**Fexinidazole**
	Not approved if < 6 years of age or < 20 kg
	Any stage, but
	Not for severe second-stage disease with CSF white cell count
	>100 cells mL-1
	Oral agent
	Pentamidine
	Early stage
	Intravenous or intramuscular
	Can cause drop in blood pressure, and changes in renal and hepatic function and blood glucose levels
	Nifurtimox plus Eflornithine
	Late stage
	Oral (nifurtimox) and intravenous (eflornithine)
	Can cause gastrointestinal upset, headache, joint and muscle pains, bone marrow suppression
	Melarsoprol
	Intravenous
	Can cause encephalopathy, with convulsions and coma, headache, thrombophlebitis, rash
T. brucei rhodesiense	**Suramin**
	Early stage
	Can cause reversible nephrotoxicity, bone marrow suppression, and rash
	Melarsoprol (see above)
	Late
T. cruzi	Benznidazole
	Oral
	Can cause gastrointestinal upset, headache, joint pains, itch, and rash
	Nifurtimox
	Oral
	Can cause gastrointestinal upset, headache, joint and muscle pains

SUMMARY

1. WHAT IS THE CAUSATIVE AGENT, HOW DOES IT ENTER THE BODY, AND HOW DOES IT SPREAD A) WITHIN THE BODY AND B) FROM PERSON TO PERSON?

- Trypanosomes are flagellated protozoan parasites.
- African trypanosomiasis (HAT) is caused by *Trypanosoma brucei rhodesiense* or *T. brucei gambiense*.
- South American trypanosomiasis is caused by *Trypanosoma cruzi*.
- Tsetse flies transmit infection in Africa from other humans or an animal, ungulate reservoir.
- Triatomine bugs transmit infection in South America from an animal reservoir.
- After an insect bite, trypanosomes spread from local tissue to lymph nodes, then enter the bloodstream, and finally invade tissues.
- In HAT, invasion of the central nervous system occurs after weeks to several months.
- In South-American trypanosomiasis, an aflagellate, intracellular life form, the amastigote, forms a pseudocyst within cells.

Continued...

...continued

2. WHAT IS THE HOST RESPONSE TO THE INFECTION AND WHAT IS THE DISEASE PATHOGENESIS?

- Nonpathogenic trypanosomes are killed by normal human serum through the action of apolipoprotein L-1. Pathogenic trypanosomes neutralize this through a serum resistance-associated protein (SRA) and other mechanisms.
- In the bloodstream, *T. brucei* species are killed by the action of antibody and complement.
- *T. brucei* species undergo antigenic variation and evade the antibody response.
- Antigenic stimulation by *T. brucei* causes polyclonal antibody production by B cells.
- Trypanosome-derived lymphocyte triggering factor (TLTF) and trypanosomal macrophage activating factor (TMAF) stimulate CD8+ T cells and macrophages, respectively.
- In HAT, various cytokines are produced, and among these TNF-α can cause weight loss.
- Invasion of the central nervous system and leptomeningeal inflammation cause a "sleeping sickness".
- *T. cruzi* sheds glycosylphosphatidylinositol (GPI), which is a potent macrophage activator.
- Inflammation occurs in *T. cruzi*-infected tissue and this may lead to long-term pathology, particularly of cardiac muscle and gastrointestinal smooth muscle.
- Autoantibodies also appear in *T. cruzi* infection but their relative contribution to pathology is unclear.

3. WHAT IS THE TYPICAL CLINICAL PRESENTATION AND WHAT COMPLICATIONS CAN OCCUR?

- Swelling may occur at the site of the insect bite.
- As trypanosomes enter the bloodstream, there are fevers, headache, malaise, and loss of appetite.

- Once the central nervous system is invaded in HAT, there is progression to coma and death with a variety of intervening neurologic problems.
- After clinical features from acute *T. cruzi* infection up to one third of patients suffer complications in the heart or gastrointestinal tract.
- Cardiac muscle can become weak and aneurysmal.
- Weakening of intestinal smooth muscle leads to dilatation and megaesophagus or megacolon.

4. HOW IS THE DISEASE DIAGNOSED, AND WHAT IS THE DIFFERENTIAL DIAGNOSIS?

- The simplest test for HAT due to *T. brucei gambiense* is a card agglutination test for antibody.
- Trypanosomes may be seen on a blood film or in a tissue sample.
- South-American trypanosomiasis may be diagnosed by a serologic test.

5. HOW IS THE DISEASE MANAGED AND PREVENTED?

- Drugs for trypanosomes can be toxic.
- For *T. brucei gambiense*, treatment has advanced with the use of oral fexinidazole or the combination of nifurtimox plus eflornithine.
- For chronic South-American trypanosomiasis, treatment with benznidazole is probably beneficial in children but of debated value in adults.
- Tsetse fly numbers can be reduced by outdoor insecticide-impregnated traps.
- Improving housing conditions reduces the population of triatomine bugs, which would otherwise live in cracks in walls or thatched roofs.

FURTHER READING

Burri C, Chappuis F, Brun R. Human African trypanosomiasis. In: Farrar J, Hotez P, Junghanns T, et al., editors. Manson's Tropical Diseases, 23rd edition. Elsevier/Saunders, London, 622–621, 2014.

Franco-Paredes C. American trypanosomiasis: Chagas Disease. In: Farrar J, Hotez P, Junghanns T, et al., editors. Manson's Tropical Diseases, 23rd edition. Elsevier/Saunders, London, 606–630, 2014.

Murphy K, Weaver C. Janeway's Immunobiology, 9th edition. Garland Science, New York/London, 2016.

REFERENCES

Bern C. Chagas Disease. N Engl J Med, 2015, 373: 456–466.

Buscher P, Cecchi G, Jamonneau V. et al. Human African trypanosomiasis. Lancet, 390: 2397–2409, 2017.

Lindner AK, Lejon V, Chappuis F, et al. New WHO Guidelines for Treatment of Gambiense Human African trypanosomiasis Including Fexinidazole: Substantial Changes for Clinical Practice. Lancet Infect Dis, 20: e38–e46, 2020.

Mesu VKBK, Kalonji WM, Bardonneau C, et al. Oral Fexinidazole for Late-Stage African Trypanosoma brucei gambiense trypanosomiasis: A Pivotal Multicentre,

Randomized, Non-Inferiority Trial. Lancet, 391: 144–154, 2018.

Pays E, Vanhollebeke B, Uzureau P, et al. The Molecular Arms Race Between African trypanosomes and Humans. Nat Rev Microbiol, 12: 575–584, 2014.

WEBSITES

Centers for Disease Control, Parasites – African Trypanosomiasis (also known as Sleeping Sickness), 2022: https://www.cdc.gov/parasites/sleepingsickness/index.html

Centers for Disease Control, Parasites – American Trypanosomiasis (also known as Chagas Disease): https://www.cdc.gov/parasites/chagas/

World Health Organization, Chagas disease (American trypanosomiasis): https://www.who.int/health-topics/chagas-disease

World Health Organization, Human African trypanosomiasis (sleeping sickness): https://www.who.int/health-topics/human-african-trypanosomiasis

Students can test their knowledge of this case study by visiting the Instructor and Student Resources: [www.routledge.com/cw/lydyard] where several multiple choice questions can be found.

Varicella-zoster virus

39

A young girl was taken to see her doctor by her mother. She had developed an itchy blistering **rash** that first appeared as red spots on her forehead 2 days previously and had now spread over her whole body. She had been previously well and was not on any medications.

On examination, the rash was indeed extensive. The girl had a temperature of 38.1°C. The doctor made a clinical diagnosis of chickenpox.

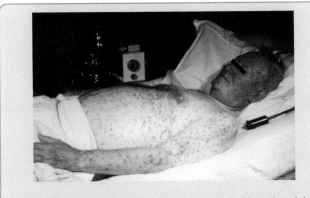

Figure 39.1 An adult male with chickenpox. Note the global distribution of the rash all over the body. *From the Centers for Disease Control & Prevention, Atlanta, Georgia. Image is found in the Public Health Image Library #10169. Additional photographic credit notes that the photo was taken by Dr Alexander D Langmuir in 1960.*

1. WHAT IS THE CAUSATIVE AGENT, HOW DOES IT ENTER THE BODY AND HOW DOES IT SPREAD A) WITHIN THE BODY AND B) FROM PERSON TO PERSON?

CAUSATIVE AGENT

The causative agent of chickenpox is varicella-zoster virus (VZV). This belongs to the *Herpesviridae* family of viruses, the characteristics of which are that the genome is double-stranded DNA, surrounded by a protein coat or **capsid** with icosahedral symmetry, and the **nucleocapsid** in turn is surrounded by a loosely fitting, irregular **lipid envelope** (Figure 39.2). The double-barreled name of the virus – VZV – is because initial infection with this virus gives rise to the disease varicella (the proper name for chickenpox), while reactivation of infection (explained below) gives rise to the disease zoster (also known as **shingles**). There are currently eight herpesviruses that can cause human disease (see Table 9.1 in the Epstein-Barr virus case, Case 9).

ENTRY AND SPREAD WITHIN THE BODY

VZV infection is acquired through entry of droplets containing viral particles into the upper respiratory tract. After an initial phase of replication at the site of entry, the virus spreads to the regional lymph nodes and organs of the **mononuclear phagocyte system** where further replication takes place, with transfer of virus to T cells. About 10–12 days after initial

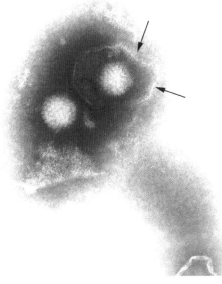

Figure 39.2 Electron micrograph of a herpesvirus, showing the geometrically regular nucleocapsid, which contains the viral DNA, and the loosely fitting irregular lipid envelope (arrows).

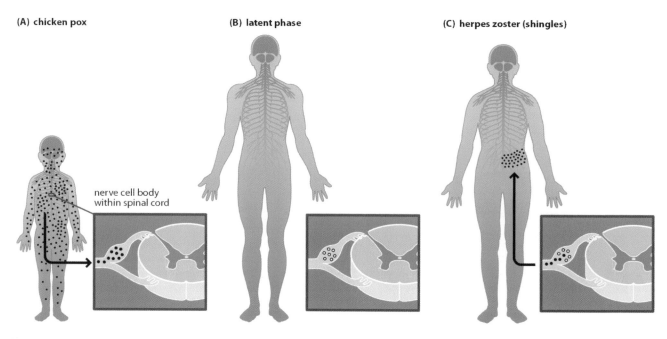

(A) chicken pox

nerve cell body
within spinal cord

(B) latent phase

(C) herpes zoster (shingles)

Figure 39.3 VZV latency. (A) During primary infection, bloodstream spread of virus results in the characteristic rash of chickenpox appearing all over the body. Some virus enters the nerve terminals of the nerves supplying the skin and ascends up the nerve axon to reach the nerve cell body, the site of latency for this virus. (B) Latent virus sits within the nerve cell bodies. In this state, there is no viral replication, and no damage to the cell. (C) This latent virus may become reactivated out of latency and start to replicate. Mature virus particles then descend down the nerve axon to reach that area of skin supplied by that particular nerve, thereby resulting in the characteristic dermatomal distribution of the zoster (or shingles) rash. *Adapted with permission from The Wellcome Trust, London, UK.*

infection, VZV-infected T cells are released from the lymph nodes into the bloodstream (**viremia**), transporting the virus to the skin, where the final phase of replication occurs, giving rise to the characteristic vesicular rash of chickenpox, the clinical manifestation of primary infection with this virus.

Like all herpesviruses, VZV establishes latency (see also Case 17: HSV). Replicating virus within the skin gains entry to the nerve terminals supplying the skin, and travels in a retrograde direction up the nerve axons to the nerve-cell bodies, the site of **latency** for this virus. In latently infected cells, the viral genome is present, but there is no replication or production of virus particles, and few, if any, viral proteins are synthesized (Figure 39.3). As there has been viremic spread of

the virus around the body, all nerve-cell bodies in the spinal cord and those of the cranial nerves may be latently infected.

Once an individual is infected with VZV, they remain infected for life. The virus remains in a latent state and does not cause any damage to the infected nerve-cell bodies. However, under certain circumstances, latent virus may become reactivated, resulting in viral replication and production of new virus particles. The precise molecular mechanisms whereby virus establishes or breaks out of latency are not fully understood, but a number of clinical factors that predispose to reactivation of latent virus are well recognized (Table 39.1). Newly formed virus particles travel back down the axon to reach the skin, resulting in a vesicular rash similar to that of chickenpox, but characterized by having a distribution limited to the area of skin supplied by the nerve in which reactivation has taken place – a so-called **dermatomal** distribution. This reactivated infection (also referred to as secondary or recurrent infection) is known as herpes zoster, zoster, or shingles.

PERSON-TO-PERSON SPREAD

An individual with chickenpox sheds large amounts of virus in droplet form from the nose and throat, and VZV is transmitted via an airborne route. The clinical attack rate among exposed susceptible individuals is in the range 70–90%, indicating that an individual with chickenpox is highly infectious.

It is possible to catch chickenpox from someone who has an attack of shingles, as there is plenty of viable virus in the vesicular rash of shingles. However, such individuals are much less infectious than patients with chickenpox, as they are not shedding large amounts of virus from the nasopharynx.

Table 39.1 Factors predisposing to reactivation of latent VZV

Increasing age

Trauma (including surgery, dentistry)

Stress (including illness)

Sunburn

Immunocompromise

    HIV infection

    Posttransplantation

    Underlying malignancy*

    Radiotherapy, chemotherapy

    High-dose steroids

*Particularly of the lymphoid and myeloid systems, e.g. leukemia, lymphoma.

INCIDENCE OF INFECTION

Infection with VZV is extremely common worldwide, with some geographic variations in epidemiology. In temperate countries, infection is usually acquired in childhood, with a peak age of infection less than 5 years old. In the UK, by adulthood, over 90% of the population have antibodies to VZV, and are therefore latently infected (and at risk of developing shingles). In tropical regions, infection usually occurs at a much older age, with a mean age of infection above 20 years old. Reasons for this are not clear but may reside around different patterns of social mixing between infected cases and susceptible contacts.

The average annual incidence of herpes zoster, the reactivated or secondary form of infection, is 0.2%. Disease is very much age-related, with rates of disease in the over 65s around 10 times higher than in the under 50s – reflecting declining cellular immune function with increasing age.

2. WHAT IS THE HOST RESPONSE TO THE INFECTION AND WHAT IS THE DISEASE PATHOGENESIS?

IMMUNE RESPONSE

Innate Immunity

Pattern-recognition receptors (PRR) that are able to sense VZV pathogen-associated molecular patterns include Toll-like receptors (TLRs) and nucleotide-binding oligomerization domain like receptor NLRP3. Viral double-stranded (ds) RNA intermediates and single-stranded RNA are also detected by cytosolic RNA receptors such as melanoma differentiation-associated protein 5 (MDA-5) and retinoic acid inducible gene (RIG-I), and viral dsDNA binds to RNA polymerase III and cyclic GMP-AMP synthase (cGAS). The end result of all these interactions is activation of transcription factors such as NFKappaB, and induction of type 1 interferons and other pro-inflammatory cytokines.

Natural killer (NK)-cell deficiencies have been reported to be associated with enhanced VZV pathogenicity, suggesting these cellular components of the innate immune system are important in control of this infection.

Adaptive Immunity

In an immunocompetent host, VZV infection stimulates both humoral and cellular arms of the adaptive immune response. Antibodies to VZV, initially of the **IgM** class, but then of the **IgG** class, can be detected about 14 days after initial infection. These antibodies have neutralizing activity, and therefore prevent the access of virus into uninfected cells. They may also mediate antibody-dependent cellular cytotoxicity, and inhibition of cell-to-cell spread. Passively administered anti-VZV immunoglobulin can prevent the development of primary varicella in exposed individuals. However, the development of

cellular immune responses is more important in controlling initial (and reactivated) infection. This involves both NK cells and CD4+ helper and **CD8**+ cytotoxic T cells. Patients with deficiencies in this arm of the adaptive response may suffer severe, if not fatal, infection.

The adaptive immune responses are important in maintaining latency of VZV – as evidenced by the fact that immunosuppressed individuals, especially those with defects in cellular, rather than humoral, immune responses, are more likely to develop shingles.

As might be expected, VZV has evolved mechanisms to evade the above host defenses. At least three VZV proteins (ORF47, ORF61, and ORF62) inhibit PRR signaling pathways; the virus is able to block IFN-1 signaling by inhibiting phosphorylation of STAT-1 and STAT-2 and by reducing Janus kinase 2 and IRF9 protein levels; VZV can modulate activation receptor ligands on NK cells to escape NK-cell killing.

PATHOGENESIS

The pathogenesis of chickenpox has been described above. Herpes zoster represents reactivation of virus in a nerve-cell body within a single sensory ganglion (a collection of nerve cell bodies), with subsequent transport of the virus down the axon to infect the skin supplied by that nerve. Zoster can occur anywhere on the body. As indicated in Table 39.1, herpes zoster is much more common in patients with underlying immunodeficiency, which may arise through a number of different causes. Malignant disorders of the reticuloendothelial system such as **lymphoma**, **leukemia**, and **multiple myeloma**, are associated with abnormalities in T-cell function and hence development of zoster, which may be unusually prolonged or aggressive. An attack of zoster in a younger-than-expected patient may be the presenting feature of underlying HIV infection. On a more subtle level, it is not unusual for patients to present with zoster rashes at times of stress, for example, bereavement or divorce, suggesting the existence of as yet ill-defined links between central nervous system (CNS) function and the immune system.

3. WHAT IS THE TYPICAL CLINICAL PRESENTATION AND WHAT COMPLICATIONS CAN OCCUR?

Many primary infections with VZV do not result in clinical disease. When symptomatic, primary infection results in varicella, also known as chickenpox. After an incubation period of 13–17 days, disease usually presents with its most striking feature, the vesicular rash (Figure 39.1), in a centripetal distribution (i.e. more evident centrally than peripherally), with lesions usually on the trunk and face (including within the mouth). Prodromal symptoms, for example, a flu-like illness, are relatively unusual, especially in children. The lesions evolve from **macules** (flat red patches in the skin) to **papules** (raised lesions) to **vesicles** (fluid-filled blisters) on a red (**erythematous**) base in a few hours (Figure 39.4). The vesicles become pustular, and then dry out and scab over the next 2–4 days, followed by regeneration of the underlying epithelium. Cropping occurs over several days, so that lesions of differing ages may be present. The patient is also usually **febrile** once the rash appears, although constitutional symptoms are usually mild, particularly in children. The rash is itchy in some, but not all patients.

Complications of chickenpox are unusual, especially in an immunocompetent child. Secondary bacterial infection of the vesicles may occur, which very occasionally may lead to **septicemia**, and also may result in scarring once the lesions heal. The disease is much more debilitating in adults, especially smokers, who are also at significant risk of varicella **pneumonia**, the commonest life-threatening complication. This presents with shortness of breath and cough about 2–3 days after onset of the rash. **Encephalitis** is uncommon but may present about 7–10 days after onset. The most likely pathogenesis of this is a post-infectious encephalitis, that is, not due to the presence of virus itself within the brain substance but arising through aberrant immune responses to infection that damage the brain tissue. A cerebellar syndrome is most often seen in children, but **hemiplegia** is also described. Recovery is usual, but long-term sequelae and even fatalities may occur.

Chickenpox in an immunosuppressed individual, whether child or adult, is altogether a more serious and life-threatening disease. Varicella pneumonia is much more common in this patient group, as is hemorrhagic varicella, a severe complication often associated with **thrombocytopenia**. Bleeding occurs into the skin lesions, from mucous membranes, and into unaffected skin, and the prognosis is poor.

Chickenpox in pregnancy is of particular concern. Adult pregnant females are more likely to die from varicella pneumonia than adult nonpregnant females. In addition, virus may cross the placenta and damage the developing fetus, giving rise to congenital varicella or varicella embryopathy, the characteristic features of which are extensive areas of skin scarring, and failure of limb bud development. This occurs in about 1% of pregnancies where maternal varicella occurs within the first 20 weeks of pregnancy. If the mother acquires infection late in pregnancy, then there is a risk that the baby will be born and acquire infection before the mother has had time to generate protective IgG anti-VZV, which can cross the placenta. Neonatal chickenpox may be very severe, with mortality rates of up to 30% reported.

SHINGLES

The clinical features of shingles are protean and depend, to a large extent, on the location of the nerve cell body in which reactivation arises. The shingles rash is acutely painful, and occurs in a dermatomal distribution (see above, and Figure 39.5). The commonest and most distressing complication is **post-herpetic neuralgia** (**PHN**), defined as pain persisting for at least 3 months (although the precise time period may vary between different studies) in the area of a zoster rash after the rash itself has disappeared. The pathogenesis is poorly understood, but it can persist for months or even years, and can have a significantly detrimental effect on the patient's quality of life. The risk of PHN increases with increasing age of the patient and is also related to the intensity of the shingles rash (i.e. the number of lesions). Other complications depend on the anatomic location of the rash. Ophthalmic zoster (occurring in the ophthalmic branch of the trigeminal nerve) is particularly serious, as virus may gain access to the cornea, giving rise to

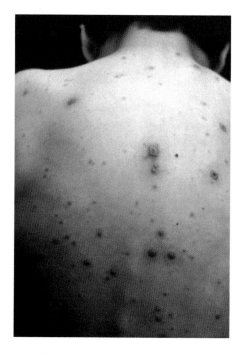

Figure 39.4 Close-up of the varicella-zoster rash. Note the presence of lesions at different stages of development and the fluid-filled vesicles on an erythematous base. *From the Centers for Disease Control & Prevention, Atlanta, Georgia. Image is found in the Public Health Image Library #6121. Additional photographic credit notes that the photo was taken by Dr Alexander D Langmuir in 1960.*

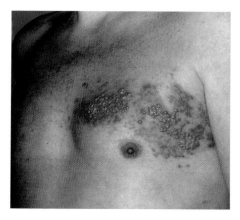

Figure 39.5 A zoster rash. This arises due to reactivation of VZV from its site of latency within the nerve cell body. The rash is dermatomal in distribution, i.e. affects only that area of skin supplied by the affected nerve. In this patient, the rash continued onto his back, stopping in the midline.

an ulcerating zoster **keratitis**. While mostly thought of as a disease of sensory nerves, motor nerve damage can also arise in zoster, giving rise to lower motor neuron paralysis, for example, zoster of the facial nerve (seventh cranial nerve) may give rise to a **Bell's palsy** (facial nerve paralysis).

Both primary and secondary VZV infections can be life-threatening in patients with underlying immunodeficiencies, particularly involving the T-cell arm of the immune response, for example, patients with HIV infection, or transplant recipients on potent immunosuppressive drugs. The presence of virus within the bloodstream (which is a routine stage of the pathogenesis of primary infection and may arise in secondary infection) means that virus may seed visceral organs, and the absence of an appropriate immune response to contain this infection may result in, for instance, fatal varicella **pneumonitis**, **hepatitis**, or encephalitis. CNS complications of zoster may also arise from a virus-induced vasculitis.

4. HOW IS THE DISEASE DIAGNOSED, AND WHAT IS THE DIFFERENTIAL DIAGNOSIS?

Diagnosis of primary infection is usually made on clinical grounds, as the vesicular rash of chickenpox is so distinctive. There are very few differential diagnoses, especially since the elimination of smallpox. The smallpox rash, in contradistinction to chickenpox, is usually centrifugal, and the lesions are rounder and deeper than the more irregular and superficial chickenpox ones. More importantly, the lesions of smallpox evolve at a similar rate and, therefore, are not present in all stages as occurs with chickenpox.

Zoster can be a difficult diagnosis to make, especially in the early stages where the patient may present with acute pain before the rash has appeared. However, once the rash appears, again, a clinical diagnosis is usually reliable, the main differential diagnosis being reactivated herpes simplex virus (HSV) infection.

Laboratory confirmation of the diagnosis can be made in a number of ways. A vesicle swab should be sent to the laboratory, where the presence of VZV DNA can be demonstrated by genome detection techniques such as the **polymerase chain reaction** (**PCR**). This has replaced other diagnostic modalities, such as virus isolation in tissue culture, immunofluorescent detection of viral antigens in cells shaken off the swab (antigen detection), or even visualization of virus particles in vesicle fluid by electron microscopy. A serum sample from a patient with chickenpox will be IgM anti-VZV positive – however, this serologic approach is not as reliable in the diagnosis of zoster, as this is a reactivated infection and IgM responses may not be as clear-cut.

5. HOW IS THE DISEASE MANAGED AND PREVENTED?

MANAGEMENT

VZV is treated with aciclovir and its derivatives. VZV is able to phosphorylate aciclovir using the virally encoded enzyme thymidine kinase, and therefore replication of VZV is sensitive to aciclovir, and derivatives of aciclovir such as valaciclovir (a valine ester) or famciclovir (see HSV case, Case 17, for more detail on the action of aciclovir). However, aciclovir therapy is not usually indicated in otherwise uncomplicated childhood chickenpox. In the US, aciclovir is licensed for that indication, but the justification for its use is an economic, not a medical one – antiviral therapy results in a speedier recovery of the child, which allows the adult carers to go back to work sooner. It is reasonable to treat chickenpox in an adult with aciclovir, as the disease is inherently more severe in this age group. Any immunodeficient patient developing primary VZV infection should be treated with high-dose aciclovir, as this is potentially life-threatening. Although aciclovir is not licensed for use in pregnancy, it can be lifesaving if varicella pneumonia develops.

It is not necessary to treat every case of shingles. Most guidelines recommend treatment if the patient has any underlying immunodeficiency, regardless of the cause (risk of internal organ dissemination and death), if the patient is over 50 years old (increased risk of PHN), or if the rash is in the territory of the ophthalmic branch of the trigeminal nerve (risk of zoster keratitis) or the facial nerve (risk of motor nerve damage). If therapy is indicated, it must be started within 72 hours of onset of the rash to be most effective.

PREVENTION

There is a licensed vaccine against primary infection with VZV – this consists of a live attenuated virus known as the Oka strain. It was initially developed for use in children with severe underlying disease such as leukemia, who would otherwise be at risk of life-threatening chickenpox. Thus, it has an excellent safety record, even in patients with impaired immune systems. As this is a live virus, occasionally vaccinees may develop vaccine-associated chickenpox, although this is usually mild. Vaccine virus may also go latent in the spinal cord and reactivate in the form of shingles, but this is rare and not usually severe. In the US, this is recommended as a routine vaccination of childhood. However, in the UK, this vaccine is currently only licensed for use in healthcare workers shown to be susceptible to infection (i.e. anti-VZV-negative). There is a theoretical risk that vaccination in childhood may paradoxically increase the burden of disease associated with

VZV infection. If vaccine-induced immunity is not long-lasting, then vaccination in childhood may result in older adolescents and adults acquiring primary infection, with a consequent increase in severity of disease. Also, it has been suggested that exposure to children with chickenpox serves to boost VZV immunity in adults which, in turn, allows continued suppression of VZV reactivation. Elimination of disease in small children may therefore lead to an increase in herpes zoster in adults.

There are also two vaccine preparations available for the prevention of reactivated or secondary infection with VZV, that is shingles or herpes zoster. The first of these (marketed in the UK as Zostavax) is comprised of the same Oka strain as above, and is therefore live attenuated, but with a much higher antigen dose. It is given as a single intramuscular (IM) injection. The second (Shingrix) is a recombinant vaccine containing VZV glycoprotein E, adjuvanted with ASO1B. This is given as a course of two injections separated by at least 2 months. Vaccination with Zostavax is recommended for immunocompetent adults over the age of 70 years, while Shingrix is advised for patients of similar age but who are severely immunosuppressed.

SUMMARY

1. WHAT IS THE CAUSATIVE AGENT, HOW DOES IT ENTER THE BODY, AND HOW DOES IT SPREAD A) WITHIN THE BODY AND B) FROM PERSON TO PERSON?

- VZV, a herpesvirus.
- Infection acquired in the throat, initial viral replication in the local lymph nodes, followed by viremic (i.e. blood-borne) spread taking virus to the skin, giving the characteristic widespread vesicular rash of varicella or chickenpox.
- Virus replicating in the skin gains access to sensory nerve terminals and travels up the nerve axon to the nucleus of the nerve cell body in the sensory ganglion.
- Nerve cell body is the site of virus latency.
- Reactivation from latency results in virus particles traveling down to the nerve terminals, giving rise to the rash of herpes zoster, or shingles.
- Patients with chickenpox shed large amounts of virus from the throat, as well as from the skin rash.
- Patients with herpes zoster are also infectious as the skin lesions contain viable virus.

2. WHAT IS THE HOST RESPONSE TO THE INFECTION AND WHAT IS THE DISEASE PATHOGENESIS?

- Innate immune responses, triggered by pattern-recognition receptors binding to VZV pathogen-associated molecular patterns, generate potent interferon and pro-inflammatory cytokine responses.
- An antibody response develops that is first detectable about 7 days after onset of the rash, initially IgM only, followed by IgG.
- T-cell responses are also key to preventing the development of life-threatening internal organ dissemination of virus.

3. WHAT IS THE TYPICAL CLINICAL PRESENTATION AND WHAT COMPLICATIONS CAN OCCUR?

- Primary infection, when symptomatic, presents as chickenpox (varicella), a widespread vesicular rash, with fever and generalized lymphadenopathy.
- The rash evolves from macule to papule to vesicle to crust in 4 days.
- Crops of lesions appear, firstly on face and trunk, and then spread to scalp and limbs.
- Complications in childhood are unusual.
- In adults, varicella is a more severe disease, and varicella pneumonia is a life-threatening complication.
- Maternal varicella in the first 20 weeks of pregnancy may result in fetal damage in 1% of cases. Maternal varicella in late pregnancy may give rise to neonatal varicella. Pregnant women with chickenpox are also at increased risk of varicella pneumonia.
- Herpes zoster (shingles) is a painful vesicular rash occurring in a dermatomal distribution.
- Complications of zoster include post-herpetic neuralgia, zoster keratitis, motor nerve damage, and dissemination in immunosuppressed hosts.

4. HOW IS THE DISEASE DIAGNOSED, AND WHAT IS THE DIFFERENTIAL DIAGNOSIS?

- Diagnosis of both chickenpox and shingles can be made on clinical grounds, on the basis of the characteristics of the rash.
- Laboratory confirmation can be made by genome amplification techniques, antigen detection, virus isolation or electron microscopy.
- Antibody tests are only of use in primary infection, where an IgM anti-VZV response can be demonstrated.

Continued...

...continued

5. HOW IS THE DISEASE MANAGED AND PREVENTED?

- Treatment is possible with aciclovir (an antiviral drug that is a deoxyguanosine analog) or derivatives such as valaciclovir or famciclovir.
- Treatment is not usually indicated in otherwise uncomplicated chickenpox in children.
- Treatment is recommended for chickenpox in adults and in any immunodeficient patient.

- In zoster, treatment is recommended if the patient is immunodeficient, aged over 50 years, or zoster is in ophthalmic or facial nerves. It must be initiated within 72 hours of onset of the rash.
- Prevention is by means of a live attenuated vaccine.
- In the UK, varicella vaccine is only licensed for VZV-susceptible healthcare workers.
- In the US, varicella vaccine is licensed for routine use in childhood.
- A high-dose live attenuated vaccine and a subunit glycoprotein E vaccine are now licensed for use for the prevention of reactivated VZV infection in the elderly.

FURTHER READING

Barer M, Irving W, Swann A, Perera N. Medical Microbiology, 19th edition. Elsevier, Philadelphia, 2019.

Goering RV, Dockrell HM, Zuckerman M, Chiodini PL. Mims' Medical Microbiology and Immunology, 6th edition. Elsevier, Philadelphia, 2018.

Humphreys H, Irving WL, Atkins BL, Woodhouse AF. Oxford Case Histories in Infectious Diseases and Microbiology, 3rd edition. Oxford University Press, Oxford, 2020.

Murphy K, Weaver C. Janeway's Immunobiology, 9th edition. Garland Science, New York/ London, 2016.

Richman DD, Whitley RJ, Hayden FG. Clinical Virology, 4th edition. ASM Press, Washington DC, 2016.

REFERENCES

Cunningham AL, Levin MJ. Herpes Zoster Vaccines. J Infect Dis, 218: S127–S133, 2018.

Gerada C, Campbell TM, Kennedy JJ, et al. Manipulation of the Innate Immune Response by Varicella Zoster Virus. Front Immunol, 11: 1, 2020.

Johnson RW, Levin MJ. Herpes Zoster and its Prevention by Vaccination. Interdiscip Top Gerontol Geriatr, 43: 131–145, 2020.

Kennedy PGE, Gershon AA. Clinical Features of Varicella-Zoster Virus Infection. Viruses, 10: 609, 2018.

Laing KJ, Ouwendijk WJD, Koelle DM, Verjans GMGM. Immunobiology of Varicella-Zoster Virus Infection. J Infect Dis, 281: S68–S74, 2018.

WEBSITES

All the Virology on the WWW Website, Tulane University, 1995: http://www.virology.net/garryfavweb12.html#Herpe

Centers for Disease Control and Prevention, Chickenpox (Varicella): https://www.cdc.gov/chickenpox/

GOV.UK, Shingles (herpes zoster): the green book, chapter 28a, 2013: https://www.gov.uk/government/publications/shingles-herpes-zoster-the-green-book-chapter-28a

GOV.UK, Varicella: the green book, chapter 34, 2013: https://www.gov.uk/government/publications/varicella-the-green-book-chapter-34

Virology Online, Herpesviruses Slide Set: http://virology-online.com/viruses/VZV.htm

Students can test their knowledge of this case study by visiting the Instructor and Student Resources: [www.routledge.com/cw/lydyard] where several multiple choice questions can be found.

Wuchereria bancrofti

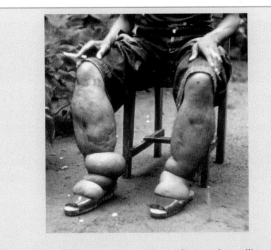

40

Figure 40.1 This man has more advanced swelling of his legs than in the case history. His bilateral lymphedema is more advanced and will not resolve with treatment. *From the Centers for Disease Control & Prevention, Atlanta, Georgia. Image is found in the Public Health Image Library #373.*

A 24-year-old man from Southern India presented with swelling of his left foot. The swelling diminished overnight. He was otherwise well with no fevers or systemic upset. Initially, there was concern that he had a deep vein **thrombosis**, but his D-dimers were not raised. There were a few enlarged lymph nodes in his groin and, on closer questioning, he reported previous episodes with the nodes becoming painful and swollen. In his hometown in India he recalled seeing individuals with grossly swollen legs. His full blood count revealed an **eosinophilia** of 1.4×10^9 L^{-1} (normal range $<0.45 \times 10^9$ L^{-1}). His filarial **serology** was positive. He was treated with a prolonged course of doxycycline. The eosinophil count fell to 0.4×10^9 L^{-1} and his swelling subsided (Figure 40.1).

1. WHAT IS THE CAUSATIVE AGENT, HOW DOES IT ENTER THE BODY AND HOW DOES IT SPREAD A) WITHIN THE BODY AND B) FROM PERSON TO PERSON?

CAUSATIVE AGENT

Wuchereria bancrofti is the causative agent of Bancroftian filariasis. It is a round worm, or nematode which, in the adult form, lives in lymphatic vessels. Females are about 4–10 cm long and males about 2–4 cm. However, they are very slender, being 0.25 and 0.3 mm in width, respectively. Other filarial nematodes that cause a similar illness are *Brugia malayi* and *B. timori*, which are confined to South-East Asia.

ENTRY AND SPREAD WITHIN THE BODY

Infection occurs with the bite of various species of mosquito vector (see below). Larvae enter the tissues and are thought to migrate along the lymphatics. They commonly reach the lymphatics draining to lymph nodes in the groin and sometimes the armpit. Adult worms mature, mate and, after

about 8 months, females release ensheathed microfilariae from the ova in their uterine bodies. These circulate in the bloodstream and are available to be ingested when a mosquito takes a blood meal (Figure 40.2). The life cycle of *W. bancrofti*

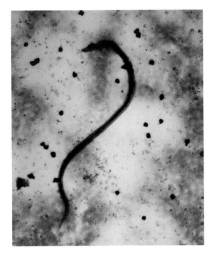

Figure 40.2 *Wuchereria bancrofti* microfilariae in a blood film. *From the Centers for Disease Control & Prevention, Atlanta, Georgia. Image is found in the Public Health Image Library #3008. Additional photographic credit is given to Dr Mae Melvine who took the photo in 1978.*

is shown in Figure 40.3. In areas where mosquitoes bite at night microfilariae appear in the bloodstream in maximum numbers around midnight. In other areas, mainly the Pacific Islands, mosquitos bite during the day and night, and microfilarial levels remain relatively constant in the bloodstream throughout the day. Microfilariae have a lifespan of about 1 year. Adult worms are capable of living up to 15 years, but usually for about 5 years.

PERSON-TO-PERSON SPREAD

Depending on geographic location, *W. bancrofti* is transmitted by *Anopheles*, *Aedes*, *Culex* or *Mansonia* mosquitoes. When mosquitoes ingest microfilariae-laden blood, the microfilariae pass into the midgut, shed their sheath, and penetrate the wall of the midgut to enter the body cavity. They migrate to muscles in the thorax and undergo two molts. The third stage larvae are the infective form and pass onto the mouth parts of the mosquito. Upon blood feeding, the larvae are deposited onto the skin surface and enter the host via the puncture made by the mosquito. Humans are the only host for *W. bancrofti* and the reservoir for infection is the local microfilaremic human population.

EPIDEMIOLOGY

Infections occur in Asia, Africa, Pacific Islands, South America, and the Caribbean. In 2000, the World Health Organization (WHO) estimated that there were 120 million cases of lymphatic filariasis in 80 countries, with 40 million individuals being severely incapacitated. Through elimination efforts, the estimated prevalence in 2018 was 51 million cases with about 38 million of these in South-East Asia. In endemic areas, infections establish from childhood and rates rise with increasing age.

2. WHAT IS THE HOST RESPONSE TO THE INFECTION AND WHAT IS THE DISEASE PATHOGENESIS?

Repeated mosquito bites expose individuals to the third stage larvae (L3). These go on to mature into adult worms, which mate and, in turn, release microfilariae. The immune response may be directed against infecting L3 larvae, adult worms or microfilariae and may involve antibody, lymphocytes, and other mononuclear cells. The immune response to each of

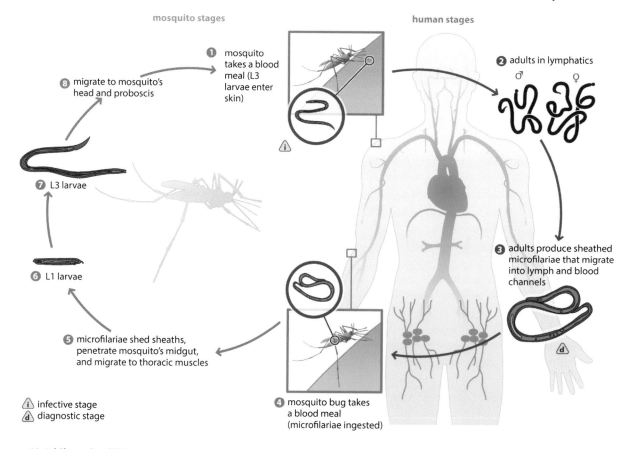

Figure 40.3 Life cycle of *Wuchereria bancrofti*. L3 larvae escape from the mosquito onto the skin surface at the time of taking a blood meal from humans (1). The larvae then enter the human host through the puncture made by the mosquito. The larvae mature into adults, which settle in lymphatics (2). Male and female adult worms mate and then females release ensheathed microfilariae larvae, which circulate in the blood (3). These may be ingested by mosquitoes feeding on humans (4). In the mosquito mid-gut, the sheaths are shed (5). Larvae then mature into the L3 form, which passes to the mosquito mouthparts (6-8). *From the Centers for Disease Control & Prevention, Atlanta, Georgia. Image is found in the Public Health Image Library #3425. Additional photographic credit is given to Alexander J da Silva, PhD, and Melanie Moser who created the image in 2003.*

these stages seems to be segregated. Some individuals will progressively stimulate a protective immune response to the L3 larvae. This may become so effective by adulthood that new infections can no longer occur. Adult worms present within lymphatics will eventually die and so, in an endemic community, there will be some who are immune and clear of infections.

In vitro, microfilariae can be killed by **IgE**-mediated degranulation of eosinophils and mast cells. This killing mechanism may be modified. In endemic areas, there will be mothers who have filariasis at the time of pregnancy. Microfilariae have been seen in cord blood and newborns have **IgM** and IgE antibodies to microfilarial antigen, as well as IgG. The IgM and IgE **antibody isotypes** do not cross the placenta and their production implies *in utero* microfilarial antigen exposure. It is postulated that this early antigen exposure modifies the subsequent immune response when children are exposed to microfilariae. In some studies in children, anti-microfilarial IgE-mediated degranulation of eosinophils and mast cells does not occur. It is thought that high titers of **IgG**4 block anti-microfilarial IgE in an antigen-specific way. This enables the survival of microfilariae. Peripheral blood lymphocytes taken from teenagers born to microfilaremic mothers do not proliferate *in vitro* and produce **interferon-γ (IFN-γ)** to the same extent as controls. There is a dominance of **Th-2** responses over **Th-1** responses, with potentially **interleukin 10** (IL-10) and **transforming growth factor-β (TGF-β)** down-regulating Th-1 responses. At the molecular level, there is down-regulation of the expression of **Toll-like receptors** on T cells (TLR1, TLR2, TLR4, and TLR9), which means that they are less responsive to stimulation through these receptors. One contributor to immunomodulation is the release by adult worms of the phosphorylcholine containing molecule ES-62, which has a variety of immunosuppressive effects. Thus, by various possible means, the immune response to infection is down-regulated through childhood and adolescence. This results in persistent infection and microfilaremia.

Immune responses do not appear to be effective in killing adult worms. However, cumulative exposure to adult worms, alive or dying, stimulates inflammatory immune responses to adult worm antigens. Adult worms also carry an endosymbiont bacterial organism called *Wolbachia*. Antigens from *Wolbachia* also stimulate an immune response. These may reach a stage where intense inflammatory reactions occur around adult worms in lymphatics. Microfilarial sheath antigen also triggers an inflammatory reaction via TLR4, dendritic cell involvement, and a Th-1 response. Inflammatory reactions contribute to lymphatic pathology. This will not be the only contributory mechanism. Adult worms themselves can cause dilatation of the lymphatics, which is called **lymphangiectasia**, without an accompanying inflammatory response. Secondary bacterial infections have been shown to contribute to the maintenance of lymphatic pathology.

A small minority of individuals experience an immune response to microfilariae in the lung. The resulting inflammatory process results in the clinical condition of tropical pulmonary eosinophilia.

There is variation between individuals and populations in the pathogenesis of disease. There is a genetic component to immune responses, with familial aggregation of certain forms of the clinical disease. People who migrate into endemic areas do not have the possible benefit of a modified immune response from childhood. They experience stronger immune responses and more lymphatic pathology. The same applies to those in endemic areas not born to microfilaremic mothers.

3. WHAT IS THE TYPICAL CLINICAL PRESENTATION AND WHAT COMPLICATIONS CAN OCCUR?

There is a spectrum of clinical manifestations. The majority of individuals, about 70%, remain asymptomatic while others experience complications related to the adult worms or the microfilariae.

A common clinical manifestation of lymphatic filariasis is swelling in the scrotum, a **hydrocele**, whereby adult *W. bancrofti* localize in the scrotal lymphatics and lymphatics of the spermatic cord. Adult worms can also cause acute inflammation of lymph nodes and lymphatics, **adenolymphangitis**, accompanied by fever. Episodes of acute adenolymphangitis probably relate to the death of adult worms and settle over a week but can recur. The lymphatic channels become damaged and dilate (lymphangiectasia). Poor lymphatic drainage predisposes to secondary bacterial infection of the skin. When bacteria inflame afferent lymphatics, red streaks appear along their course (**lymphangitis**) and further lymphatic damage occurs. Eventually **lymphedema** can develop in the limbs or genitals, which can lead to **elephantiasis** (Figures 40.1 and 40.4). Lymphedema stretches the skin and causes it to thicken, with underlying **fibrosis**. Pushing a finger into the swollen leg does not create an indented pit. Cracks in the skin can lead to further

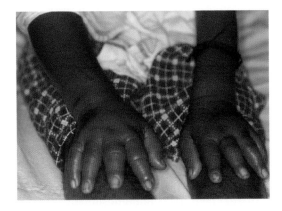

Figure 40.4 Lymphedematous arms. *From the World Health Organization, Special Programme for Research and Training in Tropical Diseases, http://www.who.int/tdr/index.html. Image #01021639. Additional photographic credit is given to Andy Crump who took the photograph in 2001 in India.*

secondary bacterial infection, aggravating the damage already done to the lymphatics. Sometimes blocked lymphatics within the abdomen can result in lymph discharging into the urinary tract. Urine takes on a milky appearance due to the fat content of lymph. This is called **chyluria**.

The presence of microfilariae in the bloodstream is in general asymptomatic but may rarely be associated with intermittent, nocturnal fevers. A strong immune response may be manifest in the lung. As microfilariae pass through the lung they cause an inflammatory reaction, with fever, cough, and wheeze and widespread inflammatory infiltrates on a chest X-ray. This is most pronounced at night. The eosinophil count is raised and there is a high titer of antifilarial antibodies. This respiratory manifestation is called tropical pulmonary eosinophilia (TPE) and is usually observed in Asia rather than Africa or South America. TPE is actually quite rare and affects about 1% of infected subjects. In TPE, microfilariae are actually scanty in the peripheral blood and are not found on blood films. Recurrent bouts of TPE damage the lungs with fibrosis. Bouts can be curtailed with a micro-filaricidal drug, diethylcarbamazine (see Section 5).

4. HOW IS THE DISEASE DIAGNOSED, AND WHAT IS THE DIFFERENTIAL DIAGNOSIS?

The diagnosis of lymphatic filariasis due to *W. bancrofti* may be based on visualizing the parasite, detecting antigen, serology or clinical features.

Adult worms are small and are rarely found in tissue. Careful ultrasound of the scrotum may sometimes reveal wriggling adults in lymphatic vessels. Microfilariae may be found in chyluria, but parasitologic diagnosis is usually based on finding microfilariae in a peripheral blood film. This requires nocturnal blood sampling in areas where lymphatic filariasis is nocturnally periodic. One can provoke microfilariae to appear in the bloodstream during the day by giving the drug diethylcarbamazine (DEC) and taking blood 30–60 minutes later. The DEC provocation test is as sensitive as a nocturnal blood film. As filariasis is a chronic infection, microfilariae may be present in the blood over prolonged periods. In endemic areas, microfilariae may be present in the blood when subjects succumb to another infection, such as malaria. Thus, a positive microfilarial blood film does not necessarily mean that the current illness is due to filariasis.

There are now tests for filarial antigen. If positive, they indicate current infection. Antigen is present in the blood in the day and the night and there is no need for nocturnal blood sampling. Current tests are highly sensitive and specific for Bancroftian filariasis. There are two commercial antigen tests. One is an **enzyme-linked immunosorbent assay** (**ELISA**) and the other an immunochromato-graphic card test. Serologic tests for antifilarial antibodies have suffered from poor sensitivity and specificity. In endemic areas, it is difficult to distinguish between current and past infection. Tests have suffered from cross-reactions with other parasites.

Eosinophilia is common and reflects the responses to all life stages. Elephantiasis, once established, is unmistakable, but earlier stages of infection have differential diagnoses and can warrant specific investigation.

Across the clinical spectrum of filariasis, there are differentials for different manifestations, including intermittent fevers, **lymphadenopathy**, **edema**, cough, wheeze, and eosinophilia. **Febrile** illness may be due to a long list of infections. Those that do not settle spontaneously include malaria, tuberculosis, and HIV. The differential diagnosis of lymphadenopathy is discussed with the *Toxoplasma* case. A swollen limb may be due to local **cellulitis**, a deep vein thrombosis, heart failure or low protein states from **nephrotic syndrome** or **cirrhosis**. Edema once it has reached the stage of elephantiasis in most endemic areas is unmistakable. In certain parts of Africa, walking barefoot introduces silicates and minerals into the skin. When these pass along the lymphatics and to lymph nodes there is a fibrotic reaction that can cause lymphatic obstruction and gross swelling. This is called **podoconiosis** and can be as severe as elephantiasis. Intermittent cough, wheeze, and eosinophilia have to be differentiated from asthma, some connective tissue diseases, **allergic alveolitis**, and lung migratory phases of intestinal nematode infections. Eosinophilia is discussed with the *Schistosoma* case.

5. HOW IS THE DISEASE MANAGED AND PREVENTED?

MANAGEMENT

There are four key drugs that have been used to treat filariasis, and each has differing effects on microfilariae and on adult worms. Drug efficacy can be monitored by blood films, ultrasound examination for adult worms, and antigen tests.

DEC has been used for several decades. It kills about 70% of microfilariae and about 50% of adult worms. It has been the backbone of attempts to eradicate filariasis. It has to be used in repeated annual cycles to reduce transmission and adult worm numbers. Ultimately, eradication of infection requires death of all adult worms, either naturally or through treatment. DEC can cause headaches and gastrointestinal (GI) upset. Dying adult worms excite episodes of adenolymphangitis. DEC is used to abort episodes of TPE due to microfilariae in the lung. However, DEC should not be used in co-infections with onchocerciasis and used with caution with loiasis due to adverse inflammatory reactions.

Albendazole can be used with DEC and ivermectin. Ivermectin is very good at killing microfilariae in the short term but has no effect on adult worms. Microfilariae therefore reappear in the blood rapidly.

Living within adult *W. bancrofti* are bacterial endosymbionts called *Wolbachia*. *Wolbachia* can be killed with the antibiotic doxycycline and if this antibiotic is used in patients with filariasis adult *W. bancrofti* die. In a trial of an 8-week course of doxycycline, there was an 80% reduction in adult worms and a loss of microfilariae in the blood 14 months after treatment. Doxycycline is avoided in pregnancy, breast-feeding mothers, and children under 12 years because of its effects on bones and teeth. These features preclude the use of doxycycline in mass treatment programs. The use of doxycycline in the case history is not typical of treatment worldwide.

Beyond specific antifilarial drugs, subjects with lymphedema need a lot of help and support in the care of their swollen limb(s) (Figure 40.5). Secondary bacterial infections need to be avoided. This requires attention to skin care, elevation of the leg to reduce swelling if possible, and prompt use of antibiotics if infection occurs. Sometimes surgery has been employed, but with limited success for swollen legs. The most effective treatment for patients suffering from hydrocele is surgery.

PREVENTION

Since 2000, the Global Progam to Eliminate Lymphatic Filariasis (GPELF) has continued elimination efforts. The mainstay of elimination efforts is periodic mass drug administration (MDA). Entire, defined populations are given chemotherapy irrespective of whether they are microfilaremic or amicrofilaremic. This is simpler than testing every individual by blood films. Repeated MDA at annual intervals is intended to reduce levels of microfilaremia so that it is less likely for mosquitoes to transmit infection. A high level of population coverage is required for this to work. MDA must also be repeated for a number of years, usually 4–6 years to cover the average lifespan of adults and their release of microfiliariae. However, in recent years, trials have shown that a single dose, three-drug combination of DEC, albendazole, and ivermectin cleared microfilariae for three years in the majority of recipients.

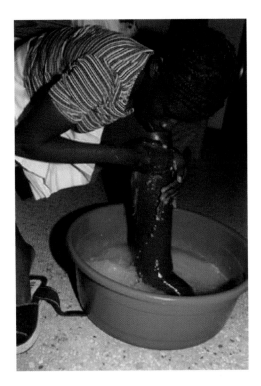

Figure 40.5 Woman carefully washing her lymphedematous leg to reduce the chances of secondary bacterial infection. *From the World Health Organization, Special Programme for Research and Training in Tropical Diseases, http://www.who.int/tdr/index.html. Image #01021471. Additional photographic credit is given to Andy Crump who took the photograph in 2001 in Haiti.*

In areas where **onchocerciasis** is present, DEC causes severe reactions in individuals co-infected with *Onchocerca volvulus*. Therefore, in these areas, ivermectin is used in combination with albendazole as part of MDA.

MDA has to work in conjunction with mosquito control in the elimination of lymphatic filariasis. This has a broader context in the prevention of malaria and other mosquito-borne diseases. Efforts in mosquito control and MDA work better combined than in isolation.

SUMMARY

1. WHAT IS THE CAUSATIVE AGENT, HOW DOES IT ENTER THE BODY, AND HOW DOES IT SPREAD A) WITHIN THE BODY AND B) FROM PERSON TO PERSON?

- Bancroftian, lymphatic filariasis is caused by *Wuchereria bancrofti*, a tissue-dwelling round worm or nematode.
- Male and female adult worms live in the lymphatics.
- Female worms release larvae called microfilariae which enter the bloodstream.
- Various species of mosquitoes ingest microfilariae in the blood and pass infection onto others.
- *Wuchereria* is only found in humans, and there is no animal reservoir.

2. WHAT IS THE HOST RESPONSE TO THE INFECTION AND WHAT IS THE DISEASE PATHOGENESIS?

- In endemic areas, pregnant mothers may have chronic filarial infection and the fetus is exposed to filarial antigen. This early exposure may modify the subsequent immune response in children.
- IgG4 antibodies may block the effects of IgE in an antigen-specific manner.
- Th-1 and Th-2 responses can be down-regulated in children and adolescents.
- Repeated exposure to third-stage larvae stimulates immunity, which may prevent further infection in older subjects.

Continued...

...continued

- Some adult individuals also develop immune responses to worms that cause lymphatic pathology.
- In Asia, there are occasional individuals who mount a strong anti-microfilaremic response, which causes intermittent pulmonary pathology with a marked peripheral eosinophilia.

3. WHAT IS THE TYPICAL CLINICAL PRESENTATION AND WHAT COMPLICATIONS CAN OCCUR?

- Infection is often asymptomatic.
- Adult worm death can cause intermittent inflammatory episodes of lymph nodes and lymphatics.
- Lymphatic damage leads to lymphedema and potentially gross swelling of dependent parts.
- Microfilariae in the blood can cause intermittent fevers.
- Tropical pulmonary eosinophilia is the clinical manifestation of a strong immune response to microfilariae with fever, cough, wheeze, and infiltrates on a chest X-ray.

4. HOW IS THE DISEASE DIAGNOSED, AND WHAT IS THE DIFFERENTIAL DIAGNOSIS?

- The presence of microfilariae in a blood film indicates current infection.
- Microfilariae may be present in the blood over long periods and therefore care must be taken in attributing the current illness to filariasis.
- Filarial antigen tests, if positive, also indicate current infection.
- In endemic areas, gross lymphedema and hydrocele is usually diagnostic of lymphatic filariasis.

5. HOW IS THE DISEASE MANAGED AND PREVENTED?

- Diethylcarbamazine (DEC) and ivermectin kill microfilariae. These are often used in conjunction with albendazole to maximize killing of adult worms.
- Doxycycline can kill adult worms through its effect on the endosymbiont bacteria, *Wolbachia*. When these bacteria are killed, the adult worms die.
- Patients complicated with gross lymphedema require careful attention to their swollen limbs and prevention of secondary infection.
- Control programs have involved repeated, annual community-wide use of combinations of drugs to reduce microfilarial numbers for transmission by mosquitoes.

FURTHER READING

Simonsen PE, Fischer PU, Hoerauf A, Weil GJ. The Filariases. In: Farrar J, Hotez P, Junghanns T, et al. editors. Manson's Tropical Diseases, 23rd edition. Elsevier Saunders, London, 737–765, 2014.

REFERENCES

Babu S, Nutman TB. Immunology of Lymphatic Filariasis. Parasite Immunol, 36: 338–346, 2014.

King CL, Suamani J, Sanuku N, et al. A Trial of Triple-Drug Treatment for Lymphatic Filariasis. New Eng J Med, 379: 1801–1810, 2018.

Local Burden of Disease 2019 Neglected Tropical Diseases Collaborators. The Global Distribution of Lymphatic Filariasis, 2000–18: A Geospatial Analysis. Lancet Glob Health, 8: e1186–1194, 2020.

Mukherjee S, Karnam A, Das M, et al. Wuchereria bancrofti Filaria Activates Human Dendritic Cells and Polarizes T Helper 1 and Regulatory T Cells Via Toll-Like Receptor 4. Commun Biol, 2: 169, 2019.

Prodjinotho UF, von Horn C, Debrah AY, et al. Pathological Manifestations in Lymphatic Filariasis Correlate with Lack of Inhibitory Properties of IgG4 Antibodies on IgE-Activated Granulocytes. PLoS Negl Trop Dis, 11: e0005777, 2017.

Taylor MJ, Makunde WH, McGarry HF, et al. Macrofilaricidal Activity After Doxycycline Treatment of Wuchereria bancrofti: A Double-Blind, Randomised Placebo Controlled Trial. Lancet, 365: 2116–2121, 2005.

WEBSITES

Centers for Disease Control and Prevention, Parasites – Lymphatic Filariasis: https://www.cdc.gov/parasites/lymphaticfilariasis/index.html

GAELF, Global Programme to Eliminate Lymphatic Filariasis: www.filariasis.org/

World Health Organization, Lymphatic filariasis (elephantiasis): https://www.who.int/health-topics/lymphatic-filariasis

Students can test their knowledge of this case study by visiting the Instructor and Student Resources: [www.routledge.com/cw/lydyard] where several multiple choice questions can be found.

Glossary

α-defensins

see also defensins

α-hemolysis

the reduction of the red blood cell hemoglobin to methemoglobin in the medium surrounding the colony of bacteria resulting in a green or brown discoloration in the medium. This is caused by hydrogen peroxide produced by the bacterium, oxidizing the hemoglobin

abscess

a localized collection of pus in any part of the body that is surrounded by swelling (inflammation)

acidophile/acidophiles

a bacterium/bacteria that grow(s) well under acid conditions

acrodermatitis chronica atrophicans

unusual, progressive, fibrosing skin process due to the effect of continuing active infection with Borrelia afzelii

acute rheumatic fever (ARF)

an inflammatory disease of the connective tissue, particularly the heart, joints, blood vessels, subcutaneous tissue, and central nervous system that occurs about three weeks following streptococcus pharyngitis in susceptible subjects

acute-phase proteins

a group of proteins (produced largely by the liver) that appear in the blood in increased amounts shortly after the onset of infection or tissue damage; they include C-reactive protein, serum amyloid, fibrinogen, mannose-binding protein, proteolytic enzyme inhibitors, and transferrin

Addison's disease

a disorder that occurs when the adrenal glands produce less of their hormones; caused by damage to the adrenal cortex through infection, autoimmunity and tumors

adenitis

inflammation of a gland

adenocarcinoma

a malignant tumor originating in glandular tissue: carcinoma – derived from epithelial cells

ADE

Antibody-dependent enhancement of dengue virus infection is identified as the main risk factor of severe dengue diseases. This is through opsonization by subneutralizing or non-neutralizing antibodies where virus infection suppresses innate cell immunity to facilitate viral replication. It is largely unknown whether suppression of type-I IFN is necessary for a successful ADE

adenolymphangitis

inflammation of the lymph nodes and the lymphatics

adenopathy

large or 'swollen' lymph nodes: see also lymphadenopathy

adhesins

bacterial proteins that promote adherence to host-cell membranes. Some are simply anchored in the bacterial membrane, whereas others are placed at the tips of pili. Adhesion is a prerequisite for entry into the body and effective adhesion contributes to virulence

adult respiratory distress syndrome (ARDS)

a diffuse pulmonary parenchymal injury associated with non-cardiogenic pulmonary edema and resulting in severe respiratory distress and hypoxemic respiratory failure. The pathologic hallmark is diffuse alveolar damage (DAD)

allergic alveolitis

inflammation of the small airways or alveoli caused by an immunological reaction to an inhaled bioaerosol (a mist or dust of biological particles) or some organic chemicals

alopecia

baldness

alpha/beta T cells (αβ T cells)

the main T cell type that recognises peptide antigens in association with major histo-compatibility antigens (HLA) class I or HLA class II; uses two chains for the T cell receptor – α and β chains

alternative complement pathway

complement activation pathway not initiated by antibody but directly through a variety of antigens including bacterial lipopolysaccharide and some virus components; see also complement pathway

alveolar macrophages

macrophages in the lung alveoli; part of the mononuclear phagocyte system

amyloidosis

a disorder that results from the abnormal deposition of a particular protein (amyloid), in various tissues of the body; amyloid proteins include immunoglobulin light chains in patients with myeloma

anaphylaxis

an immediate (within 30 min of allergen challenge) severe immune reaction mediated primarily by IgE antibodies, resulting in vasodilation and constriction of smooth muscle that may lead to death

anemia

strictly defined as a decrease in red blood cell (RBC) mass; three main classes include excessive blood loss (acutely such as a hemorrhage or chronically through low-volume loss), excessive blood cell destruction (hemolysis) or deficient red blood cell production in the bone marrow (ineffective hemopoiesis). In clinical practice, anemia is diagnosed by measuring hemoglobin levels

aneurysm

occurs when part of an artery swells. It is caused by a damaged blood vessel or a weakness in the blood vessel wall. The pressure of blood in the artery causes it to 'balloon' out at the weak point. The swelling can be small and spherical (berry-sized), normally occurring near blood vessel branches, or it can be larger and balloon-like

anhidrosis

the inability to sweat in response to heat

antibody isotypes

can be heavy chain or light chain isotypes; often used synonymously for classes of antibody (IgM, IgG, IgA, IgD and IgE; each isotype is encoded by separate immunoglobulin constant region genes μ, γ, α, δ and ε for heavy chains and κ and λ for light chains

antigenic drift

the evolutionary accumulation of amino acid substitutions in viral proteins selected by host adaptive immune systems as the virus circulates in a population. Antigenic drift can substantially limit the duration of immunity conferred by infection and vaccination

antigenic shift

a sudden shift in the antigenicity of a virus resulting from the genetic reassortment of the segmented genomes of two viral strains; seen only with influenza A viruses it may result in replacement of the hemagglutinin with a novel subtype that has not been present in human influenza viruses for a long time; the consequences of the introduction of a new hemagglutinin into human viruses is usually a pandemic, or a worldwide epidemic

antigenic variation

changes in cell surface antigenic determinants of a microorganism (by shift or drift) in order to avoid the host immune response

antimicrobial peptides (AMP)

small molecular weight proteins with a broad-spectrum antimicrobial activity; produced by a variety of cell types, as part of the innate immune response

aortic coarctation

narrowing of the aorta between the upper-body artery branches and the branches to the lower body (at the insertion of the ductus arteriosus)

aortic stenosis

narrowing of the aortic valve in the heart

aplastic anemia

a syndrome of bone marrow failure characterized by peripheral pancytopenia (reduced red blood cells but also white blood cells and platelets) and marrow hypoplasia

apoptosis

programmed death of a cell by 'suicide': involves enzymatic digestion of the DNA into packages: governed by complex signaling pathways and marked by a well-defined sequence of morphological changes, resulting in small bodies that are phagocytosed by other cells

arbovirus

a virus that is spread by the bite of infected arthropods (insects) such as mosquitoes and ticks. e.g. dengue virus, yellow fever virus, Chikungunya, yellow fever and Zika

artesunate

a derivative of Artemisinin, recommended as the first-line treatment of severe malaria

arthralgia

joint pain

arthritis

a term used to describe conditions affecting the joints; includes degeneration of the cartilage (osteoarthritis), inflammatory autoimmune arthritis (rheumatoid arthritis) and directly induced by microbes (septic arthritis)

Arthus reaction

a Type III hypersensitivity reaction; arises from deposition of immune complexes mainly in the vascular walls, serosa (pleura, pericardium, synovium), and glomeruli

Aseptic meningitis

an umbrella term for all causes of inflammation of the brain meninges that have negative cerebrospinal fluid (CSF) bacterial cultures

aspergilloma

a clump of aspergillus which populates a lung cavity (also known as a mycetoma or fungus ball)

asymptomatic

showing no symptoms of a particular disease

asthma

also called hyper-reactive airway disease; a chronic inflammatory disorder characterized by generalized but reversible bronchial airways obstruction; can be precipitated by non-specific factors, irritants including physical exertion, cold, smoke, scents and pollution; specific factors that can precipitate asthma include allergens in the form of pollen, dust, animal fur, mold, dander and some kinds of food. Some viruses and bacteria, chemical fumes or other substances at the workplace and certain medicines, e.g. aspirin and other NSAIDs

atelectasis

the collapse of part, or all, of a lung; caused by a blockage of the bronchus or bronchioles or by pressure on the lung

atopic asthma

asthma associated with specific IgE antibodies

atopic dermatitis

a common chronic skin disease (also called eczema) usually occurring in individuals with a hereditary predisposition to all atopic disorders; frequently seen in association with a high serum IgE level

atrophicus

a chronic skin disease that is characterized by the eruption of flat white hardened papules with central hair follicles often having black keratotic plugs

atrophy

wasting of tissues, organs, or the entire body, as from death and reabsorption of cells, diminished cellular proliferation, decreased cellular volume, pressure, ischemia, malnutrition, lessened function, or hormonal changes

auscultation

listening for sounds made by internal organs, such as the heart and lungs, to aid in the diagnosis of certain disorders

autophagy

the process by which cells digest parts of their own cytoplasm; allows for both recycling of macromolecular constituents under conditions of cellular stress and remodeling the intracellular structure for cell differentiation

azoospermia

absence of live spermatozoa in the semen

B lymphocytes

are a type of lymphocyte originating in the bone marrow that produces antibodies and some cytokines

bacteremia

the presence of bacteria in the bloodstream

bacteriophage

a virus that infects bacteria, also called a phage

bacteriuria

bacteria in the urine: a number of automated laboratory tests are available to detect/quantitate bacteriuria, these include turbido-metric, bioluminescence, electrical impedance, flow cytometry, and radiometric tests

Behcet's syndrome

a rare autoimmune disorder that causes ulcers and skin lesions

Bell's palsy

unilateral facial nerve paralysis, usually self-limiting

bilirubin

a product that results from the breakdown of hemoglobin: hemoglobin is broken down to heme and globin; heme is converted to bilirubin

Blackwater fever

a complication of malaria characterized by intravascular hemolysis, hemoglobinuria, and kidney failure

Bornholm's disease

infection of the chest wall musculature with a coxsackie B virus (an enterovirus) giving rise to severe chest pain

bradycardia

a resting heart rate of under 60 beats per minute in adults, though it is seldom symptomatic until the rate drops below 50 beat/min; resting heart rate under 100 beats per minute for children

Brill-Zinsser disease

reappearance of epidemic typhus years after the initial attack. The agent that causes epidemic typhus (*Rickettsia prowazekii*) remains viable for many years and then when host defenses are down, it is reactivated causing recurrent typhus

bronchiectasis

a chronic lung disease characterized by permanent, irreversible progressive, abnormal dilatation and destruction of the bronchi that results in poor clearance and pooling of mucus. This disorder also makes a person more susceptible to infections caused by bacteria

bronchiolitis

inflammation of the bronchioles, the airways that extend beyond the bronchi and terminate in the alveoli. Bronchiolitis is due to viral infections such as parainfluenza, influenza, adenovirus and, especially, respiratory syncytial virus (RSV)

bronchoalveolar lavage

a diagnostic procedure of washing a sample of cells and secretions from the alveolar and bronchial airspaces

bronchoscopy

the examination of the bronchi (the main airways of the lungs) using a flexible tube (bronchoscope). Bronchoscopy helps to evaluate and diagnose lung problems, assess blockages, obtain samples of tissue and/or fluid, and/or to help remove a foreign body

bruit

a sound, especially an abnormal one: a bruit may be heard over an artery reflecting turbulence of flow. Listening for a bruit in the neck is a simple, safe, and inexpensive way to screen for stenosis (narrowing) of the carotid artery

bullus/bullae

a blister of more than 5 mm diameter with thin skin and full of fluid

Burkitt's lymphoma

an uncommon type of Non-Hodgkin Lymphoma (NHL); commonly affects children; it is a highly aggressive type of B-cell lymphoma; associated with Epstein-Barr virus

cachexia

weight loss, wasting of muscle, loss of appetite, and general debility that can occur during a chronic disease

candidiasis

an infection caused by the fungus Candida (also called thrush)

capnophilic

term used to describe bacteria that grow best in an atmosphere of increased carbon dioxide (5–10% CO_2 in air)

capsid

the outer protein coat of a virus surrounding and protecting the nucleic acid; made up of subunits called capsomeres; in non-enveloped viruses, the outer proteins of the capsid also mediate attachment of the virus to specific receptors on the host cell surface

capsule (bacterial and fungal)

the outmost structure of bacterial and fungal cells that protects microbial cells from immune recognition and killing during infection of mammalian hosts mostly made from polysaccharides

cardia

the opening of the esophagus into the stomach; also the part of the stomach adjoining this opening

cardiomegaly

enlargement of the heart

carditis

inflammation of the heart; can be endocarditis (inflammation of the endocardium), myocarditis (inflammation of the heart muscle) or pericarditis (inflammation of the pericardium)

caseation

necrosis with conversion of damaged tissue into a soft cheesy substance

catalase

an enzyme that catalyzes the decomposition of hydrogen peroxide to water and oxygen

cathelicidin (human)

also called LL-37, is a 37 residue 18KDa cationic anti microbial peptide produced by cells that are in contact with the environment (respiratory and gastrointestinal epithelium, keratinocytes) and cells of the innate immune system including polymorphs

CATT

Card Agglutination Test for trypanosomiasis

cavitation

the formation of cavities in a body tissue or an organ, especially those cavities that form in the lung as a result of tuberculosis

CD11/CD18

leukocyte adhesion molecules (β 2 integrins)

CD4+ T cells

helper T cells; T cells that recognize specific antigenic peptides when presented to them on MHC class II molecules on antigen presenting cells

CD59

also called lymphocyte function antigen - 3 (LFA-3); it binds to CD2

CD8+ T cells

cytotoxic T cells; T cells that recognize specific antigenic peptides when presented to them on MHC class I molecules on antigen presenting cells/target cells

cellulitis

an infection of the deep subcutaneous tissue of the skin

cerebellar ataxia

a disorder of co-ordination due to a lesion affecting the cerebellum; a symptom or syndrome rather than a diagnosis, with many possible causes

cerebritis

inflammation of the brain

cervicitis

inflammation of the cervix; a common infection of the lower genital tract

chagoma

a swelling resembling a tumor that appears at the site of infection in Chagas' disease

chancre

painless ulceration formed during the primary stage of syphilis; also associated with sleeping sickness

chancroid

sexually transmitted infectious disease characterized by painful ulcers, bubo formation, and painful inguinal lymphadenopathy; caused by *Haemophilus ducreyi*, a gram-negative coccoid-bacillary rod

Charcot-Leyden crystals

crystals in the shape of elongated double pyramids, formed from eosinophils, found in the sputum in bronchial asthma and in other exudates or transudates containing eosinophils

chemokines

a class of cytokines; the main families of the four described thus far include the CC and the CXC group; they have both chemo-tactic and activating properties

cholecystitis

inflammation of the gallbladder that occurs most commonly because of an obstruction of the cystic duct

chorioretinitis

an exudative inflammatory process involving the retinal vessels usually caused by congenital viral, bacterial, or protozoan infections in neonates; congenital toxoplasmosis is the most common cause of infectious chorioretinitis in immunocompetent children; congenital cytomegalovirus (CMV) infection is the second most common etiology; fungal infections are more frequently identified, and emergent pathogens, including West Nile virus and lymphocytic choriomeningitis virus, have been described. In rare instances, chorioretinitis is part of a systemic noninfectious process

choroid plexus

a vascular proliferation of the cerebral ventricles that serves to regulate intraventricular pressure by secretion or absorption of cerebrospinal fluid

chronic fatigue syndrome

also called myalgic encephalomyelitis (or encephalopathy; ME); a disorder of unknown etiology that probably has an infectious basis

chronic granulomatous disease

an inherited disorder of phagocytic cells, resulting from an inability of phagocytes to undergo a respiratory burst to produce bactericidal superoxide anions (O–2); leads to recurrent life-threatening bacterial and fungal infections

chronic interstitial lung disease

a heterogeneous group of disorders that distort the architecture of the lung parenchyma with inflammation and fibrosis; includes idiopathic pulmonary fibrosis, sarcoidosis, hypersensitivity pneumonitis, eosinophilic pneumonias, inorganic pneumoconioses, drug-induced lung diseases, connective tissue diseases affecting the lungs, pulmonary vasculitides, the sequelae of acute respiratory distress syndrome, and bronchiolitis, among others

chronic interstitial nephritis

a large and heterogeneous group of disorders characterized primarily by interstitial fibrosis with mononuclear leukocyte infiltration and tubular atrophy; there are many causes including urinary obstruction, chronic bacterial infections, drugs (lithium, Chinese herbs) and radiation

chronic obstructive pulmonary disease (COPD)

a disease state characterized by the presence of airflow obstruction due to chronic bronchitis or emphysema; airflow obstruction is usually progressive, may be accompanied by airway hyperreactivity, and may be partially reversible

chyluria

the presence of lymph and fat in the urine, giving it a milky appearance. Due to obstruction of lymph flow, which causes rupture of lymph vessels

cirrhosis

scarring of the liver involving the formation of fibrous (scar) tissue associated with the destruction of the normal architecture of the organ

clade

a group of biological taxa (as species) that includes all descendants of one common ancestor

classical pathway (of complement)

antibody mediated activation of complement; *see also* complement pathway

clinical trials

a type of research that studies new tests and treatments e.g. drugs, surgical procedures and evaluates their effects on human health outcomes. There are 4 phases of clinical trials: *Phase I* studies usually test new drugs for the first time in a small group of people to evaluate a safe dosage range and identify side effects. *Phase II* studies test treatments that have been found to be safe in phase I but now need a larger group of human subjects to monitor for any adverse effects. *Phase III* studies are conducted on larger populations and in different regions and countries, and are often the step right before a new treatment is approved. *Phase IV* studies take place after country approval and there is a need for further testing in a wide population over a longer timeframe

colectomy

the surgical removal of the colon or part of the colon (partial colectomy)

colitis

inflammation of the large intestine (the colon). There are many forms of colitis, including ulcerative, Crohn's, infectious and pseudomembranous

collagen vascular disease (CVD)

a heterogeneous group of autoimmune disorders of unknown etiology; many CVDs involve the lungs directly or involve them as a complication of treatment of the CVD. CVDs include systemic lupus erythematosus, rheumatoid arthritis, progressive systemic sclerosis or scleroderma, dermatomyositis and polymyositis, ankylosing spondylitis, Sjögren syndrome and mixed connective-tissue disease

commensal

the normal bacterial microbiota of the body that, unless displaced from the skin or mucosal surfaces, generally do not cause disease and may actually help protect against colonization with pathogenic bacteria. If commensals transgress the epithelial barriers they may become opportunistic pathogens

complement pathway

pathway leading to activation of complement; three pathways defined: the 'classical' involving antibody, the 'alternative' pathway that activates complement without the use of antibody (i.e. via innate immune system) and the 'lectin' pathway which activates complement via mannose binding protein

conjugate vaccine

vaccines that are made by coupling the protective non-protein antigen (e.g. pneumococcal carbohydrate antigens) with a strong antigen (carrier protein – that can be recognized by the T helper cells) e.g. diptheria toxoid

conjunctivitis

an inflammation of the conjunctivae

corticosteroids

a class of drugs that have potent anti-inflammatory effects; derivatives of the steroid hormones that are produced in the adrenal cortex

CR2/CD21

complement receptor 2; found on follicular dendritic cells (where it attaches immune complexes) and B cells where it amplifies activation via the surface antibody receptor

C-reactive protein (CRP)

an acute phase protein produced in the liver; quantity found in the serum gives a measurement of systemic inflammation

Crohn's disease

a chronic inflammation of the digestive tract that can affect any part of the digestive tract from the mouth to the anus but usually involves the terminal part of the small intestine, the beginning of the large intestine (cecum), and the area around the rectum

croup

also termed laryngotracheitis or laryngotracheobronchitis, is a viral respiratory tract infection; manifests as hoarseness, a seal-like barking cough and a variable degree of respiratory distress

CSF

cerebral spinal fluid

CT scan

computed tomography scan, uses X-rays and a computer to create cross-sectional images of body organs and tissues

cyanosis

bluish discoloration of the skin and mucous membranes due to the presence of deoxygenated hemoglobin in blood vessels near the skin surface

cyst

an abnormal closed sac-like structure containing liquid, gaseous or semisolid substance that can occur in any part of the body, e.g. sebaceous, ovarian, kidney or liver

cystic fibrosis

a disease of exocrine gland function involving multiple organ systems that mainly results in chronic respiratory infections, pancreatic enzyme insufficiency, and associated complications in untreated patients. Pulmonary involvement occurs in 90% of patients surviving the neonatal period. End-stage lung disease is the principal cause of death

cystic fibrosis transmembrane conductance receptor (CFTM)

a protein that functions as a chloride channel and is regulated by cyclic adenosine monophosphate (cAMP). Mutations in the *CFTR* gene give rise to cystic fibrosis that results in abnormalities of chloride transport across epithelial cells on mucosal surfaces. This leads to decreased secretion of chloride and increased reabsorption of sodium and water across epithelial cells. The resulting reduced height of epithelial lining fluid and decreased hydration of mucus results in mucus that is stickier to bacteria, which results in infection and inflammation

cystitis

inflammation of the bladder

cytokine storm

unregulated release of proinflammatory cytokines, such as TNF-α and IL-1, into the blood by monocytes and macrophages in response to lipopolysaccharide (LPS) or lipooligosaccharide (LOS) from gram-negative bacteria

cytokines

small molecular weight proteins produced by different cell types; families classified in different ways but include the interleukins, colony-stimulating factors, tumor necrosis factors, chemokines and interferons

cytolethal distending toxin

a bacterial toxin produced by a variety of pathogenic bacteria; it enters eukaryotic cells and breaks double-stranded DNA resulting in the arrest of the cell cycle and finally leads to apoptosis of the target cell

cytolytic

destruction of a cell by lysis (breaking it open)

cytopathic

pertaining to or characterized by pathological changes in cells, e.g. cytopathic viruses

DC-SIGN

Dendritic Cell-Specific Intercellular adhesion molecule-3-Grabbing Non-integrin is a type II C-type lectin (CD209) expressed by dendritic cells and macrophages and binds to N-linked high-mannose oligosaccharides on proteins. Recognition of these MAMPS on viruses, bacteria, and fungi causes activation of phagocytosis which can also be a way of enhancing infection of DC and macrophages; *see also* MAMPS

decay accelerating factor

also called CD55; a 70kD protein present on the surface of nucleated body cells; prevents complement mediated lysis and acts as a receptor for some bacteria, e.g. *Helicobacter pylori* on stomach epithelial cells

defensins

small cationic antimicrobial proteins that insert in lipid bilayers and form pores. Alpha defensins are produced by neutrophils, macrophages and Paneth cells, and β-defensins are produced by many kinds of leucocytes and epithelial cells

dendritic cells (DCs)

several families of cells with multiple processes (resembling dendrite processes of a neuron) to enable them to interact with several other cells at the same time

dengue

caused by an arbovirus. Dengue fever (DF) and dengue hemorrhagic fever (DHF) are primarily diseases of tropical and sub-tropical areas and are caused by four closely related but antigenically distinct serotypes of the genus Flavivirus (DEN1–4); transmitted to humans by the mosquito *Aedes aegypti*; produces a spectrum of clinical illness ranging from a nonspecific viral syndrome to severe and fatal hemorrhagic disease

dermatome/dermatomal

localized area of skin that has its sensation via a single nerve from a single nerve root of the spinal cord

desmosomes

specialized cell junction characteristic of epithelial tissues into which intermediate filaments (tonofilaments of cytokeratin) are inserted; particularly present in tissues such as the skin that have to withstand mechanical stress

diathesis/diasthesia

a hereditary predisposition of the body to a disease, a group of diseases, or another disorder

disseminated intravascular coagulation (DIC)

not an illness in its own right but rather a complication or an effect of progression of other illnesses; a complex systemic thrombohemorrhagic disorder involving the generation of intravascular fibrin and the consumption of procoagulants and platelets resulting in intravascular coagulation and hemorrhage; most commonly observed in severe sepsis and septic shock; *see also* entry for septic shock

diuresis

increased formation of urine by the kidney

diverticulitis

inflammation around a diverticulum; diverticula are more commonly seen in the small or large intestine and consist of herniation of the mucosa and submucosa through the muscularis, usually at the site of a nutrient artery

Donovan body

intracytoplasmic, nonflagellated form of certain intracellular flagellates (e.g. Leishmania); also called amastigotes

Duffy binding proteins

parasite proteins that bind to Duffy antigens on erythrocyte surfaces and facilitate the invasion process in Plasmodium infections

dysesthesia

an abnormal unpleasant sensation felt when touched, caused by damage to peripheral nerves (neuropathy)

dysentery

diarrhea with blood

dyspepsia

indigestion

dysphagia

difficulty in swallowing

dyspnea

shortness of breath

dysrhythmias

irregularities in heart beats or brain waves

dysuria

burning micturition

EBV envelope

a protective outer membrane that surrounds the virus and contains glycoproteins. The virus acquires these proteins and envelope when it leaves a host cell. EBV infection is initiated by the binding of the viral envelope glycoprotein (gp350) to the cell surface receptor CR2

ECHO scan

scan using ultrasound

echocardiogram

an ultrasound heart scan

ectopic pregnancy

also known as tubal pregnancy; when the fertilized egg implants in any tissue other than the uterine wall

eczema

an inflammatory skin condition that causes itchiness, dry skin, rashes, scaly patches, blisters and skin infections

edema

abnormal accumulation of fluid outside the blood vessels; may be a manifestation of cardiac failure

EDTA

ethylenediaminetetraacetic acid; a chelating agent that sequesters Ca++ and Mg++ ions

effusion

escape of fluid from the blood vessels or lymphatics into the a cavity

egophony

increased resonance of voice sounds

electrocardiogram (ECG, EKG)

measures and records the electrical activity of the heart allowing the diagnosis of a wide range of heart conditions

electroencephalogram (EEG)

test that measures and records the electrical activity of the brain

elephantiasis

the gross (visible) enlargement of the arms, legs, genitals, or breasts to elephantoid size

ELISpot test

enzyme-linked immunosorbent spot test; an enzyme linked antibody test to enumerate antigen-specific cells usually by measuring a cytokine produced

embolism

an obstruction of a blood vessel, usually by a blood clot but can also be by an air bubble, amniotic fluid, a clump of bacteria, a chemical substance, e.g. talc, and drugs

emphysema/emphysematous

a progressive lung disease in which there is abnormal permanent enlargement of air spaces distal to the terminal bronchioles, accompanied by the destruction of the walls and without obvious fibrosis

empyema

an inflammatory fluid (pus) and debris in the pleural space resulting from an untreated pleural-space infection that progresses from free-flowing pleural fluid to a complex collection in the pleural space

encephalitis

an inflammation of the brain parenchyma; presents as diffuse and/or focal neuropsychological dysfunction; caused by infection and autoimmune mechanisms

endemic

regularly found in a particular population or geographical area

endocarditis

inflammation or infection of the endocardium, which is the inner lining of the heart muscle and, most commonly, the heart valves usually caused by bacterial infection

endocervicitis

inflammation of the mucous lining of the uterine cervix

endocytosis

the process by which a cell engulfs a large particle (bacteria, viruses, etc.) and contains it in a vesicle in the cytoplasm

endometriosis

the presence of normal endometrial mucosa (glands and stroma) that implants at sites other than the uterine cavity; this tissue responds as normal endometrium resulting in microscopic internal bleeding, with the subsequent inflammatory response, neovascularization, and fibrosis formation

endometritis

infection of the endometrium or decidua, with extension into the myometrium and parametrial tissues; the most common cause of fever during the postpartum period

endophthalmitis

inflammation of the intraocular cavities (i.e. the aqueous or vitreous humor); usually caused by infection

endoscopy

a procedure where the inside of the body is investigated using an endoscope

endosomes

membrane-bound vesicles found in the cytoplasm of virtually every animal cell. They function in the sorting and delivery of internalized material from the cell surface and the transport of materials from the Golgi to the lysosome or vacuole

endotoxin

a toxin that is part of the outer cell wall (outer membrane) of gram-negative bacteria; also called lipopolysaccharide

enteral

the route by which food or drugs are given via the mouth

enterocolitis

inflammation of the small intestine and colon

enterotoxin

a bacterial toxin that exerts its effects on the intestine, e.g. staphylococcal enterotoxin B (SEB)

enzyme-linked immunosorbent assay (ELISA)/enzyme immunoassay (EIA)

an enzyme-based system used to quantitate antigens or antibodies; enzymes coupled to antibodies include alkaline phosphatase, and peroxidase

eosinophilia

an abnormal increase in the number of eosinophils in the blood

Eotaxin

a potent chemoattractant for eosinophils. It is a member of the CC chemokine family. CCL11 is eotaxin 1, CCL24 is eotaxin 2 and eotaxin 3 is CCL26. Eotaxin 1 plays a central role in the pathogenesis of allergic airway diseases (asthma and rhinitis), in inflammatory bowel disease and gastrointestinal allergic hypersensitivity. Eotaxins are produced by IFNγ-stimulated endothelial cells, complement-activated eosinophils, TNF-activated monocytes and dermal fibroblasts.

epididymitis

inflammation of the epididymis can be acute (<6 wk) or chronic and is most commonly caused by infection

epididymoorchitis

acute inflammatory reaction of the epididymis and testis secondary to infection

epiglottitis

inflammation of the epiglottis

episome

a portion of genetic material that can exist independent of the main body of genetic material (i.e. chromosome) at some times, while at other times is able to integrate into the chromosome

epitopes

an alternative name for antigenic determinants that are the smallest parts of antigens detected by antibodies or T cell receptors

erysipelas

an acute bacterial infection of the skin accompanied by lymphadenopathy, fever, chills and leukocytosis

erythema infectiosum

in children is a normal response to infection by human parvovirus-B19; acute infection in a host who is immunocompetent results in formation of immune complexes that are deposited in the skin and joints of individuals with this condition; also known as 'slapped cheek' disease or Fifth disease

erythema migrans

a bulls-eye rash classically following infection with *Borrelia burgdorferi* delivered via a tick bite

erythema multiforme

a rash that results from an allergic response, most often secondary to a drug

erythema nodosum

a disorder characterized by the formation of tender, red nodules on the front of the legs; has been associated with certain infections and is thought to be due to a hypersensitivity reaction

erythema/erythematous

skin redness produced by congestion of the capillaries, which may result from a variety of causes; *see also* erythema nodosum

erythrocyte sedimentation rate (ESR)

measures the time taken for erythrocytes to settle in blood plasma; test used to estimate the level of inflammation in patients with some autoimmune diseases

E-selectin

also called ELAM-1 and CD62E; a member of the selectin family of adhesion molecules; recognizes sialylated carbohydrate groups related to Lewis x or Lewis a family; used by T cells to enter inflamed skin

esophageal varices

abnormally enlarged veins in the lower esophagus and upper part of the stomach; they develop when normal blood flow in the hepatic portal vein is blocked; separate blood backs up into smaller, more fragile blood vessels in the esophagus, and often stomach or rectum, causing the vessels to swell; esophageal varices can rupture, causing a life-threatening condition

exocytosis

the process by which cellular products within vesicles are secreted from a cell when the vesicle membrane fuses with the plasma membrane

exotoxin

bacterial protein toxin that is secreted by pathogenic bacteria and contributes to infectious disease: examples include diptheria toxin, tetanus toxin

extracellular traps

produced by neutrophils (called NETS) are networks of extracellular fibers that are primarily composed of DNA which bind pathogens

facultative anaerobe

an organism that can grow well in both the absence and the presence of oxygen equivalent to that of normal air (21%)

facultative (intracellular lifestyle)

means that an organism can proliferate both inside and outside host cells

farmer's lung

also called extrinsic allergic alveolitis; an allergic disease resulting from deposition of immune complexes (type III hypersensitivity) in the lung following repeated inhalation of antigenic materials, for example, from actinomycete fungi (found in moldy hay)

fasciitis

inflammation of the fascia (lining tissue under the skin)

febrile

pertaining to or marked by fever; feverish

feco-oral route

transmission of infectious organisms in feces via the mouth. Organisms transmitted via this route include hepatitis A, hepatitis E, cholera, adenovirus, and *E. coli*

fibromyalgia

a common disorder and a syndrome composed of different signs and symptoms; now recognized as one of many central pain-related common syndromes; pain associated with fibrous tissue

fibrosis

formation of excessive fibrous tissue as the result of repair or a reactive process (e.g. during inflammation which becomes chronic); a synonym for scar formation

fimbriae

thin, hair-like appendages, 1 to 20 microns in length and present on the cells of gram-negative bacteria; unlike flagella, they are non-motile and possess antigenic and hemagglutinating properties; some fimbriae mediate the attachment of bacteria to cells via adhesins

fits

the clinical expression of an abnormal cerebral discharge – a dysrhythmia; may be convulsive (stiffening or jerking) or non-convulsive (as in absence or complex partial seizures)

flagella typing

typing of flagellar H antigens by serology or PCR

Flavivirus

a genus of the family *Flaviviridae*; genus includes the West Nile virus, dengue virus, tick-borne encephalitis virus, yellow fever virus, and several other viruses which may cause encephalitis

fomites

inanimate objects that carry microbes leading to spread of infections, e.g. cups, spoons, surgical instruments, etc.

fundoscopy

examination of the optic disc, retina and blood vessels using an ophthalmoscope

G6PD deficiency

glucose-6-phosphatase dehydrogenase (G6PD) deficiency is an enzyme deficiency of erythrocytes; inherited as an X-linked disorder, G6PD deficiency affects 400 million people worldwide (about 10% of the world population); leads to lysis of erythrocytes – hemolytic anemia; it confers protection against malaria, which probably accounts for its high gene frequency; affects all races

gamma/delta T cells (γδT cells)

T cells that use γ and δ chains for their receptors; these cells do not recognize processed peptides presented on classical MHC molecules but have been reported to recognize whole protein structures, stress proteins, small molecules produced by some bacteria and even lipids

gastritis

inflammation or irritation of the stomach mucosa; can be acute or chronic; can be caused by alcohol, aspirin and *Helicobacter pylori* (chronic)

gastroenteritis

a nonspecific term for various pathological states of the gastrointestinal tract; primary manifestation is diarrhea, but may be accompanied by nausea, vomiting, and abdominal pain

G-CSF

granulocyte colony-stimulating factor; a cytokine produced by macrophages, fibroblasts and other cells; has an effect on stem cells; used to enhance the development of neutrophils when given with stem cell transplants; also injections used to drive CD34+ stem cells into the bloodstream for collection for stem cell donation

Glasgow Coma Scale score

the measurement used to quantify the level of consciousness following traumatic brain injury

glaucoma

increased intraocular pressure (IOP) and decreased visual acuity resulting from the effect of the pressure on the optic nerve

glomerulonephritis

inflammation of the capillary loops in the glomeruli of the kidney

GM-CSF

granulocyte-monocyte colony-stimulating factor; a cytokine produced by T cells, macrophages and endothelial cells in response to infection/inflammation; results in specific increase in granulocytes and monocytes in the bone marrow from their stem cells to aid in fighting the infection

Gortex tube

Gortex is a biomaterial; tubes made of Gortex are used to carry blood in heart surgery

gout

a common disorder of uric acid metabolism (found in 1% of the population) that can lead to deposition of monosodium urate (MSU) crystals in soft tissue, recurrent episodes of debilitating joint inflammation, and, if untreated, joint destruction and renal damage

granulocytopenia

a reduced number of granulocytes in the circulation – neutrophils, eosinophils and basophils

granuloma

an accumulation of macrophages, epithelioid cells, with or without lymphocytes; can result from macrophages being activated by chronic bacterial infections such as tuberculosis, leprosy and syphilis; similar but not identical structures are induced in some parasitic infections such as schistosomiasis and by non-infectious agents such as asbestos

granulysin

a protein present in the acidic granules of human NK cells and cytotoxic T cells with antimicrobial activity against a broad spectrum of microbial pathogens; when released acts directly on the microbial cell wall/membrane (e.g. of mycobacteria) leading to increased permeability and lysis

granzymes

granule-associated enzymes in cytotoxic T cells and NK cells that are involved in inducing apoptosis of virus infected cells

guarding (abdominal)

tensing of the abdominal wall muscles to guard inflamed organs within the abdomen from the pain when abdominal wall is pressed; characteristic finding in the physical examination for appendicitis or diverticulitis

Guillain-Barré syndrome

a heterogeneous group of immune-mediated processes generally characterized by motor, sensory, and autonomic dysfunction. Classically an acute inflammatory demyelinating polyneuropathy characterized by progressive symmetric ascending muscle weakness, paralysis, and hyporeflexia with or without sensory or autonomic symptoms; in severe cases, muscle weakness may lead to respiratory failure

gynecomastia

a benign enlargement of the male breast resulting from a proliferation of the glandular component of the breast

hemagglutinin

a substance that causes agglutination of erythrocytes; an integral envelope protein of influenza virus; responsible for binding of the virus to the host cell and subsequent fusion of viral and host membranes in the endosome after the virus has been taken up by endocytosis

hematuria

the presence of erythrocytes in the urine

hemiparesis

muscle weakness on one side of the body

hemiplegia

caused by damage to the brain resulting in a varying degree of weakness and lack of control on one side of the body

hemoglobinuria

the presence of free hemoglobin in the urine

hemolysin

exotoxins produced by some bacteria that damage or destroy cells; alpha hemolysin is a sphingomyelinase produced by staphylococci. It produces partial hemolysis of sheep and cattle erythrocytes and appears to have little pathogenic effect; β hemolysin (produced by streptococcus) results in complete lysis of the erythrocytes around the bacteria colonies

hemolysis

used in the empirical identification of microorganisms based on the ability of bacterial colonies grown on agar plates to break down red blood cells in blood agar plates; on growing the organism it can be classified as to whether or not it has caused hemolysis in the red blood cells (RBCs) incorporated in the medium; of particular importance in the classification of streptococcal species that produce β hemolysin

hemolytic anemia

anemia arising from abnormal destruction of red blood cells; there are many causes including an autoimmune disease where autoantibodies bind to the patient's erythrocytes and cause lysis (mediated by complement) and removal via opsonization

hemolytic uremic syndrome (HUS)

a disease mainly of infancy and early childhood; characterized by the triad of microangiopathic hemolytic anemia, thrombocytopenia, and acute renal failure; the most common cause of acute renal failure in children; diarrhea and upper respiratory infection are the most common precipitating factors of HUS. In many cases, HUS is caused by infection with certain strains of *E. coli*

hemophagocytosis

pathologic finding of activated macrophages, engulfing erythrocytes, leukocytes, platelets, and their precursor cells

hemoptysis

coughing up blood

hemozoin

produced during the intra-erythrocytic phase of Plasmodium infection by digestion of hemoglobin; may play a role in malaria-associated immunosuppression, affecting both the antigen processing and immunomodulatory functions of macrophages

Henoch-Schölein purpura

a small-vessel vasculitis characterized by immunoglobulin A (IgA), C3, and immune complex deposition in arterioles, capillaries, and venules; affects mostly children and involves the skin and connective tissues, gastrointestinal tract, joints, and scrotum as well as the kidneys

hepatitis

inflammation of the liver; may be acute or chronic; causative agents of acute hepatitis include hepatitis viruses A–E, CMV, some bacterial infections, some medicines, e.g. paracetamol and toxins such as alcohol; chronic hepatitis can also be caused by hepatitis B, C and D viruses, some drugs, toxins and can be autoimmune

hepatocellular carcinoma

a primary malignant proliferation of hepatocytes

hepatoma

primary liver cancer beginning in the liver; most are hepatocellular carcinomas (HCC)

hepatomegaly

enlargement of the liver

hereditary spherocytosis

a familial hemolytic disorder with marked heterogeneity of clinical features, ranging from an asymptomatic condition to fulminant hemolytic anemia; the morphological hallmark of HS is the abnormal osmotic fragility *in vitro*

herpangina

painful oral ulceration usually arising from an enterovirus infection

heterophile antibodies

antibodies that react with an antigen entirely different from and phylogenetically not related to the antigen responsible for their production, e.g. antibodies that agglutinate sheep red blood cells in patients with infectious mononucleosis; caused by EBV that polyclonally activates B cells independent of their specificity; usually IgM

highly active antiretroviral treatment (HAART)

a combination of three or more active drugs to effectively treat patients with HIV; more commonly referred to now as cART – combination antiretroviral therapy

histiocytes

tissue macrophages

hydatid cyst

a cyst formed in the liver, or, less frequently, elsewhere, by the larval stage of Echinococcus

hydrocele

a type of swelling in the scrotum that occurs when fluid collects in the thin sheath surrounding a testicle

hydrocephalus

a disturbance in the formation, flow, or absorption of cerebrospinal fluid (CSF) that leads to an increase in fluid in the central nervous system

hydronephrosis

defined as a dilation of the renal pelvis and calyces; physiological response to interruption of the flow of urine that can be caused by infection

hydrops fetalis (fetal hydrops)

a serious condition of the fetus defined as abnormal accumulation of fluid in two or more fetal compartments, including ascites, pleural effusion, pericardial effusion, and skin edema

hypercholesterolemia

a condition where there is a high level of cholesterol in the blood

hyperemia

an increase in blood flow to any part of the body ususally due to vasodilation

hypergastrinemia

excessive secretion of gastrin

hypersplenism

increased activity of the spleen; can be caused by tumors, anemia, malaria, tuberculosis, and various connective tissue and inflammatory diseases

hypertension

abnormally high blood pressure

hypertrophy

an increase in cell size

hypoalbuminemia

decreased levels of albumin in the blood; can be caused by various conditions, including nephrotic syndrome, hepatic cirrhosis, heart failure, and malnutrition; however, most cases of hypoalbuminemia are caused by acute and chronic inflammatory responses

hypochlorhydria

a condition characterized by an abnormally low level of hydrochloric acid in the stomach

hypochondrium

the upper lateral region of the abdomen on either side of the epigastrium and below the lower ribs

hypoesthesia

partial loss of sensitivity to sensory stimuli; diminished sensation

hypogammaglobulinemia

an immunodeficiency disease marked by abnormally low levels of all classes of immunoglobulins, with increased susceptibility to infectious diseases; may be primary (congenital), or secondary (acquired)

hypokalemia

lower than normal levels of potassium in the blood

hyponatremia

an abnormally low concentration of sodium in the blood

hypovolemia

an abnormal decrease in blood volume

hypoxemia

decreased amount of oxygen in the blood

hypoxia

deficiency in the amount of oxygen reaching tissues

ICAM-1

intercellular adhesion molecule 1; an adhesion molecule important in many immune processes; the receptor for rhinoviruses that cause the common cold

icterus

another term for jaundice; yellowing of the skin and the whites of the eyes caused by an accumulation of bilirubin in the blood; can be a symptom of gallstones, liver infection, or anemia

idiopathic thrombocytopenic purpura (ITP)

also known as primary immune thrombocytopenic purpura and autoimmune thrombocytopenic purpura; defined as isolated thrombocytopenia with normal bone marrow and the absence of other causes of thrombocytopenia

IFN-γ

a type of interferon produced by NK cells and T cells; involved mostly in signaling between cells of the immune system

IgA/sIgA

immunoglobulin A: found in the blood and the secreted form (sIgA) at mucosal surfaces, e.g. gastrointestinal epithelium; protects the body by prevention of attachment of bacteria and viruses; it is monomeric (mostly in blood) or dimeric (mostly at mucosal surfaces); there are two subclasses classes, IgA1 and IgA2

IgA nephropathy

also known as IgA nephritis and Berger's disease; kidney disease characterized by IgA deposition in the glomerular mesangium; now the most common cause of glomerulonephritis in the world

IgA1 protease

a proteolytic enzyme that cleaves specific peptide bonds in the hinge region sequence of immunoglobulin A1 (IgA1); secreted by several species of pathogenic bacteria at mucosal sites of infection to destroy the structure and function of human IgA1 thereby eliminating an important aspect of host defense; also cleaves lysosomal-associated membrane protein-1 (LAMP-1) which appears to reduce the ability of lysosomal-endosome/phagosome to mediate killing of gonococci and thus facilitate transcytosis of the gonococcus across polarized epithelial cells

IgE

immunoglobulin E; an antibody involved in combating parasite infection and the cause of type 1 hypersensitivity (allergy) reactions

IgG

immunoglobulin G dominates the secondary humoral immune response; the most common class of antibody in the blood; only antibody class to pass across the placenta to protect the neonate; there are 4 subclasses IgG1, IgG2, IgG3 and IgG4

IgM

immunoglobulin M dominates the primary humoral immune response; the largest antibody molecule and because of its size is found mainly in the bloodstream

immortalization (through EBV)

the ability of a cell line (human B cells in the case of EBV) to reproduce indefinitely. The cells escape from the normal limitation on growth of a finite number of division cycles (the Hayflick limit), by infection with EBV

immune complexes

antibodies and antigens bound in various ratios; may also contain complement components; responsible for type III hypersensitivity reactions

immune-paresis

defined as reduction in the levels of all immunoglobulin classes or reduction in all classes of immunoglobulins other than that of the paraprotein in multiple myeloma

immunofluorescence/immunofluorescence assay (IFA)

a method of determining the location of antigen (or antibody) in a tissue section or smear using a specific antibody (or antigen) labeled with a fluorochrome. The tissues or cells are examined under a fluorescence microscope which excites the dyes to give off light. Different fluorochromes emit light of different wavelengths and thus different colors

incubation period

the period between the infection of an individual by a pathogen and the manifestation of the disease it causes

in situ hybridization

a technique for localizing and detecting nucleic acid (RNA or DNA) sequences in morphologically preserved tissues sections or cell preparations by hybridizing the complementary strand of a nucleotide probe to the sequence of interest

indoleamine-2 3dioxygenase (IDO)

an enzyme that breaks down tryptophan; may play a role in immunosuppression since T cell function is dependent on this amino acid

infarction

an area of tissue death due to the local lack of oxygen

inflammatory bowel disease

collective term for ulcerative colitis (UC) and Crohn's disease (CD); inflammatory diseases of the large and small intestines

iNOS

inducible nitric oxide synthase is one of the reactive oxygen and nitrogen metabolite-metabolizing enzymes

interleukin (IL)

a family of cytokines that are themselves divided into several families e.g. IL6 family, IL10 family etc.; they are communication molecules produced by immune cells but also by other cells in the body. To date, 42 different ILs have been described (IL1-IL42)

interstitial nephritis

also called tubulointerstitial disease; inflammatory disease of the renal interstitial cells between the nephrons and sometimes tubules (as opposed to glomeruli – glomerulonephritis)

intrathecal

introduced into the space under the brain arachnoid membrane or the spinal cord

intravenous immune globulin (IVIG)

pooled serum immunoglobulins given originally to treat primary antibody deficiencies; now used to treat a variety of diseases including some secondary immunodeficiencies and autoimmune diseases

introitus

an entrance that normally goes into a canal or hollow organs (e.g. vagina)

ischemia

deprivation of oxygen to a tissue leading to damage; usually caused by vasoconstriction, thrombosis, or embolism

ischemic necrosis

necrosis caused by hypoxia; *see also* entry for hypoxia

isoenzyme typing

a technique involving electrophoresis of enzymes to type strains of bacteria within a particular species, e.g. multilocus enzyme electrophoresis (MLEE)

Jarisch-Herxheimer reaction

a transient inflammatory reaction following treatment of syphilis with antibiotics; includes fever, chills, and worsening of the skin rash or chancre

jaundice

a yellow discoloration of the skin, most easily seen by looking at the sclera of the eyes; due to excess of circulating bilirubin

kala-azar

synonymous with leishmaniasis

kallikrein

a serine protease; produced by renal tubules; causes dilation of blood vessels; part of the kinin system that plays a role in inflammation, blood pressure, coagulation and pain

Kaposi sarcoma

a connective tissue tumor caused by Human Herpes Virus 8 (HHV8), also known as Kaposi's Sarcoma-associated Herpes Virus (KSHV). It can affect people with a weakened immune system, including people with HIV (Human Immunodeficiency Virus) and AIDS

Katayama syndrome

a hypersensitivity reaction to schistosomula antigens with immune complex formation

Kato-Katz technique

a (semi) quantitative assay using thick stool smears for microscopic examination

keratitis

infection of the cornea

keratoconjunctivitis

inflammation of the cornea and conjunctiva

Kupffer cells

part of the mononuclear phagocyte system, macrophages of the liver

kyphosis

an abnormal rounding forward of the shoulders and upper back; also refers to the type of curve formed by the sacrum and thoracic spine

lamina propria

a vascular layer of connective tissue under the basement membrane lining a layer of epithelium, e.g. in the intestine where it contains many kinds of immune cells

langerin (CD207)

a surface receptor belonging to the C-type lectin receptor family present on a subset of immature dendritic cells - Langerhans cells (LC) in the skin. This major pathogen binding receptor regulates both innate and adaptive immune responses. Reported as an antiviral receptor that is able to bind and internalize incoming Human Immunodeficiency Virus (HIV) virions in Birbeck granules for degradation. The receptor is up-regulated following IFN α treatment of monocyte derived LC

latent/latency

the ability of a pathogenic virus to lie dormant within a cell. In a latent viral infection, there is no production of new virus particles

Legionnaires disease

caused by *Legionella* species that are obligate, aerobic, gram-negative bacilli: pneumonia is the most common presenting problem

Lepromin test

used to determine what kind of leprosy people have. A sample of inactivated *M. leprae* is injected just under the skin, often on the forearm, so that a small lump pushes the skin up. This is looked at 3 days and 28 days later. People who don't have leprosy will have little or no skin reaction to the antigen while people with lepromatous leprosy will also have no skin reaction to the antigen – compared with someone with tuberculoid leprosy

leptomeninges

the two innermost layers of the tissue that cover the brain and spinal cord, called the arachnoid mater and pia mater

leukemia

a cancer of the blood or bone marrow characterized by an abnormal proliferation of blood cells, usually white blood cells

leukopenia

an abnormally low white blood cell count

lichen sclerosus

a chronic inflammatory dermatosis that results in white plaques with epidermal atrophy; more commonly seen on the vulva or penis but can occur extragenitally; has recently been associated with autoantibodies to the glycoprotein extracellular matrix protein 1 (ECM1)

lipid envelope

the host-derived outer lipid layer of enveloped viruses

lipopolysaccharide (LPS)

the 'endotoxin' in the outer membrane of gram-negative bacteria made of a sugar backbone with lipid A attached; potent activator of the immune system

lymphadenitis

inflammation of a lymph node

lymphadenopathy

swollen or enlarged lymph nodes; may be generalized or local

lymphangiectasia

dilatation of lymphatic vessels

lymphangitis

inflammation of the lymphatic vessels

lymphedema

the result of accumulation of lymph in interstitial spaces; usually occurs where the lymphatic system becomes damaged following surgery and/or radiotherapy (as in cancer treatment) or as a result of infection, severe injury, burns or trauma

lymphocytosis

an increase in the number of lymphocytes in the blood caused mainly by infection or inflammation

lymphoma

a cancer in which lymphocytes are the cells of origin; over 40 kinds of lymphoma with the two major subgroups being Hodgkins lymphoma (also called Hodgkins disease) and non-Hodgkins lymphoma of which there are about 30 subtypes; treatment is different for the different patient subgroups

lysis

disintegration of a cell resulting from destruction of its membrane; e.g. bacterial cells by antibody and complement; certain viruses cause lysis; also some chemical substances e.g. enzymes

lysosomes

cytoplasmic, membrane-bound organelles that contain hydrolytic enzymes

macules

flat red patches on the skin

maculopapular rash

a description of how a rash looks. A macule is a flat, reddened area of skin present in a rash. A papule is a raised area of skin in a rash, i.e. it has both flat and raised areas

maculopathy

disease of the macula (of the retina), such as age-related macular degeneration

madarosis

loss of eyelashes or eyebrows

magnetic resonance imaging (MRI) scan

a computer-based imaging system using a powerful magnetic field and radio frequency pulses to produce detailed pictures of organs, soft tissues, bone and virtually all other internal body structures

MALDI-TOF

Matrix-Assisted Laser Desorption/Ionization Time-of-Flight mass spectrometer; mass spectrometry based analytical tool to help aid in identification using a microorganism's unique protein profile

mannose-binding lectin

also called mannose-binding protein; a calcium dependent acute phase protein; member of the collectin family; activates the complement system when it binds to pathogen surfaces

mastocytosis

a disorder characterized by mast cell proliferation and accumulation within various organs, most commonly the skin

megacolon

dilatation of the colon that is not caused by mechanical obstruction; can be acute or chronic

megaesophagus

chronic dilitation of the esophagus; usually associated with asynchronous function of the esophagus and the caudal esophageal sphincter; usually a congenital condition, causing accumulation of food and saliva and regurgitation and aspiration pneumonia from an early age; may also be secondary to systemic disease, particularly general neuromuscular disorders such as myasthenia gravis

membrane attack complex (C5–C9)

complement components C5-C9 result in the lysis of a bacterium (pore formation and membrane damage) following complement activation; *see also* complement

meninges

the three membranes (pia mater, arachnoid mater, and dura mater) that surround the brain and spinal cord

meningitis

inflammation of the meninges; can be acute or chronic; infectious and non-infectious causes e.g. bacteria and viruses and NSAIDs and antibiotics

meningoencephalitis

inflammation affecting both the brain parenchyma and the meninges; often caused by virus infection

metaplasia

cellular adaptation to physiological or mechanical stress whereby one type of differentated tissue is converted into another type of differentiated tissue; benign and reversible

microaerophilic/microaerobic

related to microorganisms requiring oxygen for growth but at lower concentration than is present in the atmosphere

microbe-associated molecular patterns (MAMPS)

evolutionarily conserved stuctures (patterns) of microorganisms but not humans that are the danger signals detected by Pattern Recognition Receptors (PRRs) such as Toll-like receptors on the surface of epithelial cells, phagocytes and antigen-presenting cells; also called Pathogen-Associated Molecular Patterns (PAMPS)

microbiota

the microorganisms that typically inhabit a particular region of the body

microcephaly

a neurodevelopmental disorder in which the circumference of the head is smaller than normal

microglia

microglia cells are the innate immune cells of the central nervous system and consequently play important roles in brain infections and inflammation. They are the phagocytic cells of the nervous system. Recent, imaging studies have revealed that in the resting healthy brain, microglia are highly dynamic, moving constantly to actively survey the brain parenchyma. Microglia can also protect the CNS by promoting neurogenesis, clearing debris, and suppressing inflammation in diseases such as stroke, autism, and Alzheimer's

miliary

characterized by lesions resembling millet seeds; miliary tuberculosis marked by appearance of many small lesions

Miller Fisher syndrome

a rare variant of Guillain-Barré syndrome that typically presents with the classic triad of ataxia, areflexia, and ophthalmoplegia

MIP-1β

macrophage inflammatory protein 1β; a C-C chemokine (class of cytokines); potent chemoattractant for monocytes

Mollaret's meningitis

a rare form of meningitis; benign recurrent aseptic meningitis

monoclonal antibodies

antibodies produced by a B-cell clone and therefore are all identical with regard to specificity; used as laboratory reagents for cell phenotyping and in immunotherapy, e.g. herceptin for treating breast cancer

mononuclear phagocytic system

also called the reticuloendothelial system; term for all the phagocytic cells of myeloid origin that are found in the different organs and bloodstream, e.g. Kupffer cells of liver, alveolar macrophages of the lung, microglial cells of the brain etc.

MSCRAMMs

microbial surface components that recognize adhesive matrix molecules

muco-ciliary escalator

the co-ordinated movement of the cilia, hair-like structures on the surface of respiratory epithelial cells

mucocutaneous leishmaniasis (MCL)

there are three forms of leishmaniasis – visceral, cutaneous and mucocutaneous. Mucosal involvement can coincide with cutaneous involvement or occur after the clearance of cutaneous lesions, sometimes even one year later. The nasal and oral mucosa are the most frequently affected areas. MCL lesions are ulcerated and can cause disfigurement

mucopurulent

pertaining to a discharge that combines pus and mucus

multi-locus enzyme electrophoresis

a molecular sub-typing technique; *see also* enzyme typing above

multi-locus sequence typing

procedure for characterizing isolates of bacterial species using the sequences of internal fragments of (usually) seven housekeeping genes

multiple myeloma

monoclonal plasma cell malignancy; multiple because there can be several growth foci of abnormal plasma cells (myeloma cells) in the bone marrow; can lead to bone erosion, anemia, kidney damage and immunosuppression

myalgia

muscle pain

mycosis

a disease caused by fungi

myelitis

inflammation of the spinal cord

myeloid

refers to the nonlymphocytic groups of white blood cells, including the granulocytes, monocytes and platelets

myeloid differentiation primary response 88 (MYD88)

is the canonical adaptor for inflammatory signaling pathways downstream of members of the Toll-like receptor (TLR) and interleukin-1 (IL-1) receptor families

myeloid leukemia

a leukemia characterized by proliferation of myeloid tissue (as of the bone marrow and spleen) and an abnormal increase in the number of granulocytes, myelocytes, and myeloblasts in the circulating blood; can be acute or chronic

myelopathy

a disorder in which the tissue of the spinal cord is diseased or damaged; can also mean pathological change in the bone marrow

myiasis

infestation of the body by fly larvae (usually through a wound or other opening) or a disease resulting from the infestation

myocarditis

collection of diseases with toxic, infectious and autoimmune etiologies characterized by inflammation of the myocardium

myopericarditis

inflammation of both the myocardium and pericardium

myositis

inflammation of the skeletal muscles

NADPH oxidase-dependent killing

killing of microorgansims through the oxygen-dependent antimicrobial mechanism of macrophages and neutrophils; relies on activation of NADPH oxidase that catalyzes the transfer of electrons from NADPH to molecular oxygen, resulting in the formation of superoxide anion and a number of downstream toxic oxidants

natural killer (NK) cell

a cell of the innate immune system that kills virus infected cells; morphologically, large granular lymphocytes

necrosis

death in response to cell or tissue damage, which ends in the release of the intracellular content and the onset of inflammation; there are different kinds of necrosis, e.g. hemorrhagic etc.

necrotizing fasciitis

a deep bacterial infection of the connective tissue that spreads along fascial planes and destroys muscle and fat

neoplasms

abnormal masses of tissue that are produced when cells grow and divide more than they should or do not die when they should. They can be benign (not cancer) or malignant (cancer)

nephrotic syndrome

a clinical entity characterized by massive loss of urinary protein (primarily albuminuria) leading to hypoproteinemia (hypoalbuminemia) and its result, edema

neuropathy

a functional disturbance or pathological change in the peripheral nervous system

neurotensin

a 13-residue peptide transmitter, sharing significant similarity in its 6 C-terminal amino acids with several other neuropeptides; distributed throughout the central nervous system; highest levels in the hypothalamus, amygdala and nucleus accumbens. It induces a variety of effects, including: analgesia, hypothermia and increased locomotor activity. It is also involved in regulation of dopamine pathways. In the periphery, neurotensin is found in endocrine cells of the small intestine, where it leads to secretion and smooth muscle contraction

neutropenia

an abnormally low neutrophil count in the blood

neutrophilia

an increased number of neutrophils in the blood

nitric oxide synthase (NOS)

produces nitric oxide (NO) and citrulline from arginine, molecular oxygen, and NADPH. NO plays an important role as a mechanism of cytotoxicity for macrophages and as a signaling molecule involved in neurotransmission

nonsteroidal anti-inflammatory drugs (NSAIDs)

used to treat inflammation in a number of diseases; includes aspirin, ibuprofen and indomethacin

nontuberculous mycobacteria (NTM)

mycobacteria other than *M. tuberculosis* (the cause of tuberculosis) and *M. leprae* (the cause of leprosy). NTM are also referred to as atypical mycobacteria

normoblasts

myeloid cell precursors of erythrocytes found in the bone marrow

nosocomial (infection)

relates to an infection acquired or taking place in a hospital

nucleic acid amplification tests (NAATs)

tests to identify a pathogen by amplifying the genetic material-nucleic acids and thus detecting small amounts. The polymerase chain reaction PCR) is one such test. NAATs for SARS-CoV-2 are thus highly sensitive tests. NAATs have 4 main roles: a) qualitative or quantitative parasite detection; b) determination of the multiplicity of infection; c) genotyping to distinguish recrudescence from re-infection; and d) detection of drug resistance mutations

nucleocapsid

the nucleic acid core and surrounding capsid of a virus

oliguric

defined as having a urine output of less than 1 mL/kg/h in infants, less than 0.5 mL/kg/h in children, and less than 400 mL/d in adults; one of the clinical hallmarks of renal failure

onchocerciasis

a parasitic disease caused by the filarial worm *Onchocerca volvulus*; transmitted through the bites of infected female blackflies of *Simulium* species, which carry immature larval forms of the parasite from human to human; larvae die causing a variety of conditions, including blindness, skin rashes, lesions, intense itching and skin depigmentation

Oncomelania

a genus of very small tropical freshwater snails, aquatic gastropod mollusks in the family Pomatiopsidae

onychomycosis

a fungal infection of the toenails or fingernails that may involve any component of the nail unit, including the matrix, bed, or plate. It can cause pain, discomfort, and disfigurement and may produce serious physical and occupational limitations, as well as reducing quality of life

ophthalmia neonatorum

conjunctivitis that is contracted by newborns during delivery; usually caused by the gonococcus

opportunistic infections

infections that take advantage of weakness in the immune defenses, e.g. those that occur in AIDS patients

opsonization

the coating of microbes with antibodies or complement (called opsonins) to enhance phagocytosis

orchitis

testicular inflammation: it may occur idiopathically or it may be associated with infections such as mumps, gonorrhoeae, filarial disease, syphilis, or tuberculosis

oro-oral

oral–oral route of transmission of infection, e.g. EBV

oropharynx

the back of the mouth where it meets the throat

Osler node

a small, raised, and tender cutaneous lesion characteristic of subacute bacterial endocarditis, usually appearing in the finger pads

osteomyelitis

acute or chronic infection of the bone usually caused by pyogenic bacteria or mycobacteria; can result in destruction of the bone and deformity if not treated

otitis media

a build-up of fluid or mucus in the middle ear: mucus can become infected with bacteria resulting in inflammation

otomycosis

a fungal infection of the external auditory canal characterized by inflammation, pruritus and scaling

oxygen-free radicals

reactive oxygen species (ROS) that are ions or very small molecules that include oxygen ions, free radicals, and peroxides, both inorganic and organic. They are highly reactive due to the presence of unpaired valence shell electrons

pancreatitis

acute or chronic inflammation of the pancreas

pancytopenia

an abnormal reduction in all three types of cells in the blood, including erythrocytes (anemia), white blood cells (neutrophils; neutropenia) and platelets (thrombocytopenia); often the result of changes in the bone marrow

pandemic

widespread occurrence of an infectious disease over a whole country or the world at a particular time

papilledema

an optic disc swelling, secondary to elevated intracranial pressure with underlying infectious, infiltrative, or inflammatory etiologies

papular

referring to one or more small solid rounded lesions rising from the skin that are each usually 0.5 cm or less in diameter (less than a half-inch across)

papule

a skin lesion that is small, solid, and raised, usually not more than 0.5 cm in diameter

paralytic ileus

functional non-mechanical 'obstruction' of intestinal flow

paraplegia

the loss of movements and/or sensation following damage to the nervous system

parasitophorous vacuole

a bubble-like compartment made of plasma membrane containing an intracellular parasite which may serve to isolate the parasite and enclose it for lysozymal attack

parasthesiae

abnormal skin sensations (as tingling or tickling or itching or burning), with no apparent physical cause; usually associated with peripheral nerve damage

parenteral

administration of a drug etc. through any route other than the digestive tract; usually via injection

paronychia

an infection of the skin which is just next to a nail (the nail-fold): the nail itself may become infected or damaged if a nail-fold infection is left untreated

pathogen-associated molecular patterns (PAMPs)

evolutionarily conserved structures (patterns) of microorganisms but not humans that are the danger signals detected by Pattern Recognition Receptors (PRRs) such as Toll-like receptors on the surface of epithelial cells, phagocytes and antigen-presenting cells; also called Microbe-Associated Molecular Patterns (MAMPs)

pathogenicity islands (PAIs)

discrete genetic units flanked by direct repeats, insertion sequences or tRNA genes, which are sites for recombination into the DNA. They are incorporated in the genome of pathogenic microorganisms but are usually absent from those nonpathogenic organisms of the same or closely related species. PAIs carry genes encoding one or more virulence factors: adhesins, toxins, invasins, etc. The G+C content of PAIs often differs from that of the rest of the genome

pattern recognition receptors (PRRs)

a class of germ line-encoded receptors that recognize specific molecular structures on the surface of pathogens, apoptotic host cells, and damaged senescent cells: Pathogen/Microbial-Associated Molecular Patterns (PAMPs/MAMPS). PRRs bridge nonspecific immunity and specific immunity. Through the recognition and binding of ligands, PRRs can produce nonspecific anti-infection, antitumor, and other immunoprotective effects. Most PRRs in the innate immune system of vertebrates can be classified into the following five types based on protein domain homology: Toll-like receptors (TLRs), nucleotide oligomerization domain (NOD)-like receptors (NLRs), retinoic acid-inducible gene-I (RIG-I)-like receptors (RLRs), C-type lectin receptors (CLRs), and absent in melanoma-2 (AIM2)-like receptors (ALRs)

pelvic inflammatory disease (PID)

infection of the womb and fallopian tubes

peptidoglycan

the rigid, shape-determining complex polymer that forms the cell wall of gram-positive bacteria. It is made up of chains of heteropolysaccharides linked by tetrapeptides in gram-negative bacteria peptidoglycan is only a few layers thick and with very few tetrapeptide cross links

perforin

a protein that can polymerize to form the membrane pores that are an important part of the killing mechanism of NK and cytotoxic T cells

pericarditis

inflammation of the outer lining of the heart

periodontal disease

a group of chronic inflammatory diseases that affect the tissues that support and anchor the teeth. If untreated, periodontal disease would result in the destruction of the gums, alveolar bone and the outer layer of the tooth root

peripheral nerve neuropathy

disorders of the peripheral nerves; in its most common form it causes pain and numbness in hands and feet, and tingling burning pain; caused by traumatic injuries, infections, metabolic problems and exposure to toxins; one of the most common causes of the disorder is diabetes

persistent generalized lymphadenopathy (PGL)

defined as enlarged lymph nodes involving at least two non-contiguous sites, other than inguinal nodes, persisting for more than 3 months. Common in patients in the early stages of HIV infection

petechial/petechiae

pinpoint-sized spots of bleeding under the skin or mucous membranes of the mouth or eyelids. The purple, red or brown dots are not raised or itchy, and are not a rash. They are caused by broken capillaries under the skin. Tiny petechiae of the face, neck and chest can be caused by prolonged straining during activities such as coughing, vomiting, giving birth and weightlifting

phage typing system

infection of bacteria by bacteriophages is highly specific; this system uses a panel of bacteriophages (also called phages) to identify unknown bacterial strains

phagocytosis

the process of ingesting large particles such as bacteria and viruses by specialized immune cells of the mononuclear phagocyte system

phagosome

the vacuole containing the large particle(s) ingested by the process of phagocytosis

pharyngitis

painful inflammation of the pharynx (sore throat)

photophobia

sensitivity or intolerance to light

picornavirus

viruses in the family *Picornaviridae*; non-enveloped, genome of single stranded, positive-sense RNA viruses with an icosahedral capsid; important human pathogens

pilin

the protein building blocks of bacterial pili

pilus/pili

smaller than flagella; used in the exchange of genetic material during bacterial conjugation; *see also* fimbriae

pinocytosis

the ingestion of soluble materials by endocytosis

placentitis

inflammation of the placenta

plasmacytoid dendritic cells (pDC)

a specialized cell type within the innate immune system that is specialized in sensing viral RNA and DNA by Toll-like receptor-7 and -9 and have the ability to rapidly produce massive amounts of type 1 IFNs upon viral encounter

pleocytosis

an abnormal increase in the number of cells (as lymphocytes) in a bodily fluid, e.g. the cerebrospinal fluid; indicative of an inflammatory, infectious, or malignant condition

pleural thickening

thickening and hardening of the pleura – two layered protective membrane surrounding the lung

pleurisy/pleuritic pain

inflammation of the pleura leading to pain on breathing (pleuritic pain)

pneumocytes

alveoli are lined by two types of pneumocytes – Type 1 and Type 2; type 1 responsible for gaseous exchange and type 2 for production and secretion of surfactant, a molecule that reduces the surface tension of pulmonary fluids and contributes to the elastic properties of the lungs

pneumonia

inflammation of the lung parenchyma; usually caused by several species of bacteria, viruses, fungi but can have a non-infectious etiology

pneumonitis

a general term referring to inflammation of the lung tissue

pneumotropic

the affinity of a microbe for lung tissue; infectious agents that cause pneumonia generally have specific adhesion molecules that allow them to bind to lung epithelia

podoconiosis

nonfilarial, non-infective 'elephantiasis', usually crystalline blockage of the limb lymphatics, almost always affecting the lower limbs and especially the feet

polycistronic

mRNA is said to be polycystronic when it contains the coding sequences for multiple gene products, each of which is independently translated from the mRNA; usual for prokaryotes but not eukaryotes

polymerase chain reaction (PCR)

a technique used to copy and amplify specific DNA sequences over a millionfold in a simple enzyme reaction; DNA corresponding to genes, or gene fragments can be amplified from chromosomal DNA; the amplified DNA can be used for the analysis and manipulation of genes

polymorphisms

silent mutations that have no effect on the encoded protein that tend to accumulate in the DNA

polymyalgia rheumatica

a disorder of the muscles and joints characterized by severe aching and stiffness of the neck, shoulder girdle and pelvic girdle; classified as a rheumatic disease but has unknown etiology

portal hypertension

elevated blood pressure in the portal vein and the smaller veins of the portal venous system. The extra blood flow makes the veins expand and makes their walls stretch and weaken. They may leak fluids into the abdomen, and they can also break and bleed. The most common cause in the western world is cirrhosis of the liver

post-herpetic neuralgia

pain arising or persisting in areas affected by herpes zoster at least 3 months after the healing of the skin lesions

Pott's disease

partial destruction of the vertebrae, usually caused by a tuberculous infection and often producing curvature of the spine

preauricular lymphadenopathy

swollen/enlarged lymph nodes draining the eyelids and conjunctivae, temporal region and pinna

preemptive therapy

therapy given to prevent development of a microbial disease or its progression, e.g. for CMV in transplantation and SARS treatment

proctitis

inflammation of the rectal mucosa; many causes

proctocolitis

inflammation of the rectum and the colon

prophylaxis

treatment given or action taken to prevent disease. Prophylactic medicine includes vaccines and medications that prevent an illness or a recurrence of a condition (migraines, seizures, etc.). Action might involve surgery, e.g. a woman at high risk of breast cancer may elect to have a mastectomy

pro-inflammatory mediators

molecules that initiate inflammatory reactions

prostaglandins

unsaturated carboxylic acids, synthesized from the fatty acid, arachidonic acid, via the cyclooxygenase pathway. They are extremely potent mediators of a diverse group of physiological processes

prostatitis

inflammation of the prostate gland; may be bacterial or non-bacterial and in the case of prostatodynia the complaints are consistent with prostatitis but no evidence of prostatic inflammation

prosthesis/prosthetic

an artificial body part, such as a limb, a heart, or a breast implant

protean

very variable, easily changeable or continually changing

protein-losing enteropathy

abnormal loss of protein from the gastrointestinal tract, or the inability of the digestive tract to take in proteins

provirus

retroviral DNA integrated into its host cell genome, which serves as a template for production of viral genomic and mRNA

pruritis

itch

pseudomembranous colitis

sometimes called antibiotic-associated colitis or *C. difficile* colitis; inflammation of the colon that occurs in some people who have received antibiotics; almost always associated with an over-growth of the bacterium *Clostridium difficile* (*C. difficile*), although in rare cases, other organisms can be involved

psoriatic arthritis

an inflammatory joint disease associated with psoriasis

pulmonary eosinophilia

infiltration of eosinophils into the lung compartments – airways, interstitium and alveoli

pulse field gel electrophoresis

a technique used to separate especially long strands of DNA by length in order to tell differences among samples. It operates by alternating electric fields to run DNA through a flat gel matrix of agarose

purpuric

having small hemorrhages in the skin, mucous membranes, or serosal surfaces

pus

contains fluid, living and dead leucocytes, dead tissue, and bacteria or other foreign substances

pyelonephritis

infection of the upper part of the urinary tract; infection of the renal pelvis and often the renal parenchyma

pyoderma (impetigo)

a highly contagious, superficial infection of exposed skin by *S. pyogenes* or *Staph. aureus*, typically the face, arms and legs; seen most frequently in young children

pyogenic (bacteria)

those bacteria that cause pus

pyogenic infection

infection characterized by severe local inflammation, usually with pus formation, generally caused by one of the pyogenic bacteria

pyrexia

fever, when the body's temperature rises above normal

pyrexia of unknown origin (PUO)

defined as a temperature greater than 38.3°C on several occasions, more than 3 weeks' duration of illness, and failure to reach a diagnosis despite one week of inpatient investigation; caused by infections (30–40%), neoplasms (20–30%), collagen vascular diseases (10–20%) and miscellaneous diseases (5–10%)

pyuria

the presence of pus in the urine; abnormal numbers of white blood cells in the urine indicative of infection

quasispecies

describes a type of population structure in which collections of closely related genomes are subjected to continuous genetic variation, competition and selection. It is very important in virology because it provides an interpretation for the extensive plasticity, both genetic and phenotypic, displayed by many viruses, in particular RNA viruses

Quellung test

bacteria are mixed into a drop of capsule-specific antiserum and observed under the microscope. Capsule swelling and increased refractility result from the interaction of the antibody with the capsule for which it is specific

radiculomyelopathy

irritation of the sacral nerve roots

radionuclide brain scanning

an imaging technique that uses a small dose of a radioactive chemical (isotope) called a tracer that can detect cancer, trauma, infection, or other disorders – in this case in the brain

rales

wet, crackly lung noises heard on inspiration

rashes

macular, petechial, scarlatiniform, urticarial, or erythema multiforme

RAST

radioallergosorbent test; test for specific IgE antibodies to a variety of allergens including cow's milk, nuts, eggs, etc.

reactive oxygen species (ROS)/reactive oxygen intermediates (ROI)

ROS form as a natural by-product of the normal metabolism of oxygen and have important roles as bacteriocidal agents and in cell signaling: release by neutrophils and macrophages are thought to play a role in tissue damage

recto-vaginal fistulae

an abnormal connection between the rectum and vagina; may result from injury during childbirth, complications following surgery, cancer or inflammatory bowel disease such as Crohn's disease; may resolve spontaneously but most often requires surgery

regulatory T cells (Tregs)

T cells that regulate cellular and antibody responses and are thought to prevent autoimmunity; mostly of the helper phenotype, i.e. CD4+ but also expressing high levels of CD25 and FoxP3

Reiter's syndrome

now known as reactive arthritis; refers to acute nonpurulent arthritis complicating an infection elsewhere in the body; can be triggered following enteric or urinogenital infections and is often associated with HLA-B27

restriction fragment length polymorphism (RFLP)

a technique by which organisms may be clonotyped by analysis of patterns derived from cleavage of their DNA

reticulocytes

newly differentiated erythrocytes without a nucleus: they retain a fine network of endoplasmic reticulum that stains with reticulocyte stains

retrovirus

an enveloped virus possessing an RNA genome. It contains an enzyme reverse transcriptase which is used to perform the reverse transcription of its genome from RNA into DNA. This can then be integrated into the host's genome with an integrase enzyme. The virus then replicates as part of the cell's DNA

Reye's syndrome

sudden (acute) brain damage (encephalopathy) and liver function problems of unknown cause, often associated with the use of aspirin to treat chickenpox or influenza in children

rheumatoid arthritis

a chronic inflammatory autoimmune disease affecting the synovial joints that can manifest itself systemically. Systems affected can include the lung, etc.

rheumatoid factor (RF)

antibodies with a specificity for the Fc region of IgG; they can be of the IgM, IgG, or IgA class; found in around 75% of patients with rheumatoid arthritis but also in some patients having a number of other autoimmune diseases such as systemic lupus erythematosus; IgM RF commonly found in low levels in normal individuals

rhinitis

a group of conditions that are caused by inflammation of the lining of the nose; can be allergic (mediated by IgE) or non-allergic through bacterial or viral infections

rhinorrhoea

a discharge from the nasal mucous membrane, especially if excessive

ribotyping

a technique used to determine genetic and evolutionary relationships between organisms; oligonucleotide probes targeted to highly conserved domains of coding sequences of ribosomal RNA are amplified and the products are visualized by gel electrophoresis banding patterns and compared with known species and strains to determine organism relatedness

Romana's sign

conjunctivitis and swelling around the eye resulting from infection via the eye with *T. cruzi*

rosaceae

a chronic (long-term) disease that affects the skin and sometimes the eyes; characterized by redness, pimples, and, in advanced stages, thickened skin; usually affects the face with other parts of the upper body only rarely involved

rose spots

red macular lesions, 2–4 millimeters in diameter occurring in patients suffering from *Salmonella typhi* and *Salmonella paratyphi*; rose spots can also occur following invasive non-typhoid salmonellosis

Roth spot

a round white retina spot surrounded by hemorrhage in bacterial endocarditis, and in other retinal hemorrhagic conditions

Sabin-Feldman dye test

a serological test for the diagnosis of toxoplasmosis; measures antibodies to toxoplasma

salivarian transmission

transmission of the African trypanosome via the saliva of the fly vector

salpingitis

an infection and inflammation in the fallopian tubes

saprophytic

feeding or growing on decaying organic matter

sarcoidosis

a disease characterized by non-caseating granulomas; associated with a number of immune system aberrations; commonly involves the lungs and the lymph nodes (especially the intrathoracic nodes), the skin, the eyes, the liver, the heart, and the nervous, musculoskeletal, renal, and endocrine systems; of unknown etiology

scotoma

an area of diminished vision within the visual field

secondary bacterial pneumonia

pneumonia caused by a secondary bacterial infection following viral infections; commonly due to *S. pneumoniae* or *S. aureus*

seizure

a common, nonspecific manifestation of neurologic injury and disease

septic arthritis

also known as infectious arthritis; direct infection of the joint, mainly bacterial

septic shock

sepsis is a systemic inflammatory response to an infectious agent that releases endotoxin or superantigen into the blood; likely mediated by TNFα, IL-1 and IFNγ; life-threatening condition

septicemia

the active multiplication of bacteria in the bloodstream that results in an overwhelming infection: can be caused by emboli

seroconversion

the development of antibodies in response to infection or immunization

serology

the study of serum, often for the presence of specific antibodies to microbes

seropositive

the presence of specific antibodies to a microorganism following infection or immunization

serovar

a sub-division of microorganisms on the basis of antigenic structure

severe combined immunodeficiency

a primary immunodeficiency that results from any of a heterogenous group of genetic conditions affecting the immune system, leading to severe T- and B-cell dysfunction

severe (as in malaria)

typically occurs due to delayed treatment of uncomplicated malaria. It is is defined by clinical or laboratory evidence of vital organ dysfunction

shingles

also known as herpes zoster; a viral disease caused by varicella-zoster virus and characterized by a painful skin rash with blisters in a limited area on one side of the body, in a dermatoma; *see also* Case 39

shock

physiologic shock is when there is not enough blood to support the organs and tissues

sickle cell disease

a group of genetic disorders that affects hemoglobin

sigmoidoscopy

procedure used to examine the rectum and the sigmoid colon (using a sigmoidoscope)

sinusitis

an inflammation of the paranasal sinuses, which may or may not be as a result of infection, from bacterial, fungal, viral, allergic, or autoimmune issues

spirochetes

motile, unicellular, spiral-shaped organisms which are morphologically distinct from other bacteria; they range from obligate anaerobes to aerobes and from free-living forms to obligate parasites

splenomegaly

enlargement of the spleen

stercorarian transmission

transmission of the American trypanosome *T. cruzi* (causing Chagas disease) via the feces of the insect vector

Stevens-Johnson syndrome

an immune-complex–mediated hypersensitivity that is a severe expression of erythema multiforme; typically involves the skin and the mucous membranes; caused by many drugs, viral infections and malignancies; cell death results causing separation of the epidermis from the dermis

Still's disease

a form of juvenile idiopathic arthritis that affects large joints and bone growth may be retarded; characterized by inflammation with high fever spikes, fatigue and salmon-colored rash

stroke

the rapidly developing loss of brain functions due to a disturbance in the blood vessels supplying blood to the brain

substance P

a small neuropeptide involved in regulation of the pain threshold

subungual

under a finger or toe nail

superantigen

antigens, mostly from bacterial toxins, that interact with a subset of T lymphocytes expressing products of the VβT cell receptor genes; as a consequence, superantigens activate large numbers of T cells independent of their specificity

suppurative infections

pus-producing infections

sylvatic

referring to diseases only occurring in wild animals

syncope

a brief loss of consciousness caused by a temporary deficiency of oxygen in the brain

systemic lupus erythematosus (SLE)

a chronic inflammatory disease that affects multiple organ systems; an autoimmune disease characterized by antinuclear antibodies especially those to double stranded DNA

tachycardia

a rapid heart rate, especially one above 100 beats per minute in an adult

tachypnea

abnormally rapid breathing or respiration

Tamm-Horsfall protein (THP)

also called uromodulin; a mucoprotein derived from the secretion of renal tubular cells and a normal constituent of urine; may play a role in preventing attachment of type 1 fimbriated *Escherichia coli* to the urinary tract thus contributing to innate immunity in that tract

tegument

a natural outer covering; in relation to many enveloped viruses is a cluster of non-essential and essential proteins that line the space between the envelope and nucleocapsid

tenesmus

the constant feeling of a need to empty the bowels

tenosynovitis

inflammation of the sheath that surrounds a tendon

terminal ileitis

inflammation of the lower section of the small intestine

TFH

T follicular helper cells are a distinct population of CD4-positive T cells found in the B cell follicles of secondary lymphoid organs. They are identified by their constant expression of the B-cell follicle homing receptor CXCR5. They regulate multiple stages of antigen-specific B cell immunity through cognate cell contact and cytokine production

Th-1

T helper 1 cells expressing CD4 are a subset of T cells that help cytotoxic CD8+ precursors to differentiate into mature cytotoxic CD8+ cells (i.e. cell mediated immunity) and activate B cells, macrophages, PMNs and NK cells. They produce pro-inflammatory cytokines like IFN-γ, IL-2, and TNF-α

Th-2

T helper 2 cells are a subset of T cells expressing CD4 that regulate the activation and maintenance of the humoral, or antibody-mediated immune responses by producing various cytokines such as IL-4, IL-5, IL-6, IL-9, IL-10 and IL-13, These cytokines are also responsible for eosinophil activation, and inhibition of several macrophage functions. They also counteract Th-1 responses

Th-17

a T helper cell subset expressing CD4 that is pro-inflammatory defined by its production of IL-17 but also produces IL-21 and IL-22. Th-17 cells play a role in host defense against extracellular pathogens, particularly at the mucosal and epithelial barriers through accumulation of immune cells, such as macrophages, neutrophils and lymphocytes, at inflammatory sites

thrombocytopenia

a disorder where the platelet count is below the normal lower limit (usually 150 × 109/L)

thrombosis

the formation of a blood clot (thrombus) inside a blood vessel that obstructs the flow of blood through the circulatory system

thyroiditis

inflammation of the thyroid gland; causes include autoimmune, e.g. Hashimoto's thyroiditis and thyrotoxicosis, infections (bacterial or viral) and drugs

tinnitus

the name given to the condition of noises 'in the ears' and/or 'in the head' with no external source. Tinnitus noises are described variously as ringing, whistling, buzzing and humming

TNF-α (tumor necrosis factor-α)

a major pro-inflammatory cytokine mainly produced by macrophages and Th-1 cells, but other cells can produce it, including NK cells, B cells, endothelial and muscle cells, fibroblasts, and osteoclasts. The primary role of TNF is in the regulation of immune cells. It activates macrophages and acts on a variety of other cells. TNF, as an endogenous pyrogen, is able to induce fever, apoptotic cell death, cachexia, and inflammation, inhibit tumorigenesis and viral replication, and respond to sepsis via IL-1 and IL-6-producing cells. Dysregulation of TNF production has been implicated in a variety of human diseases including rheumatoid arthritis, Alzheimer's disease, cancer, major depression and psoriasis

TNF-β

also known as lymphotoxin alpha, is a member of the TNF superfamily

TLRs (Toll-like receptors)

a group of pattern recognition receptors that are found on immune cells and other cell types, that play major roles in early recognition of invading microbes. They are important mediators of inflammatory pathways resulting from recognition of a wide variety of pathogen-derived ligands and link adaptive immunity with innate immunity

TORCH

Toxoplasmosis, Rubella, Cytomegalovirus and Herpes virus

toxic megacolon

an acute toxic colitis with dilatation of the colon; may complicate any number of colitides, including inflammatory, ischemic, infectious, radiation, and pseudomembranous

toxic shock syndrome

usually caused by an enterotoxin from *Staphylococcus aureus* and other infections of the vagina entering the circulation; this acts as a superantigen (see entry above) and stimulates a subset of VβT cells releasing large amounts of cytokines resulting in a life-threatening condition

tracheostomy

a surgical procedure to create an opening (stoma) in the trachea; the opening itself can be called a tracheostomy

trachoma

an infection of the eye caused by *Chlamydia trachomatis*; the most common cause of blindness in the developing world after cataracts

transformation (viral)

the change in growth, phenotype, or indefinite reproduction of cells caused by the introduction of inheritable material. Through this process, a virus causes harmful transformations of an *in vivo* cell or cell culture

transforming growth factor (TGF)

one member of a group of cytokines that were identified by their ability to promote the growth of fibroblasts but are also generally immunosuppressive

trans-stadially

the passage of a microbial parasite, such as a virus or rickettsia, from one developmental stage (stadium) of the host to its subsequent stage or stages, particularly as seen in mites

transthoracic echocardiography (TTE)

also known as a standard echocardiography where an echocardiogram is made; *see also* echocardiogram

trichiasis

an uncomfortable condition in which the eye lashes are misdirected toward the eye ball and scratch its surface, the cornea; can be caused by infection, inflammation, autoimmune conditions, and trauma such as burns

truncated

shortened

tularemia

an acute, febrile, granulomatous, infectious zoonosis caused by the aerobic gram-negative pleomorphic bacillus *Francisella tularensis*; infection of humans is through introduction of the bacillus by inhalation, intradermal injection, or oral ingestion

Type 1 hypersensitivity

is IgE-mediated hypersensitivity (or immediate type hypersensitivity). An inappropriate exaggerated IgE response can result in anaphylaxis and even death. It is mediated by activation of mast cells and basophils releasing pharmacological mediators that can result in vasodilation and bronchoconstriction

Type II hypersensitivity

Is mediated by non-IgE antibodies usually IgM or IgG and/or complement; leads to tissue damage including enhanced phagocytosis and lysis

Type III hypersensitivity

is immune-complex mediated hypersensitivity; tissue damage caused by enzymes released from neutrophils and complement activation, e.g. streptococcal glomerulonephritis

Type III secretion apparatus

one of 4 types of secretion systems of gram-negative bacteria that require ATP and mediate transport and injection of pathogenicity molecules into target eukaryotic cells, e.g. shigella

Type IV hypersensitivity

T-cell-mediated hypersensitivity; tissue damage caused by granuloma formation in mycobacterial infections

Type IV secretion apparatus

as well as delivering toxic molecules into cells, type IV secretion systems are also used to transport DNA or protein-DNA complexes. One such process is bacterial conjugation whereby two mating bacteria exchange genetic material; used in pathogenicity by *Helicobacter pylori*

typhus

an infectious disease that is spread by lice or fleas resulting in acute febrile illness; caused by rickettsial organisms

uremic

pertaining to uremia which is the accumulation in the blood of nitrogenous waste products (urea) that are usually excreted in the urine

urethritis

inflammation of the urethra

urolithiasis

the process of forming stones in the kidney, bladder and/or urethra (urinary tract)

uropathogenic *E. coli* (UPEC)

a causative agent in the vast majority of urinary tract infections (UTIs), including cystitis and pyelonephritis, and infectious complications, which may result in acute renal failure in healthy individuals as well as in renal transplant patients

urticaria

commonly referred to as hives; raised areas of erythema and edema involving the dermis and epidermis; may be acute (often due to type I hypersensitivity) or chronic; acute caused by drugs, foods, insect venoms and parasites; also physical causes include pressure and cold or light

uveitis

inflammation of one or all parts of the uveal tract. Components of the uveal tract include the iris, the ciliary body, and the choroid

uvula

a pendant, fleshy mass: palatine uvula is the small, fleshy mass hanging from the soft palate above the root of the tongue

vaginitis

inflammation of the vagina characterized by discharge, odor, irritation, and/or itching

vasculitis

inflammation of the blood vessel wall; may be caused by the immune system

VCAM-1

vascular cell adhesion molecule 1; present on endothelial cells; ligand for VLA-4, an integrin present on lymphocytes and some phagocytes; involved in adhesion of leukocytes to inflamed endothelium and therefore their movement into sites of inflammation

VEGF

Vascular endothelial growth factors are a family of secreted polypeptides with a highly conserved receptor-binding cystine-knot structure similar to that of the platelet-derived growth factors. It is a signal protein produced by many haemopoietic cells that stimulates the formation of blood vessels

vesicles

fluid filled blisters on the skin

vesicoureteric reflux

the abnormal flow of urine from the bladder back into the ureters

Vi antigen

Salmonella enterica serovar typhi is the etiological agent of typhoid fever. It produces a capsular polysaccharide known as Vi antigen, which is composed of non-stoichiometrically O-acetylated – 1,4-linked N-acetylgalactosaminuronic acid residues

viremia

the presence of viruses in the blood

virion

a complete virus particle with its DNA or RNA core and protein coat as it exists outside a cell; also called viral particle

virulent/virulence

is the relative ability of an infectious agent to cause disease. Virulence determinants or factors are the genes and proteins that play key roles in the development of disease

virus envelope

a lipid membrane derived from the host cell, which surrounds the viral nucleocapsid surrounding the nucleocapsid; composed of two lipid layers interspersed with protein molecules; contains lipids of host origin as well as proteins of viral origin; many viruses also develop glycoprotein spikes on their envelope to help them attach to specific cell surfaces

visceral leishmaniasis

also known as kala-azar, characterized by irregular bouts of fever, weight loss, enlargement of the spleen and liver, and anaemia. Most serious type of leishmaniasis and is fatal if left untreated in over 95% of cases

Weil's disease

synonymous with leptospirosis; a bacterial zoonotic disease caused by spirochaetes of the genus *Leptospira*

Weil-Felix test

an agglutination test for rickettsial infection based on production of nonspecific agglutinins in the blood of infected patients, and using various strains of *Proteus vulgaris* as antigen

Western blot

also Western immunoblot, a technique for identification of antigens in a mixture by electrophoresis, blotting onto nitrocellulose and labeling with enzyme or radio-labeled antibodies

white cell casts

casts in urine made up of white blood cells; called casts because they acquire the shape of the renal tubule

whole genome sequencing (WGS)

the analysis of the entire genomic DNA sequence of a cell at a single time, providing the most comprehensive characterization of the genome. It allows for rapid identification and characterization of microorganisms, providing a) information on the relatedness of strains, where they have come from and how they have evolved, b) identification of key virulence factors – specific characteristics that aid the microorganism in causing infection, c) antibiotic resistance profiling – compared with traditional culture methods, d) determining which antibiotics microorganisms are resistant to much more quickly, e) detection, mapping and analysis of outbreaks

Widal test

a serological test for the diagnosis of typhoid and other salmonella infections

Zollinger-Ellison syndrome

the triad of non-beta islet cell tumors of the pancreas (gastrinomas), hypergastrinemia, and severe ulcer disease

zoonosis/zoonotic disease

any infection naturally transmissible from vertebrate animals to humans

Index